T0237884

Lecture Notes in Computer Scie

Commenced Publication in 1973
Founding and Former Series Editors:
Gerhard Goos, Juris Hartmanis, and Jan van Leeuwen

Klaus Ambos-Spies   Benedikt Löwe
Wolfgang Merkle (Eds.)

Mathematical Theory and Computational Practice

5th Conference on Computability in Europe, CiE 2009
Heidelberg, Germany, July 19-24, 2009
Proceedings

 Springer

Volume Editors

Klaus Ambos-Spies
Wolfgang Merkle
Ruprecht-Karls-Universität Heidelberg
Institut für Informatik
Im Neuenheimer Feld 294, 69120, Heidelberg, Germany
E-mail: {ambos,merkle}@math.uni-heidelberg.de

Benedikt Löwe
Universiteit van Amsterdam
Plantage Muidergracht 24, 1018 TV Amsterdam, The Netherlands
E-mail: bloewe@science.uva.nl

Library of Congress Control Number: 2009930216

CR Subject Classification (1998): F.1, F.2.1-2, F.4.1, G.1.0, I.2.6, J.3

LNCS Sublibrary: SL 1 – Theoretical Computer Science and General Issues

ISSN   0302-9743

ISBN   978-3-642-03072-7 Springer Berlin Heidelberg New York

springer.com

© Springer-Verlag Berlin Heidelberg 2009

Typesetting: Camera-ready by author, data conversion by Scientific Publishing Services, Chennai, India
Printed on acid-free paper   SPIN: 12718504   06/3180   5 4 3 2 1 0

Preface

CiE 2009: Mathematical Theory and Computational Practice
Heidelberg, Germany, July 19–24, 2009

After several years of research activity, the informal cooperation "Computability in Europe" decided to take a more formal status at their meeting in Athens in June 2008: the *Association for Computability in Europe* was founded to promote the development, particularly in Europe, of computability-related science, ranging over mathematics, computer science, and applications in various natural and engineering sciences such as physics and biology, including the promotion of the study of philosophy and history of computing as it relates to questions of computability. As mentioned, this association builds on the informal network of European scientists working on computability theory that had been supporting the conference series CiE-CS over the years, and now became its new home.

The aims of the conference series remain unchanged: to advance our theoretical understanding of what can and cannot be computed, by *any* means of computation. Its scientific vision is broad: computations may be performed with discrete or continuous data by all kinds of algorithms, programs, and machines. Computations may be made by experimenting with any sort of physical system obeying the laws of a physical theory such as Newtonian mechanics, quantum theory or relativity. Computations may be very general, depending on the foundations of set theory; or very specific, using the combinatorics of finite structures. CiE also works on subjects intimately related to computation, especially theories of data and information, and methods for formal reasoning about computations. The sources of new ideas and methods include practical developments in areas such as neural networks, quantum computation, natural computation, molecular computation, and computational learning. Applications are everywhere, especially, in algebra, analysis and geometry, or data types and programming. Within CiE there is general recognition of the underlying relevance of computability to physics and a broad range of other sciences, providing as it does a basic analysis of the causal structure of dynamical systems.

This volume, *Mathematical Theory and Computational Practice*, comprises the proceedings of the fifth in a series of conferences of CiE, that was held at the Ruprecht-Karls-Universität Heidelberg, Germany.

The first four meetings of CiE were at the University of Amsterdam in 2005, at the University of Wales Swansea in 2006, at the University of Siena in 2007, and at the University of Athens in 2008. Their proceedings, edited in 2005 by S. Barry Cooper, Benedikt Löwe and Leen Torenvliet, in 2006 by Arnold Beckmann, Ulrich Berger, Benedikt Löwe and John V. Tucker, in 2007 by S. Barry Cooper, Benedikt Löwe and Andrea Sorbi, and in 2008 by Arnold Beckmann, Costas Dimitracopoulos, and Benedikt Löwe were published as *Springer Lecture Notes in Computer Science*, volumes 3526, 3988, 4497 and 5028, respectively.

CiE and its conferences have changed our perceptions of computability and its interface with other areas of knowledge. The large number of mathematicians and computer scientists attending those conferences had their view of computability theory enlarged and transformed: they discovered that its foundations were deeper and more mysterious, its technical development more vigorous, its applications wider and more challenging than they had known. The annual CiE conference has become a major event, and is the largest international meeting focused on computability theoretic issues. Future meetings in Ponta Delgada, Açores (2010, Portugal), Sofia (2011, Bulgaria), and Cambridge (2012, UK) are in planning. The series is coordinated by the CiE Conference Series Steering Committee consisting of Arnold Beckmann (Swansea), Paola Bonizzoni (Milan), S. Barry Cooper (Leeds), Benedikt Löwe (Amsterdam, Chair), Elvira Mayordomo (Zaragoza), Dag Normann (Oslo), and Peter van Emde Boas (Amsterdam).

The conference was based on invited tutorials and lectures, and a set of special sessions on a range of subjects; there were also many contributed papers and informal presentations. This volume contains 17 of the invited lectures and 34% of the submitted contributed papers, all of which have been refereed. There will be a number of post-conference publications, including special issues of *Annals of Pure and Applied Logic*, *Journal of Logic and Computation*, and *Theory of Computing Systems*.

The tutorial speakers were Pavel Pudlák (Prague) and Luca Trevisan (Berkeley).

The following invited speakers gave talks: Manindra Agrawal (Kanpur), Jeremy Avigad (Pittsburgh), Mike Edmunds (Cardiff, Opening Lecture), Peter Koepke (Bonn), Phokion Kolaitis (San Jose), Andrea Sorbi (Siena), Rafael D. Sorkin (Syracuse), Vijay Vazirani (Atlanta).

Six special Sessions were held:

Algorithmic Randomness. *Organizers:* Elvira Mayordomo (Zaragoza) and Wolfgang Merkle (Heidelberg).
  Speakers: Laurent Bienvenu, Bjørn Kjos-Hanssen, Jack Lutz, Nikolay Vereshchagin.
Computational Model Theory. *Organizers:* Julia F. Knight (Notre Dame) and Andrei Morozov (Novosibirsk).

Speakers: Ekaterina Fokina, Sergey Goncharov, Russell Miller, Antonio Montalbán.

Computation in Biological Systems — Theory and Practice.
Organizers: Alessandra Carbone (Paris) and Erzsébet Csuhaj-Varjú (Budapest).
Speakers: Ion Petre, Alberto Policriti, Francisco J. Romero-Campero, David Westhead.

Optimization and Approximation. *Organizers:* Magnús M. Halldórsson (Reykjavik) and Gerhard Reinelt (Heidelberg).
Speakers: Jean Cardinal, Friedrich Eisenbrand, Harald Räcke, Marc Uetz.

Philosophical and Mathematical Aspects of Hypercomputation.
Organizers: James Ladyman (Bristol) and Philip Welch (Bristol).
Speakers: Tim Button, Samuel Coskey, Mark Hogarth, Oron Shagrir.

Relative Computability. *Organizers:* Rod Downey (Wellington) and Alexandra A. Soskova (Sofia)
Speakers: George Barmpalias, Hristo Ganchev, Keng Meng Ng, Richard Shore.

The conference CiE 2009 was organized by Klaus Ambos-Spies (Heidelberg), Timur Bakibayev (Heidelberg), Arnold Beckmann (Swansea), Laurent Bienvenu (Heidelberg), Barry Cooper (Leeds), Felicitas Hirsch (Heidelberg), Rupert Hölzl (Heidelberg), Thorsten Kräling (Heidelberg), Benedikt Löwe (Amsterdam), Gunther Mainhardt (Heidelberg), and Wolfgang Merkle (Heidelberg).

The Program Committee was chaired by Klaus Ambos-Spies and Wolfgang Merkle:

Klaus Ambos-Spies
  (Heidelberg, Chair)
Giorgio Ausiello (Rome)
Andrej Bauer (Ljubljana)
Arnold Beckmann (Swansea)
Olivier Bournez (Palaiseau)
Vasco Brattka (Cape Town)
Barry Cooper (Leeds)
Anuj Dawar (Cambridge)
Jacques Duparc (Lausanne)
Pascal Hitzler (Karlsruhe)
Rosalie Iemhoff (Utrecht)
Margarita Korovina (Novosibirsk)
Hannes Leitgeb (Bristol)
Daniel Leivant (Bloomington)
Benedikt Löwe (Amsterdam)

Giancarlo Mauri (Milan)
Elvira Mayordomo (Zaragoza)
Wolfgang Merkle (Heidelberg, Chair)
Andrei Morozov (Novosibirsk)
Dag Normann (Oslo)
Isabel Oitavem (Lisbon)
Luke Ong (Oxford)
Martin Otto (Darmstadt)
Prakash Panangaden (Montréal)
Ivan Soskov (Sofia)
Viggo Stoltenberg-Hansen (Uppsala)
Peter van Emde Boas (Amsterdam)
Jan van Leeuwen (Utrecht)
Philip Welch (Bristol)
Richard Zach (Calgary)

We are delighted to acknowledge and thank the following for their essential financial support: Deutsche Forschungsgemeinschaft (German Research Foundation), Ruprecht-Karls-Universität Heidelberg (Bioquant and Department of Mathematics and Computer Science), The Elsevier Foundation.

We were proud to offer the program "Women in Computability" funded by the Elsevier Foundation as part of CiE 2009. The Steering Committee of the conference series CiE-CS is concerned with the representation of female researchers in the field of computability. The series CiE-CS has actively tried to increase female participation at all levels in the past years. Starting in 2008, our efforts are being funded by a grant of the Elsevier Foundation under the title *"Increasing representation of female researchers in the computability community."* As part of this program, we had another workshop, a grant scheme for female researchers, a mentorship program, and free childcare.

The high scientific quality of the conference was possible through the conscientious work of the Program Committee, the special session organizers, and the referees. We are grateful to all members of the Program Committee for their efficient evaluations and extensive debates, which established the final program. We also thank the following referees:

Pavel Alaev
Eric Allender
Luis Antunes
Argimiro Arratia
Matthias Aschenbrenner
George Barmpalias
Verónica Becher
Ulrich Berger
Daniela Besozzi
Laurent Bienvenu
Stephen Binns
Achim Blumensath
Volker Bosserhoff
Patricia Bouyer
Thomas Brihaye
Dan Browne
Marius Bujorianu
Andrés Caicedo
Wesley Calvert
Riccardo Camerlo
Alessandra Carbone
Lorenzo Carlucci
Douglas Cenzer
Peter Cholak
Robin Cockett
Alessandro Colombo
Erzsébet Csuhaj-Varjú
Ali Dashti
Jan de Gier
Gianluca Della Vedova

Marina De Vos
Pietro Di Gianantonio
Kenny Easwaran
Leah Epstein
Javier Esparza
Peter Fejer
Fernando Ferreira
Ekaterina B. Fokina
Lance Fortnow
Willem L. Fouché
Pierre Fraigniaud
Peter Gács
Nicola Galesi
Hristo Ganchev
Christine Gaßner
Lucas Gerin
Guido Gherardi
Ilaria Giordani
Rob Goldstone
Erich Grädel
Daniel Graça
Noam Greenberg
Marek Gutowski
Amit Hagar
Magnús Halldórsson
Valentina Harizanov
Monika Heiner
Peter Hertling
Stefan Hetzl
Denis Hirschfeldt

John Hitchcock
Mark Hogarth
Mathieu Hoyrup
Rupert Hölzl
Tseren-Onolt Ishdorj
Emil Jeřábek
Herman Jervell
Jan Johannsen
Iskander Kalimullin
Anna Kasprzik
Klaus Keimel
Viv Kendon
Gabriele Kern-Isberner
Bjørn Kjos-Hanssen
Leszek Kołodziejczyk
Thorsten Kräling
Oleg V. Kudinov
Oliver Kullmann
Geoffrey LaForte
Branimir Lambov
Dominique Larchey-Wendling
Dominique Lecomte
Gyesik Lee
Alexander Leitsch
Steffen Lempp
Alberto Leporati
Davorin Lesnik
Andrew Lewis
Angsheng Li
María López
Gunther Mainhardt
Pierre McKenzie
Carlo Mereghetti
Joseph S. Miller
Russell Miller
Malika More
Markus Müller
Norbert Müller
Thomas Müller
André Nies
Stela Nikolova
István Németi
Paulo Oliva
Antonio Porreca
Valeria de Paiva

Jun Pang
Arno Pauly
Gheorghe Paun
Thanases Pheidas
Thomas Piecha
Sergei Podzorov
Lucia Pomello
Petrus Potgieter
R. Ramanujam
Ramyaa Ramyaa
Rolf Rannacher
Sidney Redner
Jan Reimann
Gerhard Reinelt
Klaus Reinhardt
Fred Richman
Beatrice Riviere
Vladimir Rybakov
Dov Samet
Matthias Schröder
Peter Schuster
Monika Seisenberger
Pavel Semukhin
Stephen G. Simpson
Alla Sirokofskich
Andrea Sorbi
Alexandra Soskova
Dieter Spreen
Frank Stephan
Thomas Strahm
Alexey Stukachev
Zhuldyz Talasbaeva
Paul Taylor
Sebastiaan Terwijn
Tristan Tomala
Hideki Tsuiki
John Tucker
Nikolas Vaporis
Panayot S. Vassilevski
Giuseppe Vizzari
Paul Voda
Heribert Vollmer
Nicolai Vorobjov
Peter B. M. Vranas
Andreas Weiermann

Thomas Wilke	Karim Zahidi
Jon Williamson	Claudio Zandron
Liang Yu	Martin Ziegler

We thank Andrej Voronkov for his EasyChair system, which facilitated the work of the Program Committee and the editors considerably.

May 2009

Klaus Ambos-Spies
Benedikt Löwe
Wolfgang Merkle

Table of Contents

First-Order Universality for Real Programs

Thomas Anberrée

Division of Computer Science, University of Nottingham,
199 Taikang East Road, Ningbo 315100, China
`thomas.anberree@nottingham.edu.cn`

Abstract. J. Raymundo Marcial–Romero and M. H. Escardó described
a functional programming language with an abstract data type `Real` for
the real numbers and a non-deterministic operator `rtest: Real → Bool`.
We show that this language is universal at first order, as conjectured
by these authors: all computable, first-order total functions on the real
numbers are definable. To be precise, we show that each computable
function $f: \mathbb{R} \to \mathbb{R}$ we consider is the extension of the denotation $[\![M_f]\!]$
of some program $M_f: \texttt{Real} \to \texttt{Real}$, in a model based on powerdomains, described in previous work. Whereas this semantics is only an
approximate one, in the sense that programs may have a denotation
strictly below their true outputs, our result shows that, to compute a
given function, it is in fact always possible to find a program with a
faithful denotation. We briefly indicate how our proof extends to show
that functions taken from a large class of computable, first-order *partial*
functions in several arguments are definable.

Keywords: computability, real number computation, simply typed
lambda-calculus, denotational semantics.

1 Introduction

We prove that all computable total functions $f: \mathbb{R} \to \mathbb{R}$ are definable in a certain extension of PCF with a type for real numbers. The language possesses a
denotational semantics in the category of bounded-complete domains and our
proof proceeds by showing that every computable function under consideration
has a representative in the semantics that is the denotation of some program. In
this sense, *definable* should be understood as *definable relatively to our semantics*, which is a stronger property than the existence of a program computing a
function.

The value of this result is at least twofold. Firstly, it confirms a conjecture by
J. Raymundo Marcial–Romero and M. H. Escardó, who first described the language [8], that the language is operationally expressive enough to compute all
computable, first-order and total functions. Secondly, it shows that the denotational semantics proposed for this language by the present author [3,2], which
is the one we consider here, is good enough to expose the expressivity of the
language at first-order. This was not obvious, as our semantics does not enjoy
full adequacy with the language, in the sense that there are programs whose

K. Ambos-Spies, B. Löwe, and W. Merkle (Eds.): CiE 2009, LNCS 5635, pp. 1–10, 2009.

denotation only approximates their behaviour. That is, the denotation of a program may be strictly below its operational semantic. Therefore, it could have been the case that some function were computed by a program but that no program computing this function had a denotation representing the function in the model. The present result shows that it is not the case.

Organization. For the sake of conciseness, we leave both the language and its semantics largely unspecified. Rather, we start by describing a few properties they satisfy (Section 2), from which the universality result is derived (Section 3). Some basic background in domain theory is assumed [1].

We base our proof on the fact that the supremum operator which takes a function $f: [0,1] \to [0,1]$ and returns its supremum $\sup(f)$ is definable in pure PCF, using sequences of integers to represent real numbers [10]. We use this supremum operator to provide finer and finer intervals in which lies the desired value of a given computable function (Section 3.2). This involves translating digital representations into the native type for real numbers of the extended language (Section 3.1).

2   The Language and Its Semantics

As just mentioned, we do not fully describe the language nor its denotational semantics. Rather, we emphasize in this section the properties from which we will derive the universality result. A complete definition of the language and the semantics are in the preceding publication [3], as well as in the author's dissertation [2] which also includes the proofs of properties we assume here.

The language is based on PCF, a simply-typed lambda-calculus with a recursion operator and ground types Nat and Bool for the natural numbers and the boolean values {true, false} [9]. In addition, we assume a type Real for the real numbers. To avoid inessential technicalities in representing discrete spaces of numbers and finite products, we further assume, in the present paper, that the language has types Z and Q for the integers and the rational numbers, as well as product types and their associated projections and pairing functions. Hence, the types for the language are given by

$$\sigma = \texttt{Real} \mid \texttt{Bool} \mid \texttt{Nat} \mid \texttt{Z} \mid \texttt{Q} \mid \sigma \times \sigma \mid \sigma \to \sigma.$$

A program is a closed term of the language. The model lies in the cartesian closed category of bounded-complete, continuous, directed-complete posets, which we call bc-domains, or just domains in the present paper [1]. Each type σ is interpreted by a domain $[\![\sigma]\!]$ and every program $M: \sigma$ has a denotation $[\![M]\!]$ in $[\![\sigma]\!]$. We write R, B, N, Z and Q for $[\![\texttt{Real}]\!]$, $[\![\texttt{Bool}]\!]$, $[\![\texttt{Nat}]\!]$, $[\![\texttt{Z}]\!]$ and $[\![\texttt{Q}]\!]$, respectively. The real line, that is the set $\mathbb{R}$ endowed with the standard Euclidian topology, embeds in R via a continuous injection $\eta: \mathbb{R} \to \mathsf{R}$. The other sets {true, false}, $\mathbb{N}$, $\mathbb{Z}$ and $\mathbb{Q}$ are seen as discrete spaces which embeds in B, N, Z and Q and we also call η any of these embeddings. All functions considered in the semantics are Scott-continuous and, for example, the notation $g \colon \mathsf{R} \to \mathsf{R}$ subsumes the fact that g is continuous.

A key aspect of the language is that it possesses a non-deterministic operator rtest: Real $\to$ Bool, whose purpose is to exhibit which of $r < 1$ and $0 < r$ hold, for any input representing some real number r. In cases where $r \in [0, 1]$, either true or false is returned, non-deterministically. Via the usual if...then ...else... conditional, this non-determinism propagates to all types and leads one to seeing all programs as sets of values, and hence to interpreting types by sets of sets. So as to remain in a tamed framework for models of non-deterministic simply-typed lambda-calculi, we use Smyth powerdomains over bc-domains [3]. In particular, the denotation of a program M : Real whose only possible outputs are real numbers $r_1, \ldots, r_n$ is $\eta (r_0) \cup \cdots \cup \eta (r_n)$. In general, Smyth powerdomains are closed under finite unions.

Usefulness of the Smyth semantics. It is important to bear in mind that, due to the presence of rtest, the semantics is only an approximate one, because the denotation of even a total program may consist of more values than the program actually outputs. For example, for any program 0 for the number 0, the program rtest(0) only outputs true, but its denotation is η (true) $\cup$ η (false). Indeed, it can be proved that rtest possesses no faithful denotation in the Smyth semantics, for continuity reasons [8, Lemma 4.3]. However, if a program of ground type has a maximal denotation, then this denotation faithfully provides the unique output of the program. In particular, if $[\![M]\!] = \eta (r)$, we know that program M can only output the real number r. This is why we can still rely on the semantics to show our universality result. For each function $f \colon \mathbb{R} \to \mathbb{R}$ under consideration, we will define a function $g \colon \mathsf{R} \to \mathsf{R}$ which is the denotation of a program $F \colon$ Real $\to$ Real and such that $g(\eta (r)) = \eta (f(r))$ for all $r \in \mathbb{R}$, hence ensuring that F computes f, in the sense that FM outputs $f(r)$ whenever M outputs r.

In what follows, we are not making direct use of the rtest construct, but rather of a derived operator whose denotation is the subject of the next lemma [3,2].

Lemma 1. *There exists a continuous function* case $\colon \mathsf{R}^5 \to \mathsf{R}$ *which is the denotation of some program and satisfies the following property. For all real numbers p, q, r such that $p < q$ and all elements x, y of the domain R,*

$$\mathrm{case}(\eta (p), \eta (q), \eta (r), x, y) = \begin{cases} x & \text{if } r < p \\ x \cup y & \text{if } r \in [p, q] \\ y & \text{if } q < r. \end{cases}$$

The language also possesses operators bound$_{[a,b]}$: Real $\to$ Real, one for each pair of rational numbers (a, b) with $a < b$. A program of the form bound$_{[a,b]}(M)$ can only reduce to programs of the form bound$_{[a',b']}(M)$ with $[a', b'] \subseteq [a, b]$, thus providing the mechanism through which real numbers are represented and outputted, as sequences of finer and finer nested rational intervals. The denotations bound$_{[a,b]}$: $\mathsf{R} \to \mathsf{R}$ of the bound$_{[a,b]}$ operators satisfy the following properties.

1. For every real number $r \in [a, b]$, we have $\text{bound}_{[a,b]}(\eta(r)) = \eta(r)$.
2. For every $a, b, c, d \in \mathbb{Q}$ with $a < b$ and $c < d$ and for every $x \in \mathsf{R}$,

$$\text{bound}_{[a,b]}(\text{bound}_{[c,d]}(x)) = \begin{cases} \text{bound}_{[a,b] \cap [c,d]}(x) & \text{if } [c,d] \cap [a,b] \neq \varnothing \\ \eta(a) & \text{if } d < a \\ \eta(b) & \text{if } b < c. \end{cases}$$

An operator can be defined, which can be seen as a parameterized version of the $\text{bound}_{[a,b]}$ operators, where a and b are taken as parameters. It enjoys a useful convergence property stated in the next lemma [2, Lemma 5.13.1].

Lemma 2. *There exists a continuous function* $\text{bound}'' \colon \mathsf{R} \times \mathsf{R} \times \mathsf{R} \to \mathsf{R}$ *which is the denotation of some program and which satisfies the following property. For all real numbers a and b such that $a \leq b$ and all $x \in \mathsf{R}$, there exist finitely many intervals $[c_0, d_0], \ldots, [c_K, d_K]$ such that*

$$\text{bound}''(\eta(a), \eta(b), x) = \bigcup_{0 \leq k \leq K} \text{bound}_{[c_k, d_k]}(x)$$

and $[a - 3(b - a), b + 3(b - a)] \supseteq [c_k, d_k] \supseteq [a, b]$ for each k. In particular, if $r \in [a, b]$, one has that $\text{bound}''(\eta(a), \eta(b), \eta(r)) = \eta(r) = \eta(\text{bound}_{[a,b]}(r))$.

Of course, the four basic operators $\times$, $\div$, $+$ and $-$ are definable, in the sense of the following lemma, which we will use implicitly throughout the next two sections.

Lemma 3. *Let f be any of the four basic binary operators over $\mathbb{R}$. There exists a function $g \colon \mathsf{R} \times \mathsf{R} \to \mathsf{R}$, which is the denotation of some program of type* `Real × Real → Real`*, such that for all real numbers r and s for which $f(r, s)$ is defined, one has $g(\eta(r), \eta(s)) = \eta(f(r, s))$. As a consequence, the functions* min*,* max *and the absolute value operator are definable.*

The last ingredient we need to proceed is a limit operator that returns the limit of numerical sequences whose rate of convergence is bounded [2, Theorem 5.17.1].

Lemma 4. *There exists a continuous function* $\text{limit} \colon (\mathbb{N} \to \mathsf{R}) \to \mathsf{R}$ *which is the denotation of some program of type* `(Nat → Real) → Real` *and satisfies the following property. For every sequence $(r_n)_{n \in \mathbb{N}}$ of real numbers such that $|r_n - r_{n+1}| \leq \frac{1}{2^n}$ and for every function $f \in (\mathbb{N} \to \mathsf{R})$ such that $f(\eta(n)) = \eta(r_n)$ for all $n \in \mathbb{N}$, we have that* $\text{limit}(f) = \eta(\lim_{n \to \infty}(r_n))$.

From now on, we will often identify a number $r \in \mathbb{R}$ with its representation $\eta(r)$ in R, and consider $\mathbb{R}$ to be a subset of R. With this convention, we say that a function $g \colon \mathsf{R} \to \mathsf{R}$ *extends* a function $f \colon \mathbb{R} \to \mathbb{R}$ if $f(r) = g(r)$ for all $r \in \text{dom}(f)$. We use similar conventions for other types.

3   Universality at First Order

Using the lemmas of the previous section, we prove that any total, computable function $f: \mathbb{R} \to \mathbb{R}$ is definable in the language, in the sense that there exists a program $M: \texttt{Real} \to \texttt{Real}$ whose denotation $[\![M]\!]: \mathsf{R} \to \mathsf{R}$ extends f. To compute $f(r)$, we work with the signed-digit representation of real numbers [6], encoded in the PCF fragment of the language. Using this representation, local suprema and infima of f near r are computed to provide approximations of $f(r)$ (Section 3.2). A first step is to provide a translation from signed digit representations into native representations for real numbers in the full language (Section 3.1), making use of the limit operator mentioned in Lemma 4.

3.1   Signed-Digit Representation

Let 3 be the set $\{-1, 0, 1\}$. The function $\rho: 3^{\mathbb{N}} \times \mathbb{N} \longrightarrow \mathbb{R}$ mapping (w, n) to $2^n \sum_{i=0}^{\infty} w(i) 2^{-i-1}$ is a surjection, called the *signed digit representation* of real numbers. Each element (w, n) of $3^{\mathbb{N}} \times \mathbb{N}$ is also called a signed digit representation of $\rho(w, n)$. In the context of our semantics, a *signed digit representation* is an ordered pair (v, m) in
$$\tilde{\mathsf{R}} = [(\mathsf{N} \to \mathsf{Z}) \times \mathsf{N}]$$
that agrees with some $(w, n) \in 3^{\mathbb{N}} \times \mathbb{N}$ in the sense that $m = \eta(n)$ and $v(\eta(i)) = \eta(w(i))$ for all $i \in \mathbb{N}$. We also define the type $\widetilde{\texttt{Real}}$ as the type interpreted by $\tilde{\mathsf{R}}$, namely $\widetilde{\texttt{Real}} = (\texttt{Nat} \to \mathsf{Z}) \times \texttt{Nat}$.

Let us define the *partial* surjection
$$e: \tilde{\mathsf{R}} \to 3^{\mathbb{N}} \times \mathbb{N}$$
$$e(v, n) = \text{the unique } (w, n) \text{ that agrees with } (v, n),$$

defined at each (v, n) such that $v(\eta(i)) \in \{\eta(-1), \eta(0), \eta(1)\}$ for all $i \in \mathbb{N}$ and such that $n = \eta(m)$ for some $m \in \mathbb{N}$.

The following lemma provides a function $\rho^*: \tilde{\mathsf{R}} \to \mathsf{R}$ that translates digital representations of decimal numbers into R. Notice that there is no continuous retract in the reverse direction. This is why proving universality is not just a matter of transposing functions $\tilde{\mathsf{R}} \to \tilde{\mathsf{R}}$ to functions $\mathsf{R} \to \mathsf{R}$.

Lemma 5. *There exists a continuous function $\rho^*: \tilde{\mathsf{R}} \to \mathsf{R}$, which is the denotation of some program, such that the following diagram commutes on the domain of e.*

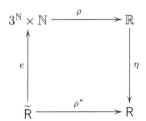

Proof. We have to show that there exists a program whose denotation $\rho_* \colon \widetilde{\mathsf{R}} \to \mathsf{R}$ extends the signed digit representation ρ in the following sense: for all $(v, m) \in \widetilde{\mathsf{R}}$ such that $m = \eta(n)$ for some $n \in \mathbb{N}$ and $v(\eta(i)) = w(i)$ for some $w \in 3^{\mathbb{N}}$ and all $i \in \mathbb{N}$, it holds that $\rho_*(v, m) = \eta(\rho(w, n))$. Let $f \colon \mathbb{Z}^{\mathbb{N}} \times \mathbb{N} \to \mathbb{R}$ be the function defined by $f(w, n) = 2^n \times \lim_{k \to \infty} (u_k)$ where the sequence $(u_k)_{k \in \mathbb{N}}$ is defined by $u_k = \sum_{i=0}^{k} w'(i) \times 2^{-i-1}$ and $w'(i) = \max(-1, \min(w(i), 1))$. Notice that $w'(i) \in \{-1, 0, 1\}$ and that $w'(i) = w(i)$ if $w(i) \in \{-1, 0, 1\}$. Hence the sequence u_k is always convergent and $f(w, n) = \rho(w, n)$ whenever $w \in 3^{\mathbb{N}}$. It is easy to see that $|u_k - \lim_{k \to \infty}(u_k)| \leq \frac{1}{2^k}$, for all $k \in \mathbb{N}$ and all $w \in \mathbb{Z}^{\mathbb{N}}$. From the properties of the language described in Section 2, in particular from Lemma 4, it follows that there is a program M_f whose denotation $[\![M_f]\!]$ extends f. We choose $\rho_* = [\![M_f]\!]$. □

Computability. The signed digit representation is commonly used to perform real number computation as well as to provide a definition of computable numerical functions. There are many definitions of computability for real numbers and numerical functions in the literature. However, most of these definitions are equivalent to the one we consider here.

Definition 1. *A function $\mathbb{R} \to \mathbb{R}$ is computable if it is computable by a Turing machine with alphabet $\Sigma = \{-1, 0, 1, ., \ldots\}$, using the signed digit representation on its input and output tapes (for more details, see [11,5,7]).*

This definition is equivalent to PCF definability at first order using the signed-digit representation.

3.2   Total Computable Functions from $\mathbb{R}$ to $\mathbb{R}$ Are Definable

Our strategy is to obtain a definition $\mathsf{R} \to \mathsf{R}$ of a total function $f \colon \mathbb{R} \to \mathbb{R}$ from a PCF definition $\widetilde{\mathsf{R}} \to \widetilde{\mathsf{R}}$ of f. We use the fact that the supremum and infimum of any computable total function from $[-1, 1]$ to $[-1, 1]$ can be computed in PCF, using the signed-digit representation [10]. We proceed in two steps. Given a computable function $f \colon \mathbb{R} \to \mathbb{R}$, our first step is to show that the functions $\sup_f, \inf_f \colon \mathbb{Q} \times \mathbb{Q} \to \mathbb{R}$ that compute the supremum and infimum of f at *any* rational interval $[p, q]$ are definable (Lemma 6). In a second step, we define a continuous function $\phi \colon \mathsf{R} \to \mathsf{R}$ which represents f (Theorem 1). The idea is to approximate $\phi(x)$ with intervals

$$\left[\inf\nolimits_f (q - \epsilon, q + \epsilon), \sup\nolimits_f (q - \epsilon, q + \epsilon)\right]$$

where q and ϵ are rational numbers such that $|x - q| \leq \epsilon$. The following lemma is a consequence of a result by Simpson [10].

Lemma 6. *Let $f \colon \mathbb{R} \to \mathbb{R}$ be a computable function. Then the functions*

$$\sup\nolimits_f \colon \mathbb{Q} \times \mathbb{Q} \to \mathbb{R} \qquad \text{and} \qquad \inf\nolimits_f \colon \mathbb{Q} \times \mathbb{Q} \to \mathbb{R}$$
$$\sup\nolimits_f (p, q) = \sup\nolimits_{r \in [p,q]} f(r) \qquad\qquad \inf\nolimits_f (p, q) = \inf\nolimits_{r \in [p,q]} f(r)$$

defined on $\{(p, q) \mid p \leq q\}$ are definable by PCF programs of type $\mathsf{Q} \times \mathsf{Q} \to \widetilde{\mathsf{Real}}$.

Proof. We only consider the case of the supremum operator; the case of the infimum operator is dual. A. Simpson proved that the operator

$$\sup_{[-1,1]}\colon ([-1,1] \to [-1,1]) \to [-1,1]$$
$$\sup_{[-1,1]}(g) = \sup \{g(r) \mid r \in [-1,1]\},$$

defined on total continuous functions, is definable in PCF, using the signed digit representation [10]. Given an interval $[p,q]$ and a continuous, total function $g'\colon [p,q] \to \mathbb{R}$, the function $g\colon [-1,1] \to [-1,1]$, $r \longmapsto g'\left(p + \frac{q-p}{2}(r+1)\right)$ is also continuous and total. Furthermore, its supremum over $[-1,1]$ is the same as the supremum of g' over $[p,q]$. It is easy to see that the operator

$$\sup'\colon (\mathbb{R} \to [-1,1]) \times \mathbb{Q} \times \mathbb{Q} \to \mathbb{R}$$
$$\sup'(g,p,q) = \sup_{[-1,1]}\left(\lambda r.g\left(p + \frac{q-p}{2}(r+1)\right)\right),$$

which takes the supremum at $[p,q]$ of computable total functions from $\mathbb{R}$ to $[-1,1]$, can be defined by some PCF program, with respect to signed binary representation, using a definition of $\sup_{[-1,1]}$. Let us use the same name for such a PCF program $\sup'\colon (\widetilde{\mathtt{Real}} \to \widetilde{\mathtt{Real}}) \times \mathtt{Q} \times \mathtt{Q} \to \widetilde{\mathtt{Real}}$.

In order to now compute the supremum of functions whose range is not restricted to $[-1,1]$, we remark that, for any continuous function $h\colon [p,q] \to \mathbb{R}$, one has that

1. $\sup_{[p,q]}(h) = 2\sup_{[p,q]}\left(\frac{1}{2}h\right)$ and
2. if $\sup_{[p,q]}(\lambda r.\max(-1,\min(h(r),1))) \in (-1,1)$ then $\sup_{[p,q]}(h) \in (-1,1)$.

Based on these remarks, we define the PCF program

$$\sup''\colon (\widetilde{\mathtt{Real}} \to \widetilde{\mathtt{Real}}) \times \mathtt{Q} \times \mathtt{Q} \to \widetilde{\mathtt{Real}}$$

$$\sup''(h,p,q) = \text{if } \text{firstdigit}(2 \times s) = 0 \text{ then } s \text{ else } 2 \times \sup''\left(\frac{1}{2}h,p,q\right)$$

where $s = \sup'(g_h,p,q)$ and $g_h = \lambda x.\max(-1,\min(h(x),1))$.

The program firstdigit$\colon \widetilde{\mathtt{Real}} \to \mathbb{Z}$ is some program that takes a signed-digit representation s of a real number and must satisfy the following: it evaluates to 0 if s is of the form $2^m \times 0.d_0 d_1 d_3 \ldots$ with $m < 0$ or $d_0 = d_1 \cdots = d_m = 0$; otherwise, for inputs s representing a real number but not of that form, firstdigit(s) evaluates to a number different from 0. The point is that, if firstdigit$(2 \times s) = 0$, then s represents a real number belonging to the interval $(-1,1)$. The program $\sup''$ works on the fact that, for $n \in \mathbb{N}$ large enough, the supremum over $[p,q]$ of the function $\frac{1}{2^n}h$ is small enough to warrant that all of its signed-digit representations s satisfy firstdigit$(s) = 0$. Since the function $f\colon \mathbb{R} \to \mathbb{R}$ is computable, it is defined by a PCF program $\tilde{f}\colon \widetilde{\mathtt{Real}} \to \widetilde{\mathtt{Real}}$. The PCF program $\sup_f\colon \mathtt{Q} \times \mathtt{Q} \to \widetilde{\mathtt{Real}}$, $\sup_f(p,q) = \sup''(\tilde{f},p,q)$ defines the function $\sup_f\colon \mathbb{Q} \times \mathbb{Q} \to \mathbb{R}$.

Theorem 1. *Any total computable function* $f\colon \mathbb{R} \to \mathbb{R}$ *is definable by a program of type* Real → Real.

Proof. We define a continuous function $\phi\colon \mathbb{R} \to \mathbb{R}$ that extends f and is the denotation of a program of type Real → Real. For each real number x, the function ϕ works by finding some rational number q such that $|x - q| \le \epsilon$ and by approximating $f(x)$ with some interval close to $[\inf_f(q - \epsilon, q + \epsilon), \sup_f(q - \epsilon, q + \epsilon)]$, for ϵ smaller and smaller, using the function bound″ : $\mathbb{R} \times \mathbb{R} \times \mathbb{R} \to \mathbb{R}$ mentioned in Lemma 2. We fix a computable enumeration $(q_i)_{i \in \mathbb{N}}$ of $\mathbb{Q}$ and also call $q\colon \mathbb{N} \to \mathbb{Q}$ some corresponding extension in the model. The function $\phi\colon \mathbb{R} \to \mathbb{R}$ is defined by $\phi(x) = \psi(x, 1, 0)$ from the auxiliary function $\psi\colon \mathbb{R} \times \mathbb{Q} \times \mathbb{N} \to \mathbb{R}$,

$$\psi(x, \epsilon, i) = \text{cases abs}\,(x - q_i) \le \epsilon \quad \to \quad \text{bound}''\left(a, b, \psi\left(x, \frac{\epsilon}{2}, 0\right)\right)$$

$$\text{abs}\,(x - q_i) \ge \frac{\epsilon}{2} \quad \to \quad \psi(x, \epsilon, i + 1)$$

where $a = \inf_f(q_i - \epsilon, q_i + \epsilon)$, $b = \sup_f(q_i - \epsilon, q_i + \epsilon)$ and

$$\text{cases } r \le p \quad \to \quad x$$
$$q \le r \quad \to \quad y$$

is a notation for case(p, q, r, x, y) (Lemma 1). Let r be a real number and let us convince ourselves that $\phi(r) = f(r)$. For any real number $\epsilon > 0$, let a_ϵ and b_ϵ be the infimum and supremum of f at $[r - 2\epsilon, r + 2\epsilon]$, respectively.

Let i_ϵ be the smallest natural number such that $|r - q_{i_\epsilon}| < \frac{\epsilon}{2}$. By definition of $\psi(r, \epsilon, 0)$, there exist $a_0, b_0, \ldots, a_J, b_J \in \mathbb{R}$ satisfying

$$a_\epsilon \le a_j \le f(r) \le b_j \le b_\epsilon \tag{1}$$

for each j, and such that

$$\bigcup_{0 \le j \le J} \text{bound}''\left(a_j, b_j, \psi\left(r, \frac{\epsilon}{2}, 0\right)\right) \subseteq \psi^{i_\epsilon}(r, \epsilon, 0). \tag{2}$$

By Lemma 2, for each j, there exist finitely many intervals $[c_{j,0}, d_{j,0}], \ldots,$ $[c_{j,K_j}, d_{j,K_j}]$, such that

$$\text{bound}''\left(a_j, b_j, \psi\left(r, \frac{\epsilon}{2}, 0\right)\right) = \bigcup_{0 \le k \le K_j} \text{bound}_{[c_{j,k}, d_{j,k}]}\left(\psi^{i_\epsilon}(r, \epsilon, 0)\right)$$

and $[a_j - 3(b_j - a_j), b_j + 3(b_j - a_j)] \subseteq [c_{j,k}, d_{j,k}] \subseteq [a_j, b_j] \subseteq [f(r), f(r)]$. From this and equations 1 and 2, we obtain that there exist some intervals $[c_0, d_0], \ldots, [c_K, d_K]$ such that

$$\bigcup_{0 \le k \le K} \text{bound}_{[c_k, d_k]}\left(\psi\left(r, \frac{\epsilon}{2}, 0\right)\right) \subseteq \psi^{i_\epsilon}(r, \epsilon, 0)$$

with $[a_\epsilon - 3(b_\epsilon - a_\epsilon), b_\epsilon + 3(b_\epsilon - a_\epsilon)] \subseteq [c_k, d_k] \subseteq [f(r), f(r)]$. It is easy to prove from there that $\psi(r, 1, 0) = f(r)$, that is $\phi(r) = f(r)$.

3.3   Generalisation to Partial Functions of Several Arguments

Using the same ideas, it is relatively easy to generalise Theorem 1 so as to encompass functions of several arguments and even many common partial functions, such as the inverse and logarithm functions. In order to apply our proof technique to partial functions $f \colon \mathbb{R}^n \to \mathbb{R}$, we need to be able to find smaller and smaller compact neighbourhoods of $(r_1, r_2, \ldots, r_n) \in \mathrm{dom}(f)$ (Such compact neighbourhoods were the intervals $[q - \epsilon, q + \epsilon]$ in the case of total functions in one argument). A sufficient condition to proceed in this manner is that $\mathrm{dom}(f)$ be recursively open in the sense of the following definition, which roughly says that there exists a recursive enumeration of compact neighbourhoods such that each point of $\mathrm{dom}(f)$ is the intersection of some of these neighbourhoods.

Definition 2 (Recursively open sets). *Let $q_1, \ldots, q_n$ and $\epsilon > 0$ be rational numbers. The closed rational ball of $\mathbb{R}^n$ of centre $(q_1, \ldots, q_n)$ and radius ϵ is the set $B(q_1, \ldots, q_n, \epsilon) = \{(r_1, \ldots, r_n) \in \mathbb{R}^n \mid \max(|r_1 - q_1|, \ldots, |r_n - q_n|) \le \epsilon\}$. A subset S of $\mathbb{R}^n$ is recursively open if there exists a computable function $\nu \colon \mathbb{N} \to \mathbb{Q}^n \times \mathbb{Q}$ such that (1) for all $k \in \mathbb{N}$, we have $B(\nu(k)) \subseteq S$ and (2) for all real $s \in S$ and all $\epsilon > 0$, there exists $k \in \mathbb{N}$ such that $B(\nu(k))$ contains s and has a radius smaller than ϵ.*

The following generalises Theorem 1 [2, Theorem 6.3.5].

Theorem 2. *Let f be a computable partial function from $\mathbb{R}^n$ to $\mathbb{R}$ such that $\mathrm{dom}(f)$ is recursively open. There exists a program whose denotation extends f.*

4   Conclusion

To prove first-order universality of the language, we based our approach on one specific characterization of computable functions (Definition 1). There are at least two other characterizations which might have seemed a more natural choice in our context but are in fact not applicable, at least directly, in the context of our approximate semantics. Brattka [4] characterized computable real-valued functions as those belonging to a certain set of *recursive relations*. Among other things, this set of recursive relations must contain the relation $\mathrm{Ord}_\mathbb{R} := \{(x, 0) \mid x < 1\} \cup \{(x, 1) \mid x > 0\} \subseteq \mathbb{R} \times \mathbb{N}$. However, the relation $\mathrm{Ord}_\mathbb{R}$ cannot be represented in our semantic model. The other characterization, by Brattka and Hertling [5], makes use of "Feasible Real Random Access Machine" manipulating registers for real numbers and natural numbers with some basic operations. But again, a family $\{<_k\}_{k \in \mathbb{N}}$ of multi-valued test operators required among the basic operations, defined by $x <_k y = \{\mathrm{true} \mid x < y\} \cup \{\mathrm{false} \mid x > y - 1/(k+1)\}$, makes it impossible to simulate those RAM machines directly in our semantic model. What would certainly be possible, however, is to use either of these two characterizations to show that all computable functions are computed by some program of the language. But the possibility is open, in principle, that such a program has a denotation strictly below its operational behaviour in our approximate semantics. Thus, without either inspecting or modifying the constructions of the above two papers, it is not possible to apply their results to obtain our definability result, which not only states that each computable function is computed by some program, but moreover that each computable function is computed by some program whose denotation extends the function. In this sense, our denotational semantics is "close enough" to the operational semantics, albeit being only an approximate one.

Notice that we could more directly derive our results without any detour in signed-digit representations, from a supremum operator sup: $(\mathsf{R} \to \mathsf{R}) \to \mathsf{R}$ that would be the denotation of some program. However, we do not know whether such an operator exists.

Acknowledgement. This work was originally part of the author's PhD dissertation, written under the supervision of Martín Escardó at the University of Birmingham [2]. All my thanks to him.

References

1. Abramsky, S., Jung, A.: Domain theory. In: Abramsky, S., Gabbay, D.M., Maibaum, T.S.E. (eds.) Handbook of Logic in Computer Science, vol. 3, pp. 1–168. Clarendon Press, Oxford (1994)
2. Anberree, T.: A Denotational Semantics for Total Correctness of Sequential Exact Real Programs. PhD thesis, School of Computer Science, The University of Birmingham, U.K (2007), http://www.cs.bham.ac.uk/~mhe/anberree-thesis.pdf
3. Anberrée, T.: A denotational semantics for total correctness of sequential exact real programs. In: Agrawal, M., Du, D.-Z., Duan, Z., Li, A. (eds.) TAMC 2008. LNCS, vol. 4978, pp. 388–399. Springer, Heidelberg (2008)
4. Brattka, V.: Recursive characterization of computable real-valued functions and relations. Theoretical Computer Science 162, 4577 (1996)
5. Brattka, V., Hertling, P.: Feasible real random access machines. Journal of Complexity 14(4), 490–526 (1998)
6. Di Gianantonio, P.: A Functional Approach to Computability on Real Numbers. PhD thesis, University of Pisa, Udine (1993)
7. Grzegorczyk, A.: On the definitions of computable real continuous functions. Fund. Math. 44, 61–71 (1957)
8. Marcial-Romero, J.R., Escardó, M.H.: Semantics of a sequential language for exact real-number computation. Theoretical Computer Science 379(1-2), 120–141 (2007)
9. Plotkin, G.D.: LCF considered as a programming language. Theor. Comput. Sci. 5(3), 225–255 (1977)
10. Simpson, A.K.: Lazy functional algorithms for exact real functionals. In: Brim, L., Gruska, J., Zlatuška, J. (eds.) MFCS 1998. LNCS, vol. 1450, pp. 456–464. Springer, Heidelberg (1998)
11. Weihrauch, K.: Computable analysis: an introduction. Springer-Verlag New York, Inc., Secaucus (2000)

Skolem + Tetration Is Well-Ordered

Mathias Barra and Philipp Gerhardy[*]

Dept. of Mathematics, University of Oslo, P.B. 1053, Blindern, 0316 Oslo, Norway
georgba@math.uio.no, philipge@math.uio.no
http://folk.uio.no/georgba
http://folk.uio.no/philipge

Abstract. The problem of whether a certain set of number-theoretic functions – defined via *tetration* (i.e. iterated exponentiation) – is *well-ordered* by the *majorisation relation*, was posed by Skolem in 1956. We prove here that indeed it is a *computable well-order*, and give a *lower bound* τ_0 on its ordinal.

1 Introduction

In this note we solve a problem posed by Thoralf Skolem in [Sko56] regarding the *majorisation relation* on $\mathbb{N}^{\mathbb{N}}$ restricted to a certain subset S_*.

Definition 1 (Majorisation). *Define the* majorisation relation '$\preceq$' *on* $\mathbb{N}^{\mathbb{N}}$ *by:*

$$f \preceq g \overset{\text{def}}{\Leftrightarrow} \exists_{N \in \mathbb{N}} \forall_{x \geq N} \left(f(x) \leq g(x) \right) \ .$$

We say that g majorises f *when* $f \preceq g$, *and as usual* $f \prec g \overset{\text{def}}{\Leftrightarrow} f \preceq g \wedge g \npreceq f$. *We say that* f *and* g *are* comparable *if* $f \prec g$ *or* $f = g$ *or* $g \prec f$.

Hence g majorises f when g *is almost everywhere (a.e.) greater than* f. The relation $\preceq$ is transitive and 'almost' anti-symmetric on $\mathbb{N}^{\mathbb{N}}$; that is, we cannot have both $f \prec g$ and $g \prec f$, and $f \preceq g \wedge g \preceq f \Rightarrow f \overset{\text{a.e.}}{=} g$.

Given $A \subseteq \mathbb{N}^{\mathbb{N}}$, one may ask whether $(A, \preceq)$ is a *total order?* if it is a *well-order?* – and if so – what is its *ordinal?*

In his 1956-paper *An ordered set of arithmetic functions representing the least ϵ-number* [Sko56], Skolem introduced the class of functions S, defined by:

$$0, 1 \in S \quad \text{and} \quad f, g \in S \Rightarrow f + g, x^f \in S \ .$$

In his words (our italics): 'we use the two *rules of production* [which] from [...] functions $f(x)$ and $g(x)$ we build $f(x) + g(x)$'. That is, S is a typical *inductively defined class*, or an *inductive closure*.

In [Sko56] the set S is stratified into the hierarchy $\bigcup_{n \in \mathbb{N}} S_n$ in a natural way: For $f \in S$ define the *Skolem-rank* $\rho_S(f)$ of f inductively by:

$$\rho_S(0) \overset{\text{def}}{=} \rho_S(1) \overset{\text{def}}{=} 0; \ \rho_S(f + g) \overset{\text{def}}{=} \max(\rho_S(f), \rho_S(g)) \quad \text{and} \quad \rho_S(x^f) \overset{\text{def}}{=} \rho_S(f) + 1 \ ,$$

[*] Both authors are supported by a grant from the Norwegian Research Council.

and define $S_n \stackrel{\text{def}}{=} \{f \in S \mid \rho_S(f) \leq n\}$. So e.g. $S_0 = \{0, 1, 2, \ldots\}$ and $S_1 = \mathbb{N}[x]$.

Skolem next defined functions $\phi_0 \stackrel{\text{def}}{=} 1$, $\phi_{n+1} \stackrel{\text{def}}{=} x^{\phi_n}$, and it is immediate that $\rho_S(\phi_n) = n$ and $\rho_S(f) < n \leq \rho_S(g) \Rightarrow f \prec \phi_n \preceq g$.

When $(A, \preceq)$ is a well-order and $f \in A$, we let $\mathrm{O}\,(A, \preceq)$ denote the ordinal/order-type of the well-order and $\mathrm{O}\,(f)$ denotes the ordinal of f w.r.t. $(A, \preceq)$.

The main results from [Sko56] are summarised in a theorem below:

Theorem A (Skolem [Sko56])
1. *If $f \in S_{n+1}$, then f can be uniquely written as $\sum_{i=1}^{k} a_i x^{f_i}$ where $a_i \in N$, $f_i \in S_n$ and $f_i \succ f_{i+1}$;*
2. *$f, g \in S \Rightarrow f \cdot g \in S$, i.e. S is closed under multiplication;*
3. *$\phi_{n+1} \in \overline{S}_{n+1} \stackrel{\text{def}}{=} S_{n+1} \setminus S_n$;*
4. *$(S, \preceq)$ is a well-order;*
5. *$\mathrm{O}\,(S_n, \preceq) = \omega^{\cdot^{\cdot^{\cdot^{\omega}}}} \Big\}{\scriptstyle n+1} = \mathrm{O}\,(\phi_{n+1})$;*
6. *$\mathrm{O}\,(S, \preceq) = \sup_{n<\omega} \mathrm{O}\,(\phi_n) = \epsilon_0 \stackrel{\text{def}}{=} \min\{\alpha \in \mathbf{ON} \mid \omega^\alpha = \alpha\}$.*

Above, $\mathbf{ON}$ is the class of all ordinals. □

In the final paragraphs of [Sko56] the following problems are suggested:

Problem 1 (Skolem [Sko56]). Define S^* by $0, 1 \in S^*$, and, if $f, g \in S^*$ then $f + g, f^g \in S^*$. Is $(S^*, \preceq)$ a well-order? If so, what is $\mathrm{O}\,(S^*, \preceq)$?

For the accurate formulation of the 2. problem, we define the number-theoretic function $t(x, y)$ of *tetration* by: $t(x, y) \stackrel{\text{def}}{=} x_y \stackrel{\text{def}}{=} \begin{Bmatrix} 1 & , y = 0 \\ x^{(x_{y-1})} & , y > 0 \end{Bmatrix} = x^{\cdot^{\cdot^{\cdot^{x}}}} \Big\}{\scriptstyle y}$.

Problem 2 (Skolem [Sko56]). Define S_* by $0, 1 \in S_*$, and, if $f, g \in S_*$ then $f + g, f \cdot g, x^f, x_f \in S_*$. Is $(S_*, \preceq)$ a well-order? If so, what is $\mathrm{O}\,(S_*, \preceq)$?

We remark here that with respect to asymptotic growth, S^* is a 'horizontal extension' of S, while S_* is a 'vertical extension'. More precisely, let $\mathcal{E}^n$ be the nth Grzegorczyk-class, and set $\overline{\mathcal{E}}^{n+1} \stackrel{\text{def}}{=} \mathcal{E}^{n+1} \setminus \mathcal{E}^n$. Then:

$$S^* \subseteq \mathcal{E}^3 \quad \text{while} \quad S_* \subseteq \mathcal{E}^4 \quad \text{and} \quad \overline{\mathcal{E}}^4 \cap S_* \neq \emptyset \ .$$

Also $2^x \in S^* \setminus S_*$, and $x_x \in S_* \setminus S^*$, so the classes are incomparable.

Problem 1. has been subjected to extensive studies, leading to a 'Yes!' on the well-orderedness, and to estimates on the ordinal. We shall briefly review the relevant results below. Problem 2. – to our best knowledge – is solved for the first time here.

On Problem 1. In [Ehr73], A. Ehrenfeucht provides a positive answer to Problem 1. by combining results by J. Kruskal [Kru60] and D. Richardson [Ric69]. Richardson uses analytical properties of certain inductively defined sets of functions $A \subset \mathbb{R}^{\mathbb{R}}$ to show (as a corollary) that $(S^*, \preceq)$ is a *total order*. Ehrenfeucht then gives a very basic well-partial-order $\subseteq$ on S^*, and invokes a deep combinatorial result by Kruskal which ensures that the total extensions of $\subseteq$ – which include $(S^*, \preceq)$ – are necessarily well-orders.

Later, in a series of papers, H. Levitz has isolated sub-orders of $(S^*, \preceq)$ with ordinal ϵ_0 [Lev75, Lev77], and he has provided the upper bound τ_0 on O $(S^*, \preceq)$ [Lev78]. Here $\tau_0 \stackrel{\text{def}}{=} \inf \{\alpha \in \mathbf{ON} \mid \epsilon_\alpha = \alpha\}$ Hence $\epsilon_0 \leq$ O $(S^*, \preceq) \leq \tau_0$; a rather large gap.

On Problem 2. Since Skolem did not precisely formulate his second problem, below follows the last paragraphs of [Sko56] verbatim:

> It is natural to ask whether the theorems [of [Sko56]] could be extended to the set S^* of functions that may be constructed from 0 and x by addition, multiplication and the general power $f(x)^{g(x)}$. However, I have not yet had the opportunity to investigate this.
>
> It seems probable that we will get a representation of a higher ordinal by taking the set of functions of one variable obtained by use of not only $x + y$, xy, and x^y but also $f(x, y)$ defined by the recursion
>
> $$f(0, y) = y, \ \ f(x + 1, y) = x^{f(x,y)}$$
>
> The difficulty will be to show the general comparability and that the set is really well ordered by the relation [$\preceq$].

> - Skolem [Sko56]

Exactly what he meant here is not clear, and at least two courses are suggested: One is to study the class obtained by *general tetration* – allowing from f and g the formation of f_g – or to consider the class most analogous to S – allowing the formation of x_f only. This paper is concerned with the second interpretation.

Finally, we have included multiplication of functions as a basic *rule of production* for S_*, since we feel this best preserves the analogy with S. Whereas in S multiplication is derivable, in S_* it is not: e.g. $x_x \cdot x$ is not definable without multiplication as a primitive.

2   Main Results and Proofs

It is obvious that $f \preceq g \Leftrightarrow x^f \preceq x^g \wedge x_f \preceq x_g$. Secondly, an S_*-function f belongs to S iff no *honest* application of the rule $f \stackrel{\text{def}}{=} x_g$ has been used, where *honest* means that $g \notin \mathbb{N}$ (identically a constant). For, if $g \equiv c$, then $x_g = x_c$ which belongs in S, and in the sequel we will tacitly assume that in the expression 'x_g' the g is not a constant. It is straightforward to show that x_x majorises all functions of S, and that x_x is $\preceq$-minimal in $\overline{S_*} \stackrel{\text{def}}{=} S_* \setminus S$.

2.1   Pre-Normal- and Normal-Forms

We next prove that all functions are represented by a unique *normal-form (NF)*.

Strictly speaking we need to distinguish between functions in S_* and the *terms* which represent them, as different terms may define the same function. We will write $f \equiv g$ to denote syntactical identity of terms generated from the definition of S_*, write $f = g$ to denote extensional identity (e.g. $x_{x+1} \not\equiv x^{x_x}$ but $x_{x+1} = x^{x_x}$), and blur this distinction when convenient. In this spirit, we will also refer to arithmetic manipulations of S_*-functions as *rewriting*.

Definition 2 (Pre-normal-form). *For $f \in S$, we call the unique normal-form for f from [Sko56] the Skolem normal-form (SNF) of f.*

Let s, t range over S. We say that a function $f \in S_*$ is in ($\Sigma\Pi$-) pre-normal-form (($\Sigma\Pi$-) PNF) if $f \in S$ and f is in SNF, or if $f \in \overline{S_*}$ and f is of the form $f = \sum_{i=1}^{n} \prod_{j=1}^{n_i} f_{ij}$ where either

$$f_{ij} \equiv x_g \quad \text{where } g \text{ is in } \Sigma\Pi\text{-PNF;}$$
$$f_{ij} \equiv x^g \quad \text{where } g \text{ is in PNF, } g \notin S, g \equiv (\prod_{i=1}^{n_g} g_i) \not\equiv x_h, \text{ and } g_{n_g} \not\equiv (s+t);$$
$$f_{ij} \equiv s \quad \text{where } s \text{ is in SNF, and } j = n_i.$$

An f_{ij} on one of the above three forms is an S_*-factor, a product Πf_i of S_*-factors is an S_*-product, also called a Π-PNF. We say that x_h is a tetration-factor, that $x^{\Pi g_j}$ is an exponent-factor with exponent-product Πg_j, and that $s \in S$ is a Skolem-factor.

The requirement on exponent-factors can be reformulated as: the exponent-product is not a single tetration-factor, nor is its Skolem-factor a sum.

Proposition 1. *All $f \in S_*$ have a $\Sigma\Pi$-PNF.*

Proof. By induction on the build-up of f. The induction start is obvious, and the cases $f = g + h$, $f = gh$ and $f = x_g$ are straightforward. Let $f = x^g$, and let $g = \Sigma\Pi g_{ij}$. Then $f = x^{\Sigma\Pi g_{ij}} = \Pi_i x^{\Pi^{n_i} g_{ij}}$. By hypothesis, this is a PNF for f except when some exponent-product $P = \Pi^{n_i} g_{ij}$ is either a single factor or when $\Pi^{n_i} g_{ij} = P' \cdot (s+t)$. Such exponent factors can be rewritten as x_{h+1} in the first case, and as $x^{P' \cdot s} \cdot x^{P' \cdot t}$ in the second case.   □

Definition 3 (Normal-form). *Let $f \in S_*$. We say that the PNF $\Sigma\Pi f_{ij}$ is a normal-form (NF) for f if*

(NF1) $f_{ij} \succeq f_{i(j+1)}$, and $f_{ij} \succ f_{i(j+1)} \Rightarrow \forall_{\ell \in \mathbb{N}} \left(f_{ij} \succ (f_{i(j+1)})^{\ell} \right)$;
(NF2) $\forall_{s \in S} \left(\Pi_j f_{ij} \succ (\Pi_j f_{(i+1)j}) \cdot s \right)$;
(NF3) *If f_{ij} is on the form x_h or x^h, then h is in NF.*

Informally **NF1–NF3** says that NF's are inherently ordered PNF's. Proving uniqueness is thus tantamount to showing that two terms in NF are syntactically identical, lest they define different functions.

The property marked **FPP** below we call the *finite power property*.

Lemma 1 (and definition of FPP). *Let $F \subseteq S_*$ be a set of comparable S_*-factors in NF such that*

$$\forall_{f_1, f_2 \in F} \forall_{\ell \in \mathbb{N}} \left(f_1 \in \overline{S_*} \ \wedge \ f_1 \succ f_2 \ \Rightarrow \ f_1 \succ (f_2)^{\ell} \right) . \qquad \textbf{(FPP)}$$

Then all NF's $f \equiv \Sigma\Pi f_{ij}$, $g \equiv \Sigma\Pi g_{ij}$ composed of factors from F are comparable. In particular $f \succ g \Leftrightarrow f_{i_0 j_0} \succ g_{i_0 j_0}$ for the least index $(i_0 j_0)$ such that $f_{ij} \not\equiv g_{ij}$.

Proof. Let $f = \Sigma_{i=1}^{n_f} \Pi_{j=1}^{m_i} f_{ij} \not\equiv \Sigma_{i=1}^{n_g} \Pi_{j=1}^{k_i} g_{ij} = g$, i.e. f and g have distinct NF's. Let $(i_0 j_0)$ be as prescribed above, and assume w.l.o.g. that $f_{i_0 j_0} \succ g_{i_0 j_0}$ (comparable by hypothesis). Since $\Sigma\Pi g_{ij}$ is a NF all summands majorise later summands. Hence, for $\ell = \max\{k_i \mid i \leq n_g\}$ and $c = n_g$ we have $(g_{i_0 j_0})^{\ell} \cdot c \succeq g_{i_0 j_0} \cdots g_{ik_i} + g_{(i+1)1} \cdots g_{(i+1)k_{i+1}} + \cdots + g_{n_g 1} \cdots g_{n_g k_{n_g}}$. Clearly $f_{i_0 j_0} \overset{\text{FPP}}{\succ} (g_{i_0 j_0})^{\ell \cdot c} \succeq (g_{i_0 j_0})^{\ell} \cdot c$ implies $f \succ g$. $\qquad \square$

Lemma 2. $f \prec g \Rightarrow^1 \forall_{\ell \in \mathbb{N}} \left((x_f)^{\ell} \prec x_g \right)$. $\qquad \square$

I.e. (honest) tetration-factors satisfy the **FPP**. We skip the proof.

2.2   The Well-Order $(S_*, \preceq)$

In this section we prove our main theorem:

Main Theorem 1. $(S_*, \preceq)$ *is a well-order.*

We establish this through the following lemmata:

Definition 4 (Tetration rank). *The* tetration rank, *denoted $\rho_T(f)$, of a function $f \in S_*$ is defined by induction as follows:*

$$\rho_T(0) \overset{def}{=} \rho_T(1) \overset{def}{=} 0 \ , \rho_T(f+g) \overset{def}{=} \rho_T(fg) \overset{def}{=} \rho_T(x^f) \overset{def}{=} \mathsf{max}(\rho_T(f), \rho_T(g)) \ ,$$

$$\rho_T(x_f) \overset{def}{=} \rho_T(f) + 1 \quad (f \text{ not } constant).$$

For all $n \in \mathbb{N}$, define $S_{*,n} \overset{def}{=} \{f \in S_* \mid \rho_T(f) \leq n\}$, and $\overline{S_{*,n+1}} \overset{def}{=} S_{*,n+1} \setminus S_{*,n}$.

Clearly $S_* = \bigcup_{n \in \mathbb{N}} S_{*,n}$, and $f, g \in S_{*,n}$ implies $f + g, fg, x^f \in S_{*,n}$. Calculating the tetration-rank of any $f \in S_*$ is straightforward, and terms with different tetration-rank cannot define the same function:

Theorem 2. *Let $\psi_n \in S_*$ be defined by $\psi_0 \overset{def}{=} x_x$, and $\psi_{n+1} \overset{def}{=} x_{\psi_n}$. Then ψ_n is comparable to all functions in S_*, and $\rho_T(f) < n \leq \rho_T(g) \Rightarrow f \prec \psi_n \preceq g$.* $\quad \square$

---

[1] Actually, the assertion remains true when $s \in S$ is substituted for $\ell \in \mathbb{N}$, but we shall not need this here.

We omit the proof for lack of space. The theorem above states that ψ_{n+1} is $\preceq$-minimal in $S_* \setminus S_{*,n}$, and that ψ_{n+1} majorises all $g \in S_{*,n}$.

The next definition is rather technical, but we will depend upon some way of 'dissecting' exponent-factors in order to facilitate comparison with tetration-factors.

Definition 5 (Tower height). Let $f = \Sigma\Pi f_{ij}$ and $\rho_T(f) = n$. Define the n-tower height $\tau_n(\Sigma\Pi f_{ij})$ inductively by $\tau_n(\Sigma\Pi f_{ij}) \overset{def}{=} \max_{ij}(\tau_n(f_{ij}))$, where τ_n is defined for factors by:

$$\tau_n(f_{ij}) \overset{def}{=} \begin{cases} 0 \ , & \text{if } \rho_T(f_{ij}) < n \quad \text{or} \quad f_{ij} \equiv x_h, \\ \tau_n(\Pi g_k) + 1 \ , & \text{if } \rho_T(f_{ij}) = n \quad \text{and} \quad f_{ij} \equiv x^{\Pi g_k} \ . \end{cases}$$

Lemma 3. *Assume that each $f \in S_{*,n}$ has a unique NF (satisfying **NF1**–**NF3**), and that $S_{*,n}$ satisfies **FPP** (so that $(S_{*,n}, \preceq)$ is a well-order). Then:*
*(1) Any two $S_{*,n+1}$-factors are comparable, and $S_{*,n+1}$ has the **FPP**;*
(2) All $S_{*,n+1}$-products are comparable and have a unique NF;
(3) If Πg_j is a $\overline{S_{*,n+1}}$-product, then

$$\exists_{h \in S_{*,n}} \exists_{0 < c, d \in \mathbb{N}} \forall_{0 < \ell \in \mathbb{N}} \left(x_{h+c} \prec x^{\Pi g_j} \preceq (x^{\Pi g_j})^\ell \prec x^{\cdot^{\cdot^{\cdot^{x^{(x^d)}}}}} \Big\}_{h+c-1} \right) \ .$$

Proof. The proof is by induction on the maximum of the tower heights of the involved terms. More precisely, since all functions have a PNF, the number

$$\min_{\tau_{n+1}}(f) \overset{def}{=} \min\left\{ m \in \mathbb{N} \mid \exists_{\Sigma\Pi f_{ij}} (f = \Sigma\Pi f_{ij} \wedge \tau_{n+1}(\Sigma\Pi f_{ij}) = m) \right\}$$

is well-defined for any $f \in S_{*,n+1}$. Given two factors, or a product of factors, when we want to prove one of the items (1)–(3), we proceed by induction on $m = \max_i (\min_{\tau_{n+1}}(f_i))$, where $f_1, \ldots f_k$ are the involved factors.

In light of THEOREM 2 – which immediately yields the **FPP** for pairs of factors of different tetration rank – the proof need only deal with $\overline{S_{*,n+1}}$-products. We note first that such products may be written in the Π-PNF $f_1 \cdots f_m \cdot P$, where $\rho_T(f_i) = n+1$ and $\rho_T(P) \leq n$. By assumption, the part P of the product has a unique NF, and is comparable to all other $S_{*,n}$-products and $S_{*,n+1}$-factors.

Induction start $(\tau_{n+1} = 0)$: If $\tau_{n+1}(f) = 0$, each f_j is a tetration factor x_{h_j} for some $h_j \in S_{*,n}$ with h_j in NF. Ordering the h_j in a decreasing sequence yields a unique NF for $f_1 \cdots f_m \cdot P$ satisfying **NF1**–**NF3** (by invoking LEMMA 2). Since all factors compare, all products compare and so (1) and (2) hold.

With regard to (3), $\tau_{n+1}(\Pi_{j=1}^k g_j) = 0$ means that $\Pi_{j=1}^k g_j$ may be assumed to be a NF by (1) and (2). Because $\rho_T(g_1) = n+1$ and $\tau_{n+1}(g_1) = 0$, we must have $g_1 \equiv x_{g_1'}$ for some $g_1' \in \overline{S_{*,n}}$. We have

$$x_{g_1'+1} = x^{x_{g_1'}} \equiv x^{g_1} \prec x^{\Pi g_j} = f \preceq x^{(x_{g_1'})^k} \preceq \left(x^{(x_{g_1'})^k} \right)^\ell \prec x^{(x_{g_1'})^{k+1}} \overset{\dagger}{\prec} x^{\cdot^{\cdot^{\cdot^{x^{(x^2)}}}}} \Big\}_{g_1'} \ .$$

Above the '$\overset{\dagger}{\prec}$' is justified by a generalisation of the inequalities $(x^y)^k = x^{yk} \prec x^{y^2}$, the proof of which we omit for lack of space. Thus, setting $h = g_1'$, $c = 1$ and $d = 2$ we have the desired inequalities.

Induction step $(\tau_{n+1} = m + 1)$: As the induction start contains all the important ideas we only include a sketch. First we prove comparability of factors (1), and only the case x_f vs. $x^{\Pi g_j}$ is involved: here we rely on the IH(3) to obtain the **FPP**. Items (2) and (3) follow more or less as for the induction start.   □

We can now prove THEOREM 1:

Proof (of THEOREM 1). That $S_{*,0} = S$ is well-ordered and have unique NF's is clear, and it vacuously satisfies the **FPP**.

LEMMA 3 furnishes $S_{*,n+1}$ with the **FPP**. If we can produce unique $\Sigma\Pi$-NF's for $f \in S_{*,n+1}$ we are – by LEMMA 1 and induction on n – done, as an increasing union of well orders is itself well ordered.

Now, let $\Sigma\Pi f_{ij} = f \in S_{*,n+1}$. Since all products have a unique NF, by rewriting each of the PNF-products Πf_{ij} to their respective NF's, and then rewriting $P \cdot s + P \cdot t$ to $P \cdot (s + t)$ where necessary – clearly an easy task given that all products are unique – before reordering the summands in decreasing order produces the desired NF.   □

It is interesting to note that the above proofs are completely constructive, and an algorithm for rewriting functions to their normal forms for comparison can be extracted easily from the proof of LEMMA 1:

Theorem 3. $(S_*, \preceq)$ *is computable.*   □

This stands in stark contrast to the Ehrenfeucht-Richardson-Kruskal proof(s) that S^* is well-ordered. Indeed, substituting S^* for S_* in the above theorem turns it into an open problem.

In the next section we will see that NF's are a clear advantage when searching for the order type $O(S_*, \preceq)$ of the well-order.

2.3   On the Ordinal of Normal-Forms

In this section the reader is assumed to be familiar with ordinals and their arithmetic. See eg. Sierpiński [Sie65]; α, β range over ordinals, γ over limits.

That $O(x_x) = \epsilon_0$ follows from THEOREMS A&2. It is also obvious that $O(f + 1) = O(f) + 1$ for all $f \in S_*$, and that if $O(f) = \alpha + 1$, then $f = f' + 1$ for some $f' \in S_*$. It follows that $\Sigma\Pi f_{ij}$ correspond to a limit – except when $f_{nn_n} \equiv c \in \mathbb{N}$.

When $f \preceq g$, we let $[f, g)$ denote the segment $\{h \in S_* \mid f \preceq h \prec g\}$, and we write $O([f, g))$ for $O([f, g), \preceq)$. In particular $O(f) = O([0, f))$. Also $f \preceq g$ implies $[0, g) = [0, f) \cup [f, g)$, viz. $O(g) = O([0, f)) + O([f, g))$.

Lemma 4. *Let* $f \equiv \Pi_i^m f_i$. *Then* $f \preceq h \preceq f \cdot 2 \Rightarrow \exists_{f' \preceq f}(h = f + f')$.   □

We omit the proof, and remark that this lemma *cannot* be generalised to the case where f is a general function.

Lemma 5. *Let* $g \preceq f \equiv \Pi_i f_i$. *Then (1)* $O(f + g) = O(f) + O(g)$. *Moreover,* *(2)* $O(\Sigma\Pi f_{ij}) = \Sigma O(\Pi f_{ij})$, *and (3)* $O(f \cdot n) = O(f) \cdot n$.

Proof. Clearly (2) $\Rightarrow$ (3). Since the majorising product of a NF succeeds the remaining sum, (2) follows by iterated applications of (1), which we prove by induction on O (g). The induction start is trivial.

Case O $(g) = \alpha' + 1$: Then, for some g' we have $g' + 1 = g$, and so

$$O(f + g) = O(f + g' + 1) = O(f + g') + 1 \overset{\text{IH}}{=} O(f) + O(g') + 1 = O(f) + O(g) \ .$$

Case O $(g) = \gamma$: As noted above, we have O $(f + g) = O(f) + O([f, f + g))$. Since $g \preceq f \Rightarrow f + g \preceq f \cdot 2$, by LEMMA 4 $[f, f + g) = \{f + g' \mid g' \prec g\}$ whence the map $\Theta : [f, f + g) \to \gamma$, defined by $\Theta(f + g') = O(g')$, is an order-preserving bijection.   □

Lemma 6. *Let* $f \equiv \Pi_k^m f_k$. *Then* $g \preceq f_m \Rightarrow O(f \cdot g) = O(f) \cdot O(g)$.

Proof. By induction on O (g) for all $\Pi_k^m f_k$ simultaneously, where $g \equiv \Sigma_i^n \Pi_j^{n_i} g_{ij}$. The induction start is trivial.

Case O $(g) = \alpha' + 1$: Set $\alpha_i = O\left(\Pi_j^{n_i} g_{ij}\right)$ and obtain that O $(g) \overset{\text{L. 5(2)}}{=} \Sigma_i^n O\left(\Pi_j^{n_i} g_{ij}\right) = \Sigma_i^n \alpha_i$. Since $g \preceq f_m$, the NF for $f \cdot g$ has the form $\Sigma_i^n f \cdot (\Pi_j^{n_i} g_{ij})$, since $\Pi_j^{n_i} g_{ij} \prec g$ for each i (since g is a successor). Hence

$$O(f \cdot g) = \Sigma_i^n O\left(f \cdot (\Pi_j^{n_i} g_{ij})\right) \overset{\text{IH}}{=} \Sigma_i^n O(f) \cdot O\left(\Pi_j^{n_i} g_{ij}\right) = \Sigma_i^n O(f) \cdot \alpha_i \ ,$$

which – by right-distributivity of ordinal arithmetic – completes this case.

Case O $(g) = \gamma$: If $n \geq 2$ the result follows from the IH by considering the NF of $f \cdot g$. If $n = 1$, and $n_1 \geq 2$, again we may invoke the IH wrt. the product $f \cdot \Pi_j^{n_1 - 1} g_{1j}$ and the function g_{1n_1}. In fact, the only remaining case is when g is a constant, which is resolved by LEMMA 5(3).   □

Theorem 4. $O\left(\Sigma \Pi f_{ij}\right) = \Sigma \Pi O(f_{ij})$.   □

Lemma 7. *Let* $g = \Pi_k^{m+1} g_k$, $m \geq 1$. *Then* $\{x^{(\Pi_i^m g_i) \cdot f}\}_{f \prec g_{m+1}}$ *is cofinal in* x^g.

Proof. We must prove that $\forall_{h \prec x^g} \exists_{f \prec g_{m+1}} (h \preceq x^{(\Pi_i^m g_i) \cdot f})$. Let $h \equiv \Sigma \Pi h_{ij} \prec x^g$, and fix $\ell \in \mathbb{N}$ such that $h \preceq (h_{11})^\ell \prec x^g$. If $\rho_T(h_{11}) < \rho_T(x^g)$ the result is immediate. Indeed, unless $x^{\Pi_i^m g_i} \prec h_{11}$ we are done.

If $h_{11} \equiv x^{h'}$ – either an exponent- or a Skolem-factor – then $h' \cdot \ell \prec g$. Choosing $f' = 1$ is sufficient unless $\Pi_i^m g_i \preceq h' \cdot \ell$, which may occur exactly when for some $h'' \prec g_{m+1}$, we have $h' = g \cdot h''$. If $g_{m+1} \equiv x$, then $h'' = c$, and since h' is an exponent-product, $c = 1$. Hence $f' = \ell + 1$ witnesses cofinality. Since g_{m+1} is a factor in an exponent-product, we have $x \prec g_{m+1} \Rightarrow x^2 \preceq g_{m+1}$; now $h'' \prec g_{m+1} \Rightarrow h'' \cdot c \prec g_{m+1}$ for all $c \in \mathbb{N}$, so that $f' = h'' \cdot (\ell + 1)$ suffices.

If $h_{11} \equiv x_{h'}$, then comparing h' to $x_{g_0 + c}$ and d satisfying the condition of LEMMA 3(3) completes the case, since those inequalities are easily seen to hold w.r.t. x^{g_1} as well.   □

Theorem 5. *Let* $g \equiv \Pi_k^m g_k$, *and let* $h \preceq g_m$. *Then* $O\left(x^{g \cdot h}\right) = O\left(x^g\right)^{O(h)}$.

Proof. By induction on $O(h)$, the induction start is trivial.

Case $O(h) = \alpha' + 1$: Then, for $h \equiv \Sigma_i^n \Pi_j^{n_i} h_{ij}$ set $h_i \stackrel{\text{def}}{=} \Pi_j^{n_i} h_{ij}$. We obtain $O\left(x^{g \cdot h}\right) = O\left(\Pi_i^n x^{g \cdot h_i}\right) \stackrel{\text{T.\,4}}{=} \Pi_i^n O\left(x^{g \cdot h_i}\right)$. Next, $n \geq 2$ since h is a successor, whence $h_i \prec h$. Hence $\Pi_i^n O\left(x^{g \cdot h_i}\right) \stackrel{\text{IH}}{=} \Pi_i^n O\left(x^g\right)^{O(h_i)} = O\left(x^g\right)^{O(h)}$, by basic ordinal arithmetic.

Case $O(h) = \gamma$: Let h be as above. If $n \geq 2$, the proof is as above, For $n = 1$, if $n_1 \geq 2$, the identity $(\mu^\alpha)^\beta = \mu^{\alpha \cdot \beta}$ combined with the IH is sufficient. The remaining possibility – h is a single factor – is resolved by $O\left(x^{g \cdot h}\right) \stackrel{\text{L.\,7}}{=} O\left(\sup_{h' \prec h} x^{g \cdot h'}\right) \stackrel{\text{IH}}{=} \sup_{h' \prec h} O\left(x^g\right)^{O(h')} \stackrel{\text{def}}{=} O\left(x^g\right)^{O(h)}$. □

2.4   The Ordinal $O\left(S_*, \preceq\right)$

An *epsilon number* ϵ satisfies $\omega^\epsilon = \epsilon$, and can alternatively be characterised as ordinals above ω closed under exponentiation [Sie65, p. 330]. Naive tetration of ordinals – which we denote by $\alpha_{(n)}$ – is well-defined, even though $\alpha_{(\omega)} = \alpha_{(\beta)}$ for all $\beta \geq \omega$. Thus $\alpha_{(\omega)}$ is an epsilon number for $\alpha \geq \omega$, and $(\epsilon_\alpha)_{(\omega)} = \epsilon_{\alpha+1}$. We follow the notation from [Lev78] and let τ_0 denote the smallest solution to the equation[2] $\epsilon_\alpha = \alpha$. In this section we prove our second main result:

Main Theorem 6. $\tau_0 \leq O\left(S_*, \preceq\right)$.

Lemma 8. $O\left(x_{f+x}\right) \geq O\left(x_f\right)_{(\omega)}$.

Proof. Set $O(x_f) = \gamma$. Then $\forall_{k \in \mathbb{N}} \left((x_f)^k \prec x^{x_f} = x_{f+1}\right)$. Hence $O\left((x_f)^k\right) \stackrel{\text{T.\,4}}{=} O(x_f)^k$, which right side is cofinal in γ^ω. Hence (†): $\gamma^\omega \leq O\left(x_{f+1}\right)$. Since $(x_{f+1})^{x_f} = x^{x_f \cdot x_f} \prec x_{f+2}$ we obtain: $O\left(x_{f+2}\right) > O\left(x^{x_f \cdot x_f}\right) \stackrel{\text{T.\,5}}{=} O\left(x^{x_f}\right)^{O(x_f)} \stackrel{\dagger}{\geq} (\gamma^\omega)^\gamma \geq \gamma^\gamma$, whence (‡): $\forall_f \left(O\left(x_{f+2}\right) \geq O\left(x_f\right)^{O(x_f)}\right)$.

This constitutes induction start $(n = 1)$ in a proof of $O\left(x_{f+2n}\right) \geq \gamma_{(n+1)}$. For the induction step, assuming $O\left(x_{f+2n}\right) \geq \gamma_{(n+1)}$, immediately yields:

$O\left(x_{f+2(n+1)}\right) = O\left(x_{(f+2n)+2}\right) \stackrel{\ddagger}{\geq} (\gamma_{(n+1)})^{\gamma_{(n+1)}} \geq \gamma_{(n+2)}$.

Finally, since $x_{f+n} \prec x_{f+x}$, we see that $\gamma_{(n+1)} \leq O\left(x_{f+x}\right)$, for arbitrary n, whence the conclusion is immediate. □

Lemma 9. *Let* $x \preceq h \in S_*$. *Then* $O\left(x_{h \cdot x}\right) \geq \epsilon_{O(h)}$.

Proof. By induction on $O(h)$. As $O(x_x) = \epsilon_0$, LEMMA 8 implies $O\left(x_{(n+1) \cdot x}\right) \geq \epsilon_n$ for $n < \omega$. Hence $O\left(x_{x^2}\right) \geq \sup_{n < \omega} \epsilon_n = \epsilon_\omega$. When $h = x$ this is exactly induction start. Both cases $\alpha + 1$ and γ follows immediately:

$(\alpha + 1)$: $O\left(x_{(h+1)x}\right) = O\left(x_{(h \cdot x + x)}\right) \stackrel{\text{L.\,8}}{\geq} O\left(x_{h \cdot x}\right)_{(\omega)} \stackrel{\text{IH}}{\geq} \left(\epsilon_{O(h)}\right)_{(\omega)} = \epsilon_{O(h)+1}$.

(γ): $O\left(x_{h \cdot x}\right) \geq \sup_{h' \prec h} O\left(x_{h' \cdot x}\right) \stackrel{\text{IH}}{\geq} \sup_{h' \prec h} \epsilon_{O(h')} = \epsilon_{O(h)}$. □

---

[2] A more graphic notation for τ_0 is $\epsilon_{\epsilon_{\epsilon_{\cdot_{\cdot_{\cdot_0}}}}} \Big\} \omega$.

Proof (of THEOREM *6).* Define $\epsilon_{0,0} \overset{\text{def}}{=} \epsilon_0$, $\epsilon_{n+1,0} \overset{\text{def}}{=} \epsilon_{\epsilon_{n,0}}$, For the functions ψ_n (as defined in THEOREM 2) we have $\psi_{n+2} \overset{\text{def}}{=} x_{\psi_{n+1}} \succ x_{\psi_n \cdot x}$ so that, for all n, we have $O(\psi_{n+2}) \overset{\text{L. 9}}{\geq} \epsilon_{O(\psi_n)}$. But then (by induction on n) $O(\psi_{2n+2}) \geq \epsilon_{O(\psi_{2n})} \overset{\text{IH}}{\geq} \epsilon_{\epsilon_{n,0}} \overset{\text{def}}{=} \epsilon_{n+1,0}$. That clearly $\tau_0 = \sup_{n<\omega} \epsilon_{n,0}$ concludes the proof. □

Corollary 1. $O(S^*, \preceq) \leq O(S_*, \preceq)$. □

3   Concluding Remarks

First observe that if

$$\forall_{f \in S_*} \exists_{n<\omega} \left(O(x_{f+1}) \leq O(x_f)_{(n)} \right) \qquad (\dagger)$$

holds, then $O(S_* \preceq) = \tau_0$, and we hope to settle ($\dagger$) in the near future.

Secondly, this article represents just an 'initial segment' of our research on the majorisation relation: why stop at tetration? *Pentration* (iterated tetration), *hexation* (iterated pentration) etc. yield a hierachy majorisation-wise cofinal in the primitive recursive functions. If $\mathcal{A}_n^-$ allows n-tration with base x, and $\mathcal{A}_n$ general n-tration (i.e. $S = \mathcal{A}_3^-$, $S^* = \mathcal{A}_3$, $S_* = \mathcal{A}_4^-$), one may ask what kind of order $(\mathcal{A}_n^{[-]}, \preceq)$ is. We plan to extend the work presented here to $\mathcal{A}^- \overset{\text{def}}{=} \bigcup_{n \in \mathbb{N}} \mathcal{A}_n^-$, and study related questions for $\mathcal{A} \overset{\text{def}}{=} \bigcup_{n \in \mathbb{N}} \mathcal{A}_n$ in a forthcoming paper. In particular, as $\epsilon_0 = \phi(1,0)$ and – if ($\dagger$) holds – $\tau_0 = \phi(2,0)$ (where ϕ is the *Veblen-function*), we would be pleased to discover that $O(\mathcal{A}_{n+2}^-) = \phi(n,0)$.

References

[Ehr73]   Ehrenfeucht, A.: Polynomial functions with Exponentiation are well ordered. Algebra Universalis 3, 261–262 (1973)

[Kru60]   Kruskal, J.: Well-quasi-orderings, the tree theorem, and Vazsonyi's conjecture. Trans. Amer. Math. Soc. 95, 261–262 (1960)

[Lev75]   Levitz, H.: An ordered set of arithmetic functions representing the least ϵ-number. Z. Math. Logik Grundlag. Math. 21, 115–120 (1975)

[Lev77]   Levitz, H.: An initial segment of the set of polynomial functions with exponentiation. Algebra Universalis 7, 133–136 (1977)

[Lev78]   Levitz, H.: An ordinal bound for the set of polynomial functions with exponentiation. Algebra Universalis 8, 233–243 (1978)

[Ric69]   Richardson, D.: Solution to the identity problem for integral exponential functions. Z. Math. Logik Grundlag. Math. 15, 333–340 (1978)

[Sie65]   Sierpiński, W.: Cardinal and Ordinal Numbers. PWN-Polish Scientific Publishers, Warszawa (1965)

[Sko56]   Skolem, T.: An ordered set of arithmetic functions representing the least ϵ-number. Det Kongelige Norske Videnskabers selskabs Forhandlinger 29(12), 54–59 (1956)

Structures of Some Strong Reducibilities

David R. Bélanger[*]

University of Waterloo
Waterloo, Ontario N2L 3G1, Canada
dbelange@math.uwaterloo.ca

Abstract. Recent developments in computability theory have given rise to new tools, among them the ibT and cl reducibilities, which, applications aside, are somewhat mysterious and deserve to be studied *per se*. This paper aims to throw some light on the little-explored degree structures induced by these new reducibilities.

1 Introduction

The cl reducibility was devised by Downey, Hirschfeldt, and LaForte as a possible tool (or definition) for the study and quantification of relative randomness. This plan suffered a setback when Lewis and Barmpalias [5] eventually showed that 1-random sets are not cl-maximal. The cl reducibility is a very natural one, however, and it is worthwhile to examine, catalogue, and admire its intrinsic properties. We also study the closely-related ibT reducibility, which was introduced by Soare.

Definition 1 (Downey-Hirschfeldt-LaForte [4]). *We say that A is* **computably Lipschitz reducible** *to B, written $A \leq_{cl} B$, if there is a Turing functional Γ and a constant $c \in \omega$ such that for all n, $\mathrm{use}_\Gamma(n) \leq n + c$.*

Definition 2 (Soare [8]). *Let $A, B \subseteq \omega$ be two sets of natural numbers. We say that A is* **identity-bounded Turing reducible** *to B, written $A \leq_{ibT} B$, if there is a Turing functional Γ with use function use_Γ such that $A = \Gamma^B$ and, for all n, $\mathrm{use}_\Gamma(n) \leq n$.*

The computably Lipschitz reducibility ($\leq_{cl}$) was first named the **strong weak truth-table reducibility** ($\leq_{sw}$), and has also been called **linear reducibility** ($\leq_\ell$) by Lewis and Barmpalias [5]; the term *computably Lipschitz* was introduced by Barmpalias and Lewis [2] after its behaviour on the topology of the Cantor space 2^ω, and this notation is adopted in Downey-Hirschfeldt [3]. That the ibT and cl reducibilities are inherently related is obvious: after all, an ibT reduction is exactly a cl reduction with constant $c = 0$. One of the goals of this paper is to elucidate the similarities, differences, and interdependencies of these two notions.

---

[*] I am deeply indebted to my Master's advisor, Barbara F. Csima, for her patient support and advice.

K. Ambos-Spies, B. Löwe, and W. Merkle (Eds.): CiE 2009, LNCS 5635, pp. 21–30, 2009.

Each of ($\leq_{ibT}$) and ($\leq_{cl}$) represents a preordering on the subsets 2^ω of the natural numbers and, as we are slowly discovering, induces a rich degree structure on its equivalence classes. As we investigate their properties, it is expedient to look to the body of known results on Turing degrees—the best-known and most-studied such structure in computability—for inspiration or technique. For example, a classical result of Spector [9] states that there is a minimal (non-computable) Turing degree, i.e., a noncomputable degree with no noncomputable degree strictly Turing below it. We might ask the analogous question: Are there minimal ibT or cl degrees? The answer, after some reflection, is negative. Here we use a natural pair of counter-examples noted in Barmpalias and Lewis [2]:

Proposition 1. *If A is a non-computable set, and if we write $A + 1 = \{x + 1 : x \in A\}$ and $2A = \{2x : x \in A\}$, then $\emptyset <_{ibT} A + 1 <_{ibT} A$ and $\emptyset <_{cl} 2A <_{cl} A$.*

We might similarly wish to seek *maximal* or *complete* degrees. There is no maximal Turing degree, since any degree a is strictly below its jump a'. When we restrict the ordering to those degrees containing a particular class of sets, however, we may be left with maximal degrees; the well-known Shoenfield Limit Lemma [6], for example, tells us that the degree $\mathbf{0}'$ of the halting set is Turing maximal and complete among the Δ_2^0 sets. Restriction along these lines will yield a host of more-or-less natural questions to ask about a degree structure: Yu and Ding [11] settled one such by showing there is no cl-complete element among the left-c.e. reals; Barmpalias [1] solved another by proving the non-existence of a cl-maximal element among the c.e. sets. Lewis-Barmpalias later showed in [5] that there are no cl-maximal sets in general. We shall, in Section 2, offer a new proof of this last result, and obtain corollaries on certain classes, including the Δ_2^0 and the Martin-Löf random sets.

Changing our focus slightly, we might examine the structure under one reducibility within a single degree of another. For example, the wtt reducibility is stronger than the Turing, and so each wtt degree is contained entirely within a Turing degree; on the other hand, there exist non-trivial *contiguous* Turing degrees, in which the reverse holds, that is, the wtt and Turing degrees co-incide. There is no exact analogue of contiguity between the cl and ibT degrees, or the wtt and cl degrees, as evidenced by the constructions in Proposition 1. These constructions, however, will generate only linear orderings. In Section 3 we investigate what sort of linear orderings can be thus embedded, and in Section 4 we do what we can to salvage some weakened notion of contiguity by searching for a degree in which all sets are totally ordered with respect to another reducibility.

We always equate a set A with its characteristic function as an infinite binary sequence:

$$A = (A(0), A(1), A(2), \cdots) .$$

An initial segment of A will be denoted by $A \upharpoonright x = A \cap [0, x]$. We use the symbols $\exists^\infty$ and $\forall^\infty$ to say "for infinitely many" and "for all but finitely many," respectively.

2 cl-Maximality

We begin with a proof of a relatively simple fact. The reader will notice that it is nothing but one of the transformations of Proposition 1 performed backward. We show its correctness in full to motivate and illuminate our strategy against more exciting problems.

Proposition 2. *There are no ibT-maximal sets.*

Proof. Suppose we have some $A \subseteq \omega$. If A is computable, then A is not maximal since, for example, $A <_{ibT} \emptyset'$. So, assume that A is not computable. Let the set W be A shifted one space to the left as a binary string, namely,

$$W = A \dot{-} 1 = \{a \dot{-} 1 : a \in A\} \ .$$

We shall now see that $A <_{ibT} W$.

Claim. $A \leq_{ibT} W$.
    Construct a computable functional Γ such that, on oracle W,

$$\Gamma^W(x) = \begin{cases} A(0) & \text{if } x = 0 \\ W(x-1) & \text{otherwise} \end{cases}$$

Then $A = \Gamma^W$ and this reduction is identity-bounded.

Claim. $W \nleq_{ibT} A$.
    Suppose, to the contrary, that $W \leq_{ibT} A$ via computable functional Γ_1. Then, for all $x \geq 0$, Γ_1 computes $A(x+1) = W(x)$ using $A \restriction x$. By induction, then, Γ_1 can compute any $A(x+1)$ using only an oracle to $A(0)$. This implies that A is computable, a contradiction.

In almost any other context, A and W would be indistinguishable—that is, the sets A and W occur simultaneously in just about any class that we could want. It is clear, for example, that A and B are share the same Turing degree. With little effort, we can get a small army of small results.

Corollary 1. *There are no ibT-maximal sets when restricted to the classes of c.e. sets or Δ_2^0 sets.*

Proof. If $(A_s)_{s\in\omega}$ is a computable approximation for A, then $(A_s \dot{-} 1)_{s\in\omega}$ is a computable approximation for $A \dot{-} 1$.

To achieve our goal here, we dropped a finite amount of information—namely, $A(0)$—from the set A, shifting the rest over to take its place, and then argued that, if there were a machine that could always tell what the next element of our new set would be based only on previous elements, then A must have been computable. Next, we wish to apply the same sort of strategy to the cl-ordering. Because in a cl-reduction the machine may skip any finite number of entries, we must toss out an infinite amount of information—which, in general, we might not be able to recover. In certain cases, however, we can remove a *well-behaved* infinite subset and not feel the loss. We say that a set A is **bi-immune** if neither A nor $\overline{A}$ contains an infinite computable subset.

Lemma 1. *If a set A is not bi-immune, then A is not cl-maximal.*

Proof. If A is computable, then $A <_{cl} \emptyset'$. So, assume that A is not computable.

First, say $B \subseteq A$ is a computable subset. Then, we may define a set W by:

$$W(x) = A\left(x + |B \cap [0,x]|\right),$$

yielding $A \leq_{cl} W$ by the Turing functional:

$$\Gamma^W(x) = \begin{cases} 1 & \text{if } x \in B \\ W\left(x - |B \cap [0,x]|\right) & \text{otherwise} \end{cases}$$

Claim. $W \nleq_{cl} A$.

Suppose that $W = \Gamma_1^A$ for some Γ_1 with use bounded by identity plus a constant c. Let N be the $(c+1)$-th smallest element of B. For each $n \geq N$, $A(n)$ can be computed by Γ using oracles to only $W \upharpoonright (n - c - 1)$, which can, in turn, be computed by Γ_1 using only $A \upharpoonright (n - 1)$. By induction, then, any element of A can be computed by Γ and Γ_1 using only the finite information in $A \upharpoonright N$. So A is computable, a contradiction.

If $\overline{A}$ has a computable subset C, then we may construct a W_0 as above such that $\overline{A} <_{cl} W_0$. Hence $A <_{cl} W_0$.

This gives immediately an alternate proof of a theorem of Barmpalias:

Corollary 2 (Barmpalias [1]). *There are no cl-maximal c.e. sets. That is, for every c.e. set A, there exists a c.e. set W such that $A <_{cl} W$.*

Proof. If A is computable, we are done; otherwise, we can effectively find a computable subset $B \subseteq A$. By Lemma 1, we can construct a set W such that $A <_{cl} W$.

Claim. W is c.e..

Take a c.e. approximation $(A_s)_{s \in \omega}$ of A. We can construct a c.e. approximation for W:

$$W_s(x) = A_s\left(x + |B \cap [0,x]|\right).$$

This can be further generalised to other classes not containing bi-immune sets, for example:

Corollary 3. *There are no cl-maximal n-c.e. sets for any $n \geq 1$.*

Sadly, this precise method can only take us so far: there *do* exist bi-immune sets, even among such relatively tame classes as the Δ_2^0 sets. We shall redress the proof by creating two sets U and V such that if $A \nleq_{cl} U$ then $A <_{cl} V$. U will resemble the W from Proposition 2, except that rather than shifting the whole sequence, we shift a certain *subsequence* whose entries become more and more spread out. V will be made from A by removing this subsequence entirely and shifting the rest to compensate, in the manner of W from Lemma 1. This will give a new proof of a result of Lewis and Barmpalias:

Proposition 3 (Lewis-Barmpalias [5]). *There are no cl-maximal sets.*

Proof. Take a set A and consider it as a binary string. Assume, once again, that A is not computable.

For $a = 0, 1, 2, 3, \ldots$ let

$$n_a = \frac{a(a+1)}{2} - 1 = -1 + (0 + 1 + 2 + 3 + \cdots + a) .$$

This gives $n_0 = 0$, $n_1 = 2$, $n_2 = 5$, and so on, with $n_a = n_{a-1} + a + 1$.

Construct a new set U from A by moving back successive n_a:

$$U(x) = \begin{cases} A(n_{a+1}) & \text{if } x = n_a \text{ for some } a \\ A(x) & \text{otherwise} \end{cases}$$

Then $A \leq_{cl} U$. If $U \not\leq_{cl} A$, then we are done; suppose that $U \leq_{cl} A$, via some functional Γ with use bounded by identity plus constant $c \geq 1$. Construct a new set V from A by deleting each n_a-th entry:

$$V(x) = A\left(x + |\{n_a : a \in \omega\} \cap [0, x]|\right)$$

Claim. $A \leq_{cl} V$.

For any $m \notin \{n_a : a \in \omega\}$, we can just read off

$$A(m) = V\left(m - |\{n_a : a \in \omega\} \cap [0, x]|\right) .$$

Given $m = n_a$ with $a \geq c$, we may use Γ to decide $A(n_a) = U(n_{a-1})$ from only the initial segment

$$A{\upharpoonright}(n_{a-1} + c) \subseteq A{\upharpoonright}(n_a - 1) ,$$

so, by induction, all of A can be solved-for using Γ, an identity-bounded oracle to V, and the finite information of $A{\upharpoonright}n_c$.

Claim. $V \not\leq_{cl} A$.

Suppose that $V \leq_{cl} A$ by Turing functional Γ_1 with use bounded by identity plus d. Suppose that $m > n_{\max\{c,d\}}$. If $m = n_a$ for some a, then we can compute $A(m) = U(n_{a-1})$ using Γ from

$$A{\upharpoonright}(n_{a-1} + c) \subseteq A{\upharpoonright}(m - 1) .$$

For all other m, we can apply Γ_1 to obtain $A(m) = V(m - |\{a \in \omega : n_a \leq m\}|)$ from

$$A{\upharpoonright}\left(m - |\{a \in \omega : n_a \leq m\}| + d\right) \subseteq A{\upharpoonright}(m - 1) .$$

By induction on m we conclude that Γ and Γ_1, using the finite information in $A{\upharpoonright}n_{\max\{c,d\}}$, can compute A—a contradiction.

We may use this construction, as we did with Lemma 1, to build from any computable approximation $(A_s)_{s \in \omega}$ two new computable approximations $(U_s)_{s \in \omega}$ and $(V_s)_{s \in \omega}$ for U and V, respectively. Further, any bound on the number of changes of an element of $(A_s)_{s \in \omega}$ can be transformed computably into a bound for $(U_s)_{s \in \omega}$ and $(V_s)_{s \in \omega}$. Hence:

Corollary 4. *There are no cl-maximal elements among the Δ_2^0 sets or the ω-c.e. sets.*

We can generalise in another direction by performing the same computable transformation on any computable relation:

Corollary 5. *There are no cl-maximal sets among any of the Σ_n^0, Π_n^0, or Δ_{n+1}^0 sets, for $n \geq 1$.*

There is no reason to stop here, of course. The cl-reducibility leaves something to be desired as a comparison of randomness—each 1-random set is strictly cl-below some other set. The cl reducibility caught the interest of Downey, Hirschfeldt, and LaForte [4] when they noticed that it preserves randomness: if A is 1-random and $A \leq_{cl} B$, then B is 1-random. For if not, we should have a machine M that, with B as an oracle, can cl-compute A with constant c, and a prefix-free machine N such that $(\forall d \exists n)[K_N(B \restriction n) < n - d]$ holds; so, composing the machines so that M acts on the output of N, $(\forall d \exists n)[K_{MN}(A \restriction n) < (n + c) - d]$ holds, contradicting the randomness of A. Hence, by Proposition 3, every 1-random set is strictly cl-below another 1-random set.

This can be further sharpened to deal with n-random sets. Van Lambalgen in [10] proved a theorem characterising A and B such that $A \oplus B$ is random. We give here a generalisation roughly as it appears in Downey-Hirschfeldt [3]:

Theorem 1 (van Lambalgen). *Given infinite sets C and D, and an infinite, coinfinite, computable set X, with $X = \{x_1 < x_2 < \cdots\}$ and $\overline{X} = \{y_1 < y_2 < \cdots\}$, let $C \oplus_X D = \{x_n : n \in C\} \cup \{y_n : n \in D\}$.*
$C \oplus_X D$ is n-random $\iff$ C and D are relatively n-random.

This implies, in particular, that if $C \oplus_X D$ is n-random, then C and D are each n-random.

Corollary 6. *Every n-random set A is strictly cl-below another n-random set B, and this B can be found uniformly.*

Proof. Take any n-random set A. Construct U and V from A as in Proposition 3. Now, taking the set of all triangular numbers $X = \{n_1, n_2, \ldots\} = \{1, 3, 6, 10, \ldots\}$, there exist sets C and D such that $A = C \oplus_X D$. It is easy to see that $U = (C \dot{-} 1) \oplus_X D$ and $V = D$. By Theorem 1, both U and V are n-random. We can conclude already that A is strictly cl-below another n-random set.

To get uniformity, we verify that $A <_{cl} U$ in all cases. By our construction in Proposition 3, it will suffice to show that $A \not\leq_{cl} V$. Suppose that $A \leq_{cl} V$ through some functional Γ. Then, in particular, Γ computes each $A(n_\alpha)$ using V as an oracle. Hence, $C \leq_T D$—but from here we can easily see, through the Kolmogorov definition, that C and D are not relatively 1-random, much less relatively n-random, contradicting Theorem 1. Therefore, $A \not\leq_{cl} V$, giving $A \not\leq_{cl} V$; so $A <_{cl} U$.

3   Linear Orderings

Proposition 4. *The rationals $\mathbb{Q}$ can be embedded as a linear ordering into the cl-degrees. In fact, this can be done within any non-computable wtt-degree $\mathbf{a}$, and if $\mathbf{a}$ is c.e. then so is each cl degree in the image of $\mathbb{Q}$.*

Proof. Recall Proposition 1, wherein we multiplied a given noncomputable set A by a factor of 2 to *stretch* it and get $A >_{cl} 2A$. We could repeat this method for successive powers of two to get an infinite sequence $A >_{cl} 2A >_{cl} 4A >_{cl} 8A >_{cl} \cdots$, but we can see immediately that every positive natural number can be included: $A >_{cl} 2A >_{cl} 3A >_{cl} 4A >_{cl} \cdots$. The obvious way, then, to get the order type of $\mathbb{Q}$ would be to multiply A by rational coefficients. While it does not make sense to multiply these sets by a negative number, say, the intuition is largely correct, and, other than some technicalities with rounding, the construction we shall use is essentially that in Proposition 1 repeated infinitely many times.

It suffices to embed the open interval $(0,1) \cap \mathbb{Q}$, which is isomorphic to $\mathbb{Q}$ as a linear ordering. By reversing the relation we may, in fact, embed it backwards, and shall do just that in order to simplify the arithmetic. For each $q \in (0,1) \cap \mathbb{Q}$, define the corresponding function $f_q : \omega \to \omega$ by $f_q(x) = \lfloor (1+q) \cdot x \rfloor$. This has a left-inverse $g_q(x) = \left\lceil \frac{x}{1+q} \right\rceil$ (to see this, note that $0 \leq (1+q) \cdot x - f_q(x) < 1$, and hence $0 \leq x - \frac{f_q(x)}{1+q} < 1$). These f_q and g_q are computable, so for any set A, $A \equiv_T f_q(A)$.

Now, if $q \leq s$ with $q, s \in (0,1) \cap \mathbb{Q}$, then $f_s(A) = f_s(g_p(f_p(A)))$; thus, $f_s(A) \leq_{ibT} f_q(A)$. On the other hand, if $q < s$ and $f_q(A) \leq_{cl} f_s(A)$, then, as usual, we can conclude that A was computable: simply take Γ and c such that $f_q(A) = \Gamma^{f_s(A)}$ and $\gamma^{f_s(A)} \leq id + c$, let $d = \left\lceil \frac{c}{s-q} \right\rceil + 1$, and notice that, for all $n \geq d$, $A{\upharpoonright}n$ is computable via g_q from $f_q(A){\upharpoonright}f_q(n)$, which is computable via Γ from $f_s(A){\upharpoonright}f_q(n) + c$, which is computable via f_s from $A{\upharpoonright}g_s(f_q(n) + c)$—but $f_s(n) > f_q(n) + c$ by our choice of n, so this is, in fact, a proper initial segment of $A{\upharpoonright}n$. As in previous proofs, by iterating this method we can compute all of A from the initial segment $A{\upharpoonright}d$. Therefore, if A is not computable, then we can embed $\mathbb{Q}$ into its wtt degree.

Corollary 7. *Any countable linear ordering can be so embedded.*

4   Incomparable c.e. Sets

We wish to continue our examination of the relationships between the ibT, cl, and wtt degrees. We already know a fair bit about linear orderings embedded between them. What about nonlinear ones? Which can be embedded into some cl degree, and which can be embedded into *any* non-trivial cl degree? In order to decide whether this second question even makes sense, we first must decide: Does there exist a cl degree in which *all* sets are ibT-comparable? In this section, we grope towards a definitive answer.

As usual, it is interesting and useful to consider the restriction to the c.e. sets. In order to construct c.e. sets from other c.e. sets, we make heavy use of the classical technique of permitting.

Proposition 5. *Each nonzero c.e. wtt-degree contains a pair U, V of c.e. sets such that $U|_{ibT}V$.*

Proof. We shall use a combination of priority and permitting strategies.

Suppose A is a non-computable c.e. set with c.e. approximating sequence $(A_s)_{s\in\omega}$. Assume that, for all s, $|A_{s+1} \setminus A_s| \leq 1$, and $\max\{n : n \in A_s\} \leq s$. We shall construct two new sets U, V such that $U \equiv_T A \equiv_{wtt} V$ but $U|_{ibT}V$.

Because we are dealing mainly with ibT reductions, we let $(\Phi_e)_{e\in\omega}$ denote a computable enumeration of ibT functionals. To obtain this, we take an enumeration $(\hat{\Phi}_e)_{e\in\omega}$ of Turing fuctionals and define, for each e:

$$\Phi_e^X(n) = \begin{cases} \hat{\Phi}_e^X(n) & \text{if } \hat{\Phi}_e^X(n)\downarrow \text{ with use} \hat{\Phi}_e^X(n) \leq n \\ \uparrow & \text{otherwise} \end{cases}$$

We may assume for convenience' sake, as in, for example, Soare [7], that each functional appears infinitely many times in both the odd and the even indices.

Requirements

$$\begin{array}{rl} \mathcal{P}: & A\leq_{wtt}U,V \\ \mathcal{R}: & U,V\leq_{wtt}A \\ \mathcal{N}_{2e}: & U \neq \Phi_e^V \\ \mathcal{N}_{2e+1}: & V \neq \Phi_e^U \end{array}$$

We achieve $\mathcal{P}$ by coding a spread-out version of A directly into U and V; we prevent $U\leq_{ibT}V$ and $V\leq_{ibT}U$ by diagonalising against each such ibT-reduction. We assign priorities to the $\mathcal{N}_e$ in the usual, descending order; $\mathcal{R}$ will be guaranteed through permitting. Note that our construction differs somewhat from so-called *standard permitting* in that, rather than waiting for any $m \leq n$ to enter A before allowing n to enter U or V, we wait for one of a very specific set of m.

To govern the permitting argument, we create, for each $\mathcal{N}_e$ and each stage s, a restraint $\alpha_{e,s}$ that shall bound any diagonalisation against $\mathcal{N}_e$ at stage s. If, at stage s, $\mathcal{N}_e$ appears to be satisfied, and $\mathcal{N}_i$ is satisfied and has stopped acting for all $i < e$, subsequent $\alpha_{e,t}$ shall be less than or equal to $\alpha_{e,s}$. Because we are constructing c.e. approximations, this means that, once satisfied, $\mathcal{N}_e$ will act on its witnesses only a finite number of times, and hence, by induction, any requirement will be injured only a finite number of times.

Our construction is further complicated by the need to provide an adequate number of witnesses, in the appropriate positions, to each $\mathcal{N}_e$. We are forced to introduce them slowly. Denote the coding position of $A(n)$ in U and V by $\lambda(n)$, so that for all n $U(\lambda(n)) = A(n) = V(\lambda(n))$. We insert one witness position (for $\mathcal{N}_0$) before $\lambda(0)$, two (one for $\mathcal{N}_0$ and one for $\mathcal{N}_1$) between $\lambda(0)$ and $\lambda(1)$, ..., and $n + 2$ (for each $\mathcal{N}_e$, $e \leq n + 1$) between $\lambda(n)$ and $\lambda(n + 1)$. A simple calculation gives the closed form $\lambda(n) = \frac{n^2 + 5n + 2}{2}$; for convenience, we also assign

$\lambda(-1) = -1$. As for the witnesses, we just allotted one to $\mathcal{N}_e$ between $\lambda(n-1)$ and $\lambda(n)$ for every $n \geq e$. We write the coding positions for $\mathcal{N}_e$ as the partial function $\Lambda_e : \omega^{\geq e} \to \omega$, $\Lambda_e(n) = \lambda(n-1) + e$. Note that $\Lambda_0(0) < \lambda(0) < \Lambda_0(1) < \Lambda_1(1) < \lambda(1) < \Lambda_0(2) < \Lambda_1(2) < \Lambda_2(2) < \lambda(2) < \cdots$.

Here are the necessary restraints:

$$\alpha_{-1,s} = -1$$

$$\alpha_{2e,s} = \max\{n \leq s : (\forall k \text{ s.t. } \alpha_{2e-1,s} \leq k < n \text{ and } 2e \leq k)[\Phi_{2e,s}^{V_s}(\Lambda_{2e}(k))\downarrow = U_s(\Lambda_{2e}(k))]\}$$

$$\alpha_{2e+1,s} = \max\{n \leq s : (\forall k \text{ s.t. } \alpha_{2e,s} \leq k < n \text{ and } 2e+1 \leq k)[$$
$$\Phi_{2e+1,s}^{U_s}(\Lambda_{2e+1}(k))\downarrow = V_s(\Lambda_{2e+1}(k))]\}$$

Strategy

When $n \in A_s \setminus A_{s-1}$, enumerate $U_s(\lambda(n)) = V_s(\lambda(n)) = 1$.

When such an n enters with $\alpha_{2e-1}, 2e \leq n < \alpha_{2e}$, define $U_s(\Lambda_{2e}(n)) = 1 \dot{-} \Phi_{2e,s}^{V_s}(\Lambda_{2e}(n))$.

When such an n enters with $\alpha_{2e}, 2e+1 \leq n < \alpha_{2e+1}$, let $V_s(\Lambda_{2e+1}(n)) = 1 \dot{-} \Phi_{2e+1,s}^{U_s}(\Lambda_{2e+1}(n))$.

Verification

Because of the direct coding of A into U and V by λ, $\mathcal{P}$ is immediate. Because n enters U or V only if some $m \leq n$ enters A, $\mathcal{R}$ is also satisfied. U and V are c.e. because they are built by stages, and each entry, being in the image of exactly one of λ and $\{\Lambda_e\}$, is changed during at most one stage. We establish $\mathcal{N}_e$ by induction.

For each i, let $\alpha_i = \lim_s \alpha_{i,s}$. We claim that this limit exists for all i. Suppose, for contradiction, that e is the smallest i such that $\lim_s \alpha_{i,s}$ does not exist. Assume further that e is even, the odd case being symmetrical. Let s_0 be such that $s \geq s_0$ implies $(\forall i < e)[\alpha_{i,s} = \alpha_e]$. Define the **settling time** or **least modulus** of $(A_s)_{s \in \omega}$ to be the function:

$$\text{set}(n) = (\mu t)(\forall s \geq t)\,[A\!\restriction\! n = A_s\!\restriction\! n]\ .$$

Then choose some $s_1 \geq s_0$ such that s_1 is greater than $\text{set}(\max\{e-1, \alpha_{e-1}-1\})$. Suppose there is a $t_0 \geq s_1$ such that $\Phi_{e,t_0}^{V_{t_0}}(\Lambda_e(\alpha_{e,t_0}))\downarrow \neq U_{t_0}(\Lambda_e(\alpha_{e,t_0}))$. The only way $\alpha_{e,t}$ can then differ, with $t > t_0$, is if, at stage t, some new element $n \leq \Lambda_e(\alpha_{e,t_0})$ enters $V_t \setminus V_{t-1}$. By our construction, then, there must also have been an $m \leq \alpha_{e,t}$ entering $A_t \setminus A_{t-1}$. We cannot have $m < e$ or $m < \alpha_{e-1} = \alpha_{e-1,t}$, by our choice of s_1. We thus have $e, \alpha_{e-1,t} \leq m < \alpha_{e,t}$. Our construction dictates that, at stage t, we diagonalise at $U_t(\Lambda_e(m))$; thus, at the next stage, $\alpha_{e,t+1} = m < \alpha_{e,t}$. This means that the sequence $(\alpha_{e,s})_{s \geq t_0}$ is decreasing and bounded below by α_{e-1}; therefore, the limit α_e exists.

Suppose next that there is no such t_0—that is, for all $s \geq s_1$, $\Phi_{e,s}^{V_s}(\Lambda_e(\alpha_e, s))\uparrow$. First, say there is some stage $t_1 \geq s_1$ where $(\forall t \geq t_1)[\Phi_{e,t}^{V_t}(\Lambda_e(\alpha_{e,t_1}))\uparrow]$. In this case, for all $t \geq t_1$, $\alpha_{e,t} \leq \alpha_{e,t_1}$. Then, for all $t \geq \max\{t_1, \text{set}(\alpha_{e,t_1})\}$, we have $\alpha_{e,t+1} = \alpha_{e,t}$, so that the limit α_e exists. If there is no such t_1, then we can define a computable function $f(n) = (\mu s \geq s_1)[\alpha_{e,s} \geq n]$ (defined on

$n \geq \max\{e, \alpha_{e-1}\}$). Then, $(\forall^\infty n)[\mathrm{set}(n) \leq f(n)]$, giving $(\forall^\infty n)[A{\restriction}n = A_{f(n)}{\restriction}n]$, so that A is computable—a contradiction.

That each sequence $(\alpha_{e,s})_{s\in\omega}$ converges means that—by definition of $\alpha_{e,s}$ and the c.e. approximating sequences $(U_s)_{s\in\omega}$ and $(V_s)_{s\in\omega}$—each requirement $\mathcal{N}_e$ is eventually satisfied.

By varying this construction we can produce variations in the result, such as:

Corollary 8. *Within any nonzero c.e. wtt-degree there exist two c.e. sets U, V such that $U|_{cl}V$.*

Corollary 9. *Within any nonzero c.e. cl-degree there exist two c.e. reals U, V such that $U|_{ibT}V$.*

References

1. Barmpalias, G.: Computably enumerable sets in the solovay and the strong weak truth table degrees. In: Cooper, S.B., Löwe, B., Torenvliet, L. (eds.) CiE 2005. LNCS, vol. 3526, pp. 8–17. Springer, Heidelberg (2005)
2. Barmpalias, G., Lewis, A.E.M.: The ibT degrees of computably enumerable sets are not dense. Ann. Pure Appl. Logic 141(1-2), 51–60 (2006)
3. Downey, R., Hirschfeldt, D.R.: Algorithmic Randomness and Complexity. Unpublished monograph
4. Downey, R.G., Hirschfeldt, D.R., LaForte, G.: Randomness and reducibility. J. Comput. System Sci. 68(1), 96–114 (2004)
5. Lewis, A.E.M., Barmpalias, G.: Randomness and the linear degrees of computability. Ann. Pure Appl. Logic 145(3), 252–257 (2007)
6. Shoenfield, J.R.: On degrees of unsolvability. Ann. of Math. 69(3), 644–653 (1959)
7. Soare, R.I.: Computability Theory and Applications. Unpublished monograph
8. Soare, R.I.: Computability theory and differential geometry. Bull. Symbolic Logic 10(4), 457–486 (2004)
9. Spector, C.: On degrees of recursive unsolvability. Ann. of Math. 64(2), 581–592 (1956)
10. van Lambalgen, M.: Random sequences. PhD thesis, University of Amsterdam (1987)
11. Yu, L., Ding, D.: There is no SW-complete c.e. real. J. Symbolic Logic 69(4), 1163–1170 (2004)

Complexity of Existential Positive First-Order Logic

Manuel Bodirsky, Miki Hermann, and Florian Richoux

LIX (CNRS, UMR 7161), École Polytechnique, 91128 Palaiseau, France
{bodirsky,hermann,richoux}@lix.polytechnique.fr

Abstract. Let Γ be a (not necessarily finite) structure with a finite relational signature. We prove that deciding whether a given existential positive sentence holds in Γ is in LOGSPACE or complete for the class $\mathrm{CSP}(\Gamma)_{\mathrm{NP}}$ under deterministic polynomial-time many-one reductions. Here, $\mathrm{CSP}(\Gamma)_{\mathrm{NP}}$ is the class of problems that can be reduced to the *constraint satisfaction problem* of Γ under *non-deterministic* polynomial-time many-one reductions.

Keywords: Computational Complexity, Existential Positive First-Order Logic, Constraint Satisfaction Problems.

1 Introduction

We study the computational complexity of the following class of computational problems. Let Γ be a structure with finite or infinite domain and with a finite relational signature. The model-checking problem for existential positive first-order logic, parametrized by Γ, is the following problem.

Problem: ExPos(Γ)
Input: An existential positive first-order sentence Φ.
Question: Does Γ satisfy Φ?

A first-order sentence is *existential positive* if it does not contain universal quantifiers and negation symbols, that is, if the only logical connectives are existential quantifiers, disjunction, conjunction, and equality. The sentence does not need to be in prenex normal form; however, every existential positive first-order sentence can be transformed in an equivalent one in this form without an exponential blowup, thanks to the absence of universal quantifiers and negation symbols.

The *constraint satisfaction problem* $\mathrm{CSP}(\Gamma)$ for Γ is defined similarly, but its input consists of a *primitive positive* sentence, that is, a sentence without universal quantifiers, negation, and disjunction. Constraint satisfaction problems frequently appear in many areas of computer science, and have attracted a lot of attention, in particular in combinatorics, artificial intelligence, and finite model theory; we refer to the recent monograph with survey articles on this subject [7]. The class of constraint satisfaction problems for infinite structures Gamma is a rich class of problems; it can be shown that for every computational problem there exists a relational structure Gamma such that CSP(Gamma) is equivalent to that problem under polynomial-time Turing reductions [1].

K. Ambos-Spies, B. Löwe, and W. Merkle (Eds.): CiE 2009, LNCS 5635, pp. 31–36, 2009.
© Springer-Verlag Berlin Heidelberg 2009

In this paper, we show that the complexity classification for existential positive first-order sentences over infinite structures can be reduced to the complexity classification for constraint satisfaction problems. For finite structures Γ, our result implies that $\text{ExPos}(\Gamma)$ is in LOGSPACE or NP-complete.

For finite Γ, the polynomial-time solvable cases of $\text{ExPos}(\Gamma)$ are precisely those relational structures Γ with an element a where all relations in Γ contain the tuple $(a, \ldots, a)$ composed only from the element a; in this case, $\text{ExPos}(\Gamma)$ is called a-*valid*. Interestingly, this is no longer true for infinite structures Γ.

Consider the structure $\Gamma := (\mathbb{N}, \neq)$, which is clearly not a-valid. However, $\text{ExPos}(\Gamma)$ can be reduced to the Boolean formula evaluation problem (which is known to be in LOGSPACE) as follows: atomic formulas in Φ of the form $x \neq y$ are replaced by *true*, and atomic formulas of the form $x \neq x$ are replaced by *false*. The resulting Boolean formula is equivalent to true if and only if Φ is true in Γ.

A universal-algebraic study of the model-checking problem for finite structures Γ and various other syntactic restrictions of first-order logic (for instance positive first-order logic) can be found in [6].

2   Result

We write $L \leq_m L'$ if there exists a deterministic polynomial-time many-one reduction from L to L'.

Definition 1 (from [4]). *A problem A is* non-deterministic polynomial-time many-one reducible *to a problem B ($A \leq_{\text{NP}} B$) iff there is a nondeterministic polynomial-time Turing machine M such that $x \in A$ if and only if there exists a y computed by M on input x with $y \in B$. We denote by A_{NP} the smallest class that contains A and is closed under $\leq_{\text{NP}}$.*

Observe that $\leq_{\text{NP}}$ is transitive [4]. To state the complexity classification for existential positive first-order logic, we need the following concepts. Let Φ be an existential positive τ-sentence for a finite relational signature τ. We construct a Boolean formula $F_\Gamma(\Phi)$ as follows. We first remove all existential quantifiers from Φ. Then we replace each atomic formula in Φ of the form $R(x_1, \ldots, x_k)$ where R denotes the empty k-ary relation over Γ by *false*. All other atomic formulas in Φ will be replaced by *true*. We write $F_\Gamma(\Phi)$ for the resulting Boolean formula. Note that if Φ is true in Γ then $F_\Gamma(\Phi)$ must be logically equivalent to *true*.

Definition 2. *We call a relational structure Γ* locally refutable *if every existential positive sentence Φ is true in Γ if and only if the Boolean formula $F(\Phi)$ (as described above) is logically equivalent to* true.

In Section 3, we will show the following result.

Theorem 3. *Let Γ be a structure with a finite relational signature τ. If Γ is locally refutable then the problem $\text{ExPos}(\Gamma)$ to decide whether an existential positive sentence is true in Γ is in LOGSPACE. If Γ is not locally refutable, then $\text{ExPos}(\Gamma)$ is complete for the class $\text{CSP}(\Gamma)_{\text{NP}}$ under polynomial-time many-one reductions.*

In particular, $\text{EXPOS}(\Gamma)$ is in P or is NP-hard (under deterministic polynomial-time many-one reductions). If Γ is finite, then $\text{EXPOS}(\Gamma)$ is in P or NP-complete, because finite domain constraint satisfaction problems are clearly in NP. The observation that $\text{EXPOS}(\Gamma)$ is in P or NP-complete has previously been made in [3] and independently in [5]. However, our proof remains the same for finite domains and is simpler than proofs in these previous works.

3   Proof

Before we prove Theorem 3, we start with the following simpler result.

Theorem 4. *Let Γ be a structure with a finite relational signature τ. If Γ is locally refutable, then the problem $\text{EXPOS}(\Gamma)$ to decide whether an existential positive sentence is true in Γ is in LOGSPACE. If Γ is not locally refutable, then $\text{EXPOS}(\Gamma)$ is NP-hard (under polynomial-time many-one reductions).*

To show Theorem 4, we first prove the following lemma.

Lemma 5. *If Γ is not locally refutable, then there are existential positive τ-formulas ψ_0 and ψ_1 with the property that*

– ψ_0 *and* ψ_1 *define non-empty relations over* Γ;
– $\psi_0 \wedge \psi_1$ *defines the empty relation over* Γ.

Proof. Because Γ is not locally refutable, there is an unsatisfiable instance Φ of $\text{EXPOS}(\Gamma)$ such that the Boolean formula $F(\Phi)$ described above is logically equivalent to *true*. Among all formulas with this property, let Φ be the one that is of minimal length.

If Φ is of the form $\Phi_1 \vee \Phi_2$ then both Φ_1 and Φ_2 are unsatisfiable over Γ, and one of the Boolean formulas $F_\Gamma(\Phi_1)$ or $F_\Gamma(\Phi_2)$ must be true; this contradicts the assumption that Φ is minimal.

If Φ is of the form $\Phi_1 \wedge \Phi_2$, then both $F_\Gamma(\Phi_1)$ and $F_\Gamma(\Phi_2)$ are true. If both Φ_1 and Φ_2 are satisfiable, then we are done, because we have found two satisfiable existential positive formulas such that their conjunction is unsatisfiable. If Φ_1 or Φ_2 is unsatisfiable over Γ, say Φ_i is unsatisfiable for $i \in \{1, 2\}$, then this contradicts the assumption that Φ is minimal, because Φ_i is smaller than Φ, unsatisfiable over Γ, and $F_\Gamma(\Phi_i)$ is true. If Φ is of the form $\exists x.\Phi'$ then this contradicts obviously the assumption that Φ is minimal. Note that Φ cannot be atomic, because in this case Φ is either unsatisfiable or $F_\Gamma(\Phi)$ is true (but not both). □

Proof of Theorem 4. If Γ is locally refutable, then $\text{EXPOS}(\Gamma)$ can be reduced to the positive Boolean formula evaluation problem, which is known to be LOGSPACE-complete. We only have to construct from an existential positive τ-sentence Φ a Boolean formula $F := F(\Phi)$ as described before Definition 2. Clearly, this construction can be performed with logarithmic work-space. We evaluate F, and reject if F is false, and accept otherwise.

If Γ is not locally refutable, we show NP-hardness of $\mathrm{EXPOS}(\Gamma)$ by reduction from 3-SAT. Let I be a 3-SAT instance. We construct an instance Φ of $\mathrm{EXPOS}(\Gamma)$ as follows. Let ψ_0 and ψ_1 be the formulas from Lemma 5 (suppose they are d-ary). Let $v_1, \ldots, v_n$ be the Boolean variables in I. For each v_i we introduce d new variables $\bar{x}_i = x_i^1, \ldots, x_i^d$. Let Φ be the instance of $\mathrm{EXPOS}(\Gamma)$ that contains the following conjuncts:

- For each $1 \leq i \leq n$, the formula $\psi_0(\bar{x}_i) \vee \psi_1(\bar{x}_i)$
- For each clause $l_1 \vee l_2 \vee l_3$ in I, the formula $\psi_{i_1}(\bar{x}_{j_1}) \vee \psi_{i_2}(\bar{x}_{j_2}) \vee \psi_{i_3}(\bar{x}_{j_3})$ where $i_p = 0$ if l_p equals $\neg x_{j_p}$ and $i_p = 1$ if l_p equals x_{j_p}, for all $p \in \{1, 2, 3\}$.

It is clear that Φ can be computed in deterministic polynomial time from I, and that Φ is true in Γ if and only if I is satisfiable.   □

Note that, applied to finite domain constraint languages Γ, we obtain again the dichotomy from [3] and [5].

Proof of Theorem 3. If Γ is locally refutable then the statement has been shown in Theorem 4. Suppose that Γ is not locally refutable. To show that $\mathrm{EXPOS}(\Gamma)$ is contained in $\mathrm{CSP}(\Gamma)_{\mathrm{NP}}$, we construct a non-deterministic Turing machine T which takes as input an instance Φ of $\mathrm{EXPOS}(\Gamma)$, and which outputs an instance $T(\Phi)$ of $\mathrm{CSP}(\Gamma)$ as follows.

On input Φ the machine T proceeds recursively as follows:

- if Φ is of the form $\exists x.\varphi$ then return $\exists x.T(\varphi)$;
- if Φ is of the form $\varphi_1 \wedge \varphi_2$ then return $T(\varphi_1) \wedge T(\varphi_2)$;
- if Φ is of the form $\varphi_1 \vee \varphi_2$ then non-deterministically return either $T(\varphi_1)$ or $T(\varphi_2)$;
- if Φ is of the form $R(x_1, \ldots, x_k)$ then return $R(x_1, \ldots, x_k)$.

The output of T can be viewed as an instance of $\mathrm{CSP}(\Gamma)$, since it can be transformed to a primitive positive τ-sentence (by moving all existential quantifiers to the front). It is clear that T has polynomial running time, and that Φ is true in Γ if and only if there exists a computation of T on Φ that computes a sentence that is true in Γ.

We now show that $\mathrm{EXPOS}(\Gamma)$ is hard for $\mathrm{CSP}(\Gamma)_{\mathrm{NP}}$ under $\leq_m$-reductions. Let L be a problem with a non-deterministic polynomial-time many-one reduction to $\mathrm{CSP}(\Gamma)$, and let M be the non-deterministic Turing machine that computes the reduction. We have to construct a deterministic Turing machine M' that computes for any input string s in polynomial time in $|s|$ an instance Φ of $\mathrm{EXPOS}(\Gamma)$ such that Φ is true in Γ if and only if there exists a computation of M on s that computes a satisfiable instance of $\mathrm{CSP}(\Gamma)$.

Say that the running time of M on s is in $O(|s|^e)$ for a constant e. Hence, there are constants s_0 and c such that for $|s| > s_0$ the running time of M and hence also the number of constraints in the input instance of $\mathrm{CSP}(\Gamma)$ produced by the reduction is bounded by $t := c|s|^e$. The non-deterministic computation of M can be viewed as a deterministic computation with access to non-deterministic advice bits as shown in [2]. We also know that for $|s| > s_0$, the machine M can access at most t non-deterministic bits. If w is a sufficiently long bit-string, we write M_w for the deterministic Turing machine obtained from M by using the bits in w as the non-deterministic bits, and $M_w(s)$ for the instance of $\mathrm{CSP}(\Gamma)$ computed by M_w on input s.

If $|s| \leq s_0$, then M' returns $\exists x.(x = x)$ if there is an $w \in \{0,1\}^*$ such that $M_w(s)$ is a satisfiable instance of $\mathrm{CSP}(\Gamma)$, and M' returns $\exists \bar{x}.\psi_0(\bar{x}) \wedge \psi_1(\bar{x})$ otherwise (i.e., it returns a false instance of $\mathrm{ExPos}(\Gamma)$; ψ_0 and ψ_1 are defined in Lemma 5). Since s_0 is a fixed finite value, M' can perform these computations in constant time.

It is convenient to assume that Γ has just a single relation R (we can always find a CSP which is deterministic polynomial-time equivalent and where the template is of this form[1]). Let l be the arity of R. Then instances of $\mathrm{CSP}(\Gamma)$ with variables $x_1, \ldots, x_n$ can be encoded as sequences of numbers that are represented by binary strings of length $\lceil \log t \rceil$ as follows: The i-th number m in this sequence indicates that the $(((i-1) \bmod l) + 1)$-st variable in the $(((i - 1) \, div \, l) + 1)$-st constraint is x_m. For $|s| > s_0$, the sentence Φ computed by M' has the form

$$\exists x_1^1, \ldots, x_l^t. \left(\bigwedge_{i=1}^{t} R(x_1^i, \ldots, x_l^i) \wedge \Psi \right). \tag{1}$$

where Ψ is an $\mathrm{ExPos}(\Gamma)$ formula defined below.

The **idea** is that *any* instance of $\mathrm{CSP}(\Gamma)$ computed by the machine M can be obtained by contracting variables in $\bigwedge_{i \leq t} R(x_1^i, \ldots, x_l^i)$. The way this is done is controlled by a Boolean formula that can be computed from the input s of M in polynomial time. The Boolean formula also contains Boolean variables for the non-deterministic advice bits of M. Each Boolean variable v in the formula is simulated by a d-tuple $\bar{x}_v$ of variables in Ψ that is forced to satisfy $\psi_0(\bar{x}_v) \vee \psi_1(\bar{x}_v)$ (ψ_0 and ψ_1 are defined in Lemma 5), similarly as in the proof of Theorem 4, such that $v = 0$ corresponds to falsity of $\psi_0(\bar{x}_v)$, and $v = 1$ corresponds to truth of $\psi_1(\bar{x}_v)$ in Γ. The formula Φ will be such that there exists a computation of M that produces a satisfiable instance I of $\mathrm{CSP}(\Gamma)$ if and only if there exists an assignment to $x_1^1, \ldots, x_l^t$ that satisfies Ψ and such that $\bigwedge_{i \leq t} R(x_1^i, \ldots, x_l^i)$ is equivalent to I.

We now provide the details of the definition of the machine M' that computes Φ. We use a construction from the proof of Cook's theorem given in [2]. In this proof, a computation of a non-deterministic Turing machine T accepting a language L is encoded by Boolean variables that represent the state and the position of the read-write head of T at time r, and the content of the tape at position j at time r. The tape content at time 0 consists of the input x, written at positions 1 through n, and the non-deterministic advice bit string w, written at positions -1 through $-|w|$. The proof in [2] specifies a deterministic polynomial-time computable transformation f_L that computes for a given string s a SAT instance $f_L(s)$ such that there is an accepting computation of T on s if and only if there is a satisfying truth assignment for $f_L(s)$.

In our case, the machine M computes a reduction and thus computes an output string. Recall our binary representation of instances of the CSP: M writes on the output tape a sequence of numbers represented by binary strings of length $\lceil \log t \rceil$. It is straightforward to modify the transformation f_L given in the proof of Theorem 2.1 in [2] to obtain for all positive integers a, a', b, b', c, c' where $a, a' \leq t$, $b, b' \leq l$, $c, c' \leq \lceil \log t \rceil$

---

[1] If $\Gamma = (D; R_1, \ldots, R_n)$ where R_i has arity r_i and is not empty, then $\mathrm{CSP}(\Gamma)$ is equivalent to $\mathrm{CSP}(D; R_1 \times \cdots \times R_n)$ where $R_1 \times \cdots \times R_n$ is the $\sum_{i=1}^{n} r_i$-ary relation defined as the Cartesian product of the relations $R_1, \ldots, R_n$. Similarly, $\mathrm{ExPos}(\Gamma)$ is equivalent to $\mathrm{ExPos}(D; R_1 \times \cdots \times R_n)$.

a deterministic polynomial-time transformation $g_{a,a',b,b',c,c'}$ that computes for a given string s a SAT instance $g_{a,a',b,b',c,c'}(s)$ with distinguished variables $z_1, \ldots, z_t$ (for the non-deterministic bits in the computation of M) such that the following are equivalent:

- $g_{a,a',b,b',c,c'}(s)$ has a satisfying assignment where z_i is set to $w_i \in \{0, 1\}$ for $1 \leq i \leq t$;
- the c-th bit in the b-th variable of the a-th constraint in $M_w(s)$ equals the c'-th bit in the b'-th variable of the a'-th constraint in $M_w(s)$.

We use the transformations $g_{a,a',b,b',c,c'}$ to define M' as follows. The machine M' first computes the formulas $g_{a,a',b,b',c,c'}(s)$. For every Boolean variable v in these formulas we introduce a new conjunct $\varphi_0(\overline{x}_v) \vee \varphi_1(\overline{x}_v)$ where $\overline{x}_v$ is a d-tuple of fresh variables. Then, every positive literal $l = x_j$ in the original conjuncts of the formula is replaced by $\varphi_1(\overline{x}_j)$, and every negative literal $l = \neg x_j$ by $\varphi_0(\overline{x}_j)$. We then existentially quantify over all variables except for $x_{z_1}, \ldots, x_{z_t}$. Let $\psi_{a,a',b,b',c,c'}(s)$ denote the resulting existential positive formula. It is clear that the formula

$$\exists x_{z_1}, \ldots, x_{z_t}. \bigwedge_{a,a',b,b'} \left(\left(\bigwedge_{c,c'} \psi_{a,a',b,b',c,c'} \right) \rightarrow x_b^a = x_{b'}^{a'} \right)$$

can be re-written in existential positive form Ψ without blow-up (we can replace implications $\alpha \rightarrow \beta$ by $\neg\alpha \vee \beta$, and then move the negation to the atomic level, where we can remove it by exchanging the role of φ_0 and φ_1), and hence Ψ can be computed by M' in polynomial time. The formula Ψ indeed has the properties required for the formula Ψ mentioned in Equation 1.   □

References

1. Bodirsky, M., Grohe, M.: Non-Dichotomies in Constraint Satisfaction Complexity. In: Aceto, L., Damgård, I., Goldberg, L.A., Halldórsson, M.M., Ingólfsdóttir, A., Walukiewicz, I. (eds.) ICALP 2008, Part II. LNCS, vol. 5126, pp. 184–196. Springer, Heidelberg (2008)
2. Garey, M.R., Johnson, D.S.: Computers and Intractability: A Guide to the Theory of NP-Completeness. W.H. Freeman and Co., New York (1979)
3. Hermann, M., Richoux, F.: On the Computational Complexity of Monotone Constraint Satisfaction Problems. In: Das, S., Uehara, R. (eds.) WALCOM 2009. LNCS, vol. 5431. Springer, Heidelberg (2009)
4. Ladner, R.E., Lynch, N.A., Selman, A.L.: A Comparison of Polynomial-Time Reducibilities. Theoretical Computer Science 1(2), 103–124 (1975)
5. Martin, B.: Dichotomies and Duality in First-order Model Checking Problems. CoRR abs/cs/0609022 (2006)
6. Martin, B.: First-Order Model Checking Problems Parameterized by the Model. In: Beckmann, A., Dimitracopoulos, C., Löwe, B. (eds.) CiE 2008. LNCS, vol. 5028, pp. 417–427. Springer, Heidelberg (2008)
7. Vollmer, H.: Complexity of Constraints (A collection of survey articles). LNCS, vol. 5250. Springer, Heidelberg (2008)

Stochastic Programs and Hybrid Automata for (Biological) Modeling

Luca Bortolussi[1] and Alberto Policriti[2,3]

[1] Dept. of Mathematics and Informatics, University of Trieste, Italy
[2] Dept. of Mathematics and Informatics, University of Udine, Italy
[3] Istituto di Genomica Applicata, Udine, Italy

Abstract. We present a technique to associate to stochastic programs written in stochastic Concurrent Constraint Programming a semantics in terms of a lattice of hybrid automata. The aim of this construction is to provide a framework to approximate the stochastic behavior by a mixed discrete/continuous dynamics with a variable degree of discreteness.

1   Introduction

In this paper we perform a further step in a program whose high-level aim is that of developing methods to replace stochastic approaches in (biological) modeling, by exploiting techniques for modeling a mixed (hybrid) continuous/discrete behavior. In other words, what we are exploring is the possibility of systematically eliminating the "stochastic ingredient", by the restriction to a discrete domain of some of the system's variables.

Our approach has been introduced and discussed in a sequence of papers (see(BP07; BP08c; BP08b)) and it is based on the following assumptions:

- there are situations in which stochastic models are—sometimes much—more realistic/precise;
- reducing the amount of the "stochastic ingredient" (however introduced) in a model, highly reduces the computational costs;
- in order to state precisely the scope and potential of a proposal like ours, two formalisms that will play the role of stochastic and hybrid reference languages, respectively, must be clearly fixed *a-priori*.

As far as the last point is concerned, our choices for the two formalisms have been the following:

1. Stochastic Concurrent Constraint Programming (**sCCP**);
2. Hybrid Automata (HA).

The above choices are, clearly, not the only possible ones and an extension of our proposal(s) to other pairs of languages, for the respective roles, is certainly possible. However, we believe such an extension is a non-trivial one, as the technicalities involved touch delicate aspects of the entire program: a language change does not seem to be a "syntactic sugar substitution". More on this point throughout the paper.

K. Ambos-Spies, B. Löwe, and W. Merkle (Eds.): CiE 2009, LNCS 5635, pp. 37–48, 2009.
© Springer-Verlag Berlin Heidelberg 2009

A general method to pass from a model written in Stochastic Concurrent Constraint Programming to a Hybrid Automata, has been introduced and discussed in (BP08a) and will be reviewed here. The Hybrid Automaton obtained is observed, in interesting cases, to retain significant properties of the stochastic simulator. Moreover, it is also shown how the level of discreteness introduced is an important parameter for evaluating the overall results of the approach. As an example, we mention the *stable oscillations* guaranteed for non-stochastic hybrid model of circadian clock, proposed in (BP08b) and obtained as counterpart of a **sCCP**-based model.

The specific contribution we put forward here is exploring the possibility of generalizing and deepening the above line of research, by trying to answer the following question: is it possible to *control* the amount of discreteness characterizing the HA proposed to simulate a given **sCCP** program?

We tackled the above problem with a refinement of our original proposal. We defined a mapping for **sCCP**-programs that instead of having a single HA as target, is designed to produce a *lattice* of HA's, whose level of discreteness determines the proximity to the top/bottom of the lattice.

As with previous experiences within our program, we had to begin from apparently secondary problems—i.e. a "massage" to the classical definition of HA— that, instead, turned out to address rather deep points. For example, the definition of Transition Driven Hybrid Automata we put forward, even though natural (discussed in some detail in Section (2)) was crucial for the successful completion of our task, as was addressing the basic problem of assigning a discrete state to the HA while the system is moving along a continuous path. We solved this problem by a mechanism assigning a *tendency* to be in a state among the possible ones, reminiscent of quantum-mechanics. Such a tendency must then be used to decode from transition's guard (our computational counterpart) the probabilistic choices to be made.

It can be argued that the stochastic dimension plays, in Systems Biology, a rather important role that can be seen as a way to render *in-silico* unknown—or computationally too complex—biological processes. This is one of the reasons for which we are developing our program in the context of Systems Biology, the other being that the kind of *computation* carried out by biological systems is *the* paradigmatic example of hybrid (discrete/continuous) computation.

A more comprehensive presentation of the material of this paper, including proofs and examples, can be found in (BP09).

2   Transition-Driven Hybrid Automata

In this section we will introduce a non-standard formalization of Hybrid Automata (HA) which we will call *Transition-Driven Hybrid Automata* (TDHA). They can be quite straightforwardly mapped into standard HA, as hinted below. More details on the standard HA can be found in (Hen96). The basic idea in definition of TDHA is to prepare for a manipulation (a.k.a. *merge/split*) of discrete and continuous states. We focus on *transitions*, which can be either discrete (corresponding to the "standard" edges of HA) or continuous (representing flows acting on system's variables). At this stage, for the sake of definiteness, we flag as *urgent* (i.e. to be taken as soon as activated) all discrete transitions: just a matter of convenience, or rather an artefact of our interest in simulation of HA. Managing standard non-deterministic guards is, however, simply a matter of slightly modifying the definition of the standard HA associated to a TDHA. Formally:

Definition 1. *A Transition-Driven Hybrid Automaton is a tuple* $\mathcal{T} = (Q, \mathbf{X}, inv, \mathfrak{TC}, \mathfrak{TD}, init)$, *where:*

- Q *is a finite set of* control modes, $\mathbf{X} = \{X_1, \ldots, X_n\}$ *is a set of real valued system's variables[1], and for each $u \in Q$, $inv[u]$ is a quantifier-free first-order formula with free variables among $\mathbf{X}$, denoting the invariant of u.*
- $\mathfrak{TC}$ *is the set of* continuous transitions or flows, *whose elements τ are triples $(u, stoich, rate)$, where: $u \in Q$ is a mode, $stoich$ is a vector of size $|\mathbf{X}|$, and $rate : \mathcal{X}_u \to \mathbb{R}$ is a function. The elements of a triple τ are indicated by* $\mathbf{cmode}[\tau]$, $\mathbf{stoich}[\tau]$, *and* $\mathbf{rate}[\tau]$, *respectively.*
- $\mathfrak{TD}$ *is the set of* discrete or instantaneous transitions, *whose elements δ are quadruples of the form $(u, v, guard, reset)$, where: u is the* exit-mode, *v is the* enter-mode, *$guard$ is a quantifier-free first-order formula with free variables among $\mathbf{X}$, and $reset$ is a quantifier-free first-order formula over $\mathbf{X}, \mathbf{X}'$. The elements of a quadruple δ are indicated by* $\mathbf{exit}[\delta]$, $\mathbf{enter}[\delta]$, $\mathbf{guard}[\delta]$, *and* $\mathbf{reset}[\delta]$, *respectively.*
- *For each $u \in Q$, $init[u]$ is a quantifier-free first-order formula over $\mathbf{X}$, defining the* initial conditions *of the system.*

For each $u \in Q$ he formula $inv[u]$ characterizes the set of admissible values that can be taken by system's variables while in mode u. We will also denote by $\mathcal{X}_u$ the set $\mathcal{X}_u = \{\mathbf{x} \mid inv[u, \mathbf{x}]\}$—the *states space* of mode u—, and by $\mathcal{X} = \bigcup_{u \in Q}\{u\} \times \mathcal{X}_u$ the *hybrid states space*.

For each $\tau \in \mathfrak{TC}$, $\mathbf{stoich}[\tau]$ and $\mathbf{rate}[\tau]$ give the *magnitude* and the *form* of the action of τ on each variable $X \in \mathbf{X}$, respectively (see below).

For each $\delta \in \mathfrak{TD}$, $\mathbf{guard}[\delta]$ represents the region of $\mathcal{X}_u$ in which the transition δ is active and $\mathbf{reset}[\delta]$ expresses the relation intervening among variables *(X)*—before the transition—and *(X')*—after.

The spirit behind the above definition is quite simple. We explicitly define automaton's modes introducing two kind of edges. Continuous edges self-loop on a mode and represent a flow acting locally on that mode. Their mathematical form is given by the rate, while their effect on system's variables is given by the stoichiometry vector. Discrete edges, instead, connect two modes (they can also self-loop), are active only in specific regions of the state space (identified by their guards), and can change the value of some variables when executed (as predicated by their reset). TDHA can be encoded quite simply into classical hybrid automata (Hen96). The control graph of the HA has as vertices the states

---

[1] Notation: the time derivative of X_j is denoted by $\dot{X}_j$, while the value of X_j after a change of mode is indicated by X'_j.

of Q and ad edges the discrete transitions in $\mathfrak{TD}$. Guards and resets associated to edges are the ones of the corresponding elements of $\mathfrak{TD}$. Urgency of discrete transitions is imposed in the standard way, i.e. by excluding from the invariant region (the allowable values of continuous variables) the sets in which guards are false. The only non-trivial part in this mapping is the flows' construction from the set $\mathfrak{TC}$ of continuous transitions. Essentially, for each transition $\tau \in \mathfrak{TC}$ active in a mode v, we multiply the function $\mathbf{rate}[\tau][\mathbf{X}]$ by the magnitude $\mathbf{stoich}[\tau][X]$ for each variable $X \in \mathbf{X}$ and add these terms to obtain the ODE's.[2] For instance, consider a system with two variables X and Y, a single mode u, and the following set $\mathfrak{TC} = \{(u, \langle 0, 1 \rangle, Y^2), (u, \langle 1, 0 \rangle, X^2), (u, \langle -1, -1 \rangle, X \cdot Y)\}$. Then the ODE for X is $flow[u, X] = 0 \cdot Y^2 + 1 \cdot X^2 + (-1) \cdot XY = X^2 - XY$, while for Y we get $flow[u, Y] = Y^2 - XY$. Further examples can be found in (BP09).

2.1   Product of TDHA

We will define now a notion of product for TDHA. As a preliminary notion, we extend the embedding of states into pairs of states to continuous and discrete transitions. Let $\mathcal{T} = (Q, \mathbf{X}, inv_1, \mathfrak{TC}, \mathfrak{TD}, init_1)$ be a TDHA, $\tau \in \mathfrak{TC}$ be a continuous transition, u be a mode in a set Q', $Q' \cap Q = \emptyset$, and $\mathbf{Y}$ be a set of variables, with $\mathbf{X} \subseteq \mathbf{Y}$. The embedding i_1 of τ, w.r.t. u and $\mathbf{Y}$ is defined as $i_1(\tau, u, \mathbf{Y}) = ((\mathbf{cmode}[\tau], u), \mathbf{stoich}[\tau]^{\mathbf{Y}}, \mathbf{rate}[\tau])$, with $\mathbf{stoich}[\tau]^{\mathbf{Y}}$ a vector on $\mathbf{Y}$ coinciding with $\mathbf{stoich}[\tau]$ for each $X \in \mathbf{X}$, and equal to zero for $X \notin \mathbf{X}$. Similarly $i_2(\tau, u, \mathbf{Y})$ is defined as $((u, \mathbf{cmode}[\tau]), \mathbf{stoich}[\tau]^{\mathbf{Y}}, \mathbf{rate}[\tau])$. An analogous notion of embedding can be defined for discrete transitions: for $\delta \in \mathfrak{TD}$ we set

$$i_1(\delta, u, \mathbf{Y}) = ((\mathbf{exit}[\delta], u), (\mathbf{enter}[\delta], u), \mathbf{guard}[\delta], \mathbf{reset}[\delta])$$

and

$$i_2(\delta, v, \mathbf{Y}) = ((v, \mathbf{exit}[\delta]), (v, \mathbf{enter}[\delta]), \mathbf{guard}[\delta], \mathbf{reset}[\delta]).$$

Definition 2. *Given two TDHA $\mathcal{T}_i = (Q_i, \mathbf{X}_i, inv_1, \mathfrak{TC}_i, \mathfrak{TD}_i, init_1)$ for $i \in \{1, 2\}$, their product $\mathcal{T} = \mathcal{T}_1 \otimes \mathcal{T}_2$, $\mathcal{T} = (Q, \mathbf{X}, inv_1, \mathfrak{TC}, \mathfrak{TD}, init_1)$, is defined by:*

- $Q = Q_1 \times Q_2$, $\mathbf{X} = \mathbf{X}_1 \cup \mathbf{X}_2$, and $inv[(u, v)] = inv_1[u] \wedge inv_2[v]$;
- $\mathfrak{TC} = \{i_1(\tau, v, \mathbf{X}) \mid \tau \in \mathfrak{TC}_1, v \in Q_2\} \cup \{i_2(\tau, u, \mathbf{X}) \mid \tau \in \mathfrak{TC}_2, u \in Q_1\}$;
- $\mathfrak{TD} = \{i_1(\delta, v, \mathbf{X}) \mid \delta \in \mathfrak{TD}_1, v \in Q_2\} \cup \{i_2(\delta, u, \mathbf{X}) \mid \delta \in \mathfrak{TD}_2, u \in Q_1\}$;
- $init[(u, v)] = init_1[u] \wedge init_2[v]$.

The definition is quite natural, given the absence of synchronization mechanisms for TDHA—examples in (BP09)—, yet we wish to focus the attention of the reader on the effect on flows. In fact, for each variable shared between $\mathcal{T}_1$ and $\mathcal{T}_2$, the result obtained is the addition of the right-hand sides of the corresponding differential equations. Hence, this construction is equivalent to the flux product defined in (BP08a).

---

[2] Formally, $flow[v][\mathbf{X}] = \sum_{\tau \in \mathfrak{TC} \mid \mathbf{cmode}[\tau] = v} \mathbf{stoich}[\tau] \cdot \mathbf{rate}[\tau]$.

3   Stochastic Concurrent Constraint Programming

In this section we briefly present the (computational/programming) language we chose to introduce the stochastic dimension in our picture. We decided to use (a simplified version of) stochastic Concurrent Constraint Programming (sCCP (Bor06), a stochastic extension of CCP (Sar93)), as it seems to be sufficiently expressive, compact, and especially suitable in many other respects—to be discussed—, for our purposes[3] . In the following we just sketch the basic notions and the concepts needed in the rest of the paper. More details on the language can be found in (Bor06; BP08c; BP09).

Definition 3. *A sCCP program is a tuple $\mathcal{A} = (A, \mathcal{D}, \mathbf{X}, dom, init(\mathbf{X}))$, where*

1. *The initial agent A and the set of definitions $\mathcal{D}$ are given by the following grammar*

$$\mathcal{D} = \emptyset \mid \mathcal{D} \cup \mathcal{D} \mid \{C \stackrel{def}{=} M\}$$
$$\pi = [g(\mathbf{X}) \rightarrow u(\mathbf{X}, \mathbf{X}')]_{\lambda(\mathbf{X})} \quad M = \pi.C \mid M + M \quad A = M \mid A \parallel A$$

2. $\mathbf{X}$ *is the set of variables of the store (with global scope);*
3. *dom is is a conjunction of quantifier-free formulae on a single variable $X \in \mathbf{X}$, with atoms consisting of equality and inequality predicates, specifying the domain of each variable.*
4. *$init(\mathbf{X})$ is a predicate on $\mathbf{X}$ of the form $\mathbf{X} = \mathbf{x_0}$, assigning an initial value to store variables.*

In the previous definition, basic actions are *guarded updates* of (some of the) variables: $g(\mathbf{X})$ is a propositional formula whose atoms are inequality predicates on variables $\mathbf{X}$ and $u(\mathbf{X}, \mathbf{X}')$ is a conjunction of predicates on $\mathbf{X}, \mathbf{X}'$ of the form $\mathbf{X}' = \mathbf{f}(\mathbf{X})$ ($\mathbf{X}'$ denotes variables of $\mathbf{X}$ after the update), for some function $\mathbf{f} : \mathbb{R}^n \rightarrow \mathbb{R}^n$. Each such action has a *stochastic duration*, specified by associating an exponentially distributed random variable to actions, whose rate depends on the state of the system through a function $\lambda : \mathbf{X} \rightarrow \mathbb{R}^+$.

All agents definable in **sCCP**, i.e. all agents $C \stackrel{def}{=} M \in \mathcal{D}$,[4] are *sequential*, i.e. they do not contain any occurrence of the parallel operator, whose usage is restricted at the upper level of the network.

sCCP sequential agents can be seen as automata synchronizing on store variables and they can be conveniently represented as labeled graphs, called *Reduced Transition Systems* (RTS) in (BP07). The steps to obtain an object suitable for the following treatment are:

---

[3] There are other probabilistic extensions of CCP studied in literature, like (DW98; GJS97; GJP99). (DW98) provides CCP with a semantics based on discrete time Markov Chains, while in (GJS97; GJP99) the stochastic ingredient is introduced by extending the store with random variables and adding a primitive for sampling. These approaches, however, are not suited for our purposes, as we need a model in which events happen probabilistically in continuous-time, as customary in biochemical modeling.

[4] In the following, with a slight abuse of notation, we sometimes write $C \in \mathcal{D}$ for $C \stackrel{def}{=} M \in \mathcal{D}$.

1. define the collection of all possible states—the *derivative set* $Der(C)$—and actions—$action(C)$—of any sequential agent appearing in a **sCCP** *simple program* (i.e. without multiple copies of the same agent).
2. on the ground of the above definition the following LTS multi-graph $RTS(C) = (S(C), E(C), \ell)$ is introduced[5]:
   - $S(C) = Der(C)$,
   - $E(C) = \{(exit(\pi), enter(\pi)) \mid \pi \in action(C)\}$,
   - $\ell(e) = (guard(\pi), update(\pi), rate(\pi))$, where π is the action defining the edge $e \in E(C)$;
3. the notion of *extended* **sCCP** program $\mathcal{A}^+ = (A^+, \mathcal{D}^+, \mathbf{Y}, dom^+, init^+(\mathbf{Y}))$, in which a counter for run-time recording the number of parallel copies of each agent, is introduced and proved isomorphic to $\mathcal{A}$. Essentially, this is a technical trick that simplifies the overall treatment: we introduce a new variable P_C for each component $C \in \mathcal{D}$ of a **sCCP** program. These variables will count the number of copies of C present in parallel in the system at a certain time. To take into account the effects of transitions on agents, we modify updates and rate functions, by increasing (resp. decreasing) the counter P_C for actions adding (resp. removing) a copy of the agent P_C. The formal definition and the proof the the following proposition, can be found in (BP09).

Proposition 1. *Let $\mathcal{A}$ be a simple sCCP program and $\mathcal{A}^+$ be its extension. The LTS of $\mathcal{A}$ and $\mathcal{A}^+$ are isomorphic as labeled multi-graphs.*

Remark 1. State variables in simple program may seem useless: they can only take values 0 or 1. We will see in the next section that, instead, their continuous approximation in the interval $[0, 1]$ will provide the mechanism for controlling a cluster of discrete states (continuously approximated). Basically, the real value of a state variable will indicate the "tendency" of the system to be in that particular state. The sum of such tendencies for discrete states merged in a continuous cluster, will always be constrained to be equal to 1.

4   From sCCP to TDHA

In this section we touch the core of our proposal, specifying our mapping in the case of a **sCCP** program associated to a *Transition-Driven* Hybrid Automaton, in such a way to deal with a *lattice* of different TDHA's.

First of all recall that transitions in **sCCP** are all discrete and stochastic. The idea behind the translation is simply to approximate some transition as continuous, leaving the others discrete. Stochasticity is mimicked by the introduction of suitable timing conditions for the discrete transitions.

---

[5] $exit(\pi), enter(\pi), guard(\pi), update(\pi), rate(\pi)$ extract resp. the executing agent, the target agent, the guard, the update and the rate of an action π.

The mapping proceeds in two steps. First we convert into TDHA's each sequential component of a **sCCP** program, then all these TDHA's are composed together using the product construction of Section (2.1).

Given a $\mathcal{A}^+ = (A^+, \mathcal{D}^+, \mathbf{Y}, dom^+, init^+(\mathbf{Y}))$, let $C \in \mathcal{D}^+$ be an agent in parallel in the initial agent A and let $RTS(C) = (S(C), E(C), \ell)$ be its RTS. The basic idea of the mapping is to partition the set of transitions $E(C)$ according to a choice of a specific continuous/discrete scheme of approximation. In order to formalize this, we use a boolean vector $\kappa \in \{0,1\}^m$, $m = |E(C)|$, indexed by edges in $E(C)$: for $e \in E(C)$, $\kappa[e] = 1$ means that the transition will be approximated as continuous, while $\kappa[e] = 0$ implies that it will remain discrete. Technically, we partition $E(C)$ in two disjoint subsets: $E(\kappa, C) = \{e \in E(C)|\ \kappa[e] = 1\}$ (to be approximated as continuous) and $E(\neg\kappa, C) = \{e \in E(C)|\ \kappa[e] = 0\}$ (to be approximated as discrete).

We need the following restriction to guarantee the existence of solutions to ODE's that will be introduced subsequently:

Definition 4. *A **sCCP** action π is called* continuously approximable *iff $rate(\pi)$ is differentiable and $rate(\pi)[\mathbf{X}] = 0$ whenever $guard(\pi)[\mathbf{X}]$ is false.*

Hence, for continuously approximable actions guards are, essentially, irrelevant. In fact, when a guard is false the corresponding transition is already deadlocked due to its rate having value zero. We call *consistent* a vector κ which is equal to 1, $\kappa[e] = 1$, only for edges e that are continuously approximable. In the following, we suppose to work only with consistent κ.

At this point we are ready to introduce the basic components of our target TDHA.

Discrete Modes. The modes of the TDHA will be, as first approximation, essentially the states $S(C)$ of $RTS(C)$. However, clearly, since continuous transitions cannot change mode, we need to consider as equivalent those states that can be reached by a path of edges that κ indicates to maintain continuous[6].

Definition 5. *Let $RTS(C)$ be the RTS of agent C, let $E(\kappa, C) \subseteq E(C)$ be the subset of continuous transitions, and let $s_1, s_2 \in S(C)$. $s_1 \sim_\kappa s_2$ iff there is a path of edges in $E(\kappa, C)$ from s_1 to s_2 in the non-oriented version of $RTS(C)$.*

Clearly $\sim_\kappa$, being a reachability relation in a non-oriented graph, is an equivalence relation. Hence we can define the set $S_\kappa(C) = S(C)/\sim_\kappa$ of RTS-states modulo $\sim_\kappa$. For each state $[s] \in S_\kappa(C)$, we define the following objects:

- The *local interaction matrix* $\nu_{\mathbf{Y}, E(\kappa, C, [s])}$ w.r.t. variables $\mathbf{Y}$ and continuous edges $E(\kappa, C, [s]) = \{e = (s_1, s_2) \in E(\kappa, C) \mid s_1, s_2 \in [s]\}$, that is edges self-looping on $[s]$. $\nu_{\mathbf{Y}, E(\kappa, C, [s])}$ has rows indexed by variables in $\mathbf{Y}$ and columns indexed by transitions in $E(\kappa, C, [s])$: $\nu_{\mathbf{Y}, E(\kappa, C, [s])}[X, e] = h$ if and only if variable X is updated by transition e according to the formula $X' = X + h$.
- The *local rate vector* $\phi_{E(\kappa, C, [s])}$ w.r.t. transitions in $E(\kappa, C, [s])$ has $|E(\kappa, C, [s])|$ entries, and it is defined by $\phi_{E(\kappa, C, [s])}[e] = rate(e)$.

---

[6] Actually, a distinction among those collapsed states will be maintained through the dependence on state variables $\mathbf{P}$, added in the extension $\mathcal{A}^+$.

Continuous flow. The continuous evolution for TDHA is given, in each mode, by the set $\mathfrak{TC}$ of continuous transitions defined as follows. To each edge $e \in E(\kappa, C, [s])$ looping on $[s]$ we associate the continuous transition

$$([s], \nu_{\mathbf{Y}, E(\kappa, C, [s])}[\cdot, e], \phi_{E(\kappa, C, [s])}[e]),$$

where the stoichiometric vector is given by the e-column of the local interaction matrix $\nu_{\mathbf{Y}, E(\kappa, C, [s])}$, and the rate of the transition is the e-entry of the local rate vector $\phi_{E(\kappa, C, [s])}$.

Therefore, summarizing:

$$\mathfrak{TC} = \{([exit(e)], \nu_{\mathbf{Y}, E(\kappa, C, [s])}[\cdot, e], \phi_{E(\kappa, C, [s])}[e]) \mid e \in E(\kappa, C)\}.$$

Invariant conditions. The space of admissible values is restricted by the domain constraints of the **sCCP** program. As we work with simple **sCCP** programs, we know that state variables P_C always belong to the interval $[0, 1]$. Hence, we can enforce the following invariant for each mode $v \in S_\kappa(C)$: $inv[v] = dom \wedge (\bigwedge_{C' \in Der(C)} P_{C'} \leq 1)$.

Discrete transitions. In order to deal with discrete transitions $e \in E(\neg\kappa, C)$, we need to take into account the *timing* of the corresponding events. As we need to remove the stochasticity, we *fire the transition at its expected time*. As some variables of the system evolve continuously, the stochastic transition becomes a non-homogeneous Poisson process (Ros96), with rate $\lambda(\mathbf{Y}(t)) = \lambda(t)$. In this case, given the cumulative rate at time t, $\Lambda(t) = \int_0^t \lambda(\mathbf{Y}(s))ds$, the probability of firing within time t is $F(t) = 1 - e^{-\Lambda(t)}$. The random variable T with this distribution can be simulated by an uniform random variable U on $(0, 1)$ solving for t the equation $\Lambda(t) = -\log U$. Taking the average on both sides, we obtain $E[\Lambda(t)] = 1$, hence we can fire the transition whenever $\Lambda(t) \geq 1$.[7] In order to describe this condition within the TDHA framework, we introduce a new variable Z_e, with initial value 0, evolving deterministically according to the equation $\dot{Z}_e = \lambda(\mathbf{Y})$. The definition of this equation can be added to $\mathfrak{TC}$ in the obvious way, i.e. by adding the continuous edge $([exit(e)], id_{Z_e}, \lambda(\mathbf{Y}))$, with id_{Z_e} the vector equal to one for the component associated to the variable Z_e and zero elsewhere[8]. When the transition fires, we also reset *all* the variables Z_e to zero, in order to mimic the memoryless property of CTMCs.

In the definition of discrete transitions, we also need to take into account guards and updates of the corresponding **sCCP** edge, the $guard(e)$ and $update(e)$ formulae, which will be part of guards and resets of the discrete transition. There are, however, two complications:

---

[7] $\Lambda(t)$ is a monotonic increasing function, due to non-negativity of λ, hence $\lambda(t)$ has a unique solution provided, for instance, $\Lambda(t) \to \infty$, a condition usually satisfied. If this does not happen, then there is a non-null probability that the transition never fires, hence the guard of the urgent transition may never become true.

[8] The stoichiometric vectors of all other continuous transitions need to be updated consequently. This will be done considering the local interaction matrix with respect to the extended set of variables.

- It may happen that the reset on a variable, say X, brings its value out of the allowed domain. Such transitions should not be permitted: they can be avoided by adding suitable conditions to the guard. Specifically, if $update(e) := \mathbf{X}' = \mathbf{f}(\mathbf{X}) \wedge \mathbf{P}' = \mathbf{P} + \mathbf{h}$, we add to $guard(e)$ the formula $dom[\mathbf{X}/\mathbf{f}(\mathbf{X})]$ obtained by replacing each system's variable X with the term $f_X(\mathbf{X})$ in the formula expressing domain constraints. This forces the values of $\mathbf{X}$ after the transition to satisfy domain constraints.
- The updates on state variables $\mathbf{P}$ reflect the change of states of the components. In particular, as the **sCCP** program is simple, they can assume just two values, 0 and 1. Actually, in the TDHA they can take values in the whole $[0, 1]$ real interval, in the case in which we are currently in a clustered state $[s]$, collapsing states $s_1, \ldots, s_n$ of $RTS(C)$. In this case, the state variables $P_{s_1}, \ldots, P_{s_n}$ sum up to 1, and generally they have values $0 < P_{s_i} < 1$, representing how likely is the s_i state in the cluster $[s]$. In order to deal with those variables correctly, we need to insure that
  1. when a state $[s]$ is left, all its state variables are set to zero and
  2. when a discrete transition looping in $[s]$ happens, then the variable of its target state s_i becomes equal to 1, while all other variables of $[s]$ are reset to 0.

  We can obtain this in the following way. Consider an **sCCP** edge connecting states s_1 and s_2, with update $update(e) := \mathbf{X}' = \mathbf{f}(\mathbf{X}) \wedge P'_{s_1} = P_{s_1} - 1 \wedge P'_{s_2} = P_{s_2} + 1$, and fix the vector κ. We will substitute the update of state variables $P'_{s_1} = P_{s_1} - 1 \wedge P'_{s_2} = P_{s_2} + 1$ by the formula $update\_statevars(\mathbf{P})$, defined by
  1. When $[s_1] \neq [s_2]$, $update\_statevars(\mathbf{P}) = \bigwedge_{s \in [s_1]} P'_s = 0 \wedge P'_{s_2} = 1$.
  2. When $[s_1] = [s_2]$, $update\_statevars(\mathbf{P}) = \bigwedge_{s \in [s_1], s \neq s_2} P'_s = 0 \wedge P'_{s_2} = 1$.

Putting all this discussion together, we have that the discrete transition associated to $e \in E(\neg \kappa, C)$ with $e = (s_1, s_2)$ and $update(e) := \mathbf{X}' = \mathbf{f}(\mathbf{X}) \wedge P'_{s_1} = P_{s_1} - 1 \wedge P'_{s_2} = P_{s_2} + 1$ is

$$([s_1], [s_2], guard(e) \wedge dom[\mathbf{X}/\mathbf{f}(\mathbf{X})] \wedge Z_e \geq 1, \mathbf{X}' = \mathbf{f}(\mathbf{X}) \wedge$$
$$\wedge update\_statevars(\mathbf{P}) \wedge (\bigwedge_{e \in E(C)} Z'_e = 0)) \in \mathfrak{TD}.$$

We can now collect all our considerations into a global definition.

Definition 6. Let $\mathcal{A} = (A, \mathcal{D}, \mathbf{X}, dom, init_0)$ be a simple **sCCP** program and $\mathcal{A}^+ = (A^+, \mathcal{D}^+, \mathbf{Y}, dom^+, init_0^+)$ be its extended version. Let C be a sequential component in parallel in A^+, with $RTS(C) = (S(C), E(C), \ell)$. Fix a boolean vector $\kappa \in \{0, 1\}^m$, $m = |E(C)|$. The Transition-Driven Hybrid Automaton associated to C w.r.t. κ is $\mathcal{T}(C, \kappa) = (Q, \mathbf{Z}, inv, \mathfrak{TC}, \mathfrak{TD}, init)$, where

- $Q = S_\kappa(C) = S(C)/\sim_\kappa$, $\mathbf{Z} = \mathbf{Y} \cup \{Z_e \mid e \in E(C)\}$, and $inv[v] = dom \wedge (\bigwedge_{C' \in Der(C)} P_{C'} \leq 1)$;
- $\mathfrak{TC} = \{([exit(e)], \nu_{\mathbf{Y}, E(\kappa, C, [s])}[\cdot, e], \phi_{E(\kappa, C, [s])}[e]) \mid e \in E(\kappa, C)\} \cup \\ \cup \{([exit(e)], id_{Z_e}, \lambda(\mathbf{Y})) \mid e \in E(\neg \kappa, C)\};$

- $\mathfrak{TD} = \{[s_1], [s_2], guard(e) \wedge dom[\mathbf{X}/\mathbf{X} + \mathbf{k}] \wedge Z_e \geq 1, \mathbf{X}' = \mathbf{X} + \mathbf{k} \wedge$
  $\wedge update\_statevars(\mathbf{P}) \wedge (\bigwedge_{e \in E(C)} Z'_e = 0) \mid e = (s_1, s_2) \in E(\neg\kappa, C)\};$
- $init = init_0^+ \wedge \left(\bigwedge_{e \in E(C)} Z_e = 0\right).$

The previous definition gives a way to associate a TDHA to a sequential agent of a **sCCP** program. In order to define the TDHA for the whole program, we simply need to combine the TDHA for its components using the product construction (cf. Def. 2). An example is presented in (BP09).

Definition 7. *Let $\mathcal{A} = (A, \mathcal{D}, \mathbf{X}, dom, init_0)$ be a simple **sCCP** program and $\mathcal{A}^+ = (A^+, \mathcal{D}^+, \mathbf{Y}, dom^+, init_0^+)$ be its extended version, with $A^+ = C_1 \parallel \ldots \parallel C_n$. Fix a boolean vector κ_i for each sequential agent C_i. The Transition-Driven Hybrid Automaton for the **sCCP** program $\mathcal{A}$, w.r.t. $\kappa = (\kappa_i)_{i=1,\ldots n}$ is*

$$\mathcal{T}(\mathcal{A}, \kappa) = \mathcal{T}(C_1, \kappa_1) \otimes \cdots \otimes \mathcal{T}(C_n, \kappa_n).$$

4.1 Lattice of TDHA

Definition 7 associates a TDHA to a **sCCP** agent for a fixed partitioning of transitions into discrete and continuous, given by the vector κ. Clearly, different choices of κ correspond to different TDHA's, with a different degree of approximation w.r.t. the original **sCCP** program. The different TDHA's that can be associated to a program $\mathcal{A}$ can be arranged into a lattice, according to the following ordering:

Definition 8. *Let $\mathcal{A}$ be a **sCCP** agent, then $\mathcal{T}(\mathcal{A}, \kappa_1) \sqsubseteq \mathcal{T}(\mathcal{A}, \kappa_2)$ iff $\kappa_1[e] = 1 \Rightarrow \kappa_2[e] = 1$, for each transition $e \in E(A) = E(C_1) \cup \ldots \cup E(C_n)$, with $A = C_1 \parallel \ldots \parallel C_n$ the initial agent of $\mathcal{A}$.*

The bottom element of this lattice is obtained for $\kappa \equiv 0$, while the top element is obtained for $\kappa[e] = 1$ iff e is continuously approximable.[9] These two choices correspond to two particular TDHA, as shown in the following propositions.

Proposition 2. *Let $\mathcal{A}$ be a **sCCP** program. The TDHA $\mathcal{T}(\mathcal{A}, 0)$ is equivalent to a timed-automata with skewed clock variables Z_e, $e \in E(A^+)$, and with system and state variables modified by resets only.*

Proof. The property holds for the TDHA associated to a sequential component C because:

1. all transitions of $\mathcal{T}(C, 0)$ are discrete;
2. all state variables $\mathbf{P}$ and system variables $\mathbf{X}$ are governed by the ODE's $\dot{X} = 0$, hence they can be modified only by resets of discrete transitions;
3. between two discrete transitions, the value of variables $\mathbf{X}$ and $\mathbf{P}$ is constant, hence the ODE's for Z_e are of the form $\dot{Z}_e = const$ and each Z_e is a skewed clock.

---

[9] We remind to the reader that transitions not continuously approximable *must* be kept discrete.

The property is clearly preserved by TDHA product.

Proposition 3. *Let A be a **sCCP** program with initial agent $A = C_1 \parallel \ldots \parallel C_n$. If e is continuously approximable for each $e \in E(A) = E(C_1) \cup \ldots \cup E(C_n)$, then $\mathcal{H}_{\mathcal{T}(A,\mathbf{1})}$, the hybrid automaton derived from $\mathcal{T}(A, \mathbf{1})$ coincides with the system of ODE's associated to A by its fluid-flow approximation (BP07).*

5   Conclusion and Further Directions

In this work we presented a method to map a given **sCCP** program to a *lattice* of Hybrid Automata, partially ordered by the level of discreteness maintained. The natural question to ask at this point if and how a sensible level of discreteness can be (semi) automatically determined, in order to maximize both expressiveness and efficiency. In our opinion this issue could be addressed syntactically by working on classes of formulae expressing significant properties for which a maximum level of discreteness can be maintained without loosing satisfiability.

Another line of work, related with expressiveness/efficiency ratio, consist in designing a *dynamic* version of the technique proposed here, by turning the vector κ introduced in Section 4 into a function dependent on variables and on parameters chosen by the user, whose value can change at run-time.

Finally, we observe that the overall technique starts from the assumption that the stochastic ingredient should be eventually fully eliminated. It is easy to argue that there are situations in which this is not a good policy and, therefore, it would be interesting to study a variant of the method presented here capable to map on *stochastic* Hybrid Automata.

References

[Bor06] Bortolussi, L.: Stochastic concurrent constraint programming. In: Proceedings of 4th International Workshop on Quantitative Aspects of Programming Languages (QAPL 2006). ENTCS, vol. 164, pp. 65–80 (2006)

[BP07] Bortolussi, L., Policriti, A.: Stochastic concurrent constraint programming and differential equations. In: Proceedings of Fifth Workshop on Quantitative Aspects of Programming Languages, QAPL 2007. ENTCS 16713 (2007)

[BP08a] Bortolussi, L., Policriti, A.: Hybrid approximation of stochastic concurrent constraint programming. In: Proceedings of IFAC 2008, Seoul (2008)

[BP08b] Bortolussi, L., Policriti, A.: The importance of being (a little bit) discrete. In: Proceedings of FBTC 2008. ENTCS 17346 (2008)

[BP08c] Bortolussi, L., Policriti, A.: Modeling biological systems in concurrent constraint programming. Constraints 13(1) (2008)

[BP09] Bortolussi, L., Policriti, A.: Stochastic Programs and Hybrid Automata for (Biological) Modeling. Technical Report,
http://www.dmi.units.it/~bortolu/sccp.htm

[DW98] Di Pierro, A., Wiklicky, H.: An operational semantics for probabilistic concurrent constraint programming. In: Proceedings of IEEE Computer Society International Conference on Computer Languages (1998)

[GJS97] Gupta, V., Jagadeesan, R., Saraswat, V.A.: Probabilistic Concurrent Constraint Programming. In: Mazurkiewicz, A., Winkowski, J. (eds.) CONCUR 1997, vol. 1243. Springer, Heidelberg (1997)

[GJP99] Gupta, V., Jagadeesan, R., Panangaden, P.: Stochastic processes as concurrent constraint programs. In: Proceedings of POPL 1999 (1999)

[Hen96] Henzinger, T.A.: The theory of hybrid automata. In: LICS 1996: Proceedings of the 11th Annual IEEE Symposium on Logic in Computer Science (1996)

[Nor97] Norris, J.R.: Markov Chains. Cambridge University Press, Cambridge (1997)

[Ros96] Ross, S.M.: Stochastic Processes. Wiley, New York (1996)

[Sar93] Saraswat, V.A.: Concurrent Constraint Programming. MIT Press, Cambridge (1993)

Numberings and Randomness

Paul Brodhead and Bjørn Kjos-Hanssen

Department of Mathematics, University of Hawai'i at Mānoa, Honolulu HI 96822
brodhead@math.hawaii.edu, bjoern@math.hawaii.edu
http://www.math.hawaii.edu/~brodhead
http://www.math.hawaii.edu/~bjoern

Abstract. We prove various results on effective numberings and Friedberg numberings of families related to algorithmic randomness. The family of all Martin-Löf random left-computably enumerable reals has a Friedberg numbering, as does the family of all Π_1^0 classes of positive measure. On the other hand, the Π_1^0 classes contained in the Martin-Löf random reals do not even have an effective numbering, nor do the left-c.e. reals satisfying a fixed randomness constant. For Π_1^0 classes contained in the class of reals satisfying a fixed randomness constant, we prove that at least an effective numbering exists.

1 Introduction

The general theory of numberings was initiated in the mid-1950s by Kolmogorov, and continued by Mal'tsev and Ershov [2]. A *numbering*, or *enumeration*, of a collection C of objects is a surjective map $F : \omega \to C$. In one of the earliest results, Friedberg [3, 1958] constructed an injective numbering ψ of the Σ_1^0 or computably enumerable (*c.e.*) sets such that the relation "$n \in \psi(e)$" is itself Σ_1^0. In a more general and informal sense, a numbering ψ of a collection of objects all having complexity $\mathcal{C}$ (such as n-c.e., Σ_n^0, or Π_n^0) is called *effective* if the relation "$x \in \psi(e)$" has complexity $\mathcal{C}$. If in addition the numbering is injective, then it is called a *Friedberg numbering*.

Brodhead and Cenzer [1] showed that there is an effective Friedberg numbering of the Π_1^0 classes in Cantor space 2^ω. They showed that effective numberings exist of the Π_1^0 classes that are homogeneous, and decidable, but not of the families consisting of Π_1^0 classes that are of measure zero, thin, perfect thin, small, very small, or nondecidable, respectively.

In this article we continue the study of existence of numberings and Friedberg numberings for subsets of ω and 2^ω. Many of our results are related to algorithmic randomness and in particular Martin-Löf randomness; see the books of Li and Vitányi [4] and Nies [6].

We now outline some notation and definitions used throughout. A subset T of $2^{<\omega}$ is a *tree* if it is closed under prefixes. The set $[T]$ of infinite paths through T is defined by $X \in [T] \leftrightarrow (\forall n) X \upharpoonright n \in T$, where $X \upharpoonright n$ denotes the initial segment $\langle X(0), X(1), \ldots, X(n-1) \rangle$. Next, P is a Π_1^0 class if $P = [T]$ for some computable tree T. Let $\sigma^\frown \tau$ denote the concatenation of σ with τ and let $\sigma^\frown i$

K. Ambos-Spies, B. Löwe, and W. Merkle (Eds.): CiE 2009, LNCS 5635, pp. 49–58, 2009.
© Springer-Verlag Berlin Heidelberg 2009

denote $\sigma^\frown\langle i\rangle$ for $i \in \omega$. The prefix ordering of strings is denoted by $\preceq$, so we have $\sigma \preceq \sigma^\frown\tau$. The string $\sigma \in T$ is a *dead end* if no extension $\sigma^\frown i$ is in T. For any $\sigma \in 2^{<\omega}$, $[\sigma]$ is the cone consisting of all infinite sequences extending σ. For a set of strings W, $[W]^{\preceq} = \bigcup_{\sigma \in W}[\sigma]$.

2   Families of Left-c.e. Reals

2.1   Basics

For our definition of left-c.e. reals we will follow the book of Nies [6]. Let $\mathbb{Q}_2$ be the set of dyadic rationals $\{\frac{a}{2^b} \leq 1 : a, b \in \omega\}$. For a dyadic rational q and real $x \in 2^\omega$, we say that $q < x$ if q is less than the real number $\sum_{i \in \omega} x(i)2^{-(i+1)}$.

Definition 1. *A real $x \in 2^\omega$ is left-c.e. if $\{q \in \mathbb{Q}_2 : q < x\}$ is c.e.*

Let $\leq_L$ denote lexicographic order on 2^ω. A dyadic rational may be written in the form $q = \sum_{i=1}^n a_i 2^{-i}$ where $a_n = 1$, and each $a_i \in \{0,1\}$. The *associated binary string* of q is $s(q) = \langle a_1, \ldots, a_n\rangle$. (If $q = 0$ then $n = 0$ and the associated string is the empty string.) Conversely, the associated dyadic rational of $\sigma \in 2^{<\omega}$ is $\sum_{i=0}^{|\sigma|-1} \sigma(i)2^{-(i+1)}$.

Lemma 1. *For each $x \in 2^\omega$, we have that*

$$\{q \in \mathbb{Q}_2 : q < x\} \text{ is c.e.} \Leftrightarrow \{\sigma \in 2^{<\omega} : \sigma^\frown 0^\omega <_L x\} \text{ is c.e.}$$

Proof. We have that $\sigma^\frown 0^\omega <_L x$ iff the associated dyadic rational of σ is less than x, and $q < x$ iff the associated binary string σ of q satisfies $\sigma^\frown 0^\omega <_L x$. In fact, $\{s(q) : q \in \mathbb{Q}_2, q < x\} = \{\sigma : \sigma^\frown \omega <_L x\}$.

Definition 2. *An effective numbering of a family of left-c.e. reals $\mathcal{R}$ is an onto map $r : \omega \mapsto \mathcal{R}$ such that*

$$\{(q, e) \in \mathbb{Q}_2 \times \omega \mid q < r(e)\}$$

is c.e. If r is also injective then r is called a Friedberg numbering *of $\mathcal{R}$.*

Theorem 1. *The family of all left-c.e. reals has an effective numbering.*

Proof. Let $W_{e,s}$ be the eth c.e. subset of $\mathbb{Q}_2$ as enumerated up to stage s. Let $r_{e,s}$ be the greatest element of $W_{e,s}$ and let $r(e) = \lim_{s\to\infty} r_{e,s}$. It is easy to check that r is an effective numbering of $\mathcal{R}$.

Some notions from algorithmic randomness will be needed repeatedly below. Ω is any fixed Martin-Löf random left-c.e. real with computable approximation $\Omega_s \leq_L \Omega_{s+1}$, $s \in \omega$. Let K denote prefix-free Kolmogorov complexity. Schnorr's Theorem states that a real $x \in 2^\omega$ is Martin-Löf random if and only if there is a constant c such that for all n, $K(x \restriction n) \geq n - c$.

Theorem 2. *The family of all Martin-Löf random left-c.e. reals has an effective enumeration.*

Proof. Let K_t a uniformly computable approximation to Kolmogorov complexity at stage t, satisfying $K_{t+1} \leq K_t$. To obtain an enumeration of the Martin-Löf random left-c.e. reals, it suffices to enumerate all Martin-Löf random left-c.e. reals y such that $K(y \upharpoonright n) \geq n - c$ for all n, uniformly in c.

Initially our m^{th} ML-random left-c.e. real m_e will look like $r_e = r(e)$ from Theorem 1, i.e. $m_{e,s} = r_{e,s}$ unless otherwise stated. Let $r_{e,t}[n]$ be the associated string, restricted or appended with zeroes if necessary to obtain length n. If at some stage t, for some $n = n_t \in \omega$,

$$K_t(r_{e,t}[n]) < n - c,$$

then let $m_{e,s} = r_{e,t}[n] \frown \Omega_s$ at all stages $s > t$ until, if ever, $K_s(r_{e,s}[n]) \geq n - c$ at some stage $s > t$. At this point, $r_{e,t}[n] < r_{e,s}[n]$, since r_e is a left-c.e. real. Resume where we left off in defining $m_e = r_e$, starting immediately at stage s with $m_{e,s} = r_{e,s}$. This process continues for the entire construction of each m_e.

This enumeration contains all left-c.e. reals which are Martin-Löf random with respect to the constant c, and only Martin-Löf left-c.e. random reals. Thus the merger of these enumerations over all c is an enumeration of all Martin-Löf random left-c.e. reals.

2.2   Kummer's Method

Kummer [5, 1990] gave a priority-free proof of Friedberg's result. The conditions set forth in the proof provide a method of obtaining Friedberg numberings.

A *c.e. class* is a uniformly c.e. collection of subsets of ω (or equivalently, of $2^{<\omega}$ or $\mathbb{Q}_2$).

Theorem 3 (Kummer [5]). *If a c.e. class can be partitioned into two disjoint c.e. subclasses L_1 and L_2 such that L_1 is injectively enumerable and contains infinitely many extensions of every finite subset of any member of L_2, then the class is injectively enumerable.*

Theorem 4. *There is a Friedberg numbering of the left-c.e. reals.*

Proof. Let $C(x) = \{\tau : \tau \frown 0^\omega <_L x\}$. Let

$$\mathcal{L} = \{C(x) : x \text{ is left-c.e.}\},$$

$$L_1 = \{C(x) : x(n) = 1 \text{ for an odd finite number of } n\},$$

and $L_2 = \mathcal{L} \setminus L_1$. It is clear that L_1 is injectively enumerable, and each finite subset F of a member of L_2 is contained in infinitely many members of L_1. The non-trivial part is to see that L_2 is c.e. Briefly, the idea is that we modify an enumeration $\{r_e\}_{e \in \omega}$ of all left-c.e. reals to only allow 1s to be added and removed in pairs of two. That is, we let $r_{e,s}^*$ be the longest prefix σ of the string associated with $r_{e,s}$ such that the number of 1s in σ is even. If in the end there are infinitely many 1s in r_e then $r_e^* = r_e$, and it is clear that $r_{e,s}^* \leq r_{e,s+1}^*$.

Theorem 5. *There is a Friedberg numbering of the Martin-Löf random left-c.e. reals.*

Proof. Let

$$\mathcal{R} = \{C(x) : x \text{ is ML-random and left-c.e.}\},$$

$L_1 = \{C(1^n{}^\frown \Omega) : n \in \omega\}$, and $L_2 = \mathcal{R} \backslash L_1$. Again, it is clear that L_1 is injectively enumerable and each finite subset of a member of L_2 can be extended to infinitely many members of L_1. We will argue that L_2 is c.e. Note that $1^n{}^\frown \Omega <_L 1^{n+1}{}^\frown \Omega$ for each n. Thus

$$L_2 = \bigcup_{n \in \omega} \{C(y) \in \mathcal{R} : 1^n{}^\frown \Omega <_L y <_L 1^{n+1}{}^\frown \Omega\}.$$

so it suffices to show that the sets

$$\{C(y) \in \mathcal{R} : y <_L 1^n{}^\frown \Omega\}, \tag{1}$$

$$\{C(y) \in \mathcal{R} : 1^n{}^\frown \Omega <_L y\} \tag{2}$$

are uniformly c.e.

Notice that $y <_L 1^n{}^\frown \Omega$ iff there is some k such that $y \restriction k <_L (1^n{}^\frown \Omega) \restriction k$, so for (1) it suffices to show that $\{C(y) \in \mathcal{R} : y \restriction k <_L (1^n{}^\frown \Omega) \restriction k\}$ is c.e., uniformly in n and k. This is non-trivial only if $k > n$, and in fact it suffices to show that a suitable subfamily $\mathcal{F}_k$ of $\{C(y) \in \mathcal{R} : y <_L \Omega\}$ containing

$$\{C(y) \in \mathcal{R} : y \restriction k <_L \Omega \restriction k\} \tag{1'}$$

is uniformly c.e. for $k \in \omega$.

We modify the enumeration $\{m_e\}_{e \in \omega}$ of the left-c.e. random reals from Theorem 2, producing a new enumeration $\{\widehat{m}_e\}_{e \in \omega}$. Initially, as long as $\Omega \restriction k$ looks like the constant-zero string 0^k then $\widehat{m}_e$ is made to look like $0^\frown \Omega$. Note that since $\Omega \neq 0^\omega$, $0^\frown \Omega <_L \Omega$.

If at any stage it looks like $\Omega \restriction k \neq 0^k$ then thereafter we let $\widehat{m}_e = m_e$ as long as $m_e \restriction k <_L \Omega \restriction k$. If at some stage s, $m_{e,s} \restriction k \geq_L \Omega_s \restriction k$, then we say that we are in an undesirable state, and we let $\widehat{m}_{e,t} = m_{e,s-1} \restriction k^\frown \Omega_t$ for all $t \geq s$ until a possible later stage where we are in a desirable state again.

Thus, if m_e really satisfies $m_e \restriction k <_L \Omega \restriction k$ then we will have $\widehat{m}_e = m_e$, and if not then $\widehat{m}_e$ will be a finite string $\sigma <_L \Omega \restriction k$ followed by Ω, so in any case it will be a Martin-Löf random real. Thus $\{C(\widehat{m}_e)\}_{e \in \omega}$ is an effective enumeration of a family $\mathcal{F}_k$ as stated. The argument for (2) is analogous.

2.3   Specifying Randomness Constants

Recall that Schnorr's Theorem states that a real $x \in 2^\omega$ is Martin-Löf random if and only if there is a constant c such that for all n, $K(x \restriction n) \geq n - c$. The optimal randomness constant of x is the least c such that this holds. For each interval $I \subseteq \omega$ we let $\mathcal{A}_I$ ($\mathcal{R}_I$) denote the set of all Martin-Löf random (and left-c.e., respectively) reals whose optimal randomness constant belongs to I. Let μ denote the fair-coin Cantor-Lebesgue measure on 2^ω. By the proof of Schnorr's Theorem we have

$$\mu(\{x : (\forall n)K(x \restriction n) \geq n - c\}) \geq 1 - 2^{-(c+1)}.$$

Consequently, if $c \geq 0$, then $\mu\mathcal{A}_{[0,c]} > 0$ and $\mathcal{A}_{[0,c]} \neq \varnothing$.

Theorem 6. *Let $c \geq 0$. There is no effective enumeration of $\mathcal{R}_{[0,c]}$.*

Proof. Suppose that $\{\alpha_e\}_{e \in \omega}$ is such an enumeration, with a uniformly computable approximation $\alpha_{e,s}$ such that $\alpha_e = \lim_{s \to \infty} \alpha_{e,s}$ and $\alpha_{e,s} \leq \alpha_{e,s+1}$. Note that

$$\mathcal{A}_{[0,c]} = \{x : (\forall n)K(x \restriction n) \geq n - c\}$$

is a Π^0_1 class. Let $\beta_s = \max\{\alpha_{e,s} : e \leq s\}$. Then $\beta = \lim_{s \to \infty} \beta_s$ is left-c.e., and since the left-c.e. members of $A_{[0,c]}$ are dense in $A_{[0,c]}$, β is the rightmost path of $A_{[0,c]}$. However the rightmost path of a Π^0_1 class is also *right-c.e.*, defined in the obvious way. Thus β is a Martin-Löf random real that is computable, a contradiction.

Theorem 7. *For each c there is an effective numbering of $\mathcal{R}_{[c+1,\infty)}$.*

Proof sketch. Let $\{m_e\}_{e \in \omega}$ be an effective enumeration of all left-c.e. random reals, with the additional property that for each e there are infinitely many e' such that for all s, $m_{e,s} = m_{e',s}$. We will define an effective numbering $\{\alpha_e\}_{e \in \omega}$ of $\mathcal{R}_{[c+1,\infty)}$.

We say that a string σ *satisfies randomness constant c at stage t* if

$$K_t(\sigma) \geq |\sigma| - c;$$

otherwise, we say that σ fails randomness constant c at stage t.

We proceed in stages $t \in \omega$, monitoring each $m_{e,t}$ for $e \leq t$ at stage t. If for some t_0, n, e, we observe that $m_{e,t_0}[n]$ fails randomness constant c, then we want to assign a place for m_e in our enumeration of $\mathcal{R}_{[c+1,\infty)}$. So we let d be minimal so that α_d has not yet been mentioned in the construction, and let $\alpha_{d,s} = m_{e,s}$ for all stages $s \geq t_0$ until further notice. If $m_{e,t_1}[n]$ at some stage $t_1 \geq t_0$ satifies randomness constant c, then we *regret* having assigned m_e a place in our enumeration $\{\alpha_e\}_{e \in \omega}$. To compensate for this regret, we choose a large number $p = p_{c,n}$ and for all stages $s \geq t_1$ let $\alpha_{d,s} = m_{e,s}[n]^\frown 0^p {}^\frown \Omega_s$. The largeness of p guarantees that $m_{e,s}[n]^\frown 0^p$ does not satisfy randomness constant c. [1] If m_e actually does fail randomness constant c, but at a larger length $n' > n$, then because there are infinitely many e' with $m_{e'} = m_e$ we will eventually assign some $\alpha_{d'}$ to some such $m_{e'}$ at a stage t_2 that is so large that $m_{e',t_2}[n'] = m_{e'}[n']$. Thus, each real in $\mathcal{R}_{[c+1,\infty)}$ will eventually be assigned a permanent $\alpha_{d'}$.

---

[1] To be precise, if $|\sigma| = n$ then there are universal constants $\hat{c}$ and $\tilde{c}$ such that, thinking of p sometimes as a string, $K(\sigma^\frown 0^p) \leq K(\sigma) + K(p) + \hat{c} \leq 2|\sigma| + 2|p| + \tilde{c} = 2n + 2\log p + \tilde{c} \leq n + p - c$ provided $p - 2\log p \geq n + \tilde{c} + c$, which is true for $p = p_{n,c}$ that we can find effectively.

Remark 1. We believe that one can even show that there is a Friedberg numbering of $\mathcal{R}_{[c+1,\infty)}$. The idea is to modify L_1 so that the strings 1^n are replaced by 1^{d_c+n} for a sufficiently large d_c, as in the footnote on page 53.

Remark 2. Theorems 6 and 7 indicate perhaps that the left-c.e. members of Σ_2^0 classes are generally easier to enumerate than those of Π_1^0 classes; this may be due to the "Σ_n^0 nature" of left-c.e. reals (for $n = 1$).

Family	Enumeration?	Friedberg?
All Π_1^0 classes		Yes, by Theorem 10
All left-c.e. reals		Yes, by Theorem 4
Π_1^0 classes C, $\mu C > 0$		Yes, by Theorem 10
Left-c.e. reals in MLR		Yes, by Theorem 5
Π_1^0 classes $\subseteq \mathcal{A}_{[0,c]}$	Yes, by Proposition 2	Open problem
Π_1^0 classes $\subseteq$ MLR	No, by Theorem 8	
Left-c.e. reals in $\mathcal{A}_{[0,c]}$	No, by Theorem 6	

Fig. 1. Existence of effective numberings and Friedberg numberings, where $\mathrm{MLR} = \bigcup_{c \in \omega} \mathcal{A}_{[0,c]}$

Whether a set of the form $\mathcal{A}_{[c_1,c_2]}$ for $0 \leq c_1 \leq c_2 < \infty$ is nonempty appears to depend on the universal prefix machine on which Kolmogorov complexity is based.

Question 1. Does there exist $0 \leq c_1 \leq c_2 < \infty$ and a choice of universal machine underlying Kolmogorov complexity such that $\mathcal{A}_{[c_1,c_2]}$ has no effective enumeration?

3   Families of Π_1^0 Classes

Definition 3 ([1]). *Let $\mathcal{C}$ be a family of closed subsets of 2^ω. We say that $\mathcal{C}$ has a* computable enumeration *if there is a uniformly computable collection $\{T_e\}_{e \in \omega}$ of trees $T_e \subseteq 2^{<\omega}$ (that is, $\{\langle \sigma, e \rangle : \sigma \in T_e\}$ is computable, and $\sigma^\frown \tau \in T_e$ implies $\sigma \in T_e$) such that $\mathcal{C} = \{[T_e] : e \in \omega\}$.*

Definition 4 ([1]). *Let $\mathcal{C}$ be a family of closed subsets of 2^ω. We say that $\mathcal{C}$ has an* effective enumeration *if there is a Π_1^0 set $S \subseteq 2^\omega \times \omega$, such that $\mathcal{C} = \{\{X : (X,e) \in S\} : e \in \omega\}$.*

Proposition 1. *Let $\mathcal{C}$ be a family of closed subsets of 2^ω. The following are equivalent:*

(1) $\mathcal{C}$ has a computable enumeration;
(2) $\mathcal{C}$ has an effective enumeration.

Proof. (1) implies (2): Let $\{T_e\}_{e \in \omega}$ be given, and define

$$S = \{(X, e) : \forall n \; X \upharpoonright n \in T_e\}.$$

(2) implies (1): Let S be given, let Φ_a be a Turing functional such that $(X, e) \in S \Leftrightarrow \Phi_a^X(e) \uparrow$, and let $T_e = \{\sigma \in 2^{<\omega} : \Phi_{a,|\sigma|}^\sigma(e) \uparrow\}$.

In light of Proposition 1, we may use either notion. Note that if C belongs to a family as in Proposition 1 then C is a Π_1^0 class.

3.1 Existence of Numberings

Theorem 8. *Let $P \subseteq 2^\omega$, let $\mathcal{C}_P$ be the collection of all Π_1^0 classes contained in P, and let $\mathcal{N}_P$ be the collection of all nonempty Π_1^0 classes contained in P. Assume P has the following properties:*

(i) P is co-dense: no cone $[\sigma]$, $\sigma \in 2^{<\omega}$, is contained in P;
(ii) P is closed under shifts: if $x \in P$ then $\sigma ^\frown x \in P$;
(iii) $\mathcal{N}_P \neq \emptyset$.

Then there is no effective numbering of either $\mathcal{C}_P$ or $\mathcal{N}_P$.

Proof. If there is a numbering of $\mathcal{N}_P$ then there is one of $\mathcal{C}_P$, because if $\emptyset \in \mathcal{C}_P$ (as is always the case) we may simply add an index of $\emptyset$ to the numbering. Thus it suffices to show that there is no effective numbering of $\mathcal{C}_P$. Suppose to the contrary that $e \mapsto [T_e]$ enumerates the family of Π_1^0 classes in $\mathcal{C}_P$. By (iii), we may assume $[T_0] \neq \emptyset$. By (i), T_0 has infinitely many dead ends. Let the dead ends of T_0 be listed in a computable way (for instance, by length-lexicographic order), as σ_n, $n \in \omega$. By (i) again, we may let τ_n be the least extension of σ_n which extends a dead end of T_n. Define a computable tree T by putting T_0 above τ_n. That is, let $[T] \cap [\tau_n] = [\tau_n T_0]$ and $[T] = [T_0] \cup \bigcup_n [\tau_n T_0]$. By (ii), the resulting class $[T]$ belongs to $\mathcal{C}_P$. Since $[T_0] \neq \emptyset$, $[T] \cap [\tau_n] \neq \emptyset = [T_n] \cap [\tau_n]$, so $[T]$ is not contained in or equal to any $[T_n]$.

All assumptions (i), (ii), (iii) of Theorem 8 are necessary: consider $P = 2^\omega$, $P = \{x\}$, where x is a single computable real, and $P = \emptyset$, respectively.

Corollary 1. *The following families of Π_1^0 classes have no effective numbering:*

1. *Π_1^0 classes containing only Martin-Löf random reals;*
2. *special Π_1^0 classes (those containing only non-computable reals);*
3. *Π_1^0 classes containing only reals x such that the Muchnik degree [7] of $\{x\}$ is above a fixed nonzero Muchnik degree;*
4. *Π_1^0 classes containing only finite (or only co-finite) subsets of ω.*

Proposition 2. *(1) The family of all Π_1^0 classes containing only reals that are Martin-Löf random with respect to a fixed randomness constant is effectively enumerable.*

(2) The family of all Σ_2^0 classes containing only Martin-Löf random reals is effectively enumerable.

Proof. (1). We enumerate all Π_1^0 classes as $\{P_i\}_{i \in \omega}$ and let

$$Q_i = P_i \cap \{x : \forall n \ K(x \upharpoonright n) \geq n - c\}.$$

Then $\{Q_i\}_{i \in \omega}$ is an enumeration of all Π_1^0 classes containing only reals that are Martin-Löf random with randomness constant c. Part (2) is analogous.

We may sum up the situation by stating that it is only the mixture of Π_1^0 and Σ_2^0 classes that leads to the negative result of Corollary 1(1). The proof of Theorem 8 for the case in Corollary 1(1) proves the following basic property of Martin-Löf tests.

Corollary 2. *For each Martin-Löf test $\{U_n\}_{n \in \omega}$ there is a Σ_1^0 class V containing all non-Martin-Löf random reals but containing no set U_n, $n \in \omega$.*

As is well-known, all Π_1^0 classes containing Martin-Löf random reals have positive measure. In contrast to Corollary 1(1), such classes can be effectively enumerated:

Theorem 9. *There is an effective numbering of the Π_1^0 classes of positive measure.*

Proof. It suffices to enumerate, uniformly in $n \in \omega$, all Π_1^0 classes of measure at least $r := \frac{1}{n}$. To accomplish this, let $e \mapsto W_e$ be an effective numbering of Σ_1^0 sets of strings, which gives rise to all Π_1^0 classes. That is, if P is a Π_1^0 class, then $P = 2^\omega \setminus [W_e]^{\preceq}$ for some e. Modify this enumeration so that strings enumerate into each W_e so long as the overall measure never surpasses $1 - r$. More precisely, if, at some stage $s > 0$, some σ is supposed to enter $W_{e,s}$ but this causes the measure of $[W_e]^{\preceq}$ to surpass $1 - r$, then we hereafter discontinue to enumerate strings into W_e; call this modified set $\widehat{W}_{e;n}$. It follows that $e \mapsto \widehat{W}_{e;n}$ is a numbering that gives rise to all Σ_1^0 classes of measure at most $1 - \frac{1}{n}$. Then the sequence of sets $\left\{ \left[\widehat{W}_{e;n} \right]^{\preceq} \right\}$ for $\langle e, n \rangle \in \omega \times \omega$ is an effective enumeration of the Σ_1^0 classes of measure less than 1.

This contrasts with the result of [1] that there is no effective numbering of the Π_1^0 classes of measure zero. We next show that any effectively enumerable family of Π_1^0 classes containing all the clopen classes has a Friedberg numbering. In fact, we show something slightly stronger.

Definition 5. *The optimal covering of $S \subseteq 2^{<\omega}$ is*

$$O = O_S = \{\sigma : [\sigma] \subseteq [S]^{\preceq} \ \& \ \neg(\exists \tau \prec \sigma)([\tau] \subseteq [S]^{\preceq})\}.$$

Let $\mathfrak{A}$ be the family of all sets O that have odd cardinality and are optimal coverings of sets S.

Theorem 10. *Any effectively enumerable family of Σ_1^0 classes $\mathcal{F}$ with $\mathcal{F} \supseteq \{[O]^{\preceq} : O \in \mathfrak{A}\}$ has a Friedberg numbering.*

Proof. For a set $Z \subseteq 2^{<\omega}$, we say that Z is *filter closed* if Z is closed under extensions ($\sigma \in Z \Rightarrow \sigma^\frown \tau \in Z$) and such that whenever both $\sigma^\frown 0$ and $\sigma^\frown 1$ are in Z then $\sigma \in Z$. The *filter closure* of Y is the intersection of all filter closed sets containing Y and is denoted by $Y^\uparrow$.

Since $\mathcal{F}$ is effectively enumerable, we may let $e \mapsto Y_e$ be a numbering of all filter closed sets of strings with $[Y_e]^{\preceq} \in \mathcal{F}$. Since $Y_e \neq Y_{e'}$ implies $[Y_e]^{\preceq} \neq [Y_{e'}]^{\preceq}$, it suffices to injectively enumerate these sets Y_e. Let

$$L_1 = \{O^\uparrow : O \in \mathfrak{A}\} \text{ and } L_2 = \{Y_e : Y_e \notin L_1\}.$$

It is clear that L_1 is injectively enumerable. By the assumption of the theorem, each $[O^\uparrow]^{\preceq} \in \mathcal{F}$. It is also clear that each finite subset of any $Y \in L_2$ is contained in infinitely many $O^\uparrow \in L_1$.

We claim that L_2 has an effectively enumeration $\{Y_e^*\}_{e \in \omega}$, to be constructed below. Fix e and let $Y_e = \{\sigma_n\}_{n \in \omega}$ in order of enumeration.

$S \subseteq 2^{<\omega}$ is an *acceptable family* if its optimal covering O has finite even cardinality. In particular $O \notin \mathfrak{A}$. We say that stage n is *good* if σ_n has greater length than any member of $O_n = O_{S_n}$ for $S_n = \{\sigma_0, \ldots, \sigma_{n-1}\}$ and does not extend any member of O_n.

Construction. We will construct Y_e^* as $Y_e^* = \bigcup_{n \in \omega} Y_{e,n}$ for uniformly computable sets $Y_{e,n}$. We set $Y_{e,-1} = \varnothing$. Suppose $n \geq 0$. If stage n is not good, we keep $Y_{e,n} = Y_{e,n-1}$.

If stage n is good, there are two cases.

Case a. S_n is an acceptable family. Then let $Y_{e,n}$ be the filter closure of O_n.

Case b. Otherwise. Then let $Y_{e,n}$ be the filter closure of $O_n \cup \{\sigma_n\}$.

We separately enumerate all sets generated from any acceptable family whose optimal covering has finite even cardinality. (*)

End of Construction.

Verification. Note that in both Case a and Case b, $Y_{e,n}$ is the filter closure of an acceptable family, so we do not enumerate any member of L_1. By (*), it therefore suffices to show that we enumerate all sets generated from an infinite family, i.e. non-clopen sets, and that each Y_e^* is some $Y_{e'}$.

If Y_e is not clopen then there are infinitely many good stages. Then in the end $Y_e^* = Y_e$, because σ_n is covered either right away (case b) or at the next good stage (case b).

Corollary 3. *The family of all Σ_1^0 classes of measure less than one, or equivalently Π_1^0 classes of positive measure, has a Friedberg numbering.*

Acknowledgments

The second author was partially supported by NSF grant DMS-0652669.

References

[1] Brodhead, P., Cenzer, D.: Effectively closed sets and enumerations. Arch. Math. Logic 46(7-8), 565–582 (2008); MR 2395559 (2009b:03119)

[2] Ershov, Y.L.: Theory of numberings. In: Handbook of computability theory. Stud. Logic Found. Math., vol. 140, pp. 473–503. North-Holland, Amsterdam (1999); MR 1720731 (2000j:03060)

[3] Friedberg, R.M.: Three theorems on recursive enumeration. I. Decomposition. II. Maximal set. III. Enumeration without duplication. J. Symb. Logic 23, 309–316 (1958); MR 0109125 (22 #13)

[4] Li, M., Vitányi, P.: An introduction to Kolmogorov complexity and its applications, 2nd edn. Graduate Texts in Computer Science. Springer, New York (1997); MR 1438307 (97k:68086)

[5] Kummer, M.: An easy priority-free proof of a theorem of Friedberg. Theoret. Comput. Sci. 74(2), 249–251 (1990); MR 1067521 (91h:03056)

[6] Nies, A.: Computability and randomness. Oxford University Press, Oxford (2009)

[7] Simpson, S.G.: An extension of the recursively enumerable Turing degrees. J. Lond. Math. Soc. 75(2), 287–297 (2007); MR 2340228 (2008d:03041)

The Strength of the Grätzer-Schmidt Theorem

Paul Brodhead and Bjørn Kjos-Hanssen

Department of Mathematics, University of Hawai'i at Mānoa, Honolulu HI 96822
brodhead@math.hawaii.edu, bjoern@math.hawaii.edu
http://www.math.hawaii.edu/~brodhead
http://www.math.hawaii.edu/~bjoern

Abstract. The Grätzer-Schmidt theorem of lattice theory states that each algebraic lattice is isomorphic to the congruence lattice of an algebra. A lattice is algebraic if it is complete and generated by its compact elements. We show that the set of indices of computable lattices that are complete is Π_1^1-complete; the set of indices of computable lattices that are algebraic is Π_1^1-complete; and that there is a computable lattice L such that the set of compact elements of L is Π_1^1-complete. As a corollary, there is a computable algebraic lattice that is not computably isomorphic to any computable congruence lattice.

Keywords: lattice theory, computability theory.

Introduction

The Grätzer-Schmidt theorem [2], also known as the *congruence lattice representation theorem*, states that each algebraic lattice is isomorphic to the congruence lattice of an algebra. It established a strong link between lattice theory and universal algebra. In this article we show that this theorem as stated fails to hold effectively in a very strong way.

We use notation associated with partial computable functions, φ_e, $\varphi_{e,s}$, $\varphi_{e,s}^\sigma$, φ_e^f as in Odifreddi [3]. Following Sacks [5] page 5, a Π_1^1 subset of ω may be written in the form

$$C_e = \{n \in \omega \mid \forall f \in \omega^\omega \ \varphi_e^f(n) \downarrow\}.$$

A subset $A \subseteq \omega$ is Π_1^1-*hard* if each Π_1^1 set is m-reducible to A; that is, for each e, there is a computable function f such that for all n, $n \in C_e$ iff $f(n) \in A$. A is Π_1^1-*complete* if it is both Π_1^1 and Π_1^1-hard. It is well known that such sets exist. Fix for the rest of the paper a number e_0 so that C_{e_0} is Π_1^1-complete. With each n, the set C_{e_0} associates a tree T_n' defined by

$$T_n' = \{\sigma \in \omega^{<\omega} \mid \varphi_{e_0,|\sigma|}^\sigma(n) \uparrow\}.$$

Note that T_n' has no infinite path iff $n \in C_e$.

A *computable lattice* $(L, \preceq)$ has underlying set $L = \omega$ and an lattice ordering $\preceq$ that is formally a subset of ω^2.

K. Ambos-Spies, B. Löwe, and W. Merkle (Eds.): CiE 2009, LNCS 5635, pp. 59–67, 2009.

We will use the symbol $\preceq$ for lattice orderings, and reserve the symbol $\leq$ for the natural ordering of the ordinals and in particular of ω. Meets and joins corresponding to the order $\preceq$ are denoted by $\wedge$ and $\vee$. Below we will seek to build computable lattices from the trees T'_n; since for many n, T'_n will be finite, and a computable lattice must be infinite according to our definition, we will work with the following modification of T'_n:

$$T_n = T'_n \cup \{\langle i \rangle : i \in \omega\} \cup \{\varnothing\}$$

where $\varnothing$ denotes the empty string and $\langle i \rangle$ is the string of length 1 whose only entry is i. This ensures that T_n has the same infinite paths as T'_n, and each T_n is infinite. Moreover the sequence $\{T_n\}_{n \in \omega}$ is still uniformly computable.

1 Computational Strength of Lattice-Theoretic Concepts

1.1 Completeness

Definition 1. *A lattice $(L, \preceq)$ is* complete *if for each subset $S \subseteq L$, both $\sup S$ and $\inf S$ exist.*

Lemma 1. *The set of indices of computable lattices that are complete is Π_1^1.*

Proof. The statement that $\sup S$ exists is equivalent to a first order statement in the language of arithmetic with set variable S:

$$\exists a [\forall b(b \in S \to b \preceq a) \ \& \ \forall c((\forall b(b \in S \to b \preceq c) \to a \preceq c)].$$

The statement that $\inf S$ exists is similar, in fact dual. Thus the statement that L is complete consists of a universal set quantifier over S, followed by an arithmetical matrix.

Example 1. In set-theoretic notation, $(\omega + 1, \leq)$ is complete. Its sublattice $(\omega, \leq)$ is not, since $\omega = \sup \omega \notin \omega$.

Proposition 1. *The set of indices of computable lattices that are complete is Π_1^1-hard.*

Proof. Let L_n consist of two disjoint copies of T_n, called T_n and T_n^*. For each $\sigma \in T_n$, its copy in T_n^* is called σ^*. Order L_n so that T_n has the prefix ordering

$$\sigma \preceq \sigma^\frown \tau,$$

T_n^* has the reverse prefix ordering, and $\sigma \prec \sigma^*$ for each $\sigma \in T_n$. We take the transitive closure of these axioms to obtain the order of L_n; see Figure 1.

Next, we verify that L_n is a lattice. For any $\sigma, \tau \in T_n$ we must show the existence of (1) $\sigma \vee \tau$, (2) $\sigma \wedge \tau$, (3) $\sigma \vee \tau^*$, and (4) $\sigma \wedge \tau^*$; the existence of $\sigma^* \vee \tau^*$ and $\sigma^* \wedge \tau^*$ then follows by duality.

We claim that for any strings $\alpha, \sigma \in T_n$, we have $\alpha^* \succeq \sigma$ iff α is comparable with σ; see Figure 1. In one direction, if $\alpha \succeq \sigma$ then $\alpha^* \succeq \alpha \succeq \sigma$, and if $\sigma \succeq \alpha$

then $\alpha^* \succeq \sigma^* \succeq \sigma$. In the other direction, if $\alpha^* \succeq \sigma$ then by the definition of $\preceq$ as a transitive closure there must exist ρ with $\alpha^* \succeq \rho^* \succeq \rho \succeq \sigma$. Then $\alpha \preceq \rho$ and $\sigma \preceq \rho$, which implies that α and ρ are comparable.

Using the claim we get that (1) $\sigma \vee \tau$ is $(\sigma \wedge \tau)^*$, where (2) $\sigma \wedge \tau$ is simply the maximal common prefix of σ and τ; (3) $\sigma \vee \tau^*$ is $\sigma^* \vee \tau^*$ which is $(\sigma \wedge \tau)^*$; and (4) $\sigma \wedge \tau^*$ is $\sigma \wedge \tau$.

It remains to show that $(L_n, \preceq)$ is complete iff T_n has no infinite path. So suppose T_n has an infinite path S. Then $\sup S$ does not exist, because S has no greatest element, S^* has no least element, each element of S^* is an upper bound of S, and there is no element above all of S and below all of S^*.

Conversely, suppose T_n has no infinite path and let $S \subseteq L_n$. If S is finite then $\sup S$ exists. If S is infinite then since T_n has no infinite path, there is no infinite linearly ordered subset of L_n, and so S contains two incomparable elements σ and τ. Because T_n is a tree, $\sigma \vee \tau$ is in T_n^*. Now the set of all elements of L_n that are above $\sigma \vee \tau$ is finite and linearly ordered, and contains all upper bounds of S. Thus there is a least upper bound for S. Since L_n is self-dual, i.e. $(L_n, \preceq)$ is isomorphic to $(L_n, \succeq)$ via $\sigma \mapsto \sigma^*$, infs also always exist. So L_n is complete.

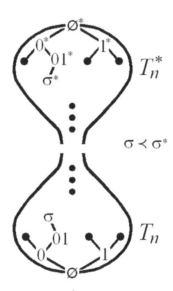

Fig. 1. The lattice L_n from Proposition 1

1.2 Compactness

Definition 2. *An element $a \in L$ is* compact *if for each subset $S \subseteq L$, if $a \preceq \sup S$ then there is a finite subset $S' \subseteq S$ such that $a \preceq \sup S'$.*

Lemma 2. *In each computable lattice L, the set of compact elements of L is Π_1^1.*

Proof. Similarly to the situation in Lemma 1, the statement that a is compact consist of a universal set quantifier over S followed by an arithmetical matrix.

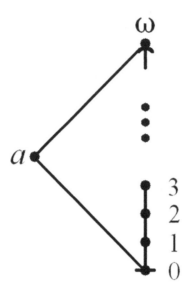

Fig. 2. The lattice $L[a]$ from Example 2

Example 2. Let $L[a] = \omega + 1 \cup \{a\}$ be ordered by $0 \prec a \prec \omega$, and let the element a be incomparable with the positive numbers. Then a is not compact, because $a \preceq \sup \omega$ but $a \not\preceq \sup S'$ for any finite $S' \subseteq \omega$.

The following result will be useful for our study of the Grätzer-Schmidt theorem.

Proposition 2. *There is a computable algebraic lattice L such that the set of compact elements of L is Π_1^1-hard.*

Proof. Let L consist of disjoint copies of the trees T_n, $n \in \omega$, each having the prefix ordering; least and greatest elements 0 and 1; and elements a_n, $n \in \omega$, such that $\sigma \prec a_n$ for each $\sigma \in T_n$, and a_n is incomparable with any element not in $T_n \cup \{0, 1\}$ (see Figure 3).

Suppose T_n has an infinite path S. Then $a_n = \sup S$ but $a_n \not\preceq \sup S'$ for any finite $S' \subseteq S$, since $\sup S'$ is rather an element of S. Thus a_n is not compact.

Conversely, suppose T_n has no infinite path, and $a_n \preceq \sup S$ for some set $S \subseteq L$. If S contains elements from $T_m \cup \{a_m\}$ for at least two distinct values of m, say $m_1 \neq m_2$, then $\sup S = 1 = \sigma_1 \vee \sigma_2$ for some $\sigma_i \in S \cap (T_{m_i} \cup \{a_{m_i}\})$, $i = 1, 2$. So $a_n \preceq \sup S'$ for some $S' \subseteq S$ of size two. If S contains 1, there is nothing to prove. The remaining case is where S is contained in $T_m \cup \{a_m, 0\}$ for some m. Since $a_n \preceq \sup S$, it must be that $m = n$. If S is finite or contains a_n, there is nothing to prove. So suppose S is infinite. Since T_n has no infinite path, there must be two incomparable elements of T_n in S. Their join is then a_n, since T_n is a tree, and so $a_n \preceq \sup S'$ for some $S' \subseteq S$ of size two.

Thus we have shown that a_n is compact if and only if T_n has no infinite path. There is a computable presentation of L where a_n is a computable function of

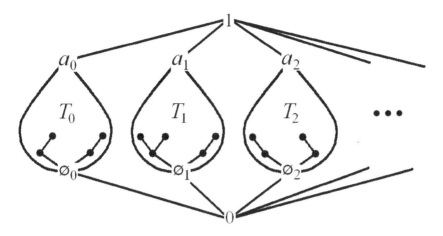

Fig. 3. The lattice L from Proposition 2

n, for instance we could let $a_n = 2n$. Thus letting $f(n) = 2n$, we have that T_n has no infinite path iff $f(n)$ is compact, i.e. $\{a \in L : a \text{ is compact}\}$ is Π^1_1-hard.

1.3 Algebraicity

Definition 3. *A lattice* $(L, \preceq)$ *is* compactly generated *if* $C = \{a \in L : a \text{ is compact}\}$ *generates* L *under* sup, *i.e., each element is the supremum of its compact predecessors. A lattice is* algebraic *if it is complete and compactly generated.*

Lemma 3. *The set of indices of computable lattices that are algebraic is* Π^1_1.

Proof. L is algebraic if it is complete (this property is Π^1_1 by Lemma 1) and *each element is the least upper bound of its compact predecessors*, i.e., any element that is above all the compact elements below a is above a:

$$\forall b(\forall c(c \in C \ \& \ c \preceq a \rightarrow c \preceq b) \rightarrow a \preceq b)$$

Equivalently,

$$\forall b(\exists c(c \in C \ \& \ c \preceq a \ \& \ c \not\preceq b) \text{ or } a \preceq b)$$

This is equivalent to a Π^1_1 statement since, by the Axiom of Choice, any statement of the form $\exists c \ \forall S \ A(c, S)$ is equivalent to $\forall (S_c)_{c \in \omega} \ \exists c \ A(c, S_c)$.

Example 3. The lattice $(\omega + 1, \leq)$ is compactly generated, since the only non-compact element ω satisfies $\omega = \sup \omega$. The lattice $L[a]$ from Example 2 and Figure 2 is not compactly generated, as the non-compact element a is not the supremum of $\{0\}$.

Proposition 3. *The set of indices of computable lattices that are algebraic is* Π_1^1*-hard.*

Proof. Let the lattice $T_n[a]$ consist of T_n with the prefix ordering, and additional elements $0 \prec a \prec 1$ such that a is incomparable with each $\sigma \in T_n$, and 0 and 1 are the least and greatest elements of the lattice. Note that $T_n[a]$ is always complete, since any infinite set has supremum equal to 1. We claim that $T_n[a]$ is algebraic iff T_n has no infinite path.

Suppose T_n has an infinite path S. Then $a \preceq \sup S$, but $a \not\preceq \sup S'$ for any finite $S' \subseteq S$. Thus a is not compact, and so a is not the sup of its compact predecessors (0 being its only compact predecessor), which means that $T_n[a]$ is not an algebraic lattice.

Conversely, suppose $T_n[a]$ is not algebraic. Then some element of $T_n[a]$ is not the join of its compact predecessors. In particular, some element of $T_n[a]$ is not compact. So there exists a set $S \subseteq T_n[a]$ such that for all finite subsets $S' \subseteq S$, $\sup S' < \sup S$. In particular S is infinite. Since each element except 1 has only finitely many predecessors, we have $\sup S = 1$. Notice that $T_n[a] \setminus \{1\}$ is actually a tree, so if S contains two incomparable elements then their join is already 1, contradicting the defining property of S. Thus S is linearly ordered, and infinite, which implies that T_n has an infinite path.

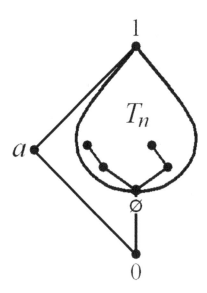

Fig. 4. The lattice $T_n[a]$ from Proposition 3

2   Lattices of Equivalence Relations

Let $\mathrm{Eq}(A)$ denote the set of all equivalence relations on A. Ordered by incusion, $\mathrm{Eq}(A)$ is a complete lattice. In a sublattice $L \subseteq \mathrm{Eq}(A)$, we write $\sup_L$ for the supremum in L when it exists, and sup for the supremum in $\mathrm{Eq}(A)$, and note that $\sup \leq \sup_L$.

A *complete sublattice of* $\mathrm{Eq}(A)$ is a sublattice L of $\mathrm{Eq}(A)$ such that $\sup_L = \sup$ and $\inf_L = \inf$. A sublattice of $\mathrm{Eq}(A)$ that is a complete lattice is not necessarily a complete sublattice in this sense. The following lemma is well known. A good reference for lattice theory is the monograph of Grätzer [1].

Lemma 4. *Suppose A is a set and $(L, \subseteq)$ is a complete sublattice of $\mathrm{Eq}(A)$. Then an equivalence relation E in L is a compact member of L if and only if E is finitely generated in L.*

Proof. One direction only uses that L is a sublattice of $\mathrm{Eq}(A)$ and L is complete as a lattice. Suppose E is not finitely generated in L. Let $C_{(a,b)}$ denote the infimum of all equivalence relations in L that contain (a, b). Then $E \subseteq \sup_L\{C_{(a,b)} : aEb\}$, but E is not below any finite join of the relations $C_{(a,b)}$. So E is not compact.

Suppose E is finitely generated in L. So there exists an n and pairs $(a_1, b_1), \ldots,$ (a_n, b_n) such that $a_i E b_i$ for all $1 \leq i \leq n$, and for all equivalence relations F in L, if $a_i F b_i$ for all $1 \leq i \leq n$ then $E \subseteq F$. Suppose $E \subseteq \sup_L\{E_i : 1 \leq i < \infty\}$ for some $E_1, E_2, \ldots \in L$. Since L is a complete sublattice of $\mathrm{Eq}(A)$, $\sup_L = \sup$, so $E \subseteq \sup\{E_i : 1 \leq i < \infty\}$. Note that $\sup\{E_i : 1 \leq i < \infty\}$ is the equivalence relation generated by the relations E_i under transitive closure. So there is some $j = j_n < \infty$ such that $\{(a_i, b_i) : 1 \leq i \leq n\} \subseteq \bigcup_{i=1}^{j} E_i$ and hence $E \subseteq \bigcup_{i=1}^{j} E_i$. Thus E is compact.

A *computable complete sublattice of* $\mathrm{Eq}(\omega)$ is a uniformly computable collection $\mathcal{E} = \{E_i\}_{i \in \omega}$ of distinct equivalence relations on ω such that $(\mathcal{E}, \subseteq)$ is a complete sublattice of $\mathrm{Eq}(\omega)$. We say that the lattice $L = (\omega, \preceq)$ is *computably isomorphic* to $(\mathcal{E}, \subseteq)$ if there is a computable function $\varphi : \omega \to \omega$ such that for all i, j, we have $i \preceq j \leftrightarrow E_{\varphi(i)} \subseteq E_{\varphi(j)}$.

Lemma 5. *The indices of compact congruences in a computable complete sublattice of $\mathrm{Eq}(\omega)$ form a Σ_2^0 set.*

Proof. Suppose the complete sublattice is $\mathcal{E} = \{E_i\}_{i \in \omega}$. By Lemma 4, E_k is compact if and only if it is finitely generated, i.e.,

$$\exists n \, \exists a_1, \ldots, a_n \, \exists b_1, \ldots, b_n \left[\bigwedge_{i=1}^{n} a_i E_k b_i \ \& \ \forall j \left(\bigwedge_{i=1}^{n} a_i E_j b_i \to E_k \subseteq E_j \right) \right].$$

Here $E_k \subseteq E_j$ is Π_1^0: $\forall x \forall y \, (x E_k y \to x E_j y)$, so the formula is Σ_2^0.

Theorem 1. *There is a computable algebraic lattice that is not computably isomorphic to any computable complete sublattice of $\mathrm{Eq}(\omega)$.*

Proof. Let L be the lattice of Proposition 2, and let f be the m-reduction of Proposition 2. Suppose φ is a computable isomorphism between L and a computable complete sublattice of $\mathrm{Eq}(\omega)$, $(\mathcal{E}, \subseteq)$. Since being compact is a lattice-theoretic property, it is a property preserved under isomorphisms. Thus an

element $a \in L$ is compact if and only if $E_{\varphi(a)}$ is a compact congruence relation. This implies that T_n has no infinite path if and only if $f(n)$ is a compact
element of L, if and only if $E_{\varphi(f(n))}$ is a compact congruence relation. By Lemma
5, this implies that $C_{e_0} = \{n : T_n$ has no infinite path$\}$ is a Σ_2^0 set, contradicting
the fact that this set is Π_1^1-complete.

3   Congruence Lattices

An *algebra* $\mathfrak{A}$ consists of a set A and functions $f_i : A^{n_i} \to A$. Here i is taken
from an index set I which may be finite or infinite, and n_i is the arity of f_i.
Thus, an algebra is a purely functional model-theoretic structure. A *congruence
relation* of $\mathfrak{A}$ is an equivalence relation on A such that for each unary f_i and all
$x, y \in A$, if xEy then $f_i(x)Ef_i(y)$, and the natural similar property holds for f_i
of arity greater than one.

The congruence relations of $\mathfrak{A}$ form a lattice under the inclusion (refinement)
ordering. This lattice $\mathrm{Con}(\mathfrak{A})$ is called the *congruence lattice* of $\mathfrak{A}$.

The following lemma is well-known and straight-forward.

Lemma 6. *If $\mathfrak{A}$ is an algebra on A, then $\mathrm{Con}(\mathfrak{A})$ is a complete sublattice of
$Eq(A)$.*

Thus, may define a *computable congruence lattice* to be a computable complete
sublattice of $\mathrm{Eq}(\omega)$ which is also $\mathrm{Con}(\mathfrak{A})$ for some algebra $\mathfrak{A}$ on ω.

Theorem 2. *There is a computable algebraic lattice that is not computably isomorphic to any computable congruence lattice.*

Proof. By Theorem 1, there is a computable algebraic lattice that is not even
computably isomorphic to any computable complete sublattice of $\mathrm{Eq}(\omega)$.

Thus, we have a failure of a certain effective version of the following theorem.

Theorem 3 (Grätzer-Schmidt [2]). *Each algebraic lattice is isomorphic to
the congruence lattice of an algebra.*

Remark 1. Let A be a set, and let L be a complete sublattice of $\mathrm{Eq}(A)$. Then
L is algebraic [1], and so by Theorem 3 L is *isomorphic* to $\mathrm{Con}(\mathfrak{A})$ for some
algebra $\mathfrak{A}$ on some set, but it is not in general possible to find $\mathfrak{A}$ such that L is
equal to $\mathrm{Con}(\mathfrak{A})$. Thus, Theorem 1 is not a consequence of Theorem 2.

Remark 2. The proof of Theorem 2 shows that not only does the Grätzer-
Schmidt theorem not hold effectively, it does not hold arithmetically. We conjecture that within the framework of reverse mathematics, a suitable form of
Grätzer-Schmidt may be shown to be equivalent to the system Π_1^1-CA$_0$ (Π_1^1-
comprehension) over the base theory ACA$_0$ (arithmetic comprehension). On the
other hand, W. Lampe has pointed out that the Grätzer-Schmidt theorem is normally proved as a corollary of a result which may very well hold effectively: each
upper semilattice with least element is isomorphic to the collection of compact
congruences of an algebra.

Conjecture 1. An upper semilattice with least element has a computably enumerable presentation if and only if it is isomorphic to the collection of compact congruences of some computable algebra.

The idea for the *only if* direction of Conjecture 1 is to use analyze and slightly modify Jónsson and Pudlák's construction [4]. The *if* direction appears to be straightforward.

Remark 3. The lattices used in this paper are not modular, and we do not know if our results can be extended to modular, or even distributive, lattices.

Acknowledgments

The authors thank Bakh Khoussainov and the participants of the University of Hawai'i at Mānoa logic and lattice theory seminar for useful suggestions. The second author was partially supported by NSF grant DMS-0652669.

References

[1] Gratzer, G.: General lattice theory. In: Davey, B.A., Freese, R., Ganter, B., Greferath, M., Jipsen, P., Priestley, H.A., Rose, H., Schmidt, E.T., Schmidt, S.E., Wehrung, F., Wille, R. (eds.), 2nd edn., Birkhauser Verlag, Basel (2003); MR 2451139

[2] Gratzer, G., Schmidt, E.T.: Characterizations of congruence lattices of abstrac algebras. Acta Sci. Math (Szeged) 24, 34–59 (1963); MR 0151406 (27 #1391)

[3] Odifreddi, P.: Classical recursion theory. Studies in Logic and the Foundations of Mathematics, vol. 125. North-Holland Publishing Co., Amsterdam (1989); The theory of functions and sets of natural numbers; With a foreword by G. E. Sacks. MR 982269 (90d:03072)

[4] Pudlak, P.: A new proof of the congruence lattice representation theorem. Algebra Universalis 6(3), 269–275 (1976); MR 0429699 (55#2710)

[5] Sacks, G.E.: Higher recursion theory, Perspectives in Mathematical Logic. Springer, Berlin (1990); MR 1080970 (92a:03062)

Hyperloops Do Not Threaten the Notion of an Effective Procedure

Tim Button[*]

Cambridge University
button@cantab.net

Abstract. This paper develops my (forthcoming) criticisms of the philosophical significance of a certain sort of infinitary computational process, a *hyperloop*. I start by considering whether hyperloops suggest that
"effectively computable" is vague (in some sense). I then consider and
criticise two arguments by Hogarth, who maintains that hyperloops undermine the very idea of effective computability. I conclude that hyperloops, on their own, cannot threaten the notion of an effective procedure.

It has been suggested that a particular kind of infinitary hypercomputer threatens the notion of an effective procedure. The infinitary computer in question is
able to perform a *hyperloop*. I start by explaining what hyperloops are, and how
they might be implemented in a general-relativistic spacetime (Sect. 1).

I consider whether hyperloops suggest that "effectively computable" is *polyvague*. I give reason to doubt this, but I also suggest that this kind of vagueness
cannot seriously threaten the notion of an effective procedure (Sect. 2).

Next, I consider Hogarth's (2009b) claim that ordinary computers and hyperloopers, running the same algorithm, will give different outputs. Hogarth
concludes that the output of an algorithm is intrinsically relativised to the machine that implements it. However, I show that hyperloopers only give different
outputs because they implement *different* algorithms (Sect. 3).

Hogarth (2004, 2009a, 2009b) also maintains that there is a deep analogy between computing and geometry. He thinks that this shows that the very idea of an
effective procedure is bankrupt. I suggest that Hogarth's attack on the notion of
an effective procedure rests on a premise that is independent from considerations
concerning hyperloops. This premise is independently interesting, but hyperloops,
as such, tell us nothing about effective procedures (Sect. 4).

1 Hyperloops: From Python to Ouroboros

Let `fun()` be an arbitrary computable function, which takes a single natural
number as an argument, and returns a boolean. If we ignore the fact that every
actual computer only has a only finite storage capacity, then, for any value of

---

[*] Thanks to Mark Hogarth, Philip Welch, Sharon Berry, Michael Potter, Peter Smith,
and five anonymous referees for this journal.

x, an ordinary computer can test whether `fun(x)`. But an ordinary computer cannot be expected to determine whether there *is* a value of x such that `fun(x)`. We can set an ordinary computer searching for a value of x such that `fun(x)`; but if `fun(x)` is not true for any value of x, the computer will loop through an ω-sequence of calculations, without ever terminating.

But suppose we had a computer which could *complete* ω-sequences of calculations. We could use this *hyperlooper* (a hyperlooping-computer) to test whether there is some value of x such that `fun(x)`. Programme 1 captures the idea:

─────────── **Programme 1** ───────────

```
xFound = 0                                    1.1
hyperloop with x = 0:                         1.2
        if fun(x):                            1.3
                xFound = 1                    1.4
                break                         1.5
        x = x+1                               1.6
if xFound: print "Yes"                        1.7
else: print "No"                              1.8
```

This programme is written in a hypothetical programming language, Ouroboros. Ouroboros extends the Python language by adding the command "`hyperloop with...`". Executing Programme 1, the computer initialises a new variable, x, to 0, and then starts looping through lines 1.3–1.6. If there is a value of x such that `fun(x)`, the computer discovers this, sets `xFound` to 1, and **breaks** out of the hyperloop (to line 1.7). If there is no value of x such that `fun(x)`, the computer loops through all possible values of x without ever changing `xFound` from its initial value of 0. Having checked every value of x, the computer proceeds to line 1.7. So the computer is guaranteed to reach line 1.7, and when it does, `xFound` has the value 1 iff there is a value of x such that `fun(x)`.

Hogarth has noted that general relativity may supply a physical framework in which to implement Programme 1. The idea is to find a spacetime in which a human user, Dave, can survey the entire *infinite* worldline of some ordinary computer, Hal.[1] In particular, in the diagram to the right, a finite part of Dave's wordline is ϱ, and Hal's entire infinite worldline is γ. On γ, Hal executes the following Python programme:

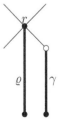

─────────── **Programme 2a** ───────────

```
x = 0                                         a.1
while fun(x) == 0:                            a.2
        x = x+1                               a.3
sendASignalToDave()                           a.4
```

---

[1] The required spacetime, the spacetime diagram, and this implementation are discussed in Hogarth (1992, 1994, 2004, 2009a); Earman & Norton (1993); Etesi & Németi (2002); Németi & Dávid (2006); Welch (2008).

If Hal finds a value of x such that fun(x), Hal sends a signal to Dave (perhaps a light-beam). Dave carries a receiving device with him, which stores a variable, xFound, initialised to a default value of 0; this receiver sets xFound to 1 on (and only on) receipt of a signal from Hal. Once Dave reaches point r, Hal's entire *infinite* history lies in Dave's past, and any signal that Hal might have sent to Dave will have arrived. So from Dave's perspective, by point r, Hal has completed a hyperloop. Dave then executes the following Python programme:

――――――――――― **Programme 2b** ―――――――――――

```
if xFound: print "Yes"                                            b.1
else: print "No"                                                  b.2
```

The overall effect is a complete implementation of Programme 1.

With the formal and physical notion of a hyperloop on the table, several questions concerning hyperloops arise.

First: *What functions can be computed with hyperloops?* Welch (2008) has answered this technical question.[2] For the purposes of this paper, the only result we require is that hyperloopers are much more powerful than Turing computers.

Second: *Are hyperloopers genuine physical possibilities?* Much interesting foundational work has addressed this question, focussing primarily on the relativistically-realised-hyperloops just discussed.[3] But in this paper, I shall avoid all questions about physical computability.

The question I wish to consider is: *What consequences do hyperloops have for the notion of effective computability?* Note that, although some authors identify "effectively computable" with "physically computable",[4] this is *not* what I have in mind. I equate the *semi-formal* notion of an algorithm with that of an effective procedure, and I say that a function is effectively computable if and only if there is some effective procedure (algorithm) for computing it.[5]

2   Polyvagueness

I want to start by considering whether the discussion of hyperloops shows that "effectively computable" exhibits a certain kind of vagueness. I shall suggest that it probably does not, but that even if it does, there is no reason to worry.

――――――――――――

[2] A computer which executes a single hyperloop can complete an ω-sequence of computations, and so can decide membership of any Π_1 or Σ_1 set. By nesting n-many hyperloops inside each other, we can write Ouroboros programmes for computers that complete ω^n-sequences of computations, and so decide any Π_n or Σ_n set. Beyond the programming capabilities of Ouroboros, but retaining the basic idea of a hyperloop, we can define hyperloopers to decide membership up through the hyperarithmetical sets. This is the limit of computation with hyperloops.

[3] e.g. Etesi & Németi (2002); Németi & Dávid (2006); and consult references in their papers. Even if relativistically-realised-hyperloops ultimately fail, some *other* physical model of a hyperloop may be feasible. For instance, Davies (2001) describes another realisation of hyperloops (which requires the infinite divisibility of matter).

[4] e.g. Etesi & Németi (2002, p.348).

[5] See Smith (2007, pp.315–23), Button (forthcoming, Sect. 3.1).

Smith (2007, p.327) calls a predicate *polyvague* "if it ambiguously locates more than one mathematical kind." To illustrate the idea of polyvagueness, consider Euler's conjecture: for all polyhedra, Vertices − Edges + Faces = 2. Lakatos (1976) investigates various potential counterexamples to the conjecture, including: a picture frame; a cube with a hollow cubic interior; and a cylinder. But before admitting these *as* counterexamples to the conjecture, we must determine whether these objects are even polyhedra. Perhaps that is (or was) not clear in every case; in which case "polyhedron" is (or was) polyvague.

Shapiro (2006) has argued that "effectively computable" began life as a polyvague predicate, but that it has been gradually sharpened over time.[6] To show this, he gives examples of procedures for which there was no unanimous consensus (at one time or other) as to whether they were effective. I want to ask whether relativistically-realised-hyperloops, as described in Sect. 1, constitute a further example to show that "effectively computable" was polyvague.

An affirmative answer might be motivated thus. In early discussions of effective procedures, we might have suggested that an effective procedure is one which takes only a finite duration of our time to complete. Alternatively, we might have suggested that an effective procedure is one which takes only a finite number of discrete stages to complete. These characterisations might have seemed to be necessarily co-extensive. However, thinking about relativistically-realised-hyperloops shows that they are not. On the former characterisation (finite duration), relativistically-realised-hyperloops are effective procedures. On the latter characterisation (finite stages), any hyperloop fails to count as effective. Since there are at least two "natural kinds" here, maybe this shows that "effectively computable" was polyvague.

For now, I will concede that this shows that "effectively computable" was polyvague. It does *not* follow that the notion of an effective procedure is bankrupt. Discovering that a predicate is polyvague rarely undermines the notion at hand. For instance, even if Lakatos is right that "polyhedron" is polyvague, the notion of a polyhedron may well remain useful; but we may want to sharpen it, for some purposes. And generally, discovering that a predicate is polyvague forces us to ask two questions. First: why do we need to use the predicate? Second: which way(s) of precisifying the predicate best serves those needs?

It is relatively easy to state why we need the predicate "effectively computable". We need to think about effective computability and effective procedures whenever we talk about what can be done using algorithms (recall my explicit characterisation of "effective computability" in Sect. 1).

So, which way of sharpening "effectively computable" best serve our need to talk about algorithms? In the present case, we must choose whether or not to count relativistically-realised-hyperloops as effective procedures. Once we focus on the question of *physical realisation*, the choice is quite clear. Suppose we decide to treat relativistically-realised-hyperloops as effective procedures, on the grounds that they take only a finite duration (from our perspective). Other

---

[6] More accurately, Shapiro (2006) says that "idealized (human) computable" is "open textured". Both Shapiro and Smith invoke Lakatos to illustrate the idea at hand.

physical realisations of hyperloops will require infinite periods of time (from our perspective). So given a single procedure involving a hyperloop, we will have to treat some physical realisations of it as effective, and treat other realisations as ineffective. Physicists may be happy to draw a distinction on these lines, but this is not the kind of distinction that seem relevant to logical and mathematical reasoning.[7] From the latter perspective – which is the home of algorithms – it would be better to abstract the notion of an effective procedure from specifics relating to the various *physical realisations* of those procedures. This abstraction favours the definition of an effective procedure in terms of its number of stages, rather than in terms of its duration.

Accordingly, I think it is clear that relativistically-realised-hyperloops ought not (now) to count as effective procedures. But, more importantly: even if it was once an open question whether relativistically-realised-hyperloops ought to count as effective procedures, that does not by itself undermine the very idea of an effective procedure.[8]

3   The Absolute Nature of Algorithms

With this background fixed, I wish to consider Hogarth's (2009b) argument that the output of an algorithm is relative to the machine that implements the algorithm. To fix ideas, Hogarth considers a simple algorithm:

Algorithm 3

```
counter = 0                          3.1
loop:                                3.2
        print counter                3.3
        counter = counter + 1        3.4
```

We are to consider implementing Algorithm 3 on two different machines, with a standard realisation of the `print` function:[9]

A. *An ordinary computer.* This machine will output an ascending sequence of arabic numerals, "0", "1", "2", ... representing finite numbers.
B. *A relativistically-realised-hyperlooper.* In this case, Hal (on γ) is to send a signal to Dave's receiver (on ϱ) every time Hal increases the value of `counter`.

---

[7] cf. Hamkins & Lewis (2000, p.568). I revisit this, emphasising *reasoning*, in Sect. 4.3.

[8] In fact, I suspect that it never was a genuinely open question. The founders of computability theory were well aware of supertasks and oracles, and so I conjecture that they would have declared that relativistically-realised-hyperloops are not effective procedures. But I shall not push this. Note also that I have no objection to our describing *hyper-effective* procedures, if we so wish. (See my discussion (forthcoming) of "effective SAD procedures".)

[9] Hogarth also considers implementing Algorithm 3 on a computer that nests a hyperloop within a hyperloop, and so completes an $\omega^2 + n$-sequence of computations.

Dave's receiver decodes the signal so as to output all the arabic numerals, "0", "1", "2", ... After this (at point r) it will output "ɯ", representing ω, and will then continue, "ɯ+1", "ɯ+2", ...

The two implementations give different outputs, and Hogarth concludes that the output of an algorithm is generally relative to the machine that implements the algorithm.[10] As Hogarth notes, if this is correct, it is extremely significant: if algorithms are essentially relative, then *all* those areas of logic and mathematics that utilise effective procedures or algorithms are likewise essentially relative.

But before discussing Hogarth's claim, it will help to recall a point from the literature on supertasks. We shall focus on Thomson's Lamp (1954). At time $t = 0$, Thomson's Lamp is off. At $t = \frac{1}{2}$, it is flicked on. At $t = \frac{3}{4}$, it is flicked off. At $t = \frac{7}{8}$, it is flicked on. . . . Is the lamp on or off at $t = 1$? No answer seems warranted, and this is sometimes thought paradoxical. But it need not be so conceived. As Benacerraf (1962) notes, we have described the state of the lamp *only* for times $0 \leq t < 1$, so *any* state of the lamp at $t = 1$ is compatible with all its previous states. Paradox dissolved![11]

Let us now return to Hogarth's argument. By analogy with Thomson's Lamp, we ask the following question: in Implementation B, what is the value of the stored variable `counter` at stage ω? As for Thomson's Lamp, no answer seems warranted. Hogarth wants to say that the value is ɯ, but why not 0, or 783, or simply `Error`? Given a value of `counter` at some stage in the computation, we know the value of `counter` at the next stage: Algorithm 3 tells us that its value increases by 1 at each successor stage. But nothing in Algorithm 3 determines `counter`'s value at limit stages (i.e. at stages numbered by some limit ordinal). Any value of `counter` at stage ω is consistent with all its previous states.

One might think that Algorithm 3's output is essentially relative *precisely because* Algorithm 3 does not define the computer's behaviour at limit stages, and so because anything could happen at limit stages. This observation saves Implementation B from *paradox*, if it saves Thomson's Lamp from the same charge. But it does not save Hogarth's argument. Consider an "algorithm" which merely initialises a variable, then prints it. Should we say that the output of this algorithm is interestingly relative to its implementation, because anything could happen at successor stages? Hardly; but then we should not say that the output of Algorithm 3 is interestingly relative, just because it leaves a computer's behaviour undefined at limit stages.

The point can be put starkly as follows. Algorithm 3 defines a function by recursion. Hogarth wants to extend that definition into the transfinite. This

---

[10] Hogarth also maintains that this raises a "new problem for rule-following" since, when someone asks us to count and never stop, we do not know whether they want us to keep counting *through* ω and beyond.

[11] Allegedly; some philosophers are not content with this resolution. For example, Dummett (2005, pp.143–4) maintains that a complete description of a lamp's states for $0 \leq t < 1$ *ought* to entail its state at $t = 1$.

plainly requires an explicit transfinite recursion clause.[12] It is easy to provide one within the framework of Ouroboros:

```
                           ── Programme 4 ──
hyperloop with counterA = 0:                              4.1
     print counterA                                      4.2
     counterA = counterA + 1                             4.3
counterB = 0                                             4.4
while counterB > -1:                                     4.5
     print "ա+" + str(counterB)                          4.6
     counterB = counterB + 1                             4.7
```

An implementation of Programme 4 seems to be what Hogarth has in mind in Implementation B. But line 4.4 explicitly defines the machine's behaviour at stage ω. So a hyperlooper running Programme 4 has a different output from an ordinary computer implementing Algorithm 3, precisely because they implement *different* algorithms. In short, hyperloops do not suggest that the output of an algorithm is essentially relativised.[13]

4   Hyperloops, Computability, and Geometry

Hogarth (2004, pp.689–690, 2009a, 2009b, Sect.2) presents a rather different argument against the idea of an effective procedure, which draws a connection between computability and geometry.

Consider a geometrical question, such as "what is the sum of a triangle's interior angles?" This might be a question about the spatiotemporal structure of our world. Or it might be a purely formal question within a particular formal geometric theory. But, Hogarth claims, there are *only* physical and formal questions here. In particular, there is no "intuitively true" geometry to ask about.

Hogarth believes that the same is true of computing. When we ask whether a function is computable, we might be asking whether some physical machine could compute that function. Or we might be asking a purely formal question within a purely formal theory of computability. But we cannot ask any *other* question. Just as there is no "intuitively true" geometry, there is no room for a notion of effective computability that is neither formal nor physical.

In what follows, I shall consider three ways to read this analogy. My main aim is to show that hyperloops should play no part in it.

---

[12] Indeed, Hogarth provides one himself, saying that "ա... is the interpretation of the absence of a signal" from Hal (2009b, Appendix). This is essentially an *extension* of the algorithm into the transfinite. It is probably worth noting that this extension is not quite right. ա is the interpretation of the absence of a signal *at point r*; indeed, ա is just the interpretation of *reaching point r*.

[13] Hogarth's "new problem for rule-following" (see fn. 10, above) dissolves in the same breath. Having added explicit rules to govern limit stages, we cannot pretend that we are "going on exactly as before".

4.1   A Circular Argument

In Button (forthcoming, Sect. 3.3), I reconstruct the analogy thus:

1. Which functions can be computed by a (configuration of) machine(s) depends upon whether the machine(s) can execute hyperloops.
2. Whether machines can execute hyperloops can only either:
   (a) depend upon the laws of physics (and so concerns physical machines); or
   (b) be stipulated within some formal theory (and so concerns the formal definition of a "machine").
3. So which functions can be computed by a (configuration of) machine(s) is either a purely formal or a purely physical question. There are no interesting questions about effective computability.

Strikingly, any non-Turing theory of computability can be used in place of hyperlooping in this argument. To demonstrate this, consider a parallel argument. The tape of a Turing machine can be in any of infinitely many possible states. By contrast, a *finite-state-machine* has only finitely many possible states, with transition rules between them. Finite-state-machines are formally much weaker than Turing machines. Now:

1*. Which functions can be computed by a (configuration of) machine(s) depends upon whether there are infinitely many, or only finitely many, possible total-machine-states.[14]
2*. The number of possible total-machine-states can only either depend upon the laws of physics, or be stipulated within some formal theory.
3. So, as above, there are no interesting questions about effective computability.

This parallel argument is obviously irrelevant: when we think about effective computability, finite-state-machines are simply too weak to concern us. But this parallel argument and the original argument have exactly the same structure. So the argument structure must *itself* be too strong. We must determine why.

To deliver an analogue of premise 1, we need two technical results. First, we must describe a formal computer and prove that it is not equivalent to a Turing computer. Second, we must describe a system of physical laws that favours this non-Turing computer (we may do this by describing spatiotemporal laws, but we could in principle describe laws of any sort). These two technical results may arise from, or give rise to, important work in the foundations of computability and the physical sciences. However, we have a philosophical argument against effective computability only when we assert, in an analogue of premise 2, that *only* physical or formal issues could decide between a Turing computer and this non-Turing computer. This assertion carries all the philosophical burden of the argument, and it is obviously contentious. So what could justify it? It is certainly entailed by what I call "Hogarth's Thesis":

---

[14] A *total-machine-state* is meant to be fully encompassing; e.g., in the case of a Turing machine, it incorporates both the machine's *internal* state and the state of its *tape*.

Hogarth's Thesis. There are questions about what is physically computable; there are questions about what is formally computable (in various different formal systems); but there is no third side to computability.

However, Hogarth's Thesis is just a restatement of the desired *conclusion* of the original argument. What we need is an *independent* argument for Hogarth's Thesis, which arises from considering hyperloops. But none is forthcoming from the reading of Hogarth's analogy that we have been considering.[15]

4.2 The (Non-)Spatiotemporality of Computation

There is a much stronger reading of Hogarth's analogy between geometry and computability. According to the strong reading, "intuitive" computability is "intuitive" physical computing in some "intuitive" spacetime. But there is no such thing as the "intuitively true" geometry or spacetime. So we cannot dismiss as "unintuitive" the (relativistic) geometries which make hyperloops possible. So we must give up on the idea of "intuitive" computability.

This strong reading draws a tight link between geometry and computability, and thereby reinstates the role of hyperloops in the attack on effective computability. Indeed, it is misleading to call this link between geometry and computability a mere "analogy". On this reading, computation is intrinsically spatiotemporal, so computation and geometry *must* share the same fate.[16]

However, as stressed in Sect. 2, there is a good notion of effective computability which is *not* spatiotemporal. That notion does not concern computation in some "intuitive" spacetime. Rather, it concerns the *abstract* notion of a procedure of arbitrary finite length, where length is measured by number of stages and not temporal duration. Hyperloops – procedures of infinite length, in these terms – cannot speak much against this notion. Indeed, *hyperloops can threaten the atemporal notion of an algorithm only if they can threaten the semi-formal notion of an arbitrary finite sequence of stages.*

So, *can* hyperloops threaten the notion of an arbitrary finite sequence? Their only shot is as follows. Suppose we define a *finite number* as one that can be counted to in a finite duration of our time. Then there is undoubtedly a sense in which we can count through ω using a relativistically-realised-hyperloop (see Sect. 3). So this might show that what counts as a finite number is relativised to a physical machine, or to a formal system. By association, the notion of a finite sequence would also be relativised.

We already know how to respond to this argument. First, the procedure for counting through ω is obviously a *new* procedure, so cannot threaten our grasp of the old procedure (compare Sect. 3). Second, we should contest any definition of "finite number" in terms of what can be counted to in a particular *duration*: finite numbers are rather more abstract (compare Sect. 2).

---

[15] To be fair to Hogarth, note again that this was *my* reconstruction of his analogy!

[16] This reading is suggested by Hogarth (2009a, p.277, 2009b, fn.2).

4.3   An Invitation to Accept Hogarth's Thesis

Consequently, hyperloops simply cannot threaten the non-spatiotemporal notion of effective computability. But this observation is likely to bring us simply to an impasse. One side of the impasse accepts Hogarth's Thesis and regards the very idea of effective computability as an outdated "relic". The other side rejects Hogarth's Thesis, so regards relativistically-realised-hyperloops as "obviously irrelevant". Both sides will be equally frustrated with the other.

The only way way to break the impasse is to consider Hogarth's Thesis directly, without talking about hyperloops. And this suggests a final, weaker reading of Hogarth's analogy. On this weak reading, Hogarth is not attempting to *argue* against the notion of effective computability. He is simply *inviting* us to let go of effective computability, and accept his Thesis.

I am grateful for the invitation, but I am also wary of it. We can all agree that actual reasoning takes place in spacetime. But I doubt that *what follows from something by reason alone* is either a spatiotemporal or purely formal question. Similarly, we can all agree that actual computing takes place in spacetime. But I doubt that *what may be effectively computed – what follows by "unimaginative" reasoning alone –* is either a spatiotemporal or a formal question. To accept Hogarth's Thesis would be to abandon this doubt.

This is not to say that Hogarth's Thesis is false. Indeed, everything I have said is compatible with the veracity of Hogarth's Thesis. I have simply attempted to establish the following: Hogarth's Thesis cannot be established by considering hyperloops. Hyperloops do threaten the notion of an effective procedure.

References

Benacerraf, P.: Tasks, Super-Tasks, and the Modern Eleatics. Journal of Philosophy 59, 765–784 (1962)

Button, T.: SAD Computers and two versions of the Church-Turing Thesis. British Journal for the Philosophy of Science (forthcoming)

Davies, E.B.: Building Infinite Machines. British Journal for the Philosophy of Science 52, 671–682 (2001)

Dummett, M.: Hume's Atomism about Events: a response to Ulrich Meyer. Philosophy 80, 141–144 (2005)

Earman, J., Norton, J.D.: Forever is a Day: Supertasks in Pitowsky and Malament-Hogarth Spacetimes. Philosophy of Science 60, 22–42 (1993)

Etesi, G., Németi, I.: Non-Turing Computations via Malament-Hogarth space-times. International Journal of Theoretical Physics 41, 341–370 (2002)

Hamkins, J.D., Lewis, A.: Infinite Time Turing Machines. Journal of Symbolic Logic 65, 567–604 (2000)

Hogarth, M.: Does General Relativity Allow an Observer to View an Eternity in a Finite Time? Foundations of Physics Letters 5, 173–181 (1992)

Hogarth, M.: Non-Turing Computers and Non-Turing Computability. In: Proceedings of the Philosophy of Science Association 1994, vol. 1, pp. 126–138 (1994)

Hogarth, M.: Deciding Arithmetic using SAD computers. British Journal for the Philosophy of Science 55, 681–691 (2004)

Hogarth, M.: Non-Turing Computers are the New Non-Euclidean Geometries. International Journal of Unconventional Computing 5, 277–291 (2009a)

Hogarth, M.: A New Problem for Rule Following. Natural Computing (2009b)

Lakatos, I.: Proofs and Refutations. Cambridge University Press, Cambridge (1976)

Németi, I., Dávid, G.: Relativistic Computers and the Turing Barrier. Journal of Applied Mathematics and Computation 178, 118–142 (2006)

Smith, P.: An Introduction to Gödel's Theorems. Cambridge University Press, Cambridge (2007)

Shapiro, S.: Computability, Proof, and Open-Texture. In: Olszewski, A., Wolenski, J., Janusz, R. (eds.) Church's Thesis After 70 Years, pp. 420–455. Ontos Verlag (2006)

Thomson, J.: Tasks and Supertasks. Analysis 15, 1–10 (1954)

Welch, P.D.: The Extent of Computation in Malament-Hogarth Spacetimes. British Journal for the Philosophy of Science 59, 659–674 (2008)

Minimum Entropy Combinatorial Optimization Problems

Jean Cardinal, Samuel Fiorini, and Gwenaël Joret*

Université Libre de Bruxelles (ULB)
B-1050 Brussels, Belgium
{jcardin,sfiorini,gjoret}@ulb.ac.be

Abstract. We survey recent results on combinatorial optimization problems in which the objective function is the entropy of a discrete distribution. These include the minimum entropy set cover, minimum entropy orientation, and minimum entropy coloring problems.

1 Introduction

Set covering and graph coloring problems are undoubtedly among the most fundamental discrete optimization problems, and countless variants of these have been studied in the last 30 years. We discuss several coloring and covering problems in which the objective function is a quantity of information expressed in bits. More precisely, the objective function is the Shannon entropy of a discrete probability distribution defined by the solution. These problems are motivated by applications as diverse as computational biology, data compression, and sorting algorithms.

Recall that the entropy of a discrete random variable X with probability distribution $\{p_i\}$, where $p_i := P[X = i]$, is defined as:

$$H(X) = -\sum_i p_i \log p_i.$$

Here, logarithms are in base 2, thus entropies are measured in bits; and $0 \log 0 := 0$. From Shannon's theorem [29], the entropy is the minimum average number of bits needed to transmit a random variable on an error-free communication channel.

We give an overview of the recent hardness and approximability results obtained by the authors on three minimum entropy combinatorial optimization problems [4,5,6]. We also provide new approximability results on the minimum entropy orientation and minimum entropy coloring problems, see Theorems 3, 7, and 8. At the end of the paper, we present a recent result that quantifies how well the entropy of a perfect graph is approximated by its chromatic entropy, with an application to a sorting problem [7].

---

* Postdoctoral Researcher of the Fonds National de la Recherche Scientifique (F.R.S. – FNRS).

K. Ambos-Spies, B. Löwe, and W. Merkle (Eds.): CiE 2009, LNCS 5635, pp. 79–88, 2009.

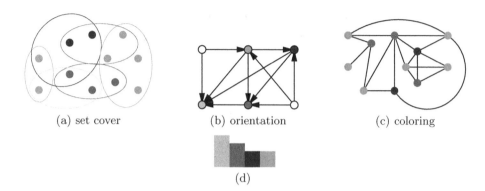

(a) set cover   (b) orientation   (c) coloring

(d)

Fig. 1. Instances of the minimum entropy combinatorial optimization problems studied in this paper, together with feasible solutions. (d) The resulting probability distribution for the given solutions : $\{4/11, 3/11, 2/11, 2/11\}$.

2 Minimum Entropy Set Cover

In the well-known minimum set cover problem, we are given a ground set U and a collection $\mathcal{S}$ of subsets of U, and we ask what is the minimum number of subsets from $\mathcal{S}$ such that their union is U. A famous heuristic for this problem is the greedy algorithm: iteratively choose the subset covering the largest number of remaining elements. The greedy algorithm is known to approximate the minimum set cover problem within a $1 + \ln n$ factor, where $n := |U|$. It is also known that this is essentially the best approximation ratio achievable by a polynomial time algorithm, unless NP has slightly super-polynomial time algorithms [12].

In the minimum entropy set cover problem, the cardinality measure is replaced by the entropy of a partition of U compatible with a given covering. The function to be minimized is the quantity of information contained in the random variable that assigns to an element of U chosen uniformly at random, the subset that covers it. This is illustrated in Figure 1(a). A formal definition of the minimum entropy set cover problem is as follows [19,6].

INSTANCE: A ground set U and a collection $\mathcal{S} = \{S_1, \ldots, S_k\}$ of subsets of U
SOLUTION: An assignment $\phi : U \to \{1, \ldots, k\}$ such that $x \in S_{\phi(x)}$ for all $x \in U$
OBJECTIVE: Minimize the entropy $-\sum_{i=1}^{k} p_i \log p_i$, where $p_i := |\phi^{-1}(i)|/|U|$

Intuitively, we seek a covering of U yielding part sizes that are either large or small, but somehow as nonuniform as possible. Also, an arbitrarily small entropy can be reached using an arbitrarily large number of subsets, making the problem quite distinct from the minimum set cover problem.

Applications. The original paper from Halperin and Karp [19] was motivated by applications in computational biology, namely haplotype reconstruction. In an abstract setting, we are given a collection of objects (that is, the set U) and, for each object, a collection of classes to which it may belong (that is, the collection of sets S_i containing the object). Assuming that the objects are selected at random from a larger population, we wish to assign to each object the most likely class it belongs to. They show that this essentially amounts to minimizing the entropy of the covering, as defined above. Experimental results derived from this work were proposed by Bonizzoni *et al.* [3], and Gusev, Măndoiu, and Paşaniuc [17].

Results. We proved that the greedy algorithm performs much better for the minimum entropy version of the set cover problem. Furthermore, we gave a complete characterization of the approximability of the problem under the $P \neq NP$ hypothesis.

Theorem 1. *[6] The greedy algorithm approximates the minimum entropy set cover problem within an additive error of* $\log e$ *(≈ 1.4427) bits. Moreover, for every $\epsilon > 0$, it is NP-hard to approximate the problem within an additive error of* $\log e - \epsilon$ *bits.*

Our analysis of the greedy algorithm for the minimum entropy set cover problem is an improvement, both in terms of simplicity and approximation error, over the first analysis given by Halperin and Karp [19]. The hardness of approximation is shown by adapting a proof from Feige, Lovász, and Tetali on the minimum sum set cover problem [13], which itself derives from the results of Feige on minimum set cover [12].

3   Minimum Entropy Orientations (Vertex Cover)

The minimum entropy orientation problem is the following [5]:

INSTANCE: An undirected graph $G = (V, E)$
SOLUTION: An orientation of G
OBJECTIVE: Minimize the entropy $-\sum_{v \in V} p_v \log p_v$, where $p_v := \rho(v)/|E|$, and
   $\rho(v)$ is the indegree of vertex v in the orientation

An instance of the minimum entropy orientation problem together with a feasible solution are given in Figure 1(b). Note that the problem is a special case of minimum entropy set cover in which every element in the ground set U is contained in exactly two subsets. Thus it can be seen as a minimum entropy vertex cover problem.

Results. We proved that the minimum entropy orientation problem is NP-hard [5]. Let us denote by $OPT(G)$ the minimum entropy of an orientation of G (in bits). An orientation of G is said to be biased if each edge vw with $\deg(v) > \deg(w)$ is oriented towards v. Biased orientations have an entropy that is provably closer to the minimum than those obtained via the greedy algorithm.

Theorem 2. *[5] The entropy of any biased orientation of G is a most $OPT(G)+1$ bits. It follows that the minimum entropy orientation problem can be approximated within an additive error of 1 bit, in linear time.*

Constant-time Approximation Algorithm for Bounded Degree Graphs. By making use of the fact that the computation of a biased orientation is purely local [25], we show that we can randomly sample such an approximate solution to guess $OPT(G)$ within an additive error of $1 + \epsilon$ bits. The complexity of the resulting algorithm does not (directly) depend on n, but only on Δ and $1/\epsilon$. This is a straightforward application of ideas presented by Parnas and Ron [27].

We consider a graph G with n vertices, m edges, and maximum degree Δ. Pick any preferred biased orientation $\overrightarrow{G}$ of G. (For instance, we may order the vertices of G arbitrarily and orient an edge vw with $\deg(v) = \deg(w)$ towards the vertex that appears last in the ordering.) The following algorithm returns an approximation of $OPT(G)$ (below, s is a parameter whose value will be decided later).

1. **For** $i = 1$ **to** s
   (a) **pick vertex** v_i **uniformly at random**
   (b) **compute the indegree** $\rho(v_i)$ **of** v_i **in** $\overrightarrow{G}$
2. **return** $H := \log m - \frac{n}{sm} \sum_{i=1}^{s} \rho(v_i) \log \rho(v_i)$

The worst-case complexity of the algorithm is $O(s\Delta^2)$.

Theorem 3. *There is an algorithm of worst-case complexity $O(\Delta^4 \log^2 \Delta/\epsilon^2)$ that, when given a graph G with maximum degree Δ and at least as many edges as vertices, returns a number H satisfying, with high probability,*

$$OPT(G) \leq H \leq OPT(G) + (1 + \epsilon).$$

Proof. Let $V := V(G)$ and $OPT := OPT(G)$. From the previous theorem we get:

$$-\sum_{v \in V} \frac{\rho(v)}{m} \log \frac{\rho(v)}{m} = \log m - \frac{1}{m} \sum_{v \in V} \rho(v) \log \rho(v) \leq OPT + 1.$$

For $i = 1, \ldots, s$, let v_i denote a uniformly sampled vertex of G. By linearity of expectation, we have:

$$E\left[\sum_{i=1}^{s} \rho(v_i) \log \rho(v_i)\right] = \frac{s}{n} \sum_{v \in V} \rho(v) \log \rho(v).$$

Noting $0 \leq \rho(v_i) \log \rho(v_i) \leq \Delta \log \Delta$ (for all i), Hoeffding's inequality then implies:

$$P\left[\left|\sum_{i=1}^{s} \rho(v) \log \rho(v) - E\left[\sum_{i=1}^{s} \rho(v) \log \rho(v)\right]\right| \geq \epsilon s\right] \leq 2\exp\left(-\frac{2s^2\epsilon^2}{s(\Delta \log \Delta)^2}\right)$$

$$\Rightarrow P\left[\left|\frac{n}{sm} \sum_{i=1}^{s} \rho(v) \log \rho(v) - \frac{1}{m} \sum_{v \in V} \rho(v) \log \rho(v)\right| \geq \epsilon \frac{n}{m}\right] \leq 2\exp\left(-\frac{2s\epsilon^2}{(\Delta \log \Delta)^2}\right)$$

$$\Rightarrow P\left[|H - OPT| \geq 1 + \epsilon\right] \leq P\left[|H - OPT| \geq 1 + \epsilon\frac{n}{m}\right] \leq 2\exp\left(-\frac{2s\epsilon^2}{(\Delta \log \Delta)^2}\right).$$

By letting $s = \Theta((\Delta \log \Delta)^2/\epsilon^2)$, with arbitrarily high probability, we conclude that the above algorithm provides an approximation of OPT within $1 + \epsilon$ bits in time $O(s\Delta^2) = O(\Delta^4 \log^2 \Delta/\epsilon^2)$. Note that this approximation can be either an under- or an over-approximation. To make the approximation one-sided, we can simply return $H + \epsilon$. □

4   Minimum Entropy Coloring

A proper coloring of a graph assigns colors to vertices such that adjacent vertices have distinct colors. We define the entropy of a proper coloring as the entropy of the color of a random vertex. An example is given in Figure 1(c). The minimum entropy coloring problem is thus defined as follows:

INSTANCE: An undirected graph $G = (V, E)$
SOLUTION: A proper coloring $\phi : V \to \mathbb{N}$ of G
OBJECTIVE : Minimize the entropy $-\sum_i p_i \log p_i$, where $p_i := |\phi^{-1}(i)|/|V|$

Note that any instance of the minimum entropy coloring problem can be seen as an implicit instance of the minimum entropy set cover problem, in which the ground set is the set of vertices of the graph, and the subsets are all stable (or independent) sets, described implicitly by the graph structure.

The problem studied by the authors in [4] was actually slightly more general: the graph G came with nonnegative weights $w(v)$ on the vertices $v \in V$, summing up to 1. The weighted version of the minimum entropy coloring problem is defined similarly as the unweighted version except now we let $p_i := \sum_{v \in \phi^{-1}(i)} w(v)$. (The unweighted version is obtained for the uniform weights $w(v) = 1/|V|$.)

Applications. Minimum entropy colorings have found applications in the field of data compression, and are related to several results in zero-error information theory [23]. The minimum entropy of a coloring is called the chromatic entropy by Alon and Orlitsky [2]. It was introduced in the context of coding with side information, a source coding scenario in which the receiver has access to an information that is correlated with the data being sent (see also [24]). Minimum entropy colorings are also instrumental in several coding schemes introduced by Doshi *et al.* [11,9,10] for functional data compression. It was also proposed for the encoding of segmented images [1].

Results. Unsurprisingly, the minimum entropy coloring problem is hard to solve even on restricted instances, and hard to approximate in general. We proved the following two results [4].

Theorem 4. *Finding a minimum entropy coloring of a weighted interval graph is strongly NP-hard.*

The following hardness result is quite strong because the trivial coloring assigning a different color to each vertex has an entropy of at most $\log n$. Actually, the approximation status of the problem is very much comparable that of the maximum stable set problem [20]. (This is not a coincidence, see for instance the discussion below and in particular Corollary 1.)

Theorem 5. *For any positive real ϵ, it is NP-hard to approximate the minimum entropy coloring problem within an additive error of $(1 - \epsilon) \log n$.*

A positive result was given by Gijswijt, Jost, and Queyranne:

Theorem 6. *[16] The minimum entropy coloring problem can be solved in polynomial time on (unweighted) co-interval graphs.*

Greedy Coloring. The greedy algorithm can be used to find approximate minimum entropy colorings. In the context of coloring problems, the greedy algorithm involves iteratively removing a maximum stable set in the graph, assigning a new color to each removed set. This algorithm is polynomial for families of graphs in which a maximum (size or weight) stable set can be found in polynomial time, such as perfect graphs. We therefore have the following result.

Corollary 1. *The minimum entropy coloring problem can be approximated within an additive error of $\log e$ (≈ 1.4427) bits when restricted to perfect graphs.*

We now show an example of an "approximate greedy" algorithm yielding a bounded approximation error on any bounded-degree graphs. The following lemma is straightforward from the proof of Theorem 1 (see [6]).

Lemma 1. *The algorithm that iteratively removes a β-approximate maximum stable set yields an approximation of the minimum entropy of a coloring within an additive error of $\log \beta + \log e$ bits.*

Theorem 7. *Minimum entropy coloring is approximable in polynomial time within an additive error of $\log(\Delta + 2) - 0.1423$ bits on graphs with maximum degree Δ.*

Proof. Using a greedy algorithm for the maximum stable set, we have $\rho = (\Delta + 2)/3$ (See Halldórsson and Radhakrishnan [18]). This ratio is valid for each step of the algorithm, as the maximum degree of the graph cannot increase. From (2), the error term for the minimum entropy problem is at most

$$\log \rho + \log e = \log(\Delta + 2) + \log e - \log 3 < \log(\Delta + 2) - 0.1423. \qquad \square$$

Coloring Interval Graphs. We give the following simple polynomial-time algorithm for approximating the minimum entropy coloring problem on (unweighted) interval graphs. This algorithm is essentially the online coloring algorithm proposed by Kierstead and Trotter [21], but in which the intervals are given in a specific order, and the intervals in S_i are 2-colored offline.

```
1. sort the intervals in increasing order of their right
   endpoints; let (v₁, v₂, ..., vₙ) be this ordering
```

2. **for** $j \leftarrow 1$ to n **do**
   - insert v_j in S_i, where i is the smallest index such that $S_1 \cup S_2 \cup \dots S_i \cup \{v_j\}$ does not contain an $(i+1)$-clique
3. color the intervals in S_1 with color 1
4. **for** $i \leftarrow 2$ to k **do**
   - color the intervals in S_i with colors $2i - 2$ and $2i - 1$

This algorithm is also similar to the BETTER-MCA algorithm introduced by Pemmaraju, Raman, and Varadarajan for the max-coloring problem [28], but instead of sorting the intervals in order of their weights, we sort them in order of their right endpoints. It achieves the same goal as the algorithm of Nicoloso, Sarrafzadeh, and Song [26] for minimum sum coloring.

Lemma 2. *At the end of the algorithm, the graph induced by the vertices in $S_1 \cup S_2 \cup \dots \cup S_i$ is a maximum i-colorable subgraph of G.*

Proof. If we consider the construction of the set $S_1 \cup S_2 \dots \cup S_i$, it matches exactly with an execution of the algorithm of Yannakakis and Gavril [31] for constructing a maximum i-colorable subgraph.   □

Kierstead and Trotter [21] proved the following lemma, that shows that step 4 of the algorithm is always feasible.

Lemma 3. *[21] The graphs induced by the sets S_i are bipartite.*

Let H' be the entropy of the probability distribution $\{|S_i|/n\}$. The proof is omitted, and relies on Lemma 2.

Lemma 4. *H' is a lower bound on the minimum entropy of a coloring of the interval graph.*

We deduce the following.

Theorem 8. *On (unweighted) interval graphs, the minimum entropy coloring problem can be approximated within an additive error of 1 bit, in polynomial time.*

Proof. The entropy of the coloring produced by the algorithm is at most that of the distribution $\{|S_i|/n\}$ plus one bit, since the intervals in S_i are colored with at most two colors. From Lemma 4 the former is a lower bound on the optimum, and we get the desired approximation.   □

Note that we do not know whether the minimum entropy coloring problem can be solved exactly in polynomial time on (unweighted) interval graphs. The hardness result of Theorem 4 only applies to probabilistic interval graphs. Also, improved approximation results could be obtained on special classes of graphs, using for instance the methods developed by Fukunaga, Halldórsson, and Nagamochi [15,14].

5   Graph Entropy and Partial Order Production

The notion of graph entropy was introduced by Körner in 1973 [22], and was initially motivated by a source coding problem. It has since found a wide range of applications (see for instance the survey by Simonyi [30]).

The entropy of a (in our case, unweighted) graph $G = (V, E)$ can be defined in a number of ways. The following is among the easiest. Let us consider a probability distribution $\{q_S\}$ on the stable sets S of G, and denote by p_v the probability that v belongs to a stable set drawn at random from this distribution (that is, $p_v := \sum_{S \ni v} q_S$). Then the entropy of G is the minimum over all possible distributions $\{q_S\}$ of

$$-\frac{1}{n} \sum_{v \in V} \log p_v. \tag{1}$$

The feasible vectors $(p_v)_{v \in V}$ form the stable set polytope $STAB(G)$ of the graph G, defined as the following convex combination:

$$STAB(G) := \text{conv}\{\mathbf{1}^S : S \text{ stable set of } G\},$$

where $\mathbf{1}^S$ is the characteristic vector of the set $S \subseteq V$, assigning the value 1 to vertices in S, and 0 to the others. Thus, the entropy of G can be written as

$$H(G) := \min_{p \in STAB(G)} -\frac{1}{n} \sum_{v \in V} \log p_v.$$

The relation between the graph entropy and the chromatic entropy (the minimum entropy of a coloring) can be made clear from the following observation: if we restrict the vector p in the definition of $H(G)$ to be a convex combination of characteristic vectors of *disjoint* stable sets, then the minimum of (1) is equal to the chromatic entropy (see [4] for a rigorous development). Hence, the mathematical program defining the graph entropy is a relaxation of that defining the chromatic entropy. It follows that the graph entropy is a lower bound on the chromatic entropy.

Recently, we showed that the greedy coloring algorithm, that iteratively colors and removes a maximum stable set, yields a good approximation of the graph entropy, provided the graph is perfect.

Theorem 9. *[7] Let G be a perfect graph, and g be the entropy of a greedy coloring of G. Then*

$$g \leq H(G) + \log H(G) + O(1).$$

The proof of Theorem 9 is a "dual fitting argument" based on a min-max relation linking the entropy of a perfect graph and its complement, due to Czisar, Körner, Lovász, Marton and Simonyi [8]. In particular, Theorem 9 implies that the chromatic entropy of a perfect graph never exceeds its entropy by more than $\log \log n$. It turns out that this is essentially tight [7].

Theorem 9 is the key tool in a recent algorithm for the partial order production problem [7]. In this problem, we want to bijectively map a set T of objects coming with an unknown total order to a vector equipped with partial order on its positions, such that the relations of the partial order are satisfied by the mapped elements. The problem admits the selection, multiple selection, sorting, and heap construction problems as special cases. Until recently, it was not known whether there existed a polynomial-time algorithm performing this task using an near-optimal number of comparisons between elements of T. By applying (twice) the greedy coloring algorithm and the approximation result of Theorem 9, we could reduce, in polynomial time, the problem to a well studied multiple selection problem and solve it with a near-optimal number of comparisons. The reader is referred to [7] for further details.

Acknowledgment. This work was supported by the Communauté Française de Belgique (projet ARC).

References

1. Agarwal, S., Belongie, S.: On the non-optimality of four color coding of image partitions. In: ICIP 2002: Proceedings of the IEEE International Conference on Image Processing (2002)
2. Alon, N., Orlitsky, A.: Source coding and graph entropies. IEEE Trans. Inform. Theory 42(5), 1329–1339 (1996)
3. Bonizzoni, P., Della Vedova, G., Dondi, R., Mariani, L.: Experimental analysis of a new algorithm for partial haplotype completion. Int. J. Bioinformatics Res. Appl. 1(4), 461–473 (2005)
4. Cardinal, J., Fiorini, S., Joret, G.: Minimum entropy coloring. J. Comb. Opt. 16(4), 361–377 (2008)
5. Cardinal, J., Fiorini, S., Joret, G.: Minimum entropy orientations. Op. Res. Lett. 36, 680–683 (2008)
6. Cardinal, J., Fiorini, S., Joret, G.: Tight results on minimum entropy set cover. Algorithmica 51(1), 49–60 (2008)
7. Cardinal, J., Fiorini, S., Joret, G., Jungers, R., Munro, J.I.: An efficient algorithm for partial order production. In: STOC 2009: 41st ACM Symposium on Theory of Computing (2009) (to appear)
8. Csiszár, I., Körner, J., Lovász, L., Marton, K., Simonyi, G.: Entropy splitting for antiblocking corners and perfect graphs. Combinatorica 10(1), 27–40 (1990)
9. Doshi, V., Shah, D., Médard, M.: Source coding with distortion through graph coloring. In: ISIT 2007: Proceedings of the IEEE International Symposium on Information Theory, pp. 1501–1505 (2007)
10. Doshi, V., Shah, D., Médard, M., Jaggi, S.: Graph coloring and conditional graph entropy. In: ACSSC 2006: Proceedings of the Fortieth Asilomar Conference on Signals, Systems and Computers, pp. 2137–2141 (2006)
11. Doshi, V., Shah, D., Médard, M., Jaggi, S.: Distributed functional compression through graph coloring. In: DCC 2007: Proceedings of the IEEE Data Compression Conference, pp. 93–102 (2007)
12. Feige, U.: A threshold of $\ln n$ for approximating set cover. J. ACM 45(4), 634–652 (1998)
13. Feige, U., Lovász, L., Tetali, P.: Approximating min sum set cover. Algorithmica 40(4), 219–234 (2004)

14. Fukunaga, T., Halldórsson, M.M., Nagamochi, H.: Rent-or-buy scheduling and cost coloring problems. In: Arvind, V., Prasad, S. (eds.) FSTTCS 2007. LNCS, vol. 4855, pp. 84–95. Springer, Heidelberg (2007)
15. Fukunaga, T., Halldórsson, M.M., Nagamochi, H.: Robust cost colorings. In: SODA 2008: Proceedings of the nineteenth annual ACM-SIAM symposium on Discrete algorithms, pp. 1204–1212 (2008)
16. Gijswijt, D., Jost, V., Queyranne, M.: Clique partitioning of interval graphs with submodular costs on the cliques. RAIRO – Operations Research 41(3), 275–287 (2007)
17. Gusev, A., Măndoiu, I.I., Paşaniuc, B.: Highly scalable genotype phasing by entropy minimization. IEEE/ACM Trans. Comput. Biol. Bioinformatics 5(2), 252–261 (2008)
18. Halldórsson, M.M., Radhakrishnan, J.: Greed is good: approximating independent sets in sparse and bounded-degree graphs. In: STOC 1994: Proceedings of the twenty-sixth annual ACM symposium on Theory of computing, pp. 439–448 (1994)
19. Halperin, E., Karp, R.M.: The minimum-entropy set cover problem. Theoret. Comput. Sci. 348(2-3), 240–250 (2005)
20. Håstad, J.: Clique is hard to approximate within $n^{1-\epsilon}$. Acta Mathematica 182, 105–142 (1999)
21. Kierstead, H.A., Trotter, W.A.: An extremal problem in recursive combinatorics. Congr. Numer. 33, 143–153 (1981)
22. Körner, J.: Coding of an information source having ambiguous alphabet and the entropy of graphs. In: Transactions of the Sixth Prague Conference on Information Theory, Statistical Decision Functions, Random Processes (Tech. Univ., Prague, 1971; dedicated to the memory of Antonín Špaček), pp. 411–425. Academia, Prague (1973)
23. Körner, J., Orlitsky, A.: Zero-error information theory. IEEE Trans. Inform. Theory 44(6), 2207–2229 (1998)
24. Koulgi, P., Tuncel, E., Regunathan, S., Rose, K.: On zero-error source coding with decoder side information. IEEE Trans. Inform. Theory 49(1), 99–111 (2003)
25. Linial, N.: Locality in distributed graph algorithms. SIAM J. Comput. 21(1), 193–201 (1992)
26. Nicoloso, S., Sarrafzadeh, M., Song, X.: On the sum coloring problem on interval graphs. Algorithmica 23(2), 109–126 (1999)
27. Parnas, M., Ron, D.: Approximating the minimum vertex cover in sublinear time and a connection to distributed algorithms. Theoret. Comput. Sci. 381(1-3), 183–196 (2007)
28. Pemmaraju, S.V., Raman, R., Varadarajan, K.: Buffer minimization using max-coloring. In: SODA 2004: Proceedings of the fifteenth annual ACM-SIAM symposium on Discrete algorithms, pp. 562–571 (2004)
29. Shannon, C.E.: A mathematical theory of communication. Bell. System Technical Journal 27, 379–423, 623–656 (1948)
30. Simonyi, G.: Graph entropy: a survey. In: *Combinatorial Optimization.* DIMACS Series in Discrete Mathematics and Theoretical Computer Science, vol. 20, pp. 399–441 (1995)
31. Yannakakis, M., Gavril, F.: The maximum k-colorable subgraph problem for chordal graphs. Inf. Process. Lett. 24(2), 133–137 (1987)

Program Self-reference in Constructive Scott Subdomains

John Case and Samuel E. Moelius III

Department of Computer & Information Sciences
University of Delaware
101 Smith Hall
Newark, DE 19716
{case,moelius}@cis.udel.edu

Abstract. Intuitively, a *recursion theorem* asserts the existence of *self-referential programs*. Two well-known recursion theorems are Kleene's Recursion Theorem (krt) and Rogers' Fixpoint Recursion Theorem (fprt). Does one of these two theorems better capture the notion of program self-reference than the other? In the context of the partial computable functions over the natural numbers ($\mathcal{PC}$), fprt is strictly *weaker* than krt, in that fprt holds in any effective numbering of $\mathcal{PC}$ in which krt holds, but *not* vice versa. It is shown that, in this context, the existence of *self-reproducing programs* (a.k.a. *quines*) is assured by krt, but *not* by fprt. Most would surely agree that a self-reproducing program is self-referential. Thus, this result suggests that krt is better than fprt at capturing the notion of program self-reference *in* $\mathcal{PC}$.

A generalization of krt to arbitrary *constructive Scott subdomains* is then given. (For fprt, a similar generalization was already known.) Surprisingly, for some such subdomains, the two theorems turn out to be *equivalent*. A precise characterization is given of those constructive Scott subdomains in which this occurs. For such subdomains, the two theorems capture the notion of program self-reference equally well.

Keywords: numberings, recursion theorems, Scott domains, self-reference, self-reproducing programs.

1 Introduction: krt and fprt

Intuitively, a *recursion theorem* asserts the existence of *self-referential programs*. Two well-known recursion theorems are Kleene's Recursion Theorem (krt) and Rogers' Fixpoint Recursion Theorem (fprt). Does one of these two theorems better capture the notion of program self-reference than the other?

The two theorems are normally stated in the context of the partial computable functions over the natural numbers ($\mathcal{PC}$). We give the formal statements of the two theorems following some necessary definitions.

K. Ambos-Spies, B. Löwe, and W. Merkle (Eds.): CiE 2009, LNCS 5635, pp. 89–98, 2009.

Let $\mathbb{N}$ be the set of natural numbers, $\{0, 1, 2, ...\}$. Let $\perp$ denote the divergent computation.[1] $\mathbb{N}_\perp \stackrel{\text{def}}{=} \mathbb{N} \cup \{\perp\}$. Let $\langle \cdot, \cdot \rangle$ be any fixed pairing function.[2] A function $f : \mathbb{N} \to \mathbb{N}_\perp$ is *effective* $\stackrel{\text{def}}{\Leftrightarrow}$ it is partial computable [Rog67]. A function $\psi : \mathbb{N} \to \mathcal{PC}$ is *effective* $\stackrel{\text{def}}{\Leftrightarrow} \lambda p, x . \psi(p)(x)$ is partial computable.[3] For any set X, a function $\psi : \mathbb{N} \to X$ is a *numbering* of X [Rog58, Ric80, Roy87, Spr90, BGS03] $\stackrel{\text{def}}{\Leftrightarrow} \psi$ is onto.[4] We will often write ψ_p as shorthand for $\psi(p)$.

An effective numbering of type $\mathbb{N} \to \mathcal{PC}$ can be thought of as a *programming language*, in the following sense. If one were to take the programs in some programming language for $\mathcal{PC}$, and number those programs, e.g., length-lexicographically, then the function that sends p to the semantics of the pth program would be an effective numbering of type $\mathbb{N} \to \mathcal{PC}$.

The following are the formal statements of Kleene's Recursion Theorem (krt) and Rogers' Fixpoint Recursion Theorem (fprt).

Definition 1. For each effective numbering $\psi : \mathbb{N} \to \mathcal{PC}$, (a) and (b) below.

(a) (**Kleene [Rog67, page 214, problem 11-4]**) krt *holds in* $\psi \Leftrightarrow$

$$(\forall g \in \mathcal{PC})(\exists e)[\psi_e = g(\langle e, \cdot \rangle)]. \tag{1}$$

(b) (**Rogers [Rog67, Theorem 11-I]**) fprt *holds in* $\psi \Leftrightarrow$

$$(\forall \text{ computable } t : \mathbb{N} \to \mathbb{N})(\exists e)[\psi_e = \psi_{t(e)}]. \tag{2}$$

krt can be interpreted as follows. The partial computable function g in (1) is an arbitrary, algorithmic task to perform with a self-copy; the program e is one that creates a copy of itself (external to itself) and then performs that task using this self-copy. In an important sense, this self-copy provides e *complete, low-level self-knowledge*. Thus, e can reflect upon its own *intensional* (synonym: *connotational*) characteristics, e.g., its size, runtime, memory usage, etc. Of course, e can *run* its self-copy, and thereby reflect upon its *extensional* (synonym: *denotational*) characteristics as well [Roy87].

fprt can be interpreted as follows. The function t in (2) is a transformation on programs; the program e is one whose semantics remain fixed under this transformation [Roy87].

The following constructive variant of fprt is also sometimes considered.[5]

Definition 2 (Riccardi [Ric80, Definition 2.1]). For each effective numbering $\psi : \mathbb{N} \to \mathcal{PC}$, FPRT *holds in* $\psi \Leftrightarrow$ there exists a computable function $n : \mathbb{N} \to \mathbb{N}$ such that, for each p,

$$\begin{aligned} &[(\psi_p \circ n)(p) \neq \perp \Rightarrow \psi_{n(p)} = \psi_{(\psi_p \circ n)(p)}] \\ \wedge\; &[(\psi_p \circ n)(p) = \perp \Rightarrow \psi_{n(p)} = \lambda x . \perp]. \end{aligned} \tag{3}$$

---

[1] Thus, $\perp$ may be thought of as the *value* of an *infinite loop*.

[2] A *pairing function* is computable, 1-1, onto, and of type $\mathbb{N}^2 \to \mathbb{N}$ [Rog67, page 64].

[3] In Section 2, the notion of *effective* is generalized to other types of functions.

[4] In this paper, we shall generally use lowercase Greek letters (e.g., γ, ψ) for numberings, and lowercase Roman letters (e.g., f, g, h) for other functions.

[5] krt similarly has constructive variants, which are considered, e.g., in [Roy87, CM09a].

```
#include <stdio.h>
int main(){char*s[]={"#include <stdio.h>%cint main(){char*s[]={",
"};printf(s[0],10);int i=0;while(i<3)printf(%c%%c%%s%%c,%%c%c,34,%c",
"s[i++],34,10);printf(s[1],34,34,10);printf(s[2],10);return 0;}%c",
};printf(s[0],10);int i=0;while(i<3)printf("%c%s%c,%c",34,
s[i++],34,10);printf(s[1],34,34,10);printf(s[2],10);return 0;}
```

$$\lambda e.e$$

| ```#include <stdio.h>
int main() { printf("75");
          return 0;   }``` | ```#include <stdio.h>
int main() { printf("#");
          return 0;   }``` |
|---|---|
| $\lambda e.\lceil \log_{256}(e+1) \rceil$ | $\lambda e.(e \bmod 256)$ |

Fig. 1. Top: A C-program that outputs its own source code. **Bottom-Left (BL)**: A C-program that outputs the *length* of its source code (in bytes). **Bottom-Right (BR)**: A C-program that outputs the *first character* of its source code. Also depicted are functions of type $\mathbb{N} \to \mathbb{N}_\perp$ corresponding to each C-program.

In (3), ψ_p and $n(p)$ play the roles played by t and e, respectively, in (2). In this sense, the function n *finds* a program $n(p)$ whose semantics remain fixed under the transformation ψ_p. Such an n is called an *effective instance of* FPRT *in* ψ.

fprt was popularized by Rogers' classic textbook [Rog67]. Therein, he writes [Rog67, page 182]: "Kleene's formulation differs slightly, and inessentially, from ours." It is true that, in any *standard* numbering of $\mathcal{PC}$, both theorems will hold. More broadly, in any effective numbering (standard or otherwise) in which krt holds, fprt holds as well [Ric80, Theorem 5.1]. However, there *do* exist non-standard, effective numberings of $\mathcal{PC}$ in which fprt holds, but in which krt does *not* hold [Ric80, Theorem 5.3].[6] In this sense, krt is strictly *stronger* than fprt.

Given that krt is stronger in this sense, are there programs (for $\mathcal{PC}$) that one would reasonably call self-referential, and whose existence is assured by krt, but *not* by fprt?

Consider the three C-programs in Figure 1. Figure 1(Top) is a program that outputs its own source code (such a *self-reproducing* program is sometimes called a *quine* [Tho99]);[7] Figure 1(BL) is a program that outputs the *length* of its source code (in bytes); and Figure 1(BR) is a program that outputs the *first character* of its source code.

Most would surely agree that a self-reproducing program, like that of Figure 1(Top), is self-referential. For the programs in Figures 1(BL) and 1(BR),

---

[6] The effective numbering of $\mathcal{PC}$ used in the proof of [Ric80, Theorem 5.3] is the same as that used in the proof of [MWY78, Theorem 3.6].

[7] Early examples of such programs are due to Lee [Lee63] and Thatcher [Tha63]. The term "quine" appears to have been inspired by Hofstadter [Hof79]. Therein, Hofstadter refers to the operation of *preceding a phrase by its own quotation* as "quining", in honor of Willard Van Orman Quine [Qui66].

however, it is less clear. For example, many C-programs begin with the character '#'. Thus, one could argue that the program in Figure 1(BR) is more *opportunistic* than self-referential.

Common among these three C-programs is that each computes a function of the form $\lambda x.f(e)$, for some effective function $f : \mathbb{N} \to \mathbb{N}_\perp$; that is, each outputs the result of applying a (partial) computable function to its source code e, whilst ignoring any possible input x.[8] These functions are also depicted in Figure 1.[9] For example, for the self-reproducing program in Figure 1(Top), the corresponding function f is simply $\lambda e.e$.[10]

Clearly, for each effective numbering $\psi : \mathbb{N} \to \mathcal{PC}$ in which krt holds, and for each effective $f : \mathbb{N} \to \mathbb{N}_\perp$, there exists an e such that

$$\psi_e = \lambda x.f(e). \tag{4}$$

But what if merely fprt holds in ψ? Then, can the same still be said *for each f*?

The answer is "no". Our first main result, Theorem 3, says that fprt assures the existence of a program e as in (4) for precisely those effective $f : \mathbb{N} \to \mathbb{N}_\perp$ with finite range.[11]

Theorem 3. Suppose that $f : \mathbb{N} \to \mathbb{N}_\perp$ is effective. Then, (a)-(c) below are equivalent.

(a) For every effective numbering $\psi : \mathbb{N} \to \mathcal{PC}$ in which fprt holds, there exists an e such that $\psi_e = \lambda x.f(e)$.
(b) For every effective numbering $\psi : \mathbb{N} \to \mathcal{PC}$ in which FPRT holds, there exists an e such that $\psi_e = \lambda x.f(e)$.
(c) The range of f is finite.

Thus, fprt does *not* assure the existence of a program like that of Figure 1(Top), *nor* even like that of Figure 1(BL) (though fprt does assure the existence of a program like that of Figure 1(BR)). Given that most would call the program of Figure 1(Top) self-referential, Theorem 3 suggests that krt is better than fprt at capturing the notion of program self-reference *in $\mathcal{PC}$*.

One can consider self-reproducing programs in effective numberings other than those of type $\mathbb{N} \to \mathcal{PC}$. For example, consider an effective numbering $\psi : \mathbb{N} \to \mathbb{N}_\perp$. (Thus, programs in this ψ code elements of $\mathbb{N}_\perp$, instead of $\mathcal{PC}$.) In such a numbering, one could reasonably call a program e *self-reproducing* $\Leftrightarrow \psi(e) =$

---

[8] In the terminology of [Cas71], $f(e)$ is the computable distortion of e wrought by f.
[9] ISO-8859-1 is a character encoding commonly used on the Internet (see, for example, [Moo96]). This encoding associates a character with each value in $\{0, ..., 255\}$. One can treat each program in Figure 1 as a base-256 number, e.g., where each character is a digit whose value is determined by the ISO-8859-1 encoding, and where the leading character is *least* significant. In this way, "`#include` ..." mod 256 = "#".
[10] Thus, $\psi_e = \lambda x.f(e) = \lambda x.(\lambda e.e)(e) = \lambda x.e$.
[11] The proof of Theorem 3 employs, among other things, techniques of Machtey, Winklmann, and Young [MWY78, proof of Theorem 3.6]. The proofs of Theorem 3 and all subsequent results can be found in [CM09b].

e. Then, one could ask the following. Does fprt assure the existence of a self-reproducing program in an effective numbering *of this type*? More generally, is fprt still *weaker* than krt in an effective numbering *of this type*?

To state such questions formally, one must first make precise what one *means* by krt and fprt in such numberings. To do so, we first introduce the notion of a *constructive Scott subdomain*. Then, in Section 3 we generalize krt to effective numberings of arbitrary constructive Scott subdomains. (For fprt, a similar generalization was already known [Ers77].) Finally, in Section 4, we revisit the aforementioned questions.

2   Constructive Scott Subdomains

In this section, we introduce *constructive Scott subdomains*. Intuitively, a constructive Scott subdomain is a collection of objects with the property that: each object can be broken up into *pieces*. $\mathcal{PC}$ is an example of such a collection, in which case the pieces are the finite functions over $\mathbb{N}$.

Having such a collection can be useful, as it can often be easier to work with each object *in pieces*, than to work with each object *as a whole*. For example, consider the effective numberings of type $\mathbb{N} \to \mathcal{PC}$. In such a numbering ψ, the predicate $\lambda p, q . [\psi_p \subseteq \psi_q]$ is *never* computable [JST09, Proposition 14]. However, if $\lambda i . F_i$ is a canonical enumeration of the finite functions over $\mathbb{N}$ [Rog67, MY78], then the predicate $\lambda i, j . [F_i \subseteq F_j]$ *is* computable.

As the reader will be able to see following the formal definitions, many collections of objects can be viewed in this way. Unless otherwise noted, concepts not explained below can be found in [SHLG94].

As is customary, we use $\sqsubseteq$ to denote the ordering relation of any given partial order. A partial order X is *flat* $\stackrel{\text{def}}{\Leftrightarrow}$ X has a least element $\bot$, and X is ordered such that, for each $x, y \in X$, $x \sqsubseteq y \Leftrightarrow [x = \bot \lor x = y]$.[12] Such is the standard ordering of $\mathbb{N}_\bot$. Thus, $\mathbb{N}_\bot$ is an example of a flat partial order.

For a subset A of a partial order X, $\bigsqcup A$ denotes the least-upper-bound of A, which may or may not *exist*, or may or may not exist *in X*. For the special case of a two element set: $x \sqcup y \stackrel{\text{def}}{=} \bigsqcup \{x, y\}$. A subset A of a partial order is *directed* $\stackrel{\text{def}}{\Leftrightarrow}$ A is *non*-empty and, for each $x, y \in A$, $\{x, y\}$ has some (not necessarily least) upper-bound in A. For each partial order X, and each $x \in X$, x is *compact* in X $\stackrel{\text{def}}{\Leftrightarrow}$ $(\forall$ directed $A \subseteq X)[[\bigsqcup A$ exists in $X \land x \sqsubseteq \bigsqcup A] \Rightarrow (\exists y \in A)[x \sqsubseteq y]]$. For each partial order X, $\mathsf{K}(X) \stackrel{\text{def}}{=} \{x \in X \mid x$ is compact in $X\}$. For example, $\mathsf{K}(\mathbb{N}_\bot) = \mathbb{N}_\bot$. Similarly, if one lets $\mathcal{P}$ be the set of all functions of type $\mathbb{N} \to \mathbb{N}_\bot$ ordered pointwise (i.e., for each $f, g \in \mathcal{P}$, $f \sqsubseteq g \Leftrightarrow (\forall x)[f(x) \sqsubseteq g(x)]$), then $\mathsf{K}(\mathcal{P})$ is exactly the set of finite functions over $\mathbb{N}$.

A *Scott domain* is a partial order D satisfying (a)-(d) below.

(a) D has a least element (denoted by $\bot$).
(b) D is *complete*: for every directed $A \subseteq D$, $\bigsqcup A$ exists in D.

---

[12] An anonymous referee attributes this notion to Scott. Flat partial orders are also considered in [SHLG94, Spr98].

(c) D is *algebraic*: for each $y \in D$, the set $A = \{x \in K(D) \mid x \sqsubseteq y\}$ is directed and $y = \bigsqcup A$.

(d) D is a *conditional upper semi-lattice*: for each $x, y \in D$, if $\{x, y\}$ has an upper-bound in D, then $x \sqcup y$ exists in D.

Suppose that D is a Scott domain, γ is a numbering of $K(D)$, and S is such that $K(D) \subseteq S \subseteq D$. A function $\psi : \mathbb{N} \to S$ is *effective via* $\gamma \stackrel{\text{def}}{\Leftrightarrow}$ the set $\{\langle i, p \rangle \mid \gamma(i) \sqsubseteq \psi(p)\}$ is computably enumerable (ce) [Rog67].[13] Recall that Section 1 defined *effective* for functions of type $\mathbb{N} \to \mathbb{N}_\perp$ and $\mathbb{N} \to \mathcal{PC}$. In the former case, the definition is equivalent to *effective via* $\text{pred} : \mathbb{N} \to K(\mathbb{N}_\perp)$, where, for each x,

$$\text{pred}(x) = \begin{cases} \perp, & \text{if } x = 0; \\ x - 1, & \text{otherwise.} \end{cases} \tag{5}$$

In the latter case, the definition is equivalent to *effective via* $\lambda i. F_i : \mathbb{N} \to K(\mathcal{P})$.[14]

A *constructive Scott subdomain* [SHLG94, Spr98, Spr07] is a tuple (D, γ, S) satisfying (a)-(g) below.

(a) D is a Scott domain.
(b) γ is a numbering of $K(D)$.
(c) The predicate $\lambda i, j. [\gamma(i) \sqsubseteq \gamma(j)]$ is computable.
(d) The predicate $\lambda i, j. [\{\gamma(i), \gamma(j)\}$ has an upper-bound in $D]$ is computable.
(e) There exists a computable function $u : \mathbb{N}^2 \to \mathbb{N}$ such that, if $\{\gamma(i), \gamma(j)\}$ has an upper-bound in D, then $(\gamma \circ u)(i, j) = \gamma(i) \sqcup \gamma(j)$.
(f) $K(D) \subseteq S \subseteq D$.
(g) There exists an effective numbering $\psi : \mathbb{N} \to S$ via γ.

We shall sometimes refer to a ψ as in (g) just above as an *effective numbering of* (D, γ, S).

A constructive Scott subdomain (D, γ, S) is *recursively complete* [SHLG94, page 270, Definition 4.8] $\stackrel{\text{def}}{\Leftrightarrow}$ $(\forall \text{ ce } A \subseteq \mathbb{N})[\gamma(A)$ is directed $\Rightarrow$ $\bigsqcup \gamma(A) \in S]$. For example, of the two constructive Scott subdomains $(\mathcal{P}, \lambda i. F_i, \mathcal{PC})$ and $(\mathcal{P}, \lambda i. F_i, K(\mathcal{P}))$, the former is recursively complete, while the latter is *not*.

3   krt in Constructive Scott Subdomains

In this section, we generalize krt to arbitrary constructive Scott subdomains. In order to give some intuition for the generalized definition, let us first return attention to effective numberings of type $\mathbb{N} \to \mathcal{PC}$. Note that, for such numberings,

---

[13] A function $\psi : \mathbb{N} \to S$ can be effective via γ, but *not* via γ'. The following is an example. Let K be the diagonal halting problem [Rog67, page 62]. Let $\{k_0 < k_1 < \cdots\} = K$ and $\{\bar{k}_0 < \bar{k}_1 < \cdots\} = \bar{K} \stackrel{\text{def}}{=} \mathbb{N} - K$. Let $f : \mathbb{N} \to \mathbb{N}$ be such that, for each i, $f(2i) = k_i$ and $f(2i + 1) = \bar{k}_i$. Let $\wp(\mathbb{N})$ be the set of all subsets of $\mathbb{N}$ ordered by $\subseteq$, and let $\lambda i. D_i : \mathbb{N} \to K(\wp(\mathbb{N}))$ be a canonical enumeration of the finite subsets of $\mathbb{N}$ [Rog67, MY78]. Note that, for each i, $f(D_i) \subseteq \bar{K} \Leftrightarrow D_i \subseteq 2\mathbb{N} + 1$. Thus, $\lambda p. \bar{K} : \mathbb{N} \to \{\bar{K}\}$ is effective via $\lambda i. f(D_i)$. However, $\lambda p. \bar{K}$ is *not* effective via $\lambda i. D_i$, as this would imply $\{i \mid D_i \subseteq \bar{K}\}$ is computably enumerable.

[14] In neither case is the choice of γ *unique*.

the following definition of krt is *equivalent* to Definition 1(a). For each effective numbering $\psi : \mathbb{N} \to \mathcal{PC}$, krt holds in $\psi \Leftrightarrow$

$$(\forall \text{ effective } f : \mathbb{N} \to \mathcal{PC})(\exists e)[\psi_e = f(e)]. \tag{6}$$

To see that this is equivalent, note that, for each $g \in \mathcal{PC}$, there exists an effective $f : \mathbb{N} \to \mathcal{PC}$ such that $(\forall p, x)[f(p)(x) = g(\langle p, x \rangle)]$, and vice versa.

Taking a cue from (6), we define krt for arbitrary constructive Scott subdomains as follows.

Definition 4. Suppose that (D, γ, S) is a constructive Scott subdomain. Then, for each effective numbering $\psi : \mathbb{N} \to S$ via γ, krt *holds in* $\psi \Leftrightarrow$

$$(\forall \text{ effective } f : \mathbb{N} \to S \text{ via } \gamma)(\exists e)[\psi(e) = f(e)]. \tag{7}$$

This definition preserves many of the intuitive properties held by krt in effective numberings of type $\mathbb{N} \to \mathcal{PC}$. For example, as will be seen in Section 4, the fact that krt *entails* fprt in such numberings carries over to effective numberings of arbitrary constructive Scott subdomains.

Suppose that (D, γ, S) is a constructive Scott subdomain. Natural questions to ask are the following. Do there exist effective numberings of (D, γ, S) in which krt holds? Do there exist such numberings in which krt does *not* hold? Our next main result, Theorem 5, gives conditions for the existence of each kind of numbering.

Theorem 5. Suppose that (D, γ, S) is a constructive Scott subdomain. Then, (a) and (b) below.

(a) There exists an effective numbering $\psi : \mathbb{N} \to S$ via γ in which krt holds $\Leftrightarrow$ (D, γ, S) is recursively complete.
(b) There exists an effective numbering $\psi : \mathbb{N} \to S$ via γ in which krt does *not* hold $\Leftrightarrow D \neq \{\bot\} \Leftrightarrow S \neq \{\bot\}$.

Definition 6 just below introduces the notion of a *regular* partial order.

Definition 6 (Spreen [Spr00]). A partial order D is *regular* $\Leftrightarrow$ for each $x, y \in \mathsf{K}(D)$,

$$y \not\sqsubseteq x \Rightarrow (\exists z \in \mathsf{K}(D))[x \sqsubseteq z \wedge \{y, z\} \text{ has } no \text{ upper-bound in } \mathsf{K}(D)].^{15} \tag{8}$$

An example of a regular partial order is $\mathcal{P}.^{16}$ To see this, let f and g be finite functions over $\mathbb{N}$ such that $g \not\sqsubseteq f$. If $f \cup g$ is not single valued, then already

---

[15] We shall generally be interested in regular *Scott domains*. It can be shown that, for each Scott domain D, and each $x, y \in \mathsf{K}(D)$, $\{x, y\}$ has an upper-bound in $\mathsf{K}(D) \Leftrightarrow \{x, y\}$ has an upper-bound in D. (See, for example, [SHLG94, page 57, Lemma 1.9].) Thus, if D is a Scott domain, then D is regular $\Leftrightarrow$ for each $x, y \in \mathsf{K}(D)$,

$$y \not\sqsubseteq x \Rightarrow (\exists z \in \mathsf{K}(D))[x \sqsubseteq z \wedge \{y, z\} \text{ has } no \text{ upper-bound in } D].$$

[16] Recall that $\mathcal{P}$ is the set of all functions of type $\mathbb{N} \to \mathbb{N}_\bot$ ordered pointwise.

$\{f, g\}$ has no upper-bound in $\mathcal{P}$. On the other hand, if there exists an $x_0 \in \mathbb{N}$ such that $f(x_0) = \bot \neq g(x_0)$, then $f \sqsubseteq h$ and $\{g, h\}$ has no upper-bound in $\mathcal{P}$, where, for each x,

$$h(x) = \begin{cases} g(x_0) + 1, & \text{if } x = x_0; \\ f(x), & \text{otherwise.} \end{cases} \tag{9}$$

An example of a *non*-regular partial order is $\wp(\mathbb{N})$, the set of all subsets of $\mathbb{N}$ ordered by $\subseteq$.

Our next main result, Theorem 7, *characterizes* the effective numberings in which krt holds, for the constructive Scott subdomains (D, γ, S) for which D is regular. Note that, in the definition of krt for arbitrary constructive Scott subdomains (Definition 4), one could write (7) as

$$(\forall \text{ effective } f : \mathbb{N} \to S \text{ via } \gamma)(\exists e)[\psi(e) \sqsubseteq f(e) \ \wedge \ f(e) \sqsubseteq \psi(e)]. \tag{10}$$

Theorem 7 says that, if (D, γ, S) is a constructive Scott subdomain and D is regular, then having just the latter half of (10) suffices to have full krt.[17]

Theorem 7. Suppose that (D, γ, S) is a constructive Scott subdomain and that D is regular. Then, for each effective numbering $\psi : \mathbb{N} \to S$ via γ, (a) and (b) below are equivalent.

(a) krt holds in ψ.
(b) $(\forall \text{ effective } f : \mathbb{N} \to S \text{ via } \gamma)(\exists e)[f(e) \sqsubseteq \psi(e)]$.

$\mathcal{P}$ is an example of a regular Scott domain. Thus, Theorem 7's characterization holds in the effective numberings of $(\mathcal{P}, \lambda i.F_i, \mathcal{PC})$. (In some sense, $(\mathcal{P}, \lambda i.F_i, \mathcal{PC})$ represents the *standard* way of formulating $\mathcal{PC}$ as a constructive Scott subdomain.)

Our next main result, Theorem 8, says that, if (D, γ, S) is a constructive Scott subdomain and D is *not* regular, then there exists an effective numbering of (D, γ, S) in which Theorem 7's characterization *fails*. Theorem 7 and Theorem 8 together say that Theorem 7's characterization holds in the effective numberings of a constructive Scott subdomain (D, γ, S) iff D is regular.

Theorem 8. Suppose that (D, γ, S) is a constructive Scott subdomain and that D is *not* regular. Then, there exists an effective numbering $\psi : \mathbb{N} \to S$ via γ satisfying (a) and (b) below.

(a) $(\forall \text{ effective } f : \mathbb{N} \to S \text{ via } \gamma)(\exists e)[f(e) \sqsubseteq \psi(e)]$.
(b) krt does *not* hold in ψ.

$\wp(\mathbb{N})$ is an example of a *non*-regular Scott domain. Thus, if one lets $\lambda i.D_i : \mathbb{N} \to \mathrm{K}(\wp(\mathbb{N}))$ be a canonical enumeration of the finite subsets of $\mathbb{N}$ [Rog67, MY78], and if one lets $\mathcal{CE}$ be the collection of all computably enumerable sets, then Theorem 7's characterization fails in the effective numberings of $(\wp(\mathbb{N}), \lambda i.D_i, \mathcal{CE})$. (In some sense, $(\wp(\mathbb{N}), \lambda i.D_i, \mathcal{CE})$ represents the *standard* way of formulating $\mathcal{CE}$ as a constructive Scott subdomain.)

---

[17] The proof of Theorem 7 bears some resemblance to Royer's proof of [Roy87, Theorem 4.2.15].

4   krt and fprt in Constructive Scott Subdomains

In this section, we compare krt and fprt in arbitrary constructive Scott subdomains. The generalization of fprt to arbitrary constructive Scott subdomains is essentially due to Ershov.[18]

Definition 9 (Ershov [Ers77]). Suppose that (D, γ, S) is a constructive Scott subdomain. Then, for each effective numbering $\psi : \mathbb{N} \to S$ via γ, fprt *holds in* $\psi \Leftrightarrow$

$$(\forall \text{ computable } t : \mathbb{N} \to \mathbb{N})(\exists e)[\psi(e) = (\psi \circ t)(e)]. \tag{11}$$

Recall from Section 1 that krt *entails* fprt in effective numberings of type $\mathbb{N} \to \mathcal{PC}$. Does this entailment relationship carry over to effective numberings of arbitrary constructive Scott subdomains? Proposition 10 just below says that the answer is "yes".[19]

Proposition 10. Suppose that (D, γ, S) is a constructive Scott subdomain. Then, for each effective numbering $\psi : \mathbb{N} \to S$ via γ, if krt holds in ψ, then fprt holds in ψ as well.

Also recall from Section 1 that there exist effective numberings of $\mathcal{PC}$ in which fprt holds, but in which krt does *not* hold [Ric80, Theorem 5.3]. Do there exist such numberings for arbitrary constructive Scott subdomains?

   The answer, as it turns out, is that such numberings exist for *some* constructive Scott subdomains, but *not* for others. Note that, for subdomains of the latter kind, krt and fprt are *equivalent*, by Proposition 10. Our final main result, Theorem 11, says that, for a constructive Scott subdomain (D, γ, S), this equivalence occurs precisely when D is flat. In such a subdomain, krt and fprt capture the notion of program self-reference equally well.

Theorem 11. Suppose that (D, γ, S) is a constructive Scott subdomain. Then, (a) and (b) below are equivalent.

(a) $(\forall \text{ effective } \psi : \mathbb{N} \to S \text{ via } \gamma)[\text{krt holds in } \psi \Leftrightarrow \text{fprt holds in } \psi]$.
(b) D is flat.

$\mathbb{N}_\perp$ is an example of a flat Scott domain.[20] Thus, this equivalence occurs in any constructive subdomain of $\mathbb{N}_\perp$. Recall from Section 3 that, in an effective numbering $\psi : \mathbb{N} \to \mathbb{N}_\perp$, one could reasonably call a program e *self-reproducing* $\Leftrightarrow \psi(e) = e$. Clearly, in such a numbering, krt suffices to assure the existence of such a program. Furthermore, by Theorem 11, fprt is equally sufficient.

---

[18] See, for example, [BGS03, Theorem 2.1].
[19] Proposition 10 generalizes Riccardi's [Ric80, Theorem 5.1].
[20] Of course, $\mathcal{PC}$ and $\mathcal{CE}$ are *not* flat. Thus, neither are $\mathcal{P}$ nor $\wp(\mathbb{N})$.

References

[BGS03]   Badaev, S., Goncharov, S., Sorbi, A.: Completeness and universality of arithmetical numberings. In: Computability and models, pp. 11–44. Springer, Heidelberg (2003)

[Cas71]   Case, J.: A note on the degrees of self-describing Turing machines. Journal of the ACM 18(3), 329–338 (1971)

[CM09a]   Case, J., Moelius, S.: Independence results for n-ary recursion theorems (2009) (submitted)

[CM09b]   Case, J., Moelius, S.: Program self-reference in constructive Scott subdomains (expanded version) (2009),
http://www.cis.udel.edu/~moelius/publications

[Ers77]   Ershov, Y.L.: Theory of Numberings, Nauka, Moscow (1977) (in Russian)

[Hof79]   Hofstadter, D.R.: Gödel, Escher, Bach: An Eternal Golden Braid. Basic Books, Inc., New York (1979)

[JST09]   Jain, S., Stephan, F., Teutsch, J.: Index sets and universal numberings, 2009. In: Ambos-Spies, K., Löwe, B., Merkle, W. (eds.) CiE 2009. LNCS, vol. 5635, pp. 270–279. Springer, Heidelberg (2009)

[Lee63]   Lee, C.Y.: A Turing machine which prints its own code script. In: Proc. of the Symposium on Mathematical Theory of Automata, pp. 155–164 (1963)

[Moo96]   Moore, K.: MIME (Multipurpose Internet Mail Extensions) Part Three: Message Header Extensions for Non-ASCII Text (1996), RFC2047

[MWY78]   Machtey, M., Winklmann, K., Young, P.: Simple Gödel numberings, isomorphisms, and programming properties. SIAM Journal on Computing 7(1), 39–60 (1978)

[MY78]   Machtey, M., Young, P.: An Introduction to the General Theory of Algorithms. North-Holland, Amsterdam (1978)

[Qui66]   Quine, W.V.: The Ways of Paradox and other essays. Random House (1966); reprinted Harvard University Press (1976)

[Ric80]   Riccardi, G.: The Independence of Control Structures in Abstract Programming Systems. PhD thesis, SUNY Buffalo (1980)

[Rog58]   Rogers, H.: Gödel numberings of partial recursive functions. Journal of Symbolic Logic 23(3), 331–341 (1958)

[Rog67]   Rogers, H.: Theory of Recursive Functions and Effective Computability. McGraw Hill, New York (1967); reprinted MIT Press (1987)

[Roy87]   Royer, J.: A Connotational Theory of Program Structure. LNCS, vol. 273. Springer, Heidelberg (1987)

[SHLG94]   Stoltenberg-Hansen, V., Lindström, I., Griffor, E.R.: Mathematical Theory of Domains. Cambridge University Press, Cambridge (1994)

[Spr90]   Spreen, D.: Computable one-to-one enumerations of effective domains. Information and Computation 84(1), 26–46 (1990)

[Spr98]   Spreen, D.: On effective topological spaces. The Journal of Symbolic Logic 63(1), 185–221 (1998)

[Spr00]   Spreen, D.: On domains witnessing increase in information. Applied General Topology 1(1) (2000)

[Spr07]   Spreen, D.: On some problems in computable topology. In: Logic Colloquium 2005, pp. 221–254. Cambridge University Press, Cambridge (2007)

[Tha63]   Thatcher, J.M.: The construction of a self-describing Turing machine. In: Proc. of the Symposium on Mathematical Theory of Automata, pp. 165–171 (1963)

[Tho99]   Thompson, G.P.: The Quine Page (1999),
http://www.nyx.net/~gthompso/quine.htm

Σ_1^0 and Π_1^0 Equivalence Structures*

Douglas Cenzer[1], Valentina Harizanov[2], and Jeffrey B. Remmel[3]

[1] Department of Mathematics, University of Florida,
P.O. Box 118105, Gainesville, Florida 32611
Tel.: 352-392-0281; Fax: 352-392-8357
cenzer@math.ufl.edu
[2] Department of Mathematics, George Washington University,
Washington, D.C. 20052
harizanv@gwu.edu
[3] Department of Mathematics, University of California-San Diego,
La Jolla, CA 92093
jremmel@ucsd.edu

Abstract. We study computability theoretic properties of Σ_1^0 and Π_1^0 equivalence structures and how they differ from computable equivalence structures or equivalence structures that belong to the Ershov difference hierarchy. Our investigation includes the complexity of isomorphisms between Σ_1^0 equivalence structures and between Π_1^0 equivalence structures.

Keywords: computability theory, equivalence structures, effective categoricity, computable model theory.

1   Introduction

Computable model theory deals with the algorithmic properties of effective mathematical structures and the relationships between such structures. Perhaps the most basic kind of relationship between two structures is that of isomorphism. It is natural to study the isomorphism problem in the context of computable mathematics by investigating the following question: given two effective isomorphic structures, how complex must an isomorphism between them be?

In what follows, we restrict our attention to countable structures for computable languages. Hence, if a structure is infinite, we can assume that its universe is the set of natural numbers, ω. We recall some basic definitions. If $\mathcal{A}$ is a structure with universe A for a language $\mathcal{L}$, then $\mathcal{L}^A$ is the language obtained by expanding $\mathcal{L}$ by constants for all elements of A. The *atomic diagram* of $\mathcal{A}$ is the set of all quantifier-free sentences of $\mathcal{L}^A$ true in $\mathcal{A}$. The *elementary diagram* of $\mathcal{A}$ is the set of all first-order sentences of $\mathcal{L}^A$ true in $\mathcal{A}$. A structure $\mathcal{A}$ is *computable* if its atomic diagram is computable, and $\mathcal{A}$ is *decidable* if its elementary

---

* This research was partially supported by NSF grants DMS-0532644 and DMS-0554841. Cenzer was partially supported by DMS 652372, Harizanov was partially supported by DMS-0704256 and Jeffrey Remmel was partially supported by NSF grant DMS 0654060.

diagram is computable. We call two structures *computably isomorphic* if there is a computable function that is an isomorphism between them. A computable structure $\mathcal{A}$ is *relatively computably isomorphic* to a (possibly noncomputable) structure $\mathcal{B}$ if there is an isomorphism between them that is computable in (the atomic diagram of) $\mathcal{B}$. A computable structure $\mathcal{A}$ is *computably categorical* (*relatively computably categorical*, respectively) if every computable structure that is isomorphic to $\mathcal{A}$ is computably isomorphic (relatively computably isomorphic, respectively) to $\mathcal{A}$. Similar definitions arise for other naturally definable classes of structures and their isomorphisms. For example, for any $n \in \omega$, a structure is Δ_n^0 if its atomic diagram is Δ_n^0; two structures are Δ_n^0 *isomorphic* if there is a Δ_n^0 isomorphism between them; and a computable structure $\mathcal{A}$ is Δ_n^0 *categorical* if every computable structure that is isomorphic to $\mathcal{A}$ is Δ_n^0 isomorphic to $\mathcal{A}$.

Among the simplest nontrivial structures is a structure consisting of nothing besides one equivalence relation—an equivalence structure $\mathcal{A} = (\omega, E)$. The study of complexity of isomorphisms between computable equivalence structures has been recently carried out by Calvert, Cenzer, Harizanov, and Morozov in [1]. Similarly, the study of structures and functions within the Ershov difference hierarchy has been recently carried out by Khoussainov, Stephan, and Yang in [4], and by Cenzer, LaForte, and Remmel in [2] where they investigated equivalence structures in particular. In this paper, we study Σ_1^0 and Π_1^0 equivalence structures. Here, we say that an equivalence structure $\mathcal{A} = (\omega, E)$ is Σ_1^0 (or *c.e.*) if E is a c.e. set and, similarly, $\mathcal{A}$ is Π_1^0 (or *co-c.e.*) if E is a Π_1^0 set. It is also the case that Σ_1^0 and Π_1^0 structures have been studied since the beginning of modern computable model theory. For example, in [5] Metakides and Nerode studied c.e. vector spaces which consist of a structure V over the natural numbers such that the operations of vector addition and scalar multiplication are computable but where there is a c.e. equivalence relation $\equiv$ the equivalences classes of which form a vector space under the vector addition and scalar multiplication. Similarly, in [6] Remmel studied co-c.e. structures where the underlying operations are computable.

We shall see that the complexity of isomorphisms between Σ_1^0 equivalence structures and between Π_1^0 equivalence structures is different from the complexity of isomorphisms between computable equivalence structures or between equivalence structures that lie in the Ershov difference hierarchy. Before we can state our results, we need some notation and definitions. For an equivalence structure $\mathcal{A} = (A, E)$ with $A = \omega$ the equivalence class of a is $[a] = \{b \in A : aEb\}$. In computability theory, it is useful to split $\mathcal{A}$ into two pieces: $Inf^{\mathcal{A}}$ and $Fin^{\mathcal{A}}$, where $Inf^{\mathcal{A}}$ consists of those elements with infinite equivalence classes, and $Fin^{\mathcal{A}}$ consists of those elements with finite equivalence classes. This is simply because it is natural to consider the different sizes of the equivalence classes of the elements in $Fin^{\mathcal{A}}$ as coding information into the equivalence relation. The *character* of an equivalence structure $\mathcal{A}$ is the set

$$\chi(\mathcal{A}) = \{(k, n) : n, k > 0 \text{ and } \mathcal{A} \text{ has at least } n \text{ equivalence classes of size } k\}.$$

This set provides a kind of skeleton for $Fin^{\mathcal{A}}$. Any set $K \subseteq (\omega - \{0\}) \times (\omega - \{0\})$ such that $(k, n+1) \in K$ implies $(k, n) \in K$ for all $n > 0$ and k, is called a *character*. A character K is *bounded* if there is some finite k_0 such that for all $(k,n) \in K$, we have $k < k_0$. The concepts of s-functions and s_1-functions introduced by Khisamiev [3] provide a means of computably approximating the characters of equivalence relations.

Definition 1. *Let* $f : \omega^2 \to \omega$. *The function* f *is an* s-*function if the following hold:*
1. *for every* $i, s \in \omega$, $f(i,s) \le f(i, s+1)$;
2. *for every* $i \in \omega$, *the limit* $m_i = \lim_s f(i,s)$ *exists.*
   We say that f *is an* s_1-*function if, in addition:*
3. *for every* $i \in \omega$, $m_i < m_{i+1}$.

Calvert, Cenzer, Harizanov and Morozov [1] gave conditions under which a given character K can be the character of a computable equivalence structure. In particular, they observed that if K is a bounded character and $\alpha \le \omega$, then there is a computable equivalence structure with character K and exactly α infinite equivalence classes. To prove the existence of computable equivalence structures for unbounded characters K, they needed additional information given by s- and s_1-functions. They showed that if K is a Σ_2^0 character, $r < \omega$, and either
(a) there is an s-function f such that

$$(k,n) \in K \Leftrightarrow card(\{i : k = \lim_{s \to \infty} f(i,s)\}) \ge n$$

or (b) there is an s_1-function f such that for every $i \in \omega$, $(\lim_s f(i,s), 1) \in K$, then there is a computable equivalence structure with character K and exactly r infinite equivalence classes.

In addition to these positive results, in [1] the authors also constructed an infinite Δ_2^0 set D such that for any computable equivalence structure $\mathcal{A}$ with unbounded character and no infinite equivalence classes, $\{k : (k,1) \in K\}$ is not a subset of D.

In [2] Cenzer, LaForte, and Remmel obtained the following results.

(i) For any n-c.e. character K, there is an equivalence structure with character K and no infinite equivalence classes.
(ii) There is an ω-c.e. character K such that any equivalence structure with character K must have infinite equivalence classes.
(iii) For any Δ_2^0 character K, there exists a d.c.e. equivalence structure with no infinite equivalence classes and character K.

Moreover, in [2] the authors defined the notions of an α-*c.e. function* and a *graph-α-c.e. function* and established the following.

(a) Any nonempty Σ_2^0 set is the range of a 2-c.e. function.
(b) For every $n \in \omega$, there is an $(n+1)$-c.e. function which is not graph-n-c.e.
(c) There is a graph-2-c.e. function that is not ω-c.e.
(d) There is a 2-c.e. bijection f such that f^{-1} is not ω-c.e.

Cenzer, LaForte, and Remmel [2] also introduced the notions of a (*weakly*) α-*c.e. isomorphism* and of a *graph-α-c.e. isomorphism* and proved the following.

(I) For every $n \in \omega$, there exist two computable equivalence structures that are $(n + 1)$-c.e. isomorphic, but not weakly n-c.e. isomorphic.

(II) There are two computable equivalence structures that are graph-2-c.e. isomorphic, but not weakly ω-c.e. isomorphic.

In [2] the authors also showed that a computable equivalence structure is computably categorical if and only if it is weakly ω-c.e. categorical. Furthermore, they showed that any computable equivalence structure with bounded character is relatively graph-2-c.e. categorical, and that any computable equivalence structure with a finite number of infinite equivalence classes is relatively graph-ω-c.e. categorical. It then follows that a computable equivalence structure is Δ_2^0 categorical if and only if it is graph-ω-c.e. categorical.

We will prove a number of results about the complexity of isomorphisms of Σ_1^0 and of Π_1^0 equivalence structures. For example, in Section 2, we show that any Σ_1^0 equivalence structure $\mathcal{A}$ with infinitely many infinite equivalence classes is isomorphic to a computable structure, while there are Σ_1^0 equivalence structures with finitely many infinite equivalence classes which are *not* isomorphic to any computable structure. We show that if Σ_1^0 equivalence structures $\mathcal{A}_1$ and $\mathcal{A}_2$ are isomorphic to a computable structure $\mathcal{A}$ that is computably categorical or relatively Δ_2^0 categorical, then $\mathcal{A}_1$ and $\mathcal{A}_2$ are Δ_2^0 isomorphic. In Section 3, we construct a Π_1^0 equivalence structure with all equivalence classes of sizes 1 or 2, which is not Δ_2^0 isomorphic to any Σ_1^0 equivalence structure. In Section 4, we consider the *spectrum question*, which is to determine the possible sets (or degrees of sets) that can be the elements with equivalent classes of size k, for some fixed k, in a computable equivalence structure of a given isomorphism type. For example, we show that for any infinite c.e. set B, there is a computable equivalence structure with infinitely many equivalence classes of size 1, infinitely many classes of size 2, and no other classes, such that $B = \{x : card([x]) = 2\}$. In Section 5, we consider the complexity of the theory $Th(\mathcal{A})$ of a computable equivalence structure $\mathcal{A}$, as well as the complexity of its elementary diagram $FTh(\mathcal{A})$. We explore the connection between the complexity of the character $\chi(\mathcal{A})$ and the theory $Th(\mathcal{A})$. We show that if $Th(\mathcal{A})$ is decidable, then the character $\chi(\mathcal{A})$ is computable. We show that if an equivalence structure $\mathcal{B}$ has a computable character, then there is a decidable structure $\mathcal{A}$ isomorphic to $\mathcal{B}$.

2 Σ_1^0 Equivalence Structures

In this section, we consider the properties of Σ_1^0 equivalence structures, their existence and categoricity. It is easy to show that the complexity of the character for Σ_1^0 equivalence structures is at the same level of the arithmetical hierarchy as for computable equivalence structures.

Lemma 1. *For any Σ_1^0 equivalence structure $\mathcal{A}$, we have:*
(a) $\{(k,a) : card([a]^{\mathcal{A}}) \geq k\}$ is a Σ_1^0 set;
(b) $Inf^{\mathcal{A}}$ is a Π_2^0 set;
(c) $\chi(\mathcal{A})$ is a Σ_2^0 set.

Thus, if $\mathcal{A}$ is a Σ_1^0 equivalence structure with infinitely many infinite equivalence classes, then it follows from Lemma 1 and Lemma 2.3 of [1] that $\mathcal{A}$ is isomorphic to a computable equivalence structure. However, it was shown in [1] that there is a Δ_2^0 character K such that any computable equivalence structure with character K must have infinitely many infinite equivalence classes. It was shown in [2] that for *any* Δ_2^0 character K, there is a d.c.e. equivalence structure $\mathcal{A}$ with character K and with no infinite equivalence classes—hence $\mathcal{A}$ is not isomorphic to any computable equivalence structure. Now, for Σ_1^0 equivalence structures we have the following existence result.

Theorem 2. *For any Σ_2^0 character K and any finite m, there is a Σ_1^0 equivalence structure $\mathcal{A}$ with character K and with exactly m infinite equivalence classes.*

Proof. Let K be a Σ_2^0 character, and let $\mathcal{B}$ be the equivalence structure given by Lemma 2.3 in [1], so that $\mathcal{B}$ has character K and has infinitely many infinite equivalence classes, and, in addition, $Fin^{\mathcal{B}}$ is a Π_1^0 set. Simply define $\mathcal{A} = (\omega, E^{\mathcal{A}})$ by $E^{\mathcal{A}} = E^{\mathcal{B}} \cup (Inf^{\mathcal{B}} \times Inf^{\mathcal{B}})$. Then the structure $\mathcal{A}$ is Σ_1^0 since $Inf^{\mathcal{B}}$ is a Σ_1^0 set, $\mathcal{A}$ has the same character K, and the infinitely many infinite equivalence classes of $\mathcal{B}$ collapse into a single equivalence class $Inf^{\mathcal{A}}$ in $\mathcal{A}$. For $m > 1$, one can then append $m - 1$ computable infinite equivalence classes.

Corollary 1. *There exists a Σ_1^0 equivalence structure $\mathcal{A}$ that is not isomorphic to any computable equivalence structure.*

Proof. Let K be a Σ_2^0 character that does not have an s_1-function. Then, by Lemma 2.6 of [1], there is no computable structure with character K and with finitely many infinite equivalence classes.

We will next consider the effective categoricity of Σ_1^0 equivalence structures. It was shown in [1] that a computable equivalence structure $\mathcal{A}$ is computably categorical if and only if $\mathcal{A}$ is relatively computably categorical, which is if and only if one of the following conditions is satisfied:
1. $\mathcal{A}$ has only finitely many finite equivalence classes;
2. $\mathcal{A}$ has finitely many infinite equivalence classes and bounded character, and there is at most one finite k such that $\mathcal{A}$ has infinitely many equivalence classes of size k.
It is also shown in [1] that a computable equivalence structure $\mathcal{A}$ is relatively Δ_2^0 categorical if and only if either $\mathcal{A}$ has finitely many infinite equivalence classes or $\mathcal{A}$ has bounded character.

Clearly, a noncomputable Σ_1^0 structure cannot be computably isomorphic to a computable structure, but we have the following best possible result.

Theorem 3. *Let $\mathcal{A}$ be a Σ_1^0 equivalence structure, and let $\mathcal{B}$ be a computable structure isomorphic to $\mathcal{A}$ such that $\mathcal{B}$ is computably categorical or relatively Δ_2^0 categorical. Then $\mathcal{A}$ and $\mathcal{B}$ are Δ_2^0 isomorphic.*

Proof. Suppose first that $\mathcal{B}$ is computably categorical. It follows from Theorem 3.16 of [1] that $\mathcal{B}$ is relatively computably categorical, so there is an isomorphism f from $\mathcal{B}$ to $\mathcal{A}$, which is computable in $\mathcal{A}$. Since $\mathcal{A}$ is Σ_1^0, it follows that f is Δ_2^0.

Now let $\mathcal{B}$ be relatively Δ_2^0 categorical. Then either:
(i) $\mathcal{B}$ has finitely many infinite equivalence classes, or
(ii) $\mathcal{B}$ has bounded character.
In case (i), $Inf^{\mathcal{B}}$ is computable and $Inf^{\mathcal{A}}$ is Σ_1^0. In Case (ii), both $Inf^{\mathcal{B}}$ and $Inf^{\mathcal{A}}$ are Σ_1^0. Furthermore, both $\{(k,a) : card([a]^{\mathcal{A}} = k\}$ and $\{(k,b) : card([b]^{\mathcal{B}}) = k\}$ are Δ_2^0 sets. It is then easy to define a Δ_2^0 isomorphism between $\mathcal{A}$ and $\mathcal{B}$.

Corollary 2. *Let $\mathcal{A}$ and $\mathcal{B}$ be isomorphic Σ_1^0 equivalence structures with infinitely many infinite equivalence classes, which satisfy one of the following conditions:*
(i) $\mathcal{A}$ has bounded character;
(ii) for every finite k, $\mathcal{A}$ has only finitely many equivalence classes of size k.
Then $\mathcal{A}$ and $\mathcal{B}$ are Δ_2^0 isomorphic.

3 Π_1^0 Equivalence Structures

In this section, we show that even simple Π_1^0 equivalence structures do not have to be Δ_2^0 isomorphic to computable structures. Note that if $\mathcal{B}$ is a Π_1^0 equivalence structure and $\mathcal{A}$ is an isomorphic computable structure that is computably categorical, then, since $\mathcal{A}$ is also relatively computably categorical, $\mathcal{A}$ and $\mathcal{B}$ are Δ_2^0 isomorphic. Thus, we have the next result.

Theorem 4. *Let $\mathcal{A}$ and $\mathcal{B}$ be isomorphic Π_1^0 equivalence structures such that $\mathcal{B}$ satisfies one of the following conditions:*
(i) $\mathcal{A}$ has only finitely many finite equivalence classes;
(ii) $\mathcal{A}$ has finitely many infinite equivalence classes and bounded character, and there is at most one finite k such that $\mathcal{A}$ has infinitely many equivalence classes of size k.
Then $\mathcal{A}$ and $\mathcal{B}$ are Δ_2^0 isomorphic.

However, the next theorem shows that Theorem 4 does not extend to all equivalence structures that are isomorphic to computable Δ_2^0 categorical structures.

Theorem 5. *There exists a Π_1^0 equivalence structure $\mathcal{A}$ with no infinite equivalence classes and with character $\{1,2\} \times \omega$, which is not Δ_2^0 isomorphic to any computable structure. (Moreover, $\mathcal{A}$ is not Δ_2^0 isomorphic to any Σ_1^0 equivalence structure.)*

It will be interesting to extend Theorem 5 to any Π_1^0 structure that has infinitely many equivalence classes of size k_1 and infinitely many equivalence classes of size k_2 for $k_1 \neq k_2$, or has unbounded character.

4 Spectra of Equivalence Structures

In this section, we begin to examine the spectrum question for equivalence structures. For a computable (or Σ_1^0, or Π_1^0) equivalence structure $\mathcal{A}$ we consider for any cardinal k, the possible Turing degree of $\{a : card([a]) = k\}$, and similarly of $\{a : card([a]) \geq k\}$. For example, we know that for any c.e. equivalence structure $\mathcal{A}$, $Inf^{\mathcal{A}}$ is Π_2^0 and $Fin^{\mathcal{A}}$ is Σ_2^0. Thus, one may ask whether there exists for any Σ_2^0 Turing degree $\mathbf{c}$, a computable equivalence structure $\mathcal{A}$ with $Fin^{\mathcal{A}}$ of degree $\mathbf{c}$.

We now give an initial result for structures with infinitely many equivalence classes of size 1 and infinitely many equivalence classes of size 2, and with no other equivalence classes. Here, the elements in classes of size 2 form a c.e. set, and the elements in classes of size 1 form a co-c.e. set. In this case, we obtain not only every c.e. degree, but also every c.e. set.

Theorem 6. *For any infinite c.e. set B, there is a computable equivalence structure A with character $\{1, 2\} \times \omega$ and no infinite equivalence classes such that $\{a : card([a]) = 2\} = B$.*

Proof. Let $\{b_0, b_1, \ldots\}$ be a computable $1 - 1$ enumeration of B. We will first give an enumeration $\{c_0, c_1, \ldots\}$ of B such that, for every n and each $i < 2n + 1$, $c_i < c_{2n+1}$. Let $c_0 = b_0$, and for every n, do the following:

(1) $c_{2n} = b_i$, where i is the least such that $b_i \notin \{c_0, c_1, \ldots, c_{2n-1}\}$, and
(2) $c_{2n+1} = b_k$, where k is the least such that $c_i < b_k$ for all $i \leq 2n$.

Now simply let $E = (\omega \times \omega) \cup \{(c_{2n}, c_{2n+1}) : n \in \omega\}$. Then for each i, $card([c_i]) = 2$, and for $a \notin B$, $[a] = \{a\}$, so that $B = \{a : card([a]) = 2\}$, as desired. It remains to show that E is a computable relation. Observe that $c_1 < c_3 < \cdots$, so for every n, $c_{2n+1} \geq n$. Now, given $a < b$, let $n = max\{a, b\}$. Then
$$aEb \implies (\exists m \leq n)[a = c_{2m} \wedge b = c_{2m+1}].$$

The proof of Theorem 6 can be modified to obtain a similar result for structures with infinitely many equivalence classes of size k_1 and infinitely many equivalence classes of size k_2, where $k_1 \neq k_2$. For three or more different cardinalities $k_1 < k_2 < \cdots < k_n$, the elements of an intermediate size equivalence classes will form a d.c.e. set, so we can ask whether any d.c.e. set can be represented in this way. Similar questions can be asked for Σ_1^0 and Π_1^0 equivalence structures.

5 Decidability of Structures

In this section, we consider the decidability of equivalence structures and their theories. The intuitive idea is that the character of an equivalence structure (together with the number of infinite classes) determines its theory, and that the character, together with the function mapping any element to the size of its equivalence class, determines its elementary diagram. Recall that $Th(\mathcal{A})$ denotes the first-order theory of a structure $\mathcal{A}$, and $FTh(\mathcal{A})$ denotes the elementary diagram of $\mathcal{A}$.

Proposition 1. *If $Th(\mathcal{A})$ is decidable, then the character $\chi(\mathcal{A})$ is computable.*

Proof. It follows from the definition of $\chi(\mathcal{A})$ that the character is uniformly definable by first-order formulas $\psi_{n,k}$ so that $(k,n) \in \chi(\mathcal{A})$ if and only if $\mathcal{A} \models \psi_{n,k}$.

It follows from the argument above that, in fact, $\chi(\mathcal{A})$ is many-one reducible to $Th(\mathcal{A})$. Define the set $K(\mathcal{A}) \subseteq \omega \times (\omega - \{0\})$ by

$$(a,k) \in K(\mathcal{A}) \iff card([a]) \geq k, \text{ and}$$

$$K^+(\mathcal{A}) = K(\mathcal{A}) \cup \{(langlea, 0) : card([a]) = \aleph_0\}.$$

Note that $K^+(\mathcal{A})$ is Turing equivalent to $K(\mathcal{A})$.

Theorem 7. *For any equivalence structure $\mathcal{A}$, the elementary diagram of $\mathcal{A}$ is Turing reducible to the join of the set $K(\mathcal{A})$ with the atomic diagram of $\mathcal{A}$.*

Theorem 8. *For any equivalence structure $\mathcal{B}$, there is a structure $\mathcal{A}$ isomorphic to $\mathcal{B}$, such that $\mathcal{A}$ and $K(\mathcal{A})$ are computable from $\chi(\mathcal{A})$.*

Proof. We may assume, without loss of generality, that $\mathcal{B}$ has no infinite equivalence classes, since, if needed, we can simply adjoin either infinitely many or some fixed finite number of equivalence classes. We may also assume that $\mathcal{B}$ has infinitely many classes with at least two elements, since otherwise $\mathcal{B}$ certainly has a decidable copy. The structure $\mathcal{A}$ will contain a distinct equivalence class $[\langle k, n \rangle]$ for each $(k,n) \in \chi(\mathcal{B})$, where we let $\langle k, n \rangle = 2^{k+1} \cdot 3^{n+1}$. Let $\chi(\mathcal{B})$ be enumerated numerically as $\langle k_0, n_0 \rangle, \langle k_1, n_1 \rangle, \ldots$ and let $b_0, b_1, \ldots$ enumerate $\omega - \chi(\mathcal{B})$. Then $E = E^{\mathcal{A}}$ is defined by using the elements $b_0, b_1, \ldots$ to fill out the equivalence classes $[\langle k_0, n_0 \rangle], [\langle k_1, n_1 \rangle], \ldots$ in order, as needed. It is easy to see that $\mathcal{A}$ and $K(\mathcal{A})$ are computable from $\chi(\mathcal{A})$.

Putting these results together, we have the next two theorems along with some immediate corollaries.

Theorem 9. *For any equivalence structure $\mathcal{A}$, $Th(\mathcal{A})$ and $\chi(\mathcal{A})$ have the same Turing degree.*

Proof. It follows from the argument in Proposition 1 that $\chi(\mathcal{A})$ is Turing reducible to $Th(\mathcal{A})$. Conversely, let $\mathcal{B}$ be an equivalence structure and let $\mathcal{A}$, isomorphic to $\mathcal{B}$, be given by Theorem 8, so that $\mathcal{A}$ and $K(\mathcal{A})$ are both computable from $\chi(\mathcal{A})$ (which, of course, equals $\chi(\mathcal{B})$). It follows from Theorem 7 that $FTh(\mathcal{A})$ is computable from $\chi(\mathcal{B})$. Now $Th(\mathcal{B}) = Th(\mathcal{A})$ is computable from $FTh(\mathcal{A})$, and hence computable from $\chi(\mathcal{B})$, as desired.

Corollary 3. *For any equivalence structure $\mathcal{A}$, $Th(\mathcal{A})$ is decidable if and only if $\chi(\mathcal{A})$ is computable.*

Theorem 10. *For any equivalence structure $\mathcal{B}$ with computable character $\chi(\mathcal{B})$, there is a decidable structure $\mathcal{A}$ isomorphic to $\mathcal{B}$. (Hence $Th(\mathcal{B})$ is decidable.)*

Proof. Again, it suffices to assume that $\mathcal{B}$ has no infinite equivalence classes. By Theorem 8, there is a structure $\mathcal{A}$ isomorphic to $\mathcal{B}$, which is computable from $\chi(\mathcal{A})$, and hence $\mathcal{A}$ and $K(\mathcal{A})$ are also computable. It now follows from Theorem 7 that $FTh(\mathcal{A})$ is decidable, and hence $Th(\mathcal{A})$, which equals $Th(\mathcal{B})$, is decidable.

Clearly, any bounded character is computable.

Corollary 4. *If the equivalence structure $\mathcal{A}$ has bounded character, then $Th(\mathcal{A})$ is decidable.*

For computably categorical structures, we can say more.

Corollary 5. *If $\mathcal{A}$ is a computably categorical equivalence structure, then $\mathcal{A}$ is decidable.*

Proof. Let $\mathcal{A}$ be computably categorical. Then $\mathcal{A}$ has bounded character, so $\chi(A)$ is computable. Hence by Theorem 10, there is a structure $\mathcal{B}$ isomorphic to $\mathcal{A}$, which is decidable. Since $\mathcal{A}$ is computably categorical, $\mathcal{A}$ is computably isomorphic to $\mathcal{B}$ and, therefore, $\mathcal{A}$ is also decidable.

Note that there are equivalence structures that are not computably categorical, which have decidable theories. For example, fix $k_1 < k_2 \leq \omega$ and let $\mathcal{A}$ have infinitely many equivalence classes of size k_1 and infinitely many classes of size k_2 and no other classes. Then $\chi(\mathcal{A})$ is computable and, thus, $Th(\mathcal{A})$ is decidable. We note that in all considered cases of decidable theories, one could, in fact, give a complete set of axioms for the theory.

6   Conclusions and Future Research

We have examined properties of Σ_1^0 and Π_1^0 equivalence structures, and studied how they compare with and differ from computable and d.c.e. structures. We have shown that any Σ_1^0 equivalence structure $\mathcal{A}$ with infinitely many infinite equivalence classes is isomorphic to a computable structure, but there are Σ_1^0 structures with finitely many infinite equivalence classes, which are *not* isomorphic to any computable structure. We have shown that if Σ_1^0 equivalence structures $\mathcal{A}_1$ and $\mathcal{A}_2$ are isomorphic to a computable structure $\mathcal{A}$ that is either computably categorical or relatively Δ_2^0 categorical, then $\mathcal{A}_1$ and $\mathcal{A}_2$ are Δ_2^0 isomorphic.

Suppose that $\mathcal{A}$ and $\mathcal{B}$ are Σ_1^0 structures with finitely many infinite equivalence classes but without computable copies, as constructed in Theorem 2. Does it follow that $\mathcal{A}$ and $\mathcal{B}$ are Δ_2^0 isomorphic?

We have constructed a Π_1^0 equivalence structure with all equivalence classes of sizes 1 and 2, which is not Δ_2^0 isomorphic to any Σ_1^0 equivalence structure. Presumably this result can be extended to any Π_1^0 equivalence structure that has infinitely many equivalence classes of size k_1 and infinitely many equivalence classes of size k_2, where $k_1 \neq k_2$, or has unbounded character.

We have begun to study the spectrum question by showing that for any infinite c.e. set B, there is a computable equivalence structure with infinitely many equivalence classes of size 1, infinitely many classes of size 2, and no other classes, in which $B = \{x : card([x]) = 2\}$. For three or more different cardinalities $k_1 < k_2 < \cdots < k_n$, the elements of each intermediate size classes form a d.c.e. set, so we can ask whether any d.c.e. set can be represented in this way. Similar questions can be asked for Σ_1^0 and Π_1^0 equivalence structures.

We have studied for computable equivalence structures $\mathcal{A}$ the complexities of their theories $Th(\mathcal{A})$ and elementary diagrams $FTh(\mathcal{A})$. We have shown that $Th(\mathcal{A})$ is decidable if and only if the character $\chi(\mathcal{A})$ is computable. We have also shown that if $\mathcal{B}$ has computable character, then there is a decidable structure $\mathcal{A}$ isomorphic to $\mathcal{B}$, and hence with decidable $Th(\mathcal{A})$. It will be also interesting to consider the complexities of $Th(\mathcal{A})$ and $FTh(\mathcal{A})$ when $\chi(\mathcal{A})$ is not computable, but is Σ_1^0, Π_1^0, Δ_2^0, or Σ_2^0.

References

1. Calvert, W., Cenzer, D., Harizanov, V., Morozov, A.: Effective categoricity of equivalence structures. Annals of Pure and Applied Logic 141, 61–78 (2006)
2. Cenzer, D., LaForte, G., Remmel, J.B.: Equivalence structures and isomorphisms in the difference hierarchy. Journal of Symbolic Logic 74, 535–556 (2009)
3. Khisamiev, N.G.: Constructive Abelian groups. In: Ershov, Y.L., Goncharov, S.S., Nerode, A., Remmel, J.B. (eds.) Handbook of Recursive Mathematics, vol. 2, pp. 1177–1231. North-Holland, Amsterdam (1998)
4. Khoussainov, B., Stephan, F., Yang, Y.: Computable categoricity and the Ershov hierarchy. Annals of Pure and Applied Logic 156, 86–95 (2008)
5. Metakides, G., Nerode, A.: Recursively enumerable vector spaces. Annals of Mathematical Logic 11, 141–171 (1977)
6. Remmel, J.B.: Combinatorial functors on co-r.e. structures. Annals of Mathematical Logic 11, 261–287 (1976)
7. Soare, R.I.: Recursively Enumerable Sets and Degrees. Springer, Berlin (1987)

Immunity for Closed Sets*

Douglas Cenzer[1], Rebecca Weber[2], and Guohua Wu[3]

[1] Department of Mathematics, University of Florida,
P.O. Box 118105, Gainesville, Florida 32611
Tel.: 352-392-0281; Fax: 352-392-8357
cenzer@math.ufl.edu
[2] Department of Mathematics, Dartmouth College,
Hanover, NH 03755-3551
rweber@math.dartmouth.edu
[3] School of Physical and Mathematical Sciences,
Nanyang Technological University, Singapore 639798
guohua@ntu.edu.sg

Abstract. The notion of immune sets is extended to closed sets and Π_1^0 classes in particular. We construct a Π_1^0 class with no computable member which is not immune. We show that for any computably inseparable sets A and B, the class $S(A,B)$ of separating sets for A and B is immune. We show that every perfect thin Π_1^0 class is immune. We define the stronger notion of *prompt immunity* and construct an example of a Π_1^0 class of positive measure which is promptly immune. We show that the immune degrees in the Medvedev lattice of closed sets forms a filter. We show that for any Π_1^0 class P with no computable element, there is a Π_1^0 class Q which is not immune and has no computable element, and which is Medvedev reducible to P. We show that any random closed set is immune.

Keywords: Computability, Π_1^0 Classes.

1   Introduction

The notion of a simple c.e. set and the corresponding complementary notion of an immune co-c.e. set are fundamental to the study of c.e. sets and degrees. Together with variations and related notions such as *effectively immune, promptly simple, hyperimmune* and so forth, they permeate the classic text of R.I. Soare [21] and its updated version.

Many of the results on c.e. sets and degrees have found counterparts in the study of effectively closed sets (Π_1^0 classes). See the surveys [10,11] for examples. In particular, hyperhyperimmune co-c.e. sets correspond to thin Π_1^0 classes [7,8,9] and hyperimmune co-c.e. sets correspond to several different notions including *smallness* studied by Binns [5,6].

---

* This research was partially supported by NSF grants DMS-0554841 and DMS 652372.

K. Ambos-Spies, B. Löwe, and W. Merkle (Eds.): CiE 2009, LNCS 5635, pp. 109–117, 2009.

In this paper we consider the notion of immune sets as applied to Π_1^0 classes and closed sets in general. We work in $2^{\mathbb{N}}$ with the topology generated by basic clopen sets called *intervals*. For any $\sigma \in \{0,1\}^*$ the interval $I(\sigma)$ is $\{X : \sigma \subset X\}$, where here $\subset$ means initial segment. Notation is standard; we note that $\sigma \restriction n$ is the length-n initial segment of σ, and if $T \subseteq \{0,1\}^*$ is a tree (i.e., it is closed under initial segment), $[T] \subseteq 2^{\mathbb{N}}$ denotes the set of infinite paths through T. For any set $P \subseteq 2^{\mathbb{N}}$, we may define the tree $T_P = \{\sigma \in \{0,1\}^* : I(\sigma) \cap P \neq \emptyset\}$; the closed sets $P \subseteq 2^{\mathbb{N}}$ are exactly those for which $P = [T_P]$. A Π_1^0 class is a closed set for which some computable tree $T \supseteq T_P$ has $[T] = P$; in this case T_P is a Π_1^0 set. For any tree T, let $Ext(T)$ be the set of nodes of T which have an infinite extension in $[T]$, thus if $P = [T]$, then $Ext(T) = T_P$.

An infinite set $C \subseteq \omega$ is said to be *immune* if it does not include any infinite c.e. subset, or equivalently if it has no infinite computable subset. A c. e. set which is the complement of an immune set is said to be *simple*. We say that a closed set $P \subseteq 2^{\mathbb{N}}$ is immune if T_P is immune.

It is easy to see that an immune closed set has no computable member. We will construct in section 2 a Π_1^0 class which is *not* immune and still has no computable member. We will show that for any computably inseparable sets A and B, the class $S(A,B)$ of separating sets for A and B is immune. We will show that every perfect thin Π_1^0 class is immune. We define the stronger notion of *prompt immunity* and construct an example of a Π_1^0 class of positive measure which is promptly immune.

In section 3, we consider connections between immunity and Binns' notion of *smallness* [5] and also connections with the Medvedev degrees of difficulty [16,19]. We show that for closed sets P and Q, the meet $P \oplus Q$ is immune if and only if both P and Q are immune, whereas the join $P \otimes Q$ is immune if and only if at least one of P and Q are immune. We show that for any Π_1^0 class P with no computable element, there is a non-immune Π_1^0 class Q with no computable element which is Medvedev reducible to P.

In section 4, we show that any random closed set (in the sense of [2]) is immune. We also show that any random closed set is not small.

2   Immunity for Π_1^0 Classes

We begin with some basic results. The following is a useful additional characterization of immunity.

Lemma 1. *P is not immune if and only if there is a computable sequence $\{\sigma_n : n \in \omega\}$ such that $\sigma_n \in T_P \cap \{0,1\}^n$ for each n.*

Proof. The reverse implication is immediate. Now suppose that C is an infinite computable subset of T_P and enumerate C as $\{\tau_0, \tau_1, \dots\}$. Observe that C must have arbitrarily long elements and define σ_n to be $\tau_i \restriction n$, where i is the least such that $|\tau_i| \geq n$.

We often refer to a Π_1^0 class with no computable members as a *special* Π_1^0 class. The following two results shows that the immune classes are a proper subset of the special classes.

Proposition 1. *If P is immune, then P is special.*

Proof. If P has a computable member X, then $\{X \restriction n : n \in \mathbb{N}\}$ is an infinite computable subset of T_P.

Theorem 1. *There exists a special Π_1^0 class P that is not immune.*

Proof. We will build a sequence of nested computable trees T_s such that $T_P = \cap_s T_s$ and a prefix-free set A such that $A_s = \{\sigma_0, \ldots, \sigma_s\} \subseteq Ext(T_s)$ and $|\sigma_s| \geq s$. We have the following requirements:

$$N_e : \varphi_e \text{ total } \Rightarrow W_e \notin P.$$

Each N_e has an associated $m_s(e)$, the minimum length of convergence of φ_e required before we act for N_e. For all e, $m_0(e) = 2e + 1$.

To meet a single requirement N_0 we act as follows. We wait until $\varphi_0(0) \downarrow$, and if that happens at stage s we let $m_s(0) = 1 + \max\{|\sigma_i| : i < s\}$ and choose all σ_t, $t \geq s$, to be incompatible with $\varphi_0(0)$. Then if at stage $s' > s$ we see $\varphi_0 \restriction m_s(0) \downarrow$, we let $T_{s'+1} = T_{s'} - \{\varphi_0 \restriction m_s(0)^\frown \tau : \tau \in \{0,1\}^*\}$. The same module holds for all other requirements; we enumerate a set R of indices of requirements that must be avoided by A. Each $m_s(e)$ changes its value at most once, and the second value it takes on is sufficiently large that the standard measure argument shows $P \cap [\sigma_i] \neq \emptyset$ for each i.

Stage 0: $\forall e \; m_0(e) = 2e + 1$; $A_0 = R_0 = \emptyset$; $T_0 = \{0,1\}^*$.

Stage $s > 0$: For each $e \leq s$ such that $\varphi_e \restriction (2e+1) \downarrow$ newly at s, set $m_s(e) = 2e + 1 + \max\{|\sigma_i| : i < s\}$. Enumerate all such e into R_s. For the rest, let $m_s(e) = m_{s-1}(e)$.

Next, if any $e \leq s$ is such that $\varphi_e \restriction m_s(e) \downarrow = \tau_e \in T_{s-1}$, let $T_s = T_{s-1} - \{\tau_e \rho : \rho \in \{0,1\}^*, e \text{ as above}\}$. Otherwise let $T_s = T_{s-1}$.

Finally, let Q be the part of T_s uncovered by A and R. That is,

$$Q = T_s - \{\tau^\frown \rho : \tau \in A_{s-1} \cup \{\varphi_e \restriction (2e+1) : e \in R\}, \rho \in \{0,1\}^*\}.$$

Note that we would get the same Q if we replaced T_s by $\{0,1\}^*$. Choose the leftmost $\sigma \in Q$ of length at least $s + 2$ and let it be $\sigma_s \in A_s$.

To verify the construction works, first note every σ_i has an extension by a straightforward measure argument: if $\varphi_e \restriction (2e+1) \downarrow$ at or before stage i, σ_i will be chosen to avoid it; if $\varphi_e \restriction (2e+1) \downarrow$ after stage i it will be allowed to remove at most $2^{-2e-1-|\sigma_i|}$ from the tree. The sum of the measure so eliminated is bounded by $\frac{2}{3}\mu([\sigma_i])$. Second, another measure argument shows there is always enough room in Q to choose a new string in A without covering all of T_s. The measure of Q at stage s is at least

$$x = 1 - \sum_{e=0}^{s} 2^{-2e-1} - \sum_{i=1}^{s-1} 2^{-i-2},$$

which we need to be greater than (at most) 2^{-s-2}. It is easily checked that $x - 2^{-s-2}$ is

$$\frac{1}{12} + \frac{1}{3 \cdot 2^{2s+1}} + \frac{1}{2^{s+2}},$$

which is clearly positive.

Since it is clear that the requirements are met, P is a Π_1^0 class, and $A \subset T_P$ is computable, the proof is complete.

The next results show many Π_1^0 classes of interest are immune. Recall $S(A, B)$ denotes the class of separating sets for A and B (all C such that $A \subseteq C$ and $B \cap C = \emptyset$); it is a closed set, and when A and B are c.e. it is a Π_1^0 class.

Proposition 2. *If A and B are computably inseparable, then $S(A, B)$ is immune.*

Proof. Suppose that $W \subset T_{S(A,B)}$ is an infinite c.e. set, enumerated without repetition as $\sigma_0, \sigma_1, \ldots$. Note that for any $\sigma \in W$ and any $n < |\sigma|$, $n \in A \Rightarrow \sigma(n) = 1$ and $n \in B \Rightarrow \sigma(n) = 0$. Since W must have elements of arbitrary length, we may computably define $i(n)$ to be the least i such that $|\sigma_i| > n$ and let $X(n) = \sigma_{i(n)}(n)$ to compute a separating set for A and B.

The notion of a *thin Π_1^0* class corresponds to that of a hyperhyperimmune set and has been studied extensively by many researchers in articles including [7,8,9].

Since any hyperhyperimmune set is also immune, the following result is expected.

Proposition 3. *If P is a perfect thin Π_1^0 class, then P is immune.*

Proof. Let P be perfect thin (and therefore having no computable member) and suppose that some computable set $W \subseteq T_P$. Then $T_P - W$ is a Π_1^0 set, so that $[T_P - W]$ is a Π_1^0 subclass of P and hence there exists a clopen set U such that $[T_P - W] = P \cap U$. It follows that $T_{P \cap U} \subseteq T_P - W$. We claim that without loss of generality $T_P - W = T_{P \cap U}$. That is, let $Q = P - U$ and consider $T_Q - W$. $T_Q - W \subseteq T_P - W$, so that $[T_Q - W] \subseteq P \cap U$. On the other hand, $T_Q - W \subseteq T_Q$, so $[T_Q - W] \subseteq Q = P - U$. Thus $[T_Q - W]$ is empty and hence $T_Q - W$ is finite. Thus we may assume without loss of generality that $T_Q \subseteq W$ and therefore $T_P - W \subset T_P - T_Q = T_{P \cap U}$. It follows that $T_P - W = T_{P \cap U}$. This means that in fact $W = T_{P-U}$ which means that T_{P-U} is computable. But P has no computable members, so that $P - U = \emptyset$ and therefore W is finite.

For any c.e. set A and $n \in \omega$, let A_n denote as usual the elements which have been enumerated into A by stage n; A is said to be *promptly simple* if there is a computable function π such that for any infinite c.e. set $W_e \subseteq \omega$, there exist n, s such that $n \in W_{e,s+1} - W_{e,s}$ and $n \in A_{\pi(s)}$.

For P a Π_1^0 class, let T be a computable tree giving P. For each s, let T_s be the collection of nodes of T which have length-s extensions in T. Let $\{\sigma_n\}_{n \in \mathbb{N}} = \{\langle\rangle, 0, 1, 00, 01, 10 \ldots\}$ denote the length-lexicographical ordering of the elements of $\{0,1\}^*$. We say that P is *promptly immune* if there is a computable function π such that for any infinite c.e. set W, there exist n, s such that

$$n \in W_{s+1} - W_s \quad \& \quad \sigma_n \notin T_{\pi(s)}.$$

There exist Π_1^0 classes with positive measure which have no computable elements. The next result is an improvement on this.

Proposition 4. *There exists a Π_1^0 class P of positive measure which is promptly immune.*

Proof. We define the Π_1^0 class $P = [T]$ in stages T_s and let $T = \bigcap_s T_s$. P will be promptly immune via the function $\pi(s) = s + 1$. For each e, we will wait for some n such that $|\sigma_n| > 2e$ to come into W_e at stage $s + 1$ and then remove σ_n from T_{s+1} by removing σ_n and all extensions (if any) from T. Initially $T_0 = \{0,1\}^*$. After stage s, we will have satisfied some of the requirements. At stage $s + 1$, we look for the least $e \le s$ which has not yet been satisfied and such that some suitable $n \in W_{e,s+1} - W_{e,s}$. We meet this requirement by setting $T_{s+1} = T_s - \{\tau : \sigma_n \subseteq \tau\}$. Not that this action removes from $[T]$ a set of measure $\le 2^{-2e-1}$, so that the total measure removed is

$$\le \sum_e 2^{-2e-1} \le \frac{2}{3}.$$

It follows that $T_s \ne \emptyset$ for any s and therefore $P = [T]$ is not empty, and in fact has measure at least $\frac{1}{3}$.

3 Immunity and Smallness

Π_1^0 classes are often viewed as collections of solutions to some mathematical problem. Muchnik and Medvedev reducibility, defined for closed subsets of $2^{\mathbb{N}}$ and indeed $\mathbb{N}^{\mathbb{N}}$ in general, order classes based on this viewpoint. The class A is *Muchnik* (a.k.a. *weakly*) *reducible* to the class B ($A \le_w B$) if for every $X \in B$ there is $Y \in A$ such that $Y \le_T X$ [17]. The class A is *Medvedev* (a.k.a. *strongly*) *reducible* to B if there is a single Turing reduction procedure which, when given any element of B as an oracle, computes an element of A; it is exactly the uniformization of Muchnik reduction [16]. These reductions have been studied extensively by Binns (e.g., [4]) and Simpson (e.g., [20]), among others, and have connections to randomness [18].

For $X, Y \in 2^{\mathbb{N}}$, the join $X \oplus Y = Z$, where $Z(2n) = X(n)$ and $Z(2n + 1) = Y(n)$. Similarly for finite sequences σ and τ of equal length, we may define $\sigma \oplus \tau = \rho$, where $\rho(2n) = \sigma(n)$ and $\rho(2n + 1) = \tau(n)$.

The quotient structure of the Π_1^0 classes under either Muchnik or Medvedev equivalence is a lattice, and both have the same join and meet operators. The join of P and Q is given by

$$P \otimes Q = \{X \oplus Y : X \in P, Y \in Q\}.$$

If $P = [S]$ and $Q = [T]$, then $P \otimes Q = S \otimes T$, where

$$S \otimes T = \{\sigma \oplus \tau, (\sigma \oplus \tau)i : \sigma \in S, \tau \in T, i \in \{0,1\}\}.$$

The meet is given by

$$P \oplus Q = \{0^\frown X : X \in P\} \cup \{1^\frown Y : Y \in Q\}.$$

Theorem 2. *For any closed sets P and Q, $P \oplus Q$ is immune if and only if both P and Q are immune.*

Proof. Suppose first that P is not immune and let $C \subseteq T_P$ be an infinite computable set. Then $\{0^\frown \sigma : \sigma \in C\}$ is a computable subset of $T_{P \oplus Q}$. A similar argument holds if Q is not immune.

Next suppose that $P \oplus Q$ is not immune and let $C \subseteq T_{P \oplus Q}$ be an infinite computable set. Let $C_i = \{\sigma : i^\frown \sigma \in C\}$ for $i = 0, 1$. Then $C_0 \subseteq T_P$, $C_1 \subseteq T_Q$ and both sets are computable. Clearly either C_0 is infinite or C_1 is infinite, which implies that either P is not immune or Q is not immune.

Theorem 3. *For any closed sets P and Q, $P \otimes Q$ is immune if and only if at least one of P and Q are immune.*

Proof. Suppose first that $P \otimes Q$ is not immune and by Lemma 1 let $C = \{\rho_0, \rho_1, \dots\}$ be a computable subset of $T_{P \otimes Q}$ with $|\rho_n| = n$ for each n. Then for each n, $\rho_{2n} = \sigma_n \oplus \tau_n$ with $\sigma_n \in T_P$ and $\tau_n \in T_Q$, showing both P and Q are not immune.

Next suppose that both P and Q are not immune and let $C_0 = \{\sigma_0, \sigma_1, \dots\} \subset T_P$ and $C_1 = \{\tau_0, \tau_1, \dots\} \subset T_Q$ with $|\sigma_n| = |\tau_n| = n$ for each n. Then $\{\sigma_n \oplus \tau_n : n \in \mathbb{N}\}$ is an infinite computable subset of $T_{P \otimes Q}$, so $P \otimes Q$ is not immune.

We may compare immunity with other "smallness" notions for Π_1^0 classes. Binns [5] defined a *small* closed set P to be one such that there is no computable function g such that, for all n, $card(\{0, 1\}^{g(n)} \cap T_P) > n$. He showed that $P \oplus Q$ and $P \otimes Q$ are each small if and only if both P and Q are small, which immediately distinguishes immunity and smallness.

Another distinction occurs in the maximum degree of each lattice. Recall the family $S(A, B)$ of separating sets of c.e. sets A, B is a Π_1^0 class. An important example is the class DNC_2, the set of diagonally noncomputable functions; here $A = \{e : \varphi_e(e) = 0\}$ and $B = \{e : \varphi_e(e) = 1\}$. DNC_2 has maximum Medvedev and Muchnik degree, and by Proposition 2 it is immune. However, Binns has proved that DNC_2 is not small. In fact, all Medvedev complete Π_1^0 classes are immune and not small, as they are all computably homeomorphic.

We look next for a containment relationship between smallness and immunity. Smallness alone will not give immunity because a Π_1^0 singleton (i.e., a computable path) is small. Binns' original special small class [5] is $S(A, B)$ for computably inseparable A, B, such that $A \cup B$ is hypersimple; by Proposition 2 it is immune. Our construction above in Theorem 1 of a special nonimmune class produces a class of positive measure, which is therefore not small. We have the following question:

Question 4. *If P is small and special, is P necessarily immune?*

We now turn to questions of density. Let 0_M denote the least Medvedev degree, which consists of all Π_1^0 classes that have a computable member. Binns has shown there is a nonsmall class of every nonzero Medvedev degree. We have the following bounding result for nonimmune classes.

Theorem 5. *For any nonzero Π_1^0 class P, there is a Π_1^0 class Q with $0_M <_M Q \leq_M P$ which is not immune.*

Proof. Let R be the Π_1^0 class of Theorem 1 which is nonzero and also not immune. It follows from Theorem 2 that $P \oplus R$ is not immune, but it is also special and certainly $P \oplus R \leq_M P$.

Question 6. *Does every nonzero Medvedev degree contain an immune Π_1^0 class?*

Ambos-Spies et al [1] showed that any promptly simple c.e. degree cups to $\mathbf{0}'$; in fact they had the much stronger result that the promptly simple degrees are exactly the c.e. degrees that are cuppable by a low c.e. degree. We have the following more modest Cupping Conjecture.

Conjecture 7. *If P is promptly immune, then there exists Q, not Medvedev complete, such that $P \otimes Q$ is Medvedev complete.*

4   Immunity and Randomness

Finally we consider the immunity of random closed sets. A closed set P may be coded as an element of $3^{\mathbb{N}}$; P is called random if that sequence is Martin-Löf random (for background on randomness see [12]). The code of P is defined from T_P; the nodes of T_P are considered in order by length and then lexicographically, and each one is represented in the code by 0, 1, or 2 according to whether the node has only the left child, only the right child, or both children, respectively. Randomness for closed sets is defined and explored in [2,3], where it is shown among other results that no Π_1^0 class is random, and that no random closed set contains an f-c.e. path for any computable f bounded by a polynomial. The following theorem does not follow immediately but is not surprising.

Theorem 8. *If P is a random closed set, then P is immune.*

Proof. Fix a computable sequence $C = (\sigma_1, \sigma_2, \dots)$ such that $|\sigma_n| = n$ for each n. For $n > 0$, let $S_n = \{Q : (\forall i \leq n)\, \sigma_i \in T_Q\}$. Then S_n is a clopen set in the space of closed sets and the sequence $\{S_n : n \in \omega\}$ is uniformly c.e. It is clear that $C \subseteq T_P$ if and only if $P \in S_n$ for all n. Now consider the Lebesgue measure $\lambda(S_n)$. Certainly $\lambda(S_1) = 2/3$. Given $\lambda_n = \lambda(S_n)$ and σ_{n+1}, let $i \leq n$ be the largest such that $\sigma_i \subset \sigma_{n+1}$. Then $\lambda_{n+1} = (\frac{2}{3})^{n+1-i}\lambda_n \leq \frac{2}{3}\lambda_n$. Hence $\lambda(S_n) \leq (\frac{2}{3})^n$ for each n. It follows that $\{S_{2n} : n \in \omega\}$ is a Martin-Löf test and hence no random closed set can belong to every S_n. Hence if P is random, then C is not a subset of T_P. Since this holds for every such C, it follows that random closed sets are immune.

Since a random ternary sequence must contain $\frac{1}{3}$ 2s in the limit, intuitively the tree it codes must branch too much to be small. This is a straightforward consequence of the following, which is drawn from Lemma 4.5 in [2].

Lemma 2. *Let Q be a random closed set. Then there exist a constant $C \in \mathbb{N}$ and $k \in \mathbb{N}$ such that for all $m > k$,*

$$C \left(\frac{4}{3}\right)^m \left(1 - m^{-\frac{1}{4}}\right) < card(T_Q \cap \{0,1\}^m) < C \left(\frac{4}{3}\right)^m \left(1 + m^{-\frac{1}{4}}\right).$$

Corollary 9. *If Q is a random closed set, Q is not small.*

Proof. For C, k as in Lemma 2, define the function $g(n)$ as

$$g(n) = \max \left\{ k + 1, \min \left\{ m : n < C \left(\frac{4}{3}\right)^m \left(1 - m^{-\frac{1}{4}}\right) \right\} \right\}.$$

It is clear that g is computable, and by Lemma 2, for all n the number of branches at level $g(n)$ will be at least n.

References

1. Ambos-Spies, K., Jockusch, C.G., Shore, R.A., Soare, R.I.: An algebraic decomposition of the recursively enumerable degrees and the coincidence of several degree classes with the promptly simple degrees. Trans. Amer. Math. Soc. 281, 109–128 (1984)
2. Barmpalias, G., Brodhead, P., Cenzer, D., Dashti, S., Weber, R.: Algorithmic randomness of closed sets. J. Logic Comp. 17, 1041–1062 (2007)
3. Brodhead, P., Cenzer, D., Dashti, S.: Random closed sets. In: Beckmann, A., Berger, U., Löwe, B., Tucker, J.V. (eds.) CiE 2006. LNCS, vol. 3988, pp. 55–64. Springer, Heidelberg (2006)
4. Binns, S.: A splitting theorem for the Medvedev and Muchnik lattices. Math. Logic Q. 49, 327–335 (2003)
5. Binns, S.: Small Π_1^0 classes. Arch. Math. Logic 45, 393–410 (2006)
6. Binns, S.: Hyperimmunity in $2^{\mathbb{N}}$. Notre Dame J. Formal Logic 4, 293–316 (2007)
7. Cenzer, D., Downey, R., Jockusch, C.J.: Countable Thin Π_1^0 Classes. Ann. Pure Appl. Logic 59, 79–139 (1993)
8. Cenzer, D., Nies, A.: Initial segments of the lattice of Π_1^0 classes. Journal of Symbolic Logic 66, 1749–1765 (2001)
9. Cholak, P., Coles, R., Downey, R., Hermann, E.: Automorphisms of the Lattice of Π_1^0 Classes. Transactions Amer. Math. Soc. 353, 4899–4924 (2001)
10. Cenzer, D., Remmel, J.B.: Π_1^0 classes in mathematics. In: Ershov, Y., Goncharov, S., Nerode, A., Remmel, J. (eds.) Handbook of Recursive Mathematics, Part Two. Elsevier Studies in Logic, vol. 139, pp. 623–821 (1998)
11. Cenzer, D., Remmel, J.B.: Effectively Closed Sets. ASL Lecture Notes in Logic (in preparation)
12. Downey, R., Hirschfeldt, D.: Algorithmic Randomness and Complexity (in preparation), http://www.mcs.vuw.ac.nz/~downey/
13. Jockusch, C.G.: Π_1^0 classes and boolean combinations of recursively enumerable sets. J. Symbolic Logic 39, 95–96 (1974)
14. Jockusch, C.G., Soare, R.: Degrees of members of Π_1^0 classes. Pacific J. Math. 40, 605–616 (1972)

15. Jockusch, C.G., Soare, R.: Π_1^0 classes and degrees of theories. Trans. Amer. Math. Soc. 173, 35–56 (1972)
16. Medvedev, Y.T.: Degrees of difficulty of the mass problem. Dokl. Akad. Nauk SSSR (N.S.) 104, 501–504 (1955) (in Russian)
17. Muchnik, A.A.: On strong and weak reducibilities of algorithmic problems. Sibirsk. Mat. Ž. 4, 1328–1341 (1963) (in Russian)
18. Simpson, S.: Mass problems and randomness. Bull. Symb. Logic 11, 1–27 (2005)
19. Simpson, S.: Π_1^0 classes and models of WKL₀. In: Simpson, S. (ed.) Reverse Mathematics 2001. Association for Symbolic Logic: Lecture Notes in Logic, vol. 21, pp. 352–378 (2005)
20. Simpson, S.: An extension of the recursively enumerable Turing degrees. J. London Math. Soc. 75, 287–297 (2007)
21. Soare, R.: Recursively Enumerable Sets and Degrees. Springer, Berlin (1987)

Lower Bounds for Kernelizations and Other Preprocessing Procedures

Yijia Chen[1], Jörg Flum[2], and Moritz Müller[2]

[1] Shanghai Jiaotong University
BASICS, Department of Computer Science, Huashan Road 1954,
Shanghai 200030, China
yijia.chen@cs.sjtu.edu.cn
[2] Albert-Ludwigs-Universität Freiburg
Abteilung für Mathematische Logik, 79104 Freiburg, Eckerstr. 1, Germany
{joerg.flum,moritz.mueller}@math.uni-freiburg.de

Abstract. We first present a method to rule out the existence of *strong* polynomial kernelizations of parameterized problems under the hypothesis P $\neq$ NP. This method is applicable, for example, to the problem SAT parameterized by the number of variables of the input formula. Then we obtain improvements of related results in [1,6] by refining the central lemma of their proof method, a lemma due to Fortnow and Santhanam. In particular, assuming that PH $\neq \Sigma_3^p$, i.e., that the polynomial hierarchy does not collapse to its third level, we show that every parameterized problem with a "linear OR" and with NP-hard underlying classical problem does not have polynomial reductions to itself that assign to every instance x with parameter k an instance y with $|y| = k^{O(1)} \cdot |x|^{1-\varepsilon}$ (here ε is any given real number greater than zero).

Keywords: Parameterized complexity, kernelization, preprocessing.

1 Introduction

Often, if a computationally hard problem must be solved in practice, one tries, in a preprocessing step, to reduce the size of the input data. Kernelizations are a type of preprocessing procedures widely studied and applied in parameterized complexity. In parameterized complexity one measures the complexity of an algorithm not only in terms of the total input length, but also takes into account other aspects of the input codified as the *parameter* $k \in \mathbb{N}$. Intuitively, the idea is to choose the parameter in such a way that it can be assumed to be small for the instances one is interested in.

A *kernelization* $\mathbb{K}$ of a parameterized problem is a polynomial time algorithm that computes for every instance x of the problem an equivalent instance $\mathbb{K}(x)$ of a size bounded in terms of k (the "small" parameter of the instance x). This suggests a method for designing algorithms: To decide a given instance x, we compute the *kernel* $\mathbb{K}(x)$ and then decide if $\mathbb{K}(x)$ is a yes-instance by brute-force. Therefore besides efficient computability, an important quality of a good kernelization is *small kernel size*. The notion of *polynomial kernelization* is an abstract model for small kernel size. A kernelization $\mathbb{K}$ is polynomial if there is a polynomial p such that for all instances x with parameter k, the size of $\mathbb{K}(x)$ is bounded by $p(k)$.

K. Ambos-Spies, B. Löwe, and W. Merkle (Eds.): CiE 2009, LNCS 5635, pp. 118–128, 2009.

Polynomial kernelizations are known for many parameterized problems (compare the survey [10]). However, till recently, only few natural problems in FPT were known to have *no* polynomial kernelizations. This has changed, since a general method to exclude polynomial kernelizations has been developed (cf. [1,6]). It is based on a lemma due to Fortnow and Santhanam [6]: Recall that an OR (see e.g. [2]) for a classical problem Q is a polynomial time computable function that assigns to every finitely many instances $x_1, \ldots, x_t$ of Q an instance y such that ($y \in Q$ if and only if $x_i \in Q$ for some $i \in \{1, \ldots, t\}$). The Lemma of Fortnow and Santhanam tells us that no NP-hard problem can have an OR with the additional property that the length $|y|$ of y is polynomially bounded in $\max_{1 \leq i \leq t} |x_i|$ unless PH $= \Sigma_3^P$.

However there are natural parameterizations of NP-hard problems that have an OR such that the *parameter* of y is polynomially bounded in $\max_{1 \leq i \leq t} |x_i|$. If such a problem would have a polynomial kernelization, then composing it with such an OR would yield an OR with the additional property excluded by the Lemma of Fortnow and Santhanam. Various applications of this result were given in [1,6], in particular, in [6] it was shown that the problem SAT parameterized by the number of propositional variables of the input formula has no polynomial kernelizations (unless PH $= \Sigma_3^P$). This answered a question posed in [11,5] and implicitly already in [8].

We explain the contents of our paper. To the best of our knowledge all known kernelizations $\mathbb{K}$ for concrete parameterized problems are *strong* in the sense that the parameter of $\mathbb{K}(x)$ is less than or equal to the parameter of x. Moreover it is known that every parameterized problem that has a kernelization already has a strong kernelization. In Section 4 we present a result with a quite simple proof showing that parameterized problems with "parameter decreasing" self-reductions do not have strong *polynomial* kernelizations. This result only requires that P $\neq$ NP (instead of PH $\neq \Sigma_3^P$). Its method can also be used to exclude strong subexponential kernelizations assuming the exponential time hypothesis (ETH). For example we show that SAT has no strong polynomial kernelization if P $\neq$ NP and no strong subexponential kernelization if ETH holds.

It is perfectly conceivable that a parameterized problem has a useful polynomial kernelization with a slight increase of the (small) parameter. In Section 3 we prove that such a slight increase may even be necessary. Such polynomial kernelizations, which are not strong, are not only interesting from a theoretical point of view but also for practical purposes: for example, such polynomial kernelizations for SAT would be sufficient for some significant applications in cryptography [11].

In Section 5 we refine the method of [1,6] to obtain significantly better lower bounds for preprocessing procedures of parameterized problems "having a linear OR." We give various applications, e.g. for SAT we get

Let $\varepsilon > 0$. If PH $\neq \Sigma_3^P$, then there is no polynomial time algorithm that for every instance α of SAT with k variables computes an equisatisfiable instance α' with

$$|\alpha'| = k^{O(1)} \cdot |\alpha|^{1-\varepsilon}. \tag{1}$$

For problems satisfying an apparently weaker condition, namely only "having an OR for instances with constant parameter," we still get quite good lower bounds; in case of SAT it would be:

$$|\alpha'| = k^{O(1)} \cdot |\alpha|^{o(1)}.$$

Due to space limitations we have to defer some proofs to the full version of the paper.

2   Preliminaries

The set of natural numbers (that is, nonnegative integers) is denoted by $\mathbb{N}$. For a natural number n let $[n] := \{1, \ldots, n\}$. By $\log n$ we mean $\lceil \log n \rceil$ if an integer is expected. We trust the reader's common sense to interpret terms $\log n$ for $n = 0$ reasonably.

We identify (classical) problems with subsets Q of $\{0, 1\}^*$. Clearly, as done mostly, we present concrete problems in a verbal, hence uncodified form. A *reduction from a problem Q to a problem Q'* is a mapping $R : \{0, 1\}^* \to \{0, 1\}^*$ such that for all $x \in \{0, 1\}^*$ we have $\big(x \in Q \iff R(x) \in Q'\big)$.

A *parameterized problem* is a pair (Q, κ) consisting of a classical problem $Q \subseteq \{0, 1\}^*$ and a *parameterization* $\kappa : \{0, 1\}^* \to \mathbb{N}$, which is required to be polynomial time computable even if the result is encoded in unary. We exemplify our way to present parameterized problems by introducing the parameterized problem p-SAT.

---

p-SAT
>   *Instance:* A propositional formula α in conjunctive
>   normal form (CNF).
>   *Parameter:* Number of variables of α.
>   *Question:* Is α satisfiable?

A parameterized problem (Q, κ) is *fixed-parameter tractable* (or, in FPT) if $x \in Q$ is solvable in time $f(\kappa(x)) \cdot |x|^{O(1)}$ for some computable $f : \mathbb{N} \to \mathbb{N}$. If f can be chosen such that $f(k) = 2^{o^{\mathrm{eff}}(k)}$ ($f(k) = 2^{k^{O(1)}}$), then (Q, κ) is in SUBEPT (EXPT).

Here o^{eff} denotes the effective version of little oh: For computable functions $f, g : \mathbb{N} \to \mathbb{N}$ we write $f = o^{\mathrm{eff}}(g)$ if there is a *computable*, nondecreasing and unbounded function $\iota : \mathbb{N} \to \mathbb{N}$ such that $f(k) \le g(k)/\iota(k)$ for all sufficiently large $k \in \mathbb{N}$. We often write $f(k) = o^{\mathrm{eff}}(g(k))$ instead of $f = o^{\mathrm{eff}}(g)$.

3   Some Basic Questions Concerning Kernelizations

Definition 1. Let (Q, κ) be a parameterized problem and $f : \mathbb{N} \to \mathbb{N}$ be a function. An *f-kernelization* for (Q, κ) is a polynomial time reduction $\mathbb{K}$ from Q to itself such that for all $x \in \{0, 1\}^*$

$$|\mathbb{K}(x)| \le f(\kappa(x)).$$

If in addition

$$\kappa(\mathbb{K}(x)) \le \kappa(x),$$

for all $x \in \{0, 1\}^*$, then $\mathbb{K}$ is a *strong f-kernelization*. A *(strong) kernelization* is a (strong) f-kernelization for some computable function $f : \mathbb{N} \to \mathbb{N}$.

We say that (Q, κ) has a *linear, polynomial, subexponential, simply exponential,* and *exponential kernelization* if there is an f-kernelization for (Q, κ) with $f(k) = O(k)$, $f(k) = k^{O(1)}$, $f(k) = 2^{o^{\mathrm{eff}}(k)}$, $f(k) = 2^{O(k)}$, and $f(k) = 2^{k^{O(1)}}$, respectively.

The following result is well-known:

Proposition 2. *Let (Q, κ) be a parameterized problem with decidable Q. The following statements are equivalent.*

1. *(Q, κ) is fixed-parameter tractable.*
2. *(Q, κ) has a kernelization.*
3. *(Q, κ) has a strong kernelization.*

Furthermore, if f is computable and $x \in Q$ is solvable in time $f(\kappa(x)) \cdot |x|^{O(1)}$, then (Q, κ) has a strong f-kernelization.

3.1 Comparing the Different Notions of Kernelizations

The notions of polynomial kernelization and of strong polynomial kernelization are distinct, as shown by the following result proven in the full version of the paper.

Proposition 3. *There is a parameterized problem that has a polynomial kernelization but no strong polynomial kernelization.*

The recent survey [10] contains examples of natural problems whose currently best known kernelizations are polynomial, simply exponential and exponential. We show that all these distinct degrees of kernelizability are indeed distinct:

Proposition 4. *The classes of parameterized problems with a linear, a polynomial, a subexponential, a simply exponential, and an exponential kernelization are pairwise distinct.*

The claim immediately follows from the following lemma.

Lemma 5. *Let $g : \mathbb{N} \to \mathbb{N}$ be time-constructible and increasing and let $f : \mathbb{N} \to \mathbb{N}$ be such that $f(k) \leq g(k-1)$ for all sufficiently large k. Then there is a parameterized problem that has a g-kernelization but no f-kernelization.*

Proof. Let g and f be as in the statement. We choose k_0 such that $f(k) \leq g(k-1)$ for all $k \geq k_0$. We consider the "inverse function" ι_g of g given by

$$\iota_g(n) := \min\{s \in \mathbb{N} \mid g(s) \geq n\}.$$

By the time-constructibility of g the function ι_g is polynomial time computable and we have for all $n \in \mathbb{N}$:

$$n \leq g(\iota_g(n)) \qquad \text{and} \qquad \text{if } \iota_g(n) \geq 1, \text{then } g(\iota_g(n) - 1) < n. \qquad (2)$$

Let Q be a problem not in P(TIME) and define κ by $\kappa(x) := \iota_g(|x|)$. By the first inequality in (2) the identity is a g-kernelization of (Q, κ), even a strong one.

Assume that (Q, κ) has an f-kernelization $\mathbb{K}$. As ι_g is unbounded, we have $\iota_g(|x|) \geq k_0$ for sufficiently long $x \in \{0,1\}^*$. Then

$$|\mathbb{K}(x)| \leq f(\kappa(x)) = f(\iota_g(|x|)) \leq g(\iota_g(|x|) - 1) < |x|.$$

Thus applying $\mathbb{K}$ at most $|x|$ times we get an equivalent instance of length at most $g(k_0)$. Therefore, $Q \in \mathrm{P}$, a contradiction.   □

3.2   Complexity of Problems with Kernelizations

By Proposition 2 a parameterized problem is fixed-parameter tractable if and only if it has a kernelization. The next result shows that a parameterized problem (Q, κ) in FPT \ EXPT with $Q \in \mathrm{NP}$ cannot have polynomial kernelizations. We show a little bit more. Recall that EXP is the class of classical problems Q such that $x \in Q$ is solvable in deterministic time $2^{|x|^{O(1)}}$.

Proposition 6. *Assume that the problem (Q, κ) has a polynomial kernelization and that $Q \in \mathrm{EXP}$. Then $(Q, \kappa) \in \mathrm{EXPT}$.*

Proof. Let $\mathbb{K}$ be a polynomial kernelization of (Q, κ). As $Q \in \mathrm{EXP}$ there is an algorithm $\mathbb{A}$ solving $x \in Q$ in time $2^{|x|^{O(1)}}$. The algorithm that on $x \in \{0,1\}^*$ first computes $\mathbb{K}(x)$ and then applies $\mathbb{A}$ to $\mathbb{K}(x)$ solves $x \in Q$ in time $|x|^{O(1)} + 2^{|\mathbb{K}(x)|^{O(1)}} = 2^{\kappa(x)^{O(1)}} \cdot |x|^{O(1)}$.   □

The model-checking problem of monadic second-order logic on the class of trees is in EXP. By a result of [7] the corresponding parameterized problem with the length of the formula as parameter is in FPT \ EXPT unless P = NP. Hence, by the preceding proposition, it has no polynomial kernelization (unless P = NP).

4   Excluding Strong Kernelizations

In this section we present a quite simple method to exclude strong kernelizations and give some applications.

Theorem 7. *Let (Q, κ) be a parameterized problem such that the 0th slice $Q(0) := \{x \in Q \mid \kappa(x) = 0\}$ is in P. Assume further that there is a polynomial reduction R from Q to itself which is* parameter decreasing, *that is, for all $x \notin Q(0)$*

$$\kappa(R(x)) < \kappa(x). \tag{3}$$

(a) If (Q, κ) has a strong polynomial kernelization, then $Q \in \mathrm{P}$.
(b) If (Q, κ) has a strong subexponential kernelization, then $(Q, \kappa) \in \mathrm{SUBEPT}$.

Proof. Let (Q, κ) and R be as in the statement. We choose a polynomial time algorithm $\mathbb{B}$ deciding $Q(0)$. Let $\mathbb{K}$ be a strong kernelization for (Q, κ). Clearly, $S := R \circ \mathbb{K}$ is computable in polynomial time and we have for all $x \in \Sigma^*$ with $\mathbb{K}(x) \notin Q(0)$ by (3)

$$\kappa(S(x)) < \kappa(\mathbb{K}(x)) \leq \kappa(x), \tag{4}$$

The following algorithm $\mathbb{A}$ decides Q. Given $x \in \Sigma^*$, the algorithm $\mathbb{A}$ computes $S(x), S(S(x)), \ldots$; by (4) we obtain an instance y with $\kappa(\mathbb{K}(y)) = 0$ after at most $\kappa(x)$ steps; then $x \in Q$ if and only if $\mathbb{K}(y) \in Q(0)$; now $\mathbb{A}$ simulates $\mathbb{B}$ on $\mathbb{K}(y)$.

To see (a), assume that $\mathbb{K}$, in addition, is a polynomial kernelization. Then $|S(x)| = |R(\mathbb{K}(x))| \leq |\mathbb{K}(x)|^c \leq \kappa(x)^{c \cdot d}$ for some $c, d \in \mathbb{N}$. Then $|S(S(x))|, |S(S(S(x)))|, \ldots$ are all bounded by $\kappa(x)^{c \cdot d}$, again by (4). Therefore $|y| \leq \kappa(x)^{c \cdot d} \leq |x|^{O(1)}$, as κ is a parameterization. It follows that $\mathbb{A}$ runs in polynomial time. This proves (a).

For (b) assume that $\mathbb{K}$ is a strong subexponential kernelization. We choose $c, d \in \mathbb{N}$ and a computable, nondecreasing and unbounded $\iota : \mathbb{N} \to \mathbb{N}$ such that $S(x)$ is computable in time $|x|^c$ and $|S(x)| \leq 2^{d \cdot \kappa(x)/\iota(\kappa(x))}$. Then, by (4), the computation of y by the algorithm $\mathbb{A}$ needs at most

$$|x|^c + 2^{c \cdot d \cdot \kappa(x)/\iota(\kappa(x))} + 2^{c \cdot d \cdot (\kappa(x)-1)/\iota(\kappa(x)-1)} + \ldots + 2^{c \cdot d \cdot 1/\iota(1)} \qquad (5)$$

many steps. Let $k := \kappa(x)$. Then

$$\sum_{\ell=1}^{k} 2^{c \cdot d \cdot \ell/\iota(\ell)} = \sum_{\ell=1}^{\lfloor \sqrt{k} \rfloor} 2^{c \cdot d \cdot \ell/\iota(\ell)} + \sum_{\ell=\lfloor \sqrt{k} \rfloor + 1}^{k} 2^{c \cdot d \cdot \ell/\iota(\ell)}$$
$$\leq \lfloor \sqrt{k} \rfloor \cdot 2^{c \cdot d \cdot \lfloor \sqrt{k} \rfloor} + k \cdot 2^{c \cdot d \cdot k/\iota(\lfloor \sqrt{k} \rfloor)}.$$

Since the function $k \mapsto \iota(\lfloor \sqrt{k} \rfloor)$ is unbounded, the sum in (5) is bounded by $|x|^c + 2^{o^{\text{eff}}(\kappa(x))}$; hence $(Q, \kappa) \in \text{SUBEPT}$. $\qquad \square$

Examples 8. The classical problems underlying

$$p\text{-SAT}, \quad p\text{-POINTED-PATH}, \quad \text{and} \quad p\text{-BICLIQUE}$$

have parameter-decreasing polynomial reductions to themselves, where

---

p-POINTED-PATH
> *Instance:* A graph $G = (V, E)$, a vertex $v \in V$, and $k \in \mathbb{N}$.
> *Parameter:* k.
> *Question:* Does G have a path of length k starting at v?

---

p-BICLIQUE
> *Instance:* A graph $G = (V, E)$ and $k \in \mathbb{N}$.
> *Parameter:* k.
> *Question:* Does G contain a k-biclique?

Here, a k-*biclique* in G is a subgraph isomorphic to $K_{k,k}$, the complete bipartite graph with $2k$ vertices.

Proof. p-SAT: We define a parameter-decreasing polynomial reduction R from p-SAT to itself: Let α be a CNF formula. If α has no variables, we set $R(\alpha) := \alpha$. Otherwise let X be the first variable in α. We let $R(\alpha)$ be a CNF formula equivalent to $\alpha(X/\text{TRUE}) \vee \alpha(X/\text{FALSE})$, where, for example, $\alpha(X/\text{TRUE})$ is the formula obtained from α by replacing X by TRUE everywhere. Clearly R is polynomial time computable.

p-POINTED-PATH: The following is a parameter-decreasing polynomial self-reduction R for p-POINTED-PATH: Let (G, v, k) be an instance of p-POINTED-PATH and assume $k \geq 3$. For any path $P : v, v_1(P), v_2(P)$ of length 2 starting from v let G_P be the graph obtained from G by deleting the two vertices $v, v_1(P)$ (and all the edges incident with one of these vertices). Let H be the graph obtained from the disjoint union of the graphs G_P (where P ranges over all paths of length 2 starting in v) by adding a new vertex w and all edges $\{w, v_2(P)\}$. Then H has a path of length $(k-1)$ starting at w if and only if G has a path of length k starting at v. Hence we can set $R((G, v, k)) := (H, w, k-1)$.

p-BICLIQUE: Let (G, k) be an instance of p-BICLIQUE with $k \geq 1$. For every edge $e = \{u, v\}$ of G, we construct a new bipartite graph G_e as follows: the vertices on the "left" are (copies of) the neighbors of v distinct from u and the vertices on the "right" are (copies of) the neightbors of u distinct from v; there is an edge between a (copy of) a vertex on the left and one on the right if and only if there is such an edge in G.

This graph G_e contains a $(k-1)$-biclique if and only if the original graph G contains a k-biclique containing the edge e. Let G^* be the disjoint union of all G_e for all edges e of G. Then G^* contains a $(k-1)$-biclique if and only if G contains a k-biclique. $\square$

Corollary 9. *1. If* $P \neq NP$, *then* p-SAT, p-POINTED-PATH, *and* p-BICLIQUE *have no strong polynomial kernelizations.*
2. If ETH *holds, then* p-SAT *and* p-POINTED-PATH *have no strong subexponential kernelizations.*

Proof. Part (1) is immediate by Theorem 7, as all three underlying problems are NP-hard[1]. Moreover, we know by this theorem that if one of the three problems has a strong subexponential kernelization, then it is in SUBEPT. However then ETH would fail in the case of p-SAT by [12], in the case of p-POINTED-PATH by [3]. $\square$

To the best of our knowledge it is not known whether p-BICLIQUE $\in$ FPT. Of course, if p-BICLIQUE $\notin$ FPT, then it has no kernelization at all.

5   Strong Lower Bounds

In [1,6] a method is developed to exclude polynomial kernelizations. In this section we refine this method and obtain stronger lower bounds for preprocessing procedures. More precisely, we show how to exclude so-called ε self-reductions and subexponential self-reductions. Due to space limitations we have to defer the proofs to the full version of the paper.

---

[1] In the case of p-BICLIQUE, see [13].

5.1   Excluding ε Self-reductions

We first introduce the corresponding concept.

Definition 10. Let $\varepsilon > 0$. A parameterized problem (Q, κ) *has an ε self-reduction* if there is a polynomial reduction from Q to itself that assigns to every instance x of Q an instance y with

$$|y| = \kappa(x)^{O(1)} \cdot |x|^{1-\varepsilon}.$$

Note that an ε self-reduction may not be a kernelization, as we do not require that $|y|$ is bounded in terms of $\kappa(x)$; moreover, it is not required that the parameter of y is bounded in terms of $\kappa(x)$. Clearly, if (Q, κ) has a polynomial kernelization, then (Q, κ) has an ε self-reduction for every $\varepsilon > 0$. The result with the bound (1) mentioned at the end of the Introduction says that, in case PH $\neq \Sigma_3^P$, for every $\varepsilon > 0$ the problem p-SAT has no ε self-reduction. This result will be a special instance of a more general result stating similar lower bounds for problems with a linear OR in the following sense:

Definition 11. Let (Q, κ) be a parameterized problem. A *linear OR for (Q, κ)* is a polynomial time algorithm $\mathbb{O}$ that for every finite tuple $\bar{x} = (x_1, \ldots, x_t)$ of instances of Q outputs an instance $\mathbb{O}(\bar{x})$ of Q such that

1. $|\mathbb{O}(\bar{x})| = t \cdot \left(\max_{i \in [t]} |x_i| \right)^{O(1)}$;
2. $\kappa(\mathbb{O}(\bar{x})) = \left(\max_{i \in [t]} |x_i| \right)^{O(1)}$;
3. $\mathbb{O}(\bar{x}) \in Q$ if and only if for some $i \in [t]$: $x_i \in Q$.

Examples 12. Each of the parameterized problems

$$p\text{-}\text{PATH}, p\text{-}\text{SAT}, p\text{-}\text{CYCLE}, uni\text{-}\text{CLIQUE}, uni\text{-}\text{DOMINATING-SET}$$

has a linear OR. The result for p-PATH is from [1], that for p-CYCLE is due to Grohe [9]. Here:

---

p-CYCLE
    Instance: A graph G and $k \in \mathbb{N}$.
    Parameter: k.
    Question: Does G have a cycle of length k?

Similarly, the problem p-PATH asks for a path of length k in G. The problem

---

uni-CLIQUE
    Instance: A graph $G = (V, E)$ and $k \in \mathbb{N}$.
    Parameter: $k \cdot \log |V|$.
    Question: Does G have a clique of cardinality k?

---

is the so-called *canonical reparameterization* [4] of the parameterized clique problem; the problem uni-DOMINATING-SET is defined similarly.

Theorem 13. *Let $\varepsilon > 0$. Let (Q, κ) be a parameterized problem with a linear OR and with NP-hard Q. If PH $\neq \Sigma_3^P$, then the problem (Q, κ) has no ε self-reductions.*

In particular, if PH $\neq \Sigma_3^P$, then all the problems mentioned in Examples 12 do not have ε self-reductions.

5.2   Excluding Subexponential Self-reductions

Recall that a *hole* in a graph is an induced cycle of length at least 4 (see [3]). We consider the parameterized problem

---

p-ODD-HOLE$_\leq$
>   *Instance:* A graph G and $k \in \mathbb{N}$.
>   *Parameter:* k.
>   *Question:* Does G have a hole of *odd* length at most k?

Let $(G_1, k_1), \ldots, (G_t, k_t)$ be instances of p-ODD-HOLE$_\leq$. If $k_1 = \ldots = k_t =: k$, then for the disjoint union G of the G_is we have $(G, k) \in p$-ODD-HOLE$_\leq$ if and only if $(G_i, k_i) \in p$-ODD-HOLE$_\leq$ for some $i \in [t]$. However, it is not clear how to define such an instance (G, k) if $k_1, \ldots, k_t$ are distinct; more precisely, we do not know whether p-ODD-HOLE$_\leq$ has a linear OR. The following concept is tailored for such situations.

Definition 14. Let (Q, κ) be a parameterized problem and let λ be a further parameterization. An *OR for λ-constant instances of* (Q, κ) is a polynomial time algorithm $\mathbb{O}$ that for every finite tuple $\bar{x} = (x_1, \ldots, x_t)$ of instances of Q with $\lambda(x_1) = \ldots = \lambda(x_t)$ outputs an instance $\mathbb{O}(\bar{x})$ of Q such that

1. $\kappa(\mathbb{O}(\bar{x})) = (\max_{i \in [t]} |x_i|)^{O(1)}$;
2. $\mathbb{O}(\bar{x}) \in Q$ if and only if for some $i \in [t]$: $x_i \in Q$.

Examples 15. The instances of the following problems are pairs (G, k), where G is a graph and $k \in \mathbb{N}$. We let λ always be the function with $\lambda(G, k) := k$. In all examples we get the claimed OR for λ-constant instances by setting $\mathbb{O}((G_1, k), \ldots, (G_t, k)) := (G, k)$, where the graph G is the disjoint union of the G_is.

(a) The problem p-ODD-HOLE$_\leq$ has an OR for λ-constant instances.

(b) The problems *uni*-CHORDLESS-PATH and *uni*-CHORDLESS-CYCLE have an OR for λ-constant instances. Here, for example,

---

uni-CHORDLESS-CYCLE
>   *Instance:* A graph $G = (V, E)$ and $k \in \mathbb{N}$.
>   *Parameter:* $k \cdot \log |V|$.
>   *Question:* Does G have a chordless cycle of length k?

Note that in the last example $\lambda(G, k) = k$ is not the parameter of (G, k) as instance of *uni*-CHORDLESS-CYCLE.

For problems with an OR for constant instances we get a slightly weaker result than that in Theorem 13 for problems with a linear OR. To state the result we first define:

Definition 16. Let (Q, κ) be a parameterized problem. A *subexponential self-reduction of* (Q, κ) is a polynomial reduction from Q to itself that assigns to every instance x of Q an instance y with

$$|y| = \kappa(x)^{O(1)} \cdot |x|^{o(1)}.$$

Clearly if (Q, κ) has a subexponential self-reduction, then it has an ε self-reduction for every $\varepsilon > 0$.

Theorem 17. *Let* (Q, κ) *be a parameterized problem with* NP*-hard* Q. *Furthermore assume that* (Q, κ) *has an* OR *for* λ*-constant instances, where* λ *is a further parameterization. If* $\mathrm{PH} \neq \Sigma_3^P$, *then* (Q, κ) *has no subexponential self-reductions.*

This improves the corresponding result of [1] in that it assumes only NP-hardness instead of NP-completeness, it assumes a weaker notion of OR (the OR used in [1] is ours for $\lambda = \kappa$) and it excludes subexponential self-reductions instead of polynomial kernelizations.

We can apply Theorem 17 to the problems in Examples 15 (b). Let us mention that with some more effort also linear ORs can be defined for these problems. For p-ODD-HOLE$_{\leq}$, however, we cannot apply Theorem 17 to rule out subexponential self-reductions (and hence polynomial kernelizations), as to the best of our knowledge it is not known whether the underlying classical problem is NP-hard. Moreover, it is not known whether p-ODD-HOLE$_{\leq}$ is in FPT. If not, then it would not have kernelizations at all. As a further application of the theorem we will show in the full version of this paper that the problem p-PATH restricted to planar connected graphs does not have subexponential self-reductions unless $\mathrm{PH} = \Sigma_3^P$.

6   Final Remarks

We presented a method to exclude polynomial or subexponential strong kernelizations and gave some applications. Furthermore we refined the recently developed method [1,6] to exclude polynomial kernelizations and obtained stronger lower bounds for various examples. More precisely, we showed how to exclude so-called ε self-reductions and subexponential self-reductions. The relationships between these new notions and various notions of kernelizations will be studied in the full version of this paper. There we will also compare polynomial kernelizations with the conceptually similar notion of *compression* algorithms introduced in [11]. We will see that the two notions coincide for "natural" problems in EXPT.

We close with a question: Is there a *natural* parameterized problem with polynomial kernelizations but without strong ones (in contrast to the artificial problem we constructed in the proof of Proposition 3)?

Acknowledgement. The first author is partially supported by the National Nature Science Foundation of China (60673049). The third author wants to thank Mike Fellows, Danny Hermelin and Frances Rosamond for many helpful discussions.

References

1. Bodlaender, H.L., Downey, R.G., Fellows, M.R., Hermelin, D.: On problems without polynomial kernels. In: Aceto, L., Damgård, I., Goldberg, L.A., Halldórsson, M.M., Ingólfsdóttir, A., Walukiewicz, I. (eds.) ICALP 2008, Part I. LNCS, vol. 5125, pp. 563–574. Springer, Heidelberg (2008)
2. Chang, R., Kadin, Y.: On computing boolean connectives of characteristic functions. Math. Systems Theory 28, 173–198 (1995)
3. Chen, Y., Flum, J.: On parameterized path and chordless path problems. In: Proceedings of the 22nd IEEE Conference on Computational Complexity (CCC 2007), pp. 250–263 (2007)
4. Chen, Y., Flum, J.: Subexponential time and fixed-parameter tractability: exploiting the miniaturization mapping. In: Duparc, J., Henzinger, T.A. (eds.) CSL 2007. LNCS, vol. 4646, pp. 389–404. Springer, Heidelberg (2007)
5. Flum, J., Grohe, M.: Parameterized Complexity Theory. Springer, Heidelberg (2006)
6. Fortnow, L., Santhanam: Infeasibility of instance compression and succinct PCPs for NP. In: Proceedings of the 40th ACM Symposium on the Theory of Computing (STOC 2008), pp. 133–142. ACM, New York (2008)
7. Frick, M., Grohe, M.: The complexity of first-order and monadic second-order logic revisited. Annals of Pure and Applied Logic 130, 3–31 (2004)
8. Grandjean, E., Kleine-Büning, H.: SAT-problems and reductions with respect to the number of variables. Journal of Logic and Computation 7(4), 457–471 (1997)
9. Grohe, M.: Private communication (2008)
10. Guo, J., Niedermeier, R.: Invitation to data reduction and problem kernelization. ACM SIGACT News 38(1) (2007)
11. Harnik, D., Naor, M.: On the compressibility of NP instances and cryptographic applications. In: Proceedings of the 47th Annual IEEE Symposium on Foundations of Computer Science (FOCS 2006), pp. 719–728 (2006)
12. Impagliazzo, R., Paturi, R., Zane, F.: Which problems have strongly exponential complexity? Journal of Computer and System Sciences 63, 512–530 (2001)
13. Johnson, D.: Announcements, updates, and greatest hits. Journal of Algorithms 8(3), 438–448 (1987)

Infinite-Time Turing Machines and Borel Reducibility

Samuel Coskey

The City University of New York

In this document I will outline a couple of recent developments, due to Joel Hamkins, Philip Welch and myself, in the theory of infinite-time Turing machines. These results were obtained with the idea of extending the scope of the study of Borel equivalence relations, an area of descriptive set theory. I will introduce the most basic aspects of Borel equivalence relations, and show how infinite-time computation may provide insight into this area.

1 Infinite-Time Turing Machines

We begin by describing a model of transfinite computation called the infinite-time Turing machine (ITTM), invented by Hamkins and Kidder and introduced by Hamkins and Lewis [1]. Like an ordinary Turing machine, an ITTM is a finite state machine with an input tape, an output tape, and a read/write head. The key addition is that when an ITTM reaches a limit ordinal numbered step, the program enters a special limit state, the read/write head is reset to the left, and the value of each cell is set to the limit of the previous values in that cell (or else zero, if the value has flipped unboundedly many times).

Definition 1. We say that a partial function f is *ittm-computable* iff there exists an ITTM which on input x, writes $f(x)$ if it is defined and diverges otherwise. We say that f is *eventually computable* iff there exists an ITTM such that on input x, the output tape eventually converges to $f(x)$ (it need not halt) if it is defined and diverges otherwise.

It is also possible to allow an ITTM an *oracle tape*, and doing so allows one to relativize computations to a real parameter. For $x, y \in 2^{\mathbb{N}}$, we say that x is ittm-computable from y, written $x \leq_{\infty} y$, iff there exists an ITTM with oracle y which writes x. Letting $x \equiv_{\infty} y$ iff $x \leq_{\infty} y$ and $y \leq_{\infty} x$, we say that $[x]_{\equiv_{\infty}}$ is the *ittm-degree* of x. The *eventual degree* relation $\equiv_{e\infty}$ is defined analogously.

The ittm-degrees share many of the most basic properties of the classical Turing degrees. For instance, since there are still only countably many ittm programs, there are continuum many ittm-degrees. It is natural to ask for the descriptive set-theoretic strengthening, namely that there exists a *perfect set* of ittm-degrees. This was recently established by Welch, building upon his earlier work [2] which had shown that there are continuum many minimal ittm-degrees.

K. Ambos-Spies, B. Löwe, and W. Merkle (Eds.): CiE 2009, LNCS 5635, pp. 129–133, 2009.

Theorem 1 (Welch). *There exists a perfect set of pairwise ittm-computably incomparable reals. There exists a perfect set of pairwise eventually computably incomparable reals.*

We next describe the pointclasses of decidable sets that correspond to these new notions of computation. We say that a set is of class D or *ittm-decidable* iff its characteristic function is ittm-computable, and of class sD or *semi ittm-decidable* iff it is the domain of a ittm-computable function. The eventually ittm-decidable and semi ittm-decidable classes E, sE are defined analogously. These classes are all provably Δ_2^1, and fit into the usual projective hierarchy as follows.

$$
\begin{array}{ccccccccc}
\Sigma_1^1 & & & sD & & & sE & & \\
 & \subset & D & \subset & \subset & E & \subset & \subset & \Delta_2^1 \\
\Pi_1^1 & \subset & \subset & sD' & \subset & \subset & sE' & \subset &
\end{array}
$$

In fact, each of these new pointclasses is contained in the class $\mathrm{prov}(\Delta_2^1)$ of ZFC provably Δ_2^1 sets. It follows that the ittm-decidable sets and the ittm-computable functions are Lebesgue measurable, a fact which will be important later on.

2   Equivalence Relations

In this section, we investigate some of the advantages of using notions from infinite-time computability in the study of Borel equivalence relations. This is the second area of logic which has received the ITTM treatment, the first being computable model theory [3].

The theory of Borel equivalence relations revolves around the following complexity notion. If E, F are equivalence relations on $2^{\mathbb{N}}$, then following [4] and [5] we say that E is *Borel reducible* to F, written $E \leq_B F$, iff there exists a Borel function $f : 2^{\mathbb{N}} \to 2^{\mathbb{N}}$ such that

$$x \, E \, y \iff f(x) \, F \, f(y)$$

for all $x, y \in 2^{\mathbb{N}}$. The function f is said to be a Borel *reduction* from E to F.

For applications, we think of the equivalence relations E, F as representing classification problems from some other area of mathematics. For instance, since any group with domain $\mathbb{N}$ is determined by its multiplication function, studying the classification problem for countable groups amounts to studying the isomorphism equivalence relation on a suitable subspace of $2^{\mathbb{N} \times \mathbb{N} \times \mathbb{N}}$. For many more examples, see [6, Section 1.2].

Now, recall that the Borel sets correspond to the bold-face pointclass Δ_1^1. Since we wish to generalize the Borel reductions, we shall use from now on the following bold-face analog of the ittm-computable functions. Namely, we say that f is *bold-face ittm-computable* iff there exists a $z \in 2^{\mathbb{N}}$ such that f is computed by an ITTM with oracle z. The *bold-face eventually computable* functions are defined analogously.

Definition 2. We say that E is *ittm-computably reducible* to F, written $E \leq_c F$, iff there is a (bold-face) ittm-computable reduction from E to F. We say that E is *eventually reducible* to F, written $E \leq_e F$, iff there is a (bold-face) eventually computable reduction from E to F.

A key observation in the study of ittm-computable and eventual reducibility is they hardly differ from Borel reducibility on many well-studied equivalence relations. Indeed, it is still unknown whether there exist Borel equivalence relations E, F such that $E \leq_c F$ but $E \not\leq_B F$. On the other hand, there exist natural equivalence relations which are so complex that Borel reducibility does not capture their relationship, and computable reducibility does. Let $x \equiv_{CK} y$ iff x, y compute (in the ordinary sense) the same countable ordinals and $x \cong_{WO} y$ iff x and y code isomorphic well-orders. Then we have the following:

Theorem 2 (Hamkins-Coskey). *The equivalence relations $\cong_{WO}$ and $\equiv_{CK}$ are Borel incomparable but ittm-computably bireducible.*

We are presently motivated by the following two goals:

- Explore the extent of the analogy between $\leq_B$ and either $\leq_c$ or $\leq_e$.
- Use the notion of ittm-computable complexity to analyze interesting relations of high complexity, where Borel reducibility is not appropriate.

3   Countable Equivalence Relations

Some progress towards these goals has been made in the countable case. An equivalence relation E is said to be *countable* iff every E-class is countable. The collection of countable Borel equivalence relations has undergone a great deal of study; let us outline the basic complexity picture. The least complex countable Borel equivalence relation is the equality relation $=$. Its immediate successor is the *almost equality* relation E_0, defined by $x \, E_0 \, y$ iff $x_n = y_n$ for all but finitely many n. There is also a most complex countable Borel equivalence relation, called E_∞. The remaining ones lie in the interval (E_0, E_∞), and it has been shown by Adams and Kechris [7] that every Borel partial ordering embeds into $\leq_B$ on this interval.

We wish to obtain all of the same conclusions for the countable ittm-computable equivalence relations, but it is not clear that this will be possible. Instead, we must work with the following infinite-time generalizations of the countable Borel equivalence relations.

Definition 3. The equivalence relation E is *write-outable* iff there exists a (bold-face) ittm-computable function such that for all x, $f(x)$ codes an enumeration of $[x]_E$. The *eventually write-outable* equivalence relations are defined analogously.

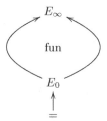

Fig. 1. The countable Borel equivalence relations

For instance, every countable Borel equivalence relation is write-outable, and the ittm-degree equivalence relation $\equiv_\infty$ is eventually write-outable but not write-outable. The write-outable equivalence relations share many properties with the countable Borel equivalence relations. The first result is the following, which essentially an immediate consequence of Theorem 1.

Corollary 1. *The equality relation $=$ is computably reducible to every write-outable equivalence relation.*

Secondly, it follows from a classical argument (from the Borel case) that E_∞ is universal for the write-outable equivalence relations. Lastly, since the arguments of Adams and Kechris are measure-theoretic, and every ittm-computable function is measurable, we also have that any Borel partial order embeds into the write-outable equivalence relations. Moreover, the analog of each of these results holds for the eventually write-outable relations under eventual reducibility. It remains only to show that E_0 is the unique immediate successor of $=$, in the sense of ittm-computable or eventually computable reducibility.

We close by mentioning an open problem in countable Borel equivalence relations. The ordinary Turing degree relation $\equiv_T$ is Borel reducible to E_∞, but it is unknown whether $\equiv_T$ is bireducible with E_∞. It is known, however, that if $\equiv_T$ is indeed universal then the Martin Conjecture must fail. Since the ittm-degree relation $\equiv_\infty$ is eventually write-outable, it is eventually reducible to E_∞, but it remains unknown whether $\equiv_\infty$ is ittm-bireducible with E_∞.

References

1. Hamkins, J.D., Lewis, A.: Infinite time Turing machines. J. Symbolic Logic 65(2), 567–604 (2000)
2. Welch, P.D.: Minimality arguments for infinite time Turing degrees. In: Sets and proofs (Leeds, 1997). London Math. Soc. Lecture Note Ser., vol. 258, pp. 425–436. Cambridge Univ. Press, Cambridge (1999)
3. Deolalikar, V., Hamkins, J.D., Schindler, R.: P $\neq$ NP$\cap$ co-NP for infinite time Turing machines. J. Logic Comput. 15(5), 577–592 (2005)

4. Friedman, H., Stanley, L.: A Borel reducibility theory for classes of countable structures. J. Symbolic Logic 54(3), 894–914 (1989)
5. Hjorth, G., Kechris, A.S.: Borel equivalence relations and classifications of countable
   models. Annals of pure and applied logic 82(3), 221–272 (1996)
6. Thomas, S.: Countable Borel equivalence relations. Online somewhere (2007)
7. Adams, S., Kechris, A.: Linear algebraic groups and countable Borel equivalence
   relations. Journal of the American mathematical society 13(4), 909–943 (2000)

Cutting Planes and the Parameter Cutwidth

Stefan Dantchev[1] and Barnaby Martin[2]

[1] Department of Computer Science, University of Durham,
Science Labs, South Road, Durham DH1 3LE, U.K.
[2] Équipe de Logique Mathématique - CNRS UMR 7056, Université Paris 7,
UFR de Mathématiques - case 7012, site Chevaleret 75205 Paris Cedex 13, France

Abstract. We introduce the parameter cutwidth for the Cutting Planes (CP) system of Gomory and Chvátal. We provide linear lower bounds on cutwidth for two simple polytopes. Considering CP as a propositional refutation system, one can see that the cutwidth of a CNF contradiction F is always bound above by the Resolution width of F. We provide an example proving that the converse fails: there is an F which has constant cutwidth, but has Resolution width $\Omega(n)$. Following a standard method for converting an FO sentence ψ, without finite models, into a sequence of CNFs, $F_{\psi,n}$, we provide a classification theorem for CP based on the sum cutwidth plus rank. Specifically, the cutwidth+rank of $F_{\psi,n}$ is bound by a constant c (depending on ψ only) iff ψ has no (infinite) models. This result may be seen as a relative of various gap theorems extant in the literature.

1 Introduction

The system of Cutting Planes (CP), introduced in [8], provides a method for solving integer linear programs (ILPs) by iteratively deriving further constraints until the problem is reduced to a general linear program (for which a polynomial algorithm is known). In terms of feasible solutions, this equates to isolating the integer hull of the solution set of the defining inequalities. It is well-known that the question as to whether a given propositional formula has a satisfying assignment can be reduced to a feasibility question for a certain ILP – the formula is therefore a contradiction iff the associated polytope has an empty integer hull. Thus CP gives rise to a refutation system for propositional CNF (conjunctive normal form) formulae, first considered in [5]. Resolution is perhaps the best-known refutation system for CNF, and width – that is the maximum number of variables in any derived clause – has shown itself to be an important parameter in its study (see [1]). We initiate an investigation of the parameter cutwidth in CP – that is, the maximum number of variables in an inequality on which a cut operation is performed. Cutwidth manifests as a more natural parameter than width as one may have a system of inequalities of large width that are inconsistent even as a linear program, i.e. can be refuted without recourse to a single cut – for example this is the case for inequalities that encode a CNF contradiction F that may be refuted by unit clause propagation. When one has large initial clauses in F, a large Resolution width is unavoidable; therefore, it becomes necessary to add 'extension' axioms to mitigate this. An advantage in our definition of cutwidth is that this need not be necessary, since we can have inequalities of arbitrary width so long as we do not perform cuts on them.

We provide linear lower bounds for two polytopes. The first has a non-empty integer hull, while the second has both an empty integer hull and is the associated polytope to a contradictory CNF (itself based on the well-known least number principle). For polytopes associated with contradictions F, we prove that Resolution width always provides an upper bound on cutwidth. But, we prove that a converse does not hold. Using similar expanders to [1], we generate a contradiction F whose Resolution width is $\Omega(n)$ but whose cutwidth is constant.

There is a standard mechanism for converting an FO sentence ψ, without finite models, into a sequence of contradictory CNFs, $F_{\psi,n}$. A series of papers in recent years has studied so-called gap phenomena in propositional refutation systems. This began with [10], and has continued with, e.g., [6,11]. A complexity gap is given for two other refutation systems based on ILP – namely those of Lovász-Schrijver and Sherali-Adams (SA) – in [6]. In each case the relevant parameter is rank and, as in [10], the separating criterion is whether or not ψ has some (infinite) model. The search for a similar gap theorem for CP is ongoing. In this paper we provide a weaker classification theorem based on the sum of cutwidth and rank. Specifically, and using the gap theorem proved for SA, we prove that the cutwidth+rank of $F_{\psi,n}$ is bound by a constant c (depending on ψ only) iff ψ has no (infinite) models.

Since our lower bounds on cutwidth, proved in Section 2, do not require an understanding of CP as a refutation system, we defer discussion of this to Section 3. In Section 3, we introduce Resolution and CP as a refutation system, and discuss the relationship between Resolution width and CP cutwidth. In Section 4, we describe the mechanism for turning FO sentences to sequences of CNFs, survey known gap theorems, and prove our classification theorem for CP – based on cutwidth+rank. For reasons of space certain proofs are omitted.

2   Cutting Planes and Lower Bounds for Cutwidth

Let $[0, 1]$ be the closed interval $0 \leq x \leq 1$ in $\mathbb{R}$.[1] Let L be a system of linear inequalities in n variables, here always with rational coefficients and bounding a polytope $P(L)$ in the $[0, 1]^n$ hypercube (the bounding inequalities $0 \leq x \leq 1$, for each variable x, should always be assumed). Cutting Planes (CP) allows for the derivation of new inequalities from a set of inequalities via the following two rules. Let $\langle , \rangle$ denote the standard dot product, and let $\bar{a}_i \in \mathbb{R}^n$, $b_i \in \mathbb{R}$ and $\bar{x}$ be an n-tuple of variables.

I. Positive linear combinations. From inequalities $\langle \bar{a}_1, \bar{x} \rangle - b_1 \geq 0, \ldots, \langle \bar{a}_k, \bar{x} \rangle - b_k \geq 0$, and for $\lambda_1, \ldots, \lambda_k$ pos. rationals, derive $\sum_{i=1}^{k} (\lambda_i \langle \bar{a}_i, \bar{x} \rangle - \lambda_i b_i) \geq 0$.

---

[1] It is conventional to consider vector spaces over the field $\mathbb{R}$, though one could equally consider $\mathbb{Q}$.

II. Cut. From $\langle \overline{a}, \overline{x} \rangle - b \geq 0$ derive $\langle \overline{a}, \overline{x} \rangle - \lceil b \rceil \geq 0$.

The twin rules allow one to generate a sequence of linear inequalities, where each inequality is either in L or is derived from previous inequalities by Rules I or II above. It is known that this sequence will converge on the integer hull of $P(L)$ [9,4]. In particular, if L has no integer solutions, one may derive the inequality $0 \geq 1$. The sequence of application of Rules I and II gives rise to a directed acyclic graph (DAG) $\mathscr{G}_L$ in which the inequalities of L are the sources and the bounding inequalities of the integer hull of $P(L)$ are the sinks. Let the rank of $\mathscr{G}_L$ be the maximum number of applications of cut (Rule II) on any path in $\mathscr{G}_L$ and let the cutwidth of $\mathscr{G}_L$ be the maximum number of variables in an inequality on which a cut was performed. Define the *rank* and *cutwidth* of L to be the minimal rank and cutwidth, respectively, of $\mathscr{G}_L$ across all possible $\mathscr{G}_L$. It is known that the rank of L is necessarily $O(n^2 \log n)$ (this requires containment in the unit hypercube) [7] and, further, is $\leq n$, if L has no integer solutions [2].

Recall L is a system of linear inequalities in n variables. Let L^* be the set of all positive linear combinations of these inequalities. Let $P^{(i)}(L)$ be the polytope whose bounding inequalities are those that may be obtained from L through at most i applications of cut. Now, if L_1 and L_2 are systems of linear inequalities such that $P(L_1) = P(L_2)$, then $L_1^* = L_2^*$. It therefore makes sense to talk about $P^{(i)}$, for some polytope P, independent of the actual inequalities used to specify P. Henceforth, we will usually drop the L. For some polytope P, let $P_{(c)}^{(i)}$ be the polytope obtainable from (the inequalities that define) P in at most i cuts of cutwidth bound by c. Let $[n]$ denote the set $\{1, \ldots, n\}$. For P a polytope in $[0,1]^n$, define

$$P' := \{\overline{x} \in P \mid \forall \overline{a} \in \mathbb{Z}^n, b \in \mathbb{R} : (\forall \overline{y} \in P \; \langle \overline{a}, \overline{y} \rangle \geq b) \Rightarrow \langle \overline{a}, \overline{x} \rangle \geq \lceil b \rceil\}.$$

If $\overline{x} \in \mathbb{R}^n$ and $J \subseteq [n]$, let $\overline{x}_{|J}$ be the $|J|$-subtuple of $\overline{x}$ given by the index set J. Now, for P a polytope in $[0,1]^n$, define

$$P_c' := \{\overline{x} \in P \mid \forall \overline{a} \in \mathbb{Z}^c, b \in \mathbb{R}, J \subseteq [n], |J| = c : \\ (\forall \overline{y} \in P \; \langle \overline{a}, \overline{y}_{|J} \rangle \geq b) \Rightarrow \langle \overline{a}, \overline{x}_{|J} \rangle \geq \lceil b \rceil\}.$$

It is not difficult to see that what we have given is simply a dual definition.

Fact 1. *For P a polytope in $[0,1]^n$, $P' = P^{(1)}$ and $P_c' = P_{(c)}^{(1)}$.*

The following is essentially a generalisation of what appears in [3]. If $\overline{x} \in (\frac{1}{2}\mathbb{Z})^n$, then let $E(\overline{x})$ be the subset of indices of $[n]$ on which $\overline{x}$ is non-integral.

Lemma 1 (Protection Lemma). *Let P be a polytope in $[0,1]^n$ and let $\overline{x} \in (\frac{1}{2}\mathbb{Z})^n$.*

- *Suppose there exists a partition $E_1, \ldots, E_t$ of $E(\overline{x})$ s.t., for each $i \in [t]$, there exists a set of points $\overline{y}_1, \ldots, \overline{y}_s \in P$ of which $\overline{x}$ is the average, that are $0-1$ on E_i but agree with $\overline{x}$ elsewhere, then $\overline{x} \in P'$.*
- *Suppose there exists a partition $E_1, \ldots, E_t$ of $E(\overline{x})$ s.t., for each $i \in [t]$ and $J \subseteq [n], |J| = c$, there exists a set of points $\overline{y}_1, \ldots, \overline{y}_s \in P$ s.t.*

- $\overline{y}_1, \ldots, \overline{y}_s$ are $0-1$ on E_i but agree with $\overline{x}$ elsewere, and
- $\overline{x}_{|J|}$ is the average of $\overline{y}_{1|J}, \ldots, \overline{y}_{s|J}$,

then $\overline{x} \in P'_c$.

Proof. The first part is proved in [3]; we prove the second part adopting a similar method. Assume for contradiction that $\overline{x}$ as in the statement is not in P'_c. I.e. choose witnessing $\overline{a} \in \mathbb{Z}^c$, b and J s.t., for all $\overline{y} \in P$, $\langle \overline{a}, \overline{y}_{|J} \rangle \geq b$, but $\langle \overline{a}, \overline{x}_{|J} \rangle < \lceil b \rceil$. Unlike the proof in [3], we can not claim here that $\overline{x} \in P$, but it is clear that $\langle \overline{a}, \overline{x}_{|J} \rangle \geq b$ (as $\overline{x}_{|J|}$ is the average of $\overline{y}_{1|J}, \ldots, \overline{y}_{s|J}$ and $\overline{y}_1, \ldots, \overline{y}_s$ are $\in P$), and hence $\langle \overline{a}, \overline{x}_{|J} \rangle \in \frac{1}{2} + \mathbb{Z}$. Thus $\sum_{e \in E(\overline{x}) \cap J} a_e$ is odd, and it follows that there exists $i \in [t]$ s.t. $\sum_{e \in E_i \cap J} a_e$ is odd. Consider the set $\overline{y}_1, \ldots, \overline{y}_s$, which are $0-1$ on E_i but agree with $\overline{x}$ elsewere, and of which $\overline{x}_{|J}$ is the average of $\overline{y}_{1|J}, \ldots, \overline{y}_{s|J}$. Since $\sum_{e \in E_i \cap J} a_e$ is odd, we have that $\langle \overline{a}, \overline{y}_{l|J} \rangle$ is integral for all $l \in [s]$. But then $\langle \overline{a}, \overline{y}_{l|J} \rangle \geq \lceil b \rceil$ and, consequently, $\langle \overline{a}, \overline{x}_{|J} \rangle \geq \lceil b \rceil$: contradiction.

We now provide a simple example of a polytope, with non-empty integer hull, whose cutwidth is linear (indeed maximal). Let $P(C_{2n+1})$ be the polytope in $[0,1]^{2n+1}$ given by the inequalities $x_i + x_{i+1 \bmod 2n+1} \geq 1$, for $i \in [2n+1]$.

Lemma 2. *The polytope $P(C_{2n+1})$ has cutwidth $> 2n$.*

Proof. We use a game argument based on the Protection Lemma. The point $(\frac{1}{2})^{2n+1}$ is plainly not in the integer hull of $P(C_{2n+1})$ (this may be shown by a single cut of cutwidth $2n+1$, through which may be derived $x_1 + \ldots x_{2n+1} \geq n+1$). We demonstrate that $(\frac{1}{2})^{2n+1}$ is in $P(C_{2n+1})^{(i)}_{(2n)}$ for all i.

We begin by proving that $\overline{x} := (\frac{1}{2})^{2n+1}$ is in $P(C_{2n+1})'_{2n}$. Consider the trivial 1-partition $E_1 := E(\overline{x})$. Let $J \subseteq [2n+1]$ s.t. $|J| = 2n$ be given, and let $l \in [2n+1]$ be the unique index not in J. Let $\overline{y}$ and $\overline{y}'$ be s.t. they each have 1 in their lth index, while one has a 0 precisely where the other has a 1 elsewhere. Further, from index $l+1$ until index $l-1 \bmod 2n+1$ one will alternate between 0 and 1, as the other alternates between 1 and 0. It is clear that both $\overline{y}$ and $\overline{y}'$ are in $P(C_{2n+1})$ and $\overline{x}_{|J}$ is the average of $\overline{y}_{|J}$ and $\overline{y}'_{|J}$. This settles the case $\overline{x} \in P(C_{2n+1})^{(1)}_{(2n)}$.

Now, $\overline{y}$ and $\overline{y}'$ are actually in the integer hull of $P(C_{2n+1})$, so we may iterate the game argument in a trivial way to show that $\overline{x} \in P(C_{2n+1})^{(i)}_{(2n)}$ for all i.

Remark 1. The system C_{2n+1} is in some way a low-width encoding of $x_1 + \ldots + x_{2n+1} \geq n + \frac{1}{2}$, which is one of its consequences. A similar argument yields that this single inequality requires cutwidth $> 2n$, but it seems preferable to prove results about the cutwidth of polytopes that may be specified by inequalities of low initial width. Now consider the system of inequalities D_{2n+1} obtained from C_{2n+1} by adding the inequalities, for $i \in [2n+1]$, $x_i + x_{i+1 \bmod 2n+1} \leq 1$. In some sense $P(D_{2n+1})$ is the natural variant of $P(C_{2n+1})$ with an empty integer hull. However, the cutwidth of $P(D_{2n+1})$ is easily seen to be 1: Sum the new inequalities of D_{2n+1} to produce $2x_1 + \ldots + 2x_{2n+1} \leq 2n + 1$. Now consider the inequalities of C_{2n+1} giving $-2x_1 - 2x_2 \leq -2$, $-2x_3 - 2x_4 \leq -2$, $\ldots$, $-2x_{2n-1} - 2x_{2n} \leq -2$. Summing gives $2x_{2n+1} \leq 1$, which cuts to $x_{2n+1} \leq 0$. A similar argument yields $x_i \leq 0$, for each i, which plainly contradicts C_{2n+1}.

We now give a cutwidth lower bound on a polytope with empty integer hull (our method closely follows the rank lower bound of [3]). Let $P(\mathrm{LNP}_n)$ denote the polytope in $[0,1]^{n(n-1)}$ specified by:

$$
\begin{aligned}
x_{i,j} + x_{j,i} &= 1 && \text{for } i,j \in [n],\ i \neq j \\
x_{i,j} + x_{j,k} - x_{i,k} &\leq 1 && \text{for } i,j,k \in [n],\ \text{pairwise distinct} \\
x_{1,j} + \ldots + x_{j-1,j} &+ x_{j+1,j} + \ldots + x_{n,j} \geq 1.
\end{aligned}
$$

We will see in the next section how this polytope is derived from a uniform sequence of propositional contradictions, expressing the first-order principle which asserts that every finite total order has a least element. We associate with a strict partial order $\prec$ on $[n]$ the vector $\overline{x}_\prec \in \{0, \frac{1}{2}, 1\}^{n(n-1)}$ given by $x_{i,j} := 0, \frac{1}{2}, 1$ when i is bigger than, incomparable to or smaller than j, respectively. A strict partial order $\prec$ is called s-scaled if there is a partition of $[n]$ into sets $A_1, \ldots, A_s$ s.t. $\prec$ is total on each of the A_is and undefined between elements in different A_is.

Lemma 3. *If $\prec$ is s-scaled with $s \geq 3$, then $\overline{x}_\prec \in P(\mathrm{LNP}_n)^{(i)}_{((n-2)/2)}$, for all i.*

Proof. By induction on i. When $i := 0$ this is plainly the case. Assume it is true for $i := m$. Let $J \subseteq [n(n-1)]$, $|J| = (n-2)/2$ be given. Let l, l' be two indices that do not appear in any of the subscripts i, j of the variables $x_{i,j}$ of J. Consider the partition of $E(\overline{x}_\prec)$ into $\binom{s}{2}$ partitions based on the possible comparisons of the sub-partial orders (e.g., the comparison of A_p and A_q would involve all variables $x_{i,j}$ s.t. either $i \in A_p$ and $j \in A_q$ or $j \in A_p$ and $i \in A_q$). Let one of these partitions be given, comparing the partial orders of A and A'. Let $\overline{y}$ and $\overline{y}'$ be s.t.

- l and l' are removed from any components of $\prec$, i.e. $y_{i,j}, y'_{i,j} := \frac{1}{2}$ if $i, j \in \{l, l'\}$.
- Excepting the previous line, in $\overline{y}$ all elements of A are put above A', and in $\overline{y}'$ vice-versa.

It may be verified that $\overline{y}$ and $\overline{y}'$ are each associated to strict partial orders that are s-scaled with $s \geq 3$, and that $\overline{x}_{|J}$ is the average of $\overline{y}_{|J}$ and $\overline{y}'_{|J}$. The result follows from the Protection Lemma.

Since the point $(\frac{1}{2})^{n(n-1)}$ is plainly not in the integer hull of $P(\mathrm{LNP}_n)$ (which is empty), but is an n-scaled order, we are able to derive the following.

Corollary 1. *The polytope $P(\mathrm{LNP}_n)$ has cutwidth $> (n-2)/2$.*

3   Resolution and Cutting Planes as a Refutation System

The refutation system *Resolution* acts on propositional formulae F in CNF. Initially, we may assume only the initial clauses of F as axioms. At any stage, if we have derived clauses of the form $(l_1 \vee \ldots \vee l_i \vee v)$ and $(l'_1 \vee \ldots \vee l'_{i'} \vee \neg v)$ (for v a variable and $l_1, \ldots, l_i, l'_1, \ldots, l'_{i'}$ literals of F) then we may deduce, via the *resolution* rule, the clause $(l_1 \vee \ldots \vee l_i \vee l'_1 \vee \ldots \vee l'_{i'})$. The aim is to derive the empty clause (boolean false). It is known that Resolution is both sound and complete, i.e. there is a legitimate derivation of the empty clause from F iff F is a contradiction. A Resolution refutation may be seen as a DAG from the initial clauses to the empty clause. The *width* (resp., *size*) of a Resolution refutation is the maximum size of a clause used in the derivation of the empty clause (resp., number of vertices in the associated DAG); the Resolution width (resp., size) of F is the minimal width (resp., size) of its Resolution refutations. A restriction of Resolution requires that the DAG be *tree-like*; tree-like Resolution is provably weaker than Resolution in general.

From a CNF formula $F := C_1 \wedge \ldots \wedge C_r$ in variables $v_1, \ldots, v_m$ we generate a system of inequalities L_F in $2m$ variables $\chi_{v_\lambda}, \chi_{\neg v_\lambda}$ ($\lambda \in [m]$). For literals $l_1, \ldots, l_t$ s.t. $(l_1 \vee \ldots \vee l_t)$ is a clause of F we have the constraining inequality

$$(1) \quad \chi_{l_1} + \ldots + \chi_{l_t} \geq 1.$$

We also have, for each $\lambda \in [m]$, the equalities of negation and bounding inequalities

$$(2) \quad \chi_{v_\lambda} + \chi_{\neg v_\lambda} = 1 \qquad (3) \quad 0 \leq \chi_{v_\lambda} \leq 1 \text{ and } 0 \leq \chi_{\neg v_\lambda} \leq 1.$$

It is clear that $P(F) := P(L_F)$ contains integral points iff F is satisfiable.

Example 1. The (negation of the) least number principle, LNP_n, may be given as the propositional CNF contradiction, on variables $v_{i,j}$, $i \neq j$, with clauses:

$$\begin{array}{ll}
\neg v_{i,j} \vee \neg v_{j,i} & \text{for } i, j \in [n], i \neq j \\
v_{i,j} \vee v_{j,i} & \text{for } i, j \in [n], i \neq j \\
\neg v_{i,j} \vee \neg v_{j,k} \vee v_{i,k} & \text{for } i, j, k \in [n], \text{ pairwise distinct} \\
v_{1,j} \vee \ldots \vee v_{j-1,j} \vee v_{j+1,j} \vee \ldots \vee v_{n,j}.
\end{array}$$

This translates, through our given scheme, to a polytope P in $[0, 1]^{2n(n-1)}$. By substituting variables of the form $\chi_{\neg v}$ by $1 - \chi_v$, one obtains the system given in the previous section, an equivalent polytope in $[0, 1]^{n(n-1)}$.

Through the translation $F \mapsto P(F)$, we see how CP may be considered as a refutation system. We define the rank and cutwidth of F to be the rank and cutwidth, respectively, of $P(F)$. While completeness follows from [9,4], we may equally argue by a standard simulation of Resolution. The resolution of clauses $(l_1 \vee \ldots \vee l_i \vee v)$ and $(l'_1 \vee \ldots \vee l'_{i'} \vee \neg v)$ is equivalent to the summing of the inequalities $\chi_{l_1} + \ldots + \chi_{l_i} + \chi_v \geq 1$ and $\chi_{l'_1} + \ldots + \chi_{l'_{i'}} + \chi_{\neg v} \geq 1$, which, via the equalities of negation, gives $\chi_{l_1} + \ldots + \chi_{l_i} + \chi_{l'_1} + \ldots + \chi_{l'_{i'}} \geq 1$. Now, some of the literals in $\{l_1, \ldots, l_i, l'_1, \ldots, l'_{i'}\}$ may be repeated (at most twice): let $l''_1, \ldots, l''_{i''}$ be the elements of the literal set without repetitions, i.e. $(l''_1 \vee \ldots \vee l''_{i''})$ is what the clauses resolve to; and this is what counts towards Resolution width. Our sum may be weakened, by adding instances $\chi_{l''_{i''}} \geq 0$, to $2\chi_{l''_1} + \ldots + 2\chi_{l''_{i''}} \geq 1$, which cuts to $\chi_{l''_1} + \ldots + \chi_{l''_{i''}} \geq 1$. It is plain to see from this argument that the CP cutwidth of F is bound above by the Resolution width of F. We will now demonstrate that it may be significantly smaller.

We consider a generalisation of the (negation of the) pigeonhole principle, as first suggested in [1]. Let $G := (U, V, E)$ be a bipartite graph s.t. $|U| = m$ and $|V| = n$ ($m > n$). For $j \in G$, let $N(j)$ be the set of neighbours of j in G. The contradiction $G\text{-PHP}_n^m$ will assert that G has a perfect matching; its clauses are

$$\neg v_{i,j} \vee \neg v_{i',j}, \text{ for } j \in V, i \neq i' \in N(j), \text{ and}$$
$$\bigvee_{j \in N(i)} v_{i,j}, \quad \text{for } i \in U.$$

For $G := K_{n+1,n}$, this is the usual pigeonhole principle. For a subset $U' \subseteq U$, define its boundary $\partial U' := \{v \in V : |N(v) \cap U'| = 1\}$. A $(m, n, d, \alpha n, \beta)$-boundary-expander is a bipartite graph $G := (U, V, E)$ with $|U| = m$ and $|V| = n$ in which the degree of the vertices in U is bound by d and, for all $U' \subseteq U$, s.t. $|U'| \leq \alpha n$, $|\partial U'| \geq \beta |U'|$. A $(m, n, d, \alpha n, \beta)$-boundary-expander is *special* if the degree bound of d also applies to the vertices in V. One major result of [1] is that the contradiction $G\text{-PHP}_n^{n+1}$ requires Resolution width $\geq (\alpha n \cdot \beta)/2$ (note that the width of the initial clauses is bound by d). Of course, central to their quest for lower bounds is that expanders of a certain kind exist. We will need the following lemma, from which we easily derive the succeeding corollary.

Lemma 4. *1.) If G is a special $(n + 1, n, d, \alpha n, \beta)$-boundary-expander, then the cutwidth of $P(G\text{-PHP}_n^{n+1})$ is $\leq d$, and 2.) There exists a special $(n + 1, n, 9, \frac{1}{972} n, \frac{9}{2})$-boundary-expander.*

Proof. (Part 1.) For each $j \in V$, and $i, i' \in N(j)$, with $i \neq i'$, we have the inequality $x_{i,j} + x_{i',j} \leq 1$. Recalling that $|N(j)| \leq d$, we claim that we can derive, in cutwidth $\leq d$, $\sum_{i \in N(j)} x_{i,j} \leq 1$. Since this is inconsistent with $\sum_{j \in N(i)} x_{i,j} \geq 1$, we are done.

Proof of claim. Let $N' \subseteq N(j)$. We prove by induction on $|N'|$ that $\sum_{i \in N'} x_{i,j} \leq 1$ may be obtained in cutwidth $\leq d$. For both $|N'| = 1$ and $|N'| = 2$, the result is trivial. Assume it is true for $|N'| = m$. Let i be arbitrary s.t. $i \in N(j)$ but $i \notin N'$. By Inductive Hypothesis, for each of the $m + 1$ subsets $N'' \subset N' \cup \{i\}$ with $|N''| = m$, we have $\sum_{i \in N''} x_{i,j} \leq 1$. Summing over all of these we obtain $m \sum_{i \in N' \cup \{i\}} x_{i,j} \leq m+1$, and the result follows from a cut of cutwidth $m+1 \leq d$. We remark that this method gives a rank of $O(d)$, yet a slightly more subtle divide-and-conquer approach could improve this to $O(\log(d))$.

(Part 2.) Omitted for reasons of space.

Corollary 2. *There is a bipartite G s.t. $G\text{-PHP}_n^{n+1}$ has CP cutwidth ≤ 9 but Resolution width $\geq \frac{1}{2 \cdot 9^{71}} n$.*

4   Sherali-Adams and a Gap for Cutting Planes

Sherali-Adams (SA) provides a static method for isolating the integer hull of a system of inequalities [12]. However, we will only be interested in it as a propositional refutation system, and we introduce it here as such (for more details see [6]). We may take a polytope $P := P(F)$, as defined from a CNF F by (1) – (3) of the previous section, and r-lift it to another polytope $P[r]$ in $\sum_{\lambda=0}^{r+1} \binom{2m}{\lambda}$ dimensions. Specifically, the variables involved in defining the polytope $P[r]$ are $\chi_{l_1 \wedge \ldots \wedge l_{r+1}}$ ($l_1, \ldots, l_{r+1}$ literals of F) and $\chi_\emptyset$. Note that we accept commutativity and idempotence of the $\wedge$-operator, e.g. $\chi_{l_1 \wedge l_2} = \chi_{l_2 \wedge l_1}$ and $\chi_{l_1 \wedge l_1} = \chi_{l_1}$. Here, $\emptyset$ represents the empty conjunct (boolean true); hence we set $\chi_\emptyset := 1$. For literals $l_1, \ldots, l_t$, s.t. $(l_1 \vee \ldots \vee l_t)$ is a clause of F, we have the constraining inequalities

$$(1')\quad \chi_{l_1 \wedge D} + \ldots + \chi_{l_t \wedge D} \geq \chi_D,$$

for D any conjunction of at most r literals of F. We also have, for each $\lambda \in [m]$ and D any conjunction of at most r literals, the equalities of negation together with the bounding inequalities

$$(2')\quad \chi_{v_\lambda \wedge D} + \chi_{\neg v_\lambda \wedge D} = \chi_D$$
$$(3')\quad 0 \leq \chi_{v_\lambda \wedge D} \leq \chi_D \quad \text{and} \quad 0 \leq \chi_{\neg v_\lambda \wedge D} \leq \chi_D.$$

The SA *rank* of the polytope $P = P(F)$ (formula F) is the minimal i such that $P[i]$ is empty.

Next we describe a translation of an FO sentence to a sequence of propositional CNF formulae. We use the language of FO logic with equality but with neither function nor constant symbols.[2] We assume that the FO sentence is given as a conjunction of FO sentences, each of which is in prenex normal form whose quantifier-free part is in CNF. The case of a purely universal sentence is easy – a sentence ψ of the form

$$\forall x_1, \ldots, x_k \; \Psi(x_1, \ldots, x_k)$$

where Ψ is quantifier-free, is translated into a sequence of propositional formulae in CNF $\langle F_{\psi,n} \rangle_{n \in \mathbb{N}}$, of which the n-th member $F_{\psi,n}$ is constructed as follows. For instantiations $x_1, \ldots, x_k \in [n]$, we can consider $\Phi(x_1, \ldots, x_k)$ to be a propositional formula over propositional variables of two different kinds: $R(x_{i_1}, \ldots, x_{i_p})$, where R is a p-ary predicate symbol, and $(x_i = x_j)$. The variables of the form $(x_i = x_j)$ evaluate to either true or false, thus we are left with variables of the form $R(x_{i_1}, \ldots, x_{i_p})$ only. The general case, a sentence ψ of the form

$$\forall x_1 \exists y_1 \forall x_2 \exists y_2 \ldots \forall x_k \exists y_k \; \Phi(x_1, \ldots, x_k, y_1, \ldots, y_k),$$

---

[2] We omit functions and constants only for the sake of a clearer exposition; note that we may simulate constants in a single FO sentence with added *outermost* existential quantification on new variables replacing those constants.

can be reduced to the previous case by Skolemization. We introduce *Skolem relations* (giving rise to *Skolem variables*) $S_i(x_1, \ldots, x_i, y_i)$ for $1 \leq i \leq k$. $S_i(x_1, \ldots, x_i, y_i)$ witnesses y_i for any given $x_1, \ldots, x_i$, so we need to add *Skolem clauses* asserting that such a witness always exists, i.e.,

$$\bigvee_{y_i=1}^{n} S_i(x_1, \ldots, x_i, y_i) \quad \text{for all } (x_1, \ldots, x_i) \in [n]^i .$$

The original sentence can be transformed into the following purely universal sentence

$$\forall x_1, \ldots x_k, y_1, \ldots y_k \bigvee_{i=1}^{k} \neg S_i(x_1, \ldots x_i, y_i) \vee \mathcal{F}(x_1, \ldots x_k, y_1, \ldots y_k).$$

By construction it is clear that, for FO sentences ψ, the CNF formula $F_{\psi,n}$ is satisfiable if and only if ψ has a model of size n. Thus satisfiability questions on the sequence $\langle F_{\psi,n} \rangle_{n \in \mathbb{N}}$ relate to questions on the existence of non-empty finite models for ψ. Note that the size of $F_{\psi,n}$ with respect to some reasonable encoding is polynomial in n.

A series of remarkable results born of the idea of generating CNF contradictions from FO principles have appeared, the so-called gap theorems. We will give two that are relevant to us – in both cases the separating criterion is the same (though this need not be so, for example there is a tetrachotomy for Nullstellensatz [11]).

Theorem 1. *Let ψ be an FO sentence with no finite models, and $F_{\psi,n}$ its translation to a propositional CNF as above. Then, either*

1. *$F_{\psi,n}$ has a tree-like Resolution refutation of size $\leq n^c$ and SA rank $\leq c$ (where the constant c depends only on ψ), or*
2. *All tree-like Resolution refutations of $F_{\psi,n}$ are of size $> 2^{\epsilon n}$ and $F_{\psi,n}$ has SA rank $> n^\epsilon$ (where $\epsilon > 0$ depends on ψ).*

Furthermore, Case 2 prevails precisely when ψ has some infinite model.

The gaps for tree-like Resolution and SA appear in [10] and [6], respectively. Our translation to CNFs does not make these results immediately applicable to width, as some of the initial clauses (the Skolem clauses) have large width. However, there is a standard mechanism for the introduction of 'extension' variables to mitigate this (see [1]). With these clauses brought down to constant size, a gap theorem for Resolution width essentially falls out of [10], exactly along the same lines: Let ψ be an FO sentence with no finite models, then $F_{\psi,n}$ has constant Resolution width if ψ has no infinite models, and at least linear Resolution width if ψ has some infinite model. It is not clear how a similar method might be employed for a gap for CP cutwidth, and a gap for CP rank is still unknown, despite the line of research in [6]. However, we can provide the following classification theorem for CP.[3]

---

[3] We resist calling it a gap theorem as the separation of constant and non-constant does not entail a gap.

Theorem 2. *Let ψ be an FO sentence with no finite models, and $F_{\psi,n}$ its translation to a propositional CNF. Then $F_{\psi,n}$ has CP rank+cutwidth bound by a constant c (which depends only on ψ) iff ψ has no infinite model.*

Proof. (Forwards.) Assume we have a CP DAG refutation for $F_{\psi,n}$ with rank+cutwidth bound by c. The crucial observation is that a single cut from $a_1\chi_{l_1} + \ldots + a_k\chi_{l_k} - \lambda \geq 0$ to $a_1\chi_{l_1} + \ldots + a_k\chi_{l_k} - \lceil\lambda\rceil \geq 0$ can be simulated by at most k lifts in SA (this follows from the general observation that one need no more lifts than there are variables to isolate the integer hull, see [12]). Therefore, we would have SA rank of $F_{\psi,n}$ of $\leq c^2$. By Case 1 of Theorem 1, we deduce that ψ has no infinite model.

(Backwards.) Omitted for reasons of space.

References

1. Ben-sasson, E., Wigderson, A.: Short proofs are narrow - resolution made simple. Journal of the ACM, 517–526 (1999)
2. Bockmayr, A., Eisenbrand, F., Hartmann, M., Schulz, A.S.: On the Chvátal rank of polytopes in the 0/1 cube. Discrete Appl. Math. 98(1-2), 21–27 (1999)
3. Buresh-Oppenheim, J., Galesi, N., Hoory, S., Magen, A., Pitassi, T.: Rank bounds and integrality gaps for cutting planes procedures. Theory of Computing 2(4), 65–90 (2006)
4. Chvátal, V.: Edmonds polytopes and a hierarchy of combinatorial problems. Discrete Math. 4, 305–337 (1973)
5. Cook, W., Coullard, C.R., Turán, G.: On the complexity of cutting-plane proofs. Discrete Appl. Math. 18(1), 25–38 (1987)
6. Dantchev, S.S.: Rank complexity gap for Lovász-Schrijver and Sherali-Adams proof systems. In: STOC 2007: Proceedings of the thirty-ninth annual ACM symposium on Theory of computing, pp. 311–317. ACM Press, New York (2007)
7. Eisenbrand, F., Schulz, S.: Bounds on the chvátal rank of polytopes in the 0/1-cube. Combinatorica 23(2), 245–261 (2003)
8. Gomory, R.E.: Solving linear programming problems in integers. In: Bellman, R., Hall, M. (eds.) Combinatorial Analysis, Proceedings of Symposia in Applied Mathematics, Providence, RI, vol. 10 (1960)
9. Gomory, R.E.: An algorithm for integer solutions to linear programs. In: Recent advances in mathematical programming, pp. 269–302. McGraw-Hill, New York (1963)
10. Riis, S.: A complexity gap for tree resolution. Computational Complexity 10(3), 179–209 (2001)
11. Riis, S.: On the asymptotic nullstellensatz and polynomial calculus proof complexity. In: Proceedings of the Twenty-Third Annual IEEE Symposium on Logic in Computer Science (LICS 2008), pp. 272–283. IEEE Computer Society Press, Los Alamitos (2008)
12. Sherali, H.D., Adams, W.P.: A hierarchy of relaxations between the continuous and convex hull representations for zero-one programming problems. SIAM J. Discrete Math. 3(3), 411–430 (1990)

Members of Random Closed Sets

David Diamondstone[1] and Bjørn Kjos-Hanssen[2]

[1] Department of Mathematics, University of Chicago, Chicago IL 60615
ded@math.uchicago.edu
http://www.math.uchicago.edu/~ded
[2] Department of Mathematics, University of Hawai'i at Mānoa, Honolulu HI 96822
bjoern@math.hawaii.edu
http://www.math.hawaii.edu/~bjoern

Abstract. The members of Martin-Löf random closed sets under a distribution studied by Barmpalias et al. are exactly the infinite paths through Martin-Löf random Galton-Watson trees with survival parameter $\frac{2}{3}$. To be such a member, a sufficient condition is to have effective Hausdorff dimension strictly greater than $\gamma = \log_2 \frac{3}{2}$, and a necessary condition is to have effective Hausdorff dimension greater than or equal to γ.

Keywords: random closed sets, computability theory.

1 Introduction

Classical probability theory studies intersection probabilities for random sets. A random set will intersect a given deterministic set if the given set is large, in some sense. Here we study a computable analogue: the question of which real numbers are "large" in the sense that they belong to some Martin-Löf random closed set.

Barmpalias et al. [2] introduced algorithmic randomness for closed sets. Subsequently Kjos-Hanssen [6] used algorithmically random Galton-Watson trees to obtain results on infinite subsets of random sets of integers. Here we show that the distributions studied by Barmpalias et al. and by Galton and Watson are actually equivalent, not just classically but in an effective sense.

For $0 \leq \gamma < 1$, let us say that a real x is a MEMBER$_\gamma$ if x belongs to some Martin-Löf (ML-) random closed set according to the Galton-Watson distribution (defined below) with survival parameter $p = 2^{-\gamma}$. We show that for $p = \frac{2}{3}$, this is equivalent to x being a member of a Martin-Löf random closed set according to the distribution considered by Barmpalias et al.

In light of this equivalence, we may state that (i) Barmpalias et al. showed that in effect not every MEMBER$_\gamma$ is ML-random, and (ii) Joe Miller and Antonio Montálban showed that every ML-random real is a MEMBER$_\gamma$; the proof of their result is given in the paper of Barmpalias et al. [2] The way to sharpen these results go via *effective Hausdorff dimension*. Each ML-random real has effective Hausdorff dimension equal to one. In Section 3 we show that (i') a MEMBER$_\gamma$

K. Ambos-Spies, B. Löwe, and W. Merkle (Eds.): CiE 2009, LNCS 5635, pp. 144–153, 2009.

may have effective Hausdorff dimension strictly less than one, and (ii′) every real of sufficiently large effective Hausdorff dimension (where some numbers strictly less than one are "sufficiently large") is a MEMBER$_\gamma$.

2   Equivalence of Two Models

We write $\Omega = 2^{<\omega}$, and 2^ω, for the sets of finite and infinite strings over $2 = \{0, 1\}$, respectively. If $\sigma \in \Omega$ is an initial substring (a prefix) of $\tau \in \Omega$ we write $\sigma \preceq \tau$; similarly $\sigma \prec x$ means that the finite string σ is a prefix of the infinite string $x \in 2^\omega$. The length of σ is $|\sigma|$. We use the standard notation $[\sigma] = \{x : \sigma \prec x\}$, and for a set $U \subseteq \Omega$, $[U]^{\preceq} := \bigcup_{\sigma \in U}[\sigma]$. Let $\mathcal{P}$ denote the power set operation. Following Kjos-Hanssen [6], for a real number $0 \leq \gamma < 1$ (so $\frac{1}{2} < 2^{-\gamma} \leq 1$), let $\lambda_{1,\gamma}$ be the distribution with sample space $\mathcal{P}(\Omega)$ such that each string in Ω has probability $2^{-\gamma}$ of belonging to the random set, independently of any other string. Let λ_γ^* be defined analogously, except that now

$$\lambda_\gamma^*(\{S : S \cap \{\sigma 0, \sigma 1\} = J\}) = \begin{cases} 1 - p & \text{if } J = \{\sigma 0\} \text{ or } J = \{\sigma 1\}, \text{ and} \\ 2p - 1 & \text{if } J = \{\sigma 0, \sigma 1\}, \end{cases}$$

independently for distinct σ, for $p = 2^{-\gamma}$.[1] For $S \subseteq \Omega$, Γ_S, the closed set determined by S, is the (possibly empty) set of infinite paths through the part of S that is downward closed under prefixes:

$$\Gamma_S = \{x \in 2^\omega : (\forall \sigma \prec x)\, \sigma \in S\}.$$

The *Galton-Watson (GW) distribution for survival parameter* $2^{-\gamma}$, also known as the $(1, \gamma)$-*induced distribution* [6], and as the distribution of a *percolation limit set* [12], is a distribution $\mathbb{P}_{1,\gamma}$ on the set of all closed subsets of 2^ω defined by   .

$$\mathbb{P}_{1,\gamma}(E) = \lambda_{1,\gamma}\{S : \Gamma_S \in E\}.$$

Thus, the probability of a property E of a closed subset of 2^ω is the probability according to $\lambda_{1,\gamma}$ that a random subset of Ω determines a tree whose set of infinite paths has property E. Similarly, let

$$\mathbb{P}_\gamma^*(E) = \lambda_\gamma^*\{S : \Gamma_S \in E\}.$$

A Σ_1^0 *subset of* $\mathcal{P}(\Omega)$ is the image of a Σ_1^0 subset of $\mathcal{P}(\omega) = 2^\omega$ via an effective isomorphism between Ω and ω.

$S \in \mathcal{P}(\Omega)$ is called $\lambda_{1,\gamma}$-*ML-random* if for each uniformly Σ_1^0 sequence $\{U_n\}_{n \in \omega}$ of subsets of $\mathcal{P}(\Omega)$ with $\lambda_{1,\gamma}(U_n) \leq 2^{-n}$, we have $S \notin \bigcap_n U_n$. In this case Γ_S is called $\mathbb{P}_{1,\gamma}$-ML-random. Similarly, $S \in \mathcal{P}(\Omega)$ is called λ_γ^*-*ML-random* if for each uniformly Σ_1^0 sequence $\{U_n\}_{n \in \omega}$ of subsets of $\mathcal{P}(\Omega)$ with $\lambda_\gamma^*(U_n) \leq 2^{-n}$, we have $S \notin \bigcap_n U_n$. In this case Γ_S is called $\mathbb{P}_\gamma^*$-ML-random.

---

[1] The notation $\lambda_{1,\gamma}$ is consistent with earlier usage [6] and is also easy to distinguish visually from λ_γ^*.

Lemma 1 (Axon [1]). *For $2^{-\gamma} = \frac{2}{3}$, $\Gamma \subseteq 2^{\omega}$ is $\mathbb{P}_{\gamma}^{*}$-ML-random if and only if Γ is a Martin-Löf random closed set under the distribution studied by Barmpalias et al.*

Thinking of S as a random variable, define further random variables

$$G_n = \{\sigma : |\sigma| = n \ \& \ (\forall \tau \preceq \sigma) \ \tau \in S\}$$

and $G = \bigcup_{n=0}^{\infty} G_n$. We refer to a value of G as a *GW-tree* when G is considered a value of the random variable under the $\lambda_{1,\gamma}$ distribution. (A *BBCDW-tree* is a particular value of the random variable analogous to G, for the distribution λ_{γ}^{*}.) We have $G \subseteq S$ and $\Gamma_G = \Gamma_S$. The set G may have "dead ends", so let

$$G_{\infty} = \{\sigma \in G : G \cap [\sigma] \text{ is infinite}\}.$$

Thus $G_{\infty} \subseteq G \subseteq S$, and values of G_{∞} are in one-to-one correspondence with values of Γ_S.

Let e be the extinction probability of a GW-tree with parameter $p = 2^{-\gamma}$,

$$e = \mathbb{P}_{1,\gamma}(\varnothing) = \lambda_{1,\gamma}(\{S : \Gamma_S = \varnothing\}).$$

For any number a let $\bar{a} = 1 - a$.

Lemma 2
$$e = \bar{p}/p.$$

Proof. Notice that we are not assuming $\langle\rangle \in S$. We have $e = \bar{p} + pe^2$, because there are two ways extinction can happen: (1) $\langle\rangle \notin S$, and (2) $\langle\rangle \in S$ but both immediate extension trees go extinct.

We use standard notation for conditional probability,

●
$$\mathbb{P}(E \mid F) = \frac{\mathbb{P}(E \cap F)}{\mathbb{P}(F)};$$

in measure notation we may also write $\lambda(E \mid F) = \lambda(E \cap F)/\lambda(F)$.

Lemma 3. *For all $J \subseteq \{\langle 0 \rangle, \langle 1 \rangle\}$,*

$$\lambda_{1,\gamma} \{G_{\infty} \cap \{\langle 0 \rangle, \langle 1 \rangle\} = J \mid G_{\infty} \neq \varnothing\} = \lambda_{\gamma}^{*}[G_1 = J].$$

Proof. By definition, $\lambda_{\gamma}^{*}[G_1 = J]$ equals

$$(2p - 1) \cdot \mathbf{1}_{J=\{\langle 0 \rangle, \langle 1 \rangle\}} + \sum_{i=0}^{1} (1 - p) \cdot \mathbf{1}_{J=\{\langle i \rangle\}},$$

so we only need to calculate $\lambda_{1,\gamma} \{G_{\infty} \cap \{\langle 0 \rangle, \langle 1 \rangle\} = J \mid G_{\infty} \neq \varnothing\}$. By symmetry, and because the probability that $G_1 = \varnothing$ is 0, it suffices to calculate this probability for $J = \{\langle 0 \rangle, \langle 1 \rangle\}$. Now if $G_1 = \{\langle 0 \rangle, \langle 1 \rangle\}$ then $\langle\rangle$ survives and both immediate extensions are non-extinct. Thus the conditional probability that $G_1 = \{\langle 0 \rangle, \langle 1 \rangle\}$ is $\frac{p(1-e)^2}{1-e} = p(1 - e)$. By Lemma 2, this is equal to $2p - 1$.

Lemma 4. *Let the number* $\mathfrak{p}_s$ *be defined by*

$$\mathfrak{p}_s = \lambda_{1,\gamma}(\langle j \rangle \in G \mid (G \cap (\langle j \rangle^\frown \Omega))_\infty = \varnothing \ \& \ \langle\rangle \in G)$$

for $j = 0$ *(or* $j = 1$, *which gives the same result). Let*

$$\lambda_f(\cdot) = \lambda_{1,\gamma}(\cdot \mid G_\infty = \varnothing \ \& \ \langle\rangle \in G).$$

Then $\lambda_f(\langle i \rangle \in G_1) = \mathfrak{p}_s$.

Proof. We have $\mathfrak{p}_s = p^2 e^2/(pe) = pe = 1 - p$. Next, $\lambda_f[G_1 = \varnothing] = \frac{p(1-p)^2}{pe^2} = p^2$ and $\lambda_f[G_1 = \{\langle 0 \rangle, \langle 1 \rangle\}] = \frac{p^3 e^4}{pe^2} = (\overline{p})^2$. Hence

$$\lambda_f[G_1 = J] = (1-p)^2 \cdot \mathbf{1}_{J=\{\langle 0 \rangle, \langle 1 \rangle\}} + \sum_{i=0}^{1} p(1-p) \cdot \mathbf{1}_{J=\{\langle i \rangle\}} + p^2 \cdot \mathbf{1}_{J=\varnothing},$$

and so $\lambda_f(\langle i \rangle \in G_1) = (1-p)^2 + p(1-p) = \mathfrak{p}_s$, as desired.

Let $\lambda_c = \lambda_{1,\gamma}(\cdot \mid G_\infty \neq \varnothing)$ be $\lambda_{1,\gamma}$ conditioned on $G_\infty \neq \varnothing$, and let λ_i be the distribution of $G_\infty \in \mathcal{P}(\Omega)$ conditional on $G_\infty \neq \varnothing$. Let μ_i, μ_f, μ_c be the distribution of the tree G corresponding to the set S under λ_i, λ_f, λ_c, respectively (so $\mu_i = \lambda_i$). We define a $\mu_i \times \mu_f \to \mu_c$ measure-preserving map $\psi : 2^\Omega \times 2^\Omega \to 2^\Omega$. The idea is to overlay two sets S_i, S_f, so that S_i specifies G_∞, and S_f specifies $G \backslash G_\infty$. Let $\psi(S_i, S_f) = G_i \cup S_f$ where G_i is the tree determined by S_i. By Lemma 4, this gives the correct probability for string $\sigma \notin G_\infty$ that is the neighbor of a string in G_∞ to be in G. By considering two cases (a string in G is in G_∞ or not) we can derive that ψ is measure-preserving.

Intuitively, a λ_i-ML-random tree may by van Lambalgen's theorem be extended to a λ_c-ML-random tree by "adding finite pieces randomly". To be precise, van Lambalgen's theorem holds in the unit interval $[0, 1]$ with Lebesgue measure λ, or equivalently the space 2^ω. If (X, μ) is a measure space then using the measure-preserving map $\varphi : (X, \mu) \to ([0, 1], \lambda)$ induced from the Carathéodory measure algebra isomorphism theorem [7], we may apply van Lambalgen as desired, and obtain

Theorem 1. *For each ML-random BBCDW-tree* H *there is a ML-random GW-tree* G *with* $G_\infty = H_\infty$.

We next prove that the live part of every infinite ML-random GW-tree is an ML-random BBCDW-tree.

Theorem 2. *For each* S, *if* S *is* $\lambda_{1,\gamma}$-ML-random *then* G_∞ *is* λ_γ^*-random.

Proof. Suppose $\{U_n\}_{n\in\omega}$ is a λ_γ^*-ML-test with $G_\infty \in \bigcap_n U_n$. Let $\Upsilon_n = \{S : G_\infty \in U_n\}$. By Lemma 3, $\lambda_{1,\gamma}(\Upsilon_n) = \lambda_\gamma^*(U_n)$. Unfortunately, Υ_n is not a Σ_1^0 class, but we can approximate it. While we cannot know if a tree will end up being infinite, we can make a guess that will usually be correct.

Let e be the probability of extinction for a GW-tree. By Lemma 2 we have $e = \frac{\bar{p}}{p}$, so since $p > 1/2$, $e < 1$. Thus there is a computable function $(n, \ell) \mapsto m_{n,\ell}$ such that for all n and ℓ, $m = m_{n,\ell}$ is so large that $e^m \leq 2^{-n}2^{-2\ell}$. Let Φ be a Turing reduction so that $\Phi^G(n, \ell)$, if defined, is the least L such that all the 2^ℓ strings of length ℓ either are not on G, or have no descendants on G at level L, or have at least $m_{n,\ell}$ many such descendants. Let

$$W_n = \{S : \text{ for some } \ell, \Phi^G(n, \ell) \text{ is undefined}\}.$$

Let $A_G(\ell) = G_\infty \cap \{0,1\}^{\leq \ell}$ be G_∞ up to level ℓ. Let the approximation $A_G(\ell, L)$ to $A_G(\ell)$ consist of the nodes of G at level ℓ that have descendants at level L. Let

$$V_n = \{S : A_G(\ell, L) \in U_n \text{ for some } \ell, \text{ where } L = \Phi^G(n, \ell)\}, \text{ and}$$

$$X_n = \{S : \text{ for some } \ell, L = \Phi^G(n, \ell) \text{ is defined and } A_G(\ell, L) \neq A_G(\ell)\}.$$

Note that $\Upsilon_n = \{S : \text{ for some } \ell, A_G(\ell) \in U_n \}$, hence $\Upsilon_n \subseteq W_n \cup X_n \cup V_n$. Thus it suffices to show that $\bigcap_n V_n$, W_n, $\bigcap_n X_n$ are all $\lambda_{1,\gamma}$-ML-null sets.

Lemma 5. $\lambda_{1,\gamma}(W_n) = 0$.

Proof. If $\Phi(\ell)$ is undefined then there is no L, which means that for the fixed set of strings on G at level ℓ, they do not all either die out or reach m many extensions. But eventually this must happen, so L must exist.

Indeed, fix any string σ on G at level ℓ. Let k be the largest number of descendants that σ has at infinitely many levels $L > \ell$. If $k > 0$ then with probability 1, above each level there is another level where actually $k+1$ many descendants are achieved. So we conclude that either $k = 0$ or k does not exist.

From basic computability theory, W_n is a Σ_2^0 class. Hence each W_n is a Martin-Löf null set.

Lemma 6. $\lambda_{1,\gamma}(X_n) \leq 2^{-n}$.

Proof. Let E_σ denote the event that all extensions of σ on level L are *dead*, i.e. not in G_∞. Let F_σ denote the event that σ has at least m many descendants on $G(L)$.

If $A_G(\ell, L) \neq A_G(\ell)$ then some $\sigma \in \{0,1\}^\ell \cap G$ has at least m many descendants at level L, all of which are dead. If a node σ has at least m descendants, then the chance that all of these are dead, given that they are on G at level L, is at most e^m (the eventual extinction of one is independent of that of another), hence writing $\mathbb{P} = \lambda_{1,\gamma}$, we have

$$\mathbb{P}(A_G(\ell, L) \neq A_G(\ell)) \leq \sum_{\sigma \in \{0,1\}^\ell} \mathbb{P}\{E_\sigma \ \& \ F_\sigma\} = \sum_{\sigma \in \{0,1\}^\ell} \mathbb{P}\{E_\sigma \mid F_\sigma\} \cdot \mathbb{P}\{F_\sigma\}$$

$$\leq \sum_{\sigma \in \{0,1\}^\ell} \mathbb{P}\{E_\sigma \mid F_\sigma\} \leq \sum_{\sigma \in \{0,1\}^\ell} e^m \leq \sum_{\sigma \in \{0,1\}^\ell} 2^{-n} 2^{-2\ell} = 2^{-n} 2^{-\ell}.$$

and hence

$$\mathbb{P} X_n \leq \sum_\ell \mathbb{P}\{A_G(\ell, L) \neq A_G(\ell)\} \leq \sum_\ell 2^{-n} 2^{-\ell} = 2^{-n}.$$

X_n is Σ_1^0 since when L is defined, $A_G(\ell)$ is contained in $A_G(\ell, L)$, and $A_G(\ell)$ is Π_1^0 in G, which means that if the containment is proper then we can eventually enumerate (observe) this fact. Thus $\cap_n X_n$ is a $\lambda_{1,\gamma}$-ML-null set.

V_n is clearly Σ_1^0. Moreover $V_n \subseteq \Upsilon_n \cup X_n$, so $\lambda_{1,\gamma}(V_n) \leq 2 \cdot 2^{-n}$, hence $\cap_n V_n$ is a $\lambda_{1,\gamma}$-ML-null set.

3 Towards a Characterization of Members of Random Closed Sets

For a real number $0 \leq \gamma \leq 1$, the γ-weight $\mathrm{wt}_\gamma(C)$ of a set of strings $C \subseteq \Omega$ is defined by

$$\mathrm{wt}_\gamma(C) = \sum_{w \in C} 2^{-|w|\gamma}.$$

We define several notions of randomness of individual reals. A *Martin-Löf (ML-)γ-test* is a uniformly Σ_1^0 sequence $(U_n)_{n < \omega}$, $U_n \subseteq \Omega$, such that for all n, $\mathrm{wt}_\gamma(U_n) \leq 2^{-n}$. A *strong ML-$\gamma$-test* is a uniformly Σ_1^0 sequence $(U_n)_{n < \omega}$ such that for each n and each prefix-free set of strings $V_n \subseteq U_n$, $\mathrm{wt}_\gamma(V_n) \leq 2^{-n}$. A real is (strongly) γ-random if it does not belong to $\cap_n [U_n]^{\preceq}$ for any (strong) ML-γ-test $(U_n)_{n < \omega}$. If $\gamma = 1$ we simply say that the real, or the set of integers $\{n : x(n) = 1\}$, is *Martin-Löf random (ML-random)*. For $\gamma = 1$, strength makes no difference. For a measure μ and a real x, we say that x is *Hippocrates μ-random* if for each sequence $(U_n)_{n < \omega}$ that is uniformly Σ_1^0, and where $\mu[U_n]^{\preceq} \leq 2^{-n}$ for all n, we have $x \notin \cap_n [U_n]^{\preceq}$. Let the ultrametric v on 2^ω be defined by $v(x, y) = 2^{-\min\{n : x(n) \neq y(n)\}}$. The γ-energy [12] of a measure μ is

$$I_\gamma(\mu) := \iint \frac{d\mu(b) d\mu(a)}{v(a, b)^\gamma}.$$

x is *Hippocrates γ-energy random* if x is Hippocrates μ-random with respect to some probability measure μ such that $I_\gamma(\mu) < \infty$.

For background on γ-energy and related concepts the reader may consult the monographs of Falconer [3] and Mattila [11] or the on-line lecture notes of Mörters and Peres [12]. The terminology *Hippocrates random* is supposed to remind us of Hippocrates, who did not consult the oracle at Delphi, but instead looked for natural causes. An almost sure property is more effective if it is possessed by all Hippocrates μ-random reals rather than merely all μ-random reals. In this sense Hippocratic μ-randomness tests are more desirable than arbitrary μ-randomness tests.

Effective Hausdorff dimension was introduced by Lutz [8] and is a notion of partial randomness. For example, if the sequence $x_0 x_1 x_2 \cdots$ is ML-random, then the sequence $x_0 0 x_1 0 x_2 0 \cdots$ has effective Hausdorff dimension equal to $\frac{1}{2}$. Let $\dim_H^1 x$ denote the effective (or constructive) Hausdorff dimension of x; then we have $\dim_H^1(x) = \sup\{\gamma : x \text{ is } \gamma\text{-random}\}$ (Reimann and Stephan [14]).

Examples of measures of finite γ-energy may be obtained from the fact that if $\dim_H^1(x) > \gamma$ then x is Hippocrates γ-energy random [6]. If x is strongly γ-random then x is γ-random and so $\dim_H^1(x) \geq \gamma$.

Theorem 3 ([6]). *Each Hippocrates γ-energy random real is a* MEMBER$_\gamma$.

Here we show a partial converse:

Theorem 4. *Each* MEMBER$_\gamma$ *is strongly γ-random.*

Proof. Let $\mathbb{P} = \lambda_{1,\gamma}$ and $p = 2^{-\gamma} \in (\frac{1}{2}, 1]$. Let $i < 2$ and $\sigma \in \Omega$. The probability that the concatenation $\sigma i \in G$ given that $\sigma \in G$ is by definition

$$\mathbb{P}\{\sigma i \in G \mid \sigma \in G\} = p.$$

Hence the absolute probability that σ survives is

$$\mathbb{P}\{\sigma \in G\} = p^{|\sigma|} = \left(2^{-\gamma}\right)^{|\sigma|} = \left(2^{-|\sigma|}\right)^\gamma.$$

Let U be any strong γ-test, i.e. a uniformly Σ_1^0 sequence $U_n = \{\sigma_{n,i} : i < \omega\}$, such that for all prefix-free subsets $U_n' = \{\sigma_{n,i}' : i < \omega\}$ of U_n, $\mathrm{wt}_\gamma(U_n') \leq 2^{-n}$. Let U_n' be the set of all strings σ in U_n such that no prefix of σ is in U_n. Clearly, U_n' is prefix-free. Let

$$[V_n]^{\preceq} := \{S : \exists i \, \sigma_{n,i} \in G\} \subseteq \{S : \exists i \, \sigma_{n,i}' \in G\}.$$

Clearly $[V_n]^{\preceq}$ is uniformly Σ_1^0. To prove the inclusion: Suppose G contains some $\sigma_{n,i}$. Since G is a tree, it contains the shortest prefix of $\sigma_{n,i}$ that is in U_n, and this string is in U_n'. Now

$$\mathbb{P}[V_n]^{\preceq} \leq \sum_{i\in\omega} \mathbb{P}\{\sigma_{n,i}' \in G\} = \sum_{i\in\omega} 2^{-|\sigma_{n,i}'|\gamma} \leq 2^{-n}.$$

Thus V is a test for $\lambda_{1,\gamma}$-ML-randomness. Suppose x is a MEMBER$_\gamma$. Let S be any $\lambda_{1,\gamma}$-ML-random set with $x \in \Gamma_S$. Then $S \notin \cap_n [V_n]^{\preceq}$, and so for some n, $\Gamma \cap [U_n]^{\preceq} = \varnothing$. Hence $x \notin [U_n]^{\preceq}$. As U was an arbitrary strong γ-test, this shows that x is strongly γ-random.

Corollary 1. *Let $x \in 2^\omega$. We have the implications*

$$dim_H^1(x) > \gamma \Rightarrow x \text{ is a } \text{MEMBER}_\gamma \Rightarrow dim_H^1(x) \geq \gamma.$$

Proof. Each real x with $dim_H^1(x) > \gamma$ is β-capacitable for some $\beta > \gamma$ [13]. This implies that x is γ-energy random [6, Lemma 2.5] and in particular x is Hippocrates γ-energy random. This gives the first implication. For the second implication, we use the fact that each strongly γ-random real x satisfies $dim_H^1(x) \geq \gamma$ (see e.g. Reimann and Stephan [14]).

The second implication of Corollary 1 does not reverse, as not every real with $dim_H^1(x) \geq \gamma$ is strongly γ-random [14].

The first implication of Corollary 1 fails to reverse as well:

Proposition 1. *Let $0 < \gamma < 1$. There is a γ-energy random real of effective Hausdorff dimension exactly γ.*

Proof. Consider the probability measure μ on 2^ω such that $\mu([\sigma^\frown 0]) = \mu([\sigma^\frown 1])$ for all σ of even length, and such that $\mu([\sigma^\frown 0]) = \mu([\sigma])$ for each σ of odd length $f(k) = 2k + 1$. A computation shows that $I_\gamma(\mu) = \sum_{k=0}^\infty 2^{(2k+1)\gamma} 2^{-k}$ which is finite if and only if $\gamma < 1/2$. We find that μ-almost all reals are μ-random and have effective Hausdorff dimension exactly $1/2$. By modifying $f(k)$ slightly we can get $I_\gamma(\mu) < \infty$ for $\gamma = 1/2$ while keeping the effective Hausdorff dimension of μ-almost all reals equal to $1/2$. Namely, what is needed is that $\sum_{k=0}^\infty 2^{f(k)\gamma} 2^{-k} < \infty$. This holds if $\gamma = 1/2$ and $f(k) = 2k - 2(1 - \varepsilon) \log k$ for any $\varepsilon > 0$ since $\sum_k k^{-(1+\varepsilon)} < \infty$. Since this $f(k)$ is asymptotically larger than $(2-\delta)k$ for any $\delta > 0$, the μ-random reals still have effective Hausdorff dimension $1/2$. The example generalizes from $\gamma = 1/2$ to an arbitrary $0 < \gamma < 1$.

Writing implication known to be strict as $\Rightarrow$ and other implication as $\rightarrow$, we have

$dim_H^1(x) > \gamma \Rightarrow x$ is γ-energy random $\rightarrow x$ is Hippocrates γ-energy random $\rightarrow x$ is a $\text{MEMBER}_\gamma \rightarrow x$ is strongly γ-random $\Rightarrow x$ is γ-random $\Rightarrow \dim(x) \geq \gamma$.

By Reimann's effective capacitability theorem [13] x is strongly γ-random if and only if x is γ-capacitable.

Conjecture 1. There is a strongly γ-random real which is not Hippocrates γ-energy random.

Conjecture 2. A real x is a MEMBER_γ if and only if x is Hippocrates γ-energy random.

To prove Conjecture 2, one might try to consult Lyons [10].

Proposition 2. *Let $0 < \gamma < 1$. If x is a real such that the function $n \mapsto x(n)$ is f-computably enumerable for some computable function f for which $\sum_{j<n} f(i)2^{-n\gamma}$ goes effectively to zero, then x is not γ-random.*

Proof. Suppose $n \mapsto x(n)$ is f-c.e. for some such f, and let $F(n) = \sum_{j<n} f(n)$. Let α be any computable function such that $\alpha(n, i) \neq \alpha(n, i + 1)$ for at most $f(n)$ many i for each n, and $\lim_{i\to\infty} \alpha(i, n) = x(n)$. Let $c(n, j)$ be the jth such i that is discovered for any $k < n$; so c is a partial recursive function whose domain is contained in $\{(n, j) : j \leq F(n)\}$. For a fixed i, α defines a real α_i by $\alpha_i(n) = \alpha(i, n)$. Let $V_n = \{x : \exists j \leq F(n)\ x \restriction n = \alpha_{c(n,j)} \restriction n)\}$. Since V_n is the union of at most $F(n)$ many cones $[x \restriction n]$,

$$\mathrm{wt}_\gamma(V_n) \leq \sum_{j=1}^{F(n)} 2^{-n\gamma} = F(n)2^{-n\gamma}$$

which goes effectively to zero by assumption. Thus there is a computable sequence $\{n_k\}_{k\in\mathbb{N}}$ such that $\mathrm{wt}_\gamma(V_{n_k}) \leq 2^{-k}$. Let $U_k = V_{n_k}$. Then U_k is Σ_1^0 uniformly in k, and $x \in \cap_k U_k$. Hence x is not γ-random.

Corollary 2 ([2]). *No member of a ML-random closed set under the BBCDW distribution is f-c.e. for any polynomial-bounded f.*

Proof. If f is polynomially bounded then clearly $\sum_{j<n} f(i)2^{-n\gamma}$ goes effectively to zero. Therefore if x is f-c.e., x is not γ-random, hence not a MEMBER$_\gamma$ for any $0 < \gamma < 1$, and thus not a member of a ML-random closed set under the BBCDW distribution.

Computing Brownian slow points. A function $f : \omega \to \omega$ is *diagonally nonrecursive* (DNR) if for each n, $f(n)$ is not equal to $\varphi_n(n)$, the value (if any) of the nth partial recursive function on input n. A real A is *Kurtz random relative to an oracle B* if it does not belong to any $\Pi_1^0(B)$ subset of 2^ω of fair-coin measure zero. Furthermore, B is Low(ML, Kurtz) if each real A that is ML-random is Kurtz random relative to B.

A starting point for the present paper was the observation (*) that each non-DNR Turing degree is Low(ML, Kurtz). A proof of this result due and credited to Kjos-Hanssen is given by Greenberg and Miller [4]; they prove that the converse holds as well. This can be used to show that each *slow point* (see Mörters and Peres [12]) of any ML-random Brownian motion must be of DNR Turing degree. The *fast* points on the other hand form a dense G_δ set, so there are fast points that are 1-generic and hence do not Turing compute any slow points.

In any case, the idea was initially to use the result (*) to understand members of random closed sets. However, as it turned out one could use the work of Hawkes [5] and Lyons [9] to better effect, in the present paper and in the precursor [6].

Acknowledgments

The authors thank the Institute of Mathematical Science at Nanjing University (and in particular Liang Yu), where the research leading to Section 2 was carried out in May 2008, for their hospitality. Section 3 contains some earlier results of the second author, who was partially supported by NSF grant DMS-0652669.

References

[1] Axon, L.: Random closed sets and probability, doctoral dissertation, University of Notre Dame (2009)

[2] Barmpalias, G., Brodhead, P., Cenzer, D., Dashti, A.S., Weber, R.: Algorithmic randomness of closed sets. J. Logic Comput. 17(6), 1041–1062 (2007)

[3] Falconer, K.: Fractal geometry. John Wiley & Sons Ltd., Chichester (1990); Mathematical foundations and applications. MR 1102677 (92j:28008)

[4] Greenberg, N., Miller, J.S.: Lowness for Kurtz randomness. Journal of Symbolic Logic 74(2), 665–678 (2009)

[5] Hawkes, J.: Trees generated by a simple branching process. J. London Math. Soc. 24(2), 373–384 (1981) MR 631950 (83b:60072)

[6] Kjos-Hanssen, B.: Infinite subsets of random sets of integers. Mathematical Research Letters 16, 103–110 (2009)

[7] Kjos-Hanssen, B., Nerode, A.: Effective dimension of points visited by Brownian motion. Theoretical Computer Science 410(4–5), 347–354 (2009)

[8] Lutz, J.H.: Gales and the constructive dimension of individual sequences. In: Welzl, E., Montanari, U., Rolim, J.D.P. (eds.) ICALP 2000. LNCS, vol. 1853, pp. 902–913. Springer, Heidelberg (2000)

[9] Lyons, R.: Random walks and percolation on trees. Annals of Probability 18(3), 931–958 (1990) MR 1062053 (91i:60179)

[10] Lyons, R.: Random walks, capacity and percolation on trees. Annals of Probability 20(4), 2043–2088 (1992) MR 1188053 (93k:60175)

[11] Mattila, P.: Geometry of sets and measures in Euclidean spaces. Cambridge Studies in Advanced Mathematics, vol. 44. Cambridge University Press, Cambridge (1995); Fractals and rectifiability. MR 1333890 (96h:28006)

[12] Mörters, P., Peres, Y., Brownian Motion, http://www.stat.berkeley.edu/~peres/

[13] Reimann, J.: Effectively closed classes of measures and randomness. Annals of Pure and Applied logic 156(1), 170–182 (2008)

[14] Reimann, J., Stephan, F.: Effective Hausdorff dimension. In: Logic Colloquium 2001, pp. 369–385 (2005)

Lowness for Demuth Randomness

Rod Downey and Keng Meng Ng[*]

Victoria University of Wellington,
School of Mathematics, Statistics and Computer Science,
PO Box 600, Wellington, New Zealand

Abstract. We show that every real low for Demuth randomness is of
hyperimmune-free degree.

1 Introduction

A fundamental theme in the study of computability theory is the idea of *computational feebleness*, which might be loosely defined as properties exhibited by non-computable sets resembling computability. This is usually described in literature as a notion of lowness, and indicates weakness as an oracle. The classical example of sets exhibiting a property of this sort are the low sets, which are the sets A such that $A' \equiv_T \emptyset'$. Thus in terms of the jump operator, low sets are indistinguishable from the computable ones. There is a plethora of results in the literature which suggest that low sets resemble computable sets, particularly for the computably enumerable (c.e.) sets.

In notions of lowness, one usually considers a certain set operation and says that A satisfies the notion of lowness if it does not give any extra power to the operation. In the above example of low sets, the operation concerned was the Turing jump operator. Slaman and Solovay demonstrated in [26] a relationship between the low sets, and another seemingly unrelated lowness notion from the theory of inductive inference. In particular they showed that every set A which was low for EX learning was also low (and in fact 1-generic below $\emptyset'$). This result says that lowness for various notions of computation can be intertwined.

In a similar vein, Bickford and Mills [5] introduced the concept of a *superlow* set. A truth-table reduction is a Turing reduction which is total on every oracle string, and a set A was defined to be superlow if $A' \equiv_{tt} \emptyset'$, where the equivalence $\equiv_{tt}$ is induced by the pre-ordering of truth-table reducibility. One expects that the superlow sets would resemble the computable sets very strongly. Indeed a standard construction of a low c.e. set by the preservation of jump computations already made the constructed set superlow (as in the low basis theorem). At first blush we might be tempted to think that the low and superlow sets are very similar, or even the same. However the low and superlow sets have turned out to be not even elementarily equivalent. Recent examples have suggested that the dynamic properties of low and superlow c.e. sets are very different. For instance

---

[*] Both authors were partially supported by the Marsden Fund of New Zealand.

K. Ambos-Spies, B. Löwe, and W. Merkle (Eds.): CiE 2009, LNCS 5635, pp. 154–166, 2009.
© Springer-Verlag Berlin Heidelberg 2009

Downey and Ng [12] showed that there is a low c.e. degree which is not the join of any two superlow c.e. degrees.

Recent development in algorithmic randomness have revealed that the theory of low and superlow c.e. sets is much deeper than originally thought. Various lowness notions for Kolmogorov complexity and other operations arising in algorithmic randomness have suggested a deep connection with subclasses of the low sets. Several subclasses of the superlow sets have sprung up, and have been shown to be even better candidates for studying properties resembling computability. A central theme in these classes is the notion of *traceability*.

An order function h is a total computable, non-decreasing and unbounded function. A set A is said to be *jump traceable* with respect to an order h, if there is a computable g, such that for all x, $|W_{g(x)}| \leq h(x)$, and $J^A(x) \in W_{g(x)}$. Here, $J^A(x)$ denotes the value of the universal function $\Phi_x^A(x)$ partial computable in A. Note that the range of J^A is contained in $\mathbb{N}$, and not restricted to binary values. A set A is said to be jump traceable, if there is an order h for which it is jump traceable with respect to h. This notion was introduced by Nies [23].

A jump traceable set A differs from a superlow set in the sense that we are able to effectively enumerate finitely many candidates for each $J^A(x)$. For a superlow set A we are only able to approximate whether $J^A(x)$ converges. Since we are able to code finite information into $J^A(x)$, it might appear that being jump traceable is stronger than being superlow. Recall that a set A is n-c.e. if there is a computable approximation $A_s(x)$ to A, such that the number of changes in $A_s(x)$ is bounded by n at every x. A is ω-c.e. if the number of changes is bounded by a computable function.

Ng [19] showed that for n-c.e. sets, jump traceability and superlowness were the same. Earlier, Nies [23] showed that they coincide on the c.e. sets. However when we consider the next level on the Ershov hierarchy, these two notions separate: it is not hard to see that no jump traceable set can be Martin-Löf random, so a superlow Martin-Löf random set cannot be jump traceable. In the other direction, Nies [23] showed that there was an ω-c.e. jump traceable set which was not superlow. If we consider non-Δ_2^0 sets, the situation becomes even more bizzare. There is a perfect Π_1^0 class of sets which are jump traceable, via an exponential bound. Such a phenomenon highlights an important inherent property of being traceable; we are only able to enumerate possible values of $A \upharpoonright_n$, but beyond that we are given no additional information to suggest which one of the enumerated values is correct. Indeed Kjos-Hanssen and Nies [17] showed that jump traceable sets could even be superhigh.

Traceability plays a very important role in understanding lowness notions arising in algorithmic information theory. If R is a notion of effective randomness, then *low for R* would denote all the sets A for which $R^A = R$ (i.e. every random Z is still random relative to A). The work of Terwijn and Zambella [28], Kjos-Hanssen, Nies and Stephan [18], and Bedregal and Nies [4] has revealed an interesting interaction between "predictability" in terms of traceability, and simplicity in terms of Kolmogorov complexity. Recall that a set Z is of hyperimmune-free degree, if every function computable from Z is dominated by a computable function. A is said to be computably traceable if A is "uniformly hyperimmune-free". That is, there is a computable function h such that for each $f \leq_T A$, there exists a computable sequence of canonical finite sets $D_{g(x)}$ with $|D_{g(x)}| \leq h(x)$, and such that $f(x) \in D_{g(x)}$ for all x. They showed that

Theorem 1.1. *A is low for Schnorr randomness iff A is computably traceable.*

Hence the notion of being low for Schnorr randomness coincided with a combinatorial notion, that of being computably traceable.

A very robust class exhibiting low information content is the class of K-*trivial reals*[1]. Formally a real A is K-trivial if there is some constant c such that $K(A \mid_n) \leq K(n) + c$ for every n, where K denotes the prefix-free Kolmogorov complexity. Here $A \mid_n$ denotes the first n bits of A. Hence an initial segment of a K-trivial real contains no more information than its own length; clearly all computable reals are K-trivial. The most well-known work on the K-trivials was the work of Nies showing the coincidence of several simple classes:

Theorem 1.2 (Nies [22,24]). *A is K-trivial iff A is low for Martin-Löf randomness iff A is low for K.*

A real A is low for K if $\exists c \forall \sigma (K(\sigma) \leq K^A(\sigma) + c)$; that is, A does not help in the compression of strings when used as an oracle. The robustness of this class was further demonstrated when various other characterizations were found; for instance the reals low for weak 2-randomness, and the bases for Martin-Löf randomness [16]. Here A is a base for Martin-Löf randomness if $A \leq_T Z$ for some Z which is random relative to A. Intuitively there cannot be many possibilities for initial segments of a base for randomness, because we can use the given Turing reduction Φ (where $A = \Phi^Z$) to lower the Kolmogorov complexity of possible initial segments of Z. This was in fact the driving force behind the "hungry sets" theorem of [16]. We refer the reader to Franklin and Stephan [14] for a Schorr random version of a base. Other notions of bases which have been studied are the LR-bases for randomness [2,3], and the JT-bases for randomness [19]. These are notions obtained from a base for randomness, by replacing Turing reducibility with different weak reducibilites.

The resemblance which the K-trivial reals bear with the computable sets makes one wonder if they are related to the low sets. Is there also a combinatorial characterization in terms of traceability like the one for Schnorr lowness? Recent developments have suggested that this was the case. In [10], Downey, Hirschfeldt, Nies and Stephan showed that the K-trivial reals were natural solutions to Post's problem in the following sense:

Theorem 1.3. *Every K-trivial real is Turing incomplete.*

They used a new method widely known as the "Decanter method". This method exploited the fact that for any given K-trivial real, we could challenge its triviality very slowly. This resembles the "drip-feeding" action of a decanter, and

---

[1] We identify sets of natural numbers with real numbers. It is common to use the term "sets of natural numbers" in traditional computability theory, while in algorithmic information theory it is useful to think of these as infinite binary sequences.

hence the fanciful name. Nies [23,24] then applied a non-uniform method of the Decanter method to show:

Theorem 1.4. *Every K-trivial real is superlow.*

In fact, the same proof also shows that every K-trivial real is jump traceable at an order of $n \log^2 n$. These results suggested that jump traceability was the appropriate combinatorial notion associated with K-triviality, in the same way as computable traceability was related to lowness for Schnorr randomness. Armed with this insight, Figueira, Nies and Stephan [13] defined the notion of strong jump traceability. They defined A to be *strongly jump traceable*, if A is jump traceable with respect to all order functions. Figueira, Nies and Stephan used a cost function construction to show the existence of a promptly simple strongly jump traceable c.e. set. One can view c.e. strong jump traceability as a natural strengthening of being superlow. Unlike the case of computable traceability, strong jump traceability is different from jump traceability; in fact there is an entire hierarchy of jump traceable sets ordered by the growth rates of the bounding functions on the size of the trace. This hierarchy contains infinitely many strata in either direction. That is, there is no single maximal bound for jump traceability (Figueira Nies and Stephan [13]), and neither is there a single minimal bound (Ng [21]). In fact, Greenberg and Downey [8] showed that if one got down to a level of $\log \log n$, then every jump traceable real was Δ_2^0.

Figueira, Nies and Stephan asked if strong jump traceability was the coveted combinatorial characterization of the K-trivials. Cholak, Downey and Greenberg [6] answered this for the c.e. case by showing that the c.e. strongly jump traceable sets form a proper sub-ideal of the c.e. K-trivials. In fact, they showed that if A was c.e. and jump traceable at order $\sim \sqrt{\log n}$, then A was also K-trivial. This gave the first example of a combinatorial property which implies K-triviality. Even though neither notion of jump traceability gives us an exact characterization of the K-trivials, the associated results provide a good idea of the upper and lower bounds on the order of jump traceability which would capture K-triviality. By analyzing the proofs which give the lowerbound $\sim \sqrt{\log n}$ and upperbound $\sim n \log^2 n$, two possible characterizations had been suggested. Greenberg suggested that A is K-trivial iff A is jump traceable for all orders h with $\sum_{n \in \omega} \frac{1}{h(n)} < \infty$. This was refuted by Barmpalias, Downey and Greenberg [1], and independently by Ng [20]. The second conjecture is that the collection of orders should be the class of all orders h satisfying $\sum_{n \in \omega} 2^{-h(n)} < \infty$, and is still open.

Several other lowness notions have been studied with respect to other concepts of randomness. We list a few notable examples. Downey, Greenberg, Mihailović and Nies [9] showed that the computably traceable sets were exactly those which were low for computable measure machines. Here, a computable measure machine is a prefix-free machine with a computable halting probability, and A is low for computable measure machines (c.m.m.) if for each c.m.m. M relative to A, there is a c.m.m. N and a constant c such that $K_M^A(\sigma) \geq K_N(\sigma) - c$ for every σ.

Nies [24] showed that the only sets which were low for computable randomness[2], were the computable sets. The combined work of Greenberg, Miller, Stephan and Yu [15,27] revealed that the sets which were low for Kurtz randomness, were exactly the hyperimmune-free and non-DNR degrees. These were also the sets which were low for weak 1-genericity, which showed yet another interaction between lowness notions in classical computability, and randomness. For more examples we refer the reader to Chapter 8 of Nies' book [25].

In the next section, we contribute with another result in this direction. We consider lowness with respect to a less well-known notion of randomness, known as Demuth random. This was introduced by Demuth [7] and was originally motivated by topics in constructive analysis. This appears to be a very natural (strong) randomness notion to study, and not much work has yet been done on this class.

Definition 1.5. *A Demuth test is a sequence of c.e. open sets* $\{W_{g(x)}\}_{x \in \mathbb{N}}$ *such that* $\mu W_{g(x)} < 2^{-x}$ *for every* x, *and* g *is* ω-*c.e. We say that* Z *passes the test if* $\forall^\infty x Z \notin W_{g(x)}$. *A real is Demuth random if it passes every Demuth test.*

A Demuth test is a sequence of c.e. open sets, but the function giving the weak indices is an ω-c.e. function. Informally if we were building such a test (to try and catch some real number) we have additional power over building a ML-test because we can change the name of $W_{g(x)}$ a bounded number of times. That is, we can remove a certain part (or even all) of what we have enumerated into $W_{g(x)}$ so far, as long as we only do it a computably bounded number of times. The definition of passing a Demuth test is as in the Solovay sense, and we cannot always require that $W_{g(x)} \supseteq W_{g(x+1)}$. There is no universal Demuth test, although there is a single special test $\{W_{\hat{g}(x)}\}$ which is universal in the sense that every real passing the special test is Demuth random. However the function $\hat{g}(x)$ has to emulate every ω-c.e. function, and so $\hat{g}(x)$ is Δ_2^0. Hence the special test is not a Demuth test.

Clearly the Demuth randoms lie between 2-randomness and ML-randomness. It is not hard to construct a Δ_2^0 Demuth random using the special test, and obviously no ω-c.e. set can be Demuth random. Hence the containments are proper. Demuth randoms exhibit properties which can be found in both 1- and 2-randomn reals. For instance every Demuth random (like the 2-randoms) are GL_1 and hence of hyperimmune degree by a result of Miller and Nies (Theorem 8.1.19 of [25]). Here a real is of hyperimmune degree if it is not of hyperimmune-free degree. Since there are hyperimmune-free weakly 2-randoms, this implies that Demuth randomness and weak 2-randomness are incomparable notions. However unlike the 2-randoms, the Demuth test notion is essentially computably enumerable.

We contribute two theorems to the understanding of this notion of randomness. First, we prove that every Demuth random is array computable. This notion was introduced by Downey, Jockusch and Stob [11] to describe the class of reals below which certain multiple permitting arguments could not be carried out.

---

[2] A real is computably random if it succeeds on every computable martingale.

This again suggests that Demuth randoms are like the 2-randoms, having low computational strength. In particular, the Demuth randoms below $\emptyset'$ form an interesting class, being both low *and* array computable but not superlow.

Theorem 1.6. *Each Demuth random Z is array computable.*

Proof. We observe the proof that every Demuth random is GL_1 already does it; this can be found in Chapter 3 of Nies [25]. The proof actually produces an ω-c.e. function g which dominates the function $\Theta^Z(x) :=$ least s such that $\Phi_x^Z(x)[s] \downarrow$ (which of course implies that $Z' \leq_T Z \oplus \emptyset'$). By usual convention the output value of $\Phi_x^Z(x)$ is $< \Theta^Z(x)$. Since every function computable in Z can be coded into the diagonal, we have a computable function p such that for every e, and almost every $y > e$, we have $\Phi_e^Z(y) = \Phi_{p(e,y)}^Z(p(e,y)) < \Theta^Z(p(e,y)) < g(p(e,y)) < \tilde{g}(y)$, where $\tilde{g}(y) := \max\{g(p(0,y)), g(p(1,y)), \cdots, g(p(y,y))\}$ is ω-c.e. as well. $\square$

Next, we study the notion of lowness with respect to Demuth randomness. A relativized Demuth test involves full relativization. That is, a Demuth test relative to A is a sequence $\{W_{g(x)}^A\}$ where $\mu W_{g(x)}^A < 2^{-x}$ for every x. Here, $g(x) = \lim \tilde{g}(x, s)$ for some A-computable function $\tilde{g}$, and the number of $\tilde{g}$-mind changes is bounded by an A-computable function. A real Z is Demuth random relative to A if it passes every Demuth test relative to A. We say that A is *low for Demuth randomness* if every Demuth random is Demuth random relative to A. In the next section we prove that every real low for Demuth randomness is of hyperimmune-free degree.

However it is still unknown if there is any set which is non-computable and low for Demuth randomness. A construction of such a real will have to build a hyperimmune-free degree, and if one uses the standard forcing method then one has to address the issue of constructing the effective objects required in the proof. We conjecture that every real which is low for Demuth randomness is computable.

2   No Set of Hyperimmune Degree Can Be Low for Demuth Randomness

We work in the Cantor space 2^ω with the usual clopen topology. The basic open sets are of the form $[\sigma]$ where σ is a finite string, and $[\sigma] = \{X \in 2^\omega \mid X \supset \sigma\}$. We fix some effective coding of the set of finite strings, and identify finite strings with their code numbers. We treat W_x as a c.e. open set, consisting of basic clopen sets. We say that $[\sigma] \in W_x$ to mean that the code number of σ is in W_x, and we say that a string $\tau \in W_x$ if $\tau \supseteq \sigma$ for some $[\sigma] \in W_x$. Equivalently we say that τ is *captured by* W_x. The same definition holds if we replace τ by an infinite binary string. We prove:

Theorem 2.1. *No set of hyperimmune degree can be low for Demuth randomness.*

Proof. Suppose A is of hyperimmune degree. Let h^A be a function total computable in A and non-decreasing, which escapes domination by all total computable functions. That is, for all total computable g, $\exists^\infty x(g(x) < h^A(x))$. We build a $Z \leq_T A'$ which is Demuth random, but not Demuth random relative to A. To do this, we give an A-computable approximation $\{Z_s\}$ to Z. The construction will try to achieve two goals. The first is to make Z Demuth random by making Z avoid all Demuth tests. The second goal is to ensure that for infinitely many x, there are at most $h^A(x)$ many mind changes of $Z_s \mid_x$. Hence we can easily use the approximation Z_s to build a Demuth test relative to A capturing Z infinitely often. Hence Z cannot be Demuth random relative to A.

2.1   The Motivation

Before we describe the strategy used to prove Theorem 2.1, let us see why an attempted construction of a c.e. set A which is low for Demuth randomness fails. Let us consider a single (relativized) Demuth test $\{V_x^A\}$, played by the opponent, where the index for V_x^A can change $h^A(x)$ times. Now we have to cover $\{V_x^A\}$ with a plain Demuth test $\{U_x\}$, by making sure that $V_x^A \subseteq U_x$ for every x. If $h^A(x) = 0$ for all x, then we could just follow the construction of a c.e. set which is low for random. We would enumerate y into A (to make A non-computable), if the penalty we have to pay for making the enumeration of y is small. Even when h^A is computable, we can always arrange the enumerations so that $V_x^A \subseteq U_x$ eventually, because we could use $h^A(x)$ as the bound for the index change of U_x.

The problem is that an enumeration into A not only increases the amount we have to put into U_x, but also gives the opponent a chance to redefine $h^A(x)$. Suppose he has defined $h^A(x)$ with use b_x. At some stage we will have to commit ourselves to a number $g(x)$, and promise never to change the index for U_x more than $g(x)$ times. We would of course declare that $g(x) > h^A(x)$, but once we do that, the opponent could challenge us to change $A \mid_{b_x}$ to ensure the non-computability of A. We have to eventually change $A \mid_{b_x}$ at some x, and allow the opponent to make $h^A(x) > g(x)$, and then we are stuck.

Note that the opponent will be likely to have a winning strategy, if h^A escapes domination by all computable functions. He could then carry out the above for each e, patiently waiting for an x such that $h^A(x) > \varphi_e(x)$, and then defeat the eth Demuth test. This is the basic idea used in the following proof, where we will play the opponent's winning strategy.

2.2   Listing All Demuth Tests

In order to achieve the first goal, we need to specify an effective listing of all Demuth tests. It is enough to consider all Demuth tests $\{U_x\}$ where $\mu(U_x) < 2^{-3(x+1)}$. Let $\{g_e\}_{e \in \mathbb{N}}$ be an effective listing of all partial computable functions of a single variable. For every g in the list, we will assume that in order to output $g(x)$, we will have to first run the procedures to compute $g(0), \cdots, g(x-1)$, and wait for all of them to return, before attempting to compute $g(x)$. We may also assume that g is non-decreasing. This minor but important restriction on g ensures that:

(i) $dom(g)$ is either $\mathbb{N}$, or an initial segment of $\mathbb{N}$,
(ii) for every x, $g(x+1)$ converges strictly after $g(x)$, if ever.

By doing this, we will not miss any total non-decreasing computable function. It is easy to see that there is a total function $k \leq_T \emptyset'$ that is universal in the following sense:

1. if $f(x)$ is ω-c.e. then for some e, $f(x) = k(e, x)$ for all x,
2. for all e, the function $\lambda x k(e, x)$ is ω-c.e.,
3. there is a uniform approximation for k such that for all e and x, the number of mind changes for $k(e, x)$ is bounded by

$$\begin{cases} g_e(x) & \text{if } g_e(x) \downarrow, \\ 0 & \text{otherwise.} \end{cases}$$

Let $k(e, x)[s]$ denote the approximation for $k(e, x)$ at stage s. Denote $U_x^e = W_{k(e,x)}$, where we stop enumeration if $\mu(W_{k(e,x)}[s])$ threatens to exceed $2^{-3(x+1)}$. Then for each e, $\{U_x^e\}$ is a Demuth test, and every Demuth test is one of these. To make things clear, we remark that there are two possible ways in which $U_x^e[s] \neq U_x^e[s+1]$. The first is when $k(e, x)[s] = k(e, x)[s+1]$ but a new element is enumerated into $W_{k(e,x)}$. The second is when $k(e, x)[s] \neq k(e, x)[s+1]$ altogether; if this case applies we say that U_x^e has a *change of index at stage* $s + 1$.

2.3   The Strategy

Now that we have listed all Demuth tests, how are we going to make use of the function h^A? Note that there is no single universal Demuth test; this complicates matters slightly. The eth requirement will ensure that Z passes the first e many (plain) Demuth tests. That is,

$$\mathcal{R}_e : \text{ for each } k \leq e, \, Z \text{ is captured by } U_x^k \text{ for only finitely many } x.$$

$\mathcal{R}_e$ will do the following. It starts by picking a number r_e, and decides on $Z \mid_{r_e}$. This string can only be captured by U_x^k for $x \leq r_e$, so there are only finitely many pairs $\langle k, x \rangle$ to be considered since we only care about $k \leq e$. Let S_e denote the collection of these open sets. If any $U_x^k \in S_e$ captures $Z \mid_{r_e}$, we would change our mind on $Z \mid_{r_e}$. If at any point in time, $Z \mid_{r_e}$ has to change more than $h^A(0)$ times, we would pick a new follower for r_e, and repeat, comparing with $h^A(1), h^A(2), \cdots$ each time. The fact that we will eventually settle on a final follower for r_e, will follow from the hyperimmunity of A; all that remains is to argue that we can define an appropriate computable function *at each* $\mathcal{R}_e$, in order to challenge the hyperimmunity of A.

Suppose that $r_e^0, r_e^1, \cdots$ are the followers picked by $\mathcal{R}_e$. The required computable function P would be something like $P(n) = \sum_{k \leq e} \sum_{x \leq r_e^n} g_k(x)$, for if $P(N) < h^A(N)$ for some N, then we would be able to change $Z \mid_{r_e^N}$ enough times on the Nth attempt. There are two considerations. Firstly, we do not know which of $g_0, \cdots, g_e$ are total, so we cannot afford to wait on non converging computations when computing P. However, as we have said before, we can have a different P at each requirement, and the choice of P can be non-uniform. Thus, P could just sum over all the total functions amongst $g_0, \cdots, g_e$.

The second consideration is that we might not be able to compute $r_e^0, r_e^1, \cdots$, if we have to recover r_e^n from the construction (which is performed with oracle A). We have to somehow figure out what r_e^n is, externally to the construction. Observe that however, if we restrict ourselves to non-decreasing $g_0, g_1, \cdots$, it would be sufficient to compute an upperbound for r_e^n. We have to synchronize this with the construction: instead of picking r_e^n when we run out of room to change $Z \mid_{r_e^{n-1}}$, we could instead pick r_e^n the moment enough of the $g_k(x)$ converge and demonstrate that their sum exceeds $h^A(r_e^{n-1})$. To recover a bound for say, r_e^1 externally, we compute the first stage t such that all of the $g_k(x)[t]$ have converged for $x \leq r_e^0$ and g_k total.

2.4   Notations Used for the Formal Construction

The construction uses oracle A. At stage s we give an approximation $\{Z_s\}$ of Z, and at the end we argue that $Z \leq_T A'$. The construction involves finite injury of the requirements. $\mathcal{R}_1$ for instance, would be injured by $\mathcal{R}_0$ finitely often while $\mathcal{R}_0$ is waiting for hyperimmune permission from h^A. We intend to satisfy $\mathcal{R}_e$, by making $\mu(U_x^e \cap [Z \mid_r])$ small for appropriate x, r. At stage s, we let $r_e[s]$ denote the follower used by $\mathcal{R}_e$. At stage s of the construction we define Z_s up till length s. We do this by specifying the strings $Z_s \mid_{r_0[s]}, \cdots, Z_s \mid_{r_k[s]}$ for an appropriate number k (such that $r_k[s] = s - 1$). We adopt the convention of $r_{-1} = -1$ and $\alpha \mid_{-1} = \alpha \mid_0 = \langle \rangle$ for any string α. We let $S_e[s]$ denote all the pairs $\langle k, x \rangle$ for which $\mathcal{R}_e$ wants to make Z avoid U_x^k at stage s. The set $S_e[s]$ is specified by

$$S_e[s] = \{\langle k, x \rangle \mid k \leq e \ \wedge \ r_{k-1}[s] + 1 \leq x \leq r_e[s]\}.$$

Define the sequence of numbers

$$M_n = \sum_{j=n}^{2n} 2^{-(1+j)};$$

these will be used to approximate Z_s. Roughly speaking, the intuition is that $Z_s(n)$ will be chosen to be either 0 or 1 depending on which of $(Z_s \mid_n)^\frown 0$ or $(Z_s \mid_n)^\frown 1$ has a measure of $\leq M_n$ when restricted to a certain collection of U_x^e.

If P is an expression we append $[s]$ to P, to refer to the value of the expression as evaluated at stage s. When the context is clear we drop the stage number from the notation.

2.5   Formal Construction of Z

At stage $s = 0$, we set $r_0 = 0$ and $r_e \uparrow$ for all $e > 0$, and do nothing else. Suppose $s > 0$. We define $Z_s \mid_{r_k[s]}$ inductively; assume that has been defined for some k. There are two cases to consider for $\mathcal{R}_{k+1}$:

1. $r_{k+1}[s] \uparrow$: set $r_{k+1} = r_k[s] + 1$, end the definition of Z_s and go to the next stage.

2. $r_{k+1}[s] \downarrow$: check if $\sum_{\langle e,x \rangle \in S_{k+1}[s]} 2^{r_{k+1}} g_e(x)[s] \leq h^A(r_{k+1}[s])$. The sum is computed using converged values, and if $g_e(x)[s] \uparrow$ for any e, x we count it as 0. There are two possibilities:

   (a) $sum > h^A(r_{k+1})$: set $r_{k+1} = s$, and set $r_{k'} \uparrow$ for all $k' > k+1$. End the definition of Z_s and go to the next stage.

   (b) $sum \leq h^A(r_{k+1})$: pick the leftmost node $\sigma \supseteq Z_s \upharpoonright_{r_k[s]}$ of length $|\sigma| = r_{k+1}[s]$, such that $\sum_{\langle e,x \rangle \in S_{k+1}[s]} \mu(U_x^e[s] \cap [\sigma]) \leq M_{r_{k+1}[s]}$. We will later verify that σ exists by a counting of measure. Let $Z_s \upharpoonright_{r_{k+1}[s]} = \sigma$.

We say that $\mathcal{R}_{k+1}$ has *acted*. If 2(a) is taken, then we say that $\mathcal{R}_{k+1}$ has *failed the sum check*. This completes the description of Z_s.

2.6   Verification

Clearly, the values of the markers $r_0, r_1, \cdots$ are kept in increasing order. That is, at all stages s, if $r_k[s] \downarrow$, then $r_0[s] < r_1[s] < \cdots < r_k[s]$ are all defined. From now on when we talk about Z_s, we are referring to the fully constructed string at the end of stage s. It is also clear that the construction keeps $|Z_s| < s$ at each stage s.

Lemma 2.2. *Whenever step 2(b) is taken, we can always define $Z_s \upharpoonright_{r_{k+1}[s]}$ for the relevant k and s.*

Proof. We drop s from notations, and proceed by induction on k. Let Υ be the collection of all possible candidates for $Z_s \upharpoonright_{r_{k+1}}$, that is, $\Upsilon = \{\sigma : \sigma \supseteq Z \upharpoonright_{r_k} \wedge |\sigma| = r_{k+1}\}$. Suppose that $k \geq 0$:

$$\sum_{\sigma \in \Upsilon} \sum_{\langle e,x \rangle \in S_{k+1}} \mu(U_x^e \cap [\sigma]) = \sum_{\langle e,x \rangle \in S_{k+1}} \sum_{\sigma \in \Upsilon} \mu(U_x^e \cap [\sigma])$$

$$\leq \sum_{\langle e,x \rangle \in S_{k+1}} \mu(U_x^e \cap [Z \upharpoonright_{r_k}]) \leq \sum_{\langle e,x \rangle \in S_k} \mu(U_x^e \cap [Z \upharpoonright_{r_k}]) + \sum_{x=r_k+1}^{r_{k+1}} \sum_{e \leq k+1} \mu(U_x^e)$$

$$\leq M_{r_k} + \sum_{x=r_k+1}^{r_{k+1}} 2^{-2x} \text{ (since } k \leq r_k) \leq M_{r_k} + \sum_{x=2r_k+1}^{r_k+r_{k+1}} 2^{-(1+x)}$$

$$= \sum_{x=r_k+1}^{2r_{k+1}} 2^{-(1+x)} 2^{r_{k+1}-r_k} \text{ (adjusting the index } x) = M_{r_{k+1}} |\Upsilon|.$$

Hence, there must be some σ in Υ which passes the measure check in 2(b) for $Z \upharpoonright_{r_{k+1}}$. A similar, but simpler counting argument follows for the base case $k = -1$, using the fact that the search now takes place above $Z \upharpoonright_{r_k} = \langle \rangle$.   □

Lemma 2.3. *For each e, the follower $r_e[s]$ eventually settles.*

Proof. We proceed by induction on e. Note that once $r_{e'}$ has settled for every $e' < e$, then $\mathcal{R}_e$ will get to act at every stage after that. Hence there is a stage s_0 such that

(i) $r_{e'}$ has settled for all $e' < e$, and
(ii) r_e receives a new value at stage s_0.

Note also that $\mathcal{R}_e$ will get a chance to act at every stage $t > s_0$, and the only reason why r_e receives a new value after stage s_0, is that $\mathcal{R}_e$ fails the sum check. Suppose for a contradiction, that $\mathcal{R}_e$ fails the sum check infinitely often after s_0.

Let $q(n-1)$ be the stage where $\mathcal{R}_e$ fails the sum check for the nth time after s_0. In other words, $q(0), q(1), \cdots$ are precisely the different values assigned to r_e after s_0. Let $\mathcal{C}$ be the collection of all $k \leq e$ such that g_k is total, and d be a stage where $g_k(x)[d]$ has converged for all $k \leq e$, $k \notin \mathcal{C}$ and $x \in dom(g_k)$. We now define an appropriate computable function to contradict the hyperimmunity of A. Define the total computable function p by: $p(0) = 1 + \max\{s_0, d,$ the least stage t where $g_k(r_e[s_0])[t] \downarrow$ for all $k \in \mathcal{C}\}$. Inductively define $p(n+1) = 1 +$ the least t where $g_k(p(n))[t] \downarrow$ for all $k \in \mathcal{C}$. Let $P(n) = \sum_{k \leq e} \sum_{x \leq p(n)} 2^{p(n)} g_k(x)[p(n+1)]$, which is the required computable function.

One can show by a simple induction, that $p(n) \geq q(n)$ for every n, using the fact that $\mathcal{R}_e$ is given a chance to act at every stage after s_0, as well as the restrictions we had placed on the functions $\{g_k\}$. Let N be such that $P(N) \leq h^A(N)$. At stage $q(N+1)$ we have $\mathcal{R}_e$ failing the sum check, so that $h^A(N) < h^A(q(N)) < \sum_{\langle k, x \rangle \in S_e} 2^{q(N)} g_k(x)$, where everything in the last sum is evaluated at stage $q(N+1)$. That last sum is clearly $< P(N) \leq h^A(N)$, giving a contradiction.   □

Let $\hat{r}_e$ denote the final value of the follower r_e. Let $Z = \lim_s Z_s$. We now show that Z is not Demuth random relative to A. For each e and s, $Z_{s+1+\hat{r}_e} \upharpoonright_{\hat{r}_e}$ is defined, by Lemma 2.2.

Lemma 2.4. *For each e, $\#\{t \geq 1 + \hat{r}_e : Z_t \upharpoonright_{\hat{r}_e} \neq Z_{t+1} \upharpoonright_{\hat{r}_e}\} \leq h^A(\hat{r}_e)$.*

Proof. Suppose that $Z_{t_1} \upharpoonright_{\hat{r}_e} \neq Z_{t_2} \upharpoonright_{\hat{r}_e}$ for some $1 + \hat{r}_e \leq t_1 < t_2$. We must have $r_{e'}$ already settled at stage t_1, for all $e' \leq e$. Suppose that $Z_{t_2} \upharpoonright_{\hat{r}_e}$ is to the left of $Z_{t_1} \upharpoonright_{\hat{r}_e}$, then let e' be the least such that $Z_{t_2} \upharpoonright_{\hat{r}_{e'}}$ is to the left of $Z_{t_1} \upharpoonright_{\hat{r}_{e'}}$. The fact that $\mathcal{R}_{e'}$ didn't pick $Z_{t_2} \upharpoonright_{\hat{r}_{e'}}$ at stage t_1, shows that we must have a change of index for U_b^a between t_1 and t_2, for some $\langle a, b \rangle \in S_{e'} \subseteq S_e$. Hence, the total number of mind changes is at most $2^{\hat{r}_e} \sum_{\langle a,b \rangle \in S_e} g_a(b)$, where divergent values count as 0. $2^{\hat{r}_e}$ represents the number of times we can change our mind from left to right consecutively without moving back to the left, while $\sum_{\langle a,b \rangle \in S_e} g_a(b)$ represents the number of times we can move from right to left. Since $\mathcal{R}_e$ never fails a sum check after $\hat{r}_e$ is picked, it follows that the number of mind changes has to be bounded by $h^A(\hat{r}_e)$.   □

By asking A' appropriate 1-quantifier questions, we can recover $Z = \lim_s Z_s$. Hence Z is well-defined and computable from A'. To see that Z is not Demuth random in A, define the A-Demuth test $\{V_x^A\}$ by the following: run the construction and enumerate $[Z_s \upharpoonright_x]$ into V_x^A when it is first defined. Subsequently each time we get a new $Z_t \upharpoonright_x$, we change the index for V_x^A, and enumerate the new $[Z_t \upharpoonright_x]$ in. If we ever need to change the index $> h^A(x)$ times, we stop and do nothing. By Lemma 2.4, Z will be captured by $V_{\hat{r}_e}^A$ for every e.

Lastly, we need to see that Z passes all $\{U_x^e\}$. Suppose for a contradiction, that $Z \in U_x^e$ for some e and $x > \hat{r}_e$. Let δ be such that $Z \in [\delta] \in U_x^e$, and let $e' \geq e$ such that $\hat{r}_{e'} > |\delta|$. Go to a stage in the construction where δ appears in U_x^e and never leaves, and $r_{e'} = \hat{r}_{e'}$ has settled. At every stage t after that, observe that $\langle e, x \rangle \in S_{e'}$, and that $\mathcal{R}_{e'}$ will get to act, at which point it will discover that $\mu(U_x^e \cap [Z \restriction_{\hat{r}_{e'}}]) = 2^{-\hat{r}_{e'}} > M_{\hat{r}_{e'}}$. Thus, $\mathcal{R}_{e'}$ never picks $Z \restriction_{\hat{r}_{e'}}$ as an initial segment for Z_t, giving us a contradiction.   □

References

1. Barmpalias, G., Downey, R., Greenberg, N.: K-trivial degrees and the jump-traceability hierarchy. Proceedings of the American Mathematical Society (to appear)
2. Barmpalias, G., Lewis, A., Ng, K.M.: The importance of Π_1^0-classes in effective randomness (submitted)
3. Barmpalias, G., Lewis, A., Stephan, F.: Π_1^0 classes, LR degrees and Turing degrees. Annals of Pure and Applied Logic
4. Bedregal, B., Nies, A.: Lowness properties of reals and hyper-immunity. In: WoLLIC 2003. Electronic Lecture Notes in Theoretical Computer Science, vol. 84 (2003)
5. Bickford, M., Mills, C.: Lowness properties of r.e. sets. Theoretical Computer Science (typewritten unpublished manuscript)
6. Cholak, P., Downey, R., Greenberg, N.: Strong jump-traceability I: The computably enumerable case. Advances in Mathematics 217, 2045–2074 (2008)
7. Demuth, O.: Remarks on the structure of tt-degrees based on constructive measure theory. Commentationes Mathematicae Universitatis Carolinae 29(2), 233–247 (1988)
8. Downey, R., Greenberg, N.: Strong jump-traceability II: The general case (in preparation)
9. Downey, R., Greenberg, N., Mihailović, N., Nies, A.: Lowness for computable machines. In: Computational Prospects of Infinity. Lecture Notes Series of the Institute for Mathematical Sciences, NUS, vol. 15, pp. 79–86 (2008)
10. Downey, R., Hirschfeldt, D., Nies, A., Stephan, F.: Trivial reals. In: Proceedings of the 7th and 8th Asian Logic Conferences, pp. 103–131. World Scientific, Singapore (2003)
11. Downey, R., Jockusch, C., Stob, M.: Array nonrecursive sets and multiple permitting arguments. Recursion Theory Week 1432, 141–174 (1990)
12. Downey, R., Ng, K.M.: Splitting into degrees with low computational strength (in preparation)
13. Figueira, S., Nies, A., Stephan, F.: Lowness properties and approximations of the jump. In: Proceedings of the Twelfth Workshop of Logic, Language, Information and Computation (WoLLIC 2005). Electronic Lecture Notes in Theoretical Computer Science, vol. 143, pp. 45–57 (2006)
14. Franklin, J., Stephan, F.: Schnorr trivial sets and truth-table reducibility. Technical Report TRA3/08, School of Computing, National University of Singapore (2008)
15. Greenberg, N., Miller, J.: Lowness for Kurtz randomness. Journal of Symbolic Logic (to appear)
16. Hirschfeldt, D., Nies, A., Stephan, F.: Using random sets as oracles (to appear)
17. Kjos-Hanssen, B., Nies, A.: Superhighness (to appear)

18. Kjos-Hanssen, B., Nies, A., Stephan, F.: Lowness for the class of Schnorr random sets. Notre Dame Journal of Formal Logic 35(3), 647–657 (2005)
19. Ng, K.M.: Ph.D Thesis (in preparation)
20. Ng, K.M.: Strong jump traceability and beyond (submitted)
21. Ng, K.M.: On strongly jump traceable reals. Annals of Pure and Applied Logic 154, 51–69 (2008)
22. Nies, A.: On a uniformity in degree structures. In: Complexity, Logic and Recursion Theory. Lecture Notes in Pure and Applied Mathematics, February 1997, pp. 261–276 (1997)
23. Nies, A.: Reals which compute little. CDMTCS Research Report 202, The University of Auckland (2002)
24. Nies, A.: Lowness properties and randomness. Advances in Mathematics 197, 274–305 (2005)
25. Nies, A.: Computability And Randomness. Oxford University Press, Oxford (2006) (to appear)
26. Slaman, T., Solovay, R.: When oracles do not help. In: Fourth Annual Conference on Computational Learning Theory, pp. 379–383 (1971)
27. Stephan, F., Liang, Y.: Lowness for weakly 1-generic and Kurtz-random. In: Cai, J.-Y., Cooper, S.B., Li, A. (eds.) TAMC 2006. LNCS, vol. 3959, pp. 756–764. Springer, Heidelberg (2006)
28. Terwijn, S., Zambella, D.: Algorithmic randomness and lowness. Journal of Symbolic Logic 66, 1199–1205 (2001)

Graph States and the Necessity of Euler Decomposition

Ross Duncan[1] and Simon Perdrix[2,3]

[1] Oxford University Computing Laboratory,
Wolfson Building, Parks Road, OX1 3QD Oxford, UK
ross.duncan@comlab.ox.ac.uk
[2] LFCS, University of Edinburgh, UK
[3] PPS, Université Paris Diderot, France
simon.perdrix@pps.jussieu.fr

Abstract. Coecke and Duncan recently introduced a categorical formalisation of the interaction of complementary quantum observables. In this paper we use their diagrammatic language to study graph states, a computationally interesting class of quantum states. We give a graphical proof of the fixpoint property of graph states. We then introduce a new equation, for the Euler decomposition of the Hadamard gate, and demonstrate that Van den Nest's theorem—locally equivalent graphs represent the same entanglement—is equivalent to this new axiom. Finally we prove that the Euler decomposition equation is not derivable from the existing axioms.

Keywords: quantum computation, monoidal categories, graphical calculi.

1 Introduction

Upon asking the question "What are the axioms of quantum mechanics?" we can expect to hear the usual story about states being vectors of some Hilbert space, evolution in time being determined by unitary transformations, etc. However, even before finishing chapter one of the textbook, we surely notice that something is amiss. Issues around normalisation, global phases, etc. point to an "impedence mismatch" between the theory of quantum mechanics and the mathematics used to formalise it. The question therefore should be "What are the axioms of quantum mechanics *without Hilbert spaces'?*"

In their seminal paper [1] Abramsky and Coecke approached this question by studying the categorical structures necessary to carry out certain quantum information processing tasks. The categorical treatment provides as an intuitive pictorial formalism where quantum states and processes are represented as certain diagrams, and equations between them are described by rewriting diagrams. A recent contribution to this programme was Coecke and Duncan's axiomatisation of the algebra of a pair complementary observables [2] in terms of the

K. Ambos-Spies, B. Löwe, and W. Merkle (Eds.): CiE 2009, LNCS 5635, pp. 167–177, 2009.

red-green calculus. The formalism, while quite powerful, is known to be incomplete in the following sense: there exist true equations which are not derivable from the axioms.

In this paper we take one step towards its completion. We use the red-green language to study *graph states*. Graph states [3] are very important class of states used in quantum information processing, in particular with relation to the one-way model of quantum computing [4]. Using the axioms of the red-green system, we attempt to prove Van den Nest's theorem [5], which establishes the local complementation property for graph states. In so doing we show that a new equation must be added to the system, namely that expressing the Euler decomposition of the Hadamard gate. More precisely, we show that Van den Nest's theorem is equivalent to the decomposition of H, and that this equation cannot be deduced from the existing axioms of the system.

The paper procedes as follows: we introduce the graphical language and the axioms of the red-green calculus, and its basic properties; we then introduce graph states and and prove the fixpoint property of graph states within the calculus; we state Van den Nest's theorem, and prove our main result—namely that the theorem is equivalent to the Euler decomposition of H. Finally we demonstrate a model of the red-green axioms where the Euler decomposition does not hold, and conclude that this is indeed a new axiom which should be added to the system.

2   The Graphical Formalism

Definition 1. *A* diagram *is a finite undirected open graph generated by the following two families of vertices:*

where $\alpha \in [0, 2\pi)$, and a vertex $H = \boxed{H}$ belonging to neither family.

Diagrams form a monoidal category $\mathcal{D}$ in the evident way: composition is connecting up the edges, while tensor is simply putting two diagrams side by side. In fact, diagrams form a †-compact category [6,1] but we will suppress the details of this and let the pictures speak for themselves. We rely here on general results [7,8,9] which state that a pair diagrams are equal by the axioms of †-compact categories exactly when they may be deformed to each other.

Each family forms a *basis structure* [10] with an associated *local phase shift*. The axioms describing this structure can be subsumed by the following law. Define $\delta_0 = \epsilon^\dagger$, $\delta_1 = 1$ and $\delta_n = (\delta_{n-1} \otimes 1) \circ \delta$, and define $\delta_n^\dagger$ similarly.

Spider Law. *Let f be a connected diagram, with n inputs and m outputs, and whose vertices are drawn entirely from one family; then*

$$f = \delta_m \circ p(\alpha) \circ \delta_n^{\dagger} \qquad \text{where } \alpha = \sum_{p(\alpha_i) \in f} \alpha_i \quad \text{mod } 2\pi$$

with the convention that p(0) = 1.

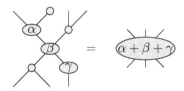

The spider law justifies the use of "spiders" in diagrams: coloured vertices of arbitrary degree labelled by some angle α. By convention, we leave the vertex empty if $\alpha = 0$.

We use the spider law as rewrite equation between graphs. It allows vertices of the same colour to be merged, or single vertices to be broken up. An important special case is when $n = m = 1$ and no angles occur in f; in this case f can be reduced to a simple line. (This implies that both families generate the same compact structure.)

$$\begin{array}{c} \text{\textbardbl} \end{array} = \text{\textbar}$$

Lemma 1. *A diagram without H is equal to a bipartite graph.*

Proof. If any two adjacent vertices are the same colour they may be merged by the spider law. Hence if we can do such mergings, every green vertex is adjacent only to red vertices, and vice versa.

We interpret diagrams in the category **FdHilb**$_{\text{wp}}$; this the category of complex Hilbert spaces and linear maps under the equivalence relation $f \equiv g$ iff there exists θ such that $f = e^{i\theta}g$. A diagram f with n inputs and m output defines a linear map $[\![f]\!] : \mathbb{C}^{\otimes 2n} \to \mathbb{C}^{\otimes 2m}$. Let

$$[\![\epsilon_Z^{\dagger}]\!] = \tfrac{1}{\sqrt{2}}(|0\rangle + |1\rangle) \qquad\qquad [\![\epsilon_X^{\dagger}]\!] = |0\rangle$$

$$[\![\delta_Z^{\dagger}]\!] = \begin{pmatrix} 1\,0\,0\,0 \\ 0\,0\,0\,1 \end{pmatrix} \qquad\qquad [\![\delta_X^{\dagger}]\!] = \tfrac{1}{\sqrt{2}}\begin{pmatrix} 1\,0\,0\,1 \\ 0\,1\,1\,0 \end{pmatrix}$$

$$[\![p_Z(\alpha)]\!] = \begin{pmatrix} 1 & 0 \\ 0 & e^{i\alpha} \end{pmatrix} \qquad [\![p_X(\alpha)]\!] = e^{-\frac{i\alpha}{2}}\begin{pmatrix} \cos\frac{\alpha}{2} & i\sin\frac{\alpha}{2} \\ i\sin\frac{\alpha}{2} & \cos\frac{\alpha}{2} \end{pmatrix}$$

$$[\![H]\!] = \frac{1}{\sqrt{2}} \begin{pmatrix} 1 & 1 \\ 1 & -1 \end{pmatrix}$$

and set $[\![f^\dagger]\!] = [\![f]\!]^\dagger$. The map $[\![\cdot]\!]$ extends in the evident way to a monoidal functor.

The interpretation of $\mathcal{D}$ contains a universal set of quantum gates. Note that $p_Z(\alpha)$ and $p_X(\alpha)$ are the rotations around the X and Z axes, and in particular when $\alpha = \pi$ they yield the Pauli X and Z matrices. The $\wedge Z$ is defined by:

The δ_X and δ_Z maps *copy* the eigenvectors of the Pauli X and Z; the ϵ maps *erase* them. (This is why such structures are called basis structures).

Now we introduce the equations[1] which make the X and Z families into *complementary* basis structures as in [2]. Note that all of the equations are also satisfied in the Hilbert space interpretation. We present them in one colour only; they also hold with the colours reversed.

Copying

Bialgebra

π-Commutation

A consequence of the axioms we have presented so far is the Hopf Law:

---

[1] We have, both above and below, made some simplifications to the axioms of [2] which are specific to the case of qubits. We also supress scalar factors.

Finally, we introduce the equations for H:

The special role of H in the system is central to our investigation in this paper.

3   Generalised Bialgebra Equations

The bialgebra law is a key equation in the graphical calculus. Notice that the left hand side of the equation is a 2-colour bipartite graph which is both a $K_{2,2}$ (i.e a complete bipartite graph) and a C_4 (i.e. a cycle composed of 4 vertices) with alternating colours. In the following we introduce two generalisations of the bialgebra equation, one for any $K_{n,m}$ and another one for any C_{2n} (even cycle).

We give graphical proofs for both generalisations; both proofs rely essentially on the primitive bialgebra equation.

Lemma 2. *For any n, m, "$K_{n,m} = P_2$", graphically:*

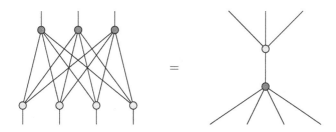

Lemma 3. *For n, an even cycle of size $2n$, of alternating colours, can be rewritten into hexagons. Graphically:*

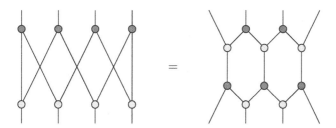

4   Graph States

In order to explore the power and the limits of the axioms we have described, we now consider the example of *graph states*. Graph states provide a good testing ground for our formalism because they are relatively easy to describe, but have wide applications across quantum information, for example they form a basis for universal quantum computation, capture key properties of entanglement, are related to quantum error correction, establish links to graph theory and violate Bell inequalities.

In this section we show how graph states may be defined in the graphical language, and give a graphical proof of the fix point property, a fundamental property of graph states. The next section will expose a limitation of the theory, and we will see that proving Van den Nest's theorem requires an additional axiom.

Definition 2. *For a given simple undirected graph G, let $|G\rangle$ be the corresponding graph state*

$$|G\rangle = \left(\prod_{(u,v)\in E(G)} \wedge Z_{u,v} \right) \left(\bigotimes_{u\in V(G)} \frac{|0\rangle_u + |1\rangle_u}{\sqrt{2}} \right)$$

where $V(G)$ (resp. $E(G)$) is the set of vertices (resp. edges) of G.

Notice that for any $u, v, u', v' \in V(G)$, $\wedge Z_{u,v} = \wedge Z_{v,u}$ and $\wedge Z_{u,v} \circ \wedge Z_{u',v'} = \wedge Z_{u',v'} \circ \wedge Z_{u,v}$, which make the definition of $|G\rangle$ does not depends on the orientation or order of the edges of G.

Since both the state $|+\rangle = \frac{|0\rangle+|1\rangle}{\sqrt{2}}$ and the unitary gate $\wedge Z$ can be depicted in the graphical calculus, any graph state can be represented in the graphical language. For instance, the 3-qubit graph state associated to the triangle is represented as follows:

More generally, any graph G, $|G\rangle$ may be depicted by a diagram composed of $|V(G)|$ green dots. Two green dots are connected with a H gate if and only if the corresponding vertices are connected in the graph. Finally, one output wire is connected to every green dot. Note that the qubits in this picture are the output wires rather than the dots themselves; to act on a qubit with some operation we simply connect the picture for that operation to the wire.

Having introduced the graph states we are now in position to derive one of their fundamental properties, namely the *fixpoint property*.

Property 1 (Fixpoint). Given a graph G and a vertex $u \in V(G)$,

$$R_x(\pi)^{(u)} R_z(\pi)^{(N_G(u))} |G\rangle = |G\rangle$$

The fixpoint property can shown in the graphical calculus by the following example. Consider a star-shaped graph shown below; the qubit u is shown at the top of the diagram, with its neighbours below. The fixpoint property simply asserts that the depicted equation holds.

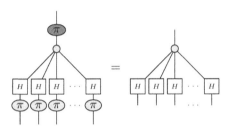

Theorem 1. *The fixpoint property is provable in the graphical language.*

5   Local Complementation

In this section, we present the Van den Nest theorem. According to this theorem, if two graphs are locally equivalent (i.e. one graph can be transformed into the other by means of local complementations) then the corresponding quantum states are LC-equivalent, i.e. there exists a local Clifford unitary[2] which transforms one state into the other. We prove that the local complementation property is true if and only if H has an Euler decomposition into $\pi/2$-green and red rotations. At the end of the section, we demonstrate that the $\pi/2$ decomposition does not hold in all models of the axioms, and hence show that the axiom is truly necessary to prove Van den Nest's Theorem.

Definition 3 (Local Complementation). *Given a graph G containing some vertex u, we define the* local complementation *of u in G, written $G * u$ by the complementation of the neighbourhood of u, i.e. $V(G * u) = V(G)$, $E(G * u) := E(G)\Delta(N_G(u) \times N_G(u))$, where $N_G(u)$ is the set of neighbours of u in G (u is not $N_G(u)$) and Δ is the symmetric difference, i.e. $x \in A\Delta B$ iff $x \in A$ xor $x \in B$.*

Theorem 2 (Van den Nest). *Given a graph G and a vertex $u \in V(G)$,*

$$R_x(-\pi/2)^{(u)} R_z^{(N_G(u))} |G\rangle = |G * u\rangle .$$

---

[2] One-qubit Clifford unitaries form a finite group generated by $\pi/2$ rotations around X and Z axis: $R_x(\pi/2), R_z(\pi/2)$. A n-qubit local Clifford is the tensor product of n one-qubit Clifford unitaries.

We illustrate the theorem in the case of a star graph:

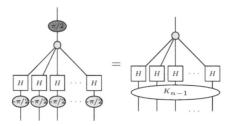

where K_{n-1} denotes the totally connected graph.

Theorem 3. *Van den Nest's theorem holds if and only if H can be decomposed into $\pi/2$ rotations as follows:*

Notice that this equation is nothing but the Euler decomposition of H:

$$H = R_Z(-\pi/2) \circ R_X(-\pi/2) \circ R_Z(-\pi/2)$$

Several interesting consequences follow from the decomposition. We note two:

Lemma 4. *The H-decomposition into $\pi/2$ rotations is not unique:*

Lemma 5. *Each colour of $\pi/2$ rotation may be expressed in terms of the other colour.*

Remark: The preceding lemmas depend only on the *existence* of a decomposition of the form $H = R_z(\alpha)\,R_x(\beta)\,R_z(\gamma)$. It is straight forward to generalise these result based on an arbitrary sequence of rotations, although in the rest of this paper we stick to the concrete case of $\pi/2$.

Most of the rest of the paper is devoted to proving Theorem 3: the equivalence of Van den Nest's theorem and the Euler form of H. We begin by proving the easier direction: that the Euler decomposition implies the local complementation property.

Triangles. We begin with the simplest non trivial examples of local complementation, namely triangles. A local complementation on one vertex of the triangle removes the opposite edge.

Lemma 6

Complete Graphs and Stars. More generally, S_n (a star composed of n vertices) and K_n (a complete graph on n vertices) are locally equivalent for all n.

Lemma 7

General case. The general case can be reduced to the previous case: first green rotations can always be pushed through green dots for obtaining the lhs of equation in Lemma 7. After the application of the lemma, one may have pairs of vertices having two edges (one coming from the original graph, and the other from the complete graph). The Hopf law is then used for removing these two edges.

Lemma 8. *Local complementation implies the H-decomposition:*

This completes the proof of Theorem 3. Note that we have shown the equivalence of two equations, both of which were expressible in the graphical language. What remains to be established is that these properties—and here we focus on the decomposition of H—are not derivable from the axioms already in the system. To do so we define a new interpretation functor.

Let $n \in \mathbb{N}$ and define $[\![\cdot]\!]_n$ exactly as $[\![\cdot]\!]$ with the following change:

$$[\![p_X(\alpha)]\!]_n = [\![p_X(n\alpha)]\!] \qquad [\![p_Z(\alpha)]\!]_n = [\![p_Z(n\alpha)]\!]$$

Note that $[\![\cdot]\!] = [\![\cdot]\!]_1$. Indeed, for all n, this functor preserves all the axioms introduced in Section 2, so its image is indeed a valid model of the theory. However we have the following inequality

$$[\![H]\!]_n \neq [\![p_Z(-\pi/2)]\!]_n \circ [\![p_X(-\pi/2)]\!]_n \circ [\![p_Z(-\pi/2)]\!]_n$$

for example, in $n = 2$, hence the Euler decomposition is not derivable from the axioms of the theory.

6   Conclusions

We studied graph states in an abstract axiomatic setting and saw that we could prove Van den Nest's theorem if we added an additional axiom to the theory. Moreover, we proved that the $\pi/2$-decomposition of H is exactly the extra power which is required to prove the theorem, since we prove that the Van den Nest theorem is true if and only if H has a $\pi/2$ decomposition. It is worth noting that the system without H is already universal in the sense every unitary map is expressible, via an Euler decomposition. The original system this contained two representations of H which could not be proved equal; it's striking that removing this ugly wart on the theory turns out to be necessary to prove a non-trivial theorem. In closing we note that this seemingly abstract high-level result was discovered by studying rather concrete problems of measurement-based quantum computation.

References

1. Abramsky, S., Coecke, B.: A categorical semantics of quantum protocols. In: Proceedings of the 19th Annual IEEE Symposium on Logic in Computer Science: LICS 2004, pp. 415–425. IEEE Computer Society, Los Alamitos (2004)
2. Coecke, B., Duncan, R.: Interacting quantum observables. In: Aceto, L., Damgård, I., Goldberg, L.A., Halldórsson, M.M., Ingólfsdóttir, A., Walukiewicz, I. (eds.) ICALP 2008, Part II. LNCS, vol. 5126, pp. 298–310. Springer, Heidelberg (2008)
3. Hein, M., Dür, W., Eisert, J., Raussendorf, R., Van den Nest, M., Briegel, H.J.: Entanglement in graph states and its applications. In: Proceedings of the International School of Physics "Enrico Fermi" on Quantum Computers, Algorithms and Chaos (July 2005)
4. Raussendorf, R., Briegel, H.J.: A one-way quantum computer. Phys. Rev. Lett. 86, 5188–5191 (2001)

5. Van den Nest, M., Dehaene, J., De Moor, B.: Graphical description of the action of local clifford transformations on graph states. Physical Review A 69, 022316 (2004)
6. Kelly, G., Laplaza, M.: Coherence for compact closed categories. Journal of Pure and Applied Algebra 19, 193–213 (1980)
7. Joyal, A., Street, R.: The geometry of tensor categories i. Advances in Mathematics 88, 55–113 (1991)
8. Selinger, P.: Dagger compact closed categories and completely positive maps. In: Proceedings of the 3rd International Workshop on Quantum Programming Languages (2005)
9. Duncan, R.: Types for Quantum Computation. PhD thesis. Oxford University (2006)
10. Coecke, B., Pavlovic, D., Vicary, J.: A new description of orthogonal bases. Math. Structures in Comp. Sci., 13 (2008), http://arxiv.org/abs/0810.0812 (to appear)

On Stateless Multicounter Machines*

Ömer Eğecioğlu and Oscar H. Ibarra

Department of Computer Science,
University of California, Santa Barbara, CA 93106, USA
{omer,ibarra}@cs.ucsb.edu

Abstract. We investigate the computing power of stateless multi-counter machines with reversal-bounded counters. Such a machine has m-counters and it operates on a one-way input delimited by left and right end markers. The move of the machine depends only on the symbol under the input head and the signs of the counters (zero or positive). At each step, every counter can be incremented by 1, decremented by 1 (if it is positive), or left unchanged. An input string is accepted if, when the input head is started on the left end marker with all counters zero, the machine eventually reaches the configuration where the input head is on the right end marker with all the counters again zero. Moreover, for a specified k, no counter makes more than k-reversals (i.e., alternations between increasing mode and decreasing mode) on any computation, accepting or not. We mainly focus our attention on deterministic realtime (the input head moves right at every step) machines. We show hierarchies of computing power with respect to the number of counters and reversals. It turns out that the analysis of these machines gives rise to rather interesting combinatorial questions.

Keywords: Stateless multicounter machines, reversal-bounded, realtime computation, hierarchies.

1 Introduction

There has been recent interest in studying stateless machines (which is to say machines with only one state), see e.g. [10], because of their connection to certain aspects of membrane computing and P systems, a subarea of molecular computing that was introduced in a seminal paper by Gheorge Păun [8] (see also [9]). A membrane in a P system consists of a multiset of objects drawn from a given finite type set $\{a_1, \ldots, a_k\}$. The system has no global state (i.e., stateless) and works on the evolution of objects in a massively parallel way. Thus, the membrane can be modeled as having counters $x_1, \ldots, x_k$ to represent the multiplicities of objects of types $a_1, \ldots, a_k$, respectively, and the P system can be thought of as a counter machine in a nontraditional form: without states, and with parallel counter increments/decrements. Here, we only consider stateless machines with sequential counter increments/decrements.

---

* This research was supported in part by NSF Grants CCF-0430945 and CCF-0524136.

K. Ambos-Spies, B. Löwe, and W. Merkle (Eds.): CiE 2009, LNCS 5635, pp. 178–187, 2009.
© Springer-Verlag Berlin Heidelberg 2009

Since stateless machines have no states, the move of such a machine depends only on the symbol(s) scanned by the input head(s) and the local portion of the memory unit(s). Acceptance of an input string has to be defined in a different way. For example, in the case of a pushdown automaton (PDA), acceptance is by "null" stack. It is well known that nondeterministic PDAs with states are equivalent to stateless nondeterministic PDAs [2]. However, this is not true for the deterministic case [6]. For Turing Machines, where acceptance is when the machine enters a halting configuration, it can be shown that the stateless version is less powerful than those with states. In [1,5,10] the computing power of stateless multihead automata were investigated with respect to decision problems and head hierarchies. For these devices, the input is provided with left and right end markers. The move depends only on the symbols scanned by the input heads. The machine can be deterministic, nondeterministic, one-way, two-way. An input is accepted if, when all heads are started on the left end marker, the machine eventually reaches the configuration where all heads are on the right end marker. In [7], various types of stateless restarting automata and two-pushdown automata were compared to the corresponding machines with states.

In this paper, we investigate the computing power of stateless multicounter machines with reversal-bounded counters. Such a machine has m-counters. It operates on a one-way input delimited by left and right end markers. The move of the machine depends only on the symbol under the input head and the signs of the counters, which indicate if the counter is zero or not. An input string is accepted if, when the input head is started on the left end marker with all counters zero, the machine eventually reaches the configuration where the input head is on the right end marker with all the counters again zero. Moreover, the machine is k-*reversal bounded*: i.e. for a specified k, no counter makes more than k alternations between increasing mode and decreasing mode (i.e. k pairs of increase followed by decrease stages) in any computation, accepting or not. In this paper, we mainly study deterministic realtime machines (the input head moves right at every step) and show hierarchies of computing power with respect to the number of counters and reversals.

2   Stateless Multicounter Machines

A deterministic stateless one-way m-counter machine operates on an input of the form $cw\$$, where c and $\$$ are the left and right end markers. At the start of the computation, the input head is on the left end marker c and all m counters are zero. The moves of the machine are described by a set of rules of the form:

$$(x, s_1, .., s_m) \rightarrow (d, e_1, ..., e_m)$$

where $x \in \Sigma \cup \{c, \$\}$, Σ is the input alphabet, s_i = sign of counter C_i (0 for zero and 1 for positive), $d = 0$ or 1 (direction of the move of the input head: $d = 0$ means don't move, $d = 1$ means move head one cell to the right), and $e_i = +, -,$ or 0 (increment counter i by 1, decrement counter i by 1, or do not change counter i), with the restriction that $e_i = -$ is applicable only if $s_i = 1$.

Note that since the machine is deterministic, no two rules can have the same left hand sides.

The input w is accepted if the machine reaches the configuration where the input head is on the right end marker $ and all counters are zero. The machine is k-reversal if it has the property that each counter makes at most k "full" alternations between increasing mode and decreasing mode in any computation, accepting or not. Thus, e.g., $k = 2$ means a counter can only go from increasing to decreasing to increasing to decreasing.

There are two types of stateless machines: in *realtime machines* $d = 1$ in every move, i.e., the input head moves right at each step. In this case all the counters are zero when the input head reaches $ for acceptance. In *non-realtime machines* d can be 0 or 1. In particular, when the input head reaches $, the machine can continue computing until all counters become zero and then accept. In this paper, we are mainly concerned with deterministic realtime machines.

3   Stateless Realtime Multicounter Machines

We will show that these devices are quite powerful, even in the unary input alphabet case of $\Sigma = \{a\}$.

Since the machine is realtime, in each rule, $(x, s_1, .., s_m) \rightarrow (d, e_1, ..., e_m)$, $d = 1$. So there is no need to specify d and we can just write the rule as $(x, s_1, ..., s_m) \rightarrow (e_1, ..., e_m)$.

We refer to a vector of signs $v = (s_1, .., s_m)$ that may arise during the computation of an m-counter machine as a *sign-vector*. Thus a sign-vector is a binary string $s_1 \cdots s_m$ of length m, or equivalently a subset S of the set $\{1, 2, \ldots, m\}$. The correspondence between a subset S and a sign vector v is given by putting $s_i = 1$ iff $i \in S$. The string 0^m as a sign-vector signifies that all counters are zero. We will use both binary strings and subsets of $\{1, 2, \ldots, m\}$ interchangeably in describing sign-vectors. For $w \in \Sigma^*$ and $a \in \Sigma$, we define $|w|_a$ as the number of occurrences of a in w.

Theorem 1. *Every language over* $\Sigma = \{a\}$ *accepted by a stateless realtime multicounter machine* M *is of the form* $a^r(a^s)^*$ *for some* $r, s \geq 0$.

Proof. We show that

1. if $(c, 0^m) \rightarrow 0^m$ is a move of M then M accepts ε or Σ^*,
2. if $(c, 0^m) \rightarrow S$ with $S \neq 0^m$ for some S is a move of M, then M accepts a singleton or an infinite language.

For 1., we consider two subcases: if M has no move of the form $(a, 0^m) \rightarrow (e_1, ..., e_m)$, then an accepting computation must proceed from c to $ directly; i.e. the accepted input is c. In this case $r = s = 0$. However if M does have the move $(a, 0^m) \rightarrow 0^m$, then M accepts Σ^*, corresponding to the values $r = 0, s = 1$.

In case 2., $(c, 0^m) \rightarrow S_1$ with $S_1 \neq 0^m$ is a rule of M and that M halts. Let r be the smallest integer such that M accepts a^r. Then $r > 0$ and when the machine is reading the a immediately before $, the sign-vector is some nonempty

S_r, and the counter values are 1 for counters corresponding to 1's in S_r, and 0 otherwise.

We consider the two subcases depending on whether or not $(a, 0^m) \to 0^m$ is a rule of M. If $(a, 0^m) \to 0^m$ is a rule of M then M accepts $a^r \Sigma^*$. If $(a, 0^m) \to S$ with $S \neq 0^m$ is a rule of M then let $s > 0$ be the smallest integer such that M accepts a^{r+s}. If there is no such integer, then M accepts the singleton a^r. Otherwise $a^{r+ks} \in L$ for $k \geq 0$. □

3.1   1-Reversal Machines

We are interested in machines accepting only a singleton language of the form $L = \{a^n\}$. We derive the precise value of the maximum such n and show that the program of the machine achieving this n is unique (up to relabeling of the counter indices). Note that we can interpret a^n as the "maximum" number that a 1-reversal m-counter machine can count.

By Theorem 1, if $(c, 0^m) \to 0^m$, then a realtime stateless machine accepts either ε or Σ^*. Therefore an accepting computation of an m-counter machine that accepts a non-null singleton can be represented in the form

$$0^m \to S_1 \to S_2 \to \cdots \to S_r \to 0^m \tag{1}$$

where $S_i \neq 0^m$ for $1 \leq i \leq r$ and the arrows indicate the sequence of sign-vectors after each move of the machine. Such a machine accepts the singleton language $\{a^r\}$, or equivalently can count up to r.

Borrowing terminology from the theory of Markov chains, we call a nonempty sign-vector S *transient* if S appears exactly once in (1), and *recurrent* if S appears more than once in (1). Thus a transient S contributes exactly one move, whereas a recurrent S is "re-entered" more than once during the course of the computation of M. We prove a few lemmas characterizing possible computations of a 1-reversal machine accepting a singleton.

Lemma 1. *If S is recurrent and S' appears between two occurrences of S in (1), then $S \subseteq S'$.*

Proof. Suppose $j \in S \setminus S'$. Then counter C_j zero at the start of the computation, it is nonzero at the first appearance of S, and is zero at or before the appearance of S' in the computation. Since M is 1-reversal, it is not possible for counter C_j to be nonzero again at the second occurrence of S. □

Lemma 2. *If S is recurrent and S' appears between two occurrences of S in (1), then $S' = S$.*

Proof. Suppose that the first occurrence of S is followed by S''; i. e.

$$S \to S'' \to \cdots \to S' \to \cdots \to S$$

We first show that $S'' \subseteq S$. By way of contradiction, assume $j \in S'' \setminus S$. Then at S the counter C_j must be incremented from 0 to 1. But this cannot happen more than once at the two S's since M is 1-reversal. Therefore $S'' \subseteq S$. By Lemma 1, $S \subseteq S''$, and therefore $S = S''$. It follows that all sign-vectors S' between two occurrences of S are equal to S.   □

The next lemma gives an upper bound on the number of distinct sign-vectors that can appear in (1).

Lemma 3. *Suppose $S_1, S_2, \ldots, S_d$ are the distinct non-null sign-vectors that appear in the computation of a 1-reversal machine M with m counters that accepts a singleton. Then $d \leq 2m - 1$.*

Proof. Put $S_0 = S_{d+1} = 0^m$ and consider the $m \times (d+2)$ binary matrix B where the j-th column is S_{j-1} for $1 \leq j \leq d+2$. Since M is 1-reversal, each row of B has at most one interval consisting of all 1's. Therefore there are a total of at most m horizontal intervals of 1's in B. These together have at most $2m$ endpoints: i.e. a pair 01 where a such an interval of 1's starts, and a pair 10 where an interval ends. Since the S_j are distinct, going from S_j to S_{j+1} for $0 \leq j \leq d$ must involve at least one bit change, i.e. at least one of the $2m$ pairs of 01 and 10's. It follows that $d+1 \leq 2m$.   □

Lemma 4. *Suppose at the first occurrence of the sign-vector S in (1), the values of the counters are $v_1(S) \geq v_2(S) \geq \cdots \geq v_m(S) \geq 0$. Then*

1. *$v_1(S) - 1$ is an upper bound to the number of times M makes a move from S back to S.*
2. *$v_1(S) + v_2(S)$ is an upper bound to the largest counter value when M makes a move from S to some S'.*

Proof. The lemma is a consequence of the fact that since M halts, some counter for any non-null sign-vector S must be decremented.   □

Lemmas 3 and 4 immediately give an upper bound on how high a 1-reversal, m-counter machine can count. We necessarily have $v_1(S_1) = 1$, and this value can at most double at each S_i. Therefore we obtain the bound

$$1 + 2 + 2^2 + \cdots + 2^{2m-2} = 2^{2m-1} - 1 .\tag{2}$$

As an example, for $m = 2$, the bound given in (2) is 7. This is almost achieved by the machine M defined by the following program with four rules:

$$(\text{¢}, 00) \rightarrow +0, \ (a, 01) \rightarrow 0-, \ (a, 10) \rightarrow ++, \ (a, 11) \rightarrow -+ , \tag{3}$$

where we have used the notation $(a, 11) \rightarrow -+$ for the move $(a, 1, 1) \rightarrow (-, +)$, and similarly for others. The computation of M proceeds as follows:

Sign-vector	Entering counter values	Move number
0 0	0 0	0
1 0	1 0	1
1 1	2 1	2
1 1	1 2	3
0 1	0 3	4
0 1	0 2	5
0 1	0 1	6
0 0	0 0	

Therefore M can count up to $n = 6$. However, we can do better than (2). The reason is that in order to be able to double $v_1(S)$, other than the trivial case of $S = S_1$, S has to be recurrent, and it is not possible to have all S_i recurrent if M is 1-reversal. Consider the machine M_m^* whose moves are defined as:

$$
\begin{aligned}
(c, 0^m) &\rightarrow +0^{m-1} , \\
(a, 1^i 0^{m-i}) &\rightarrow +^{i+1} 0^{m-i-1} \text{ for } 0 < i < m , \\
(a, 0^j 1^{m-j}) &\rightarrow 0^j - +^{m-j-1} \text{ for } 0 \le j < m .
\end{aligned}
\tag{4}
$$

The first set of sign-vectors above defined for input a are transient and the second ones are recurrent. The transient sign-vectors accumulate as much as possible in the counters, and the recurrent ones spend as much time as possible while about doubling the maximum counter value. The machine M described in (3) is M_2^*. The machine M_3^*, which can count up to $n = 19$, is as follows:

Sign-vector	Entering counter values	Move number
0 0 0	0 0 0	0
1 0 0	1 0 0	1
1 1 0	2 1 0	2
1 1 1	3 2 1	3
1 1 1	2 3 2	4
1 1 1	1 4 3	5
0 1 1	0 5 4	6
0 1 1	0 4 5	7
0 1 1	0 3 6	8
0 1 1	0 2 7	9
0 1 1	0 1 8	10
0 0 1	0 0 9	11
0 0 1	0 0 8	12
0 0 1	0 0 7	13
0 0 1	0 0 6	14
0 0 1	0 0 5	15
0 0 1	0 0 4	16
0 0 1	0 0 3	17
0 0 1	0 0 2	18
0 0 1	0 0 1	19
0 0 0	0 0 0	

Theorem 2. *The machine M_m^* described in (4) is 1-reversal and accepts the singleton $\{a^n\}$ with*

$$
n = (m-1)2^m + m .
\tag{5}
$$

The value of n given in (5) is the maximal value that a single reversal m counter machine can count. Furthermore, the program of any machine that achieves this bound is unique up to relabeling of the counters.

Proof. First we establish the bound in (5). The first m moves of the machine result in the sign-vector 1^m and the counter contents $m, m-1, \ldots, 2, 1$. After m more moves, we arrive at the sign-vector 01^{m-1} with contents of the counters

$$0, 2m-1, 2m-2, \ldots, m+1.$$

At this point the first counter has made a reversal, but all the other counters are still increasing. When first j counters are zeroed, each of them has completed a single reversal and the sign-vector becomes $0^j 1^{m-j}$ for $1 \leq j < m$. The largest counter value when we enter this sign-vector is $2^{j-1}m - 2^{j-1} + 1$. When the machine starts to decrement the last counter, its content is $2^{m-1}m - 2^{m-1} + 1$. Therefore the number of moves from 10^{m-1} to the last appearance of $0^{m-1}1$ is

$$m - 1 + (2^{m-1}m - 2^{m-1} + 1) + (2^{m-1}m - 2^{m-1}) = (m-1)2^m + m .$$

We sketch the proof of uniqueness. Since M halts, any recurrent sign-vector must decrease one or more counters. Therefore a recurrent S is followed by some subset of S. Since each such recurrent sign-vector can be used to double the maximum content of the counters whereas a transient one only contributes 1 move, the chain of subsets must be as long as possible. By relabeling if necessary, we can assume that these subsets are $0^j 1^{m-j}$ for $0 \leq j < m$. This leaves $m - 1$ distinct sign vectors, and the pattern of intervals of 1's that is necessary because M is 1-reversal, forces these to be transient and in the form $1^i 0^{m-i} \to +^{i+1} 0^{m-i-1}$ for $0 \leq i < m$. Finally the largest values of the counters when the machine enters the recurrent sign-vector 1^m is when the moves are defined as in M_m^*.   □

Next we prove that for 1-reversal machines $m + 1$ counters is better than m counters. Here we no longer assume that the language accepted is a singleton (or finite), nor a unary alphabet.

Theorem 3. *Suppose L is accepted by a realtime 1-reversal machine with m counters. Then L is accepted by a realtime 1-reversal machine with $m + 1$ counters. Furthermore the containment is strict.*

Proof. Given a realtime 1-reversal machine M with m counters that accepts M, we can view M as an $m + 1$ counter machine which behaves exactly like M on the first m counters, and never touches the $(m + 1)$-st counter. Since the acceptance of an input string is defined by entering \$ when all counters are zero, this machine is also 1-reversal and accepts L. By theorem 2, the singleton $\{a^n \mid n = m2^{m+1} + m + 1\}$ is accepted by the 1-reversal machine M_{m+1}^*. Since $(m-1)2^m + m < m2^{m+1} + m + 1$ for $m > 0$, this language is not accepted by any 1-reversal realtime machine with m-counters.   □

3.2   *k*-Reversal Machines

Now we consider k-reversal machines, $k \geq 1$. The next result gives an upper bound on the maximal n that is countable by a k-reversal m-counter machine.

Theorem 4. *If the upper bound on n for 1-reversal m-counter machine is $f(m)$, then $f((2k-1)m)$ is an the upper bound on n for a k-reversal m-counter machine.*

Proof. We sketch the proof. Let $L = \{a^n\}$ be a singleton language accepted by a k-reversal m-counter machine M. We will show how we can construct from M a 1-reversal $(2k-1)m$-counter machine M' that makes at least as many steps as M and accepts a language $L' = \{a^{n'}\}$ for some $n' \geq n$. The result then follows. The construction of M' from M is based on the following ideas:

1. Consider first the case $k = 2$. Assume for now that the counters reverse from decreasing to increasing at different times.
2. Let C be a counter in M that makes 2 reversals. We associate with C three counters C, T, C' in M'. Initially, $T = C' = 0$.
3. C in M' simulates C in M as long as C does not decrement. When C decrements, T is set to 1 (i.e., it is incremented). Then as long as C does not increment the simulation continues.
4. When C in M increments, C in M' is decremented while simultaneously incrementing C' until C becomes zero. During the decrementing process all other counters remain unchanged. But to make M' operate in realtime, its input head always reads an a during this process.
5. When the counter C of M' becomes zero, T is set to zero (i,e., it is decremented), and C' is incremented by 1.
6. Then the simulation continues with C' taking the place of C. Counters C and T remain at zero and no longer used.

So if C in M makes 2 reversals, we will need three 1-reversal counters C, T, C' in M'. If C makes 3 reversals, we will need five 1-reversal counters C, T, C', T', C'' in M'. In general, if C makes k reversals, we will need $(2k-1)$ 1-reversal counters in M'. It follows that if there are m counters where each counter makes k reversals, we will need $(2k-1)m$ 1-reversal counters. If some of the counters "reverse" (to increasing) at the same time, we handle them systematically one at a time, by indexing the counters.   □

Corollary 1. *If $L = \{a^n\}$ is accepted by a k-reversal m-counter machine, then*

$$n \leq ((2k-1)m - 1)2^{(2k-1)m} + (2k-1)m .$$

We can also prove that the number of counters matters for k-reversal machines.

Theorem 5. *For any fixed k, there is a language accepted by a k-reversal $(m+1)$-counter machine which is not accepted by any k-reversal m-counter machine.*

Proof. Consider $L = \{a^n\}$ where n is the largest number that a k-reversal m-counter machine can count. A bound for n is given in Corollary 1. Suppose M is such a machine and the sequence of sign-vectors in its calculation is

$$0^m \rightarrow S_1 \rightarrow S_2 \rightarrow \cdots \rightarrow S_t \rightarrow 0^m$$

where $S_i \neq 0^m$ for $1 \leq i \leq t$. M must have the rule $(a, S_t) \rightarrow (e_1, e_2, \ldots, e_m)$ where $e_i = -$ for $i \in S_t$ and $e_i = 0$ otherwise. We construct a k-reversal $(m+1)$-counter machine M' that accepts a longer singleton. Define M' by

1. If $(c, 0^m) \to (e_1, ..., e_m)$ is in M, then $(c, 0^{m+1}) \to (e_1, ..., e_m, +)$ is in M',
2. If $(a, s_1, ..., s_m) \to (e_1, ..., e_m)$ is in M, then $(a, s_1, ..., s_m, 1) \to (e_1, ..., e_m, +)$ is in M',
3. $(a, 0, ..., 0, 1) \to (0, ..., 0, -)$ is in M'.

Thus with every move of M the counter C_{m+1} is incremented until the last S_t clears its contents, at which point C_{m+1} starts clearing its contents, i.e. makes one reversal. It is clear that M' is k-reversal like M, and it accepts the longer singleton $L = \{a^{2n-1}\}$.   □

We can also show that for a fixed m, which may depend on k, $k + 1$ reversals are better than k.

Theorem 6. *For any fixed m and $k < 2^{m-1}/m$, there is a language accepted by a $(k+1)$-reversal m-counter machine which is not accepted by any k-reversal m-counter machine.*

Proof. We need the following generalization of Lemma 3 to k-reversal machines. Suppose $S_1, S_2, \ldots, S_d$ are the distinct non-null sign-vectors that appear in the computation of a k-reversal machine M with m counters that accepts a singleton. Then $d \leq 2km - 1$. To prove this inequality, put $S_0 = S_{d+1} = 0^m$ and as in Lemma 3, consider the $m \times (d+2)$ binary matrix B where the j-th column is S_{j-1} for $1 \leq j \leq d + 2$. Since M is k-reversal, each row of B has at most k intervals consisting of all 1's. Therefore there are a total of at most km horizontal intervals of 1's in B, which together have at most $2km$ endpoints. Since the S_j are distinct, going from S_j to S_{j+1} for $0 \leq j \leq d$ must involve at least one bit change, i.e. at least one of the $2m$ pairs of 01 and 10's. It follows that $d + 1 \leq 2km$.

Since $k < 2^{m-1}/m$, $d < 2^m - 1$. Therefore we can find a non-null set S with $S \neq S_i$ for $1 \leq i \leq d$. Now the longest singleton accepted by a k-reversal m-counter machine M is $\{a^n\}$ where n is as given in Theorem 2. We use S to construct a $(k+1)$-reversal m-counter machine M' which accepts a string longer than n. M' is constructed as follows.

1. $(c, 0^m) \to (e_1, ..., e_m)$ is in M', where $e_i = +$ for $i \in S$ and $e_i = 0$ otherwise.
2. If $(c, 0^m) \to (f_1, ..., f_m)$ is in M, then $(a, S) \to (S_1)$ is in M', where S_1 is the sign-vector defined by $i \in S_1$ if $f_i = +$ and $i \notin S_1$ if $f_i = 0$,
3. If $(a, s_1, ..., s_m) \to (g_1, ..., g_m)$ is in M, with $s_1 \cdots s_m \notin \{0^m, S\}$, then $(a, s_1, ..., s_m) \to (g_1, ..., g_m)$ is in M'.

Thus M' makes an extra initial move, and then continues as M does. The appearance of S at the beginning of the computation can only introduce one more reversal. Therefore M' is $(k+1)$-reversal, and accepts $\{a^{n+1}\}$.   □

Remark. If M is reversal bounded, then we must have $s = 0$ or $s = 1$ in Theorem 1. Note that the latter case arises only when $(a, 0^m) \to 0^m$ is a rule of M. Since there are finitely many programs for fixed m, there are finitely many sequences of sign vectors of the form (1). Let m' denote the maximum number of reversals among these sign vectors. Then for $k \geq m'$, every language language accepted by a k-reversal m-counter machine is accepted by a m'-reversal m-counter machine. Therefore some assumption on the number of reversals k is necessary for the hierarchy guaranteed in Theorem 6.

4   Stateless Non-realtime Multicounter Machines

We briefly discuss the case when in each rule, d can be 1 or 0 (i.e., the input head need not move right at each step). Clearly, any language accepted by a realtime machine can be accepted by a non-realtime machine. The latter is strictly more powerful, since it has been shown in [4] that the language $L = \{w \mid w \in \{a, b\}^*, |w|_a = |w|_b\}$ can be accepted by a non-realtime 1-reversal 2-counter machine but not by any realtime k-reversal m-counter machine for any $k, m \geq 1$. In fact, a stronger result was shown in [4]. A non-realtime reversal-bounded multicounter machine is *restricted* if it can only accept an input when the input head first reaches the right end marker \$ and all counters are zero. Hence, there is no applicable rule when the input head is on \$. However, the machine can be non-realtime (i.e., need not move at each step) when the head is not on \$. The machine can also be *nondeterministic:* two or more rules can have the same left hand side. Then L cannot be accepted by any restricted nondeterministic non-realtime reversal-bounded multicounter machine [4]. These results show that even only allowing non-realtime computation on the end marker (i.e., allowing the machine to remain on \$ and continue computing before accepting) makes the machine more powerful.

References

1. Frisco, P., Ibarra, O.H.: On stateless multihead finite automata and multihead pushdown automata. In: Proceedings of 13th International Conference on Developments in Language Theory. LNCS (July 2009) (to appear)
2. Hopcroft, J.E., Ullman, J.D.: Introduction to Automata Theory, Languages and Computation. Series in Computer Science. Addison -Wesley, Reading (1979)
3. Ibarra, O.H.: Reversal-bounded multicounter machines and their decision problems. J. Assoc. for Computing Machinery 25, 116–133 (1978)
4. Ibarra, O.H., Eğecioğlu, Ö.: Hierarchies and characterizations of stateless multicounter machines. In: Federrath, H. (ed.) Proceedings of 15th International Computing and Combinatorics Conference (COCOON 2009). LNCS (2001) (to appear)
5. Ibarra, O.H., Karhumaki, J., Okhotin, A.: On stateless multihead automata: hierarchies and the emptiness problem. In: Laber, E.S., Bornstein, C., Nogueira, L.T., Faria, L. (eds.) LATIN 2008. LNCS, vol. 4957, pp. 94–105. Springer, Heidelberg (2008)
6. Korenjak, A.J., Hopcroft, J.E.: Simple deterministic languages. In: Proceedings of IEEE 7th Annu. Symp. on Switching and Automata Theory, pp. 36–46 (1966)
7. Kutrib, M., Messerschmidt, H., Otto, F.: On stateless two-pushdown automata and restarting automata. In: Pre-Proceedings of the 8th Automata and Formal Languages (May 2008)
8. Păun, G.: Computing with Membranes. Journal of Computer and System Sciences 61(1), 108–143 (2000) (and Turku Center for Computer Science-TUCS Report 208) (November 1998), http://www.tucs.fi
9. Păun, G.: Computing with Membranes: An Introduction. Springer, Berlin (2002)
10. Yang, L., Dang, Z., Ibarra, O.H.: On stateless automata and P systems. In: Pre-Proceedings of Workshop on Automata for Cellular and Molecular Computing (August 2007)

Computability of Continuous Solutions of Higher-Type Equations

Martín Escardó

School of Computer Science, University of Birmingham, UK

Abstract. Given a continuous functional $f: X \to Y$ and $y \in Y$, we wish to compute $x \in X$ such that $f(x) = y$, if such an x exists. We show that if x is unique and X and Y are subspaces of Kleene–Kreisel spaces of continuous functionals with X exhaustible, then x is computable uniformly in f, y and the exhaustion functional $\forall_X : 2^X \to 2$. We also establish a version of the above for computational metric spaces X and Y, where is X computationally complete and has an exhaustible set of Kleene–Kreisel representatives. Examples of interest include functionals defined on compact spaces X of analytic functions.

Keywords: Higher-type computability, Kleene–Kreisel spaces of continuous functionals, exhaustible set.

1 Introduction

Given a continuous functional $f: X \to Y$ and $y \in Y$, we consider the equation $f(x) = y$ with the unknown $x \in X$. We show that if X and Y are subspaces of Kleene–Kreisel spaces [1] with X *exhaustible* [2], the solution is computable uniformly in f, y and the exhaustion functional $\forall_X : 2^X \to 2$, provided there is a unique solution (Section 3). Here exhaustibility plays the role of a computational counter-part of the topological notion of compactness (Section 2). Moreover, under the same assumptions for X and Y, it is uniformly semi-decidable whether a solution $x \in X$ fails to exist.

Recall that the Kleene–Kreisel spaces are obtained from the discrete space $\mathbb{N}$ by iterating finite products and function spaces in a suitable cartesian closed category (e.g. filter spaces, limit spaces, k-spaces, equilogical spaces or QCB spaces). For computability background, see [1] or [2]. We exploit the fact that, by cartesian closedness, computable functionals are closed under λ-definability.

The computation of unique solutions of equations of the form $g(x) = h(x)$ with $g, h: X \to Y$ is easily reduced to the previous case, because there are (abelian) computable group structures on the ground types that can be lifted componentwise to product types and pointwise to function types, and hence $x \in X$ is a solution of such an equation iff it is a solution of the equation $f(x) = 0$, where $f(x) = h(x) - g(x)$. And, by cartesian closedness, the case in which g and h computably depend on parameters, and in which the solution computably depends on the same parameters, is covered. Moreover, because the Kleene–Kreisel spaces are closed under finite products and countable powers,

K. Ambos-Spies, B. Löwe, and W. Merkle (Eds.): CiE 2009, LNCS 5635, pp. 188–197, 2009.

this includes the solution of finite and countably infinite systems of equations with functionals of finitely many or countably infinitely many variables.

We also consider a generalization to computational metric spaces that applies to computational analysis, where f can be a functional and x a function (Section 4). And we develop examples of sets of analytic functions that are exhaustible and can play the role of the space X (Section 5).

Organization. (2) Exhaustible subspaces of Kleene–Kreisel spaces. (3) Equations over Kleene–Kreisel spaces. (4) Equations over metric spaces. (5) Exhaustible spaces of analytic functions.

2   Exhaustible Subspaces of Kleene–Kreisel Spaces

In previous work we investigated exhaustible sets of total elements of effectively given domains and their connections with Kleene–Kreisel spaces of continuous functionals [2]. Here we work directly with exhaustible subspaces of Kleene–Kreisel spaces, where in this section we translate notions and results for them from that work. Denote by Y^X the space of continuous functionals from X to Y.

Definition 2.1. Let $2 = \{0, 1\}$ be discrete.

1. A space K is called *exhaustible* if the universal quantification functional

$$\forall_K : 2^K \to 2$$

   defined by $\forall_K(p) = 1$ iff $p(x) = 1$ for all $x \in K$ is computable.
2. It is called *searchable* if there is a computable selection functional

$$\varepsilon_K : 2^K \to K$$

   such that for all $p \in 2^K$, if there is $x \in K$ with $p(x) = 1$ then $p(\varepsilon_K(p)) = 1$.
3. A set $F \subseteq X$ is *decidable* if its characteristic map $X \to 2$ is computable. $\quad\square$

Equivalently, K is exhaustible iff the functional $\exists_K : 2^K \to 2$ defined by $\exists_K(p) = 1$ iff $p(x) = 1$ for some $x \in K$ is computable. If K is searchable, then it is exhaustible, because $\exists_K(p) = p(\varepsilon_K(p))$. The empty space is exhaustible, but not searchable, because there is no map $2^\emptyset \to \emptyset$.

The following results are directly adapted to our setting from [2].

Lemma 2.1

1. *The Cantor space $2^{\mathbb{N}}$ is searchable.*
2. *Any exhaustible subspace of a Kleene–Kreisel space is compact, and moreover, if it is non-empty, it is searchable, a computable retract, and a computable image of the Cantor space.*
3. *Searchable spaces are closed under computable images, intersections with decidable sets, and finite products.*
4. *A product of countably many searchable subspaces of a common Kleene–Kreisel space is searchable uniformly in the sequence of quantifiers.*

Thus, exhaustibility is a computational counter-part of the topological notion of compactness, at least for subspaces of Kleene–Kreisel spaces.

3 Equations over Kleene–Kreisel Spaces

We emphasize that in this paper, including Section 4, the terminology *uniform* is used in the sense of recursion theory, rather than metric topology.

Theorem 3.1. *If $f\colon X \to Y$ is a continuous map of subspaces of Kleene–Kreisel spaces with X exhaustible, and $y \in Y$, then, uniformly in $\forall_X$, f, and y:*

1. *It is semi-decidable whether the equation $f(x) = y$ fails to have a solution.*
2. *If $f(x) = y$ has a unique solution $x \in X$, then it is computable.*

Hence if $f\colon X \to Y$ is a computable bijection then it has a computable inverse, uniformly in $\forall_X$ and f.

The conclusion is a computational counter-part of the topological theorem that any continuous bijection from a compact Hausdorff space to a Hausdorff space is a homeomorphism.

Remark 3.1. The uniqueness assumption in the second part is essential. In fact, consider e.g. $X = 2$ (which is trivially exhaustible) and $Y = \mathbb{N}^{\mathbb{N}}$. Then a map $f\colon X \to Y$ amounts to two functions $f_0, f_1\colon \mathbb{N} \to \mathbb{N}$. Hence computing a solution to the above equation amounts to finding $i \in 2$ such that $f_i = y$ holds, that is, $f_i(n) = y(n)$ for all $n \in \mathbb{N}$. In other words, under the assumption that $f_0 = y$ or $f_1 = y$, we want to find i such that $f_i = y$. If the only data supplied to the desired algorithm are f_0, f_1, y, this is not possible, because no finite amount of information about the data can determine that one particular disjunct holds. When specialized to this example, the proof of the theorem relies on the additional information that only one of the disjuncts holds.   □

The following will be applied to semi-decide absence of solutions:

Lemma 3.1. *Let X be an exhaustible subspace of a Kleene–Kreisel space and $K_n \subseteq X$ be a sequence of sets that are decidable uniformly in n and satisfy $K_n \supseteq K_{n+1}$. Then, uniformly in the data:*

emptiness of $\bigcap_n K_n$ is semi-decidable.

Proof. Because X is compact by exhaustibility, K_n is also compact as it is closed. Because X is Hausdorff, $\bigcap_n K_n = \emptyset$ iff there is n such that $K_n = \emptyset$. But emptiness of this set is decidable uniformly in n by the algorithm $\forall x \in X.x \notin K_n$. Hence a semi-decision procedure is given by $\exists n.\forall x \in X.x \notin K_n$.   □

As a preparation for a lemma that will be applied to compute unique solutions, notice that if a singleton $\{u\} \subseteq \mathbb{N}^Z$ is exhaustible, then the function u is computable, because $u(z) = \mu m.\forall v \in \{u\}.v(z) = m$. Moreover, u is computable uniformly in $\forall_{\{u\}}$, in the sense that there is a computable functional

$$U\colon S \to \mathbb{N}^Z \quad \text{with} \quad S = \{\phi \in 2^{2^{\mathbb{N}^Z}} \mid \phi = \forall_{\{v\}} \text{ for some } v \in \mathbb{N}^Z\},$$

such that $u = U\left(\forall_{\{u\}}\right)$, namely $U(\phi)(z) = \mu m.\phi(\lambda u.u(z) = m)$. Lemma 3.2 below generalizes this, using an argument from [2] that was originally used to prove that non-empty exhaustible subsets of Kleene–Kreisel spaces are computable images of the Cantor space and hence searchable. Here we find additional applications and further useful generalizations.

Lemma 3.2. *Let X be an exhaustible subspace of a Kleene–Kreisel space and $K_n \subseteq X$ be a sequence of sets that are exhaustible uniformly in n and satisfy $K_n \supseteq K_{n+1}$. Then, uniformly in the data:*

if $\bigcap_n K_n$ is a singleton $\{x\}$, then x is computable.

Proof. By Lemma 2.1, X is a computable retract of its Kleene–Kreisel super-space. Because any Kleene–Kreisel space is a computable retract of a Kleene–Kreisel space of the form $\mathbb{N}^Z$, and because retractions compose, there are computable maps $s\colon X \to \mathbb{N}^Z$ and $r\colon \mathbb{N}^Z \to X$ with $r \circ s = \mathrm{id}_X$. It suffices to show that the function $u = s(x) \in \mathbb{N}^Z$ is computable, because $x = r(u)$. The sets $L_n = s(K_n) \subseteq \mathbb{N}^Z$, being computable images of exhaustible sets, are themselves exhaustible. For any $z \in Z$, the set $U_z = \{v \in \mathbb{N}^Z \mid v(z) = u(z)\}$ is clopen and $\bigcap_n L_n = \{u\} \subseteq U_z$. Because $\mathbb{N}^Z$ is Hausdorff, because $L_n \supseteq L_{n+1}$, because each L_n is compact and because U_z is open, there is n such that $L_n \subseteq U_z$. That is, $v \in L_n$ implies $v(z) = u(z)$. Therefore, for every $z \in Z$ there is n such that $v(z) = w(z)$ for all $v, w \in L_n$. Now, the map $n(z) = \mu n.\forall v, w \in L_n.v(z) = w(z)$ is computable by the exhaustibility of L_n. But $u \in L_{n(z)}$ for any $z \in Z$ and therefore u is computable by exhaustibility as $u(z) = \mu m.\forall v \in L_{n(z)}.v(z) = m$, as required. □

To build sets K_n suitable for applying these two lemmas, we use:

Lemma 3.3. *For every computable retract of a Kleene–Kreisel space, there is a family $(=_n)$ of equivalence relations that are decidable uniformly in n and satisfy*

$$x = x' \iff \forall n.\, x =_n x',$$
$$x =_{n+1} x' \implies x =_n x'.$$

Proof. Let X be a Kleene–Kreisel space and $s\colon X \to \mathbb{N}^Z$ and $r\colon \mathbb{N}^Z \to X$ be computable maps with Z a Kleene–Kreisel space and $r \circ s = \mathrm{id}_X$. By the density theorem, there is a computable dense sequence $\delta_n \in Z$. Then the definition

$$x =_n x' \iff \forall i < n.s(x)(\delta_i) = s(x')(\delta_i)$$

clearly produces an equivalence relation that is decidable uniformly in n and satisfies $x =_{n+1} x' \implies x =_n x'$. Moreover, $x = x'$ iff $s(x) = s(x')$, because s is injective, iff $s(x)(\delta_n) = s(x')(\delta_n)$ for every n, by density, iff $x =_n x'$ for every n, by definition. □

Proof (of Theorem 3.1). The set $K_n = \{x \in X \mid f(x) =_n y\}$, being a decidable subset of an exhaustible space, is exhaustible. Therefore the result follows from Lemmas 3.1 and 3.2, because $x \in \bigcap_n K_n$ iff $f(x) =_n y$ for every n iff $f(x) = y$ by Lemma 3.3. □

Algorithms 3.2. In summary, the algorithm for semi-deciding non-existence of solutions is

$$\exists n.\forall x \in X.f(x) \neq_n y,$$

and that for computing the solution x_0 as a function of $\forall_X$, f, and y is:

$$\forall x \in K_n.p(x) = \forall x \in X.f(x) =_n y \implies p(x),$$
$$\forall v \in L_n.q(v) = \forall x \in K_n.q(s(x)),$$
$$n(z) = \mu n.\forall v, w \in L_n.v(z) = w(z),$$
$$u(z) = \mu m.\forall v \in L_{n(z)}.v(z) = m,$$
$$x_0 = r(u).$$

Here $r\colon \mathbb{N}^Z \to X$ is a computable retraction with section $s\colon X \to \mathbb{N}^Z$, where Z is a Kleene–Kreisel space, as constructed in the proof of Lemma 3.2.   □

Of course, even in the absence of uniqueness, *approximate* solutions with precision n are trivially computable with the algorithm

$$\varepsilon_X(\lambda x.f(x) =_n y),$$

using the fact that non-empty exhaustible subsets of Kleene–Kreisel spaces are searchable. But the above unique-solution algorithm uses the quantification functional $\forall_X$ rather than the selection functional ε_X. In the next section we compute solutions as limits of approximate solutions.

4   Equations over Metric Spaces

For the purposes of this and the following section, we can work with computational spaces in the sense of TTE [3] using Baire-space representations, or equivalently, using partial equivalence relations on representatives living in effectively given domains [4]. Our development applies to both, and we can more generally assume for the former that representatives form subspaces of arbitrary Kleene–Kreisel spaces rather than just the Baire space $\mathbb{N}^{\mathbb{N}}$. We first formulate the main result of this section and then supply the missing notions in Definition 4.2:

Theorem 4.1. *Let X and Y be computational metric spaces with X computationally complete and having an exhaustible Kleene–Kreisel representation.*

If $f\colon X \to Y$ is continuous and $y \in Y$, then, uniformly in f, y and the exhaustibility condition:

1. *It is semi-decidable whether the equation $f(x) = y$ fails to have a solution.*
2. *If $f(x) = y$ has a unique solution $x \in X$, then it is computable.*

Hence any computable bijection $f\colon X \to Y$ has a computable inverse, uniformly in f and the exhaustibility condition.

Given that exhaustibility is a computational counter-part of the topological notion of compactness, and that compact metric spaces are complete, it is natural to conjecture that, at least under suitable computational conditions, the assumption of computational completeness in the above theorem is superfluous. We leave this as an open question. In connection with this, notice that this theorem is analogous to a well-known result in constructive mathematics [5], with the assumptions reformulated in our higher-type computational setting.

There is a technical difficulty in the proof of the theorem: at the intensional level, where computations take place, solutions are unique only up to equivalence of representatives. In order to overcome this, we work with pseudo-metric spaces at the intensional level and with a notion of decidable closeness for them. Recall that a *pseudo-metric* on a set X is a function $d \colon X \times X \to [0, \infty)$ such that

$$d(x, x) = 0, \quad d(x, y) = d(y, x), \quad d(x, z) \le d(x, y) + d(y, z).$$

Then d is a *metric* if it additionally satisfies $d(x, y) = 0 \implies x = y$. If d is only a pseudo-metric, then $(\sim)$ defined by

$$x \sim y \iff d(x, y) = 0$$

is an equivalence relation, referred to as *pseudo-metric equivalence*. A pseudo-metric topology is Hausdorff iff it is T_0 iff the pseudo-metric is a metric. Moreover, two points are equivalent iff they have the same neighbourhoods. Hence any sequence has at most one limit up to equivalence.

A *computational metric space* is a computational pseudo-metric space in which the pseudo-metric is actually a metric, and hence we formulate the following definitions in the generality of pseudo-metric spaces.

Definition 4.2. We work with any standard (admissible) representation of the Hausdorff space $[0, \infty)$.

1. A *computational pseudo-metric space* is a computational space X endowed with a computable pseudo-metric, denoted by $d = d_X \colon X \times X \to [0, \infty)$.
2. A *fast Cauchy sequence* in a computational pseudo-metric space X is a sequence $x_n \in X$ with $d(x_n, x_{n+1}) < 2^{-n}$. The subspace of $X^{\mathbb{N}}$ consisting of fast Cauchy sequences is denoted by Cauchy(X).
3. A computational pseudo-metric space X is called *computationally complete* if every sequence $x_n \in$ Cauchy(X) has a limit uniformly in x_n.
4. A computational pseudo-metric space X has *decidable closeness* if there is a family of relations $\sim_n$ on X that are decidable uniformly in n and satisfy:
   (a) $x \sim_n y \implies d(x, y) < 2^{-n}$,
   (b) $x \sim y \implies \forall n . x \sim_n y$.
   (c) $x \sim_{n+1} y \implies x \sim_n y$,
   (d) $x \sim_n y \iff y \sim_n x$,
   (e) $x \sim_{n+1} y \sim_{n+1} z \implies x \sim_n z$.
   The last condition is a counter-part of the triangle inequality. It follows from the first condition that if $x \sim_n y$ for every n, then $x \sim y$. Write

$$[x] = \{y \in X \mid x \sim y\}, \qquad [x]_n = \{y \in X \mid x \sim_n y\}.$$

Then the equivalence class $[x]$ is the closed ball of radius 0 centered at x. $\square$

For instance, the spaces $\mathbb{R}$ and $[0, \infty)$ are computationally complete metric spaces under the Euclidean metric, but don't have decidable closeness.

Remark 4.1. In the above definition, we don't require the representation topology of X to agree with the pseudo-metric topology generated by open balls. But notice that the metric topology is always coarser than the representation topology, because, by continuity of the metric, open balls are open in the representation topology. Hence the representation topology of any computational metric space is Hausdorff. Moreover, if X has an exhaustible Kleene–Kreisel space of representatives and the metric topology is compact, then both topologies agree, because no compact Hausdorff topology can be properly refined to another compact Hausdorff topology.   □

We are ready to prove the theorem.

Lemma 4.1. *For every computational metric space X there is a canonical computable pseudo-metric $d = d_{\ulcorner X \urcorner}$ on the representing space $\ulcorner X \urcorner$ such that:*

1. *The representation map $\rho = \rho_X : \ulcorner X \urcorner \to X$ is an isometry:*

$$d(t, u) = d(\rho(t), \rho(u)).$$

   In particular:
   (a) $t \sim u \iff d(t, u) = 0 \iff \rho(t) = \rho(u)$.
   (b) If $f : X \to Y$ is a computable map of metric spaces, then any representative $\ulcorner f \urcorner : \ulcorner X \urcorner \to \ulcorner Y \urcorner$ preserves the relation $(\sim)$.
2. *If X is computationally complete, then so is $\ulcorner X \urcorner$.*
3. *The representing space $\ulcorner X \urcorner$ has decidable closeness.*

Proof. Construct $d_{\ulcorner X \urcorner} : \ulcorner X \urcorner \times \ulcorner X \urcorner \to [0, \infty)$ as the composition of a computable representative $\ulcorner d_X \urcorner : \ulcorner X \urcorner \times \ulcorner X \urcorner \to \ulcorner [0, \infty) \urcorner$ of $d_X : X \times X \to [0, \infty)$ with the representation map $\rho_{[0,\infty)} : \ulcorner [0, \infty) \urcorner \to [0, \infty)$. A limit operator for $\ulcorner X \urcorner$ from a limit operator for X is constructed in a similar manner. For given $t, u \in \ulcorner X \urcorner$, let q_n be the n-th term of the sequence $\ulcorner d_X \urcorner(t, u) \in \ulcorner [0, \infty) \urcorner \subseteq \text{Cauchy}(\mathbb{Q})$, and define $t \sim_n u$ to mean that $[-2^{-n}, 2^{-n}] \subseteq [q_n - 2^{-n+1}, q_n + 2^{-n+1}]$.   □

Lemma 4.2. *Let Z be a subspace of a Kleene–Kreisel space with complete computational pseudo-metric structure and decidable closeness, and $K_n \subseteq Z$ be a sequence of sets that are exhaustible uniformly in n and satisfy $K_n \supseteq K_{n+1}$. Then, uniformly in the data:*

if $\bigcap_n K_n$ is an equivalence class, then it has a computable member.

Proof. Let $z \in \bigcap_n K_n$. For any m, we have $\bigcap_n K_n = [z] \subseteq [z]_{m+1}$, and hence there is n such that $K_n \subseteq [z]_{m+1}$, because the sets K_n are compact, because $K_n \supseteq K_{n+1}$, because Z is Hausdorff and because $[z]_{m+1}$ is open. Hence for every $u \in K_n$ we have $u \sim_{m+1} z$, and so for all $u, v \in K_n$ we have $u \sim_m v$. By the exhaustibility of K_n and the decidability of $(\sim_n)$, the function $n(m) = \mu n.\forall u, v \in K_n . u \sim_m v$ is computable. By the searchability of K_n, there is a computable sequence $u_m \in K_{n(m)}$. Because $n(m) \leq n(m+1)$, we have that $K_{n(m)} \supseteq K_{n(m+1)}$ and hence $u_m \sim_m u_{m+1}$ and so $d(u_m, u_{m+1}) < 2^{-m}$ and u_m is a Cauchy sequence. By completeness, u_m converges to a computable point u_∞. Because $z \in K_{n(m)}$, we have $u_m \sim_m z$ for every m, and hence $d(u_m, z) < 2^{-m}$. And because $d(u_\infty, u_m) < 2^{-m+1}$, the triangle inequality gives $d(u_\infty, z) < 2^{-m} + 2^{-m+1}$ for every m and hence $d(u_\infty, z) = 0$ and therefore $u_\infty \in \bigcap_n K_n$. □

The proof of the following is essentially the same as that of Theorem 3.1, but uses Lemma 4.2 rather than Lemma 3.2, and Lemma 4.1 instead of Lemma 3.3.

Lemma 4.3. *Let Z and W be subspaces of Kleene–Kreisel spaces with computational pseudo-metric structure and decidable closeness, and assume that Z is computationally complete and exhaustible.*

If $g \colon Z \to W$ is a computable map that preserves pseudo-metric equivalence and $w \in W$ is computable, then, uniformly in $\forall_Z$, g, and w:

1. *It is semi-decidable whether the equivalence $g(z) \sim w$ fails to have a solution $z \in Z$.*
2. *If $g(z) \sim w$ has a unique solution $z \in Z$ up to equivalence, then some solution is computable.*

Proof. The set $K_n = \{z \in Z \mid g(z) \sim_n w\}$, being a decidable subset of an exhaustible space, is exhaustible. Therefore the result follows from Lemmas 3.1 and 4.2, because $z \in \bigcap_n K_n$ iff $g(z) \sim_n w$ for every n iff $g(z) = w$. □

Algorithm 4.3. The solution $z = u_\infty$ is then computed from $\forall_Z$, g and w as follows, where we have expanded $\forall_{K_n}$ as a quantification over Z:

$$n(m) = \mu n. \forall u, v \in Z. g(u) \sim_n w \wedge g(v) \sim_n w \implies u \sim_m v,$$
$$u_\infty = \lim_m \varepsilon_K(\lambda z. g(z) \sim_{n(m)} w).$$

Thus, although there are common ingredients with Theorem 3.1, the resulting algorithm is different from 3.2, because it relies on the limit operator and approximate solutions. □

But, for Theorem 4.1, approximate solutions are computable uniformly in $\ulcorner f \urcorner$ and $\ulcorner y \urcorner$ only, as different approximate solutions are obtained for different representatives of f and y:

Proof (of Theorem 4.1.). Let $f \colon X \to Y$ and $y \in Y$ be computable. Now apply Lemma 4.3 with $Z = \ulcorner X \urcorner$, $W = \ulcorner Y \urcorner$, $g = \ulcorner f \urcorner$, $w = \ulcorner y \urcorner$, using Lemma 4.1 to fulfil the necessary hypotheses. If $f(x) = y$ has a unique solution x, then $g(z) \sim w$ has a unique solution z up to equivalence, and $x = \rho(z)$ for any solution z, and hence x is computable. Because g preserves $(\sim)$ by Lemma 4.1, if $g(z) \sim w$ has a solution z, then $x = \rho(z)$ is a solution of $f(x) = y$. This shows that $f(x) = y$ has a solution iff $g(z) = w$ has a solution, and we can reduce the semi-decision of absence of solutions of $f(x) = y$ to absence of solutions of $g(z) = w$. □

5   Exhaustible Spaces of Analytic Functions

For any $\epsilon \in (0,1)$, any $x \in [-\epsilon, \epsilon]$, any $b > 0$, and any sequence $a \in [-b,b]^{\mathbb{N}}$, the Taylor series $\sum_n a_n x^n$ converges to a number in the interval $[-b/(1+\epsilon), b/(1-\epsilon)]$. The following is proved by a standard computational analysis argument:

Lemma 5.1. *Any analytic function $f \in \mathbb{R}^{[-\epsilon,\epsilon]}$ of the form $f(x) = \sum_n a_n x^n$ is computable uniformly in any given $\epsilon \in (0,1)$, $b > 0$ and $a \in [-b,b]^{\mathbb{N}}$.*

Definition 5.1. Denote by $A = A(\epsilon, b) \subseteq \mathbb{R}^{[-\epsilon,\epsilon]}$ the subspace of such analytic functions and by $T = T_{\epsilon,b} \colon [-b,b]^{\mathbb{N}} \to A(\epsilon, b)$ the functional that implements the uniformity condition, so that $f = T(a)$.   □

The following results also hold uniformly in ϵ and b, but we omit explicit indications for the sake of brevity. The results are uniform in the exhaustibility assumptions too. Because $[-b,b]^{\mathbb{N}}$ is compact and T is continuous, the space A is compact as well. Moreover:

Theorem 5.2. *The space A has an exhaustible set of Kleene–Kreisel representatives.*

Proof. The space $[-b,b]^{\mathbb{N}}$ has an exhaustible space of representatives, e.g. using signed-digit binary representation. Because exhaustible spaces are preserved by computable images, the image of any representative $\ulcorner T \urcorner \colon \ulcorner [-b,b]^{\mathbb{N}} \urcorner \to \ulcorner A \urcorner$ of T gives an exhaustible set of representatives of A contained in the set $\ulcorner A \urcorner$ of all representatives of A.   □

Hence the solution of a functional equation with a unique analytic unknown in A can be computed using Theorem 4.1.

Lemma 5.2. *For any non-empty space X with an exhaustible set of Kleene–Kreisel representatives, the maximum- and minimum-value functionals*

$$\max{}_X, \min{}_X \colon \mathbb{R}^X \to \mathbb{R}$$

are computable.

Of course, any $f \in \mathbb{R}^X$ attains its maximum value because it is continuous and because spaces with exhaustible sets of representatives are compact.

Proof. We discuss max only. By e.g. the algorithm given in [6], this is the case for $X = 2^{\mathbb{N}}$. Because the representing space $\ulcorner X \urcorner$, being a non-empty exhaustible subspace of a Kleene–Kreisel space, is a computable image of the Cantor space, the space X itself is a computable image of the cantor space, say with $q \colon 2^{\mathbb{N}} \to X$. Then the algorithm $\max_X(f) = \max_{2^{\mathbb{N}}}(f \circ q)$ gives the required conclusion.   □

Corollary 5.1. *If K is a subspace of a metric space X and K has an exhaustible set of Kleene–Kreisel representatives, then K is computably located in X, in the sense that the distance function $d_K \colon X \to \mathbb{R}$ defined by*

$$d_K(x) = \min\{d(x,y) \mid y \in K\}$$

is computable.

Corollary 5.2. *For any metric space X with an exhaustible set of Kleene–Kreisel representatives, the max-metric $d(f,g) = \max\{d(f(x),g(x)) \mid x \in X\}$ on $\mathbb{R}^X$ is computable.*

Corollary 5.3. *For $f \in \mathbb{R}^{[-\epsilon,\epsilon]}$, it is semi-decidable whether $f \notin A$.*

Proof. Because A is computationally located in $\mathbb{R}^{[-\epsilon,\epsilon]}$ as it has an exhaustible set of representatives, and because $f \notin A \iff d_A(f) \neq 0$. □

Another proof, which doesn't rely on the exhaustibility of a set of representatives of A, uses Theorem 4.1: $f \notin A$ iff the equation $T(a) = f$ doesn't have a solution $a \in [-b,b]^{\mathbb{N}}$. But this alternative proof relies on a complete metric on $[-b,b]^{\mathbb{N}}$. For simplicity, we consider a standard construction for 1-bounded metric spaces. Because we apply this to metric spaces with exhaustible sets of representatives, this is no loss of generality as the diameter of such a space is computable as $\max(\lambda x. \max(\lambda y. d(x,y)))$ and hence the metric can be computably rescaled to become 1-bounded.

Lemma 5.3. *For any computational 1-bounded metric space X, the metric on $X^{\mathbb{N}}$ defined by $d(x,y) = \sum_n 2^{-n-1} d(x_n, y_n)$ is computable and 1-bounded, and it is computationally complete if X is.*

Proof. Use the fact that the map $[0,1]^{\mathbb{N}} \to [0,1]$ that sends a sequence $a \in [0,1]^{\mathbb{N}}$ to the number $\sum_n 2^{-n-1} a_n$ is computable. Regarding completeness, it is well known that a sequence in the space $X^{\mathbb{N}}$ is Cauchy iff it is componentwise Cauchy in X, and in this case its limit is calculated componentwise. □

Corollary 5.4. *The Taylor coefficients of any $f \in A$ can be computed from f.*

Proof. Because $[-b,b]^{\mathbb{N}}$ has an exhaustible set of representatives, the function T has a computable inverse by Theorem 4.1 and Lemma 5.3. □

References

1. Normann, D.: Recursion on the countable functionals. Lec. Not. Math., vol. 811. Springer, Heidelberg (1980)
2. Escardó, M.: Exhaustible sets in higher-type computation. Log. Methods Comput. Sci. 4(3), 3:3, 37 (2008)
3. Weihrauch, K.: Computable analysis. Springer, Heidelberg (2000)
4. Bauer, A.: A relationship between equilogical spaces and type two effectivity. MLQ Math. Log. Q. 48(suppl. 1), 1–15 (2002)
5. Bishop, E., Bridges, D.: Constructive Analysis. Springer, Berlin (1985)
6. Simpson, A.: Lazy functional algorithms for exact real functionals. In: Brim, L., Gruska, J., Zlatuška, J. (eds.) MFCS 1998. LNCS, vol. 1450, pp. 323–342. Springer, Heidelberg (1998)

Equivalence Relations on Classes of Computable Structures*

Ekaterina B. Fokina and Sy-David Friedman

Kurt Gödel Research Center for Mathematical Logic
University of Vienna
Währingerstraße 25 A-1090 Vienna Austria
efokina@logic.univie.ac.at, sdf@logic.univie.ac.at

Abstract. If $\mathcal{L}$ is a finite relational language then all computable $\mathcal{L}$-structures can be effectively enumerated in a sequence $\{\mathcal{A}_n\}_{n\in\omega}$ in such a way that for every computable $\mathcal{L}$-structure $\mathcal{B}$ an index n of its isomorphic copy $\mathcal{A}_n$ can be found effectively and uniformly. Having such a universal computable numbering, we can identify computable structures with their indices in this numbering. If K is a class of $\mathcal{L}$-structures closed under isomorphism we denote by K^c the set of all computable members of K. We measure the complexity of a description of K^c or of an equivalence relation on K^c via the complexity of the corresponding sets of indices. If the index set of K^c is hyperarithmetical then (the index sets of) such natural equivalence relations as the isomorphism or bi-embeddability relation are Σ^1_1. In the present paper we study the status of these Σ^1_1 equivalence relations (on classes of computable structures with hyperarithmetical index set) within the class of Σ^1_1 equivalence relations as a whole, using a natural notion of hyperarithmetic reducibility.

1 Introduction

Formalization of the notion of algorithm and studies of the computability phenomenon have resulted in increasing interest in the investigation of effective mathematical objects, in particular of algebraic structures and their classes. We call an algebraic structure *computable* if its universe is a computable subset of ω and all its basic predicates and operations are uniformly computable. For a class K of structures, closed under isomorphism, we denote by K^c the set of computable members of K. One of the questions of computable model theory is to study the algorithmic complexity of such classes of computable structures and various relations on these structures. In particular, we want to have a nice way to measure the complexity of a description of K^c or to compare the complexity of relations defined on different classes of computable structures. We say that K has a *computable characterization*, if we can separate computable structures in K from all other structures (not in K or noncomputable). Possible approaches to formalize the idea of computable characterizations of classes were described

---

* This work was partially supported by FWF Grant number P 19375 - N18.

K. Ambos-Spies, B. Löwe, and W. Merkle (Eds.): CiE 2009, LNCS 5635, pp. 198–207, 2009.

in [10]. One such approach involves the notion of an index set from the classical theory of numberings [7]. It will be described below. The same approach can be used to formalize the question of *computable classification* of K up to some equivalence relation, i. e. the question of existence of a description of each element of K up to isomorphism, or other equivalence relation, in terms of relatively simple invariants. In this paper we will discuss different ways to measure the complexity of equivalence relations on classes of computable structures.

This work is analogous to research in descriptive set theory, where the complexity of classes of structures and relations on these classes is studied via Borel-reducibility. In this paper we will study the questions that can be considered as computable versions of questions from [12,14,8].

First of all, we introduce the necessary definitions and basic facts from computable model theory.

2   Background

2.1   Computable Sequences and Indices of Structures

Consider a sequence $\{\mathcal{A}_n\}_{n\in\omega}$ of algebraic structures.

Definition 1. *A sequence $\{\mathcal{A}_n\}_{n\in\omega}$ is called* computable *if each structure $\mathcal{A}_n$ is computable, uniformly in n.*

In other words, there exists a computable function which gives us an index for the atomic diagram of $\mathcal{A}$ uniformly in n; equivalently, we can effectively check the correctness of atomic formulas on elements of each structure in the sequence uniformly.

Definition 2. *We call a sequence $\{\mathcal{A}_n\}_{n\in\omega}$ of computable structures* hyperarithmetical, *if there is a hyperarithmetical function which gives us, for every n, an index of the atomic diagram of $\mathcal{A}_n$.*

Let $\mathcal{L}$ be a finite relational language. A result of A. Nurtazin [15] shows that there is a universal computable numbering of all computable $\mathcal{L}$-structures, i. e. there exists a computable sequence $\{\mathcal{A}_n\}_{n\in\omega}$ of computable $\mathcal{L}$-structures, such that for every computable $\mathcal{L}$-structure $\mathcal{B}$ we can effectively find a structure $\mathcal{A}_n$ which is isomorphic to $\mathcal{B}$. Fix such a universal computable numbering.

One of the approaches from [10] involves the notion of index set.

Definition 3. *An* index set *of an $\mathcal{L}$-structure $\mathcal{B}$ is the set $I(\mathcal{B})$ of all indices of computable structures isomorphic to $\mathcal{B}$ in the universal computable numbering of all computable $\mathcal{L}$-structures. For a class K of structures, closed under isomorphism, the* index set *$I(K)$ is the set of all indices of computable members in K.*

We can use a similar idea to study the computable classification of classes of structures up to some equivalence relation. There are many interesting equivalence relations from the model-theoretic point of view. We can consider classes of structures up to isomorphism, bi-embeddability, or elementary bi-embeddability, bi-homomorphism, etc. There has been a lot of work on the isomorphism problem for various classes of computable structures (see, for example, [2,3,6,10]). There also has been some work on the computable bi-embeddability problem in [5], where the relation between the isomorphism problem and the embedding problem for some well-known classes of structures is studied. The approach used in the mentioned papers follows the ideas from [10] and makes use of the following definition.

Definition 4. *Let $\{\mathcal{A}_n\}_{n\in\omega}$ be as before the universal computable numbering of all computable $\mathcal{L}$-structures. Let K be a class of $\mathcal{L}$-structures closed under isomorphism, and let $I(K)$ be its index set. Then*

- *the* isomorphism problem *for K is the set of pairs $(a,b) \in I(K) \times I(K)$ such that $\mathcal{A}_a \cong \mathcal{A}_b$;*
- *the* bi-embeddability problem *for K is the set of pairs $(a,b) \in I(K) \times I(K)$ such that there is an embedding of $\mathcal{A}_a$ into $\mathcal{A}_b$ and an embedding of $\mathcal{A}_b$ into $\mathcal{A}_a$ (for any language L, for L-structures $\mathcal{A}$ and $\mathcal{B}$ we say that $\mathcal{A}$ embeds into $\mathcal{B}$, $\mathcal{A} \sqsubseteq \mathcal{B}$, if $\mathcal{A}$ is isomorphic to a substructure of $\mathcal{B}$).*

Generalizing this idea, every binary relation E on a class K of structures can be associated with the set $I(E, K)$ of all pairs of indices $(a, b) \in I(K) \times I(K)$ such that the structures $\mathcal{A}_a$ and $\mathcal{A}_b$ are in the relation E. We can measure the complexity of various relations on computable structures via the complexity of the corresponding sets of pairs of indices.

2.2   Computable Trees

In further constructions we will often use computable trees. Here we give some definitions useful for describing trees. Our trees are isomorphic to subtrees of $\omega^{<\omega}$. For the language, we take a single unary function symbol, interpreted as the predecessor function. We write $\emptyset$ for the top node (our trees grow down), and we think of $\emptyset$ as its own predecessor. Thus, our trees are defined on ω with their structure given by the predecessor function, but we often consider them as subtrees of $\omega^{<\omega}$ and treat their elements as finite sequences. In particular, following [14], we define a relation $\leq$ on tree nodes $s, t \in \omega^{<\omega}$ of the same length in the following way: $s \leq t$ iff for all i, the i-th coordinate of t is greater than or equal to the i-th coordinate of s. We also define the operation $s + t$ as coordinate-wise addition.

2.3 Σ_1^1 Sets

Kleene defined the *analytical hierarchy*, starting with computable relations on numbers and functions (from ω to ω) and closing under projection and complement. We need only the bottom part of this hierarchy. We use symbols $f, g, \ldots$ (possibly with indices) as function variables, and $x, y, \ldots$ (possibly, with indices) as number variables. A relation $R(\overline{x}, f)$ is *computable* if there is some e such that for all $\overline{x} \in \omega^{<\omega}$ and all $f \in \omega^{\omega}$,

$$\varphi_e^f(\overline{x}) \downarrow = \begin{cases} 1 \text{ if } R(\overline{x}, f) \\ 0 \text{ otherwise} \end{cases}$$

Definition 5. *Let $S(\overline{x})$ be a relation.*

1. $S(\overline{x})$ *is Σ_1^1 if it can be expressed in the form $(\exists f)(\forall y)\, R(\overline{x}, y, f)$, where $R(\overline{x}, y, f)$ is computable,*
2. $S(\overline{x})$ *is Π_1^1 if it can be expressed in the form $(\forall f)(\exists y)\, R(\overline{x}, y, f)$, where $R(\overline{x}, y, f)$ is computable,*
3. $S(\overline{x})$ *is Δ_1^1 if it is both Σ_1^1 and Π_1^1.*

If $S(\overline{x})$ is a k-place relation, we may consider the set S' of codes for k-tuples belonging to S. It is clear that S is Σ_1^1 iff S' is Σ_1^1. The same is true for Π_1^1 and Δ_1^1 relations. The next result gives familiar conditions equivalent to being Σ_1^1 [1,16]. We identify finite sequences with their codes.

Proposition 1 (Kleene)
   The following are equivalent:

1. S *is Σ_1^1,*
2. *there is a computable relation $R(n, u)$, on pairs of numbers, such that $n \in S$ iff $(\exists f)\,(\forall s)\, R(n, f \restriction s)$,*
3. *there is a c.e. relation $R(n, u)$, on pairs of numbers, such that $n \in S$ iff $(\exists f)\,(\forall s)\, R(n, f \restriction s)$,*
4. *there is a computable sequence of computable trees $\{T_n\}_{n \in \omega}$ such that $n \in S$ iff T_n has a path.*

2.4   Sets of Indices

We review the formal approach of [10] to the problem of determining whether or not a class of computable structures can be nicely characterized or classified relative to some natural equivalence relation.

According to [10], we say that a class K *has a computable characterization*, if its index set is hyperarithmetical. This condition is equivalent to existence of a computable infinitary sentence φ such that K^c (the set of computable structures in K) consists exactly of all computable models of φ. This definition expresses the fact that the set of all computable members of K can be nicely defined among all other structures for the same language. Note, that if $I(K)$ is hyperarithmetical and E is the isomorphism or bi-embeddability relation, then the corresponding equivalence relation $I(E, K)$ on indices is a Σ_1^1 set. In the worst case, when this equivalence relation is properly Σ_1^1, the easiest way to say that two computable structures from K are in the relation E is to say "There are functions between the structures that are isomorphisms (embeddings) between them". Often there are easier ways to verify the relation (such as counting basis elements of vector spaces to determine the isomorphism). Within this approach we say that there is a *computable classification for K up to E* if the corresponding equivalence relation $I(E, K)$ on indices of computable structures from K is a hyperarithmetical set. The standard way to measure the complexity of equivalence relations on computable models from K is the following.

Definition 6. *Let Γ be a complexity class (e.g., Σ_3^0, Π_1^1, etc.). $I(E, K)$ is m-complete Γ if $I(E, K)$ is Γ and for any $S \in \Gamma$, there is a computable function f such that*

$$n \in S \text{ iff } f(n) \in I(E, K).$$

By universality of the fixed computable numbering of $\mathcal{L}$-structures, this condition is equivalent to the condition that there is a computable sequence of pairs of computable $\mathcal{L}$-structures $\{(\mathcal{A}_n, \mathcal{B}_n)\}_{n \in \omega}$ from K for which $n \in S$ iff $\mathcal{A}_n E \mathcal{B}_n$.

3   A Complete Σ_1^1 Equivalence Relation

We can also measure the complexity of a relation E on a class of computable structures not as a set but as a relation, i. e. in a class of relations. This approach can be considered as a computable analog of the study of Borel-reducibility between analytic equivalence relations [12]. For this we need an appropriate notion of reducibility which allows us to compare relations on computable structures.

Let K be a class of structures with hyperarithmetical index set $I(K)$. As we have mentioned before, using indices we can identify every relation E on computable members of K with the corresponding relation $I(E, K)$ on natural numbers. Therefore, it is natural to restrict our attention to relations on ω.

Definition 7. *For equivalence relations E', E'' on ω we say that E' is h-reducible to E'', $E' \leq_h E''$, if there is a hyperarithmetical function f such that for all x, y, $xE'y$ iff $f(x)E''f(y)$.*

If E is an equivalence relation on K^c (the set of all computable members of K), then we often make no difference between E and $I(E, K)$ in the following sense. If E' is an arbitrary equivalence relation on ω then we say that E' *h-reduces to* E iff there exists a hyperarithmetical sequence of computable structures $\{\mathcal{A}_x\}_{x \in \omega}$ from K such that for all x, y, $xE'y$ iff $\mathcal{A}_x E \mathcal{A}_y$ (this is equivalent to $E' \leq_h I(E, K)$ in the sense of Definition 7).

Definition 8. *Let $\mathcal{R}$ be a class of relations. A relation $R \in \mathcal{R}$ is an Σ_1^1 h-complete for $\mathcal{R}$, if it is Σ_1^1 and every Σ_1^1 relation $R' \in \mathcal{R}$ h-reduces to R.*

Proposition 2. *Let $\mathcal{R}$ be the class of all isomorphism relations on classes K^c, where K is any class of structures with hyperarithmetical index set. The isomorphism relation on computable undirected graphs is Σ_1^1 h-complete for $\mathcal{R}$.*

Proof. Let K be a class of structures with hyperarithmetical index set. Using the effective transformations from [9] or [11], we get an h-reduction of the isomorphism relation on computable models of K to the isomorphism relation on computable graphs.

Theorem 1. *1. There is a class K of structures with hyperarithmetical index set and an equivalence relation on K^c which is an h-complete Σ_1^1 equivalence relation.*

2. There is a class K of structures with hyperarithmetical index set and a preorder on K^c which is an h-complete Σ_1^1 preorder.

The former statement of the theorem follows from the latter, as a complete Σ_1^1 equivalence relation results from any complete Σ_1^1 preorder. The converse is also true: the proof of the corresponding fact from [14] can be carried out effectively, and the Borel reducibility in the proof can be, in fact, substituted by h-reducibility:

Proposition 3. *Any h-complete Σ_1^1 equivalence relation on ω is induced by an h-complete Σ_1^1 preorder on ω.*

Theorem 1 can be regarded as a computable analog of results from [14]. Below we sketch the proof, which closely follows the proof of the non-effective version in [14]. First, we need a technical result on the representation of Σ_1^1 preorders.

Theorem 2. *Let R be a Σ_1^1 preorder on ω. Then there exists a computable sequence of computable trees $\{T_n^R\}_{n \in \omega}$, such that*

1. $xRy \Leftrightarrow T_{\langle x,y \rangle}^R$ *has a path;*
2. $\forall s, t \in \omega^{<\omega}$ *of the same length, such that $s \leq t$, if $s \in T_{\langle x,y \rangle}^R$ then $t \in T_{\langle x,y \rangle}^R$;*
3. $\forall x T_{\langle x,x \rangle}^R = \omega^{<\omega}$;
4. *If $s \in T_{\langle x,y \rangle}^R, t \in T_{\langle y,z \rangle}^R$ and $|s| = |t|$, then $s + t \in T_{\langle x,z \rangle}^R$.*

Proof. For every natural number m we define the function $\langle m \rangle$ by:

$$\langle m \rangle(x) = \begin{cases} 1 \text{ if } x = m+1 \\ 0 \text{ otherwise.} \end{cases}$$

We turn the preorder R on ω into a Σ_1^1 preorder R_0 on 2^ω in the following way:

$$xR_0y \Longleftrightarrow ((\exists m,n)x = \langle m \rangle \wedge y = \langle n \rangle \wedge mRn) \vee (x = y).$$

By [14], we get a tree S on $2 \times 2 \times \omega$, such that:

1. for all $x, y \in 2^\omega$ xR_0y iff for some $z \in \omega^\omega$, $(x|n, y|n, z|n) \in S$ for all n;
2. if $(u, v, s) \in S$ and $s \leq t$ then $(u, v, t) \in S$;
3. if $u \in 2^{<\omega}$ and $s \in \omega^{<\omega}$ have the same length, then $(u, u, s) \in S$;
4. if $(u, v, s) \in S$ and $(v, w, t) \in S$ then $(u, w, s + t) \in S$.

Note that the construction from [14] is highly effective and the resulting tree S is computable. Now we define a sequence of trees on ω using S. For every $s \in \omega^{<\omega}$ of length k, we let $s \in T_{\langle m,n \rangle}^R$ iff $(\langle m \rangle|k, \langle n \rangle|k, s) \in S$. The sequence $\{T_n^R\}_{n \in \omega}$ has all the necessary properties.

Proof (of Theorem 1). We define a computable structure $\mathcal{A}$ from K. Every such structure will code a computable sequence of computable trees with some additional property. The language for the class consists of one unary predicate symbol V and two unary function symbols g, h. In each model the function g is a successor function on $V = \{v_0, v_1, \ldots, v_n, \ldots\}$, and it defines on V a copy of ω. Each $v_n \in V$ is a root of a tree $T_n^{\mathcal{A}}$, with its structure given by h as a predecessor function. As we have mentioned before, we often consider $T_n^{\mathcal{A}}$ consisting of finite strings $s \in \omega^{<\omega}$. For a structure $\mathcal{A}$ of this kind to be in K we require that for all n, if $s \in T_n^{\mathcal{A}}, |s| = |t|$ and $s \leq t$ then $t \in T_n^{\mathcal{A}}$. From the definition it follows that $I(K)$ is hyperarithmetical.

Consider a preorder $\leq^*$ on K given in the following way.

$$A_1 \leq^* A_2 \Leftrightarrow \exists \varphi[\ \forall s(|s| = |\varphi(s)|) \text{ and } \forall s, t(s \preceq t \rightarrow \varphi(s) \preceq \varphi(t))$$
$$\text{and } \forall s \in \omega^{<\omega}\{z | s \in T_z^{\mathcal{A}_1}\} \subseteq \{z | \varphi(s) \in T_z^{\mathcal{A}_2}\}].$$

Then $\leq^*$ is a Σ_1^1 preorder.

Consider an arbitrary Σ_1^1 preorder R on ω and prove that R is h-reducible to $\leq^*$. That is, for every x we need to hyperarithmetically build (uniformly in x) a computable structure $\mathcal{A}_x \in K$ such that $xRy \Leftrightarrow \mathcal{A}_x \leq^* \mathcal{A}_y$.

By Theorem 2, for R there is a computable sequence of computable trees T_n^R with the properties 1–4. We define the structure $\mathcal{A}_x$ as follows. For every z, the tree $T_z^{\mathcal{A}_x}$ (under the z-th element of $V^{\mathcal{A}_x}$) equals $T_{\langle z, x \rangle}^R$. Then $\{\mathcal{A}_x\}_{x \in \omega}$ is a hyperarithmetical (even computable) sequence of computable structures from K. We check that $xRy \Leftrightarrow \mathcal{A}_x \leq^* \mathcal{A}_y$.

($\Rightarrow$): Suppose xRy. Then there is a path f in the tree $T_{\langle x, y \rangle}^R$. Define a function $\varphi(s) = s + f \upharpoonright |s|$, where $s \in \omega^{<\omega}$. Then φ has the necessary properties. Let $z \in \omega$ and $s \in \omega^k$ be such that $s \in T_z^{\mathcal{A}_x}$. By definition it means that $s \in T_{\langle z, x \rangle}^R$. As xRy, we have that $f \upharpoonright k \in T_{\langle x, y \rangle}^R$. By property 4 of the sequence $\{T_n^R\}$, $s + f \upharpoonright k \in T_{\langle z, y \rangle}^R$, which gives us $\varphi(s) \in T_z^{\mathcal{A}_y}$.

($\Leftarrow$): Suppose $\mathcal{A}_x \leq^* \mathcal{A}_y$, and φ is the function witnessing this fact. Consider the tree $T_x^{\mathcal{A}_x} = T_{\langle x, x \rangle}^R$. By property 3 of $\{T_n^R\}$, the function f, defined by $f \upharpoonright k = 0^k$ is a path in $T_x^{\mathcal{A}_x}$. Therefore, by the definition of $\leq^*$, $\varphi(0^k) \in T_x^{\mathcal{A}_y}$, for all k. Hence, by property 1, xRy.

In fact, the proof of Theorem 1 gives us an even stronger result. In the definition of the reducibility (Definition 7) we can replace "hyperarithmetical" by "computable" and still get the correct statements for the new reducibility. The following definition was first introduced in [4] as an effective analog of the Borel reducibility on classes of structures (for arbitrary structures, not necessarily computable). The universe of a structure is a subset of ω, possibly finite. As above, for a class K, the structures all have the same language, and K is closed under isomorphism (modulo the restriction on the universe).

Definition 9. *1. A* Turing computable transformation *from K' to K'' is a computable operator $\Phi = \varphi_e$ such that for each $\mathcal{A} \in K'$, there exists $\mathcal{B} \in K''$ with $\varphi_e^{D(\mathcal{A})} = \chi_{D(\mathcal{B})}$. We write $\Phi(\mathcal{A})$ for $\mathcal{B}$.*

2. *We say that* $\cong_{K'}$ *tc-reduces to* $\cong_{K''}$ *if for* $\mathcal{A}, \mathcal{A}' \in K'$,

$$\mathcal{A} \cong_{K'} \mathcal{A}' \text{ iff } \varPhi(\mathcal{A}) \cong_{K''} \varPhi(\mathcal{A}').$$

We can use the same approach to compare arbitrary equivalence relations on classes of structures. Namely,

Definition 10. *For equivalence relations* E', E'' *on* K', K'' *respectively,* E' *is tc-reducible to* E'', $E' \leq_{tc} E''$, *if there is a computable transformation* $\varPhi$ *from* K' *to* K'' *such that for* $\mathcal{A}, \mathcal{B} \in K'$, $\mathcal{A}E'\mathcal{B}$ *iff* $\varPhi(\mathcal{A})E''\varPhi(\mathcal{B})$.

As we study the relations on computable members of K, we can again restrict our attention on relations on ω.

Definition 11. *For equivalence relations* E', E'' *on* ω, E' *is tc-reducible to* E'', $E' \leq_{tc} E''$, *if there is a computable function* f *such that for all* x, y, $xE'y$ *iff* $f(x)E''f(y)$.

As before, we identify an equivalence relation E on K^c with $I(E, K)$. Then for an arbitrary equivalence relation E' on ω, we say that E' *tc*-reduces to E iff there exists a *computable* sequence of computable structures $\{\mathcal{A}_x\}_{x \in \omega}$ from K such that for all x, y, $xE'y$ iff $\mathcal{A}_x E \mathcal{A}_y$.

Corollary 1.   1. *There is a class* K *of structures with hyperarithmetical index set and an equivalence relation on* K^c *which is a tc-complete* Σ_1^1 *equivalence relation.*
 2. *There is a class* K *of structures with hyperarithmetical index set and a pre-order on* K^c *which is a tc-complete* Σ_1^1 *preorder.*
 3. *The isomorphism relation on computable undirected graphs is a tc-complete* Σ_1^1 *isomorphism relation.*

4 Σ_1^1 Complete Bi-Embeddability Relation

Theorem 3.   1. *The embeddability relation* $\sqsubseteq$ *on the class of computable trees is an h-complete* Σ_1^1 *preorder.*
 2. *The bi-embeddability relation* $\equiv$ *on the class of computable trees is an h-complete* Σ_1^1 *equivalence relation.*

Proof. We sketch the proof of the first statement of the theorem. The second statement follows from it in the obvious way. As before, we use the ideas from [14]. We have only to take care that there is enough effectiveness and uniformity of the arguments. Let K be the class of structures constructed in Theorem 1. We give a uniform effective procedure of constructing a computable tree from every computable member of K, which allows us to reduce $\leq^*$ to $\sqsubseteq$ on the class of trees.

Let G_0 be a tree defined in the following way. For every vertex x of a complete infinitely branching tree $\omega^{<\omega}$, except for the root, we add a new vertex x' between

x and its predecessor. We use G_0 as a base to construct the tree $G_{\mathcal{A}}$ corresponding to a structure $\mathcal{A} \in K^c$. To code $\mathcal{A}$ into a tree, we add to G_0 new vertices of the form $(x, s, 0^k)$ and $(x, s, 0^{2x+2}10^k)$, where $x \in \omega, s \in \omega^{<\omega}, k \in \omega$ and $s \in T_x^{\mathcal{A}}$. We connect (x, s, w) to (x, s, w') iff w' is the predecessor of w and we connect $(x, s, \emptyset)$ to s, considered as an element of G_0. The resulting tree is $G_{\mathcal{A}}$. If $\mathcal{A}$ is computable then obviously $G_{\mathcal{A}}$ is computable and the procedure is uniform.

Suppose $\mathcal{A}_1 \leq^* \mathcal{A}_2$. We prove that there is an embedding of $G_{\mathcal{A}_1}$ into $G_{\mathcal{A}_2}$. By definition of $\leq^*$, there is a function $\varphi : \omega^{<\omega} \to \omega^{<\omega}$ (which, in fact, can be taken 1-to-1 due to Property 2 of all $T_x^{\mathcal{A}_i}$). We send every element $s \in \omega^{<\omega}$ into $\varphi(s)$, and every s' into $\varphi(s)'$. Thus, we have defined an embedding of G_0 into itself. Every vertex of the form (x, s, w) we send into $(x, \varphi(s), w)$. This map will define an embedding of $G_{\mathcal{A}_1}$ into $G_{\mathcal{A}_2}$, as by definition of $\leq^*$ we have $s \in T_x^{\mathcal{A}_1} \Rightarrow \varphi(s) \in T_x^{\mathcal{A}_2}$.

On the other hand, if $G_{\mathcal{A}_1} \sqsubseteq G_{\mathcal{A}_2}$, then let g be the function witnessing the embedding. For every vertex z of $G_{\mathcal{A}_1}$, the number of its neighbors does not exceed the number of neighbors of $g(z)$, moreover the distance between any two vertices z_1, z_2 of $G_{\mathcal{A}_1}$ is the same as between $g(z_1)$ and $g(z_2)$ in $G_{\mathcal{A}_2}$. In particular, all elements of $\omega^{<\omega}$ must be sent to elements of $\omega^{<\omega}$. This map $\varphi : \omega^{<\omega} \to \omega^{<\omega}$ has the necessary properties and witnesses $\mathcal{A}_1 \leq^* \mathcal{A}_2$.

Again, all the procedures are in fact computable, thus, we get the following corollary.

Corollary 2. *1. The embeddability relation $\sqsubseteq$ on the class of trees is a tc-complete Σ_1^1 preorder.*
2. The bi-embeddability relations $\equiv$ on the class of trees is a tc-complete Σ_1^1 equivalence relation.

5   Questions

We conclude the paper with the following questions. In descriptive set theory there are many examples of Borel equivalence relations that cannot be Borel-reduced to the isomorphism relation on any class Mod_φ for any countable infinitary sentence φ. The computable analog of this statement is false, as any Δ_1^1 equivalence relation on ω is h-reducible to equality on ω. However the following remains open:

Question 1. Is there a class K with hyperarithmetical index set, such that its isomorphism relation is an h-complete (a tc-complete) Σ_1^1 equivalence relation?

We do not know if there exists a hyperarithmetical class of computable structures with Σ_1^1 (but not Δ_1^1) isomorphism relation which is not h-complete among all isomorphism relations on hyperarithmetical classes of computable structures. An affirmative answer to the following question may help solve this problem:

Question 2. Does there exist a *hyperarithmetical* class K of computable structures which contains a unique structure of non-computable Scott rank (up to isomorphism)?

If such a class exists then the isomorphism relation on the class of computable graphs cannot be h-reduced to the isomorphism relation on K. Indeed, there exist non-isomorphic graphs of high (i.e. $\geq \omega_1^{CK}$) Scott rank. They must be sent to non-isomorphic structures in K. However, no computable structure of high Scott rank can be sent to a computable structure of computable Scott rank under hyperarithmetical reducibility.

The main result of [8] shows that in the non-effective setting, every Σ_1^1 equivalence relation is Borel-equivalent to the bi-embeddability relation on the class of all countable models of an infinitary sentence φ.

Question 3. Let E be a Σ_1^1 equivalence relation on ω. Does there always exist a class K of structures with hyperarithmetical $I(K)$, such that E is h-*equivalent* to the bi-embeddability relation on computable members of K?

References

1. Ash, C.J., Knight, J.F.: Computable Structures and the Hyperarithmetical Hierarchy. Elsevier, Amsterdam (2000)
2. Calvert, W.: The isomorphism problem for computable abelian p-groups of bounded length. J. Symbolic Logic 70(1), 331–345 (2005)
3. Calvert, W.: The isomorphism problem for classes of computable fields. Arch. Math. Logic 43(3), 327–336 (2004)
4. Calvert, W., Cummins, D., Knight, J.F., Miller, S.: Comparing classes of finite structures. Algebra and Logic 43, 374–392 (2004)
5. Carson, J., Fokina, E., Harizanov, V., Knight, J., Maher, C., Quinn, S., Wallbaum, J.: Computable Embedding Problem (submitted)
6. Downey, R., Montalbán, A.: The isomorphism problem for torsion-free Abelian groups is analytic complete. Journal of Algebra 320, 2291–2300 (2008)
7. Ershov, Y.L.: Theory of numberings, Nauka, Moscow (1977)
8. Friedman, S.D., Motto Ros, L.: Analytic equivalence relations and bi-embeddability (2008) (preprint)
9. Goncharov, S.S.: Computability and Computable Models, Mathematical problems from applied logic. II. In: Gabbay, D.M., Goncharov, S.S., Zakharyaschev, M. (eds.) Logics for the XXIst century. International Mathematical Series, pp. 99–216. Springer, New York (2006)
10. Goncharov, S.S., Knight, J.F.: Computable structure and non-structure theorems. Algebra and Logic 41, 351–373 (2002) (English translation)
11. Hirschfeldt, D.R., Khoussainov, B., Shore, R.A., Slinko, A.M.: Degree spectra and computable dimensions in algebraic structures. Annals of Pure and Applied Logic 115, 71–113 (2002)
12. Kechris, A.: New directions in descriptive set theory. Bull. Symbolic Logic 5(2), 161–174 (1999)
13. Knight, J., Miller, S., Vanden Boom, M.: Turing computable embeddings. J. Symbolic Logic 72(3), 901–918 (2007)
14. Louveau, A., Rosendal, C.: Complete analytic equivalence relations. Trans. Amer. Math. Soc. 357(12), 4839–4866 (2005)
15. Nurtazin, A.T.: Computable classes and algebraic criteria for autostability, Ph.D. Thesis, Institute of Mathematics and Mechanics, Alma-Ata (1974)
16. Rogers, H.: Theory of recursive functions and effective computability. McGraw-Hill, New York (1967)

Fractals Generated by Algorithmically Random Brownian Motion

Willem L. Fouché*

Department of Decision Sciences,
University of South Africa, PO Box 392, 0003 Pretoria, South Africa
fouchwl@gmail.com

To the memory of my parents Hennie and Charlotte

Abstract. A continuous function x on the unit interval is an algorithmically random Brownian motion when every probabilistic event A which holds almost surely with respect to the Wiener measure, is reflected in x, provided A has a suitably effective description. In this paper we study the zero sets and global maxima from the left as well as the images of compact sets of reals of Hausdorff dimension zero under such a Brownian motion. In this way we shall be able to find arithmetical definitions of perfect sets of reals whose elements are linearly independent over the field of recursive real numbers.

Mathematics Subject Classification (2000): 03D20, 68Q30, 60J65, 60G05, 60G17.

Keywords: algorithmic randomness, Brownian motion, fractal geometry.

1 Introduction

In [3], the author proposed the problem of studying the fractal geometry and sample path properties of Brownian motion from the point of view of algorithmic randomness. Some progress on this problem was made in [4], [5] and by Kjos-Hanssen and Nerode in [9]. In this paper, we study the set of left-maxima and the set of zeroes of as well as the images of ultra-thin sets (closed sets of Hausdorff dimension 0) under an algorithmically random Brownian motion. We show that these images are perfect sets whose elements are linearly independent over the field of rational numbers. We also discuss the definability of these sets within the arithmetical hierarchy, a theme that will be pursued in a sequel to this paper.

The arguments in this paper rely heavily on [3], [4] as well as on the beautiful constructions of Kahane in [8].

---

* The research is based upon work supported by the National Research Foundation (NRF) of South Africa. Any opinion, findings and conclusions or recommendations expressed in this material are those of the author and therefore the NRF does not accept any liability in regard thereto.

K. Ambos-Spies, B. Löwe, and W. Merkle (Eds.): CiE 2009, LNCS 5635, pp. 208–217, 2009.
© Springer-Verlag Berlin Heidelberg 2009

2   Preliminaries

The set of non-negative integers is denoted by ω and we write $\mathcal{N}$ for the product space $\{0,1\}^\omega$. The set of words over the alphabet $\{0,1\}$ is denoted by $\{0,1\}^*$. If $a \in \{0,1\}^*$, we write $|a|$ for the length of a. If $\alpha = \alpha_0\alpha_1\ldots$ is in $\mathcal{N}$, we write $\overline{\alpha}(n)$ for the word $\prod_{j<n} \alpha_j$. We use the usual recursion-theoretic terminology Σ_r^0 and Π_r^0 for the arithmetical subsets of $\omega^k \times \mathcal{N}^l$, $k, l \geq 0$. (See, for example, [7]). We write λ for the Lebesgue probability measure on $\mathcal{N}$. For a binary word s of length n, say, we write $[s]$ for the "interval" $\{\alpha \in \mathcal{N} : \overline{\alpha}(n) = s\}$. A sequence (a_n) of real numbers converges *effectively* to 0 as $n \to \infty$ if for some total recursive $f : \omega \to \omega$, it is the case that $|a_n| \leq (m+1)^{-1}$ whenever $n \geq f(m)$.

For any binary word a we denote its (prefix-free) Kolmogorov complexity by $K(a)$. An infinite binary string α is Kolmogorov-Chaitin or Martin-Löf complex if

$$\exists_d \forall_n \, K(\overline{\alpha}(n)) \geq n - d.$$

In the sequel, we shall denote this set by KC and refer to its elements as KC-strings. (See, e.g., [2] and [10] for more background.)

The mean, or expected value of a random variable X will be denoted by $E(X)$. Two random variables X and Y on possibly different probability spaces are said to be *similar* when they have the same probability distributions. A random variable X with mean μ and variance σ^2 is normal if it has a density function

$$\frac{1}{\sqrt{2\pi}\,\sigma} \, e^{-(t-\mu)^2/2\sigma^2}.$$

A Brownian motion on the unit interval is a real-valued function $(\omega, t) \mapsto X_\omega(t)$ on $\Omega \times [0,1]$, where Ω is the underlying space of some probability space, such that $X_\omega(0) = 0$ a.s. and for $t_1 < \ldots < t_n$ in the unit interval, the random variables $X_\omega(t_1), X_\omega(t_2) - X_\omega(t_1), \cdots, X_\omega(t_n) - X_\omega(t_{n-1})$ are statistically independent and normally distributed with means all 0 and variances $t_1, t_2 - t_1, \cdots, t_n - t_{n-1}$, respectively. We say in this case that the Brownian motion is *parametrised* by Ω.

It is a fundamental fact that any Brownian motion has a "continuous version". This means the following: Write Σ for the σ-algebra of Borel sets of $C[0,1]$ where the latter is topologised by the uniform norm topology. There is a unique probability measure W on Σ such that for $0 \leq t_1 < \ldots < t_n \leq 1$ and for a Borel subset B of $\mathbf{R}^n$, we have

$$P(\{\omega \in \Omega : (X_\omega(t_1), \cdots, X_\omega(t_n)) \in B\}) = W(A),$$

where

$$A = \{x \in C[0,1] : (x(t_1), \cdots, x(t_n)) \in B\}).$$

The measure W is known as the *Wiener measure*. We shall usually write $X(t)$ instead of $X_\omega(t)$.

3 Complex Oscillations

We next survey the results from [1], [3] and [4] which will play an important role in this paper. For $n \geq 1$, we write C_n for the class of continuous functions on the unit interval that vanish at 0 and are linear with slopes $\pm\sqrt{n}$ on the intervals $[(i-1)/n, i/n]$, $i = 1, \cdots, n$. With every $x \in C_n$, one can associate a binary string $a = a_1 \cdots a_n$ by setting $a_i = 1$ or $a_i = 0$ according to whether x increases or decreases on the interval $[(i-1)/n, i/n]$. We call the sequence a the code of x and denote it by $c(x)$. The following notion was introduced by Asarin and Prokovskii in [1].

Definition 1. *A sequence (x_n) in $C[0,1]$ is complex if $x_n \in C_n$ for each n and there is a constant $d > 0$ such that $K(c(x_n)) \geq n - d$ for all n. A function $x \in C[0,1]$ is a complex oscillation if there is a* complex oscillation *complex sequence (x_n) such that $\|x - x_n\|$ converges effectively to 0 as $n \to \infty$.*

The class of complex oscillations is denoted by $\mathcal{C}$. It was shown by Asarin and Prokovskiy [1] that the class $\mathcal{C}$ has Wiener measure 1.

For the results in this paper, in analogy with Martin-Löf [10], we shall require a recursive characterisation of the almost sure events, with respect to Wiener measure, which are reflected in each complex oscillation. In order to describe this characterisation, we use, as in [3], an analogue of a $\varPi_2^0$ subset of $C[0,1]$ which is of constructive measure 0. We introduce some notation. If F is a subset of $C[0,1]$, we denote by $\overline{F}$ the topological closure of F in $C[0,1]$ with the supremum norm topology. For $\epsilon > 0$, we let $O_\epsilon(F)$ be the set $\{f \in C[0,1] : \exists_{g \in F} \|f - g\| < \epsilon\}$. (Here $\|.\|$ denotes the supremum norm.) For convenience sake, we write F^0 for the complement of F and F^1 for F.

Definition 2. *A sequence $\mathcal{F}_0 = (F_i : i < \omega)$ in Σ is an effective generating sequence if*

1. *for $F \in \mathcal{F}_0$, for $\epsilon > 0$ and $\delta \in \{0,1\}$, we have, for $G = O_\epsilon(F^\delta)$ or for $G = F^\delta$, that $W(\overline{G}) = W(G)$,*
2. *there is an efffective procedure that yields, for each sequence $0 \leq i_1 < \ldots < i_n < \omega$ and $k < \omega$ a binary rational number β_k such that*

$$|W(F_{i_1} \cap \ldots \cap F_{i_n}) - \beta_k| < 2^{-k},$$

3. *for $n, i < \omega$, a strictly positive rational number ϵ and for $x \in C_n$, both the relations $x \in O_\epsilon(F_i)$ and $x \in O_\epsilon(F_i^0)$ are recursive in x, ϵ, i and n, relative to an effective representation of the rationals.*

If $\mathcal{F}_0 = (F_i : i < \omega)$ is an effective generating sequence and $\mathcal{F}$ is the Boolean algebra generated by $\mathcal{F}_0$, then there is an enumeration $(T_i : i < \omega)$ of the elements of $\mathcal{F}$ (with possible repetition) in such a way, for a given i, one can effectively describe T_i as a finite union of sets of the form

$$F = F_{i_1}^{\delta_1} \cap \ldots \cap F_{i_n}^{\delta_n}$$

where $0 \leq i_1 < \ldots < i_n$ and $\delta_i \in \{0, 1\}$ for each $i \leq n$. We call any such sequence $(T_i : i < \omega)$ a *recursive enumeration* of $\mathcal{F}$. We say in this case that $\mathcal{F}$ is *effectively generated* by $\mathcal{F}_0$ and refer to $\mathcal{F}$ as an *effectively generated algebra* of sets. A sequence (A_n) of sets in $\mathcal{F}$ is said to be $\mathcal{F}$-*semi-recursive* if it is of the form $(T_{\phi(n)})$ for some total recursive function $\phi : \omega \to \omega$ and some effective enumeration (T_i) of $\mathcal{F}$. (Note that the sequence (A_n^c), where A_n^c is the complement of A_n, is also an $\mathcal{F}$-semirecursive sequence.) In this case, we call the union $\cup_n A_n$ a $\Sigma_1^0(\mathcal{F})$ set. A set is a $\Pi_1^0(\mathcal{F})$ set if it is the complement of a $\Sigma_1^0(\mathcal{F})$ set. It is of the form $\cap_n A_n$ for some $\mathcal{F}$-semirecursive sequence (A_n). A sequence (B_n) in $\mathcal{F}$ is a *uniform* sequence of $\Sigma_1^0(\mathcal{F})$ sets if, for some total recursive function $\phi : \omega^2 \to \omega$ and some effective enumeration (T_i) of $\mathcal{F}$, each B_n is of the form

$$B_n = \bigcup_m T_{\phi(n,m)}.$$

In this case, we call the intersection $\cap_n B_n$ a $\Pi_2^0(\mathcal{F})$ set. If, moreover, the W-measure of B_n converges *effectively* to 0 as $n \to \infty$, we say that the set given by $\cap_n B_n$ is a $\Pi_2^0(\mathcal{F})$ set of constructive measure 0.

The proof of the following theorem appears in [3].

Theorem 1. *Let $\mathcal{F}$ be an effectively generated algebra of sets. If x is a complex oscillation, then x is in the complement of every $\Pi_2^0(\mathcal{F})$ set of constructive measure 0.*

We shall also make frequent use of the following result from [3] which is a consequence of Theorem 1.

Theorem 2. *If B is a $\Sigma_1^0(\mathcal{F})$ set and $W(B) = 1$, then $\mathcal{C}$, the set of complex oscillations, is contained in B.*

Remark. It follows that if a $\Pi_1^0(\mathcal{F})$ set A contains at least one complex oscillation, then $W(A) > 0$. Consequently, if B is a $\Sigma_2^0(\mathcal{F})$ set with $W(B) = 0$, then B will contain no complex oscillation.

We introduce a class of effective generating sequences which is very useful for reflecting properties of one-dimensional Brownian motion into complex oscillations. Let $\mathcal{G}_0$ be a family of sets in Σ each having a description of the form:

$$a_1 X(t_1) + \cdots + a_n X(t_n) \leq L \tag{1}$$

or of the form (1) with $\leq$ replaced by $<$, where all the a_j, t_j $(0 \leq t_j \leq 1)$ are rational numbers, L is a recursive real number and X is one-dimensional Brownian motion. If $\epsilon > 0$ and $G \in \Sigma$ is described by (1), we have that $O_\epsilon(G)$ is described by the inequality

$$a_1 X(t_1) + \cdots + a_n X(t_n) < L + \epsilon \sum_j |a_j| \tag{2}$$

while $O_\epsilon(G^0)$ is given by

$$a_1 X(t_1) + \cdots + a_n X(t_n) > L - \epsilon \sum_j |a_j|. \tag{3}$$

We require that it be possible to find an enumeration $(G_i : i < \omega)$ of $\mathcal{G}_0$ such that, for given i, if G_i is given by (1), we can effectively compute the sign, the denominators and numerators of the rational numbers a_j, t_j and, moreover, that the recursive real L can be computed up to arbitrary accuracy. This has the implication that there is an effective procedure, Π, such that, for given i, ϵ, m with $i, m < \omega$ and ϵ a positive rational, the validity of (2) and (3) can be decided by Π when G_i is given by (1) and when $X \in C_m$.

It is shown in [4] that $\mathcal{G}_0 = (G_i : i < \omega)$ is an effective generating sequence in the sense of Definition 2. The associated effectively generated algebra of sets $\mathcal{G}$ will be referred to as a gaussian algebra.

For a closed subinterval I of the unit interval and a real number b, we write $[M(I) \geq b]$ for the event $[\sup\{X(t) : t \in I\} \geq b]$ and $[m(I) \leq b]$ for the event $[\inf\{X(t) : t \in I\} \leq b]$, where X is one-dimensional Brownian motion on the unit interval. We let $\mathcal{M}_0$ be the set of the events of the form $[M(I) \leq b]$ or $[m(I) \leq b]$ where b is an arbitrary rational number and where I is a subinterval of the unit interval with rational endpoints. It follows from the arguments on pp 434 - 438 in [3] that the elements of $\mathcal{M}_0$ can be effectively enumerated rendering $\mathcal{M}_0$ an effective generating sequence. This result will be frequently used in our study of the zero set and global left-maxima of a complex oscillation. We denote by $\mathcal{M}$ the Boolean algebra generated by $\mathcal{M}_0$.

We can also define an effective generating sequence $\mathcal{N}_0$ as we did $\mathcal{M}_0$ where we replace the Brownian motion X by the stochastic process Y where $Y = M[0, t] - X(t)$. It is well-known that Y is similar to $|X|$. (see, e.g., Freedman [6].) This has the implication that the proof in [3] that $\mathcal{M}_0$ is effectively generating can be easily adapted to show that the same holds for $\mathcal{N}_0$. The Boolean algebra generated by $\mathcal{N}_0$ is denoted by $\mathcal{N}$.

4   Hamel Sets Generated by Complex Oscillations

Recall that a subset of the reals is perfect if it is closed and has no isolated points. A perfect subset of the unit interval is called a *Hamel set*, if its elements are linearly independent over the field of rational numbers, or, equivalently, if it is a perfect subset of some Hamel basis of the reals over the rationals.

Set

$$E = \left\{ \frac{1}{2} + \sum_{k=2}^{\infty} \epsilon_k \frac{1}{2^{k^2}} : \epsilon_k \in \{-1, 1\} \text{ for all } k \right\}.$$

For fixed rational numbers $0 < a_1 < b_1 < a_2 < \ldots < a_m < b_m < 1$, set $I_n = [a_n, b_n]$ for $n = 1, \ldots, m$. It is readily seen that E is a perfect set and that it is ultrathin in the sense that its Hausdorff dimension is 0.

By using the geometric and probabilistic constructions on pp 255-257 of Kahane [8], it can be shown that, for nonzero rational numbers $r_1, \ldots, r_m$, the event

for all $1 \leq n \leq m$ there is some $t_n \in I_n \cap E$ such that $\sum_{n=1}^{m} r_n X(t_n) = 0$, has Wiener measure 0. We shall now show that no complex oscillation satisfies this event. This will enable us to prove the following

Theorem 3. *If x is a complex oscillation then the elements of the image $x(E)$ of the set E under x will be linearly independent over the field of rational numbers.*

Proof (Sketch): For $\ell \geq 2$, set

$$D_\ell = \left\{ \frac{1}{2} + \sum_{k=2}^{\ell} \epsilon_k \frac{1}{2^{k^2}} : \epsilon_k \in \{-1, 1\} \text{ for all } k = 1, \ldots, \ell \right\}.$$

Write $D = \cup_{\ell \geq 2} D_\ell$. Note that the topological closure of D is $D \cup E$.

For a continuous function X on the unit interval, define the predicates $P(X)$ and $R(X)$ by

$$P(X) \longleftrightarrow \forall_{1 \leq n \leq m} \exists_{t_n \in I_n \cap E} \left(\sum_{n=1}^{m} r_n X(t_n) = 0 \right)$$

and

$$R(X) \longleftrightarrow \exists_C \forall_{\ell \geq C} \exists_{t \in \prod_{n=1}^{m}(I_n \cap D_\ell)} \left| \sum_{n=1}^{m} r_n X(t_{n,\ell}) \right| \leq \frac{1}{\ell}.$$

Here we have denoted the component of $t \in \prod_{n=1}^{m}(I_n \cap D_\ell)$ in $I_n \cap D_\ell$ by $t_{n,\ell}$. Note that $R(X)$ defines a $\Sigma_2^0(\mathcal{G})$ set for some gaussian algebra $\mathcal{G}$.

We first show that for any continuous function X on the unit interval, it is the case that $R(X) \to P(X)$. Since $P(X)$ defines a set of Wiener measure zero, it follows from the remark following Theorem 2 that for no complex oscillation x will it be true that $R(x)$ holds. Next we shall show that if x is a complex oscillation then $P(x) \to R_1(x)$, where $R_1(X)$ is defined as $R(X)$ but in terms of a different set of intervals $J_1, \ldots J_m$ with rational endpoints. In particular, if x is a complex oscillation, then $\neg P(x)$.

Suppose X is a continuous function on the unit interval such that $R(X)$ holds. Then there is some integer C such that for all $\ell \geq C$ and all $1 \leq n \leq m$ we can find some $t_{n,\ell} \in I_n \cap D_\ell$ such that

$$|\sum_{n=1}^{m} r_n X(t_{n,\ell})| \leq \frac{1}{\ell}.$$

We can find some sequence ℓ_i of natural numbers such that, for all $1 \leq n \leq m$, the sequence t_{n,ℓ_i} converges to some number t_n. It is clear that $t_n \in E \cap I_n$ for all n. Moreover,

$$\sum_{n=1}^{m} r_n X(t_n) = 0.$$

It follows that $R(X) \to P(X)$.

Next suppose x is a complex oscillation such that $P(x)$ holds. Let $t_1, \ldots, t_m$ be such that $t_n \in E \cap I_n$ for every n and $\sum_{n=1}^{m} r_n x(t_n) = 0$. For $1 \leq n \leq m$

set $J_n = [c_n, d_n]$ where $c_n = \frac{a_n + b_{n-1}}{2}$ and $d_n = \frac{b_n + a_{n+1}}{2}$. Here $b_0 = 0$ and $a_{m+1} = 1$). Choose C so large that for all $\ell \geq C$ and all $1 \leq n \leq m$ we can find some $t_{n,\ell} \in J_n \cap D_\ell$ such that $|t_{n,\ell} - t_n| \leq \frac{1}{2^\ell}$.

Now

$$|\sum_{n=1}^{m} r_n x(t_{n,\ell})| = |\sum_{n=1}^{m} r_n (x(t_{n,\ell}) - x(t_n))|.$$

By Proposition 1 in [4], for $C_1 > 2$ and C sufficiently large, for all $\ell \geq C$, the right-hand side is bounded from above by

$$C_1 \frac{\ell \sum_n |r_n|}{2^{\frac{\ell}{2}}}$$

which is $\leq \frac{1}{\ell}$ for all $\ell \geq C$. It follows that $R_1(x)$ holds where $R_1(X)$ is defined as $R(X)$ but in terms of the intervals J_n in stead of I_n. This concludes the proof of the theorem.

By using a suitable extension of the notion of a Gaussian algebra, one can show that $x(E)$ is in fact linearly independent over the field of recursive real numbers.

By combining these results and by exploiting the recursive isomorphism ϕ in [4], one can find relativised definitions within the arithmetical hierarchy [7] of a continuum of perfect sets, each of which is linearly independent over the rationals. This result will be discussed in a sequel to this paper.

5   Zeroes and Left-Maxima of Complex Oscillations

For a compact subset A of Euclidean space $\mathbf{R}^d$ and real numbers α, ϵ with $0 \leq \alpha < d$ and $\epsilon > 0$, consider all finite or countable coverings of A by balls B_n of diameter $\leq \epsilon$ and the corresponding sums

$$\sum_n |B_n|^\alpha,$$

where $|B|$ denotes the diameter of B. All the metric notions here are to be understood in terms of the standard ℓ^2 norms on Euclidean space. The infimum of the sums over all coverings of A by balls of diameter $\leq \epsilon$ is denoted by $H_\alpha^\epsilon(A)$. When ϵ decreases to 0, the corresponding $H_\alpha^\epsilon(A)$ increases to a limit (which may be infinite). The limit is denoted by $H_\alpha(A)$ and is called the Hausdorff measure of A in dimension α.

If $0 < \alpha < \beta \leq d$, then, for any covering (B_n) of A,

$$\sum_n |B_n|^\beta \leq \sup_n |B_n|^{\beta - \alpha} \sum_n |B_n|^\alpha,$$

from which it follows that

$$H_\beta^\epsilon(A) \leq \epsilon^{\beta - \alpha} H_\alpha^\epsilon(A).$$

Hence if $H_\alpha(A) < \infty$, then $H_\beta(A) = 0$. Equivalently,

$$H_\beta(A) > 0 \implies H_\alpha(A) = \infty.$$

Therefore,

$$\sup\{\alpha : H_\alpha(A) = \infty\} = \inf\{\beta : H_\beta(A) = 0\}.$$

This common value is called the Hausdorff dimension of A and denoted by $\dim A$.
If α is such that $0 < H_\alpha(A) < \infty$, then $\alpha = \dim A$. However, if $\alpha = \dim A$, we cannot say anything about the value of $H_\alpha(A)$.

For a real-valued continuous function f on the unit interval and for $t \in [0,1]$, set

$$M_f(t) = \sup_{s \in [0,t]} f(t).$$

Write R_f for the set of zeroes of the function $M_f(t) - f(t)$ in the unit interval, i.e., R_f is the set of global maxima from the left of the function f.

Proposition 1. *Suppose f is a real-valued continuous function on the unit interval such that $M_f(b) > M_f(a)$ for some $b > a$ in the unit interval. Suppose, moreover, that for some positive constants C, α,*

$$|f(a+h) - f(a)| \leq C|h|^\alpha,$$

for all a, h such that $a, a + h$ are in the unit interval. Then $\dim R_f \geq \alpha$.

Proof: Since M_f is an increasing function, it is a distribution function of some Radon measure ν, such that $\nu(a,b] = M_f(b) - M_f(a)$. It follows from the hypotheses that ν is a non-zero measure. Moreover, it is supported on the set R_f. For $0 \leq a < b \leq 1$, we have

$$M_f(b) - M_f(a) \leq \sup_{0 \leq h \leq b-a} f(a+h) - f(a) \leq C(b-a)^\alpha.$$

The result follows from Frostman's lemma. (See eg p130 of [8]).

If x is a complex oscillation then it is easily seen that $M_x(t) > 0$ for some t. Moreover, it follows from Proposition 1 in [4] that a complex oscillation x satisfies the hypotheses of Proposition 1 of this paper. Consequently, it follows from Proposition 1 that

Corollary 1. *If x is a complex oscillation, and R_x is the set of its global maxima from the left in the unit interval, then $\dim R_x \geq \frac{1}{2}$.*

In particular, this argument shows that if X is a one-dimensional Brownian motion on the unit interval, then, almost surely, the associated random set R_X has Hausdorff dimension at least $\frac{1}{2}$.

As noted before, it is well-known that, if X is one-dimensional Brownian motion on the unit interval, then the random variable $Y(t) := M_X(t) - X(t)$ has the same distribution as $|X(t)|$. We can therefore conclude that, if we denote the (random) zero set of X by Z_X, then, almost surely

$$\dim Z_X \geq \frac{1}{2}.$$

For a compact subset A of the unit interval, denote by $N_\epsilon(A)$ the minimal number of closed intervals of radius ϵ required to cover A. Since

$$H_\alpha(A) \leq \liminf_{\epsilon \to 0^+} N_\epsilon(A)(2\epsilon)^\alpha,$$

we can conclude that if

$$\alpha > \liminf_{\epsilon \to 0^+} \frac{\log N_\epsilon(A)}{\log \frac{1}{\epsilon}},$$

then $H_\alpha(A) = 0$. Consequently,

$$\dim A \leq \liminf_{\epsilon \to 0^+} \frac{\log N_\epsilon(A)}{\log \frac{1}{\epsilon}}. \tag{4}$$

We shall use this observation to show that if x is a complex oscillation, then $\dim Z_x \leq \frac{1}{2}$.

For real numbers a, b with $b > 0$, we write $a \ll b$ (Vinogradovs notation) to signify the existence of a constant $c > 0$ independent of a such that $a \leq cb$. For a natural number m and a continuous version of one-dimensional Brownian motion over the unit interval, let $C_m(X)$ be the (random) number of $0 \leq k \leq 2^m$ such that X has a zero in the interval $I_{km} := [\frac{k-1}{2^m}, \frac{k}{2^m})$. It is well-known that

$$\mathrm{Prob}(\exists_{t \in I_{km}} X(t) = 0) \ll \frac{1}{\sqrt{k}}.$$

(See, for example [11].) Consequently,

$$E(C_m(X)) \ll \sum_{k=1}^{2^m} \frac{1}{\sqrt{k}} \ll 2^{\frac{m}{2}}.$$

It follows that that for any positive rational ϵ, we have

$$E \sum_{m=1}^\infty \frac{C_m(X)}{2^{m(\frac{1}{2}+\epsilon)}} = \sum_{m=1}^\infty E \frac{C_m(X)}{2^{m(\frac{1}{2}+\epsilon)}} \ll \sum_{m=1}^\infty \frac{2^{\frac{m}{2}}}{2^{m(\frac{1}{2}+\epsilon)}} < \infty.$$

As a very weak consequence of this inequality, we deduce that, for every positive rational ϵ, almost surely,

$$C_m(X) \leq 2^{m(\frac{1}{2}+\epsilon)}$$

infinitely often. Our next aim is to reflect this observation into every complex oscillation.

For a natural number m and a positive rational ϵ, the event $\exists_{m>t} C_m(X) < 2^{m(\frac{1}{2}+\epsilon)}$ is fully and faithfully described by the predicate

$$\exists_{m>t} \exists_{s \in \{0,1\}^{2^m}} \left(\forall_{k<2^m} s_k = 1 \leftrightarrow M(I_{km}) \geq 0 \wedge m(I_{km}) \leq 0 \right)$$

$$\wedge \sum_{k<2^m} s_k \leq 2^{m(\frac{1}{2}+\epsilon)}.$$

This defines a $\Sigma_1^0(\mathcal{M})$ set of Wiener measure 1 and therefore contains every complex oscillation x. We conclude that for every $x \in \mathcal{C}$ and every positive rational ϵ

$$\liminf_{\eta \to 0^+} \frac{\log N_\eta(Z_x)}{\log \frac{1}{\eta}} \leq \liminf_{m \to \infty} \frac{\log C_m(x)}{\log 2^m} \leq \frac{1}{2} + \epsilon.$$

It follows from (4) that, for every complex oscillation x,

$$\dim Z_x \leq \frac{1}{2}.$$

By arguing in terms of the algebra $\mathcal{N}$ instead $\mathcal{M}$, the same argument shows that $\dim R_x \leq \frac{1}{2}$. By taking Corollary 1 into account, we conclude with

Theorem 4. *If x is a complex oscillation and R_x is the set of global maxima from the left of x, then*

$$\dim R_x = \frac{1}{2}.$$

References

1. Asarin, E.A., Prokovskiy, A.V.: Use of the Kolmogorov complexity in analyzing control system dynamics. Automat. Remote Control 47, 21–28 (1986); Translated from: Primeenenie kolmogorovskoi slozhnosti k anlizu dinamiki upravlemykh sistem. Automatika i Telemekhanika (Automation Remote Control) 1, 25–33 (1986)
2. Chaitin, G.A.: Algorithmic information theory. Cambridge University Press, Cambridge (1987)
3. Fouché, W.L.: Arithmetical representations of Brownian motion I. J. Symb. Logic 65, 421–442 (2000)
4. Fouché, W.L.: The descriptive complexity of Brownian motion. Advances in Mathematics 155, 317–343 (2000)
5. Fouché, W.L.: Dynamics of a generic Brownian motion: Recursive aspects, in: From Gödel to Einstein: Computability between Logic and Physics. Theoretical Computer Science 394, 175–186 (2008)
6. Freedman, D.: Brownian motion and diffusion, 2nd edn. Springer, New York (1983)
7. Hinman, P.G.: Recursion-theoretic hierarchies. Springer, New York (1978)
8. Kahane, J.-P.: Some random series of functions, 2nd edn. Cambridge University Press, Cambridge (1993)
9. Kjos-Hanssen, B., Nerode, A.: The law of the iterated logarithm for algorithmically random Brownian motion. In: Artemov, S., Nerode, A. (eds.) LFCS 2007. LNCS, vol. 4514, pp. 310–317. Springer, Heidelberg (2007)
10. Martin-Löf, P.: The definition of random sequences. Information and Control 9, 602–619 (1966)
11. Peres, Y.: An Invitation to Sample Paths of Brownian Motion, http://www.stat.berkeley.edu/~peres/bmall.pdf

Computable Exchangeable Sequences Have Computable de Finetti Measures

Cameron E. Freer[1] and Daniel M. Roy[2]

[1] Department of Mathematics
Massachusetts Institute of Technology
freer@math.mit.edu
[2] Computer Science and Artificial Intelligence Laboratory
Massachusetts Institute of Technology
droy@csail.mit.edu

Abstract. We prove a uniformly computable version of de Finetti's theorem on exchangeable sequences of real random variables. In the process, we develop machinery for computably recovering a distribution from its sequence of moments, which suffices to prove the theorem in the case of (almost surely) continuous directing random measures. In the general case, we give a proof inspired by a randomized algorithm which succeeds with probability one. Finally, we show how, as a consequence of the main theorem, exchangeable stochastic processes in probabilistic functional programming languages can be rewritten as procedures that do not use mutation.

Keywords: de Finetti, exchangeability, computable probability theory, probabilistic programming languages, mutation.

1 Introduction

This paper examines the computable probability theory of exchangeable sequences of real-valued random variables; the main contribution is a uniformly computable version of de Finetti's theorem.

The classical result states that an exchangeable sequence of real random variables is a mixture of independent and identically distributed (i.i.d.) sequences of random variables. Moreover, there is an (almost surely unique) measure-valued random variable, called the *directing random measure*, conditioned on which the random sequence is i.i.d. The distribution of the directing random measure is called the *de Finetti measure*.

We show that computable exchangeable sequences of real random variables have computable de Finetti measures. In the process, we show that a distribution on $[0, 1]^\omega$ is computable if and only if its moments are uniformly computable.

This work is formulated in essentially the type-2 theory of effectivity (TTE) framework for computable analysis, though it is also related to domain theoretic representations of measures. Furthermore, our formalism coincides with those distributions from which we can generate exact samples on a computer.

K. Ambos-Spies, B. Löwe, and W. Merkle (Eds.): CiE 2009, LNCS 5635, pp. 218–231, 2009.
© Springer-Verlag Berlin Heidelberg 2009

In particular, we highlight the relationship between exchangeability and mutation in probabilistic programming languages. The computable de Finetti theorem can be used to uniformly transform procedures which induce exchangeable stochastic processes into equivalent procedures which do not use mutation (see Section 5).

1.1 de Finetti's Theorem

We assume familiarity with the standard measure-theoretic formulation of probability theory (see, e.g., [Bil95] or [Kal02]). Fix a basic probability space $(\Omega, \mathcal{F}, \mathbb{P})$ and let $\mathcal{B}_{\mathbf{R}}$ denote the Borel sets of $\mathbf{R}$. By a *random measure* we mean a random element in the space of Borel measures on $\mathbf{R}$, i.e., a kernel from $(\Omega, \mathcal{F})$ to $(\mathbf{R}, \mathcal{B}_{\mathbf{R}})$. An event $A \in \mathcal{F}$ is said to occur *almost surely* (a.s.) if $\mathbb{P}A = 1$. We denote the indicator function of a set B by $\mathbf{1}_B$.

Let $X = \{X_i\}_{i \geq 1}$ be an infinite sequence of real random variables. An infinite sequence X is *exchangeable* if, for any finite set $\{k_1, \ldots, k_j\}$ of distinct indices, $(X_{k_1}, \ldots, X_{k_j}) \overset{d}{=} (X_1, \ldots, X_j)$, where $\overset{d}{=}$ denotes equality in distribution.

Theorem 1 (de Finetti [Kal05, Chap. 1.1]). *Let $X = \{X_i\}_{i \geq 1}$ be an exchangeable sequence of real-valued random variables. There is a random probability measure ν on $\mathbf{R}$ such that $\{X_i\}_{i \geq 1}$ is conditionally i.i.d. with respect to ν. That is,*

$$\mathbb{P}[X \in \cdot \mid \nu] = \nu^{\infty} \quad a.s. \tag{1}$$

Moreover, ν is a.s. unique and given by

$$\nu(B) = \lim_{n \to \infty} \frac{1}{n} \sum_{i=1}^{n} \mathbf{1}_B(X_i) \quad a.s., \tag{2}$$

where B ranges over $\mathcal{B}_{\mathbf{R}}$.   □

We call ν the *directing random measure*, and its distribution, $\mu := \mathcal{L}(\nu)$, the *de Finetti measure*. As in [Kal05, Chap. 1, Eq. 3], we may take expectations on both sides of (1) to arrive at a characterization

$$\mathbb{P}\{X \in \cdot\} = \mathbb{E}\nu^{\infty} = \int m^{\infty} \mu(dm) \tag{3}$$

of an exchangeable sequence as a mixture of i.i.d. sequences.

A Bayesian perspective suggests the following interpretation: exchangeable sequences arise from independent observations from some latent random measure. Posterior analysis follows from placing a prior distribution on ν.

In 1931, de Finetti [dF31] proved the classical result for binary exchangeable sequences, in which case the de Finetti measure is simply a mixture of Bernoulli distributions; the exchangeable sequence is equivalent to repeatedly flipping a coin whose weight is drawn from some distribution on $[0, 1]$. Later, Hewitt and

Savage [HS55] and Ryll-Nardzewski [RN57] extended the result to arbitrary real-valued exchangeable sequences. We will refer to this more general version as the *de Finetti theorem*. Hewitt and Savage [HS55] provide a history of the early developments, and a history of more recent extensions can be found in Kingman [Kin78], Diaconis and Freedman [DF84], Lauritzen [Lau84], and Aldous [Ald85].

1.2   The Computable de Finetti Theorem

Consider an exchangeable sequence of $[0,1]$-valued random variables. In this case, the de Finetti measure is a distribution on the (Borel) measures on $[0,1]$. For example, if the de Finetti measure is a Dirac delta on the uniform distribution on $[0,1]$, then the induced exchangeable sequence consists of independent, uniformly distributed random variables on $[0,1]$.

As another example, let p be a random variable, uniformly distributed on $[0,1]$, and let $\nu = \delta_p$. Then the de Finetti measure is the uniform distribution on Dirac delta measures on $[0,1]$, and the corresponding exchangeable sequence is $p, p, \ldots$, i.e., a constant sequence, marginally uniformly distributed.

Clearly, sampling a measure ν, and then repeated sampling from ν, induces an exchangeable sequence; de Finetti's theorem states that all exchangeable sequences have an (a.s.) unique representation of this form.

In both examples, the de Finetti measures are *computable measures*. (In Section 2, we make this and related notions precise. For a more detailed example, see our discussion of the Pólya urn construction of the Beta-Bernoulli process in Section 5.) A natural question to ask is whether computable exchangeable sequences always arise from independent samples from computable random distributions. In fact, sampling from a computable de Finetti measure gives a computable directing random measure ν; repeated sampling from ν induces a computable exchangeable sequence (Proposition 1). Our main result is the converse: every computable real-valued exchangeable sequence arises from a computable de Finetti measure.

Theorem 2 (Computable de Finetti). *Let X be a real-valued exchangeable sequence and let μ be the distribution of its directing random measure ν. Then X is computable iff μ is computable. Moreover, μ is uniformly computable in X, and conversely.*

1.3   Outline of the Proof

Let $\mathcal{B}_{\mathbf{R}}$ denote the Borel sets of $\mathbf{R}$, let $\mathcal{I}_{\mathbf{R}}$ denote the set of finite unions of open intervals, and let $\mathcal{I}_{\mathbf{Q}}$ denote the set of finite unions of open intervals with rational endpoints. For $k \geq 1$ and $\beta \in \mathcal{B}_{\mathbf{R}}^k = \mathcal{B}_{\mathbf{R}} \times \cdots \times \mathcal{B}_{\mathbf{R}}$, we write $\beta(i)$ to denote the ith coordinate of β.

Let $X = \{X_i\}_{i \geq 1}$ be an exchangeable sequence of real random variables, with directing random measure ν. For every $\gamma \in \mathcal{B}_{\mathbf{R}}$, we define a $[0,1]$-valued random variable $V_\gamma := \nu\gamma$. A classical result in probability theory [Kal02, Lem. 1.17] implies that a Borel measure on $\mathbf{R}$ is uniquely characterized by the mass it places on the open intervals with rational endpoints. Therefore, the distribution of the stochastic process $\{V_\tau\}_{\tau \in \mathcal{I}_{\mathbf{Q}}}$ determines the de Finetti measure μ (the distribution of ν).

The *mixed moments* of the variables $\{V_\gamma\}_{\gamma \in C}$, for a subset $C \subseteq \mathcal{B}_{\mathbf{R}}$, are defined to be the expectations of the monomials $\prod_{i=1}^{k} V_{\beta(i)}$, for $k \geq 1$ and $\beta \in C^k$. The following relationship between the finite dimensional distributions of the sequence X and the mixed moments of $\{V_\beta\}_{\beta \in \mathcal{B}_{\mathbf{R}}}$ is an immediate consequence of the characterization of an exchangeable sequence as a mixture of i.i.d. sequences given by (3).

Lemma 1. $\mathbb{P}\big(\bigcap_{i=1}^{k}\{X_i \in \beta(i)\}\big) = \mathbb{E}\big(\prod_{i=1}^{k} V_{\beta(i)}\big)$ *for $k \geq 1$ and $\beta \in \mathcal{B}_{\mathbf{R}}^k$.*   □

Assume that X is computable. If ν is a.s. continuous, we can compute the mixed moments of $\{V_\tau\}_{\tau \in \mathcal{I}_{\mathbf{Q}}}$. In Section 3, we show how to computably recover a distribution from its moments. This suffices to recover the de Finetti measure.

In the general case, point masses in ν prevent us from computing the mixed moments. Here we use a proof inspired by a randomized algorithm which (a.s.) avoids the point masses and recovers the de Finetti measure. For the complete proof, see Section 4.4.

2   Computable Measures

We assume familiarity with the standard notions of computability theory and computably enumerable (c.e.) sets (see, e.g., [Rog67] or [Soa87]). Recall that $r \in \mathbf{R}$ is called a c.e. real when the set of all rationals less than r is a c.e. set. Similarly, r is a co-c.e. real when the set of all rationals greater than r is c.e. A real r is a computable real when it is both a c.e. and co-c.e. real.

The following representations for probability measures on computable T_0 spaces [GSW07, §3] are devised from more general TTE representations in [Sch07] and [Bos08], and agree with [Wei99] in the case of the unit interval. This work is also related to domain theoretic representations of measure theory (given in [Eda95], [Eda96], and [AES00]) and to constructive analysis (see [Bau05]), though we do not explore this here.

Schröder [Sch07] has connected TTE representations with *probabilistic processes* [SS06] which sample representations. In Section 5 we explore consequences of the computable de Finetti theorem for probabilistic programming languages.

2.1   Representations of Probability Measures

Let S be a T_0 second-countable space with a countable basis $\mathcal{S}$ that is closed under finite unions. Assume that there is (and fix) an enumeration of $\mathcal{S}$ for which the sets $\{C \in \mathcal{S} : C \subseteq A \cap B\}$ are c.e., uniformly in the indices of $A, B \in \mathcal{S}$.

Let $\mathcal{M}_1(S)$ denote the set of Borel probability measures on S (i.e., the probability measures on the σ-algebra generated by $\mathcal{S}$). Such measures are determined by the measure they assign to elements of $\mathcal{S}$; we define the representation of

$\eta \in \mathcal{M}_1(S)$ with respect to $\mathcal{S}$ to be the set $\{(B,q) \in \mathcal{S} \times \mathbf{Q} : \eta B > q\}$. We say that η is a *computable measure* when its representation is a c.e. set (in terms of the fixed enumeration of $\mathcal{S}$). In other words, η is computable if the measure it assigns to a basis set $B \in \mathcal{S}$ is a c.e. real, uniformly in the index of B in the enumeration of $\mathcal{S}$. This representation for $\mathcal{M}_1(S)$ is admissible with respect to the weak topology, hence computably equivalent (see [Wei00, Chap. 3]) to the canonical representation for Borel measures given in [Sch07]. (Consequences for measurable functions with respect to admissible representations can be found in [BG07, §2].) We will be interested in representing measures $\eta \in \mathcal{M}_1(S)$ where S is either $\mathbf{R}^\omega$, $[0,1]^k$, or $\mathcal{M}_1(\mathbf{R})$, topologized as described below.

Consider $\mathbf{R}^\omega$ under the product topology, where $\mathbf{R}$ is endowed with its standard topology. Fix the canonical enumeration of $\mathcal{I}_\mathbf{Q}$ and the enumeration of $\bigcup_{k \geq 1} \mathcal{I}_\mathbf{Q}^k$ induced by interleaving (and the pairing function). A suitable basis for this product topology is given by the cylinders $\{\sigma \times \mathbf{R}^\omega : \sigma \in \bigcup_{k \geq 1} \mathcal{I}_\mathbf{Q}^k\}$. Let $\vec{x} = \{x_i\}_{i \geq 1}$ be a sequence of real-valued random variables (e.g., an exchangeable sequence X, or $\{V_\tau\}_{\tau \in \mathcal{I}_\mathbf{Q}}$ under the canonical enumeration of $\mathcal{I}_\mathbf{Q}$). The representation of the joint distribution of $\vec{x}$ is $\bigcup_{k \geq 1} \{(\sigma,q) \in \mathcal{I}_\mathbf{Q}^k \times \mathbf{Q} : \mathbb{P}(\bigcap_{i=1}^k \{x_i \in \sigma(i)\}) > q\}$. We say that $\vec{x}$ is computably distributed when this set is c.e. Informally, we will call a sequence $\vec{x}$ computable when it is computably distributed.

Fix $k \geq 1$, and consider the *right order topology* on $[0,1]$ generated by the basis $\mathcal{T} = \{(c,1] : c \in \mathbf{Q}, \ 0 \leq c < 1\} \cup \{[0,1]\}$. Let $\mathcal{S} = \mathcal{T}^k$ be a basis for the product of the right order topology on $[0,1]^k$. Let $\vec{w} = (w_1, \ldots, w_k)$ be a random vector in $[0,1]^k$. The representation of the joint distribution of $\vec{w}$ with respect to $\mathcal{S}$ is $\{(\vec{c},q) \in \mathbf{Q}^k \times \mathbf{Q} : \mathbb{P}(\bigcap_{i=1}^k \{w_i > c_i\}) > q\}$. We will use this representation in Proposition 1 for $\{V_{\sigma(i)}\}_{i \leq k}$ where $\sigma \in \mathcal{I}_\mathbf{Q}^k$.

The de Finetti measure μ is the distribution of the directing random measure ν, an $\mathcal{M}_1(\mathbf{R})$-valued random variable. For $k \geq 1$, $\beta \in \mathcal{B}_\mathbf{R}^k$, and $\vec{c} \in \mathbf{Q}^k$, we define the event $Y_{\beta,\vec{c}} := \bigcap_{i=1}^k \{\nu\beta(i) > c_i\}$. The sets $\{\nu : \nu\tau > c\}$ for $\tau \in \mathcal{I}_\mathbf{Q}$ and $c \in \mathbf{Q}$ form a subbasis for the weak topology on $\mathcal{M}_1(\mathbf{R})$. We therefore represent the distribution μ by $\bigcup_{k \geq 1} \{(\sigma,\vec{c},q) \in \mathcal{I}_\mathbf{Q}^k \times \mathbf{Q}^k \times \mathbf{Q} : \mathbb{P}Y_{\sigma,\vec{c}} > q\}$, and say that μ is computable when $\mathbb{P}Y_{\sigma,\vec{c}}$ is a c.e. real, uniformly in σ and $\vec{c}$.

2.2 Computable de Finetti Measures

In the remainder of the paper, let X be a real-valued exchangeable sequence with directing random measure ν and de Finetti measure μ.

Proposition 1. *X is uniformly computable in μ.*

Proof. We give a proof that relativizes to μ; without loss of generality, assume that μ is computable. Then $\mathbb{P}(\bigcap_{i=1}^k \{\nu : \nu\sigma(i) > c_i\})$ is a c.e. real, uniformly in $\sigma \in \mathcal{I}_\mathbf{Q}^k$ and $\vec{c} \in \mathbf{Q}^k$. Therefore, the joint distribution of $\{V_{\sigma(i)}\}_{i \leq k}$ under the right order topology is computable, uniformly in σ. By Lemma 1, $\mathbb{P}(\bigcap_{i=1}^k \{X_i \in \sigma(i)\}) = \mathbb{E}(\prod_{i=1}^k V_{\sigma(i)})$. Because the monomial maps $[0,1]^k$ into $[0,1]$ and is continuous, this expectation is a c.e. real, uniformly in σ [Sch07, Prop. 3.6]. $\square$

3   The Computable Moment Problem

One often has access to the moments of a distribution, and wishes to recover the underlying distribution. Let η be a distribution on $[0,1]^\omega$, and let $\vec{x} = \{x_i\}_{i \geq 1}$ be a sequence of $[0,1]$-valued random variables with distribution η. Classically, the distribution of η is uniquely determined by the mixed moments of $\vec{x}$. We show that a distribution on $[0,1]^\omega$ is computable iff its sequence of mixed moments is uniformly computable.

To show that η is computable, it suffices to show that $\eta(\sigma \times [0,1]^\omega)$ is a c.e. real for $\sigma \in \mathcal{I}_{\mathbf{Q}}^k$, uniformly in σ. Recall that $\eta(\sigma \times [0,1]^\omega) = \mathbb{E}\mathbf{1}_{\sigma \times [0,1]^\omega}$. Using the effective Weierstrass theorem ([Wei00, p. 161] or [PR89, p. 45]), we can approximate the indicator function by a sequence of rational polynomials that converge pointwise from below:

Lemma 2 (Polynomial approximations). *There is a computable sequence* $\{p_{n,\sigma} : n \in \omega, \ \sigma \in \bigcup_{k \geq 1} \mathcal{I}_{\mathbf{Q}}^k\}$ *of rational polynomials where* $p_{n,\sigma}$ *is in* k *variables (for* $\sigma \in \mathcal{I}_{\mathbf{Q}}^k$*) and, for* $\vec{x} \in [0,1]^k$ *we have* $-1 \leq p_{n,\sigma}(\vec{x}) \leq \mathbf{1}_\sigma(\vec{x})$ *and* $\lim_n p_{n,\sigma}(\vec{x}) = \mathbf{1}_\sigma(\vec{x})$. □

By the dominated convergence theorem, the expectations of the sequence converge to $\eta(\sigma \times [0,1]^\omega)$ from below:

Lemma 3. *Fix* $k \geq 1$ *and* $\sigma \in \mathcal{I}_{\mathbf{Q}}^k$, *and let* $\vec{x} = (x_1, \ldots, x_k)$ *be a random vector in* $[0,1]^k$. *Then* $\mathbb{E}(\mathbf{1}_\sigma(\vec{x})) = \sup_n \mathbb{E}(p_{n,\sigma}(\vec{x}))$. □

Theorem 3 (Computable moments). *Let* $\vec{x} = (x_i)_{i \geq 1}$ *be a random vector in* $[0,1]^\omega$ *with distribution* η. *Then* η *is computable iff the mixed moments of* $\{x_i\}_{i \geq 1}$ *are uniformly computable.*

Proof. All monomials in $\{x_i\}_{i \geq 1}$ are bounded and continuous on $[0,1]^\omega$. If η is computable, then the mixed moments are uniformly computable by a computable integration result [Sch07, Prop. 3.6].

Now suppose the mixed moments of $\{x_i\}_{i \geq 1}$ are uniformly computable. To establish the computability of η, it suffices to show that $\eta(\sigma \times [0,1]^\omega)$ is a c.e. real, uniformly in σ. Note that $\eta(\sigma \times [0,1]^\omega) = \mathbb{E}(\mathbf{1}_{\sigma \times [0,1]^\omega}(\vec{x})) = \mathbb{E}(\mathbf{1}_\sigma(x_1, \ldots, x_k))$. By Lemma 3, $\mathbb{E}(\mathbf{1}_\sigma(x_1, \ldots, x_k)) = \sup_n \mathbb{E}(p_{n,\sigma}(x_1, \ldots, x_k))$. The sequence of polynomials $(p_{n,\sigma}(x_1, \ldots, x_k))_{n \in \omega}$ is uniformly computable by Lemma 2. The expectation of each polynomial is a $\mathbf{Q}$-linear combination of moments, hence computable. □

4   The Computable de Finetti Theorem

4.1   Continuous Directing Random Measures

For $k \geq 1$ and $\psi \in \mathcal{I}_{\mathbf{R}}^k$, we say that ψ has *no mass on its boundary* when $\mathbb{P}(\bigcup_{i=1}^k \{X_i \in \partial\psi(i)\}) = 0$.

Lemma 4. *Suppose X is computable. Then the mixed moments of $\{V_\tau\}_{\tau \in \mathcal{I}_Q}$ are uniformly c.e. reals and the mixed moments of $\{V_{\bar{\tau}}\}_{\tau \in \mathcal{I}_Q}$ are uniformly co-c.e. reals. In particular, if $\sigma \in \mathcal{I}_Q^k$ has no mass on its boundary, then the moment $\mathbb{E}(\prod_{i=1}^k V_{\sigma(i)})$ is a computable real.*

Proof. Let $k \geq 1$ and $\sigma \in \mathcal{I}_Q^k$. By Lemma 1, $\mathbb{E}(\prod_{i=1}^k V_{\sigma(i)}) = \mathbb{P}(\bigcap_{i=1}^k \{X_i \in \sigma(i)\})$, which is a c.e. real because X is computable. Similarly, $\mathbb{E}(\prod_{i=1}^k V_{\overline{\sigma(i)}}) = \mathbb{P}(\bigcap_{i=1}^k \{X_i \in \overline{\sigma(i)}\})$, which is the probability of a finite union of closed rational cylinders, and thus a co-c.e. real (as we can computably enumerate all $\tau \in \mathcal{I}_Q^k$ contained in the complement of a given closed rational cylinder). When σ has no mass on its boundary, $\mathbb{E}(\prod_{i=1}^k V_{\sigma(i)}) = \mathbb{E}(\prod_{i=1}^k V_{\overline{\sigma(i)}})$, which is both c.e. and co-c.e. $\qquad\square$

Proposition 2 (Computable de Finetti – a.s. continuous ν). *Assume that ν is continuous with probability one. Then μ is uniformly computable in X, and conversely.*

Proof. By Proposition 1, X is uniformly computable in μ. We now give a proof of the other direction that relativizes to X; without loss of generality, assume that X is computable. Let $k \geq 1$ and consider $\sigma \in \mathcal{I}_Q^k$. The (a.s.) continuity of ν implies that $\mathbb{P}(\bigcup_{i=1}^k \{X_i \in \partial\sigma(i)\}) = 0$, i.e., σ has no mass on its boundary. Therefore the moment $\mathbb{E}(\prod_{i=1}^k V_{\sigma(i)})$ is computable by Lemma 4. By the computable moment theorem (Theorem 3), the joint distribution of the variables $\{V_\tau\}_{\tau \in \mathcal{I}_Q}$ is computable. $\qquad\square$

4.2 "Randomized" Proof Sketch

The task of lower bounding the measure of the events $Y_{\sigma,\bar{c}}$ for $\sigma \in \mathcal{I}_Q^k$ and $\bar{c} \in Q^k$ is complicated by the potential existence of point masses on any of the boundaries $\partial\sigma(i)$ of $\sigma(i)$. If X is a computable exchangeable sequence for which ν is discontinuous at some rational point with non-zero probability, the mixed moments of $\{V_\tau\}_{\tau \in \mathcal{I}_Q}$ are c.e., but not co-c.e., reals (by Lemma 4). Therefore the computable moment theorem (Theorem 3) is inapplicable. For arbitrary directing random measures, we give a proof of the computable de Finetti theorem which works regardless of the location of point masses.

Consider the following sketch of a "randomized algorithm": We draw a countably infinite set of real numbers $A \subseteq R$ from a continuous distribution with support on the entire real line (e.g., Gaussian or Cauchy). Note that, with probability one (over the draw), A will be dense in R, and no point mass of ν will be in A. Let $\mathcal{I}_A$ denote the set of finite unions of intervals with endpoints in A. Because there is (a.s.) no mass on any boundary of any $\psi \in \mathcal{I}_A$, we can proceed as in the case of (a.s.) continuous ν, in terms of this alternative basis.

If X is computable, we can compute all moments (relative to the infinite representations of the points of A) by giving c.e. (in A) lower and (as in Lemma 4) upper bounds. By the computable moment theorem (Theorem 3) we can compute (in A) the joint distribution of $\{V_\psi\}_{\psi \in \mathcal{I}_A}$. This joint distribution classically determines the de Finetti measure. Moreover, we can compute (in A) all rational lower bounds on any $\mathbb{P}Y_{\sigma,\vec{c}}$ for $\sigma \in \mathcal{I}_\mathbf{Q}^k$, our original basis.

This algorithm essentially constitutes a (degenerate) probabilistic process (see [SS06]) which returns a representation of the de Finetti measure with probability one. A proof along these lines could be formalized using the TTE and probabilistic process frameworks. Here, however, we provide a construction that makes explicit the representation in terms of our rational basis, and which can be seen as a "derandomization" of this algorithm.

Weihrauch [Wei99, Thm. 3.6] proves a computable integration result via an argument that could likewise be seen as a derandomization of an algorithm which densely subdivides the unit interval at random locations to avoid mass on the division boundaries. Similar arguments are used to avoid mass on the boundaries of open hypercubes [Mül99, Thm. 3.7] and open balls [Bos08, Lem. 2.15]. These arguments also resemble the classic proof of the Portmanteau theorem [Kal02, Thm. 4.25], in which an uncountable family of sets with disjoint boundaries is defined, almost all of which have no mass on their boundaries.

4.3 Rational Refinements and Polynomial Approximations

We call $\psi \in \mathcal{I}_\mathbf{R}^k$ a *refinement* of $\varphi \in \mathcal{I}_\mathbf{R}^k$, and write $\psi \lhd \varphi$, when $\overline{\psi(i)} \subseteq \varphi(i)$ for all $i \leq k$. For $n \in \omega$ and $\vec{c} = (c_1, \ldots, c_k) \in \mathbf{Q}^k$, we will denote by $p_{n,\vec{c}}$ the polynomial $p_{n,\sigma}$ (defined in Lemma 2), where $\sigma = (c_1, 2) \times \cdots \times (c_k, 2)$. Let $\vec{x} = (x_1, \ldots, x_k)$, and similarly with $\vec{y}$. We can write $p_{n,\vec{c}}(\vec{x}) = p_{n,\vec{c}}^+(\vec{x}) - p_{n,\vec{c}}^-(\vec{x})$, where $p_{n,\vec{c}}^+$ and $p_{n,\vec{c}}^-$ are polynomials with positive coefficients. Define the $2k$-variable polynomial $q_{n,\vec{c}}(\vec{x}, \vec{y}) := p_{n,\vec{c}}^+(\vec{x}) - p_{n,\vec{c}}^-(\vec{y})$. We will denote $q_{n,\vec{c}}(V_{\psi(1)}, \ldots, V_{\psi(k)}, V_{\zeta(1)}, \ldots, V_{\zeta(k)})$ by $q_{n,\vec{c}}(V_\psi, V_\zeta)$, and similarly with $p_{n,\vec{c}}$.

Proposition 3. *Let $\sigma \in \mathcal{I}_\mathbf{Q}^k$ and let $\vec{c} \in \mathbf{Q}^k$ and $n \in \omega$. If X is computable, then $\mathbb{E}q_{n,\vec{c}}(V_\sigma, V_{\overline{\sigma}})$ is a c.e. real (uniformly in $\sigma, \vec{c}$, and n).*

Proof. By Lemma 4, each monomial of $p_{n,\vec{c}}^+(V_\sigma)$ has a c.e. real expectation, and each monomial of $p_{n,\vec{c}}^-(V_{\overline{\sigma}})$ has a co-c.e. real expectation, and so by linearity of expectation, $\mathbb{E}q_{n,\vec{c}}(V_\sigma, V_{\overline{\sigma}})$ is a c.e. real.   □

4.4 Proof of the Computable de Finetti Theorem

Proof of Theorem 2 (Computable de Finetti). The sequence X is uniformly computable in μ by Proposition 1. We now give a proof of the other direction that relativizes to X; without loss of generality, suppose that X is computable. Let $\pi \in \mathcal{I}_\mathbf{Q}^k$ and $\vec{c} \in \mathbf{Q}^k$. It suffices to show that the probability $\mathbb{P}Y_{\pi,\vec{c}}$ is a c.e. real (uniformly in π and $\vec{c}$). We do this by a series of reductions to quantities of the form $\mathbb{E}q_{n,\vec{c}}(V_\sigma, V_{\overline{\sigma}})$ for $\sigma \in \mathcal{I}_\mathbf{Q}^k$, which by Proposition 3 are c.e. reals.

Note that $\mathbb{P}Y_{\pi,\vec{c}} = \mathbb{E}1_{Y_{\pi,\vec{c}}}$. If $\psi \in \mathcal{I}_\mathbf{R}^k$ satisfies $\psi \lhd \pi$, then $Y_{\psi,\vec{c}} \subseteq Y_{\pi,\vec{c}}$. Furthermore, $\mathcal{I}_\mathbf{R}^k$ is dense in $\mathcal{I}_\mathbf{Q}^k$, and so by the dominated convergence theorem we have that $\mathbb{E}1_{Y_{\pi,\vec{c}}} = \sup_{\psi \lhd \pi} \mathbb{E}1_{Y_{\psi,\vec{c}}}$. By the definitions of Y and V we have $\mathbb{E}1_{Y_{\psi,\vec{c}}} = \mathbb{E}(1_{(c_1,2)\times\cdots\times(c_k,2)}(V_{\psi(1)}, \ldots, V_{\psi(k)}))$, which, by a Weierstrass approximation (Lemma 3) equals $\sup_n \mathbb{E}p_{n,\vec{c}}(V_\psi)$.

If $\zeta, \varphi \in \mathcal{I}_\mathbb{R}$ satisfy $\zeta \lhd \varphi$ then $V_\zeta \leq V_\varphi$ a.s. Multiplication is continuous, and so the dominated convergence theorem gives us

$$\mathbb{E}\left(\prod_{i=1}^k V_{\psi(i)}\right) = \sup_{\sigma \lhd \psi} \mathbb{E}\left(\prod_{i=1}^k V_{\sigma(i)}\right) \quad \text{and} \tag{4}$$

$$\mathbb{E}\left(\prod_{i=1}^k V_{\overline{\psi(i)}}\right) = \inf_{\tau \rhd \psi} \mathbb{E}\left(\prod_{i=1}^k V_{\overline{\tau(i)}}\right), \tag{5}$$

where σ and τ range over $\mathcal{I}_\mathbb{Q}^k$. By the linearity of expectation, $\mathbb{E}p_{n,\vec{c}}^+(V_\psi) = \sup_{\sigma \lhd \psi} \mathbb{E}p_{n,\vec{c}}^+(V_\sigma)$ and $\mathbb{E}p_{n,\vec{c}}^-(V_\psi) = \inf_{\tau \rhd \psi} \mathbb{E}p_{n,\vec{c}}^-(V_\tau)$.

If ψ has no mass on its boundary then $V_{\psi(i)} = V_{\overline{\psi(i)}}$ a.s. for $i \leq k$, and so

$$\mathbb{E}p_{n,\vec{c}}(V_\psi) = \mathbb{E}q_{n,\vec{c}}(V_\psi, V_\psi) \tag{6}$$

$$= \mathbb{E}q_{n,\vec{c}}(V_\psi, V_{\overline{\psi}}) \tag{7}$$

$$= \mathbb{E}p_{n,\vec{c}}^+(V_\psi) - \mathbb{E}p_{n,\vec{c}}^-(V_{\overline{\psi}}) \tag{8}$$

$$= \sup_{\sigma \lhd \psi} \mathbb{E}p_{n,\vec{c}}^+(V_\sigma) - \inf_{\tau \rhd \psi} \mathbb{E}p_{n,\vec{c}}^-(V_{\overline{\tau}}) \tag{9}$$

$$= \sup_{\sigma \lhd \psi \lhd \tau} \mathbb{E}q_{n,\vec{c}}(V_\sigma, V_{\overline{\tau}}). \tag{10}$$

Because there are only countably many point masses, the $\psi \in \mathcal{I}_\mathbb{R}^k$ with no mass on their boundaries are dense in $\mathcal{I}_\mathbb{Q}^k$. Therefore,

$$\sup_{\psi \lhd \pi} \mathbb{E}p_{n,\vec{c}}(V_\psi) = \sup_{\psi \lhd \pi} \sup_{\sigma \lhd \psi \lhd \tau} \mathbb{E}q_{n,\vec{c}}(V_\sigma, V_{\overline{\tau}}). \tag{11}$$

Note that $\{(\sigma, \tau) : (\exists \psi \lhd \pi)\, \sigma \lhd \psi \lhd \tau\} = \{(\sigma, \tau) : \sigma \lhd \pi \text{ and } \sigma \lhd \tau\}$. Hence

$$\sup_{\psi \lhd \pi} \sup_{\sigma \lhd \psi} \sup_{\tau \rhd \psi} \mathbb{E}q_{n,\vec{c}}(V_\sigma, V_{\overline{\tau}}) = \sup_{\sigma \lhd \pi} \sup_{\tau \rhd \sigma} \mathbb{E}q_{n,\vec{c}}(V_\sigma, V_{\overline{\tau}}). \tag{12}$$

Again by dominated convergence we have $\sup_{\tau \rhd \sigma} \mathbb{E}q_{n,\vec{c}}(V_\sigma, V_{\overline{\tau}}) = \mathbb{E}q_{n,\vec{c}}(V_\sigma, V_{\overline{\sigma}})$. In summary, we have $\mathbb{P}Y_{\pi,\vec{c}} = \sup_n \sup_{\sigma \lhd \pi} \mathbb{E}q_{n,\vec{c}}(V_\sigma, V_{\overline{\sigma}})$. Finally, by Proposition 3, $\{\mathbb{E}q_{n,\vec{c}}(V_\sigma, V_{\overline{\sigma}}) : \sigma \lhd \pi \text{ and } n \in \omega\}$ is a set of uniformly c.e. reals. $\qquad\square$

5   Exchangeability in Probabilistic Programs

Functional programming languages with probabilistic choice operators have recently been proposed as universal languages for statistical modeling (e.g., IBAL [Pfe01], $\lambda_\circ$[PPT08], and Church [GMR+08]). Although the semantics of probabilistic programs have been studied extensively in theoretical computer science in the context of randomized algorithms (e.g., [Koz81] and [JP89]), this application of probabilistic programs to universal statistical modeling has a different character which has raised a number of interesting theoretical questions (e.g., [RP02], [PPT08], [GMR+08], [RMG+08], and [Man09]).

The computable de Finetti theorem has implications for the semantics of probabilistic programs, especially concerning the choice of whether or not to allow mutation (i.e., the ability of a program to modify its own state as it runs). For concreteness, we will explore this connection using Church, a probabilistic functional programming language. Church extends Scheme (a dialect of LISP) with a boolean-valued *flip* procedure, which denotes the Bernoulli distribution. In Church, an expression denotes the distribution induced by evaluation. For example, (+ (*flip*) (*flip*) (*flip*)) denotes a Binomial($n = 3, p = \frac{1}{2}$) distribution and (λ (x) (if (*flip*) x 0)) denotes the probability kernel $x \mapsto \frac{1}{2}(\delta_x + \delta_0)$. Church is call-by-value and so (= (*flip*) (*flip*)) denotes a Bernoulli distribution, while the application of the procedure (λ (x) (= x x)) to the argument (*flip*), written ((λ (x) (= x x)) (*flip*)), denotes δ_1. (For more details, see [GMR+08].) If an expression modifies its environment using mutation it might not denote a fixed distribution. For example, a procedure may keep a counter variable and return an increasing sequence of integers on repeated calls.

Consider the Beta-Bernoulli process and the Pólya urn scheme written in Church. While these two processes look different, they induce the same distribution on sequences. We will define the procedure **sample-coin** such that calling **sample-coin** returns a new procedure which itself returns random binary values. The probabilistic program

```
(define my-coin (sample-coin))
(my-coin) (my-coin) (my-coin) (my-coin) (my-coin) ...
```

defines a random binary sequence. Fix $a, b > 0$. Consider the following two implementations of **sample-coin** (and recall that the (λ () ...) special form creates a procedure of no arguments):

(i)   *(ii)*

```
(define (sample-coin)          (define (sample-coin)
  (let ((weight (beta a b)))     (let ((red a)
    (λ () (flip weight)) ) )          (total (+ a b)) )
                                   (λ () (let ((x (flip red/total)))
                                      (set! red (+ red x))
                                      (set! total (+ total 1))
                                      x ) ) )
```

In case *(i)*, evaluating (my-coin) returns a 1 with probability **weight** and a 0 otherwise, where the shared **weight** parameter is itself drawn from a Beta(a, b) distribution on $[0, 1]$. Note that the sequence of values obtained by evaluating (my-coin) is exchangeable but not i.i.d. (e.g., an initial sequence of ten 1's leads one to predict that the next draw is more likely to be 1 than 0). However, conditioned on the **weight** (a variable within the opaque procedure my-coin) the sequence is i.i.d. A second random coin constructed by (define your-coin (sample-coin)) will have a different weight (a.s.) and will generate a sequence that is independent of my-coin's.

The sequence induced by *(i)* is exchangeable because applications of beta and flip return independent samples. The code in *(ii)* implements the Pólya urn scheme with a red balls and b black balls (see [dF75, Chap. 11.4]), and the sequence of return values is exchangeable because its joint distribution depends only on the number of red and black balls and not on their order.

Because the sequence induced by *(ii)* is exchangeable, de Finetti's theorem implies that its distribution is equivalent to that induced by i.i.d. draws from some some random measure (the directing random measure). In the case of the Pólya urn scheme, the directing random measure is a random Bernoulli whose weight parameter has a Beta(a, b) distribution. Therefore *(i)* denotes the the de Finetti measure, and *(i)* and *(ii)* induce the same distribution over sequences.

However, there is an important difference between these two implementations. While *(i)* denotes the de Finetti measure, *(ii)* does not, because samples from it do not denote fixed distributions: The state of a procedure my-coin sampled using *(ii)* changes after each iteration, as the sufficient statistics are updated (using the mutation operator set!). Therefore, each element of the sequence is generated from a different distribution. Even though the sequence of calls to such a my-coin has the same *marginal* distribution as those given by repeated calls to a my-coin sampled using *(i)*, a procedure my-coin sampled using *(ii)* could be thought of as a probability kernel which depends on the state.

In contrast, a my-coin sampled using *(i)* does not modify itself via mutation; the value of weight does not change after it is randomly initialized and therefore my-coin denotes a fixed distribution — a particular Bernoulli. Its marginal distribution is a random Bernoulli, precisely the directing random measure of the Beta-Bernoulli process.

The computable de Finetti theorem could be used to uniformly transform *(ii)* into a procedure sample-weight which does not use mutation and whose application is equivalent in distribution to the evaluation of (beta a b). In general, an implementation of Theorem 2 transforms code which generates an exchangeable sequence (like *(ii)*) into code representing the de Finetti measure (i.e., a procedure of the form *(i)* which does not use mutation). In addition to their simpler semantics, mutation-free procedures are often desirable for practical reasons. For example, having sampled the directing random measure, an exchangeable sequence of random variables can be efficiently sampled in parallel without the overhead necessary to communicate sufficient statistics.

5.1 Partial Exchangeability of Arrays and Other Data Structures

The example above involved binary sequences, but the computable de Finetti theorem can be used to transform implementations of real exchangeable sequences. For example, given a probabilistic program outputting repeated draws from a distribution with a Dirichlet process prior via the Chinese restaurant process [Ald85] representation, we can automatically recover a version of the random directing measure characterized by Sethuraman's stick-breaking construction [Set94].

More general de Finetti-type results deal with variables taking values other than reals, or weaken exchangeability to various notions of partial exchangeability. The Indian buffet process, defined in [GG05], can be interpreted as a set-valued exchangeable sequence and can be written in a way analogous to the Pólya urn in *(ii)*. A corresponding stick-breaking construction given by [TGG07] (following [TJ07]) is analogous to the code in *(i)*, but gives only a Δ_1-index for the (a.s.) finite set of indices (with respect to a random countable set of reals sampled from the base measure), rather than its canonical index (see [Soa87, II.2]). This observation was first noted in [RMG⁺08]. In the case of the Indian buffet process on a discrete base measure (where we can fix an enumeration of all finite sets of elements in the support), the computable de Finetti theorem implies the existence of a computable de Finetti measure that gives canonical indices for the sets. In the general case, the computable de Finetti theorem is not directly applicable, and the question of its computability is open.

Probabilistic models of relational data ([KTG⁺06] and [RT09]) can be viewed as random binary (2-dimensional) arrays whose distributions are *separately* (or *jointly*) exchangeable, i.e., invariant under (simultaneous) permutations of the rows and columns. Aldous [Ald81] and Hoover [Hoo79] gave de Finetti-type results for infinite arrays that satisfy these two notions of partial exchangeability. An extension of the computable de Finetti theorem to this setting would provide an analogous uniform transformation on arrays.

Data structures such as trees or graphs might also be approached in this manner. Diaconis and Janson [DJ08] and Austin [Aus08] explore many connections between partially exchangeable random graphs and the theory of graph limits.

Acknowledgements

The authors would like to thank Vikash Mansinghka, Hartley Rogers, and the anonymous referees for helpful comments.

References

[AES00] Alvarez-Manilla, M., Edalat, A., Saheb-Djahromi, N.: An extension result for continuous valuations. J. London Math. Soc. 61(2), 629–640 (2000)

[Ald81] Aldous, D.J.: Representations for partially exchangeable arrays of random variables. J. Multivariate Analysis 11(4), 581–598 (1981)

[Ald85] Aldous, D.J.: Exchangeability and related topics. In: École d'été de probabilités de Saint-Flour, XIII—1983. Lecture Notes in Math., vol. 1117, pp. 1–198. Springer, Berlin (1985)

[Aus08] Austin, T.: On exchangeable random variables and the statistics of large graphs and hypergraphs. Probab. Surv. 5, 80–145 (2008)

[Bau05] Bauer, A.: Realizability as the connection between constructive and computable mathematics. In: CCA 2005: Second Int. Conf. on Comput. and Complex in Analysis (2005)

[BG07] Brattka, V., Gherardi, G.: Borel complexity of topological operations on computable metric spaces. In: Cooper, S.B., Löwe, B., Sorbi, A. (eds.) CiE 2007. LNCS, vol. 4497, pp. 83–97. Springer, Heidelberg (2007)

[Bil95]   Billingsley, P.: Probability and measure, 3rd edn. John Wiley & Sons Inc., New York (1995)

[Bos08]   Bosserhoff, V.: Notions of probabilistic computability on represented spaces. J. of Universal Comput. Sci. 14(6), 956–995 (2008)

[dF31]   de Finetti, B.: Funzione caratteristica di un fenomeno aleatorio. Atti della R. Accademia Nazionale dei Lincei, Ser. 6. Memorie, Classe di Scienze Fisiche, Matematiche e Naturali 4, 251–299 (1931)

[dF75]   de Finetti, B.: Theory of probability, vol. 2. John Wiley & Sons Ltd., London (1975)

[DF84]   Diaconis, P., Freedman, D.: Partial exchangeability and sufficiency. In: Statistics: applications and new directions (Calcutta, 1981), pp. 205–236. Indian Statist. Inst., Calcutta (1984)

[DJ08]   Diaconis, P., Janson, S.: Graph limits and exchangeable random graphs. Rendiconti di Matematica, Ser. VII 28(1), 33–61 (2008)

[Eda95]   Edalat, A.: Domain theory and integration. Theoret. Comput. Sci. 151(1), 163–193 (1995)

[Eda96]   Edalat, A.: The Scott topology induces the weak topology. In: 11th Ann. IEEE Symp. on Logic in Comput. Sci., pp. 372–381. IEEE Comput. Soc. Press, Los Alamitos (1996)

[GG05]   Griffiths, T.L., Ghahramani, Z.: Infinite latent feature models and the Indian buffet process. In: Adv. in Neural Inform. Processing Syst., vol. 17, pp. 475–482. MIT Press, Cambridge (2005)

[GMR$^+$08] Goodman, N.D., Mansinghka, V.K., Roy, D.M., Bonawitz, K., Tenenbaum, J.B.: Church: a language for generative models. In: Uncertainty in Artificial Intelligence (2008)

[GSW07]   Grubba, T., Schröder, M., Weihrauch, K.: Computable metrization. Math. Logic Q. 53(4-5), 381–395 (2007)

[Hoo79]   Hoover, D.N.: Relations on probability spaces and arrays of random variables, Institute for Advanced Study. Princeton, NJ (preprint) (1979)

[HS55]   Hewitt, E., Savage, L.J.: Symmetric measures on Cartesian products. Trans. Amer. Math. Soc. 80, 470–501 (1955)

[JP89]   Jones, C., Plotkin, G.: A probabilistic powerdomain of evaluations. In: Proc. of the Fourth Ann. Symp. on Logic in Comp. Sci., pp. 186–195. IEEE Press, Los Alamitos (1989)

[Kal02]   Kallenberg, O.: Foundations of modern probability, 2nd edn. Springer, New York (2002)

[Kal05]   Kallenberg, O.: Probabilistic symmetries and invariance principles. Springer, New York (2005)

[Kin78]   Kingman, J.F.C.: Uses of exchangeability. Ann. Probability 6(2), 183–197 (1978)

[Koz81]   Kozen, D.: Semantics of probabilistic programs. J. Comp. System Sci. 22(3), 328–350 (1981)

[KTG$^+$06] Kemp, C., Tenenbaum, J., Griffiths, T., Yamada, T., Ueda, N.: Learning systems of concepts with an infinite relational model. In: Proc. of the 21st Nat. Conf. on Artificial Intelligence (2006)

[Lau84]   Lauritzen, S.L.: Extreme point models in statistics. Scand. J. Statist. 11(2), 65–91 (1984)

[Man09]   Mansinghka, V.K.: Natively Probabilistic Computing. PhD thesis, Massachusetts Institute of Technology (2009)

[Mül99]   Müller, N.T.: Computability on random variables. Theor. Comput. Sci. 219(1-2), 287–299 (1999)

[Pfe01]   Pfeffer, A.: IBAL: A probabilistic rational programming language. In: Proc. of the 17th Int. Joint Conf. on Artificial Intelligence, pp. 733–740. Morgan Kaufmann Publ., San Francisco (2001)

[PPT08]   Park, S., Pfenning, F., Thrun, S.: A probabilistic language based on sampling functions. ACM Trans. Program. Lang. Syst. 31(1), 1–46 (2008)

[PR89]   Pour-El, M.B., Richards, J.I.: Computability in analysis and physics. Springer, Berlin (1989)

[RMG⁺08] Roy, D.M., Mansinghka, V.K., Goodman, N.D., Tenenbaum, J.B.: A stochastic programming perspective on nonparametric Bayes. In: Nonparametric Bayesian Workshop, Int. Conf. on Machine Learning (2008)

[RN57]   Ryll-Nardzewski, C.: On stationary sequences of random variables and the de Finetti's equivalence. Colloq. Math. 4, 149–156 (1957)

[Rog67]   Rogers, Jr., H.: Theory of recursive functions and effective computability. McGraw-Hill, New York (1967)

[RP02]   Ramsey, N., Pfeffer, A.: Stochastic lambda calculus and monads of probability distributions. In: Proc. of the 29th ACM SIGPLAN-SIGACT Symp. on Principles of Program. Lang., pp. 154–165 (2002)

[RT09]   Roy, D.M., Teh, Y.W.: The Mondrian process. In: Adv. in Neural Inform. Processing Syst., vol. 21 (2009)

[Sch07]   Schröder, M.: Admissible representations for probability measures. Math. Logic Q. 53(4-5), 431–445 (2007)

[Set94]   Sethuraman, J.: A constructive definition of Dirichlet priors. Statistica Sinica 4, 639–650 (1994)

[Soa87]   Soare, R.I.: Recursively enumerable sets and degrees. Springer, Berlin (1987)

[SS06]   Schröder, M., Simpson, A.: Representing probability measures using probabilistic processes. J. Complex. 22(6), 768–782 (2006)

[TGG07]   Teh, Y.W., Görür, D., Ghahramani, Z.: Stick-breaking construction for the Indian buffet process. In: Proc. of the 11th Conf. on A.I. and Stat. (2007)

[TJ07]   Thibaux, R., Jordan, M.I.: Hierarchical beta processes and the Indian buffet process. In: Proc. of the 11th Conf. on A.I. and Stat. (2007)

[Wei99]   Weihrauch, K.: Computability on the probability measures on the Borel sets of the unit interval. Theoret. Comput. Sci. 219(1–2), 421–437 (1999)

[Wei00]   Weihrauch, K.: Computable analysis: an introduction. Springer, Berlin (2000)

Spectra of Algebraic Fields and Subfields

Andrey Frolov[1,*], Iskander Kalimullin[1,**] and Russell Miller[2,***]

[1] N.G. Chebotarev Research Institute of Mechanics and Mathematics
Kazan State University
Universitetskaya St. 17
Kazan 420008, Tatarstan Russia
Andrey.Frolov@ksu.ru
Iskander.Kalimullin@ksu.ru
[2] Queens College of CUNY
65-30 Kissena Blvd., Flushing, NY 11367 USA
and
The CUNY Graduate Center
365 Fifth Avenue, New York, NY 10016 USA
Russell.Miller@qc.cuny.edu
http://qcpages.qc.cuny.edu/~rmiller

Abstract. An algebraic field extension of $\mathbb{Q}$ or $\mathbb{Z}/(p)$ may be regarded either as a structure in its own right, or as a subfield of its algebraic closure $\overline{F}$ (either $\overline{\mathbb{Q}}$ or $\overline{\mathbb{Z}/(p)}$). We consider the Turing degree spectrum of F in both cases, as a structure and as a relation on $\overline{F}$, and characterize the sets of Turing degrees that are realized as such spectra. The results show a connection between enumerability in the structure F and computability when F is seen as a subfield of $\overline{F}$.

Keywords: Computability, computable model theory, field, algebraic, spectrum.

1 Introduction

By definition, the *spectrum of a structure* $\mathfrak{A}$, written $\mathrm{Spec}(\mathfrak{A})$, is the set of all Turing degrees of structures $\mathfrak{B} \cong \mathfrak{A}$ with domain ω. The intention is to measure the inherent complexity of the isomorphism type of $\mathfrak{A}$, by describing the set of Turing degrees capable of computing a copy of $\mathfrak{A}$. We restrict the domain to ω in order to measure only the complexity of the functions and relations of $\mathfrak{A}$, without interference from an unusual choice of domain.

Likewise, the *spectrum of a relation* R on a computable structure $\mathfrak{M}$ is the set $\mathrm{DgSp}_{\mathfrak{M}}(R)$ of Turing degrees of all images of R under isomorphisms from $\mathfrak{M}$

---

* The first two authors were partially supported by RFBR grants 05-01-00605 and 09-01-97010.
** The second author was partially supported by RF President grant MK-4314.2008.1.
*** The corresponding author was partially supported by Grant # 13397 from the Templeton Foundation, and by Grants # 69723-00 38 and 61467-00 39 from The City University of New York PSC-CUNY Research Award Program.

K. Ambos-Spies, B. Löwe, and W. Merkle (Eds.): CiE 2009, LNCS 5635, pp. 232–241, 2009.
© Springer-Verlag Berlin Heidelberg 2009

onto other computable structures $\mathfrak{B}$. (In general R will not lie in the signature of $\mathfrak{M}$, for otherwise its spectrum would contain only the degree $\mathbf{0}$.) This measures the complexity of the relation R, again by asking how simple or complex R can become under different presentations of the structure $\mathfrak{M}$. As before, restricting to computable presentations $\mathfrak{B}$ of $\mathfrak{M}$ keeps any extra complexity from creeping in when we select the underlying structure.

These two notions are compared at some length in [9]. Both are common in the study of computable model theory. In this paper we apply them to algebraic fields, by which we mean algebraic extensions of any of the prime fields $\mathbb{Q}$ and $\mathbb{Z}/(p)$ (for prime p). Theorem 2 will describe precisely the possible spectra of such fields, viewed as structures. Of course, the algebraic fields are exactly the subfields of the algebraic closures $\overline{\mathbb{Q}}$ and $\overline{\mathbb{Z}/(p)}$, so it is also natural to view such fields as unary relations on computable presentations of these algebraic closures. Theorem 4 will describe precisely the possible spectra of algebraic fields as relations in this sense. In Section 5, we will then compare the possible spectra of these fields in these two contexts and shed light on the relative usefulness of these two methods of presenting a field,

Notation and background for computability theory can be found in [19], and for fields in [10], [21], and many other sources. Many influential papers on computable fields have appeared over the years, among which we would cite in particular [7], [17], [4], [13], and [20]. For a basic introduction to the subject, [14] is useful. We will assume many standard field-theoretic results, but one is so important as to bear mention right now.

Theorem 1 (Kronecker [12]; or see [3]). *The fields $\mathbb{Q}$ and $\mathbb{Z}/(p)$, and all finite extensions of these fields, have splitting algorithms. (That is, the set of irreducible polynomials in $\mathbb{Q}[X]$ is computable, and likewise in $\mathbb{Z}/(p)[X]$.) Moreover, to determine a splitting algorithm for a computable field $E = F(x_1, \ldots, x_n)$, we need to know only a splitting algorithm for F, an enumeration of F within E, the atomic diagram of E, and (for each $i < n$) whether x_{i+1} is algebraic or transcendental over $F(x_1, \ldots, x_i)$.*

A splitting algorithm for E enables us to compute, for any $p(X) \in E[X]$, how many roots p has in E. With this information, we may then find all those roots in E (and know when we are finished!).

2   Spectra of Fields

The first part of the main theorem of this section follows by relativizing a theorem of Ershov [4, §12, Thm. 2]. Ershov's proof has only been published in German, and ours differs from his in our use of the Theorem of the Primitive Element. Moreover, the last two parts of Theorem 2 are (we believe) new. For reasons of space, we refer the reader to [8], where Appendix A includes the proof.

Theorem 2.   *1. For every algebraic field F, there exists a set $V_F \in 2^\omega$ with*

$$Spec(F) = \{d : V_F \text{ is c.e. in } d\}.$$

2. *Conversely, for every $V \in 2^\omega$, there exists an algebraic field F of arbitrary characteristic such that*

$$Spec(F) = \{ \boldsymbol{d} : V \text{ is c.e. in } \boldsymbol{d} \}.$$

3. *Finally, if Q is the prime field of any fixed characteristic c, then the set $\{V_F \upharpoonright n \ : \ F \text{ algebraic over } Q \ \& \ n \in \omega\}$ forms a computable extendible subtree $T \subset 2^{<\omega}$ with no isolated paths, and every path through T is of the form V_F for some F.*

For reference in Section 4, we give the definition of V_F here, for arbitrary algebraic extensions F of any prime field Q. First we define two c.e. sets T_F and U_F, each of which depends only on the isomorphism type of F. Let $p_0(X), p_1(X), \ldots$ list the irreducible polynomials in $Q[X]$, and set $d_i = \deg(p_i(X))$ and $D_i = \sum_{j<i} d_j$. Then, for each i and each $k < d_i$, we define

$$D_i + k \in T_F \iff F \text{ contains at least } (k+1) \text{ distinct roots of } p_i(X).$$

So the first d_0 bits of T_F tell the exact number of roots of $p_0(X)$ in F, and the next d_1 bits tell the number of roots of $p_1(X)$, etc. Obviously T_F is computably enumerable in the degree of any field isomorphic to F.

To define U_F, we need to consider pairs of polynomials. Since we have a splitting algorithm for Q, it is computable whether a pair $\langle g_0(X), g_1(X, Y) \rangle \in (Q[X] \times Q[X, Y])$ satisfies both of the following conditions.

- $Q[X]/(g_0)$ is a field. This is equivalent to demanding that g_0 be nonconstant and irreducible in $Q[X]$.
- g_1, when viewed as a polynomial in Y, is irreducible in the polynomial ring $(Q[X]/(g_0))[Y]$. (The coefficients of Y in g_1 are really in $Q[X]$, of course; here we consider their images in $Q[X]/(g_0)$.) Equivalently, if x is a root of g_0, then $g_1(x, Y)$ is irreducible in the polynomial ring $Q(x)[Y]$.

So we may computably enumerate a list $\boldsymbol{G}_0, \boldsymbol{G}_1, \ldots$ of all pairs satisfying these properties, writing $\boldsymbol{G}_j = \langle g_{j0}, g_{j1} \rangle$. Now define

$$U_F = \{ j : (\exists x, y \in F)[g_{j0}(x) = 0 = g_{j1}(x, y)] \}.$$

Again this set is c.e. in the degree of any field isomorphic to F.

In fact, $T_F \leq_1 U_F$, so it is not really necessary to take the join of T_F and U_F, but we consider it more perspicuous to do so. Define $V_F = T_F \oplus U_F$. The remainder of the proof of Theorem 2 appears in the appendix given in [8].

We regard Theorem 2 as a substantial step in classifying the possible spectra of models of standard theories of mathematics. We say this informally, because of course, the class of algebraic fields is not an EC class, nor even an EC_Δ class: there is no set T of first-order sentences for which the algebraic fields (even of a fixed characteristic) are precisely the models of T. So we must speak, non-rigorously, of standard classes $\mathcal{K}$ of mathematical structures. The goal, for a given $\mathcal{K}$, is to provide a criterion ψ such that for all subsets $\mathcal{S}$ of the set of Turing degrees,

$$\psi \text{ holds of } \mathcal{S} \iff \mathcal{S} = \mathrm{Spec}(\mathfrak{M}) \text{ for some } \mathfrak{M} \in \mathcal{K}.$$

Ideally, ψ should use only set- and degree-theoretic properties: Turing reducibility, jumps, and so forth. In our case, ψ was the property that there exists a $V \subseteq \omega$ such that $\mathcal{S}$ is the set of degrees in which V is c.e.

Now there do exist classes $\mathcal{K}$, even EC-classes, for which such a ψ is known. For example, take θ to be the conjunction of the axioms for dense linear orders: the only possible spectrum $\mathcal{S}$ of a countable model of θ is the set of all Turing degrees. The one other class $\mathcal{K}$ for which a nontrivial criterion is known is the class of torsion-free abelian groups: Coles, Downey, and Slaman examined this class in [2], mainly with an eye to studying 1-degrees, and while they did not state it specifically, it is clear from their work that parts 1 and 2 of Theorem 2 would also hold with "algebraic field" replaced by "torsion-free abelian group." (These two classes are closely linked in computable model theory, and neither is an EC_Δ class.) A few other examples can be found, in addition to dense linear orders: the class of complete graphs, for example, or the class of finite models in a given signature, but for these classes the condition ψ is quite trivial. There is no ψ known for any of the following classes: graphs, trees, linear orders, Boolean algebras, abelian groups, p-groups, fields, or rational vector spaces. Moreover, for several pairs of these classes, the conditions ψ (while unknown as yet) must be nonequivalent. For each of these classes, we would regard the discovery of a condition ψ as an important result.

3   Consequences

Several known theorems combine nicely with Theorem 2. For example, we have a result on the jump degree of an algebraic field F. Recall that the *jump degree* of a countable structure $\mathfrak{M}$ is the least degree under $\leq_T$ (if any exists) in the set $\{\boldsymbol{d}' : \boldsymbol{d} \in \mathrm{Spec}(\mathfrak{M})\}$. Jump degrees are studied in [11] and several other papers. The useful result for us was proven by Coles, Downey, and Slaman.

Theorem 3 (Coles, Downey, Slaman; see [2]). *For all $A \subseteq \omega$ there is a set B such that (1) A is c.e. in B, and (2) every set C with A c.e. in C satisfies $B' \leq_T C'$.*

Applying this result along with Theorem 2 immediately gives jump degrees.

Corollary 1. *Every algebraic field has a jump degree.*   □

In [18], Richter constructed a set A such that $\{\boldsymbol{d} : A \text{ is c.e. in } \boldsymbol{d}\}$ has no least Turing degree under $\leq_T$. Combining this with Theorem 2 yields a result already proven by other means by Calvert, Harizanov, and Shlapentokh:

Corollary 2 (Calvert, Harizanov, Shlapentokh; see [1]). *There exists an algebraic field whose spectrum contains no least Turing degree.*   □

On the other hand, given any set S, the collection of degrees which can enumerate the join $S \oplus \overline{S}$ (where $\overline{S}$ is the complement of S) is the upper cone of Turing degrees above S. This allows us to reprove another result from the above paper.

Corollary 3 (Calvert, Harizanov, Shlapentokh; see [1]). *Every upper cone of Turing degrees forms the spectrum of some algebraic field.*   □

4 Subfields of $\overline{\mathbb{Q}}$ and $\overline{\mathbb{Z}/(p)}$

Pick any prime field Q (either the rationals $\mathbb{Q}$, or $\mathbb{Z}/(p)$), and fix one computable presentation $\overline{Q}$ of the algebraic closure of Q. Since $\overline{Q}$ is computably categorical, we see that for any relation R on $\overline{Q}$, and any degree $\boldsymbol{d} \in \mathrm{DgSp}_{\overline{Q}}(R)$, there is a relation S on $\overline{Q}$ itself such that $(\overline{Q}, R) \cong (\overline{Q}, S)$ and $\boldsymbol{d} = \deg(S)$. (This appears in [9, Lemma 1.6]: if $(\overline{Q}, R) \cong (\mathfrak{A}, T)$ with $\boldsymbol{d} = \deg(T)$, take $S = g(T)$, where $g : \mathfrak{A} \to \overline{Q}$ is a computable isomorphism.)

It is well-known that any field isomorphism extends to an isomorphism of the algebraic closures of the fields; we express this in the following lemma.

Lemma 1. *For an algebraic field F, let f and g be any two embeddings of F into $\overline{Q}$. Then $(g \circ f^{-1})$ extends to an automorphism of $\overline{Q}$.*   □

Consequently, we may speak without ambiguity of the degree spectrum of F, viewed as a relation on $\overline{Q}$; the exact choice of a subfield of $\overline{Q}$ isomorphic to F is irrelevant, since any two choices have the same degree spectrum.

When an algebraic field F is viewed as a structure, its spectrum is defined by a condition of computable enumerability. When the same field is viewed as a subfield of $\overline{Q}$, its spectrum as a relation will no longer involve enumerability: instead of simply waiting for a root of a polynomial $p(X) \in Q[X]$ to appear in F, we can find all the roots of $p(X)$ in $\overline{Q}$ and check immediately (using an oracle for the subfield F) whether any of those finitely many roots actually lies in F. This is easily seen in the case of normal extensions, which is now practically trivial.

Proposition 1. *Let F be a normal algebraic extension of the prime field Q, and fix a computable copy of $\overline{Q}$. Then F has exactly one homomorphic image in $\overline{Q}$, and $\mathrm{DgSp}_{\overline{Q}}(F) = \{deg(T_F^*)\}$, where $T_F^* = \{i : \exists a \in F(p_i(a) = 0)\}$.*

Indeed, this proposition would apply not only to Q, but to any ground field E with a splitting algorithm, provided that we consider only embeddings of F into $\overline{E}$ which extend a given embedding $h : E \hookrightarrow \overline{E}$.

Proof. Fix any $g : F \hookrightarrow \overline{Q}$, and pick any $x \in \overline{Q}$. Either F contains no roots of the minimal polynomial $p(X)$ of x in $Q[X]$ (so $x \notin g(F)$), or, by normality, F contains $d = \deg(p)$ roots of $p(X)$. But $\overline{Q}$ also contains only d roots of $p(X)$, so all of them, including x, must then lie in $g(F)$.

This proves uniqueness of the image $g(F)$. Of course, there may still be many homomorphisms $g : F \hookrightarrow \overline{Q}$. Rabin proved in [17] that one of them (say h) must be computable, and so $g(F) = h(F) \leq_T T_F^*$. Conversely, given any polynomial $p(X) \in Q[X]$, we can compute from a $g(F)$-oracle whether p has any roots in F, just by finding a root of p in $\overline{Q}$ and checking whether it lies in $g(F)$. (By normality, it suffices to check this for just one root of p.) So $T_F^* \leq_T g(F)$, and $\deg(T_F^*)$ is the unique degree in $\mathrm{DgSp}_{\overline{Q}}(F)$.   □

If F is algebraic but not normal over Q, then F does have more than one possible homomorphic image in $\overline{Q}$. For us, a crucial distinction will be whether F is *almost normal* over Q.

Definition 1. *An algebraic field extension $L \subseteq F$ is almost normal if there is a finite field extension $L \subseteq E$ such that $E \subseteq F$ is a normal extension.*

Corollary 4. *If F is an almost normal field extension of the prime field Q, then $\mathrm{DgSp}_{\overline{Q}}(F)$ is a singleton.*

Proof. Let E be a finite extension of Q over which F is normal. Then by the Proposition and the subsequent remark, the image $g(F)$ of any embedding $g : F \to \overline{P}$ is determined by $g(E)$, and specifically by the values of g on a finite set B generating E. Now for any two such embeddings g_0 and g_1, we need only finitely much information – namely $g_0 \restriction B$ and $g_1 \restriction B$ – to compute an automorphism of $\overline{Q}$ which maps $g_0(b)$ to $g_1(b)$ for each $b \in B$. This automorphism must then map $g_0(F)$ onto $g_1(F)$, and since it is computable, $g_0(F) \equiv_T g_1(F)$.   □

Now we are ready to classify all spectra of algebraic subfields.

Theorem 4. *Let F be any algebraic field, with prime subfield Q, and let V_F be the set defined in Section 2. If F is almost normal over Q, then $\mathrm{DgSp}_{\overline{Q}}(F) = \{\deg(V_F)\}$. If not, then $\mathrm{DgSp}_{\overline{Q}}(F) = \{\boldsymbol{d} : \deg(V_F) \leq_T \boldsymbol{d}\}$, the upper cone of degrees above the degree of V_F.*

Proof. First we need a lemma.

Lemma 2. *For every algebraic field F, every subfield $\tilde{F} \subseteq \overline{Q}$ isomorphic to F computes V_F. Thus $\mathrm{DgSp}_{\overline{Q}}(F)$ is contained in the upper cone above $\deg(V_F)$.*

Proof. For any i, we can find all roots of $p_i(X)$ in $\overline{Q}$ and check how many of them lie in $\tilde{F}$. This computes T_F. Likewise, given any pair $\langle g_0(X), g_1(X, Y) \rangle$, we can find all roots of g_0 in $\overline{Q}$, check which ones (say $x_0, \ldots, x_k$) lie in $\tilde{F}$, and then check whether $\tilde{F}$ contains any roots of $g_1(x_j, Y)$, for each $j \leq k$. This computes U_F, so $V_F \leq_T \tilde{F}$.   □

By Theorem 2, we may take F itself to be a presentation with $F \equiv_T V_F$. With a V_F-oracle, therefore, we may compute an embedding $g : F \to \overline{Q}$. But then the image $g(F)$ is also V_F-computable, since for any $x \in \overline{Q}$ we may find its minimal polynomial $p(X) \in Q[X]$, determine from V_F (really from T_F) the number of roots of $p(X)$ in F, find that many roots of $p(X)$ in F, and determine whether g maps any of them to x. Thus $\deg(g(F)) \in \mathrm{DgSp}_{\overline{Q}}(F)$ and $g(F) \leq_T V_F$, so by Lemma 2 $g(F) \equiv_T V_F$, putting $\deg(V_F) \in \mathrm{DgSp}_{\overline{Q}}(\tilde{F})$.

If F is almost normal over Q, then Corollary 4 now proves the desired result. So we may assume that F is a non-normal extension of every finitely generated subfield of F. To complete our proof, we must show that in this case every degree $\boldsymbol{d}$ which computes V_F lies in $\mathrm{DgSp}_{\overline{Q}}(F)$.

So suppose that $D \in \boldsymbol{d}$ and $V_F \leq_T D$. The following construction is computable in a D-oracle. Let $F_0 = Q$ and let g_0 be the unique embedding of F_0 into $\overline{Q}$, and set $i_0 = -1$. We will build a computable increasing sequence $i_0 < i_1 < \cdots$ and an embedding $g = \cup_s g_s$ with $F_s = \mathrm{dom}(g_s) \subseteq F$ being the subfield of F generated by all roots in F of all polynomials $p_j(X)$ with $j \leq i_s$. Thus $\cup_s F_s$ will equal F. Moreover, the image $g_s(F_s)$ will be defined so as to code $D{\restriction}(s+1)$, with this coding being respected at subsequent stages.

It is important to note, in the following description of the procedure at stage $s + 1$, that the first part of the procedure, creating the coding opportunity, requires only a V_F-oracle, along with knowledge of g_s. (Of course, our D-oracle computes V_F.) Given g_s and i_s and the finite algebraic extension $F_s = \mathrm{dom}(g_s) \supseteq Q$ within F, we search through all roots of $p_{i_s+1}(X)$ in F, then all roots of $p_{i_s+2}(X)$, etc., until we find an $x \in F$ which is a root of some $p_i(X)$ with $i > i_s$, such that the minimal polynomial $q(X) \in F_s[X]$ of x does not have $\deg(q)$-many roots in F. Eventually we must find such a root, since F is not almost normal over Q, and Sublemma 1 below shows that with our V_F-oracle, we can recognize the root when we find it. For the least $i > i_s$ for which $p_i(X)$ has such a root, we let $i_{s+1} = i$ and let $q_s(X)$ be the minimal polynomial in $F_s[X]$ of that root. This completes the first part of the procedure.

Now we use our full D-oracle to take the coding step. Let r_0 be the least root of $q_s(X)$ in $\overline{Q}$ (under the order $<$ on the domain ω of $\overline{Q}$). If $s \in D$, then define g'_s to extend g_s by mapping the least root of $q_s(X)$ in F to r_0. If $s \notin D$, then we wish to ensure that r_0 does not lie in the image $g_{s+1}(F_{s+1})$. Let $\overline{q}_s(X) \in \overline{Q}[X]$ be the image of $q_s(X)$ under the map g_s on its coefficients. Let F'_s be the subfield of F generated by F_s and all roots of $q_s(X)$ in F. Of course, all these generators of F'_s must be mapped to roots in $\overline{Q}$ of $\overline{q}_s$, of which there are only finitely many. So we check through these finitely many possible maps until we find a map g'_s, which extends to a field embedding of F'_s into $\overline{Q}$ and satisfies $r_0 \notin \mathrm{range}(g'_s)$. To see that such an g'_s must exist, let $K_s \subset \overline{Q}$ be the splitting field of $\overline{q}_s(X)$ over $g_s(F_s)$. By irreducibility of $q_s(X)$, the Galois group $\mathrm{Gal}(K_s/g_s(F_s))$ acts transitively on the roots of $\overline{q}_s$ in K_s. We know that there exists an embedding of F'_s into K_s, and that embedding must omit at least one root r_1 of $\overline{q}_s$ from its image, since F does not contain all $\deg(q_s)$-many possible roots of q_s. By transitivity, there is an element of $\mathrm{Gal}(K_s/g_s(F_s))$ mapping r_1 to r_0, and the composition of this element with the given embedding omits r_0 from its image. So when we search, we will find the desired extension g'_s.

The coding step is now finished. To complete stage $s + 1$ (in both cases $s \in D$ and $s \notin D$), we now extend this g'_s to the remaining roots of $p_{i_{s+1}}$ in F and to all roots in F of all polynomials $p_j(X)$ with $i_s < j < i_{s+1}$. This can be done in any systematic way, and we let g_{s+1} be this extension, so g_{s+1} extends g_s to the subfield F_{s+1} generated by all roots of all $p_j(X)$ with $j \leq i_{s+1}$. This completes stage $s + 1$. Notice that if $s \notin D$, then r_0 remains outside the image of g_{s+1}, since all roots of $q_s(X)$ in F are mapped to elements $\neq r_0$. Indeed, the same reasoning shows that the embedding $\cup_s g_s$ constructed through all these stages will have r_0 in its image iff $s \in D$.

Now the union $g = \cup_s g_s$ is an embedding of F into $\overline{Q}$, computable in D. We define $\tilde{F} = g(F)$ to be its image, so that $\deg(\tilde{F}) \in \mathrm{DgSp}_{\overline{Q}}(F)$, and we claim that $\tilde{F} \equiv_T D$. First, the entire procedure is D-computable. To decide from a D-oracle whether an arbitrary $x \in \overline{Q}$ lies in $\tilde{F}$, just find the unique i for which $p_i(x) = 0$. Then use D to compute V_F and find all roots of $p_i(X)$ in F, and check whether g maps any of them to x. Thus $\tilde{F} \leq_T D$.

Conversely, our coding allows us to run the entire construction and thus compute g (and hence D) from an $\tilde{F}$-oracle, as follows. By Lemma 2, $V_F \leq_T \tilde{F}$, so given g_s, we can run the first part of stage $s + 1$, defining $q_s(X)$ and setting up the coding. Then we find the least root r_0 of $q_s(X)$ in $\overline{Q}$, and check whether it lies in $\tilde{F}$. If not, then by the construction $s \notin D$, and with this knowledge we can run the remainder of the procedure for stage $s+1$ and compute g_{s+1}. Conversely, as argued at the end of the construction, r_0 can only lie in $\tilde{F}$ if it appeared there at stage $s + 1$ under the steps followed when $s \in D$. So, if $r_0 \in \tilde{F}$, then we know that $s \in D$, and again this knowledge allows us to run the remainder of the procedure for stage $s + 1$ and compute g_{s+1}. The real point is to compute D, of course, and it is now clear that we can decide by this process, for any s, whether $s \in D$ or not. So $D \leq_T \tilde{F}$, and thus $\boldsymbol{d} = \deg(D) \in \mathrm{DgSp}_{\overline{Q}}(F)$. (Computing g_{s+1} below $\tilde{F}$ is really the inductive step which allows our computation of D to continue through as many stages as necessary.)

It remains to prove the sublemma we used.

Sublemma 1. *With a V_F-oracle and our fixed presentation of F we can determine, uniformly in i and s, the irreducible factors $q(X)$ of $p_i(X)$ in $F_s[X]$, and check which ones have their full complement of $\deg(q)$-many roots in F.*

Proof. From a V_F-oracle we know a finite set generating F_s over Q, so we may find a primitive generator x of F_s and its minimal polynomial $g_0(X) \in Q[X]$. Also, we have a splitting algorithm for F_s, which allows us to find the irreducible factors of $p_i(X)$ in $F_s[X]$. If $q(X)$ is one of these, then we may compute its splitting field in $\overline{Q}$ over the image $g_s(F_s)$. We find a primitive generator $z \in \overline{Q}$ of this splitting field and determine the minimal polynomial $p_k(X) \in Q[X]$ of z. Then we use T_F (which is part of V_F) to find all roots $z_1, \ldots, z_n \in F$ of $p_k(X)$. However, an individual $Q[z_m] \subseteq F$ may only be conjugate to the splitting field of $q(X)$ over F_s, rather than being an actual splitting field. So, for each root z_m, we check whether the subfield $Q[z_m]$ of F contains all generators of F_s (hence contains the coefficients of $q(X)$) and also contains a full complement of roots of $q(X)$. This is easy: the splitting set of $Q[z_m]$ is computable in F, by Theorem 1. If there is an m for which this holds, then we have our answer; if not, then F cannot contain a full complement of roots of $q(X)$, since such a set of roots, along with the generators of F_s, would generate a subfield containing some root of $p_k(X)$.

In this computation, the use of g_s is not necessary; we could compute a splitting field of $q(X)$ over F_s without knowing g_s. Indeed, it would be sufficient to have as oracles T_F and a presentation of F. (From these we can find a finite generating set for F_s, using the definition of F_s in the construction. If F_s were replaced by an arbitrary subfield of F, we would need to know a finite set of generators for it.) Alternatively, a V_F-oracle would suffice, since from it we could compute both T_F and a presentation of F. □

This completes the proof of Theorem 4. □

5 Conclusions

The most obvious theorem relevant to the results of Sections 2 and 4 is Rabin's Theorem, from [17]; see also [14].

Theorem 5 (Rabin). *Let F be any computable field.*

1. *There exists a Rabin embedding g of F into a computable algebraically closed field $\overline{F}$. (This means that g is a computable field homomorphism and $\overline{F}$ is algebraic over the image $g(F)$.)*
2. *For every Rabin embedding g of F into any computable ACF E, the image of g is a computable subset of E iff F has a splitting algorithm.*

A relativized version states that the image $g(E)$ is Turing-equivalent to the *splitting set* of F, i.e. the set of reducible polynomials in $F[X]$. Details are available in [15].

We view Rabin's Theorem as saying that Σ_1 questions, such as the reducibility of polynomials over F, become computable if one is given F as a subfield of $\overline{F}$, rather than simply as a freestanding field. This phenomenon is specific to algebraic fields. For example, in the language of trees with an immediate-predecessor function, all computable trees of height $\leq \omega$ embed into the computable tree $\omega^{<\omega}$, yet existential questions about a computable tree T do not become computable when one is given the ability to compute a subtree of $\omega^{<\omega}$ isomorphic to T. Indeed, in a field of infinite transcendence degree over its prime subfield Q, being algebraic over Q is a Σ_1 property which need not become computable just because one can compute the image of the field in a computable algebraic closure. For fields, algebraicity is the key: one knows exactly how many roots $p(X) \in F[X]$ has (counted by multiplicity) in its algebraic closure $\overline{F}$, and so, if one can compute the image of F within $\overline{F}$, one can find them all and check how many lie in the image of F. We conjecture that similar results may hold for other algebraic structures (in the model-theoretic sense of *algebraic*), but that some further constraints on the structure are necessary, such as the ability to determine the maximum number of possible realizations of a type. (Local finiteness and finite axiomatizability of the theory, in a finite signature, would likely suffice for this.)

The relation of these matters to the current work is that in our results again, computable enumerability in fields as freestanding structures converts to computability when we view the fields as subfields of their algebraic closures. The spectrum of the field F is defined by a Σ_1 condition: the ability of d to *enumerate* a given set V_F. The spectrum of F as a subfield of $\overline{F}$, on the other hand, is defined by the ability of d to *compute* V_F (and, for almost normal fields, the ability of V_F to compute d). Of course, this makes it harder for a degree to present F as a subfield of a computable copy of $\overline{F}$ than it is to present F as a field; conversely, a presentation of F as subfield of $\overline{F}$ gives us more power than a mere presentation of F as a field.

References

1. Calvert, W., Harizanov, V., Shlapentokh, A.: Turing degrees of isomorphism types of algebraic objects. Journal of the London Math. Soc. 73, 273–286 (2007)
2. Coles, R.J., Downey, R.G., Slaman, T.A.: Every set has a least jump enumeration. Journal of the London Mathematical Society 62(2), 641–649 (2000)
3. Edwards, H.M.: Galois Theory. Springer, New York (1984)
4. Ershov, Y.L.: Theorie der Numerierungen III. Zeitschrift für Mathematische Logik und Grundlagen der Mathematik 23(4), 289–371 (1977)
5. Ershov, Y.L., Goncharov, S.S.: Constructive Models. Kluwer Academic/Plenum Press, New York (2000)
6. Fried, M.D., Jarden, M.: Field Arithmetic. Springer, Berlin (1986)
7. Frohlich, A., Shepherdson, J.C.: Effective procedures in field theory. Phil. Trans. Royal Soc. London, Series A 248(950), 407–432 (1956)
8. Frolov, A., Kalimullin, I., Miller, R.G.: Spectra of algebraic fields and subfields, http://qcpages.qc.cuny.edu/~rmiller/CiE09.pdf
9. Harizanov, V., Miller, R.: Spectra of structures and relations. Journal of Symbolic Logic 72(1), 324–347 (2007)
10. Jacobson, N.: Basic Algebra I. W.H. Freeman & Co., New York (1985)
11. Knight, J.F.: Degrees coded in jumps of orderings. Journal of Symbolic Logic 51, 1034–1042 (1986)
12. Kronecker, L.: Grundzüge einer arithmetischen Theorie der algebraischen Größen. J. f. Math. 92, 1–122 (1882)
13. Metakides, G., Nerode, A.: Effective content of field theory. Annals of Mathematical Logic 17, 289–320 (1979)
14. Miller, R.G.: Computable fields and Galois theory. Notices of the AMS 55(7), 798–807 (2008)
15. Miller, R.G.: Is it harder to factor a polynomial or to find a root? Transactions of the American Mathematical Society (to appear)
16. Miller, R.G.: d-Computable categoricity for algebraic fields. The Journal of Symbolic Logic (to appear)
17. Rabin, M.: Computable algebra, general theory, and theory of computable fields. Transactions of the American Mathematical Society 95, 341–360 (1960)
18. Richter, L.J.: Degrees of structures. Journal of Symbolic Logic 46, 723–731 (1981)
19. Soare, R.I.: Recursively Enumerable Sets and Degrees. Springer, New York (1987)
20. Stoltenberg-Hansen, V., Tucker, J.V.: Computable Rings and Fields. In: Griffor, E.R. (ed.) Handbook of Computability Theory, pp. 363–447. Elsevier, Amsterdam (1999)
21. van der Waerden, B.L.: Algebra, vol. I. Springer, New York (Hard Cover 1970); trans., Blum, F., Schulenberger, J.R. (2003 softcover)

Definability in the Local Theory of the ω-Enumeration Degrees

Hristo Ganchev*

Sofia University, Faculty of Mathematics and Informatics
5 James Bourchier blvd. 1164 Sofia, Bulgaria
ganchev@fmi.uni-sofia.bg

Abstract. We continue the study of the local theory of the structure of the ω-enumeration degrees, started by Soskov and Ganchev [7]. We show that the classes of 1-high and 1-low ω-enumeration degrees are definable. We prove that a standard model of arithmetic is definable as well.

1 ω-Enumeration Degrees

In computability theory there is a vast variety of reducibility relations. Informally, the common feature among all of them is that an object A is reducible to an object B (denoted by $A \leq B$), if there is an algorithm for transforming the information contained in B into the information contained in A. For example $A \leq_1 B$, if there is a one-one computable function g, such that $\chi_A = \chi_B \circ g$, (here χ_A and χ_B stand for the characteristic functions of A and B respectively). As a second example consider Turing reducibility, for which $A \leq_T B$, if there is a Turing machine which transforms χ_B into χ_A. We say that A is enumeration reducible to B ($B \neq \emptyset$), $A \leq_e B$, if there is a Turing machine transforming every function enumerating B into a function enumerating A. Finally, A is c.e. in B if there is a Turing machine transforming χ_B into a function enumerating A.

Now let us try to define a reducibility relation between sequences of sets of natural numbers and sets of natural numbers. This problem may be solved in various ways thus yielding a large number of reducibilities. In this paper, we shall be concerned with the solution proposed by Soskov in [6].

Let us denote

$$\mathcal{S}_\omega = \{\mathcal{A} = \{A_n\}_{n<\omega} \mid A_n \subseteq \mathbb{N}\}.$$

Take an element $\mathcal{A} \in \mathcal{S}_\omega$ and a set $X \subseteq \mathbb{N}$. If we are to say that $\mathcal{A} \leq X$, we should be able to obtain effectively the information contained in $\mathcal{A}$ from the information contained in X. Note, that in order to do this, we should be able to restore each element of the sequence and the order in which they occur. This is so, since $\mathcal{A} = \{A_n\}_{n<\omega}$ is a mapping from the set of natural numbers to the power set $2^{\mathbb{N}}$:

---

* The author was partially supported by the European Operational programm HRD through contract BGO051PO001/07/3.3-02/53 with the Bulgarian Ministry of Education and by BNSF through contract D002-258/18.12.08.

K. Ambos-Spies, B. Löwe, and W. Merkle (Eds.): CiE 2009, LNCS 5635, pp. 242–249, 2009.
© Springer-Verlag Berlin Heidelberg 2009

$$
\begin{array}{cccccc}
0 & 1 & 2 & \ldots & n \ldots \\
\end{array}
$$
$$
\mathcal{A}: \downarrow \ \downarrow \ \downarrow \ \ldots \ \downarrow \ldots
$$
$$
A_0 A_1 A_2 \ldots A_n \ldots
$$

In order to simulate this mapping we shall use the Turing jump J_T. The Turing jump is an unary operation (definable in second order arithmetic) $J_T : 2^{\mathbb{N}} \to 2^{\mathbb{N}}$, such that for any X, $X \leq J_T(X)$ and $J_T(X) \not\leq X$. So, in some sense, X gives rise to a "natural" sequence $X, J_T(X), J_T^2(X), \ldots$, which may be regarded as copy of $\mathbb{N}$.

$$
\begin{array}{ccccc}
X & J_T(X) & J_T^2(X) & \ldots & J_X) \ldots \\
X: \uparrow & \uparrow & \uparrow & \ldots & \uparrow \ldots \\
0 & 1 & 2 & \ldots & n \ldots \\
\end{array}
$$

Combining the two mappings we arrive to the following definition:

Definition 1. *Let $\mathcal{A} \in \mathcal{S}_\omega$ and $X \subseteq \mathbb{N}$. We shall say, that $\mathcal{A}$ is uniformly reducible to X and write $\mathcal{A} \preceq_\omega X$, if*

$$
\forall n(A_n \ c.e. \ in \ J_T^n(X) \ uniformly \ in \ n).
$$

The uniformity condition is necessary, since it guarantees the existence of one algorithm which reduces the sequence $A_0, A_1, A_2, \ldots$ to the sequence $X, J_T(X), J_T^2(X), \ldots$ (recall that the existence of one algorithm is crucial to all reducibilities considered in computability theory).

The relation $\preceq_\omega$ gives a tool for comparing elements of $\mathcal{S}_\omega$, namely

Definition 2. *Let $\mathcal{A}, \mathcal{B} \in \mathcal{S}_\omega$. We shall say that $\mathcal{A} \leq_\omega \mathcal{B}$, if*

$$
\forall X \subseteq \mathbb{N}(\mathcal{B} \preceq_\omega X \Longrightarrow \mathcal{A} \preceq_\omega X).
$$

The relation $\leq_\omega$ is a preorder, so it generates a nontrivial equivalence relation on $\mathcal{S}_\omega$:

$$
\mathcal{A} \equiv_\omega \mathcal{B} \iff \mathcal{A} \leq_\omega \mathcal{B} \ \& \ \mathcal{B} \leq_\omega \mathcal{A}.
$$

We call the respective equivalence classes ω-enumeration degrees and denote

$$
\mathbf{d}_\omega(\mathcal{A}) = \{\mathcal{B} \in \mathcal{S}_\omega \mid \mathcal{A} \equiv_\omega \mathcal{B}\}.
$$

We shall denote the collection of all ω-enumeration degrees by $\mathbf{D}_\omega$. The preorder $\leq_\omega$ on $\mathcal{S}_\omega$ induces a partial order $\leq_\omega$ on $\mathbf{D}_\omega$, namely

$$
\mathbf{a} \leq_\omega \mathbf{b} \iff \exists \mathcal{A} \in \mathbf{a} \exists \mathcal{B} \in \mathbf{b}(\mathcal{A} \leq_\omega \mathcal{B}).
$$

Clearly $\mathbf{0}_\omega = \mathbf{d}_\omega(\emptyset, \emptyset, \ldots, \emptyset, \ldots)$ is the least degree in $\mathbf{D}_\omega$. Also, we have that for arbitrary $\mathcal{A}, \mathcal{B} \in \mathcal{S}_\omega$, $\mathbf{d}_\omega(\mathcal{A} \oplus \mathcal{B}) = \mathbf{d}_\omega(A_0 \oplus B_0, A_1 \oplus B_1, \ldots)$ is the l.u.b. of the set $\{\mathbf{d}_\omega(\mathcal{A}), \mathbf{d}_\omega(\mathcal{B})\}$. Thus $\mathcal{D}_\omega = (\mathbf{D}_\omega, \mathbf{0}_\omega, \leq_\omega, \vee)$ is an upper semi-lattice with least element.

The structure $\mathcal{D}_\omega$ is first introduced by Soskov [6] and is further studied in [1,7].

2   Basic Properties of the ω-Enumeration Degrees

Let $\mathcal{A}$ be an element of $\mathcal{S}_\omega$. We set the jump sequence of $\mathcal{A}$ to be the sequence $P(\mathcal{A}) = (P_0(\mathcal{A}), P_1(\mathcal{A}), \ldots, P_n(\mathcal{A}), \ldots)$, where the sets $P_i(\mathcal{A})$ are defined by

$$P_0(\mathcal{A}) = A_0;$$
$$P_{n+1}(\mathcal{A}) = J_e(P_n(\mathcal{A})) \oplus A_{n+1}.$$

Here $J_e(A)$ stands for the enumeration jump of the set A. Recall that $J_e(A) = E_A \oplus \overline{E_A}$, where $E_A = \{\langle x, i \rangle \mid x \in W_i(A)\}$[1]. The sequences $\mathcal{A}$ and $P(\mathcal{A})$ are closely related, since $\mathcal{A} \equiv_\omega P(\mathcal{A})$. Furthermore, using the jump sequences, Soskov and Kovachev [8] are able to show that the relation $\leq_\omega$ is a real reducibility relation between sequences.

Theorem 1 (Soskov, Kovachev [8])

$$\mathcal{A} \leq_\omega \mathcal{B} \iff \forall n(A_n \leq_e P_n(\mathcal{B}) \text{ uniformly in } n)$$

Another important role played by the jump sequences is in the definition of a jump operation on sequences.

Definition 3 (Soskov [6]). *Let $\mathcal{A} \in \mathcal{S}_\omega$. We define the jump of $\mathcal{A}$ to be the sequence $\mathcal{A}' = \{P_{1+n}(\mathcal{A})\}_{n \in \omega}$.*

In other words $\mathcal{A}'$ is obtained from the jump sequence of $\mathcal{A}$ by simply deleting its first element. Besides the simplicity of its definition, the jump operation has another nice property, namely:

$$\mathcal{A}' \preceq_\omega X \iff \exists Y \subseteq \mathbb{N}(\mathcal{A} \preceq_\omega Y \ \& \ J_T(Y) \equiv_T X),$$

for each $\mathcal{A} \in \mathcal{S}_\omega$ and $X \subseteq \mathbb{N}$. In other words, the set X "codes" the jump of the sequence $\mathcal{A}$, if and only if it is equivalent to the Turing jump of a set "coding" $\mathcal{A}$ itself.

The so defined jump operation is strictly monotone, i.e.

$$\mathcal{A} \leq_\omega \mathcal{B} \implies \mathcal{A}' \leq_\omega \mathcal{B}',$$

$$\mathcal{A}' \not\leq_\omega \mathcal{A}.$$

We set

$$\mathbf{a}' = \mathbf{d}_\omega(\mathcal{A}')$$

for arbitrary $\mathcal{A} \in \mathbf{a}$. The previous properties guarantee, that this definition is unambiguous.

Soskov and Ganchev [7] prove that the jump operation on the ω-enumeration degrees has a very unexpected property:

---

[1] Here W_i stands for the c.e. set with Gödel index i. Furthermore, $W_i(A) = \{x \mid \langle x, u \rangle \in W_i \ \& \ D_u \subseteq A\}$, where D_u is the finite set with canonical index u.

Theorem 2 (Soskov, Ganchev [7]). *Let* $\mathbf{a}, \mathbf{b} \in \mathbf{D}_\omega$*, be such that* $\mathbf{a}^{(n)}$ *(the n-th iteration of the jump operation on* $\mathbf{a}$*) is less or equal to* $\mathbf{b}$*. Then the set*

$$\{\mathbf{x} \in \mathcal{D}_\omega \mid \mathbf{a} \leq_\omega \mathbf{x} \,\&\, \mathbf{x}^{(n)} = \mathbf{b}\}$$

has a least element. We shall denote this element by $I_{\mathbf{a}}^n(\mathbf{b})$*.*

Note that this theorem is neither true for the structure of the Turing degrees, $\mathcal{D}_T$, nor the structure of the enumeration degrees, $\mathcal{D}_e$. This suggests that $\mathcal{D}_\omega'$ (the structure of the ω-enumeration degrees augmented by the jump operation) is rather different from the structures $\mathcal{D}_T$ and $\mathcal{D}_e$. Nevertheless, it turns out that this is not quite so. First Soskov [6], shows that $\mathcal{D}_e'$ (the structure of the enumeration degrees augmented by the jump operation) is embeddable in $\mathcal{D}_\omega'$ by the mapping $\kappa : \mathbf{D}_e \to \mathbf{D}_\omega$, acting by the rule:

$$\kappa(\mathbf{d}_e(A)) = \mathbf{d}_\omega(A, \emptyset, \emptyset, \ldots).$$

Then Soskov and Ganchev [7] are able to prove, that the set $\mathbf{D}_1 = \kappa[\mathbf{D}_e]$ is first order definable in the theory of $\mathcal{D}_\omega'$, and so $Th_1(\mathcal{D}_e')$ is interpretable within $Th_1(\mathcal{D}_\omega')$. Furthermore it is shown that that the structures $\mathcal{D}_e$ and $\mathcal{D}_\omega'$ have isomorphic automorphism groups.

So, although the structures $\mathcal{D}_\omega$ and $\mathcal{D}_e'$ are quite different, the first being far richer then the second one, they are closely related.

In the next section, we shall obtain some nice results for the local theory of the ω-enumeration degrees, using results about Σ_2^0 enumeration degrees and degrees c.e. and above a Turing degree $\mathbf{a}$.

3   The Local Theory

Let us denote

$$\mathcal{G}_\omega = ([\mathbf{0}_\omega, \mathbf{0}_\omega'], \mathbf{0}_\omega, \leq_\omega).$$

By local theory we mean the theory of the structure $\mathcal{G}_\omega$. From now on we shall restrict our considerations only to ω-enumeration degrees belonging to $\mathcal{G}_\omega$. Thus from now on, unless explicitly otherwise stated, by an arbitrary omega-enumeration degree we will mean an arbitrary omega enumeration degree below $\mathbf{0}_\omega'$.

In the local theory the classes of the n-high and n-low degrees are of particular interest. They are defined by

$$H_n = \{\mathbf{a} \mid \mathbf{a}^{(n)} = \mathbf{0}_\omega^{(n+1)}\}$$
$$L_n = \{\mathbf{a} \mid \mathbf{a}^{(n)} = \mathbf{0}_\omega^{(n)}\}$$

Further more, we set $H = \bigcup H_n$, $L = \bigcup L_n$ and $I = [\mathbf{0}_\omega, \mathbf{0}_\omega'] \backslash (L \cup H)$. The last three classes are studied in [7]. It is shown, that they have a strong connection with the class of the, so called, almost zero degrees.

In order to define the notion of an almost zero degree, for every $n \in \mathbb{N}$ set $\mathbf{o}_n$ to be the least degree, satisfying the equality

$$\mathbf{x}^{(n)} = \mathbf{0}_\omega^{(n+1)},$$

i.e., $\mathbf{o}_n = I_{\mathbf{0}_\omega}^n(\mathbf{0}_\omega^{(n+1)})$. Clearly, the degrees $\mathbf{o}_n$ form a strictly decreasing sequence

$$\mathbf{0}'_\omega = \mathbf{o}_0 >_\omega \mathbf{o}_1 >_\omega \mathbf{o}_2 >_\omega \ldots$$

The first natural question to ask is whether this sequence converges to $\mathbf{0}_\omega$, i.e., is it true that

$$\forall n(\mathbf{x} \leq_\omega \mathbf{o}_n) \Longrightarrow \mathbf{x} = \mathbf{0}_\omega.$$

The answer to this question is negative. In fact, the degrees below all $\mathbf{o}_n$ form a countable ideal, whose elements are called almost zero (a.z.) degrees. A remarkable property of this ideal is that it has no minimal upper bound (beneath $\mathbf{0}'_\omega$). In addition the a.z. degrees give a nice characterisation for the classes H and L, namely

$$\begin{aligned}
\mathbf{x} \in H &\Longleftrightarrow \forall a.z.\ \mathbf{y}(\mathbf{y} \leq_\omega \mathbf{x}) \\
\mathbf{x} \in L &\Longleftrightarrow \forall a.z.\ \mathbf{y}(\mathbf{y} \wedge \mathbf{x} = \mathbf{0}_\omega).
\end{aligned} \tag{1}$$

We can reformulate (1) in the terms of the degrees $\mathbf{o}_n$:

$$\begin{aligned}
\mathbf{x} \in H &\Longleftrightarrow \forall \mathbf{y}(\forall n(\mathbf{y} \leq_\omega \mathbf{o}_n) \Rightarrow \mathbf{y} \leq_\omega \mathbf{x}) \\
\mathbf{x} \in L &\Longleftrightarrow \forall \mathbf{y}(\forall n(\mathbf{y} \leq_\omega \mathbf{o}_n) \Rightarrow \mathbf{y} \wedge \mathbf{x} = \mathbf{0}_\omega)
\end{aligned}$$

This suggests, that for every n we can use $\mathbf{o}_n$ as a parameter to obtain a first order definition for each class H_n and L_n. Indeed, we have

$$\begin{aligned}
\mathbf{x} \in H_n &\Longleftrightarrow \mathbf{o}_n \leq_\omega \mathbf{x} \\
\mathbf{x} \in L_n &\Longleftrightarrow \mathbf{o}_n \wedge \mathbf{x} = \mathbf{0}_\omega.
\end{aligned}$$

In particular

$$\begin{aligned}
\mathbf{x} \in H_1 &\Longleftrightarrow \mathbf{o}_1 \leq_\omega \mathbf{x} \\
\mathbf{x} \in L_1 &\Longleftrightarrow \mathbf{o}_1 \wedge \mathbf{x} = \mathbf{0}_\omega.
\end{aligned} \tag{2}$$

Our next goal is to prove the following theorem:

Theorem 3. *The degree $\mathbf{o}_1$ is definable in $\mathcal{G}_\omega$.*

Proof. According to [7]

$$\mathbf{o}_1 = (\emptyset, J_e(\emptyset), J_e^2(\emptyset), \ldots) \tag{3}$$

From here, we conclude that $\mathbf{o}_1$ is a non-cuppable degree, i.e.,

$$\forall \mathbf{y}(\mathbf{y} \vee \mathbf{o}_1 = \mathbf{0}'_\omega \Longrightarrow \mathbf{y} = \mathbf{0}'_\omega).$$

Furthermore, since $\mathbf{0}'_\omega = \mathbf{d}_\omega(J_e(\emptyset), J_e^2(\emptyset), J_e^3(\emptyset)\ldots)$, we may conclude, that for arbitrary $\mathbf{x}$ the g.l.b. of $\mathbf{x}$ and $\mathbf{o}_1$ exists and is exactly $I_{\mathbf{0}_\omega}^1(\mathbf{x}')$, i.e.,

$$\mathbf{x} \wedge \mathbf{o}_1 = I_{\mathbf{0}_\omega}^1(\mathbf{x}'). \tag{4}$$

Consider the formula

$$\mathcal{K}(\mathbf{x}, \mathbf{a}_1, \mathbf{a}_2, \mathbf{a}_3) \overset{def}{\Longleftrightarrow} \bigwedge_{1 \leq i < j \leq 3} \mathbf{x} = (\mathbf{a}_i \vee \mathbf{x}) \wedge (\mathbf{a}_j \vee \mathbf{x}).$$

Kalimullin [2] shows, that for each enumeration degree $\mathbf{u}$, there are enumeration degrees $\mathbf{a}_1$, $\mathbf{a}_2$, and $\mathbf{a}_3$ such that

(K1) $\mathbf{u} \lesssim \mathbf{a}_1, \mathbf{a}_2, \mathbf{a}_3$;
(K2) $\mathbf{a}_1, \mathbf{a}_2, \mathbf{a}_3 \lesssim \mathbf{u}'$;
(K3) $\mathbf{u}' = \mathbf{a}_1' = \mathbf{a}_2' = \mathbf{a}_3'$;
(K4) $\mathcal{K}(\mathbf{x}, \mathbf{a}_1, \mathbf{a}_2, \mathbf{a}_3)$ is true for all $\mathbf{x} \in [\mathbf{u}, \mathbf{u}']$;
(K5) $\mathbf{u}' = \mathbf{a}_1 \vee \mathbf{a}_2 \vee \mathbf{a}_3$

Fix enumeration degrees $\mathbf{a}_1$, $\mathbf{a}_2$ and $\mathbf{a}_3$ satisfying (K1) – (K5) for $\mathbf{u} = \mathbf{0}_e$ and $\widetilde{\mathbf{a}}_1$, $\widetilde{\mathbf{a}}_2$ and $\widetilde{\mathbf{a}}_3$ satisfying (K1) – (K4) for $\mathbf{u} = \mathbf{0}_e'$. Set

$$\mathbf{b}_1 = \kappa(\mathbf{a}_1) \vee I_{\mathbf{0}_\omega}^1(\kappa(\widetilde{\mathbf{a}}_1));$$
$$\mathbf{b}_2 = \kappa(\mathbf{a}_1) \vee I_{\mathbf{0}_\omega}^1(\kappa(\widetilde{\mathbf{a}}_2));$$
$$\mathbf{b}_3 = \kappa(\mathbf{a}_1) \vee I_{\mathbf{0}_\omega}^1(\kappa(\widetilde{\mathbf{a}}_3)).$$

It is easy to see that $\mathbf{b}_1$, $\mathbf{b}_2$ and $\mathbf{b}_3$ satisfy (K1), (K2), (K4), (K5) for $\mathbf{u} = \mathbf{0}_\omega$. On the other hand, it is true that $\mathbf{b}_i' = \kappa(\widetilde{\mathbf{a}}_i)$, for $1 \leq i \leq 3$, and hence according to (4) and (K5)

$$(\mathbf{o}_1 \wedge \mathbf{b}_1) \vee (\mathbf{o}_1 \wedge \mathbf{b}_2) \vee (\mathbf{o}_1 \wedge \mathbf{b}_3) = \mathbf{o}_1.$$

Now suppose that $\mathbf{x}$ is a non-cuppable ω-enumeration degree for which there are degrees $\widetilde{\mathbf{b}}_1$, $\widetilde{\mathbf{b}}_2$ and $\widetilde{\mathbf{b}}_3$ satisfying (K1), (K2), (K4) and (K5) for $\mathbf{u} = \mathbf{0}_\omega$, such that

$$(\mathbf{x} \wedge \widetilde{\mathbf{b}}_1) \vee (\mathbf{x} \wedge \widetilde{\mathbf{b}}_2) \vee (\mathbf{x} \wedge \widetilde{\mathbf{b}}_3) = \mathbf{x}. \tag{5}$$

Let $\mathcal{X} \in \mathbf{x}$ and consider $P_0(\mathcal{X})$. According to Cooper, Sorbi and Yi [5], every non-trivial Δ_2^0 enumeration degree is cuppable. Hence, either $P_0(\mathcal{X})$ is enumeration equivalent to $\emptyset$ or no non-computably enumerable Δ_2^0 set is enumeration reducible to it:

$$P_0(\mathcal{X}) \equiv_e \emptyset \quad \text{or} \quad \forall Y (Y \leq_e P_0(\mathcal{X}) \,\&\, Y \text{ is } \Delta_2^0 \Longrightarrow Y \leq_e \emptyset). \tag{6}$$

Suppose that $P_0(\mathcal{X}) \not\equiv_e \emptyset$. Fix $\widetilde{\mathbf{b}}_1$, $\widetilde{\mathbf{b}}_2$ and $\widetilde{\mathbf{b}}_3$ satisfying (K1), (K3), (K4) and (5).

Fix $\widetilde{\mathcal{B}}_i \in \widetilde{\mathbf{b}}_i$, for $1 \leq i \leq 3$, and consider $\mathbf{d}_e(P_0(\widetilde{\mathcal{B}}_1))$, $\mathbf{d}_e(P_0(\widetilde{\mathcal{B}}_2))$ and $\mathbf{d}_e(P_0(\widetilde{\mathcal{B}}_3))$. We may conclude the last three satisfy (K2), (K4) and (K5) for $\mathbf{u} = \mathbf{0}_e$. Thus at least two of them are different from $\mathbf{0}_e$ (this is implied by (K5)). Suppose, that these are $\mathbf{d}_e(P_0(\widetilde{\mathcal{B}}_1))$ and $\mathbf{d}_e(P_0(\widetilde{\mathcal{B}}_2))$. According to Kalimullin [2], we may conclude that these two degrees are Δ_2^0, so that without loss of generality we may assume that $P_0(\widetilde{\mathcal{B}}_1)$ and $P_0(\widetilde{\mathcal{B}}_2)$ are Δ_2^0 sets. Now from (6), for $i = 1, 2$

$$\forall Y (Y \leq_e P_0(\mathcal{X}) \,\&\, Y \leq_e P_0(\widetilde{\mathcal{B}}_i) \Longrightarrow Y \leq_e \emptyset).$$

Since $\mathbf{x} = (\mathbf{x} \wedge \widetilde{\mathbf{b}}_1) \vee (\mathbf{x} \wedge \widetilde{\mathbf{b}}_2) \vee (\mathbf{x} \wedge \widetilde{\mathbf{b}}_3)$ it must be the case that $P_0(\mathcal{X}) \equiv_e P_0(\widetilde{\mathcal{B}}_3)$. Hence, $P_0(\widetilde{\mathcal{X}}_3) \not\equiv_e \emptyset$. Now applying Kalimullin's theorem once again (this time for $P_0(\widetilde{\mathcal{B}}_1)$ and $P_0(\widetilde{\mathcal{B}}_3)$) we obtain, that $P_0(\widetilde{\mathcal{B}}_3)$ is Δ_2^0. But this contradicts the assumption about $P_0(\mathcal{X})$ and thus $P_0(\mathcal{X}) \equiv_e \emptyset$.

So, we have proved that whenever $\mathbf{x}$ is a non-cuppable ω-enumeration degree for which there are degrees $\widetilde{\mathbf{b}}_1$, $\widetilde{\mathbf{b}}_2$ and $\widetilde{\mathbf{b}}_3$ satisfying (5), (K1), (K2), (K4) and (K5) for $\mathbf{u} = \mathbf{0}_\omega$, it is true that for each $\mathcal{X} \in \mathbf{x}$, $\mathcal{X} \equiv_e \emptyset$.

On the other hand $\mathbf{o}_1$ is the biggest degree generated by a sequence beginning with $\emptyset$. Thus $\mathbf{o}_1$ is definable in the first order language of $\mathcal{G}_\omega$.

$\square$

Corollary 1. *The classes* H_1, L_1 *and* $\mathbf{D}_1$ *are definable by a first order formula in* $\mathcal{G}_\omega$.

Proof. The degrees in $\mathbf{D}_1$ satisfy the condition:

$$\mathbf{x} \in \mathbf{D}_1 \iff \forall \mathbf{z}(\mathbf{z} \vee \mathbf{o}_1 = \mathbf{x} \vee \mathbf{o}_1 \implies \mathbf{x} \leq_\omega \mathbf{z}). \qquad \square$$

The last corollary will enable us to prove that there is a standard model of arithmetic definable in $\mathcal{G}_\omega$. Note, that this does not tell us which is the degree of the first order theory of $\mathcal{G}_\omega$, since it is not clear whether $\mathcal{G}_\omega$ is inerpretable in first order arithmetic.

Theorem 4. *FOA is interpretable in* $\mathcal{G}_\omega$.

Proof. Denote by R_ω the collection of all degrees, that are the g.l.b. of a degree in $\mathbf{D}_1[\mathbf{0}_\omega, \mathbf{0}'_\omega]$ and $\mathbf{o}_1$, i.e.,

$$R_\omega = \{\mathbf{x} \wedge \mathbf{o}_1 \mid \mathbf{x} \in \mathbf{D}_1[\mathbf{0}_\omega, \mathbf{0}'_\omega]\}.^2$$

Since for $\mathbf{x} \leq_\omega \mathbf{0}'_\omega$,

$$(\mathbf{x} \wedge \mathbf{o}_1)' = \mathbf{x}',$$

it turns out that

$$\{\mathbf{x}' \mid \mathbf{x} \in \mathbf{D}_1[\mathbf{0}_\omega, \mathbf{0}'_\omega]\} = \{\mathbf{x}' \mid \mathbf{x} \in R_\omega\}.$$

According to the results in [7] the jump operation is an isomorphism between the intervals $[\mathbf{0}_\omega, \mathbf{o}_1]_\omega$ and $[\mathbf{0}'_\omega, \mathbf{0}''_\omega]_\omega$. So, we obtain:

$$(R_\omega, \leq_\omega) \cong (\{\mathbf{x}' \mid \mathbf{x} \in \mathbf{D}_1[\mathbf{0}_\omega, \mathbf{0}'_\omega]\}, \leq_\omega) \cong (\{\mathbf{x}' \mid \mathbf{x} \in \mathbf{D}_e[\mathbf{0}_e, \mathbf{0}'_e]\}, \leq_e).$$

McEvoy [3] shows that the elements in $\{\mathbf{x}' \mid \mathbf{x} \in \mathbf{D}_e[\mathbf{0}_e, \mathbf{0}'_e]\}$ are exactly the Π_2^0 enumeration degrees above $\mathbf{0}'_e$. On the other hand these are exactly the degrees to which the c.e. in and above $J_T(\emptyset)$ degrees are mapped by the Rogers embedding ι. Thus $(R_\omega, \leq_\omega)$ is isomorphic to the structure of the c.e. in and above $J_T(\emptyset)$ degrees.

Nies, Slaman and Shore [4] prove that for every Turing degree $\mathbf{a}$, there is a standard model of arithmetic definable in the degrees c.e. in and above $\mathbf{a}$.

Now the theorem follows from the fact, that R_ω is first order definable in $\mathcal{G}_\omega$. $\qquad \square$

References

1. Ganchev, H.: Exact pair theorem for the ω-enumeration degrees. In: Cooper, S.B., Löwe, B., Sorbi, A. (eds.) CiE 2007. LNCS, vol. 4497, pp. 316–324. Springer, Heidelberg (2007)

---

[2] Reacall, that $\mathbf{x} \wedge \mathbf{o}_1$ always exists, according to (4).

2. Kalimullin, I.S.: Definability of the jump operator in the enumeration degrees. Journal of Mathematical Logic 3, 257–267 (2003)
3. McEvoy, K.: Jumps of quasi-minimal enumeration degrees. J. Symbolic Logic 50, 839–848 (1985)
4. Nies, A., Shore, R.A., Slaman, T.A.: Interpretability and definability in the recursively enumerable degrees. Proc. London Math. Soc. 77(2), 241–291 (1998)
5. Cooper, S.B., Sorbi, A., Yi, X.: Cupping and noncupping in the enumeration degrees of σ_2^0 sets. Ann. Pure Appl. Logic 82, 317–342 (1997)
6. Soskov, I.N.: The ω-enumeration degrees. Journal of Logic and Computation 17, 1193–1214 (2007)
7. Soskov, I.N., Ganchev, H.: The jump operator on the ω-enumeration degrees. Ann. Pure and Appl. Logic (to appear)
8. Soskov, I.N., Kovachev, B.: Uniform regular enumerations. Mathematical Structures in Comp. Sci. 16(5), 901–924 (2006)

Computability of Analytic Functions with Analytic Machines

Tobias Gärtner[1] and Günter Hotz[2]

[1] Max Planck Institut für Informatik, Saarbrücken
[2] Fachbereich für Informatik, Universität des Saarlandes, Saarbrücken

Abstract. We present results concerning analytic machines, a model of real computation introduced by Hotz which extends the well known Blum, Shub and Smale machines by infinite converging computations. We use the machine model to define computability of complex analytic (i.e. holomorphic) functions and examine in particular the class of analytic functions which have analytically computable power series expansions. We show that this class is closed under the basic analytic operations composition, local inversion and analytic continuation.

1 Introduction

In order to define computable functions on the real and complex numbers, several different proposals have been made. One prominent approach, introduced by Blum, Shub and Smale [1] defines a machine model that deals with real numbers as atoms. A computation on such a *BSS machine* is a finite sequence of arithmetic operations and conditional branches. Blum, Shub and Smale and others have developed an elaborate complexity theory for these machines, and e.g. posed the $P \neq NP$ problem over the real numbers. In [2], a comprehensive description of the theory is presented. A main characteristic of the functions computable within this model is that those function are, in a sense, piecewise defined rational functions. Elementary transcendental functions like the exponential function or trigonometric functions are not computable in this model.

Another successful approach, *recursive analysis*, defines computability over the real and complex numbers by allowing infinite computations on Turing machines, for instance by Grzegorczyk [3]. In the *type two theory of effectivity (TTE)*, continuous structures like the real and complex numbers are mapped to the tape of the machine by means of representations or naming systems, and the Turing machines perform infinite converging computations. An overview over the theory can be found in Weihrauch's book [4]. A main characteristic of the functions defined in this framework is that they are automatically continuous, and therefore discontinuous functions like the Heaviside function are not computable. A comprehensive overview over many different approaches to computability over the real and complex numbers can be found in [5].

There are also approaches which try to connect the different approaches, like Brattka and Hertling's feasible real RAM. In this paper, we continue the work

K. Ambos-Spies, B. Löwe, and W. Merkle (Eds.): CiE 2009, LNCS 5635, pp. 250–259, 2009.

of Hotz, Chadzelek and Vierke [6,7,8] and use the approach of the *analytic machines*. These machines are a kind of synthesis of BSS machines and recursive analysis, since they allow exact arithmetic operations with real numbers as atoms on the one hand and infinite converging computations on the other. The model can be restricted to perform rational arithmetic on rounded real inputs. A certain class, the 'quasi-strongly δ-$\mathbb{Q}$ analytic functions' subsumes both BSS computable and type 2 computable functions. One of the main results of the theory is Chadzelek's hierarchy theorem [7], which in particular states that analytic machines are not closed under composition.

The focus of this paper is the well known class of holomorphic or complex analytic functions. Complex function theory has been one of the most successful parts of mathematics and is, by many, also perceived as one of the most beautiful. The fact that each analytic function can be expressed by a power series also suggests that these functions have interesting computability properties, as a power series actually can be regarded as a specification of computation. We use the model of analytic machines in order to define computable analytic functions. We define the class of (analytically) computable analytic function and the class of coefficient computable functions, which are those functions whose power series expansion is computable by an analytic machine. We show that coefficient computable analytic functions are also computable. Furthermore, we show that the class of coefficient computable analytic functions is closed under composition, local inversion and analytic continuation. In view of the fact that analytic machines are not closed under composition, the closure properties of coefficient computable functions were not entirely expected.

The computability of analytic functions has already been studied in the framework of TTE, for example Ker-I Ko [9], Müller [10] have studied polynomial time computable analytic functions, and Müller [11] has studied constructive aspects of analytic functions. In the last work, Müller shows that in the framework of TTE, that the power series expansion of type 2 computable analytic functions is computable, and also the converse, that a function is computable if its power series is computable. He goes further and examines which information is necessary for these results to be constructive. Since in contrast to type 2 machines, the convergence speed of analytic machines is not known, his methods are not applicable to our machine model.

2   Definition and Properties of Analytic Machines

The machine model underlying this work is a register machine model, which is similar to the model of Blum, Shub and Smale (BSS machines). For finite computations, the models have the same computational power. In a sense, the analytic machines extend the BSS machines by infinite converging computations. In the following section, we will briefly present the model of analytic machines. The definition has been adapted and shortened from [7]. For a more detailed description we refer the reader to this work.

Machine Model. First, we give the definition of our mathematical machine that allows us to reason formally about computations and the functions computed by a machine. A mathematical machine over an alphabet A (not necessarily finite) is a tuple $\mathcal{M} = (K, K_i, K_f, K_t, \Delta, A, \mathrm{in}, \mathrm{out})$. The set K is the *set of configurations* of $\mathcal{M}$, $K_i, K_f, K_t \subset K$ are *initial, final* and *target configurations*. The function $\Delta : K \to K$ with $\Delta|_{K_f} = id_{K_f}$ is the *transition function*, and the functions $\mathrm{in} : A^* \to K_i$ and $\mathrm{out} : K \to A^*$ are *input* and *output functions* over A.

A sequence $b = (k_i)_{i \in \mathbb{N}} \in K^{\mathbb{N}}$ with $k_0 = \mathrm{in}(x)$ and $k_{i+1} = \Delta(k_i)$ is called *computation of $\mathcal{M}$* applied to x. The computation is called *finite* or *halting* if there is an $n \in \mathbb{N}$ such that $k_n \in K_f$, and the minimal n with this property is called the *length* of the computation and $\mathrm{out}(k_n)$ its *result*.

Assuming a metric on A^*, we can also consider infinite converging computations. If for a computation $b = (k_i)_{i \in \mathbb{N}}$ a target configuration is assumed infinitely often, and the limit $\lim_{n \to \infty} \mathrm{out}(k_{i_n})$ for the subsequence $(k_{i_n})_{n \in \mathbb{N}} \subset K_t$ of target configurations exists, we call the computation *analytic*. In this case, this limit is the *result of the analytic computation* b, and $\mathrm{out}(k_{i_n})$ is called the *n-th approximation*. If for an input $x \in A$ the resulting computation is neither finite nor analytic, then we say that $\mathcal{M}$ *does not converge* on this input, otherwise we say that the *machine converges* on this input.

The *domain* $\mathcal{D}_M$ of a machine $\mathcal{M}$ is the set of all $x \in A^*$ such that the computation of $\mathcal{M}$ with input x is analytic or finite, and its *halting set* $\mathcal{H}_M$ is the set of all $x \in A^*$ such that its computation with input x is finite.

On its domain, the machine $\mathcal{M}$ now gives rise to the definition of a function $\Phi_M : \mathcal{D}_M \to A^*$, the *function defined by $\mathcal{M}$*. For inputs x such that the computation of $\mathcal{M}$ with input x is finite or analytic with result $y \in A^*$, we define $\Phi_M(x) = y$.

The mathematical machine will now be specialized to register machines over the ring $\mathcal{R}$. These machines are random access machines that operate with elements of $\mathcal{R}$ as atoms. For generality, the definition will be made for any ring $\mathcal{R}$ containing the integers (endowed with a metric in the case of analytic machines). The machines will, however, only be used for the fields of rationa, real an complex numbers $\mathbb{Q}, \mathbb{R}$ and $\mathbb{C}$. The basic notion of the machines is depicted in Fig. 1. A machine consists of a control unit with an accumulator α, a program counter β and an index register γ. There is an infinite input tape $x = (x_0, x_1, \dots)$ that can be read, an infinite output tape $y = (y_0, y_1, \dots)$ that can be written on, and an infinite memory $z = (z_0, z_1, \dots)$. Furthermore, each machine has a finite program $\pi = (\pi_1, \dots, \pi_N)$ with instructions from the instruction set $\Omega = \Omega_{\mathcal{R}}$ shown in Fig. 1. The set of configurations of a machine with program π is given by the contents of its registers and memory cells:
$$K = \{k = (\alpha, \beta, \gamma, \pi, x, y, z) \,|\, \alpha \in \mathcal{R}, \ \beta, \gamma \in \mathbb{N}, x, y, z \in \mathcal{R}^{\mathbb{N}}\}.$$ Initial, final and target configurations are given by $K_i = \{k \in K \,|\, \alpha{=}\gamma{=}0, \beta{=}1, \forall j : y_j, z_j{=}0\}$, $K_f = \{k \in K \,|\, \pi_\beta = \mathtt{end}\}$ and $K_t = \{k \in K \,|\, \pi_\beta = \mathtt{print}\}$.

The instruction $\mathtt{print}$ is used to identify target configurations, and $\mathtt{exception}$ stops the machine in a non-halting configuration.

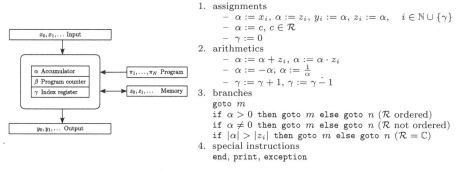

Fig. 1. Register machine and set of instructions

With the definition of our machine model, we now define computable functions.

Definition 2.1. *Let* $D \subset \mathcal{R}^*$. *We call a function* $f : D \to \mathcal{R}^*$ *analytically* $\mathcal{R}$-*computable if there exists an* $\mathcal{R}$-*machine* $\mathcal{M}$ *such that* $D \subset \mathcal{D}_{\mathcal{M}}$ *and* $f = \Phi_{\mathcal{M}}|_D$. *We call* f $\mathcal{R}$-*computable if there exists an* $\mathcal{R}$-*machine* $\mathcal{M}$ *such that* $D \subset \mathcal{H}_{\mathcal{M}}$ *and* $f = \Phi_{\mathcal{M}}|_D$.
A set $A \subset \mathcal{R}^*$ *is called* (analytically) decidable *if its characteristic function* χ_A *is (analytically) computable.*

Properties of Computable Functions. For the finitely computable functions, there is Blum, Shub and Smale's well-known representation theorem for $\mathbb{R}$-computable functions (see, e.g. [2]).

Theorem 2.2. *Suppose* $D \subset \mathbb{R}^n$ *(or* $\mathbb{C}^n$*), and let* f *be an* $\mathbb{R}$-*computable (or* $\mathbb{C}$-*computable) function on* D. *Then* D *is the union of countably many semi-algebraic sets, and on each of those sets* f *is a rational function.*

Undecidable Problems. For Turing machines, the well known halting problem is undecidable: There is no Turing machine, that accepts as inputs the description of a Turing machines and decides if that given machine halts on a given input or not. This problem is, however, easily seen to be decidable by an analytic $\mathbb{R}$-machine.
    The analogon of the halting problem for analytic machines is the *convergence problem*. That is, decide (analytically) whether a given analytic machine converges on its input or not. By diagonalization, it is easy to see that this function is not analytically computable [7,6]. If we compose several analytic machines, however, the convergence problem becomes decidable. Intuitively, the term *composition* in the context of analytic machines means that an analytic machine uses the result of another analytic machine as input. Formally it is based on the computed functions, we say that a function f is *computable by* i *composed analytic machines*, if there are i $\mathbb{R}$-analytically computable functions $f_1, \dots, f_i$ such that $f = f_i \circ \cdots \circ f_1$.

Chadzelek and Hotz [7] have shown that the convergence problem for analytic machines is computable by the composition of *three* analytic machines. Chadzelek has pointed out that it is possible to decide the problem with two analytic machines, but only if infinite intermediate results are used. The following theorem improves this result and shows that *two* machines suffice to decide the convergence problem, thus answering the open question of Chadzelek [8].

Theorem 2.3. *The convergence problem for analytic machines over the real numbers is decidable by two composed analytic machines.*

Proof. We give the description of two machines $\mathcal{M}_1$ and $\mathcal{M}_2$ such that the composition of their computed functions decides the convergence problem.

Let $(a_n)_{n \in \mathbb{N}}$ be the sequence of outputs of target configurations of the input machine $\mathcal{M}$. The goal is to check whether (a_n) is a Cauchy sequence. Let $b_k :=$ $\sup_{n > m \geq k} |a_n - a_m|$. Then, the sequence (a_n) is convergent iff $\lim_{k \to \infty} b_k = 0$. In order to decide convergence of a sequence to zero, it is sufficient to consider the sequence formed of the numbers rounded to the next higher power of two. Therefore we define $r(x, k) := \max\{j \leq k : 2^{-j} \geq x\}$ for $x < 1$ and $r(x, k) := 0$ for $x > 1$, $k \in \mathbb{N}$ and set $\tilde{b}_k := r(b_k, k)$.

Then we have $\lim_{k \to \infty} b_k = 0$ iff $\lim_{k \to \infty} \tilde{b}_k = \infty$ (note that this also covers sequences that are ultimately constantly zero). The machine $\mathcal{M}_1$ now computes approximations $\tilde{b}_{k,j}$ of $\tilde{b}_k$. Let $b_{k,j} := \max\{|a_n - a_m| \,|\, j > n > m \geq k\}$ and $\tilde{b}_{k,j} = r(b_{k,j}, k)$. Then $\tilde{b}_{k,j}$ is non-increasing with j and therefore for each k there is a j such that $\tilde{b}_{k,j} = \tilde{b}_k$, i.e. the sequence $(\tilde{b}_{k,j})_j$ becomes stationary. By parallel computation in rounds of k, $\mathcal{M}_1$ computes the approximation $\tilde{b}_{k,j}$ of $\tilde{b}_k$ for larger and larger j, and for each k, this approximation becomes stationary for sufficiently large j. The machine $\mathcal{M}_1$ stores the information about the $\tilde{b}_{k,j}$ in a single real number by a unary encoding. For $n \in \mathbb{N}$ let $u(n)$ be the unary encoding of n. Then, in each round $\mathcal{M}_1$ outputs the real number $\mathcal{M}_1^{(j)} = 0.0u(\tilde{b}_{1,j})0u(\tilde{b}_{2,j})0u(\tilde{b}_{3,j})0 \ldots 0u(\tilde{b}_{j,j})$. Since for each k the computation of $\tilde{b}_{1,j}$ will become stationary, this sequence of outputs converges to a real number $\mathcal{M}_1(\mathcal{M}) = 0.0u(\tilde{b}_1)0u(\tilde{b}_2)0u(\tilde{b}_3)0 \ldots$

Now, it follows that the original sequence (a_n) converges iff $\mathcal{M}_1(\mathcal{M})$ is an irrational number. Because if the original sequence converges, then $\tilde{b}_k$ is unbounded and the number computed by $\mathcal{M}_1$ has no periodic dual expansion. If, on the other hand, the original sequence is not convergent, then $\tilde{b}_k$ will become stationary (since it is nondecreasing with k and bounded) and the number computed by $\mathcal{M}_1$ has a periodic dual expansion.

The second machine $\mathcal{M}_2$ just has to decide whether its input is rational or not, which an analytic machine simply does by enumerating all rational numbers. □

The undecidability of the convergence problem for analytic machines and the decidability of this problem by two composed analytic machines imply that those functions cannot be closed under composition (this follows in fact already from Chadzelek's result that three composed analytic machines decide convergence):

Corollary 2.4. *The set of analytically computable functions is not closed under composition.*

3 Computability of Analytic Functions

Now we come to the main part of this paper, the application of the machine model of analytic machines to define the computability of analytic functions. By an *analytic function* on a region of the complex plane $\mathbb{C}$ we mean a function that has a power series expansion in each point of the region, or equivalently, which is complex-differentiable in every point of D. Those functions are also called holomorphic functions, but in our context we prefer the term analytic because it stresses the power series aspect of the function.

We will define two classes of computable analytic functions, first those functions which are analytic and also analytically computable (i.e. computable by an analytic machine) and then those functions which have a power series expansion that is computable by an analytic machine, the functions which we call coefficient computable. We will then show properties of the functions in the respective classes and examine the relationship between those two classes. In particular, we will show that every coefficient computable analytic function is also computable, and that the class of coefficient computable analytic functions is closed under composition, local inversion and analytic continuation.

Notation. In this section, if not mentioned otherwise, D denotes a *region* in $\mathbb{C}$, i.e. a nonempty open connected subset of $\mathbb{C}$. For a complex number z_0 and $r > 0$, $U_r(z_0)$ is the open disc with center at z_0 and radius r and $\overline{U}_r(z_0)$ its closure.

Computable Analytic Functions. First, we show that the case of finitely computable functions or, equivalently, BSS-computable functions is not very interesting in the context of analytic functions:

Lemma 3.1. *Let D be a region in $\mathbb{C}$ and $f : D \to \mathbb{C}$ be analytic and (finitely) $\mathbb{C}$-computable. Then f is a rational function on D.*

Proof. This follows by the representation theorem 2.2 and the identity theorem for analytic functions.

In light of Lemma 3.1, we will focus our attention on analytic functions which are not necessarily finitely computable. Therefore, in the following, we will freely omit the term 'analytic' when we talk about computability, i.e. instead of 'analytically computable analytic functions' we will just say 'computable analytic functions'.

Definition 3.2. *Let $D \subset \mathbb{C}$ be a region in the complex plane, and $f : D \to \mathbb{C}$ be a function. We call f a*

- computable analytic function *if f is holomorphic and analytically computable on D. We call f*

- coefficient computable in $z_0 \in D$ if f is holomorphic on D and the power series expansion of f in z_0 is an analytically computable sequence, and we call f
- coefficient computable on D if f is holomorphic on D and coefficient computable in each point $z_0 \in D$ and
- uniformly coefficient computable on D if the mapping $\mathbb{N} \times D \to \mathbb{C}, (n, z_0) \mapsto a_k(z_0)$ is analytically computable, where $a_k(z_0)$ is the k-th coefficient of the power series of f in z_0.

Now, we show that coefficient computability is the stronger notion, i.e. that each coefficient computable function is also computable.

Theorem 3.3. *Suppose* $f : D \to \mathbb{C}$ *is an analytic function on* D *with power series expansion* $f(z) = \sum_{k=0}^{\infty} a_k(z - z_0)^k$ *at* z_0, *such that the sequence* $(a_k)_{k \in \mathbb{N}}$ *is analytically computable, i.e.* f *is coefficient computable in* z_0. *Let be* $R > 0$ *such that* $\overline{U}_R(z_0) \subseteq D$. *Then* f *is analytically computable in the open disc* $U_R(z_0)$.

Proof. Let $\mathcal{K}$ be the machine that (analytically) computes the coefficients of the power series of f in z_0. We give the description of a machine $\mathcal{M}$ which analytically computes f in $U_R(z_0)$. We denote the n-th approximation of $\mathcal{K}$'s computation of the coefficient a_k with $a_k^{(n)}$. Let $z \in U_R(z_0)$ be the input for $\mathcal{M}$. The basic idea would now be to approximate the value $f(z)$ by a diagonal approximation $\sum_{k=0}^{n} a_k^{(n)}(z - z_0)^k$. The problem is that there is no information about the speed of convergence of $a_k^{(n)}$ with n, which could also be different for each k. The trick is to use upper bounds on the coefficients given by basic theorems of function theory. We recall that by Cauchy's inequalities (cf. eg. [12]) that for f there is a constant $M > 0$ such that $|a_k| \leq \frac{M}{R^k}$ for all $k \in \mathbb{N}$. The machine $\mathcal{M}$ now has information about M and R stored as constants (w.l.o.g rational bounds can be chosen), and if a current approximation $a_k^{(n)}$ exceeds the bound $\frac{M}{R^k}$, it is replaced by this bound. Summarizing, $\mathcal{M}$ computes (as n-th approximation)

$$\mathcal{M}^{(n)}(z) = \sum_{k=0}^{n} b_k^{(n)}(z - z_0)^k \quad \text{with} \quad b_k^{(n)} = \begin{cases} a_k^{(n)} & : \ |a_k^{(n)}| \leq \frac{M}{R^k} \\ \frac{M}{R^k} & : \ \text{else} \end{cases}$$

We assert that the sequence of outputs $\mathcal{M}^{(n)}(z)$ converges to $f(z)$. In order to see this, the difference $|f(z) - \mathcal{M}^{(n)}(z)| = |f(z) - \sum_{k=0}^{n} b_k^{(n)}(z - z_0)^k|$ is split into three parts, each of which becomes arbitrarily small for large n. The first part is $\sum_{k=0}^{n_0} |(a_k - b_k^{(n)})(z - z_0)^k|$ for a fixed n_0, and the second and third parts are $\sum_{k=n_0}^{n} |(a_k - b_k^{(n)})(z - z_0)^k|$ and $\sum_{k=n}^{\infty} |a_k(z - z_0)^k|$, respectively. By choosing n_0 large enough, the second and third parts can be made arbitrarily small. For n_0 then fixed, the first part can also be made arbitrarily small, since for large enough n the fixed number n_0 of the analytic computations of $a_0^{(n)}, \ldots, a_{n_0}^{(n)}$ can be made arbitrarily precise, and hence their finite sum. $\square$

A refinement of the proof of Thm. 2.2 will lead us now to an even stronger result, namely that not only a function is computable if it is coefficient computable, but also all its derivatives. For the following, recall the notation $[k]_d := \frac{k!}{(k-d)!} = k \cdot (k-1) \cdots (k-d+1)$. We observe that if the sequence (a_k) is analytically computable, then so are all sequences $(a_{(d,k)})_k = ([k+d]_d a_{k+d})_k$. Furthermore, they are *uniformly* computable, i.e. the mapping $(d,k) \mapsto a_{(d,k)}$ is analytically computable. Now, we can show that

Theorem 3.4. *Suppose* $f : D \to \mathbb{C}$ *is an analytic function on* D *which is coefficient computable in* $z_0 \in D$. *Let further be* $R > 0$ *such that* $\overline{U}_R(z_0) \subseteq D$. *Then each derivative* $f^{(d)}$ *is analytically computable in the open disc* $U_R(z_0)$, *and furthermore, all derivatives are* uniformly *computable, i.e. the mapping* $(d,z) \mapsto f^{(d)}(z)$ *is analytically computable on* $\mathbb{N} \times U_R(z_0)$.

Proof. The proof is similar to the proof of Thm. 2.2. Using Cauchy's inequalities for the derivatives we get the estimate $|a_{d,k}| = (k+d)_d a_{k+d} \leq \frac{M}{R^{k+d}}(k+d)_d$. The approximation of $f(z)$ given by the machine is again split into three parts, and with a little more technical effort, each part can be shown to become arbitrarily small. $\qquad\square$

Corollary 3.5. *Suppose* $f : D \to \mathbb{C}$ *is an analytic function on* D *which is coefficient computable in* $z_0 \in D$. *Let further be* $R > 0$ *such that* $\overline{U}_R(z_0) \subseteq D$. *Then* f *is uniformly coefficient computable on* $U_R(z_0)$.

This shows that coefficient computability in a point is the basic property for the computability of analytic functions in our sense, since it implies the computability and even uniform coefficient computability in a disc around this point.

Closure Properties of Coefficient Computable Functions. We now turn our attention to closure properties of the class of coefficient computable functions. We show that this class is closed under the basic analytic operations composition, local inversion and analytic continuation.

Composition. Since the class of generally analytically computable functions is not closed under composition (cf. Cor. 2.4), one could expect that this is also the case for the class of coefficient computable functions, which are defined by means of analytic machines. But with the theorems of the last section we can show that this class actually is closed under composition. The standard approach here would be to use a recursive description of the coefficients of the composition to compute these; since the precision of the approximations are, however, unknown, this approach is not successful. Instead, we directly express the coefficients in terms of the derivatives and apply Thm. 3.4.

Theorem 3.6. *Suppose* $g : D \to \mathbb{C}$ *and* $f : g(D) \to \mathbb{C}$ *are coefficient computable functions. Then the composition* $f \circ g : D \to \mathbb{C}$ *is also coefficient computable.*

Proof. The general form of the n-th derivative of the composition $f \circ g$ at z_0 is given by *Faa di Bruno's Formula*:

$$(f \circ g)^{(n)}(z_0) = \sum_{(k_1,\ldots,k_n) \in P_n} \frac{n!}{k_1! \cdots \cdots k_n!} \left(f^{(k_1 + \cdots + k_n)} \right)(z_1) \prod_{m=1}^{n} \left(\frac{g^{(m)}(z_0)}{m!} \right)^{k_m}$$

Here, the set $P_n = \{(k_1, \ldots, k_n) \mid k_1 + 2k_2 + \cdots + nk_n = n\}$ denotes the set of partitions of n.
In order to compute the coefficients of $f \circ g$ at z_0 observe that the coefficients of each derivative of f at z_1 and g at z_0 are uniformly computable, and that the n-th derivative of $f \circ g$ is a finite polynomial in those.   □

Local Inversion. Similarly to case of composition, we can apply Thm. 3.4 to show that the local inversion of a coefficient computable function is again coefficient computable:

Theorem 3.7. *Suppose* $f : D \to \mathbb{C}$ *is coefficient computable and* $f'(z_0) \neq 0$. *Then the local inverse* f^{-1} *of* f *is coefficient computable in a neighborhood of* $f(z_0)$.

Proof. Similarly to Thm. 3.6 one can show by induction, the coefficients of f^{-1} at $z_1 = f(z_0)$ can be expressed as rational functions in the derivatives of f at z_0 of order less or equal than n and therefore in the coefficients of the power series of f in z_0.   □

Analytic Continuation. A very important notion in the realm of analytic functions is the notion of *analytic continuation*. We briefly recall the basic concept. Let $f : D \to \mathbb{C}$ be an analytic function, $z_0 \in D$ and $f(z) = \sum_{k=0}^{\infty} a_k (z - z_0)^k$ the power series expansion of f with convergence radius $r > 0$. Let $U_r(z_0)$ be the disk with center z_0 and radius r. Then it can happen that the range of convergence $U := U_r(z_0)$ of the power series exceeds the boundary of D. If we assume further that $D \cap U$ is connected, a well-defined analytic function g is defined on $D \cup U$ by $g(z) = f(z)$ if $z \in D$ and $\sum_{k=0}^{\infty} a_k (z - z_0)^k$ if $z \in U$. It is well-defined because $D \cap U$ is connected, since then f and the function defined by the power series have to be equal on all of $D \cap U$ because of the identity theorem (note that connectedness is sufficient for the application of this theorem; simple connectedness is not necessary). The function g is called *immediate analytic continuation* of f. This process can be iterated, and an extension of f on a much larger region G may be defined. If the requirement of the connectedness of $D \cap U$ is omitted, the resulting 'function' can have multiple values. Consequently developed, this leads to the theory of *Riemann surfaces*, but this is not the subject of this article. We confine ourselves to regions that are subsets of the complex plane.
A very interesting result about coefficient computable functions is that they are closed under analytic continuation:

Theorem 3.8. *Suppose that* $f : D \to \mathbb{C}$ *is coefficient computable and that* f *has an analytic continuation* $g : G \to \mathbb{C}$ *on a region* $G \supset D$. *Then* g *is coefficient computable on* G.

Proof. This is a consequence of Cor. 3.5. In order to show that f is coefficient computable in $w \in G$, fix $z_0 \in D$. Inductively, we obtain a sequence of points $z_1, z_2, \ldots, z_n = w$ and functions $f = f_0, f_1, f_2, \ldots, f_n = w$ such that f_{i+1} is an analytic continuation of f_i, z_{i+1} is in the disc of convergence of f_i at z_i and f_i is coefficient computable at z_i.   □

This result is interesting because it relates an important notion of complex function theory to computability: In function theory, there is the notion of a *function germ*, which conveys that a holomorphic function is completely determined by the information of its power series expansion in a single point. Theorem 3.8 shows that a similar notion holds for coefficient computability: Coefficient computability of a function in a single point implies its computability in every point of its domain.

4   Conclusion and Outlook

In this paper, we have used the model of analytic machines to define two notions of computability for complex analytic functions. We have shown coefficient computability to be the stronger of the two notions. We have further shown that, in contrast to the general analytically computable functions, the class of coefficient computable functions is closed under composition, and also under local inversion and analytic continuation.

Those results have in a sense, however, not been constructive, since in order to show computability, constants containing information about the regarded functions have been assumed in the constructed machines. Open questions remain whether any of these results can, at least in part, be gained constructively. Another question that remains open is whether or to which extent analytically computable analytic functions are also coefficient computable, i.e. the converse of Thm. 3.3.

It could also be interesting to compare these results to the corresponding ones for type 2 Turing machines from the viewpoint of relativization [13].

References

1. Blum, L., Shub, M., Smale, S.: On a theory of computation and complexity over the real numbers. Bull. of the AMS 21, 1–46 (1989)
2. Blum, L., Cucker, F., Shub, M., Smale, S.: Complexity and Real Computation. Springer, New York (1998)
3. Grzegorczyk, A.: On the definition of computable real continuous functions. Fundamenta Mathematicae 44, 61–71 (1957)
4. Weihrauch, K.: Computable Analysis. Springer, Berlin (2000)
5. Ziegler, M.: Real computability and hypercomputation. TR C-07013, KIAS (2007)
6. Hotz, G., Vierke, G., Schieffer, B.: Analytic machines. ECCC 2, 25 (1995)
7. Chadzelek, T., Hotz, G.: Analytic machines. TCS 219, 151–167 (1999)
8. Chadzelek, T.: Analytische Maschinen. Dissertation, Saarland University (1998)
9. Ko, K.I.: Complexity Theory of Real Functions. Birkhäuser, Basel (1991)
10. Müller, N.T.: Uniform computational complexity of Taylor series. In: Ottmann, T. (ed.) ICALP 1987. LNCS, vol. 267, pp. 435–444. Springer, Heidelberg (1987)
11. Müller, N.T.: Constructive aspects of analytic functions. Informatik Berichte FernUniversität Hagen 190, 105–114 (1995)
12. Rudin, W.: Real and Complex Analysis. McGraw-Hill, New York (1987)
13. Ziegler, M.: Computability and continuity on the real arithmetic hierarchy and the power of type-2 nondeterminism. In: Cooper, S.B., Löwe, B., Torenvliet, L. (eds.) CiE 2005. LNCS, vol. 3526, pp. 562–571. Springer, Heidelberg (2005)

An Application of Martin-Löf Randomness to Effective Probability Theory

Mathieu Hoyrup[1] and Cristóbal Rojas[2]

[1] LORIA - 615, rue du jardin botanique, BP 239
54506 Vandœuvre-lès-Nancy, France
hoyrup@loria.fr
[2] Institut de Mathématiques de Luminy, Campus de Luminy, Case 907
13288 Marseille Cedex 9, France
rojas@iml.univ-mrs.fr

Abstract. In this paper we provide a framework for computable analysis of measure, probability and integration theories. We work on computable metric spaces with computable Borel probability measures. We introduce and study the framework of *layerwise computability* which lies on Martin-Löf randomness and the existence of a universal randomness test. We then prove characterizations of effective notions of measurability and integrability in terms of layerwise computability. On the one hand it gives a simple way of handling effective measure theory, on the other hand it provides powerful tools to study Martin-Löf randomness, as illustrated in a sequel paper.

Keywords: Algorithmic randomness, universal test, computable analysis, effective probability theory, Lebesgue integration, layerwise computability.

1 Introduction

While computability on topological spaces is now well-established (see [1,2] e.g.), the landscape for computability on measurable spaces and probability spaces is rather uneven. An effective presentation of measurable spaces is proposed in [3]. Computability on L^p-spaces has been studied in [4,5,6] for euclidean spaces with the Lebesgue measure. Computability of measurable sets has been studied, on the real line with the Lebesgue measure in [7] and on second countable locally compact Hausdorff spaces with a computable σ-finite measure in [8]. In the latter article a computability framework for bounded integrable functions is also introduced, when the measure is finite. Another approach based on probabilistic computing has been recently developed in [9]. The connection of this with the previous mentioned works remains to be established.

On the other hand, another effective approach to probability theory has already been deeply investigated, namely algorithmic randomness, as introduced by Martin-Löf in [10]. This theory was originally developed on the Cantor space, i.e. the space of infinite binary sequences, endowed with a computable probability measure. Since then, the theory has been mainly studied on the Cantor space

K. Ambos-Spies, B. Löwe, and W. Merkle (Eds.): CiE 2009, LNCS 5635, pp. 260–269, 2009.
© Springer-Verlag Berlin Heidelberg 2009

from the point of view of recursion theory, focused on the interaction between randomness and reducibility degrees. The theory has been recently extended to more general spaces in [11, 12, 13].

In this paper, we propose a general unified framework for the computable analysis of measure and integration theory, and establish intimate relations with algorithmic randomness. We first consider two natural ways (more or less already present in the literature) of giving effective versions of the notions of *measurable set*, *measurable map* and *integrable function*.

Then we develop a third approach which we call *layerwise computability* that is based on the existence of a universal randomness test. This fundamental result proved by Martin-Löf in his seminal paper is a peculiarity of the effective approach of mathematics, having no counterpart in the classical world. Making a systematic use of this has quite unexpected strong consequences: (i) it gives *topological* characterizations of effective measurability notions; (ii) measure-theoretic notions, usually defined almost everywhere, become set-theoretic when restricting to effective objects; (iii) the practice of these notions is rather light: most of the basic manipulations on computability notions on topological spaces can be straightforwardly transposed to effective measurability notions, by the simple insertion of the term "layerwise". This language trick may look suspicious, but in a sense this paper provides the background for this to make sense and being practiced.

In this way, Martin-Löf randomness and the existence of a universal test find an application in computable analysis. In [14] we show how this framework in turn provides powerful tools to the study of algorithmic randomness, extending Birkhoff's ergodic theorem for random points from *computable* functions's to *effectively measurable* ones in a simple way thanks to layerwise computability. In [14] we also show that this framework provides a general way of deriving results in the spirit of [15].

In Sect. 2 we recall the background on computable probability spaces and define the notion of *layering of the space*, which will be the cornerstone of our approach. In Sect. 3 we present two approaches to make measure-theoretical notions on computable probability space effective. Some definitions are direct adaptations of preceding works, some others are new (in particular the notions of effectively measurable maps and effectively integrable functions). In Sect. 4 we present our main contribution, namely *layerwise computability*, and state several characterizations. Being rather long, the proofs are gathered in the appendix.

2   Preliminaries

Computable metric space. Let us first recall some basic results established in [12, 13]. We work on the well-studied computable metric spaces (see [1, 16, 2, 17]).

Definition 1. *A* **computable metric space** *is a triple* $(X, d, \mathcal{S})$ *where:*

1. (X, d) *is a separable metric space,*
2. $\mathcal{S} = \{s_i : i \in \mathbb{N}\}$ *is a countable dense subset of X with a fixed numbering,*
3. $d(s_i, s_j)$ *are uniformly computable real numbers.*

$\mathcal{S}$ is called the set of *ideal points*. If $x \in X$ and $r > 0$, the metric ball $B(x, r)$ is defined as $\{y \in X : d(x, y) < r\}$. The set $\mathcal{B} := \{B(s, q) : s \in \mathcal{S}, q \in \mathbb{Q}, q > 0\}$ of *ideal balls*, which is a basis of the topology, has a canonical numbering $\mathcal{B} = \{B_i : i \in \mathbb{N}\}$. An *effective open set* is an open set U such that there is a r.e. set $E \subseteq \mathbb{N}$ with $U = \bigcup_{i \in E} B_i$. If $B_i = B(s, r)$ we denote by $\overline{B}_i$ the closed ball $\overline{B}(s, r) = \{x \in X : d(x, s) \leq r\}$. The complement of $\overline{B}_i$ is effectively open, uniformly in i. If X' is another computable metric space, a function $f : X \to X'$ is *computable* if the sets $f^{-1}(B_i')$ are effectively open, uniformly in i. Let $\overline{\mathbb{R}} := \mathbb{R} \cup \{-\infty, +\infty\}$. A function $f : X \to \overline{\mathbb{R}}$ is *lower (resp. upper) semi-computable* if $f^{-1}(q_i, +\infty]$ (resp. $f^{-1}[-\infty, q_i)$) is effectively open, uniformly in i (where $q_0, q_1, \ldots$ is a fixed effective enumeration of the set of rational numbers $\mathbb{Q}$). We remind the reader that there is an effective enumeration $(f_i)_{i \in \mathbb{N}}$ of all the lower semi-computable functions $f : X \to [0, +\infty]$.

A numbered basis $\mathcal{B}' = \{B_0', B_1', \ldots\}$ of the topology is called *effectively equivalent* to $\mathcal{B}$ if every $B_i' \in \mathcal{B}'$ is effectively open uniformly in i, and every $B_i \in \mathcal{B}$ is an effective union of elements of $\mathcal{B}'$, uniformly in i.

Computable probability space. In [3] is studied an effective version of measurable spaces. Here, we restrict our attention to metric spaces endowed with the Borel σ-field (the σ-field generated by the open sets).

Let $(X, d, \mathcal{S})$ be a computable metric space. We first recall what it means for a Borel probability measure over X to be computable. Several equivalent approaches can be found in [12, 18, 19, 13] for instance.

Definition 2. *A Borel probability measure μ is* **computable** *if $\mu(B_{i_1} \cup \ldots \cup B_{i_n})$ are lower semi-computable, uniformly in $i_1, \ldots, i_n$.*

In [18, 19] it is proved that μ is computable if and only if $\int f_i \, d\mu$ are lower semi-computable, uniformly in i (where f_i are the lower semi-computable functions).

Proposition 1. *Let μ be a computable Borel probability measure. If $f : X \to [0, +\infty)$ is upper semi-computable and bounded by M then $\int f \, d\mu$ is upper semi-computable (uniformly in a description of f and M).*

Following [13] we introduce:

Definition 3 (from [13]). *A* **computable probability space** *is a pair (X, μ) where X is a computable metric space and μ is a computable Borel probability measure on X.*

From now and beyond, we will always work on computable probability spaces.

A ball $B(s, r)$ is said to be a μ-***almost decidable ball*** if r is a computable positive real number and $\mu(\{x : d(s, x) = r\}) = 0$. The following result has been independently proved in [9] and [13].

Theorem 1. *Let (X, μ) be a computable probability space. There is a basis $\mathcal{B}^\mu = \{B_1^\mu, B_2^\mu, \ldots\}$ of uniformly μ-almost decidable balls which is effectively equivalent to the basis $\mathcal{B}$ of ideal balls. The measures of their finite unions are then uniformly computable.*

Algorithmic randomness. Here, (X, μ) is a computable probability space. Martin-Löf randomness was first defined in [10] on the space of infinite symbolic sequences. Generalizations to abstract spaces have been investigated in [20, 11, 12, 13]. We follow the latter two approaches, developed on computable metric spaces.

Definition 4. *A* **Martin-Löf test** *(ML-test) is a sequence of uniformly effective open sets U_n such that $\mu(U_n) < 2^{-n}$.*
A point x **passes** *a ML-test U if $x \notin \bigcap_n U_n$. A point is* **Martin-Löf random** *(ML-random) if it passes all ML-tests. We denote the set of ML-random points by ML_μ.*

If a set $A \subseteq X$ can be enclosed in a ML-test (U_n), i.e. $A \subseteq \bigcap_n U_n$ then we say that A is an *effective μ-null set*.
The following fundamental result, proved by Martin-Löf on the Cantor space with a computable probability measure, can be extended to any computable probability space using Thm. 1 (almost decidable balls behave in some way as the cylinders in the Cantor space, as their measures are computable).

Theorem 2 (adapted from [10]). *Every computable probability space (X, μ) admits a universal Martin-Löf test, i.e. a ML-test U such that for all $x \in X$, x is ML-random $\iff$ x passes the test U. Moreover, for each ML-test V there is a constant c (computable from a description of V) such that $V_{n+c} \subseteq U_n$ for all n.*

Definition 5. *Let (X, μ) be a computable probability space. Let $(U_n)_{n \in \mathbb{N}}$ be a universal ML-test. We call $K_n := X \setminus U_n$ the nth* **layer** *of the space and the sequence $(K_n)_{n \in \mathbb{N}}$ the* **layering** *of the space.*

One can suppose w.l.o.g. that the universal test is decreasing: $U_{n+1} \subseteq U_n$. Hence the set ML_μ of ML-random points can be expressed as an increasing union: $ML_\mu = \bigcup_n K_n$. In [14] we prove that the sets K_n are compact, in an effective way, which justifies their name.

3 Effective Versions of Measurability Notions

We now consider effective versions of the notions of measurable set, measurable map, and integrable function.

3.1 The Approach Up to Null Sets

This approach is by *equivalence classes*. As a concequence, the obtained definitions cannot distinguish between objects which coincide *up to a null set*.

Measurable sets. This approach to computability of measurable sets was first proposed by Šanin [7] on $\mathbb{R}$ with the Lebesgue measure, and generalized by Edalat [8] to any second countable locally compact Hausdorff spaces with a computable regular σ-finite measure. We present the adaptation of this approach to computable probability spaces (which are not necessarily locally compact).

Let (X, μ) be a computable probability space and $\mathfrak{S}$ the set of Borel subsets of X. The function $d_\mu : \mathfrak{S}^2 \to [0, 1]$ defined by $d_\mu(A, B) = \mu(A \Delta B)$ for all Borel sets A, B is a pseudo-metric. Let $[\mathfrak{S}]_\mu$ be the quotient of $\mathfrak{S}$ by the equivalence relation $A \sim_\mu B \iff d_\mu(A, B) = 0$ and $\mathcal{A}_\mu$ be the set of finite unions of μ-almost decidable balls from $\mathcal{B}^\mu$ with a natural numbering $\mathcal{A}_\mu = \{A_1, A_2, \ldots\}$. We denote by $[A]_\mu$ the equivalence class of a Borel set A. The following result was proved in [21] (Thm. 2.3.2.1) for computable metric spaces.

Proposition 2. $([\mathfrak{S}]_\mu, d_\mu, \mathcal{A}_\mu)$ *is a computable metric space.*

The following definition is then the straightforward adaptation of [7,8].

Definition 6. *A Borel set A is called a μ-recursive set if its equivalence class $[A]_\mu$ is a computable point of the computable metric space $[\mathfrak{S}]_\mu$.*

In other words, there is a total recursive function $\varphi : \mathbb{N} \to \mathbb{N}$ such that $\mu(A \Delta A_{\varphi(n)}) < 2^{-n}$ for all n. The measure of any μ-recursive set is computable. Observe that an ideal ball need not be μ-recursive as its measure is in general only lower semi-computable. On the other hand, μ-almost decidable balls are always μ-recursive.

Measurable maps. Here Y is a computable metric space. To the notion of μ-recursive set corresponds a natural effective version of the notion of μ-recursive map:

Definition 7. *A measurable map $T : (X, \mu) \to Y$ is called a μ-recursive map if there exists a basis of balls $\hat{\mathcal{B}} = \{\hat{B}_1, \hat{B}_2, \ldots\}$ of Y, which is effectively equivalent to the basis of ideal balls $\mathcal{B}$, and such that $T^{-1}(\hat{B}_i)$ are uniformly μ-recursive sets.*

Integrable functions. Computability on L^p spaces has been studied in [4,5,6] for euclidean spaces with the Lebesgue measure. The L^1 case can be easily generalized to any computable probability space, and a further generalization including σ-finite measures might be carried out without difficulties.

Let (X, μ) be a computable probability space. Let $\mathcal{F}$ be the set of measurable functions $f : X \to \overline{\mathbb{R}}$ which are integrable. Let $I_\mu : \mathcal{F} \times \mathcal{F} \to [0, +\infty)$ be defined by $I_\mu(f, g) = \int |f - g| \, d\mu$. I_μ is a metric on the quotient space $L^1(X, \mu)$ with the relation $f \sim_\mu g \iff I_\mu(f, g) = 0$. There is a set $\mathcal{E} = \{f_0, f_1, \ldots\}$ of uniformly computable effectively bounded functions ($|f_i| < M_i$ with M_i computable from i) which is dense in $L^1(X, \mu)$ (this is a direct consequence of Prop. 2.1 in [12]). $\mathcal{E}$ is called the set of *ideal functions*.

Proposition 3. $(L^1(X, \mu), d_\mu, \mathcal{E})$ *is a computable metric space.*

This leads to a first effective notion of integrable function:

Definition 8. *A function $f : X \to \overline{\mathbb{R}}$ is μ-recursively integrable if its equivalence class is a computable point of the space $L^1(X, \mu)$, i.e. f can be effectively approximated by ideal functions in the L^1 norm.*

If $f : X \to \overline{\mathbb{R}}$ is integrable, then f is μ-recursively integrable if and only if so are $f^+ = \max(f, 0)$ and $f^- = \max(-f, 0)$.

3.2   The Approach Up to *Effective* Null Sets

On a metric space, every Borel probability measure is *regular*, i.e. for every Borel set A and every $\varepsilon > 0$ there is a closed set F and an open set U such that $F \subseteq A \subseteq U$ and $\mu(U \setminus F) < \varepsilon$ (see [22]). It gives an alternative way to define an effective version of the notion of measurable set. We will see how to define *effectively μ-measurable* maps and *effectively μ-integrable* functions using the same idea.

Measurable sets. Edalat [8] already used regularity of measures to define μ-computable sets, a notion that is stronger than μ-recursivity. Let us consider the adaptation of this notion to computable probability spaces (for coherence reasons, we use the expression "effective μ-measurability" instead of "μ-computability").

Definition 9. *A Borel set A is* **effectively μ-measurable** *if there are uniformly effective open sets U_i, V_i such that $X \setminus V_i \subseteq A \subseteq U_i$ and $\mu(U_i \cap V_i) < 2^{-i}$.*

This is a generalization of the notion of effective μ-null set (see after Def. 4): a set of measure zero is an effective μ-null set if and only if it is effectively μ-measurable.

Example 1. The whole space X is effectively μ-measurable. More generally, an effective open set is effectively μ-measurable if and only if its measure is computable. The *Smith-Volterra-Cantor set* or *fat Cantor set*, which is an effective compact subset of $[0, 1]$ whose Lebesgue measure is $1/2$, is effectively λ-measurable where λ denotes the Lebesgue measure (see SVC(4) in [23]).

Measurable maps. To the notion of effectively μ-measurable set corresponds a natural effective version of the notion of measurable map:

Definition 10. *A measurable map $T : (X, \mu) \to Y$ is* **effectively μ-measurable** *if there exists a basis of balls $\hat{\mathcal{B}} = \{\hat{B}_1, \hat{B}_2, \ldots\}$ of Y, which is effectively equivalent to the basis of ideal balls $\mathcal{B}$, and such that $T^{-1}(\hat{B}_i)$ are uniformly effectively μ-measurable sets.*

Integrable functions. In [8] a notion of μ-*computable integrable function* is proposed: such a function can be effectively approximated from above and below by simple functions. This notion is developed on any second countable locally compact Hausdorff spaces endowed with a computable finite Borel measure. In this approach only bounded functions can be handled, as they are dominated by simple functions, which are bounded by definition. We overcome this problem, providing at the same time a framework for metric spaces that are not locally compact, as function spaces.

The following definition is a natural extension of the counterpart of Def. 9 for the characteristic function $\mathbf{1}_A$ of an effectively μ-measurable set A.

Definition 11. *A function* $f : X \to [0, +\infty]$ *is* **effectively** μ-**integrable** *if there are uniformly lower semi-computable functions* $g_n : X \to [0, +\infty]$ *and upper semi-computable functions* $h_n : X \to [0, +\infty)$ *such that:*

1. $h_n \leq f \leq g_n$,
2. $\int (g_n - h_n)\, d\mu < 2^{-n}$,
3. h_n *is bounded by some* M_n *which is computable from* n.

Observe that a set A is effectively μ-measurable if and only if its characteristic function $\mathbf{1}_A$ is effectively μ-integrable.

4   The Algorithmic Randomness Approach: Layerwise Computability

4.1   Layerwise Computability

We remind the reader that every computable probability space comes with a Martin-Löf layering $(K_n)_{n \in \mathbb{N}}$ (see Def. 5). In the following definition, $\mathcal{B} = \{B_i : i \in \mathbb{N}\}$ is the basis of ideal balls of Y.

Definition 12. *A set* A *is* **layerwise semi-decidable** *if it is semi-decidable on every* K_n, *uniformly in* n. *In other words,* A *is layerwise semi-decidable if there are uniformly effective open sets* U_n *such that* $A \cap K_n = U_n \cap K_n$ *for all* n. *A set* A *is* **layerwise decidable** *if both* A *and its complement are layerwise semi-decidable. A function* $T : (X, \mu) \to Y$ *is* **layerwise computable** *if* $T^{-1}(B_i)$ *are layerwise semi-decidable, uniformly in* i.

In the language of representations, a set A is layerwise semi-decidable (resp. layerwise decidable) if there is a machine which takes n and a Cauchy representation of $x \in K_n$ as inputs, and eventually halts if and only if $x \in A$ (resp. halts and outputs 1 if $x \in A$, 0 if $x \notin A$) (if $x \notin K_n$, nothing is assumed about the behavior of the machine).

Actually, every computability notion on computable metric spaces has in principle its layerwise version. For instance one can define *layerwise lower semi-computable* functions $f : X \to \overline{\mathbb{R}}$.

Let us state some basic properties of layerwise computable maps, when considering the push-forward measure ν defined by $\nu(A) = \mu(T^{-1}(A))$.

Proposition 4. *Let* $T : (X, \mu) \to Y$ *be a layerwise computable map.*

- *The push-forward measure* $\nu := \mu \circ T^{-1} \in \mathcal{M}(Y)$ *is computable.*
- T *preserves ML-randomness, i.e.* $T(ML_\mu) \subseteq ML_\nu$. *Moreover, there is a constant* c *(computable from a description of* T*) such that* $T(K_n) \subseteq K'_{n+c}$ *for all* n, *where* (K'_n) *is the canonical layering of* (Y, ν).
- *If* $f : (Y, \nu) \to Z$ *is layerwise computable then so is* $f \circ T$.
- *If* $A \subseteq Y$ *is layerwise decidable (resp. semi-decidable) then so is* $T^{-1}(A)$.

The first point implies that in the particular case when $Y = \mathbb{R}$, a layerwise computable function is then a computable random variable as defined in [24]: its distribution ν over $\mathbb{R}$ is computable. Observe that when ν is the push-forward of μ, layerwise computability notions interact as the corresponding plain computability ones; however, without this assumption on ν the last three points may not hold.

As shown by the following proposition, if layerwise computable objects differ at one ML-random point then they essentially differ, i.e. on a set of positive measure.

Proposition 5. *Let $A, B \subseteq X$ be layerwise decidable sets and $T_1, T_2 : (X, \mu) \to Y$ layerwise computable functions.*

- *$A = B \mod 0$ if and only if $A \cap ML_\mu = B \cap ML_\mu$.*
- *$T_1 = T_2$ almost everywhere if and only if $T_1 = T_2$ on ML_μ.*

4.2   Characterizations of Effective Measure-Theoretic Notions in Terms of Layerwise Computability

Measurable sets. The notion of effective μ-measurable set is strongly related to the Martin-Löf approach to randomness. Indeed, if A is a Borel set such that $\mu(A) = 0$ then A is effectively μ-measurable if and only if it is an effective μ-null set. If A is effectively μ-measurable, coming with C_n, U_n, then $\bigcup_n C_n$ and $\bigcap_n U_n$ are two particular representatives of $[A]_\mu$ which coincide with A on ML_μ. We can even go further, as the following result proves.

Theorem 3. *Let A be a Borel set. We have:*

1. *A is μ-recursive $\Longleftrightarrow$ A is equivalent to an effectively μ-measurable set.*
2. *A is effectively μ-measurable $\Longleftrightarrow$ A is layerwise decidable.*

The equivalences are uniform. These characterizations enable one to use layerwise computability to simplify proofs: for instance basic operations that preserve decidability of sets, as finite unions or complementation, also preserve *layerwise* computability in a straightforward way, hence they preserve μ-recursivity and effective μ-measurability.

Let A be a μ-recursive set: it is equivalent to a layerwise decidable set B. By Prop. 5 the set $A^* := B \cap ML_\mu$ is well-defined and constitutes a canonical representative of the equivalence class of A under $\sim_\mu$. If A is already layerwise decidable then $A^* = A \cap ML_\mu$. From this, the operator $*$ is idempotent, it commutes with finite unions, finite intersections and complements.

Proposition 6. *If A is a layerwise semi-decidable set then*

- *$\mu(A)$ is lower semi-computable,*
- *$\mu(A)$ is computable if and only if A is layerwise decidable.*

Measurable maps. We obtain a version of Thm. 3 for measurable maps.

Theorem 4. *Assume Y is a complete computable metric space. Let T : $(X, \mu) \to Y$ be a measurable map. We have:*

1. *T is μ-recursive $\iff$ T coincides almost everywhere with an effectively μ-measurable map.*
2. *T is effectively μ-measurable $\iff$ T is layerwise computable.*

The equivalences are uniform. Observe that while all other implications directly derive from Thm.3, the first one is not so easy as we have to carry out the explicit construction of an effectively μ-measurable function from the equivalence class of T. These characterizations show that computability, which trivially implies layerwise computability, implies μ-recursivity and effective μ-measurability.

Let T be μ-recursive: there is a layerwise computable function T' which is equivalent to T. Let T^* be the restriction of T' to ML_μ. By Prop. 5 T^* is uniquely defined.

Integrable functions. We know from Thm. 3 that A is effectively μ-measurable if and only if A is layerwise decidable, which is equivalent to the layerwise computability of $\mathbf{1}_A$. As a result, $\mathbf{1}_A$ is effectively μ-integrable if and only if $\mathbf{1}_A$ is layerwise computable. The picture is not so simple for unbounded integrable functions: although $\int f \, d\mu$ is always computable when f is effectively μ-integrable, it is only *lower semi-computable* when f is layerwise computable.

Proposition 7. *Let $f : X \to [0, +\infty]$.*

- *If f is layerwise lower semi-computable then $\int f \, d\mu$ is lower semi-computable (uniformly in a description of f).*
- *If f is bounded and layerwise computable then $\int f \, d\mu$ is computable (uniformly in a description of f and a bound on f).*

Hence, we have to add the computability of $\int f \, d\mu$ to get a characterization.

Theorem 5. *Let $f : X \to [0, +\infty]$ be a μ-integrable function. We have:*

1. *f is μ-recursively integrable $\iff$ f is equivalent to an effectively μ-integrable function.*
2. *f is effectively μ-integrable $\iff$ f is layerwise computable and $\int f \, d\mu$ is computable.*

The equivalences are uniform, but a description of $\int f \, d\mu$ as a computable real number must be provided.

We now get a rather surprising result, which is a weak version of Prop. 6 for integrable functions.

Proposition 8. *Let $f : X \to [0, +\infty]$ be a layerwise lower semi-computable function. If $\int f \, d\mu$ is computable then f is layerwise computable.*

References

1. Edalat, A., Heckmann, R.: A computational model for metric spaces. Theor. Comput. Sci. 193, 53–73 (1998)
2. Weihrauch, K.: Computable Analysis. Springer, Berlin (2000)
3. Wu, Y., Weihrauch, K.: A computable version of the daniell-stone theorem on integration and linear functionals. Theor. Comput. Sci. 359(1-3), 28–42 (2006)
4. Pour-El, M.B., Richards, J.I.: Computability in Analysis and Physics. Perspectives in Mathematical Logic. Springer, Berlin (1989)
5. Zhang, B.Y., Zhong, N.: L^p-computability. Math. Logic Q. 45, 449–456 (1999)
6. Kunkle, D.: Type-2 computability on spaces of integrable functions. Math. Logic Q. 50(4-5), 417–430 (2004)
7. Šanin, N.: Constructive Real Numbers and Constructive Function Spaces. Translations of Mathematical Monographs, vol. 21. American Mathematical Society, Providence (1968)
8. Edalat, A.: A computable approach to measure and integration theory. In: LICS 2007: Proceedings of the 22nd Annual IEEE Symposium on Logic in Computer Science, Washington, DC, USA, pp. 463–472. IEEE Computer Society, Los Alamitos (2007)
9. Bosserhoff, V.: Notions of probabilistic computability on represented spaces. Journal of Universal Computer Science 14(6), 956–995 (2008)
10. Martin-Löf, P.: The definition of random sequences. Information and Control 9(6), 602–619 (1966)
11. Hertling, P., Weihrauch, K.: Random elements in effective topological spaces with measure. Inf. Comput. 181(1), 32–56 (2003)
12. Gács, P.: Uniform test of algorithmic randomness over a general space. Theor. Comput. Sci. 341, 91–137 (2005)
13. Hoyrup, M., Rojas, C.: Computability of probability measures and Martin-Löf randomness over metric spaces. Inf. Comput (2009) (in press/arxiv)
14. Hoyrup, M., Rojas, C.: Applications of effective probability theory to Martin-Löf randomness. In: ICALP (2009)
15. Davie, G.: The Borel-Cantelli lemmas, probability laws and Kolmogorov complexity. Annals of Probability 29(4), 1426–1434 (2001)
16. Yasugi, M., Mori, T., Tsujii, Y.: Effective properties of sets and functions in metric spaces with computability structure. Theor. Comput. Sci. 219(1-2), 467–486 (1999)
17. Brattka, V., Presser, G.: Computability on subsets of metric spaces. Theor. Comput. Sci. 305(1-3), 43–76 (2003)
18. Edalat, A.: When scott is weak on the top. Mathematical Structures in Computer Science 7(5), 401–417 (1997)
19. Schröder, M.: Admissible representations for probability measures. Math. Log. Q. 53(4-5), 431–445 (2007)
20. Zvonkin, A., Levin, L.: The complexity of finite objects and the development of the concepts of information and randomness by means of the theory of algorithms. Russian Mathematics Surveys 256, 83–124 (1970)
21. Rojas, C.: Randomness and ergodic theory: an algorithmic point of view. PhD thesis, Ecole Polytechnique (2008)
22. Billingsley, P.: Convergence of Probability Measures. John Wiley, New York (1968)
23. Bressoud, D.M.: A Radical Approach to Lebesgue's Theory of Integration. Mathematical Association of America Textbooks. Cambridge University Press, Cambridge (2008)
24. Müller, N.T.: Computability on random variables. Theor. Comput. Sci. 219(1-2), 287–299 (1999)

Index Sets and Universal Numberings[*]

Sanjay Jain[1], Frank Stephan[2], and Jason Teutsch[3]

[1] Department of Computer Science,
National University of Singapore, Singapore 117417, Republic of Singapore
sanjay@comp.nus.edu.sg
[2] Department of Computer Science and Department of Mathematics,
National University of Singapore, Singapore 117543, Republic of Singapore
fstephan@comp.nus.edu.sg
[3] Center for Communications Research, 4320 Westerra Court,
San Diego, California 92121-1969, United States of America
teutsch@cs.uchicago.edu

Abstract. This paper studies the Turing degrees of various properties defined for universal numberings, that is, for numberings which list all partial-recursive functions. In particular properties relating to the domain of the corresponding functions are investigated like the set DEQ of all pairs of indices of functions with the same domain, the set DMIN of all minimal indices of sets and DMIN* of all indices which are minimal with respect to equality of the domain modulo finitely many differences. A partial solution to a question of Schaefer is obtained by showing that for every universal numbering with the Kolmogorov property, the set DMIN* is Turing equivalent to the double jump of the halting problem. Furthermore, it is shown that the join of DEQ and the halting problem is Turing equivalent to the jump of the halting problem and that there are numberings for which DEQ itself has 1-generic Turing degree.

1 Introduction

It is known that for acceptable numbering many problems are very hard: Rice [14] showed that all semantic properties like $\{e : \varphi_e$ is total$\}$ or $\{e : \varphi_e$ is somewhere defined$\}$ are non-recursive and that the halting problem K is Turing reducible to them. Similarly, Meyer [10] showed that the set $\text{MIN}_\varphi = \{e : \forall d < e \ [\varphi_d \neq \varphi_e]\}$ of minimal indices is even harder: $\text{MIN}_\varphi \equiv_T K'$. In contrast to this, Friedberg [3] showed that there is a numbering ψ of all partial-recursive functions such that $\psi_d \neq \psi_e$ whenever $d \neq e$. Hence, every index in this numbering is a minimal index: $\text{MIN}_\psi = \mathbb{N}$. One could also look at the corresponding questions for minimal indices for domains. Then, as long as one does not postulate that every function occurs in the numbering but only that every domain occurs, there

---

[*] This research was supported in part by NUS grant numbers R252-000-212-112 (all authors), R252-000-308-112 (S. Jain and F. Stephan) and R146-000-114-112 (F. Stephan). Missing proofs can be found in Technical Report TRA3/09 of the School of Computing at the National University of Singapore.

K. Ambos-Spies, B. Löwe, and W. Merkle (Eds.): CiE 2009, LNCS 5635, pp. 270–279, 2009.
© Springer-Verlag Berlin Heidelberg 2009

are numberings for which the set of minimal indices of domains is recursive and other numberings for which this set is Turing equivalent to K'. But there is a different result if one requires that the numbering is universal in the sense that it contains every partial-recursive function. Then the set $\mathrm{DMIN}_\psi = \{e : \forall d < e \ [W_d^\psi \neq W_e^\psi]\}$ is not recursive but satisfies $\mathrm{DMIN}_\psi \oplus K \equiv_T K'$, see Proposition 4 below. On the other hand, DMIN_ψ is for some universal numberings ψ not above K'. Indeed, DMIN_ψ is 1-generic for certain numberings. In the present work, various properties linked to the domains of functions for universal and domain-universal numberings are studied. In particular the complexities of these sets are compared with K, K', K'' and so on.

Schaefer [15] tried to lift Meyer's result one level up in the arithmetic hierarchy and asked whether $\mathrm{MIN}_\psi^* \equiv_T K''$; Teutsch [17] asked the corresponding question for domains: is $\mathrm{DMIN}_\psi^* \equiv_T K''$? These questions were originally formulated for Gödel numberings. In the present work, partial answers are obtained: on one hand, if the numbering ψ is a Kolmogorov numbering then DMIN_ψ^* and MIN_ψ^* are both Turing equivalent to K''; on the other hand, there is a universal numbering (which is not a Gödel numbering) such that DMIN_ψ^* and MIN_ψ^* are 1-generic and hence not above K.

Besides this, a further main result of this paper is to show that for certain universal numbering ψ the domain equality problem DEQ_ψ has 1-generic Turing degree; hence the domain-equivalence problem of ψ is not Turing hard for K'.

After this short overview of the history of minimal indices and the main results of this paper, the formal definitions are given, beginning with the fundamental notion of numberings and universal numberings. For an introduction to the basic notions of Recursion Theory and Kolmogorov Complexity, see the textbooks of Li and Vitányi [9], Odifreddi [11,12] and Soare [16].

Definition 1. Let $\psi_0, \psi_1, \psi_2, \ldots$ be a family of functions from $\mathbb{N}$ to $\mathbb{N}$. ψ is called a *numbering* iff the set $\{\langle e, x, y \rangle : \psi_e(x) \!\downarrow= y\}$ is recursively enumerable; ψ is called a *universal numbering* iff every partial-recursive function equals to some function ψ_e; ψ is called a *domain-universal numbering* iff for every r.e. set A there is an index e such that the domain W_e^ψ of ψ_e equals A.

A numbering ψ is *acceptable* or a *Gödel numbering* iff for every further numbering ϑ there is a recursive function f such that $\psi_{f(e)} = \vartheta_e$ for all e; a numbering ψ has the *Kolmogorov property* iff

$$\forall \text{ numberings } \vartheta \ \exists c \ \forall e \ \exists d < ce + c \ [\psi_d = \vartheta_e]$$

and a numbering ψ is a *Kolmogorov numbering* iff it has the Kolmogorov property effectively, that is,

$$\forall \text{ numberings } \vartheta \ \exists c \ \exists \text{ recursive } f \ \forall e \ [f(e) < ce + c \wedge \psi_{f(e)} = \vartheta_e].$$

A numbering ψ is a *K-Gödel numbering* [1] iff for every further numbering ϑ there is a K-recursive function f such that $\psi_{f(e)} = \vartheta_e$ for all e. Similarly one can define *K-Kolmogorov numberings*.

Note that a universal numbering is a weakening of an acceptable numbering while in the field of Kolmogorov complexity, the notion of a universal machine is stronger than that of an acceptable numbering; there a universal machine is a numbering of strings (not functions) with the Kolmogorov property and so this notion is stronger than the notion of an acceptable numbering of strings.

Definition 2. Given a numbering ψ, let $\text{DMIN}_\psi = \{e : \forall d < e \ [W_d^\psi \neq W_e^\psi]\}$, $\text{DMIN}_\psi^* = \{e : \forall d < e \ [W_d^\psi \neq^* W_e^\psi]\}$ and $\text{DMIN}_\psi^m = \{e : \forall d < e \ [W_d^\psi \not\equiv_m W_e^\psi]\}$. Here $A =^* B$ means that the sets A, B are finite variants and $A \neq^* B$ means that the sets A, B are not finite variants. Furthermore, $A \equiv_m B$ iff there are recursive functions f, g such that $A(x) = B(f(x))$ and $B(x) = A(g(x))$ for all x; $A \not\equiv_m B$ otherwise. The superscript "m" in DMIN_ψ^m is just referring to many-one reduction.

2 Minimal Indices and Turing Degrees

The next result is well-known and can, for example, be derived from [2, Theorem 5.7].

Proposition 3. *Let φ be any acceptable numbering. Now $K' \leq_T A \oplus K$ iff one can enumerate relative to the oracle A a set E of indices of total recursive functions such that for every total recursive f there is an $e \in E$ with $\varphi_e = f$.*

Meyer [10] showed the next result for Gödel numberings. Here the result is given for universal numberings; by a well-known result of Friedberg this is false for some domain-universal numberings.

Proposition 4. *For every universal numbering ψ, $K' \leq_T \text{DMIN}_\psi \oplus K$.*

Schaefer [15] and Teutsch [17,18] investigated the complexity of DMIN_ψ^*. The next two results generalize their findings from Gödel numberings to domain-universal numberings.

Proposition 5. *For every domain-universal numbering ψ, $K' \leq_T \text{DMIN}_\psi^* \oplus K$.*

Proposition 6. *For every domain-universal numbering ψ, $K'' \equiv_T \text{DMIN}_\psi^* \oplus K'$.*

Proof. Let φ be a Gödel numbering and note that

$$K'' \equiv_T \{e : W_e^\varphi \text{ is co-finite}\}.$$

Furthermore, let a be the unique element of DMIN_ψ^* such that W_a^ψ is co-finite. For any given e, find using K' the least d such that $W_d^\psi = W_e^\varphi$. Furthermore, let $D = \text{DMIN}_\psi^* \cap \{0, 1, 2, \ldots, d\}$ and define

$$\text{ndiff}(c, d, x) = |\{y \leq x : W_c^\psi(y) \neq W_d^\psi(y)\}|$$

to be the number of differences between W_c^ψ and W_d^ψ below x. One can find with oracle K' the unique $b \in D$ such that $\text{ndiff}(b, d, x) \leq \text{ndiff}(c, d, x)$ for all $c \in D$ and almost all x. Note that b is the unique member of DMIN_ψ^* with $W_b^\psi =^* W_e^\varphi$. Now W_e^φ is co-finite iff $b = a$; hence

$$\{e : W_e^\varphi \text{ is co-finite}\} \leq_T \text{DMIN}_\psi^* \oplus K'.$$

As $\text{DMIN}_\eta^* \leq_T K''$ for all η, $K'' \equiv_T \text{DMIN}_\psi^* \oplus K'$. $\square$

Remark 7. The following proofs make use of Owings' Cardinality Theorem [13]. This says that whenever there is an $m > 0$ and a B-recursive $\{0, 1, 2, \ldots, m\}$-valued function mapping every m-tuple $(a_1, a_2, \ldots, a_m)$ to a number in $\{0, 1, 2, \ldots, m\}$ which is different from $A(a_1) + A(a_2) + \ldots + A(a_m)$ then $A \leq_T B$. Kummer [4,7] generalized this result and showed that whenever there are an $m > 0$ and B-r.e. sets enumerating uniformly for every m-tuple $(a_1, a_2, \ldots, a_m)$ up to m numbers including $A(a_1) + A(a_2) + \ldots + A(a_m)$ then $A \leq_T B$.

Theorem 8. *For every universal numbering ψ with the Kolmogorov property, $K \leq_T \text{DMIN}_\psi^*$ and $K'' \equiv_T \text{DMIN}_\psi^*$.*

Proof. Let σ_n be the n-th finite string in an enumeration of $\mathbb{N}^*$. Due to the Kolmogorov property, one can recursively partition the natural numbers into intervals I_n such that for every n there is a number $z \in \text{DMIN}_\psi^*$ with $\min(I_n) \cdot (|\sigma_n| + 1) + |\sigma_n| < z < \max(I_n)$. For every $p \in I_n$ with $\sigma_n = a_1 a_2 \ldots a_m$ let

$$\vartheta_p(x) = \psi_{p \cdot (m+1) + K_x(a_1) + K_x(a_2) + \ldots + K_x(a_m)}(x)$$

and note that

$$\vartheta_p =^* \psi_{p \cdot (m+1) + K(a_1) + K(a_2) + \ldots + K(a_m)}$$

as the approximations $K_x(a_1), K_x(a_2), \ldots, K_x(a_m)$ coincide respectively with $K(a_1), K(a_2), \ldots, K(a_m)$ for almost all x. By the Kolmogorov property there is a constant m such that for every p there is an $e < \max\{pm, m\}$ with $\psi_e = \vartheta_p$; fix this m from now on.

Now, for any $a_1, a_2, \ldots, a_m$, choose n such that $\sigma_n = a_1 a_2 \ldots a_m$ and let $g(a_1, a_2, \ldots, a_m)$ and $h(a_1, a_2, \ldots, a_m)$ be the unique values in $\mathbb{N}$ and $\{0, 1, 2, \ldots, m\}$, respectively, such that

$$g(a_1, a_2, \ldots, a_m) \cdot (m + 1) + h(a_1, a_2, \ldots, a_m) = \max(I_n \cap \text{DMIN}_\psi^*).$$

By choice of m, $g(a_1, a_2, \ldots, a_m) \in I_n$ and $g(a_1, a_2, \ldots, a_m) > 0$. Hence

$$\vartheta_{g(a_1, a_2, \ldots, a_m)} =^* \psi_{g(a_1, a_2, \ldots, a_m) \cdot (m+1) + K(a_1) + K(a_2) + \ldots + K(a_m)}$$

and $\psi_e = \vartheta_{g(a_1, a_2, \ldots, a_m)}$ for some $e < g(a_1, a_2, \ldots, a_m) \cdot m$. So $g(a_1, a_2, \ldots, a_m) \cdot (m + 1) + K(a_1) + K(a_2) + \ldots + K(a_m)$ is not in DMIN_ψ^* and

$$h(a_1, a_2, \ldots, a_m) \in \{0, 1, 2, \ldots, m\} - \{K(a_1) + K(a_2) + \ldots + K(a_m)\}.$$

So $h \leq_T \text{DMIN}^*_\psi$ and h produces on input $a_1, a_2, \ldots, a_m$ a value in $\{0, 1, 2, \ldots, m\}$ different from $K(a_1) + K(a_2) + \ldots + K(a_m)$. Owings' Cardinality Theorem [4,13] states that the existence of such a function h implies $K \leq_T \text{DMIN}^*_\psi$.

It is well-known that $\text{DMIN}^*_\psi \leq_T K''$. On the other hand one can now apply Proposition 5 to get that $K' \leq_T \text{DMIN}^*_\psi$ and Proposition 6 to get that $K'' \leq_T \text{DMIN}^*_\psi$. □

Theorem 9. *For every universal numbering ψ with the Kolmogorov property,* $K'' \equiv_T \text{DMIN}^m_\psi$.

Remark 10. Teutsch [17,18] considered also the problem $\text{DMIN}^T_\psi = \{e : \forall d < e$ $[W^\psi_d \neq_T W^\psi_e]\}$. He showed that if ψ is an acceptable numbering then $K''' \leq_T \text{DMIN}^T_\psi \oplus K'$. The above techniques can also be used to show that if ψ is a Kolmogorov numbering then $K''' \equiv_T \text{DMIN}^T_\psi$.

One might also ask what the minimum Turing degree of MIN^*_ψ is. While Friedberg showed that MIN_ψ can be recursive, this is not true for MIN^*_ψ. Furthermore, one can easily adapt the construction from Theorem 18 to construct a K-Gödel numbering where MIN^*_ψ is 1-generic. In addition to this, Propositions 5 and 6 can be transferred to MIN^*_ψ. Furthermore, the proof from above for numberings with the Kolmogorov property works also for MIN^*_ψ instead of DMIN^*_ψ. Thus one has the following result.

Theorem 11. *For every universal numbering ψ with the Kolmogorov property,* $K' \equiv_T \text{MIN}_\psi$, $K'' \equiv_T \text{MIN}^*_\psi$ *and* $K'' \equiv_T \text{MIN}^m_\psi$.

3   Prominent Index Sets

It is known from Rice's Theorem that almost all index sets in Gödel numberings are Turing hard for K. On the other hand, in Friedberg numberings, the index set of the everywhere undefined function is just a singleton and hence recursive. So it is a natural question how the index sets depend on the chosen underlying universal numbering. In particular the following index sets are investigated within this section.

Definition 12. For a universal numbering ψ define the following notions:

- $\text{EQ}_\psi = \{\langle i, j \rangle : \psi_i = \psi_j\}$ and $\text{EQ}^*_\psi = \{\langle i, j \rangle : \psi_i =^* \psi_j\}$;
- $\text{DEQ}_\psi = \{\langle i, j \rangle : W^\psi_i = W^\psi_j\}$ and $\text{DEQ}^*_\psi = \{\langle i, j \rangle : W^\psi_i =^* W^\psi_j\}$;
- $\text{INC}_\psi = \{\langle i, j \rangle : W^\psi_i \subseteq W^\psi_j\}$ and $\text{INC}^*_\psi = \{\langle i, j \rangle : W^\psi_i \subseteq^* W^\psi_j\}$;
- $\text{EXT}_\psi = \{\langle i, j \rangle : \forall x \in W^\psi_i \ [x \in W^\psi_j \wedge \psi_j(x) = \psi_i(x)]\}$;
- $\text{CONS}_\psi = \{\langle i, j \rangle : \forall x \in W^\psi_i \cap W^\psi_j \ [\psi_i(x) = \psi_j(x)]\}$;
- $\text{DISJ}_\psi = \{\langle i, j \rangle : W^\psi_i \cap W^\psi_j = \emptyset\}$;
- $\text{INF}_\psi = \{i : W^\psi_i \text{ is infinite}\}$.

Note that although these sets come as a sets of pairs, one can also fix the index i and consider the classic index set of all j such that ψ_j is consistent with ψ_i in the way described above. But the index sets of pairs are quite natural and so some more of these examples will be investigated.

Kummer [8] obtained a breakthrough and solved an open problem of Herrmann posed around 10 years earlier by showing that there is a domain-universal numbering where the domain inclusion problem is K-recursive. He furthermore concluded that also the extension-problem for universal numberings can be made K-recursive.

Theorem 13 (Kummer [8]). *There is a domain universal numbering ψ and a universal numbering ϑ such that* $\mathrm{INC}_\psi \leq_T K$ *and* $\mathrm{EXT}_\vartheta \equiv_T K$. *The numbering ϑ can easily be obtained from ψ.*

Note that this result needs that ψ is only domain universal and not universal; if ψ would be universal then $K' \leq_T \mathrm{INC}_\psi \oplus K$ and hence $\mathrm{INC}_\psi \not\leq_T K$. It is still open whether $K \leq_T \mathrm{INC}_\psi$ for all domain universal numberings ψ. But for the function-extension problem, Kummer's result is optimal.

Proposition 14. $\mathrm{EXT}_\psi \geq_T K$ *for every universal numbering.*

As Kummer showed, this result cannot be improved. But in the special case of K-Gödel numberings, EXT_ψ takes the Turing degree of K' as shown in the next result.

Theorem 15. $\mathrm{EXT}_\psi \equiv_T K'$ *for every K-Gödel numbering ψ.*

The next result is not that difficult and proves that there is one index set whose Turing degree is independent of the underlying numbering: the index set of the consistent functions.

Proposition 16. $\mathrm{CONS}_\psi \equiv_T K$ *for all universal numberings ψ.*

Remark 17. Another example of this type is the set DISJ_ψ. Here $\mathrm{DISJ}_\psi \equiv_T K$ for every domain-universal numbering ψ. The sufficiency is easy as one can test with one query to the halting problem whether W_i^ψ and W_j^ψ intersect. The necessity is done by showing that the complement of K is r.e. relative to DISJ_ψ: Let i be an index with $W_i^\psi = K$. Then $x \notin K$ iff there is an j with $x \in W_j^\psi \wedge \langle i, j \rangle \in \mathrm{DISJ}_\psi$. Hence the complement of K is recursively enumerable relative to DISJ_ψ and so $K \leq_T \mathrm{DISJ}_\psi$.

Tennenbaum defined that A *is Q-reducible to* B [11, Section III.4] iff there is a recursive function f with $x \in A \Leftrightarrow W_{f(x)} \subseteq B$ for all x. Again let i be an index of K: $K = W_i^\psi$. Furthermore, define f such that $W_{f(x)} = \{\langle i, j \rangle : x \in W_j^\psi\}$. Now K is Q-reducible to the complement of DISJ_ψ as $x \in K$ iff $W_{f(x)}$ is contained in the complement of DISJ_ψ.

For wtt-reducibility and other reducibilities stronger than wtt, no such result is possible. Indeed, one can choose ψ such that $\{e : \psi_e$ is total$\}$ is hypersimple and $W_e^\psi = \emptyset$ iff $e = 0$. Then $\{\langle i, j \rangle : i > 0 \wedge j > 0 \wedge \langle i, j \rangle \in \mathrm{DISJ}_\psi\}$ is hyperimmune and wtt-equivalent to DISJ_ψ. Thus no set wtt-reducible to DISJ_ψ has diagonally nonrecursive wtt-degree [5]. In particular, $K \not\leq_{wtt} \mathrm{DISJ}_\psi$ for this numbering ψ.

Recall that a set A is 1-generic iff for every r.e. set B of strings there is an n such that either $A(0)A(1)A(2)\ldots A(n) \in B$ or $A(0)A(1)A(2)\ldots A(n) \cdot \{0,1\}^*$ is disjoint from B. Jockusch [6] gives an overview on 1-generic sets. Note that the Turing degree of a 1-generic set G is generalized low$_1$ which means that $G' \equiv_T G \oplus K$. Hence $G \not\geq_T K$ and this fact will be used at various places below.

Theorem 18. *There is a K-Gödel numbering ψ such that*

- *DMIN$_\psi$ and DMIN$_\psi^*$ are 1-generic;*
- *DEQ$_\psi$ and INC$_\psi$ have the Turing degree K';*
- *DEQ$_\psi^*$ and INC$_\psi^*$ have the Turing degree K''.*

Proof. The basic idea is the following: One partitions the natural numbers in the limit into two types of intervals: coding intervals $\{e_m\}$ and genericity intervals J_m. The coding intervals contain exactly one element while the genericity intervals are very large. They satisfy the following requirements:

- $|J_m| \geq c_K(m)$ where c_K is the convergence module of K, that is, where $c_K(m) = \min\{s \geq m : \forall n \leq m \,[n \in K \Rightarrow n \in K_s]\}$. In the construction, an approximation c_{K_s} of c_K from below is used.
- There is a limit-recursive function $m \mapsto \sigma_m$ such that $\sigma_m \in \{0,1\}^{|J_m|}$ and for every $\tau \in \{0,1\}^{\min(J_m)}$ and for every genericity requirement set R_n with $n \leq m$ the following implication holds: if $\tau\sigma_m$ has an extension in R_n then already $\tau\sigma_m \in R_n$. Here $R_n = \{\rho \in \{0,1\}^* : $ some prefix of ρ is enumerated into W_n^φ within $|\rho|$ steps$\}$. Note that the R_n are uniformly recursive and φ is the default Gödel numbering.
- There are infinitely many genericity intervals J_m such that for all $x \in J_m$ it holds that $\sigma_m(x - \min(J_m)) = $ DMIN$_\psi(x) = $ DMIN$_\psi^*(x)$.

All strings $\sigma_{k,0}$ are just 0 and in stage $s+1$ the following is done:

- Inductively over k define $e_{0,s} = 0$ and $e_{k+1,s} = e_{k,s} + |\sigma_{k,s}| + 1$ and $J_{k,s} = \{x : e_{k,s} < x < e_{k+1,s}\}$.
- Determine the minimal m such that one of the following three cases hold:
  (ρ_m) $m < s$ and $\exists \rho_m \in \{0,1\}^s \,\exists \tau \in \{0,1\}^{\min(J_{m,s})} \,\exists n \leq m \,[\tau\sigma_{m,s}\rho_m \in R_n \wedge \tau\sigma_{m,s} \notin R_n]$;
  (c_K) $m < s$ and $|J_{m,s}| < c_{K_s}(m) \leq s$;
  (none) $m = s$.
  Note that one of the three cases is always satisfied and thus the search terminates.
- In the case (ρ_m), update the approximations to σ_m as follows:

$$\sigma_{k,s+1} = \begin{cases} \sigma_{k,s}\rho_m & \text{if } k = m; \\ \sigma_{k,s} & \text{if } k \neq m. \end{cases}$$

- In the case (c_K) the major goal is to make the interval $J_{m,s}$ having a sufficient long length. Thus

$$\sigma_{k,s+1} = \begin{cases} \sigma_{k,s}0^s & \text{if } k = m; \\ \sigma_{k,s} & \text{if } k \neq m. \end{cases}$$

– In the case (none), no change is made, that is, $\sigma_{k,s+1} = \sigma_{k,s}$ for all k.

Let e_m, J_m, σ_m be the limit of all $e_{m,s}, J_{m,s}, \sigma_{m,s}$. One can show by induction that all these limits exists. The set $\{d : \exists m \, [d \in J_m]\}$ is recursively enumerable as whenever $e_{m,s+1} \neq e_{m,s}$ then $e_{m,s+1} \geq s$; hence $\exists m \, [d \in J_m]$ iff $\exists s > d+1 \, \exists m \, [d \in J_{m,s}]$. Now one constructs the numbering ψ from a given universal numberings φ by taking for any d, x the first case which is found to apply:

– if there are $s > x + d$ and $m \leq d$ with $d = e_{m,s}$ and $\varphi_{m,s}(x)$ defined then let $\psi_d(x) = \varphi_m(x)$;
– if there are $s > x + d$ and $m \leq d$ with $d \in J_{m,s}$ and $(\sigma_{m,s}(d - \min(J_m)) = 0)$ $\vee \, \forall y \, [x \neq \langle d, y \rangle]$ then let $\psi_d(x) = 0$;
– if none of these two cases ever applies then $\psi_d(x)$ remains undefined.

Without loss of generality it is assumed that φ_0 is total and thus 0 is the least index e with $W_e^\psi = \mathbb{N}$. It is easy to see that the following three constraints are satisfied.

– If $d = e_m$ then $\psi_d = \varphi_m$;
– If $d \in J_m$ and $\sigma_m(d - \min(J_m)) = 1$ then $W_d^\psi =^* \{\langle x, y \rangle : x \neq d\}$;
– If $d \in J_m$ and $\sigma_m(d - \min(J_m)) = 0$ then $W_d^\psi = \mathbb{N}$.

Note that the first condition is co-r.e.: Hence one can either compute from d an m with $e_m = d$ or find out that d is in $\bigcup_{m' \in \mathbb{N}} J_{m'}$. But it might be that one first comes up with a candidate m for $e_m = d$ and later finds out that actually $e_m \in \bigcup_{m' \in \mathbb{N}} J_{m'}$. So the algorithm is first to determine an m and to follow φ_m where m is correct whenever really the first case applies; later, in the case that the second or third case applies, one already fixed finitely many values of ψ_d which does not matter as the second and third case tolerate finitely many fixed values. Indeed, one knows only in the limit whether the second or the third case will apply. Therefore only $=^*$ is postulated in the second case and ψ_d is made total in the third case by making $\psi_d(x)$ defined whenever there is an $s > x + d$ such that $d \in J_{m,s}$ and $\sigma_{m,s}(d - \min(J_{m,s})) = 0$.

For each n there is at most one interval J_m and at most one $d \in J_m$ such that $d > e_n$ and $W_d^\psi =^* \{\langle x, y \rangle : x \neq d\} =^* W_{e_n}^\psi$; if d exists then let $F(n) = d$ else let $F(n) = 0$. Now for every J_m and every $d \in J_m$, $d \in \text{DMIN}_\psi^*$ iff $\sigma_m(d - \min(J_m)) = 1$ and $d \neq F(n)$ for all $n \leq m$. As there are infinitely many indices of total functions, $F(m) = 0$ infinitely often and there are infinitely many genericity intervals J_m which do not intersect the range of F. For each such interval J_m and every d not in the range of F, the construction of σ_m and ψ implies the following: if $\sigma_m(d - \min(J_m)) = 1$ then $d \in \text{DMIN}_\psi \cap \text{DMIN}_\psi^* \wedge W_d^\psi \neq^* \mathbb{N}$ else $d \notin \text{DMIN}_\psi \cup \text{DMIN}_\psi^* \wedge W_d^\psi = \mathbb{N}$. Thus if τ is the characteristic function of DMIN_ψ or DMIN_ψ^* restricted to the domain $\{0, 1, 2, \ldots, e_m\}$ and $n \leq m$ then $\tau \sigma_m \in R_n$ whenever some extension of $\tau \sigma_m$ is in R_n. Hence the sets DMIN_ψ and DMIN_ψ^* are both 1-generic.

Furthermore, let $\{a_0, a_1, a_2, \ldots\}$ be either $\{d : W_d^\psi = \{0\}\}$ or $\{d : W_d^\psi =^* \{0\}\}$. It is easy to see that $\{a_0, a_1, a_2, \ldots\} \subseteq \{e_0, e_1, e_2, \ldots\}$ and $a_n \geq e_n$. By construction $a_{n+1} \geq e_{n+1} \geq c_K(n)$ for all n and it follows that $K \leq_T \mathrm{DEQ}_\psi$, $K \leq_T \mathrm{DEQ}_\psi^*$, $K \leq_T \mathrm{INC}_\psi$ and $K \leq_T \mathrm{INC}_\psi^*$. Having the oracle K and knowing that ψ is a K-Gödel numbering, one can now use the same methods as in Gödel numberings to prove that the sets DEQ_ψ and INC_ψ (respectively, DEQ_ψ^* and INC_ψ^*), are complete for K' (respectively, complete for K''). $\square$

Proposition 19. *For any universal numbering ψ, the set MIN_ψ is never 1-generic and never hyperimmune.*

The next result shows that not only DMIN_ψ but also DEQ_ψ can have a Turing degree properly below that of K'.

Theorem 20. *There is a universal numbering ψ such that DEQ_ψ has 1-generic Turing degree.*

4   Open Problems

In the following several major open questions of the field are identified.

Open Problem 21. *Is there a universal numbering ψ such that DMIN_ψ has minimal Turing degree?*

This is certainly possible for MIN_ψ as one can code every Turing degree below K into MIN_ψ for a suitable ψ. Recall that $\mathrm{INC}_\psi = \{\langle i, j \rangle : W_i^\psi \subseteq W_j^\psi\}$ and $\mathrm{DEQ}_\psi = \{\langle i, j \rangle : W_i^\psi = W_j^\psi\}$. Obviously

$$\mathrm{DMIN}_\psi \leq_T \mathrm{DEQ}_\psi \leq_T \mathrm{INC}_\psi \leq_T K'.$$

By Theorem 18 there is a universal numbering ψ such that $\mathrm{DMIN}_\psi <_T \mathrm{DEQ}_\psi \equiv_T K'$ and Friedberg showed that there is a domain-universal numbering ϑ for which DEQ_ϑ is recursive. Theorem 20 showed that one can make DEQ_ψ to have 1-generic Turing degree as well for some universal numbering. Hence the first two Turing reductions can be made proper while the following remains unknown.

Open Problem 22. *Is there a universal numbering ψ with $\mathrm{INC}_\psi <_T K'$?*

Note that for universal numberings, this question is equivalent to asking whether $\mathrm{INC}_\psi \not\geq_T K$. The reason is that $\mathrm{INC}_\psi \oplus K \equiv_T K'$ holds for universal numberings by $\mathrm{DMIN}_\psi \leq_T \mathrm{INC}_\psi \leq_T K'$ and Proposition 4. For domain-universal numberings, one can even ask the stronger question whether there is a domain-universal numbering ϑ with $\mathrm{INC}_\vartheta <_T K$. Kummer [8] already showed that $\mathrm{INC}_\vartheta \leq_T K$ can be obtained for some domain-universal numbering ϑ, see Theorem 13 above.

In Theorem 8 above it was shown that for numberings ψ satisfying the Kolmogorov property, $\mathrm{DMIN}_\psi^* \equiv_T K''$. On the other hand, by Theorem 18 there is a universal numbering ψ with DMIN_ψ^* being 1-generic. Although these results give already much knowledge about DMIN_ψ^*, the original problem of Schaefer [15] is still not completely solved.

Open Problem 23. *Is $\mathrm{DMIN}_\psi^* \equiv_T K''$ for all Gödel numberings ψ?*

Acknowledgments. The authors would like to thank Lance Fortnow, Martin Kummer, Wei Wang and Guohua Wu for discussions on the paper.

References

1. Case, J., Jain, S., Suraj, M.: Control Structures in Hypothesis Spaces: The Influence on Learning. Theoretical Computer Science 270, 287–308 (2002)
2. Fortnow, L., Gasarch, W., Jain, S., Kinber, E., Kummer, M., Kurtz, S.A., Pleszkoch, M., Slaman, T.A., Solovay, R., Stephan, F.: Extremes in the degrees of inferability. Annals of Pure and Applied Logic 66, 231–276 (1994)
3. Friedberg, R.: Three theorems on recursive enumeration. The Journal of Symbolic Logic 23(3), 309–316 (1958)
4. Harizanov, V., Kummer, M., Owings Jr., J.C.: Frequency computation and the cardinality theorem. Journal of Symbolic Logic 57, 682–687 (1992)
5. Kjos-Hanssen, B., Merkle, W., Stephan, F.: Kolmogorov Complexity and the Recursion Theorem. In: Durand, B., Thomas, W. (eds.) STACS 2006. LNCS, vol. 3884, pp. 149–161. Springer, Heidelberg (2006)
6. Jockusch, C.G.: Degrees of generic sets. London Mathematical Society Lecture Notes 45, 110–139 (1981)
7. Kummer, M.: A proof of Beigel's cardinality conjecture. The Journal of Symbolic Logic 57, 677–681 (1992)
8. Kummer, M.: A note on the complexity of inclusion. Notre Dame Journal of Formal Logic (to appear)
9. Li, M., Vitányi, P.: An Introduction to Kolmogorov Complexity and Its Applications, 2nd edn. Springer, Heidelberg (1997)
10. Meyer, A.R.: Program size in restricted programming languages. Information and Control 21, 382–394 (1972)
11. Odifreddi, P.: Classical Recursion Theory. Studies in Logic and the Foundations of Mathematics, vol. 125. North-Holland, Amsterdam (1989)
12. Odifreddi, P.: Classical Recursion Theory II. Studies in Logic and the Foundations of Mathematics, vol. 143. Elsevier, Amsterdam (1999)
13. Owings Jr., J.C.: A cardinality version of Beigel's Nonspeedup Theorem. The Journal of Symbolic Logic 54, 761–767 (1989)
14. Gordon Rice, H.: Classes of enumerable sets and their decision problems. Transactions of the American Mathematical Society 74, 358–366 (1953)
15. Schaefer, M.: A guided tour of minimal indices and shortest descriptions. Archives for Mathematical Logic 37, 521–548 (1998)
16. Soare, R.I.: Recursively enumerable sets and degrees. In: Perspectives in Mathematical Logic. Springer, Berlin (1987)
17. Teutsch, J.: Noncomputable Spectral Sets, PhD Thesis, Indiana University (2007)
18. Teutsch, J.: On the Turing degrees of minimal index sets. Annals of Pure and Applied Logic 148, 63–80 (2007)

Ordinal Computability[*]

Peter Koepke

Mathematisches Institut, Universität Bonn
Endenicher Allee 60, D 53115 Bonn, Germany
koepke@math.uni-bonn.de

Abstract. Ordinal computability uses ordinals instead of natural numbers in abstract machines like register or TURING machines. We give an overview of the computational strengths of α-β-machines, where α and β bound the time axis and the space axis of some machine model. The spectrum ranges from classical TURING computability to ∞-∞-computability which corresponds to GÖDEL's model of constructible sets. To illustrate some typical techniques we prove a new result on Infinite Time Register Machines (= ∞-ω-register machines) which were introduced in [6]: a real number $x \in {}^{\omega}2$ is computable by an Infinite Time Register Machine iff it is TURING computable from some finitely iterated hyperjump $0^{(n)}$.

Keywords: Ordinal machines, Infinite Time Register Machines, Constructible sets.

1 Ordinal Computations

Standard computability theory is fundamentally based on the set $\mathbb{N} = \{0, 1, \ldots\}$ of natural numbers: steps of a computation are clocked by natural numbers ("time"), cells of a TURING tape are indexed by natural numbers ("space"), and read and write registers contain natural numbers ("space"). The canonical wellorder $0 < 1 < 2 < \ldots$ of natural numbers allows recursive algorithms, the arithmetic properties of the structure $(\mathbb{N}, +, \cdot, 0, 1)$ allow to code various information. A standard TURING computation can be represented by a space-time function or diagram with natural numbers axes.

↑	$n+1$	0	0	0	0	0	⋯	0	0	⋯
	n	0	0	0	0	0	⋯	0	1	⋯
S	⋮	⋮	⋮	⋮	⋮	⋮	⋰	⋮		
P	4	1	1	1	1	1	⋯	1	1	
A	3	1	1	1	1	1	⋯	1	1	
C	2	0	0	0	1	1	⋯	1	1	
E	1	0	0	1	1	0	⋯	0	0	
	0	1	0	0	0	0	⋯	0	0	
		0	1	2	3	4	⋯	m	$m+1$	⋯ ⋯
		T	I	M	E				→	

A standard TURING computation. Head positions are indicated by shading.

---

[*] The author wants to thank JOEL HAMKINS and PHILIP WELCH for a very inspiring discussion at the EMU 2008 workshop at New York in which the techniques and results bounding the strength of Infinite Time Register Machines were suggested and conjectured.

K. Ambos-Spies, B. Löwe, and W. Merkle (Eds.): CiE 2009, LNCS 5635, pp. 280–289, 2009.
© Springer-Verlag Berlin Heidelberg 2009

The ordinal numbers were introduced by GEORG CANTOR to extend the natural numbers $\mathbb{N}$ into the transfinite. There are strong analogies between the class-sized structure $(\text{Ord}, <, +, \cdot, 0, 1)$ of all ordinals and $(\mathbb{N}, <, +, \cdot, 0, 1)$:

- $(\text{Ord}, <)$ is a wellorder, i.e., it is a strict linear order and every nonempty subset has a $<$-minimal element;
- ordinal addition and multiplication satisfy recursive laws familiar from natural number arithmetic.

An obvious generalisation from the perspective of transfinite ordinal theory is to replace the natural numbers in computations by ordinal numbers. For the TURING machine model this means working on a TURING tape indexed by *ordinals* along a time indexed by *ordinals*. At successor ordinals the *ordinal machine* behaves much like a standard finite time and finite space machine. The behaviour at limit ordinals is governed by specific *limit rules*. Limit rules are derived from the usual limit operations in the ordinals: if $(\alpha_s)_{s<\lambda}$ is a sequence of ordinals whose length λ is a limit ordinal, define its *limit* and *limit inferior* by

$$\lim_{s \to \lambda} \alpha_s = \bigcup_{s < \lambda} \alpha_s$$

and

$$\liminf_{s \to \lambda} \alpha_s = \bigcup_{s < \lambda} \bigcap_{s < r < \lambda} \alpha_r.$$

An *ordinal computation* can then be visualised like

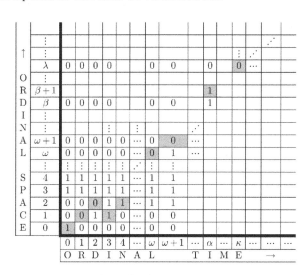

An ordinal computation

The limit rules will be applied independently of the machine state. Thus an ordinal algorithm will be given by a *standard* program for a TURING, register or similar machine.

More abstractly, ordinal computations have the following "shapes":

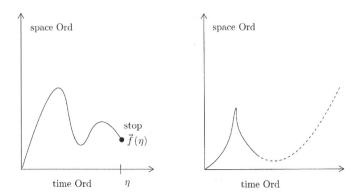

The left diagram represents a *halting* computation which computes some ordinal values $\vec{f}(\eta)$ at the halting time η. The right computation does not halt in the available time and thus *diverges*.

These diagrams suggest to consider restrictions of the available time and space by some ordinals α and β, resp. Note that in the common VON NEUMANN formalisation of ordinals, an ordinal is the set of its predecessors:

$$\alpha = \{\xi | \xi \in \alpha\} = \{\xi | \xi < \alpha\}.$$

An α-β-machine has computation diagrams of the forms:

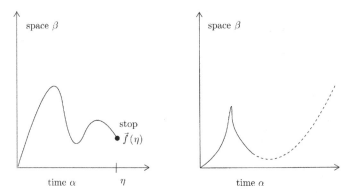

This gives a spectrum of α-β-machines for $\omega \leqslant \alpha \leqslant \infty$ and $\omega \leqslant \beta \leqslant \infty$ where the ω-ω-machines should be equal to their standard counterparts. Here we let $\infty = \mathrm{Ord}$ be the class of all ordinals.

2 Register Machines

For a concrete example we give a formal definition of register machines working on ordinals. We base our presentation of infinite time machines on the *unlimited register machines* as presented in [1]. Ordinal programs are *standard* programs.

Definition 1. *Fix limit ordinals α and β, $\omega \leqslant \alpha \leqslant \infty$, $\omega \leqslant \beta \leqslant \infty$. Let $P = I_0, I_1, \ldots, I_{s-1}$ be an URM program, i.e., a sequence of instructions as described in (i)-(v) below. Let $Z : \beta \to 2$, which will serve as an oracle. A pair*

$$I : \theta \to \omega, R : \theta \to (^\omega \beta)$$

is an α-β-(register) computation by P if the following hold:

a) *θ is an ordinal $\leqslant \alpha$; θ is the* length *of the computation;*
b) *$I(0) = 0$; the machine starts in state 0;*
c) *If $t < \theta$ and $I(t) \notin s = \{0, 1, \ldots, s-1\}$ then $\theta = t + 1$; the machine stops if the machine state is not a program state of P;*
d) *If $t < \theta$ and $I(t) \in \text{state}(P)$ then $t + 1 < \theta$; the next configuration is determined by the instruction $I_{I(t)}$:*

  i. *if $I_{I(t)}$ is the instruction $\mathtt{reset(Rn)}$ then let $I(t+1) = I(t)+1$ and define $R(t+1) : \omega \to \text{Ord}$ by*

  $$R_k(t+1) = \begin{cases} 0, \text{ if } k = n \\ R_k(t), \text{ if } k \neq n \end{cases}$$

  ii. *if $I_{I(t)}$ is the instruction $\mathtt{increase(Rn)}$ then let $I(t+1) = I(t)+1$ and define $R(t+1) : \omega \to \text{Ord}$ by*

  $$R_k(t+1) = \begin{cases} R_k(t) + 1, \text{ if } k = n \\ R_k(t), \text{ if } k \neq n \end{cases}$$

  iii. *if $I_{I(t)}$ is the oracle instruction $\mathtt{oracle(Rn)}$ then let $I(t+1) = I(t)+1$ and define $R(t+1) : \omega \to \text{Ord}$ by*

  $$R_k(t+1) = \begin{cases} Z(R_n(t)), \text{ if } k = n \\ R_k(t), \text{ if } k \neq n \end{cases}$$

  iv. *if $I_{I(t)}$ is the transfer instruction $\mathtt{Rn{=}Rm}$ then let $I(t+1) = I(t)+1$ and define $R(t+1) : \omega \to \text{Ord}$ by*

  $$R_k(t+1) = \begin{cases} R_m(t), \text{ if } k = n \\ R_k(t), \text{ if } k \neq n \end{cases}$$

  v. *if $I_{I(t)}$ is the jump instruction $\mathtt{if\ Rm{=}Rn\ then\ goto\ q}$ then let $R(t+1) = R(t)$ and*

  $$I(t+1) = \begin{cases} q, \text{ if } R_m(t) = R_n(t) \\ I(t) + 1, \text{ if } R_m(t) \neq R_n(t) \end{cases}$$

e) *If $t < \theta$ is a limit ordinal, then*

$$\forall k \in \omega (R_k(t) = \begin{cases} \liminf_{r \to t} R_k(r) \text{ if } \liminf_{r \to t} R_k(r) < \beta \\ 0 \text{ if } \liminf_{r \to t} R_k(r) = \beta \end{cases}$$
$$I(t) = \liminf_{r \to t} I(r).$$

By the second clause in the definition of $R_k(t)$ the register is reset *in case $\liminf_{r \to t} R_k(r) = \beta$.*

The register computation is obviously determined recursively by the initial register contents $R(0)$, the oracle Z and the program P. We call it the α-β-register computation by P with input $R(0)$ and oracle Z. If the computation stops then $\theta = \eta + 1$ is a successor ordinal and $R(\eta)$ is the final register content. In this case we say that P α-β-register computes $R(\eta)(0)$ from $R(0)$ and the oracle Z, and we write

$$P : R(0), Z \mapsto_{\alpha,\beta} R(\eta)(0).$$

The interpretation of programs yields associated notions of computability.

Definition 2. *An n-ary function $F : \omega^n \to \omega$ is α-β-register computable in the oracle Z if there is a register program P such that for every n-tuple $(a_0, \ldots, a_{n-1}) \in {}^\omega\omega$ holds:*

$$P : (a_0, \ldots, a_{n-1}, 0, 0, \ldots), Z \mapsto_{\alpha,\beta} F(a_0, \ldots, a_{n-1}).$$

We then denote F by P^Z. F is α-β-register computable if F is α-β-computable in the empty oracle $\emptyset$.

Obviously any standard recursive function is α-β-computable.

Definition 3. *A subset $A \subseteq \mathcal{P}(\beta)$ is α-β-register computable if there is a register program P, and an oracle $Y : \beta \to 2$ such that for all $Z \subseteq \beta$:*

$$P^{Y \times Z}(0) = \chi_A(Z)$$

where $Y \times Z$ is some appropriate pairing of Y and Z and χ_A is the characteristic function of A.

In sections 4 and 5 we shall determine the computational strength of ∞-ω-register machines. Earlier the author studied a weaker *non-resetting* type of infinitary register machine, which was there called "Infinite Time Register Machine" [5]. Those machines exactly corresponded to hyperarithmetic definitions. Since the ∞-β-machines introduced above are in closer analogy with the established Infinite Time TURING Machines (ITTM) of [2] we use the term *Infinite Time Register Machine* for the new concept. The machines of [5] could be called *Non-Resetting Infinite Time Register Machines*.

Definition 4. *Call an ∞-ω-register machine an Infinite Time Register Machine (ITRM). Correspondingly we use the terms "ITRM-computation" and "ITRM-computable" for "∞-ω-register computation" and "∞-ω-register computable".*

3 Strengths of Ordinal Machines

An obvious task is to determine the class of α-β-computable functions and sets for varying α and β. For register machines as defined above one obtains the subsequent table. Note that one should naturally have at least as much time available as there is space ($\alpha \geqslant \beta$) to be able to reach all possible register contents. There are still open positions ("?") and ongoing research. References behind the strength statements are to the bibliography.

Ordinal register computability

Register machines	space ω	space admissible α	space Ord
time ω	standard register machine computable $= \Delta_1^0$	-	-
time admissible α	?	α register machine (α recursion theory) computable $= \Delta_1(L_\alpha)$ [7]	-
time Ord	ITRM Infinite time register machine computable $= L_{\omega_\omega^{\mathrm{CK}}} \cap \mathcal{P}(\omega)$ [see sections 4 and 5]	?	Ordinal register machine computable $= L \cap \mathcal{P}(\mathrm{Ord})$ [8]

For TURING machines we get a nearly identical table with a marked difference between ITRMs and ITTMs.

Ordinal Turing computability

TURING	space ω	space admissible α	space Ord
time ω	standard TURING machine computable $= \Delta_1^0$	-	-
time admissible α	?	α TURING machine (α-recursion theory) computable $= \Delta_1(L_\alpha)$ [7]	-
time Ord	ITTM $\Delta_1^1 \subsetneq$ computable in real parameter $\subsetneq \Delta_2^1$ [2]	?	Ordinal TURING machine computable $= L \cap \mathcal{P}(\mathrm{Ord})$ [4]

4 Halting Times of ITRMs

Determinations of α-β-computability strengths typically involve the combination of techniques from recursion theory, admissibility theory, descriptive set theory, and set theoretic constructibility theory. As an example and a new result we prove the estimate for ITRMs in detail. One can view the proofs as an analysis of the computable power of "resetting registers" at infinite liminf's: a resetting register basically corresponds to one hyperjump.

Definition 5. *A limit ordinal δ is called* admissible *if there is no total $\Sigma_1(L_\delta)$-definable function from some $\beta < \delta$ cofinally into δ.*

From the perspective of this paper it is interesting that one can characterise admissibility in ordinal computability without refering to the constructible L_α-hierarchy:

Lemma 1. *A limit ordinal δ is admissible iff there is no δ-δ-register computable total function mapping some $\beta < \delta$ cofinally into δ (see [3]).*

Let $\omega_0^{\mathrm{CK}}, \omega_1^{\mathrm{CK}}, \ldots, \omega_\omega^{\mathrm{CK}}, \ldots$ be the monotone enumeration of the admissible ordinals and their limits. We shall prove:

Theorem 1. *A real number $a \in {}^\omega 2$ is computable by an ITRM iff $a \in L_{\omega_\omega^{\mathrm{CK}}}$.*

The implication from left to right, i.e., the upper bound for the set of ITRM-computable reals, follows from bounding the halting times, i.e., the lengths of halting computations, of ITRMs below $\omega_\omega^{\mathrm{CK}}$.

Theorem 2. *Consider an ITRM with register program P. Then there is some $n < \omega$ such that for arbitrary inputs $i < \omega$: if the ITRM-computation according to P with input i and empty oracle halts then it halts before ω_n^{CK}.*

For the proof fix an infinite time register computation $\mathcal{C}$

$$I : \theta \to \omega, R : \theta \to^\omega \omega$$

by P with some input $(R_0(0), R_1(0), \ldots)$ and oracle $\emptyset$. Assume that P only mentions registers among $R_0, \ldots, R_{l-1}$.

Lemma 2. *Let $\tau < \theta$ be of the form $\tau = \tau_0 + \delta$ where δ is admissible $> \omega$. Assume that*

$$\forall k < l \ R_k(\tau) = \liminf_{t < \tau} R_k(t).$$

Then $\theta = \infty$.

Proof. Since computations can be composed by concatenation, we may assume that $\tau_0 = 0$ and $\tau = \delta$. Let $I(\delta) = n$ and

$$R(\delta) = (n_0, \ldots, n_{l-1}, 0, 0, \ldots)$$

where $R_0, \ldots, R_{l-1}$ includes all the registers mentioned in the program P. Since the constellation at δ is determined by liminf's there is some $\gamma < \delta$ such that

- $\forall t \in (\gamma, \delta) I(t) \geqslant n$ and $\{t \in (\gamma, \delta) | I(t) = n\}$ is closed unbounded in δ;
- for all $k < l$: $\forall t \in (\gamma, \delta) \ R_k(t) \geqslant n_k$ and $\{t \in (\gamma, \delta) | R_k(t) = n_k\}$ is closed unbounded in δ.

These closed unbounded sets are Δ_1-definable over the set L_δ. By the admissibility of L_δ their intersection is closed unbounded in δ of ordertype δ. In particular one can choose $\bar{\delta} \in (\gamma, \delta)$ such that $(I(\bar{\delta}), R_0(\bar{\delta}), \ldots, R_{l-1}(\bar{\delta})) = (I(\delta), R_0(\delta), \ldots, R_{l-1}(\delta))$ and

$$\forall t \in [\bar{\delta}, \delta](I(\bar{\delta}) \leqslant I(t) \wedge R_0(\bar{\delta}) \leqslant R_0(t) \wedge \ldots \wedge R_{l-1}(\bar{\delta}) \leqslant R_{l-1}(t)).$$

Then one can easily show by induction, using the liminf rules: If $\sigma \geqslant \bar{\delta}$ is of the form $\sigma = \bar{\delta} + \delta \cdot \xi + \eta$ with $\eta < \delta$ then

$$(I(\sigma), R_0(\sigma), \ldots, R_{l-1}(\sigma)) = (I(\bar{\delta} + \eta), R_0(\bar{\delta} + \eta), \ldots, R_{l-1}(\bar{\delta} + \eta)).$$

In particular the computation does not stop.

Lemma 3. *Let* $n < \omega$. *Let* $\tau < \theta$ *be of the form* $\tau = \bar{\tau} + \omega_{n+1}^{\mathrm{CK}}$ *and*

$$\mathrm{card}\{k < l | R_k(\tau) = 0\} \leqslant n.$$

Then $\theta = \infty$.

Proof. Set $N = \{k < l | R_k(\tau) = 0\}$. We prove the Lemma by induction on n. If $n = 0$ then $\forall k < l \; R_k(\tau) = \liminf_{t < \tau} R_k(t)$. By Lemma 7, $\theta = \infty$.

Now consider $n = m + 1$ where the claim holds for m. Assume that $\theta < \infty$. By Lemma 5, there must be some $k_0 < l$ such that $R_{k_0}(\tau) \neq \liminf_{t < \tau} R_{k_0}(t)$. By the limit rule, $R_{k_0}(\tau) = 0$ and $\liminf_{t < \tau} R_{k_0}(t) = \omega$. Take $\delta < \tau$ such that

$$\forall \eta \in (\delta, \tau) \; \forall i \in (l \setminus N) \cup \{k_0\} : R_i(\eta) \neq 0.$$

Take $\tau_0 \in (\delta, \tau)$ of the form $\tau_0 = \overline{\tau_0} + \omega_n^{\mathrm{CK}} = \overline{\tau_0} + \omega_{m+1}^{\mathrm{CK}}$. Then

$$\mathrm{card}\{k < l | R_k(\tau_0) = 0\} \leqslant n - 1 = m.$$

But then by the induction hypothesis, $\theta = \infty$, contradiction.

We get the following corollaries:

Theorem 3. *Assume that* $\theta \geqslant \omega_{l+1}^{\mathrm{CK}} + 1$, *where* l *is the number of registers mentioned in the program* P. *Then* $\theta = \infty$.

Theorem 4. *If a real number* $a \in {}^{\omega}2$ *is computable by an ITRM then* $a \in L_{\omega_{\omega}^{\mathrm{CK}}}$.

Proof. The computations of $a(n)$ for $n < \omega$ have lengths $< \omega_{l+1}^{\mathrm{CK}} + 1$ for some fixed $l < \omega$. They can thus be carried out absolutely *inside* the structure $(L_{\omega_{\omega}^{\mathrm{CK}}}, \in)$. Hence $a \in L_{\omega_{\omega}^{\mathrm{CK}}}$.

5   Hyperjumps

The lower bound for the strengths of ITRM computability uses the characterisation of the reals in $L_{\omega_{\omega}^{\mathrm{CK}}}$ by the finite *hyperjumps* $0, 0^+, 0^{++}, \ldots, 0^{(k+1)} = (0^{(k)})^+, \ldots$ of the constant function 0.

Assume a fixed recursive enumeration $P_0, P_1, \ldots$ of all register programs and let $P_n^Z : \omega \rightharpoonup \omega$ be the partial function given by P with oracle Z. The hyperjump $Z^+ \in {}^{\omega}2$ of $Z \in {}^{\omega}2$ is defined by:

$$Z^+(n) = 1 \text{ iff } \{(i, j) \in \omega \times \omega | P_n^Z(2^i \cdot 3^j) = 1\} \text{ is a wellfounded relation.}$$

The hyperjump can be computed by the following result from [6].

Theorem 5. *The set* $\mathrm{WO} = \{Z \in {}^{\omega}2 | Z \text{ codes a wellorder}\}$ *is computable by an ITRM.*

Proof. The following program P on an ITRM outputs yes/no depending on whether the oracle Z codes a wellfounded relation. The program is a backtracking algorithm which searches for a "leftmost" infinite descending chain in Z. A stack is used to organise the backtracking. We code a stack $(r_0, \ldots, r_{m-1})$ of natural numbers by $r = 2^{r_0} \cdot 3^{r_1} \cdots p_{m-1}^{r_{m-1}+1}$. We present the program in pseudo-code and assume that it is translated into a register program according to Definition 1 so that the order of commands is kept. Also the stack commands like push are understood as *macros* which are inserted into the code with appropriate renaming of variables and statement numbers.

```
      push 1; %% marker to make stack non-empty
      push 0; %% try 0 as first element of descending sequence
      FLAG=1; %% flag that fresh element is put on stack
Loop: Case1: if FLAG=0 and stack=0 %% inf descending seq found
             then begin; output 'no'; stop; end;
      Case2: if FLAG=0 and stack=1 %% inf descending seq not found
             then begin; output 'yes'; stop; end;
      Case3: if FLAG=0 and length-stack > 1
      %% top element cannot be continued infinitely descendingly
             then begin; %% try next
             pop N;
             push N+1;
             FLAG:=1; %% flag that fresh element is put on stack
             goto Loop;
             end;
      Case4: if FLAG=1 and stack-is-decreasing
             then begin;
             push 0; %% try to continue sequence with 0
             FLAG:=0; FLAG:=1; %% flash the flag
             goto Loop;
             end;
      Case5: if FLAG=1 and not stack-is-decreasing
             then begin;
             pop N;
             push N+1; %% try next
             FLAG:=0; FLAG:=1; %% flash the flag
             goto Loop;
             end;
```

The correctness of this program is proved in [6].

From [9], Corollary VII.1.10 and the Notes for VII.5 we obtain:

Proposition 1. *A real* $a \in {}^{\omega}2$ *is an element of* $L_{\omega_{\omega}^{\mathrm{CK}}}$ *iff* a *is standard* TURING *computable from some* $0^{(n)}$ *with* $n < \omega$.

By Theorem 5 the collection of ITRM-computable reals is closed with respect to the hyperjump. It is also closed with respect to standard Turing computations. Proposition 1 then implies the converse direction of Theorem 5.

Theorem 6. *Every real number $a \in {}^\omega 2 \cap L_{\omega_\omega^{CK}}$ is computable by an ITRM.*

6 Conclusions and Further Considerations

Ordinal computability is able to characterise important classes of sets from higher recursion theory, descriptive set theory, and constructibility theory. It remains to be seen whether ordinal computability will have some further applications besides its unifying role. Several cases of the strengths of α-β-computability still need to be determined and are the subject of current research.

Concerning the specific result on ITRMs we plan to refine the above analysis to a level-by-level correspondence between numbers of registers, numbers of the admissible ordinals used to bound halting computations, and numbers of iterations of the hyperjump. One register should correspond to one admissible, and to one application of the hyperjump. Further registers will be required for auxiliary classical computations and bookkeeping.

References

[1] Cutland, N.J.: Computability: An Introduction to Recursive Function Theory. In: Perspectives in Mathematical Logic. Cambridge University Press, Cambridge (1980)

[2] Hamkins, J.D., Lewis, A.: Infinite Time Turing Machines. J. Symbolic Logic 65(2), 567–604 (2000)

[3] Irrgang, B., Seyfferth, B.: Multitape ordinal machines and primitive recursion. In: Beckmann, A., et al. (eds.) Logic and Theory of Algorithms - CiE 2008 - Local Proceedings, pp. 175–184. University of Athens (2008)

[4] Koepke, P.: Turing computations on ordinals. Bull. Symbolic Logic 11(3), 377–397 (2005)

[5] Koepke, P.: Infinite Time Register Machines. In: Beckmann, A., Berger, U., Löwe, B., Tucker, J.V., et al. (eds.) CiE 2006. LNCS, vol. 3988, pp. 257–266. Springer, Heidelberg (2006)

[6] Koepke, P., Miller, R.: An enhanced theory of Infinite Time Register Machines. In: Beckmann, A., Dimitracopoulos, C., Löwe, B. (eds.) CiE 2008. LNCS, vol. 5028, pp. 306–315. Springer, Heidelberg (2008)

[7] Koepke, P., Seyfferth, B.: Ordinal machines and admissible recursion theory. Annals of Pure and Applied Logic (2009) (in print)

[8] Koepke, P., Siders, R.: Register computations on ordinals. Archive for Mathematical Logic 47, 529–548 (2008)

[9] Simpson, S.G.: Subsystems of Second Order Arithmetic. In: Perspectives in Mathematical Logic. Springer, Heidelberg (1999)

A Gandy Theorem for Abstract Structures and Applications to First-Order Definability

Oleg V. Kudinov[1,*] and Victor L. Selivanov[2,**]

[1] S.L. Sobolev Institute of Mathematics
Siberian Division Russian Academy of Sciences
kud@math.nsc.ru
[2] A.P. Ershov Institute of Informatics Systems
Siberian Division Russian Academy of Sciences
vseliv@ngs.ru

Abstract. We establish a Gandy theorem for a class of abstract structures and deduce some corollaries, in particular the maximal definability result for arithmetical structures in the class. We also show that the arithmetical structures under consideration are biinterpretable (without parameters) with the standard model of arithmetic. As an example we show that for any $k \geq 3$ a predicate on the quotient structure of the h-quasiorder of finite k-labeled forests is definable iff it is arithmetical and invariant under automorphisms.

Keywords: Gandy theorem, definability, least fixed point, biinterpretability, labeled forest, h-quasiorder.

1 Introduction

We establish a Gandy theorem for a class of abstract structures. This theorem (stating informally that the least fixed point of any positive Σ-definition is a Σ-predicate) is interesting in its own right because most of the known versions of the Gandy theorem apply to set-theoretic structures which look different from our structures here. An exception is presented in [Er96] where a version of Gandy theorem is established for the standard model of arithmetic.

As an application we show that in any arithmetical structure **A** in our class with the additional requirement that all elements are in a sense uniformly Σ-definable a predicate is first-order definable iff it is arithmetical. Another application is the biinterpretability (without parameters) of any arithmetical structure **A** as above with the standard model of arithmetic.

We illustrate the obtained general facts by the structure of finite labeled forests. For any $k \geq 2$, let $\mathcal{F}_k$ be the collection of finite forests the elements of which are labeled by natural numbers from the set $k = \{0, 1, \ldots, k - 1\}$.

---

* Supported by DFG-RFBR Grant 06-01-04002 and by RFBR grants 08-01-00336 and 07-01-00543a.
** Supported by DFG-RFBR Grant 06-01-04002 and by RFBR grant 07-01-00543a.

K. Ambos-Spies, B. Löwe, and W. Merkle (Eds.): CiE 2009, LNCS 5635, pp. 290–299, 2009.
© Springer-Verlag Berlin Heidelberg 2009

The h-quasiorder $\leq$ on $\mathcal{F}_k$ is defined as follows: $F \leq G$, if there is a monotone mapping from F to G that preserves the labelings. By $(G\mathcal{F}_k; \leq)$ we denote the corresponding quotient partial ordering. In [KS07, KS07a, KSZ08] some results on (first-order) definability in the structure $(G\mathcal{F}_k; \leq)$ were established. This line of research is parallel to the popular ongoing study of definability in the degree structures of computability theory, because $(G\mathcal{F}_k; \leq)$ is isomorphic to a natural initial segment of the Wadge degrees of k-partitions of the Baire space [He93, Se07].

Applying the developed theory to some expansions of $(G\mathcal{F}_k; \leq)$, we deduce some best possible definability results, in particular we show that for any $k \geq 3$, a predicate P on $G\mathcal{F}_k$ is definable in $(G\mathcal{F}_k; \leq)$ iff P is arithmetical (w.r.t. the natural constructivization of $G\mathcal{F}_k$) and invariant under the automorphisms of $(G\mathcal{F}_k; \leq)$. This implies all the definability results in [KS07, KS07a, KSZ08] concerning the finite labeled forests, and also gives answers to some open questions raised in [KS07a]. Another interesting corollary is the biinterpretability (without parameters) of the structure $(G\mathcal{F}_k; \leq)$, expanded by the minimal non-smallest elements, with the standard model of arithmetic. Interestingly, for most of the degree structures in computability theory such biinterpretabilty (even with parameters) is either false or still open. By *expansion* of a structure of signature σ we mean a structure in an extended signature with the same universe where all symbols from σ are interpreted in the same way as in the given structure.

2 Σ-Definability in Abstract Structures

In [Ba75, Er96], an interesting theory of admissible sets closely related to computability theory was developed that emphasizes the role of bounded quantifiers in formulas of a finite relational signature $\{\in, \ldots\}$ containing the binary epsilon-relation symbol. Do we gain something useful if we replace the epsilon-relation symbol with a binary relation symbol $\leq$ and consider the resulting structures instead of the set-theoretic structures?

A step in this direction was made in [Er96] where some central facts like Gödel incompleteness theorem were deduced from a Gandy theorem for the standard model of arithmetic. We develop a similar theory for a class of abstract structures. In this section, we briefly recall some notions and facts mostly observed in [Er96]. We work with an arbitrary finite relational signature $\sigma = \{\leq, \ldots\}$ containing a distinguished binary relation symbol $\leq$.

RQ-Formulas of σ are constructed from the atomic formulas according to the usual rules of first-order logic concerning $\wedge, \vee, \neg, \rightarrow$, the usual rules for the (unbounded) quantification with $\forall, \exists$ and the following additional formation rules for the bounded quantification: if φ is an RQ-formula and x, y are variables then the expressions $\forall x \leq y\varphi$ and $\exists x \leq y\varphi$ are RQ-formulas. Δ_0-*Formulas* of signature σ are constructed inductively according to the following rules: any atomic formula of signature σ is a Δ_0-formula; if φ and ψ are Δ_0-formulas then $\neg\varphi$, $(\varphi \wedge \psi)$, $(\varphi \vee \psi)$ and $(\varphi \rightarrow \psi)$ are Δ_0-formulas; if x, y are variables and φ is a Δ_0-formula then $\forall x \leq y\varphi$ and $\exists x \leq y\varphi$ are Δ_0-formulas. Σ-*Formulas* of signature σ are constructed inductively according to the following rules: any Δ_0-formula is a Σ-formula; if φ and ψ are Σ-formulas then $(\varphi \wedge \psi)$ and $(\varphi \vee \psi)$ are Σ-formulas; if x, y are variables and φ is a Σ-formula then $\forall x \leq y\varphi$, $\exists x \leq y\varphi$ and $\exists x\varphi$ are Σ-formulas.

For a given structure $\mathbf{A}$ of signature σ, a predicate on A is Δ_0-*definable* (resp. Σ-*definable*) if it is defined by a Δ_0-formula (resp. by a Σ-formula). A predicate on A is Δ-*definable* if both the predicate and its negation are Σ-definable. First-order definability (called definability in this paper) is defined in the same manner.

For σ-structures $\mathbf{A}$ and $\mathbf{B}$, let $\mathbf{A} \leq_{\mathrm{end}} \mathbf{B}$ denote that $\mathbf{A}$ is a substructure of $\mathbf{B}$ and $(A; \leq)$ is an initial segment of $(B; \leq)$. Let $\mathbf{A} \leq_{\mathrm{end}} \mathbf{B}$ and let some evaluation of the free variables of a formula φ in $\mathbf{A}$ be fixed. If φ is a Δ_0-formula then $\mathbf{A} \models \varphi$ iff $\mathbf{B} \models \varphi$. If φ is a Σ-formula then $\mathbf{A} \models \varphi$ implies $\mathbf{B} \models \varphi$.

When defining more and more complicated predicates in a structure, it is natural to use predicates already defined in the definitions of new predicates. It is well-known that in defining new Δ_0-predicates (resp. Σ-predicates) we can use the Δ_0-predicates (resp. Δ-predicates)

The introduced classes of formulas have some nice properties in the so called *bounded* structures $\mathbf{A}$; these are σ-structures such that $\leq$ is a transitive directed relation (directed means that for all $x, y \in A$ there is $z \in A$ with $x, y \leq z$) and, for any Δ_0-formula φ, $\mathbf{A} \models \forall x \leq t\exists y\varphi$ implies $\mathbf{A} \models \exists v\forall x \leq t\exists y \leq v\varphi$. For a Σ-formula φ and a variable u not occurring in φ, let φ^u be the Δ_0-formula obtained from φ by substituting any occurrence $\exists x$ of unbounded existential quantifier in φ by the corresponding bounded existential quantifier $\exists x \leq u$. The next fact is known as the Σ-reflection principle: Let $\mathbf{A}$ be a bounded structure and φ a Σ-formula of signature σ. Then $\varphi \equiv_{\mathbf{A}} \exists u\varphi^u$ where $\varphi \equiv_{\mathbf{A}} \psi$ means the equivalence of formulas φ and ψ in the structure $\mathbf{A}$.

As for the usual first-order formulas, any RQ-formula is equivalent to an RQ-formula in negation normal form (i.e., a formula without implications that has negations only on the atomic formulas). Let P be an n-ary predicate symbol not in σ, φ an RQ-formula of signature $\sigma \cup \{P\}$, and $\bar{x} = x_1, \ldots, x_m$ a list of variables that includes all free variables of φ. We say that P *occurs positively in* φ if φ is in negation normal form and has no subformulas $\neg P(y_1, \ldots, y_n)$. For any n-ary predicate Q on a σ-structure $\mathbf{A}$, we denote by $(\mathbf{A}, Q)$ the expansion of $\mathbf{A}$ to the $\sigma \cup \{P\}$-structure where P is interpreted as Q. Then we can define an operator $\Gamma = \Gamma_{\varphi, \bar{x}}$ on the n-ary predicates on A that sends any Q to the predicate

$$\Gamma_{\varphi, \bar{x}}(Q) = \{(a_1, \ldots, a_n) \mid (\mathbf{A}, Q) \models \varphi(a_1, \ldots, a_n)\}.$$

Proposition 1. *Let* $\mathbf{A}$ *be a bounded* σ-*structure and* P *an* n-*ary predicate symbol that occurs positively in a* Σ-*formula* φ *with the free variables among* $\bar{x}$.

(i) The operator Γ *is monotone, i.e.,* $Q \subseteq R \subseteq A^n$ *implies* $\Gamma(Q) \subseteq \Gamma(R)$.

(ii) The operator Γ *sends* Σ-*predicates to* Σ-*predicates. If* $\psi(\bar{x})$ *is a* Σ-*formula that defines* $Q \subseteq A^n$ *then the* Σ-*formula* $\varphi[P(\bar{v}), \psi(\bar{v})]$, *obtained from* φ *by substituting any occurrence of* $P(\bar{v})$ *by* $\psi(\bar{v})$, *defines* $\Gamma(Q)$.

(iii) The symbol P *occurs positively in the* Σ-*formula* $\exists u\varphi^u$ *and* $\exists u\varphi^u$ *induces the same operator as* φ.

By Proposition 1(i) and the Tarski fixed-point theorem, the operator Γ has the least fixed point denoted by $LFP(\Gamma_{\varphi,\bar{x}})$. In general, the least fixed points defined in this way may be complicated. But for some structures $\mathbf{A}$ it turns out that the least fixed points of any Σ-formula φ as above is a Σ-predicate (in this case we say that $\mathbf{A}$ has the *Gandy property*). Many interesting examples or such structures are known in the theory of admissible sets [Ba75, Er96]. As shown in [Er96], the standard model of arithmetic has the Gandy property. In the next section we extend the last example to a class of structures.

3   Gandy Theorem for Abstract Structures

In this section we establish a version of the Gandy theorem for a class of bounded σ-structures. An important technical tool here is some coding technique.

We say that a σ-structure $\mathbf{A}$ *admits a Δ-coding of finite sets* if there is a binary Δ-predicate $E(x, y)$ on $\mathbf{A}$ such that $E(x, y)$ implies $x \leq y$ for all $x, y \in A$ and, for all $n < \omega$ and $x_1, \ldots, x_n \in A$, there is $y \in A$ with

$$\mathbf{A} \models \forall x(E(x, y) \leftrightarrow x = x_1 \vee \cdots \vee x = x_n).$$

Informally, y is considered as a code of the set $\{x \mid E(x, y)\}$. The definition requires the existence of at least one code for any finite subset of A, but does not require that the code is unique.

Observe that, by the axiom of choice, there is a sequence $\{set^n\}_{n<\omega}$ of functions $set^n : A^n \to A$ which codes the finite sets in the sense that, for all $n < \omega$ and $x, x_1, \ldots, x_n \in A$ we have: $x_1, \ldots, x_n \leq set^n(x_1, \ldots, x_n)$ and $E(x, set^n(x_1, \ldots, x_n)) \leftrightarrow x \in \{x_1, \ldots, x_n\}$ (in particular, set^0 is an element of A such that $E(x, set^0)$ is false for all $x \in A$). This implies the existence of the coding $\langle x, y \rangle = set^2(set^1(x), set^2(x, y))$ of ordered pairs of elements of A such that

$$x, y \leq \langle x, y \rangle, \ \langle x, y \rangle = \langle x_1, y_1 \rangle \leftrightarrow x = x_1 \wedge y = y_1, \ l(\langle x, y \rangle) = x, \ r(\langle x, y \rangle) = y$$

for some unary functions l and r on A. Unfortunately, it is not always easy to find such coding functions which are Δ-functions; this is a reason for a bit more complicated considerations below which aim to find more liberal multivalued codings that admit Δ-decodings.

For any $n \geq 0$, define the $(n + 1)$-ary Δ-predicate S_n by

$$S_n(x_1, \ldots, x_n, y) \leftrightarrow \forall x \leq y(E(x, y) \to x = x_1 \vee \cdots \vee x = x_n) \wedge \bigwedge_i E(x_i, y).$$

Then we have: For all $n \geq 0$ and $x_1, \ldots, x_n \in A$ there is $y \in A$ such that $S_n(x_1, \ldots, x_n, y)$; If $S_n(x_1, \ldots, x_n, y)$ and $S_n(y_1, \ldots, y_n, y)$ then $\{x_1, \ldots, x_n\} = \{y_1, \ldots, y_n\}$; $S_n(x_1, \ldots, x_n, y)$ implies $x_i \leq y$.

Define the ternary Δ-predicate P (that resembles the standard set-theoretic coding of ordered pairs defined above) by

$$P(x_1, x_2, y) \leftrightarrow \exists y_1, y_2 \leq y(S_2(y_1, y_2, y) \wedge S_1(x_1, y_1) \wedge S_2(x_1, x_2, y_2)).$$

Then we have: For any $x_1, x_2 \in A$ there is $y \in A$ with $P(x_1, x_2, y)$; If $P(x_1, x_2, y)$ and $P(x_1', x_2', y)$ then $x_1 = x_1'$ and $x_2 = x_2'$. This implies the existence of decoding Δ-functions for the coding of pairs. Namely, define the unary Δ-predicate $Pair$ by $Pair(y) \leftrightarrow \exists x_1, x_2 \leq y P(x_1, x_2, y)$ and the unary functions l, r on A as follows: if $\neg Pair(y)$ then $l(y) = r(y) = y$, otherwise $l(y) = x_1$ and $r(y) = x_2$ where x_1, x_2 are the unique elements with $P(x_1, x_2, y)$. One easily checks that l, r are Δ-functions which decode the pairing function $\langle \cdot \rangle$ fixed above.

Let $(\omega; \leq)$ be Δ-definable in a σ-structure $\mathbf{A}$, i.e., there are Δ-predicates N and $\preceq$ such that $(\omega; \leq)$ is isomorphic to $(N; \preceq)$ via a function $n \mapsto \tilde{n}$. Then it is easy to see that every element of N is Δ-definable, and there is a unary Δ-function s on $\mathbf{A}$ such that $s(\tilde{n}) = \widetilde{n+1}$ for each $n < \omega$. In the assumptions above (where for simplicity of notation we assume, as also in similar considerations below, that $N \subseteq A$), we define the Δ-predicates Seq ($Seq(n, y)$ is to mean that y "looks like" a code of $\{\langle 0, y_0 \rangle, \ldots, \langle n-1, y_{n-1} \rangle\}$ interpreted as the sequence $y_0, \ldots, y_{n-1}$ of length n) and Q ($Q(n, y, i, z)$ is to mean that z is the i-th element of the sequence coded by y) as follows:

$$Seq(n, y) \leftrightarrow n \in N \wedge \forall x \varepsilon y (Pair(x) \wedge l(x) \in N \wedge l(x) \prec n) \wedge \forall x \varepsilon y (l(x) \prec n$$
$$\rightarrow \exists x_1 \varepsilon y (l(x_1) = s(l(x))) \wedge \forall x, x_1 \varepsilon y (l(x) = l(x_1) \rightarrow r(x) = r(x_1))$$

and

$$Q(n, y, i, z) \leftrightarrow Seq(n, y) \wedge i \in N \wedge i \prec n \wedge \exists x \varepsilon y (l(x) = i \wedge r(x) = z).$$

To shorten the formulas above, we abbreviated $\exists x \leq y(E(x, y) \wedge \cdots)$ (resp. $\forall x \leq y(E(x, y) \rightarrow \cdots)$) to $\exists x \varepsilon y(\cdots)$ (resp. $\forall x \varepsilon y(\cdots)$). Thus, the "quantifications" $\exists x \varepsilon y$ and $\forall x \varepsilon y$ preserve the Δ- and Σ-predicates.

It is clear that for all $n < \omega$ and $y_0, \ldots, y_{n-1} \in A$ there is an element $y = set^n(\{\langle \tilde{0}, y_0 \rangle, \ldots, \langle \widetilde{n-1}, y_{n-1} \rangle\})$ which codes such a sequence, i.e., $Seq(\tilde{n}, y)$ is true. Note that in the assumptions above there is a ternary Δ-function q on $\mathbf{A}$ such that $Seq(n, y)$, $i \in N$ and $i \prec n$ imply $Q(n, y, i, q(n, y, i))$.

We call a σ-structure $\mathbf{A}$ *locally finite* if all the initial segments $\{x \mid x \leq y\}$, $y \in A$, are finite. The main result of this section is now formulated as follows:

Theorem 1. *Let $\mathbf{A}$ be a bounded locally finite σ-structure that admits a Δ-coding of finite sets and a Δ-definable copy of $(\omega; \leq)$. Then $\mathbf{A}$ has the Gandy property.*

Proof. We have to show that for any Σ-formula φ of signature $\sigma \cup \{P\}$ as at the end of Section 2 the least fixed point of the operator $\Gamma = \Gamma_{\varphi, \bar{x}}$ is a Σ-predicate. Simplifying notation a bit, we consider only the case when P is a unary predicate and the list $\bar{x}$ consists of only one variable v, and use the notation above concerning a Δ-definable copy $(N; \preceq)$ of $(\omega; \leq)$ such that $N \subseteq A$.

By Proposition 1(i), the operator Γ on the class $P(A)$ of subsets of A is monotone. Let us check that the operator is also continuous (i.e., $a \in \Gamma(S)$, $S \subseteq A$, implies that $a \in \Gamma(F)$ for some finite subset $F \subseteq S$). Indeed, $a \in \Gamma(S)$ implies $(\mathbf{A}, S) \models \exists u \varphi^u(a)$ by Propositions 1(iii) and the Σ-reflection principle, so $(\mathbf{A}, S) \models \varphi^u(a)$ for some $u \in A$. Choose $b \in A$ with $a, u \le b$ and let $\mathbf{B}$ be the substructure of $\mathbf{A}$ with the universe $B = \{x \mid x \le b\}$. Since $(\mathbf{B}, S \cap B) \le_{\text{end}}$ $(\mathbf{A}, S)$ and φ^u is Δ_0, $(\mathbf{B}, S \cap B) \models \varphi^u(a)$ by the remarks in the previous section. Since $(\mathbf{B}, S \cap B) \le_{\text{end}} (\mathbf{A}, S \cap B)$, $(\mathbf{A}, S \cap B) \models \varphi^u(a)$ and hence $(\mathbf{A}, S \cap B) \models \exists u \varphi^u(a)$. Therefore, $a \in \Gamma(F)$ for the finite set $F = S \cap B \subseteq S$.

By monotonicity and continuity of Γ, $LFP(\Gamma) = \bigcup\{\Gamma^n(\emptyset) \mid n < \omega\}$ where $\Gamma^n(\emptyset)$ is the n-th iterate of Γ starting with $\emptyset$, i.e., $\Gamma^0(\emptyset) = \emptyset$ and $\Gamma^{n+1}(\emptyset) = \Gamma(\Gamma^n(\emptyset))$ (of course, we have $\emptyset \subseteq \Gamma^1(\emptyset) \subseteq \Gamma^2(\emptyset) \subseteq \cdots$). By induction on n one now easily checks that, for any $a \in A$, the condition $a \in \Gamma^{n+2}(\emptyset)$ is equivalent to the following:

(*) there exist finite sets $F_0, \ldots, F_n \subseteq A$ such that $F_0 \subseteq \Gamma(\emptyset)$, $F_{i+1} \subseteq \Gamma(F_i)$ for all $i < n$, and $a \in \Gamma(F_n)$.

It suffices now to find a Σ-formula $\theta(a, m)$ of signature σ with the free variables a, m such that, for all $a \in A$ and $m \in N$, the condition (*) is equivalent to $\mathbf{A} \models \theta(a, m)$; then the Σ-formula $\exists m(m \in N \wedge \theta(a, m))$ will define $LFP(\Gamma)$ in $\mathbf{A}$. For this, we use the codings of finite sets, ordered pairs and sequences. Informally, we would like to express the existence of $F_0, \ldots, F_n$ by the existence of a code of a finite set $\{\langle \tilde{0}, F_0 \rangle, \ldots \langle \tilde{n}, F_n \rangle\}$ where $N = \{\tilde{0} \prec \tilde{1} \prec \cdots\}$. Let $\theta(a, m)$ be the formula

$$\exists y(Seq(s(m), y) \wedge \forall t \in z_0 \varphi[t; P(v), v \ne v]) \wedge$$
$$\forall x \varepsilon y(l(x) < m \rightarrow \forall t \varepsilon z' \varphi[t; P(v), E(v, r(x))]) \wedge \varphi[a; P(v), E(v, z_m)]).$$

where $z_0 = q(s(m), y, \tilde{0})$, $z' = q(s(m), y, s(l(x)))$, $z_m = q(s(m), y, m)$. The formula $\theta(a, m)$ is equivalent to a Σ-formula and has the desired property. $\quad\square$

4 Σ-Definability of Σ-Truth

In this section we derive some important corollaries of results of the previous section.

Theorem 2. *Let $\mathbf{A}$ be a bounded locally finite σ-structure that admits a Δ-coding of finite sets and a Δ-definable copy $(N; \preceq)$ of $(\omega; \le)$. Then a predicate on ω is computably enumerable (resp. computable) iff its image in N under the isomorphism $n \mapsto \tilde{n}$ is Σ-definable (resp. Δ-definable). In particular, $(\omega; +, \cdot)$ is Δ-definable in $\mathbf{A}$.*

Let now $\{\varphi_n\}_{n \ge 0}$ be the standard Gödel numbering of the RQ-formulas; w.l.o.g. we assume that $i < n$ for each free variable v_i in φ_n (where $V = \{v_0, v_1, \ldots\}$ is the countable list of all variables used in the RQ-formulas). Then all the usual syntactic notions on formulas correspond to Δ-predicates on the Gödel numbers, in particular there is a unary Δ-predicate $Sigma$ true exactly on the isomorphic images of Gödel numbers of Σ-formulas.

Next we are going to show that truth of Σ-formulas in a structure satisfying the conditions above is coded by a Σ-predicate. For this we have to code the finitary (i.e., almost constant) evaluations of variables in A by elements of A.

For this we again use the predicate Seq from the previous section. If $Seq(n, y)$ is true for some $n \in N$ and $y \in A$, define the evaluation $\gamma_y : V \to A$ by: $\gamma_y(v_i) = q(n, y, \tilde{i})$ if $i < n$ and $\gamma_y(v_i) = \tilde{0}$ otherwise. Finally, define the binary predicate Tr on A as follows: $Tr(m, y)$ iff $Sigma(m)$, $Seq(m, y)$ and $\mathbf{A} \models \varphi_n(\gamma_y)$ where n satisfies $m = \tilde{n}$.

Theorem 3. *Let* $\mathbf{A}$ *be a bounded locally finite σ-structure that admits a Δ-coding of finite sets and a Δ-definable copy of* $(\omega; \leq)$. *Then* Tr *is a Σ-predicate.*

Proof (sketch). By a straightforward rewriting of the Tarski-style recursive definition of truth of Σ-formulas one defines Tr as the least fixed point of a positive Σ-definition. By Theorem 1, Tr is a Σ-predicate.   □

5   First-Order Definability

In this section we apply the previous results to characterizing the first-order definable predicates in σ-structures satisfying some additional restrictions.

The first restriction is that the structures $\mathbf{A}$ will be assumed arithmetical, i.e., equipped with a numbering α (i.e., a surjection from ω onto A) modulo which the equality predicate and all signature predicates are arithmetical. Obviously, any first-order definable predicate on an arithmetical structure $(\mathbf{A}; \alpha)$ is arithmetical (w.r.t. α) and invariant under the automorphisms of $\mathbf{A}$; we say that $(\mathbf{A}; \alpha)$ has the *maximal definability property* if the converse is also true, i.e., any arithmetical predicate invariant under the automorphisms of $\mathbf{A}$ is first-order definable.

The second restriction concerns the definability of elements. We say that *the elements of* $(\mathbf{A}; \alpha)$ *are uniformly Σ-definable* if there is an arithmetical sequence of unary Σ-formulas $\{\psi_n(v_0)\}$ such that ψ_n defines the element $\alpha(n)$ in $\mathbf{A}$ for each $n < \omega$. Obviously, under the last restriction the structure $\mathbf{A}$ is rigid, i.e., has no non-trivial automorphisms.

Theorem 4. *Let* $(\mathbf{A}; \alpha)$ *be an arithmetical σ-structure with uniformly Σ-definable elements such that $\mathbf{A}$ is bounded, locally finite and admits a Δ-coding of finite sets and a Δ-definable copy $(N; \preceq)$ of $(\omega; \leq)$. Then $(\mathbf{A}; \alpha)$ has the maximal definability property.*

We say that a structure $\mathbf{B}$ of a finite relational signature τ is *biinterpretable* with a structure $\mathbf{C}$ of a finite relational signature ρ if $\mathbf{B}$ is first-order definable in $\mathbf{C}$ (in particular, there is a bijection $f : B \to B_1$ on a definable set $B_1 \subseteq C^m$ for some $m \geq 1$ which induces an isomorphism on the τ-structure $\mathbf{B}_1$ definable in $\mathbf{C}$), $\mathbf{C}$ is first-order definable in $\mathbf{B}$ (in particular, there is a similar bijection $g : C \to C_1$ on a definable set $C_1 \subseteq B^n$ for some $n \geq 1$), the function $g^m \circ f : B \to B^{nm}$ is definable in $\mathbf{B}$ and the function $f^n \circ g : C \to C^{mn}$ is definable in $\mathbf{C}$. This notion is in fact a particular case of a popular model-theoretic notion of biinterpretability.

Theorem 5. *Let* $(\mathbf{A}; \alpha)$ *be an arithmetical σ-structure with uniformly Σ-definable elements such that $\mathbf{A}$ is bounded, locally finite and admits a Δ-coding of finite sets and a Δ-definable copy $(N; \preceq)$ of $(\omega; \leq)$. Then $\mathbf{A}$ is biinterpretable with $(\omega; +, \cdot)$.*

6   Example: The Structure $(G\mathcal{F}_k; \leq)$

Here we apply the results obtained above to some expansions of the structure $(G\mathcal{F}_k; \leq)$. First we very briefly recall some notions and known facts, mostly from [Se04].

Let $\mathcal{F}_k$ and $\mathcal{T}_k$ be the classes of all finite k-forests and finite k-trees, respectively. For each $i < k$, let $\mathcal{T}_k^i$ be the set of finite k-trees which carry the label i on their roots. For technical reasons we consider also the empty k-forest $\emptyset \in \tilde{\mathcal{F}}_k$ (which is not a tree) assuming that $\emptyset \leq F$ for each $F \in \tilde{\mathcal{F}}_k$. A natural subset of k-trees is formed by k-chains, i.e., by words over the alphabet $k = \{0, \dots, k-1\}$. We will denote such words in the usual way, as strings of symbols. E.g., 01221 and 011022 denote some words over the alphabet $\{0, 1, 2\}$.

The h-quasiorder $\leq$ on $\mathcal{F}_k$ is defined as follows: $F \leq G$, if there is a monotone mapping from F to G that preserves the labelings. The h-equivalence is the equivalence relation induced by the h-quasiorder. Let $G\mathcal{F}_k$, $G\mathcal{T}_k$ and $G\mathcal{T}_k^i$ be the quotient sets of respectively $\mathcal{F}_k$, $\mathcal{T}_k$ and $\mathcal{T}_k^i$ modulo the h-equivalence. For $F \in \tilde{\mathcal{F}}_k$, $[F]$ denotes the h-equivalence class of F in any of the structures under consideration. The partial order $(G\mathcal{F}_k; \leq)$ is a distributive lattice, the corresponding operations are denoted by $\sqcup$ and $\sqcap$. The sets $G\mathcal{T}_k^0, \dots, G\mathcal{T}_k^{k-1}$ form a partition of $G\mathcal{T}_k$. For each $i < k$, let T_i be the unary predicate on $G\mathcal{F}_k$ which is true exactly on the elements of $G\mathcal{T}_k^i$. Any element x of $G\mathcal{F}_k$ has a unique canonical representation $x = \bigsqcup \mathcal{Y}$ where $\mathcal{Y}$ is a finite set of join-irreducible elements distinct from $[\emptyset]$; we call the elements of $\mathcal{Y}$ *components* of x. The join-irreducible elements distinct from $[\emptyset]$ coincide with $G\mathcal{T}_k$.

For any finite k-forest F and any $i < k$, let $p_i(F)$ be the k-tree obtained from F by joining a new smallest element and assigning the label i to this element. In particular, $p_i(\emptyset)$ will be the singleton tree carrying the label i. The operations p_i preserve the h-equivalence, so we use the same notation p_i to denote also the operations induced by p_i on $G\mathcal{F}_k$. These operations have the following properties: $x \leq p_i(x)$, $x \leq y \to p_i(x) \leq p_i(y)$, $p_i p_i(x) = p_i(x)$, $p_i(x) \leq p_j(y) \to p_i(x) \leq y$ for $i \neq j$, $p_i(G\mathcal{F}_k) = G\mathcal{T}_k^i$.

Any element of $G\mathcal{F}_k$ is the value of a term of signature $\{\sqcup, p_0, \dots, p_{k-1}, [\emptyset]\}$ without variables. E.g., [1210] and [2010] coincide with $p_1(p_2(p_1(p_0([\emptyset]))))$ and $p_2(p_0(p_1(p_0([\emptyset]))))$, respectively. We omit parenthesis whenever they are clear from the context, e.g. we could write the last term as $p_2 p_0 p_1 p_0([\emptyset])$.

The main result of this section is the following

Theorem 6. *For any $k \geq 3$, the structure* $\mathbf{A}_k = (G\mathcal{F}_k; \leq, T_0, \dots, T_{k-1})$ *has the Gandy and the maximal definability properties, and is biinterpretable with* $(\omega; +, \cdot)$.

Proof. It suffices to check that $\mathbf{A}_k$ (with the natural constructivization) satisfies the conditions of Theorems 1, 4 and 5. Since $(G\mathcal{F}_k; \leq)$ is a lattice, $\mathbf{A}_k$ is bounded. By [He93, Se04], $\mathbf{A}_k$ is locally finite. It is an easy exercise to show that operations $\sqcup, \sqcap$ and elements $[\emptyset], [0], \dots, [k-1]$, hence also any of the elements $\tilde{0} = [\emptyset], \tilde{1} = [0] \sqcup [1], \tilde{2} = [01] \sqcup [10], \tilde{3} = [010] \sqcup [101], \dots$ are Δ_0-definable.

Let $N = \{\tilde{n} \mid n < \omega\}$, then $(N; \leq)$ has order type ω and N is Δ_0-definable because

$$x \in N \leftrightarrow [2] \not\leq x \wedge \cdots \wedge [k-1] \not\leq x \wedge x \notin GT_k,$$
$$x \in GT_k \leftrightarrow T_0(x) \vee \cdots \vee T_{k-1}(x).$$

It remains to show that $\mathbf{A}_k$ admits a Δ-coding of finite sets (in fact, we show it admits even a Δ_0-coding). For this we first establish Δ_0-definability of some more predicates and functions. The functions $p_0, \ldots, p_{k-1}$ are Δ_0 because $p_i(x)$ is the smallest element of GT_k^i above x. The binary predicate $Com(x, y)$ meaning that y is a component in the canonical representation of x is Δ_0 because

$$Com(x, y) \leftrightarrow y \in GT_k \wedge y \leq x \wedge \neg \exists y_1 \leq x(y_1 \in GT_k \wedge y < y_1).$$

The function $c : GF_k \to GT_k^0$ defined by $c(x) = p_0(p_1(x^*), \cdots, p_{k-1}(x^*))$, where $x^* = p_0(x) \sqcup \cdots \sqcup p_{k-1}(x)$, is clearly Δ_0. Note that, for any $x \in GF_k$, $x < x^* < c(x)$ and x^* has at least $k - 1 \geq 2$ components. Let us show that there is a Δ_0-function $d : GF_k \to GF_k$ such that $dc(x) = x$ and $d(x) \leq x$ for all $x \in GF_k$ (in particular, c is injective and we can restore x from its "code" $c(x) \in GT_k^0$).
Set $f([\emptyset]) = [\emptyset]$ and $f(y) = \sqcap\{z \mid Com(y, z)\}$ for $y \in GF_k$, $y \neq [\emptyset]$. Since

$$u = f(y) \leftrightarrow u \leq y \leq n \wedge \forall z \leq y(Com(y, z) \to u \leq z) \wedge$$
$$\neg \exists u_1 \leq y(u < u_1 \wedge \forall z \leq y(Com(y, z) \to u_1 \leq z)),$$

f is a Δ_0-function. Since $p_i(x) \sqcap p_j(x) = x$ for $i \neq j$, $f(x^*) = x$ for all x. For any $y \in GF_k$, let $g(y) = \bigsqcup\{z \leq y \mid T_1(z) \vee \cdots \vee T_{k-1}(z)\}$. Then g is a Δ_0-function and $g(c(x)) = p_1(x^*) \sqcup \cdots \sqcup p_{k-1}(x^*)$ for any $x \in GF_k$. Since $ffg(c(x)) = x$ and $g(y), f(y) \leq y$, the function $d = f \circ f \circ g$ has the desired properties.
Next we define a sequence $\{set^n\}_{n < \omega}$ as follows. We set $set^0 = [\emptyset]$ and, for $n \geq 1$ and $x_1, \ldots, x_n \in GF_k$, $set^n(x_1, \ldots, x_n) = y_1 \sqcup \cdots \sqcup y_m$ where m is the cardinality of $\{x_1, \ldots, x_n\}$ and $(y_1, \ldots, y_m)$ is the sequence in GT_k^1 constructed from $(x_1, \ldots, x_n)$ as follows. First, we define $i_1, \ldots, i_m \in \{1, \ldots, n\}$ by induction on $j \geq 1$: i_j is the smallest i such that x_i is maximal in $(\{x_1, \ldots, x_n\} \setminus \{x_{i_l} \mid 1 \leq l < j\}; \leq)$. Second, define y_j by induction: $y_1 = c(x_{i_1})$ and, for $j \in \{1, \ldots, m-1\}$, $y_{j+1} = (p_1 \circ p_2)^{h_j}(c(x_{i_{j+1}}))$ where h_j is the height of the minimal (repetition-free) k-tree T with $[T] = y_j$ (see [Se04] for additional details).
Finally, let $E(x, y)$ be the Δ_0-predicate $\exists t \leq y(Com(y, t) \wedge x = d(y))$. By construction of $i_1, \ldots, i_m$, $x_{i_l} \not\leq x_{i_j}$ for $1 \leq l < j \leq m$. By properties of the functions c, d above, $c(x_{i_l}) \not\leq c(x_{i_j})$, hence also $y_l \not\leq y_j$ for $1 \leq l < j \leq m$. If $l < j$ then h_j is above the height of the minimal k-tree T with $[T] = y_l$, hence $y_j \not\leq y_l$. Therefore, $y_1, \ldots, y_m$ are pairwise incomparable. Together with the properties of functions c, d this implies the desired properties of E. This completes the proof of the Gandy property.
It remains to show that the elements of $\mathbf{A}_k$ are uniformly Σ-definable. Any such element is the value of a variable-free term t_a of signature $\{p_0, \ldots, p_{k-1}, [\emptyset]\}$. For any $n < \omega$, let $\psi_n(v_0)$ be the Δ_0-formula of signature $\{\leq, T_0, \ldots, T_{k-1}\}$

equivalent to the formula $v_0 = t_{\alpha(n)}$ of signature $\{p_0, \ldots, p_{k-1}, [\emptyset]\}$. Clearly, the sequence $\{\psi_n\}$ is computable, hence arithmetical. $\qquad\square$
Next we derive some corollaries:

Theorem 7. *(i) For any $k \geq 3$, the structure $(G\mathcal{F}_k; \leq, [0], \ldots, [k-1])$ has the maximal definability property, and is biinterpretable with $(\omega; +, \cdot)$.*
(ii) For any $k \geq 3$, a predicate P on $G\mathcal{F}_k$ is definable in $(G\mathcal{F}_k; \leq)$ if and only if P is arithmetical (w.r.t. the natural constructivization $\alpha : \omega \to G\mathcal{F}_k$) and invariant under the automorphisms of $(G\mathcal{F}_k; \leq)$.

As easy consequences, we answer some questions left open in [KS07a]:

Corollary 1. *For any $k \geq 3$, the set GC_k of the equivalence classes of finite k-chains and the binary relation "$y = f(x)$ for some automorphism f of $(G\mathcal{F}_k; \leq)$" are first-order definable in $(G\mathcal{F}_k; \leq)$.*

Proof. The set and the relation are clearly arithmetical. They are also invariant under the automorphisms of $(G\mathcal{F}_k; \leq)$ because any such automorphism is induced by a permutation of $0, \ldots, k-1$ [KS07a]. $\qquad\square$

Remark. In this paper, we illustrated the developed theory only by $G\mathcal{F}_k$. We guess that there are many more similar applications, in particular to some structures considered in [Ku06]. Up to now, we obtained similar results for the structure of words over an alphabet with at least 2 symbols equipped with the infix partial order.

References

[Ba75]   Barwise, J.: Admissible Sets and Structures. Springer, Berlin (1975)
[Er96]   Ershov, Y.L.: Definability and Computability. Plenum, New-York (1996)
[He93]   Hertling, P.: Topologische Komplexitätsgrade von Funktionen mit endlichem Bild. Informatik-Berichte 152, 34 pages, Fernuniversität Hagen (1993)
[KS07]   Kudinov, O.V., Selivanov, V.L.: Definability in the homomorphic quasiorder of finite labeled forests. In: Cooper, S.B., Löwe, B., Sorbi, A. (eds.) CiE 2007. LNCS, vol. 4497, pp. 436–445. Springer, Heidelberg (2007)
[KS07a] Kudinov, O.V., Selivanov, V.L.: Undecidability in the homomorphic quasiorder of finite labeled forests. Journal of Logic and Computation 17, 1135–1151 (2007)
[KSZ08] Kudinov, O.V., Selivanov, V.L., Zhukov, A.V.: Definability in the h-quasiorder of labeled forests. Annals of Pure and Applied Logic (2008), doi: 10.1016/ j.apal. 2008.09.026
[KSZ09] Kudinov, O.V., Selivanov, V.L., Zhukov, A.V.: Definability of closure operations in the h-quasiorder of labeled forests. In: Local volume of this conference
[Ku06]   Kuske, D.: Theories of orders on the set of words. RAIRO Theoretical Informatics and Applications 40, 53–74 (2006)
[Se04]   Selivanov, V.L.: Boolean hierarchy of partitions over reducible bases. Algebra and Logic 43(1), 44–61 (2004)
[Se07]   Selivanov, V.L.: Hierarchies of Δ_2^0-measurable k-partitions. Mathematical Logic Quarterly 53, 446–461 (2007)

Constructing New Aperiodic
Self-simulating Tile Sets

Grégory Lafitte[1,*] and Michael Weiss[2,**]

[1] Laboratoire d'Informatique Fondamentale de Marseille (LIF),
CNRS – Aix-Marseille Université,
39, rue Joliot-Curie, F-13453 Marseille Cedex 13, France
[2] Università degli Studi di Milano,
Bicocca Dipartimento di Informatica, Sistemistica e Comunicazione,
336, Viale Sarca, 20126 Milano, Italy

Abstract. Wang tiles are unit size squares with colored edges. By using a fixed-point theorem *à la* Kleene for tilings we give novel proofs of classical results of tilings problems' undecidability by way of diagonalization on tilings (made possible by this theorem). Then, we present a general technique to construct aperiodic tile sets, *i.e.*, tile sets that generate only aperiodic tilings of the plane. Our last construction generalizes the notion of self-simulation and makes possible the construction of tile sets that self-simulate *via* self-similar tilings, showing how complex the self-simulation can be.

Introduction

In the early sixties, Wang has introduced the study of tilings with colored tiles to solve mathematical logical problems [Wan61, Wan62]. In this model, a tile is a unit size square with colored edges and two tiles can be assembled if their common edge has the same color. To tile consists in assembling tiles from a tile set (a finite set of different tiles) on the grid $\mathbb{Z}^2$. The tiles can be repeated as many times as needed, but cannot be turned.

The domino problem, *i.e.*, the problem to know whether a tile set can tile the plane, was proved undecidable by Berger in [Ber66]: he constructed for any Turing machine M and any input w, a tile set $\tau_{M,w}$ such that this tile set can generate a tiling of the plane if and only if the computation of M does not stop on the input w. Therefore, tilings can be seen as a Turing-capable model of computation on the plane.

In the first part of this paper, we aim at proving classical tilings' results using a new promising theorem obtained recently in [LW08b]: this theorem is a recursion theorem *à la* Kleene and states that for any recursive function f, there exists a tile set τ_i such that τ_i exactly simulates $\tau_{f(i)}$ and thus, τ_i can be seen as a fixed-point

* This author has been supported by the French ANR grant *Sycomore*.
** This author thanks the Swiss national fund and the Commission de recherche of the University of Geneva for supporting his post-doc.

for f. The simulation, introduced in [LW08a], expresses that a tile set τ exactly simulates a tile set τ' if τ generates a set of rectangular patterns of equal sizes which are isomorphic to the tiles of τ'. The formal definition is given in section 1. Using this strong tool, we propose a new way to prove undecidability results on tilings, without complex intricate constructions but only with a diagonalization technique that we introduce in this paper.

In the second part of this paper we use our fixed-point theorem $\grave{a}$ la Kleene to obtain a general framework for constructing aperiodic tile sets, $i.e.$, tile sets that generate only non periodic tilings of the plane. Aperiodicity is one of the most challenging research on tilings. Without aperiodicity, any question on tilings would be trivial. Aperiodicity can be seen as the chore of the computability complexity of tilings. In [DRS08], Durand et $al.$ have used the classical Kleene recursion theorem to build a tile set that self-simulates with squares and thus is aperiodic. In this paper, we generalize this construction to obtain a general framework to build aperiodic tile sets that self-simulate with any shape that tiles the plane periodically.

From this, a new question is raised: how $complex$ can a self-simulation be? The previous construction shows that the self-simulation can be carried out by regular periodic structures. In the last section, we go a step further and show that self-simulations can be hidden in more complex structures such as self-similar structures. We show, with the fixed-point theorem $\grave{a}$ la Kleene on tilings, the existence of a tile set that self-simulates via a self-similar L-shape-structure.

In section 1, we recall the basic notions of tilings and simulation between tile sets. In section 2, we use our fixed-point theorem $\grave{a}$ la Kleene to prove in a natural and completely new way three of the most important results on tilings. In section 3, we define a generalization of the exact simulation with patterns that are not only rectangles and use this new notion to show how to build tile sets that self-simulate, and thus are aperiodic. Lastly, section 4 shows an example of a self-simulation based on a self-similar tiling.

1 Notions of Simulation

1.1 Tilings and Reductions

We begin with the basic notions of tilings. A tile is an oriented unit size square with colored edges from C, where C is a finite set of colors. A tile set is a finite set of tiles. To tile consists in placing the tiles of a given tile set on the grid $\mathbb{Z}^2$ such that two adjacent tiles share the same color on their common edge. Since a tile set can be described with a finite set of integers, then we can enumerate the tile sets, and τ_i designates the ith tile set. Let τ be a tile set. A tiling P generated by τ is called a τ-tiling. It is associated to a tiling function f_P where $f_P(x, y)$ gives the tile at position (x, y) in P.

Different notions of reduction have been introduced in [LW07, LW08a, DRS08]. We recall some of the notions relative to these reductions and we refer the reader to these papers for detailed explanations and exposition of their properties.

A pattern is a finite tiling. If it is generated by τ, we call it a τ-pattern. A reduction function R of size (a, b) between two tile sets τ and τ' is a function from the $a \times b$ τ-patterns to the tiles of τ. A τ-tiling P simulates a τ'-tiling Q if there exists two integers a, b and a reduction function R of size (a, b) between τ and τ' such that, if we cut P in rectangular patterns of size $a \times b$ and replace each of these patterns by their corresponding tiles given by R then we obtain Q. In such a case we write $P \trianglelefteq Q$, or $P \trianglelefteq^R Q$ when we want to specify the reduction. Therefore, with P it is possible to build Q. The reduction function R is not necessarily one-to-one; an aperiodic tiling can simulate a periodic tiling.

1.2   Simulations

This notion of reduction refers to tilings. To compare tile sets, we need to compare the sets of tilings that they can produce. In [LW08a], and slightly differently in [DRS08], the following notion of simulation has been introduced to compare tile sets:

Definition 1. *Let τ and τ' be two tile sets. We say that τ exactly simulates τ' if there exist $a, b \in \mathbb{Z}$ and a one-one reduction function R from the $a \times b$ patterns of τ to the tiles of τ' such that the two following conditions are respected:*

i) for any τ'-tiling Q, there exists a τ-tiling P such that $Q \trianglelefteq^R P$;
ii) for any τ-tiling P, there exists a τ'-tiling Q such that $Q \trianglelefteq^R P$.

We denote it by $\tau' \trianglelefteq_e \tau$ (or $\tau' \trianglelefteq_e^R \tau$ to specify the reduction R).

Therefore, a tile set τ exactly simulates a tile set τ' if it can simulate any τ'-tiling and if it cannot generate a tiling that does not simulate a τ'-tiling. Moreover, the behaviors of τ and τ' are really close since both of them are generating the same kind of tilings at different *levels*.

In this paper, we will use only the exact simulation, since it appears to be the most restrictive one and the most adapted to compare tile sets. We note that exact simulation preserves, among other properties, tilability, periodicity, self-similarity, *i.e.*, if τ exactly simulates a tile set which tiles the plane or is periodic or is self-similar then τ, *resp.*, tiles the plane or is periodic or is self-similar.

2   Proving Undecidability Results Using a Fixed-Point Theorem

In this section, we show how to use a technique of diagonalization over tilings to prove in a more intuitive way some of the main classical results on tilings. These kinds of proofs are direct with our fixed-point theorem *à la* Kleene [LW08b] since it embodies the idea of diagonalization. We recall the fixed-point theorem *à la* Kleene that has been proved in [LW08b]:

Theorem 1 (Fixed-point theorem *à la* Kleene for tilings, [LW08b])
Given a recursive function f, there exists an integer e such that τ_e exactly simulates $\tau_{f(e)}$.

When Wang introduced the model of tiling of the plane with colored tiles [Wan61, Wan62], a first conjecture was raised: the domino problem, *i.e.*, the problem to know whether a given tile set can tile the whole plane. Berger, Wang's student, proved in a long and technical proof the undecidability of the problem by simulating a Turing machine in a tiling [Ber66]. The proof was strongly improved in [Rob71, AD97] and more recently in [Oll08]. These enhancements introduced neat constructions in which the simulation of Turing machines can be carried. In this paper we present a direct proof of the undecidability of the domino problem using a diagonalization argument.

Theorem 2. *The domino problem is undecidable.*

Proof. We suppose that the problem is decidable. We build a tile set τ which tiles the plane if and only if τ does not tile the plane. Since we have supposed that the problem is decidable, we have a recursive function f such that $f(i)$ outputs 1 if τ_i tiles the plane and outputs 0 otherwise. Let j be the index of a tile set that tiles the plane and k the index of a tile set that does not tile the plane. Using f, we build a recursive function g which is defined as follows: $g(i) = j$ if $f(i) = 0$ and $g(i) = k$ otherwise.

Since g is a recursive function, by our fixed-point theorem *à la* Kleene there exists a tile set τ_e such that τ_e exactly simulates $\tau_{g(e)}$. If τ_e tiles the plane, then $\tau_{g(e)} = \tau_k$ does not tile it and if τ_e does not tile the plane, then $\tau_{g(e)} = \tau_j$ tiles it. Since τ_e exactly simulates $\tau_{g(e)}$, then τ_e tiles the plane if and only if $\tau_{g(e)}$ also tiles it. Therefore, τ_e tiles the plane if and only if τ_e does not tile it, which is a contradiction. □

This construction shows how to diagonalize over tilings and makes way for direct proofs of other undecidability results on tilings. The diagonalisation construction proves two other main classical results. These two results have been proved by Gurevich and Koryakov in [GK72]:

Corollary 1. *1. The problem to know whether a tile set can tile the plane periodically is undecidable;*
2. let A be the set of periodic tile sets and B the set of tile sets that does not tile the plane. Then A is recursively inseparable of B, i.e., there does not exist a recursive set C such that $A \subset C$ and $B \subset \overline{C}$.

This kind of proof by diagonalization can be used for any property on tile sets which is preserved by exact reduction, *i.e.*, a property is *preserved* by exact simulation if for any tile set τ_i, if τ_k exactly simulates τ_i, then τ_k satisfies the property P if and only if τ_i satisfies P. We strongly believe that a deeper generalization of these kinds of constructions will soon lead to a real Rice theorem for properties on tilings and obtain a powerful tool to understand undecidability related to tilings.

3   Building New Aperiodic Tile Sets by Way of Theorem 1

The quest for aperiodic tile sets is one of the most intriguing in the tilings field of study since these tile sets have the property to tile the plane only in an irregular way. If building a tile set that generates aperiodicity is quite easy, it is harder to show that it cannot generate periodicity. The research of the last forty years has shown different examples of aperiodic tile sets [Ber66, Rob71, Mye74, CK97] [DLS01, Oll08]. One of the most interesting construction for our purpose is the one of Durand *et al.* [DRS08] where an aperiodic tile set is built with the help of the classical Kleene's recursion theorem. In this paper we generalize this construction using our fixed-point theorem *à la* Kleene for tilings and we give a framework to build original aperiodic tile sets. The idea is to transform the tiles of a tile set in patterns of a same domain $\mathcal{A}$ and then apply our fixed-point theorem *à la* Kleene to obtain a tile set which simulates itself under some conditions that guarantee the aperiodicity of this tile set.

3.1   Regular Transformations of the Plane

First of all, we define the transformation of $\mathbb{Z}^2$ by a periodic tiling of the plane: Let $\mathcal{A}$ be a pattern of $\mathbb{Z}^2$ that generates a periodic tiling P, *i.e.*, there exists two independent vectors of $\mathbb{Z}^2$ such that the tiling P is invariant by translation of these vectors. We say that $\mathcal{A}$ *induces a regular transformation* of $\mathbb{Z}^2$ by P, or just *induces a transformation of* $\mathbb{Z}^2$, if there exists a one-one function $f : \mathbb{Z}^2 \to P$ such that, for any $x, y \in \mathbb{Z}^2$, $f(x)$ and $f(y)$ are two neighbor patterns of P if and only if x and y are two neighbors in $\mathbb{Z}^2$. The transformation f is called a *regular transformation* of the plane. A regular transformation of the plane is thus a transformation of $\mathbb{Z}^2$ which preserves the neighborhood relation.

3.2   Generalization of the Exact Simulation

The exact simulation seen previously allows only to simulate the tiles of a tile set with rectangular patterns. We need to extend this notion to allow simulations with patterns which are not only rectangles but any pattern of domain $\mathcal{A}$ where $\mathcal{A}$ induces a transformation of $\mathbb{Z}^2$. Indeed, if a pattern $\mathcal{A}$ induces a transformation of the plane $\mathbb{Z}^2$, then we can obtain a notion of simulation with patterns of domain $\mathcal{A}$ since the neighborhood relationship and, therefore, the local constraints are preserved.

Let P be a periodic tiling generated by $\mathcal{A}$. Since $\mathcal{A}$ induces a transformation of the plane by P, then there exists a regular transformation $f : \mathbb{Z}^2 \to P$ which preserves the neighborhood relationship. We define the following reduction: for two tile sets τ and τ', a *reduction function of domain* $\mathcal{A}$, denoted $R^{\mathcal{A}}$, is a function from the τ-patterns of domain $\mathcal{A}$ to the tiles of τ'.

The figure 1.1 shows a reduction from patterns of domain $\mathcal{A}$ to tiles. With this extended reduction function notion, we obtain a natural notion of simulation: let Q be a τ-tiling and Q' be a τ'-tiling. We say that Q *simulates* Q' *by* $\mathcal{A}$ if there exists a reduction function $R^{\mathcal{A}}$ from the τ-patterns of domain $\mathcal{A}$ to the tiles of τ' such that, if we cut Q in pattern of domain $\mathcal{A}$ with respect to P (where P is a periodic tiling generated by $\mathcal{A}$), and replace these patterns by their corresponding tiles given by $R^{\mathcal{A}}$, then we obtain Q'. We denote this reduction $Q' \trianglelefteq_{\mathcal{A}} Q$ or $Q' \trianglelefteq_{\mathcal{A}}^{R^{\mathcal{A}}} Q$ when we want to specify the reduction function.

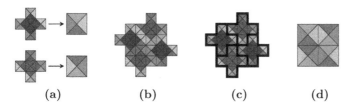

(a)   (b)   (c)   (d)

Fig. 1. If we cut the tiling of (b) in patterns of domain $\mathcal{A}$ (c) and apply the reduction of (a) we obtain the tiling of (d)

The figure 1 shows how to go from a tiling Q to a tiling Q' with a reduction function $R^{\mathcal{A}}$.

By generalization of the exact simulation we obtain the following definition:

Definition 2. *Let* τ *and* τ' *be two tile sets. Let* $\mathcal{A}$ *be a pattern that induces a transformation of* $\mathbb{Z}^2$. *We say that* τ *exactly simulates* τ' *by* $\mathcal{A}$ *if there exists a one-one reduction function* $R^{\mathcal{A}}$ *from the* τ-*patterns of domain* $\mathcal{A}$ *to the tiles of* τ' *such that the two following conditions are respected:*

i) for any τ'-*tiling* Q' *there exists a* τ-*tiling* Q *such that* $Q' \trianglelefteq^{R^{\mathcal{A}}}_{\mathcal{A}} Q$;

ii) for any τ-*tiling* Q *there exists a* τ'-*tiling* Q' *such that* $Q' \trianglelefteq^{R^{\mathcal{A}}}_{\mathcal{A}} Q$.

We denote this reduction by $\tau \trianglelefteq_{e,\mathcal{A}} \tau'$.

One can see that if we restrict the patterns $\mathcal{A}$ to rectangles, then we obtain the *exact* simulation.

3.3   Self-similar Aperiodic Tile Sets

From this definition, we want to build tile sets that exactly self-simulate by a pattern $\mathcal{A}$. This can be done by building a mapping that transforms a tile set τ in a tile set τ' that exactly simulates τ by $\mathcal{A}$, and then applying our fixed-point theorem *à la* Kleene to obtain a tile set that simulates itself.

One can note that the trivial tile set composed of only one unicolor tile produces a unique tiling that self-simulates in any possible way. Thus, since we want our fixed-point to be aperiodic, we need to characterize tile sets that self-simulate:

Lemma 1. *Let* $\mathcal{A}$ *be a pattern that induces a regular transformation of the plane. If* τ *is a tile set that exactly self-simulates by* $\mathcal{A}$, *then:*

i) either τ *generates a deterministic periodic pattern, i.e., there exists* $A \subseteq \tau$ *such that* τ *generates a periodic pattern using once and only once each tile of* A^1;

ii) or τ *is aperiodic.*

---

[1] We note that it is recursive to know whether a tile set τ generates or not a deterministic periodic pattern since it is enough to check this property for the finitely many subsets of τ.

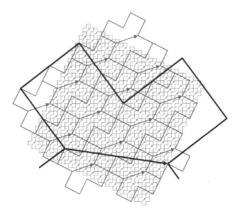

Fig. 2. The self-similarity structure of our aperiodic tile set

3.4   Building Some New Aperiodic Tile Sets

With the help of lemma 1, we can obtain a panel of aperiodic tile sets. Indeed, for any pattern $\mathcal{A}$ that induces a regular transformation of $\mathbb{Z}^2$, we can build an aperiodic tile set. We give an example of such a construction.

For our example, we use a pattern with the shape of a cross ⌗. Since this pattern induces a regular transformation of $\mathbb{Z}^2$ then we can build, for any tile set τ, a tile set τ' such that τ' exactly simulates τ by $\mathcal{A}$ (figure 1.a) We called f the function that takes as input the index of a tile set τ_i and outputs the index of the tile set $\tau_{g(i)}$ which exactly simulates τ_i by $\mathcal{A}$. Using f, we build the following recursive function g: it takes as input the code i of a tile set. If the tile set generates a deterministic periodic pattern, then g outputs the code of a tile set that does not tile the plane. The function g outputs $f(i)$ otherwise.

Since g is recursive, by the theorem à la Kleene there exists a fixed-point τ_e such that τ_e exactly simulates $\tau_{g(e)}$. Since $\tau_{g(e)}$ exactly simulates τ_e by $\mathcal{A}$, then τ_e exactly simulates itself by $\mathcal{A}'$ where $\mathcal{A}'$ corresponds to a pattern that has the same shape than $\mathcal{A}$ but enlarged by a factor corresponding to the size of the reduction between τ_e and $\tau_{g(e)}$. Indeed, we can see that each tile of τ_e is represented by a pattern of domain $\mathcal{A}$ in $\tau_{g(e)}$ (*exact* simulation by $\mathcal{A}$) and each tile of $\tau_{g(e)}$ is represented by a rectangular pattern in τ_e (*exact* simulation).

By the lemma 1, τ_e is either aperiodic, or it generates a deterministic periodic pattern. Because of the construction of g which outputs the code of a tile set that does not tile the plane when given in input a tile set that generates a deterministic periodic pattern, we have that such a tile set cannot be the fixed-point. Therefore, the fixed-point τ_e is an aperiodic tile set.

The figure 2 shows the structure of the tilings generated by the fixed-point. We note that this tiling is self-similar and has the property that the simulation turns at each step. We note that if we restrict this construction to rectangular patterns $\mathcal{A}$, then we obtain the construction given in [DRS08]. Our construction is more general because it allows the construction of aperiodic tile sets for any pattern $\mathcal{A}$ that induces a transformation of the plane. In the following section, we show how to generalize the construction of tile sets that self-simulate using a self-similar tile set.

4   Pushing the Construction a Step Further

The definition 2 restricted the patterns by which a self-simulation can be done since the neighborhood relationship has to be preserved. With this construction, the self-simulation can be made only *via* periodic patterns. In this section, we show that the structure in which the self-simulation is made can be more complicated. For this purpose, we use self-similar tilings. Our main example builds a self-similar tile set based on L-shapes[2] where the L-shape structure is aperiodic and is represented in figure 3.b.

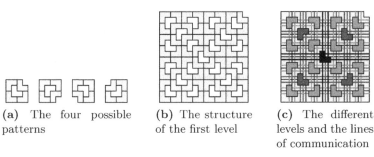

(a) The four possible patterns

(b) The structure of the first level

(c) The different levels and the lines of communication

Fig. 3

The first step consists in making L-shapes with tiles. For that, we need three tiles for any of the four possible L-shapes. These tiles can assemble in an unique way. We call these L-shapes, L of level 1. The second step consists in assembling four L of level 1. Four L of level 1 are assembled in such a way that they surround an L-shape. Therefore there exist four different ways to assemble four L of level 1, given in figure 3.a. The central L-shape is called a L of level 2. The pattern composed of four L of level 1 and completed with an L of level 2 in its center is called a pattern of level 2. By recursion, four patterns of level i are assembled to define an L-shape of level i in its center. The pattern obtained is called a pattern of level $i + 1$. For instance, the figure 3.a represents the four possible assemblages of four L of level 1 that define an L of level 2 in their center; the figure 3.c represents a pattern of level 4 and the black L-shape in its middle is an L of level 4. We now show that this recursion process can be done with tiles.

To be sure that any tiling generated by this tile set respects the hierarchy of levels of patterns, we add communication corridors between all the L's of same level. All the L's of level 1 are directly in contact with their neighbors. Therefore the colors on the sides of the tiles can guarantee that the L's of level 1 are placed

---

[2] This construction includes corridors of communication to make possible the self-simulation which will be explained later. The self-similarity based on L-shapes can also be made without them.

in the right way. One can note that an L of a certain level i lays on the intersection of two lines and two columns; These two lines and two columns contain only L's of level i (and parts of L's of level 1). By generalization, for any i, there exists a regular subgrid composed of lines and columns of width 2 where the vertex define the L's of level i. Therefore, any tile of the tiling belongs to one column and one line of a subgrid which means that any tile has to transmit two, and only two, kinds of information: one vertical and one horizontal. To achieve this goal, it suffices to superimpose to any tile the different communications that it can send. The figure 3.c shows the different levels of L's and the communication corridors between them.

We obtain a tiling where two neighbors L's of level i can communicate and guarantee that they are put in the right way, as it is done at level 1. At the end, we obtain a tiling where any level of L's respects the recursion process defined previously and obtain with Wang tiles the well-known self-similar L-tiling.

The goal is to use this particular self-similar tile set to simulate other tile sets. To a tile set τ it is possible to build a tile set μ such that μ generates tilings with the L-self-similarity and such that the L-shapes generated by μ correspond to tiles of τ. To do this, we superimpose to any possible L-shape the code of the different tiles of τ. Therefore, a particular L-shape simulates a tile of τ. We now want that the L's of a certain level to simulate a τ-tiling. We already know that any L simulates a tile. We need to guarantee that the local constraint is respected. To achieve this goal, we use the communication corridors. An L of level i has exactly four neighbors L's on its north, south, east and west side. These four L's have to represent tiles that match with the tile represented by L. The communication corridor is used to send also the information of the colors of the sides of the tiles that the L's represents. The corridor between two neighbor L's of same level is completed if and only if both of the L's have sent the same color. Therefore, we guarantee that all the L's of a level i represent the tiles of a τ-tiling. By recursion, this is done at any level, and therefore, any μ-tiling simulates a countable infinite number of τ-tilings.

The transformation defined previously can be made by a recursive function f that transforms a tile set τ_i in the tile set $\tau_{f(i)}$ that simulates τ_i with the L-structure. Therefore, by application of the fixed-point theorem à la Kleene, there exists a tile set τ_e which is a fixed-point for f which means that τ_e self-simulates: a τ_e-tiling P has the L-self-similarity, and any level of L's of P represents a τ_e-tiling.

Therefore, we have shown that there exist tile sets that self-simulate via self-similar tile sets. This self-simulation can be extended to a whole lot of other self-similar tile sets, such as, for example, the ones of Amman [AGS92].

References

[AD97]   Allauzen, C., Durand, B.: Appendix A: Tiling problems. In: The classical decision problem, Perspectives of Mathematical Logic, pp. 407–420. Springer, Heidelberg (1997)

[AGS92]   Ammann, R., Grünbaum, B., Shephard, G.C.: Aperiodic tiles. Discrete Comput. Geom. 8, 1–25 (1992)

[Ber66]   Berger, R.: The undecidability of the domino problem. Mem. Amer. Math Soc. 66, 1–72 (1966)

[CK97]   Culik, K., Kari, J.: On aperiodic sets of Wang tiles. In: Freksa, C., Jantzen, M., Valk, R. (eds.) Foundations of Computer Science. LNCS, vol. 1337, pp. 153–162. Springer, Heidelberg (1997)

[DLS01]   Durand, B., Levin, L.A., Shen, A.: Complex tilings. In: STOC 2001: Proceedings of the thirty-third annual ACM symposium on Theory of computing, pp. 732–739 (2001)

[DRS08]   Durand, B., Romashchenko, A.E., Shen, A.: Fixed point and aperiodic tilings. Developments in Language Theory, 276–288 (2008)

[GK72]   Gurevich, Y.S., Koryakov, I.O.: Remarks on berger's paper on the domino problem. Siberian Math. Jour. 13(2), 319–321 (1972)

[LW07]   Lafitte, G., Weiss, M.: Universal tilings. In: Thomas, W., Weil, P. (eds.) STACS 2007. LNCS, vol. 4393, pp. 367–380. Springer, Heidelberg (2007)

[LW08a]   Lafitte, G., Weiss, M.: Simulations between tilings. In: Beckmann, A., Dimitracopoulos, C., Löwe, B. (eds.) Logic and Theory of Algorithms, 4th Conference on Computability in Europe, CiE 2008, Athens, Greece. University of Athens (June 2008)

[LW08b]   Lafitte, G., Weiss, M.: Computability of tilings. In: International Federation for Information Processing, Fifth IFIP International Conference on Theoretical Computer Science, vol. 273, pp. 187–201 (2008)

[Mye74]   Myers, D.: Nonrecursive tilings of the plane. II. J. Symb. Log 39(2), 286–294 (1974)

[Oll08]   Ollinger, N.: Two-by-two substitution systems and the undecidability of the domino problem. In: Beckmann, A., Dimitracopoulos, C., Löwe, B. (eds.) CiE 2008. LNCS, vol. 5028, pp. 476–485. Springer, Heidelberg (2008)

[Rob71]   Robinson, R.M.: Undecidability and nonperiodicity for tilings of the plane. Inv. Math 12, 117–209 (1971)

[Wan61]   Wang, H.: Proving theorems by pattern recognition II. Bell Systems Journal 40, 1–41 (1961)

[Wan62]   Wang, H.: Dominoes and the ∀∃∀ -case of the decision problem. In: Proc. Symp. on Mathematical Theory of automata, pp. 23–55 (1962)

Relationship between Kanamori-McAloon Principle and Paris-Harrington Theorem

Gyesik Lee

ROSAEC center, Seoul National University
599 Gwanak-ro, Gwanak-gu Seoul, 151-742 Korea
gslee@ropas.snu.ac.kr

Abstract. We give a combinatorial proof of a tight relationship between the Kanamori-McAloon principle and the Paris-Harrington theorem with a number-theoretic parameter function. We show that the provability of the parametrised version of the Kanamori-McAloon principle can exactly correspond to the relationship between Peano Arithmetic and the ordinal ε_0 which stands for the proof-theoretic strength of Peano Arithmetic. Because A. Weiermann already noticed the same behaviour of the parametrised version of Paris-Harrington theorem, this indicates that both propositions behave in the same way with respect to the provability in Peano Arithmetic.

Keywords: Kanamori-McAloon principle, Paris-Harrington theorem, Peano Arithmetic, independence.

1 Introduction

In combinatorics, the finite Ramsey theorem [1] shows that unavoidable structure exists in colourings of finite hypergraphs. In any colouring of the edges of a sufficiently large complete graph, we will always find monochromatic complete subgraphs. Below is a formal definition.

Given $X \subseteq \mathbb{N}$ and $n \in \mathbb{N}$, let $[X]^n$ denote the set of all subsets of X with n elements. Each $m \in \mathbb{N}$ is identified with the set of its predecessors $\{0, \ldots, m-1\}$. If C is a colouring defined on $[X]^n$, we write $C(x_1, \ldots, x_n)$ for $C(\{x_1, \ldots, x_n\})$ assuming $x_1 < \cdots < x_n$. A set $H \subseteq X$ is called C-*homogeneous* if C is constant on $[H]^n$. We write

$$X \to (k)^n_c$$

if for all $C : [X]^n \to c$ there exists C-homogeneous $H \subseteq X$ s.t. $card(H) \geq k$. Let $R(n, c, k)$ be the least natural number ℓ such that $\ell \to (k)^n_c$.

The Paris-Harrington theorem [2] adds the requirement that H be *relatively large*, i.e., $card(H) \geq \min(H)$. If for all $C : [X]^n \to c$ there exists a relatively large C-homogeneous $H \subseteq X$ s.t. $card(H) \geq k$, we denote this fact by

$$X \to^* (k)^n_c. \tag{1}$$

Then put PH $:= (\forall\, n, c, k)(\exists\, \ell)[\ell \to^* (k)^n_c]$. The independence of PH from Peano Arithmetic is obviously caused by the *relative largeness* condition because the

K. Ambos-Spies, B. Löwe, and W. Merkle (Eds.): CiE 2009, LNCS 5635, pp. 310–323, 2009.
© Springer-Verlag Berlin Heidelberg 2009

finite Ramsey theorem itself is provable in a very weak system. Indeed, Erdős and Rado [3] showed the totality of R in $I\Delta_0 + (\exp)$:

Theorem 1 (Erdős and Rado [3]). *(In $I\Delta_0 + (\exp)$) Let $n, c, k > 0$ and $k \geq n$. Then*

$$R(n, c, k) \leq c * (c^{n-1}) * \cdots * (c^2) * (c(k - n) + 1).$$

Here $a * b := a^b$, and $*$ is right-associative.

That is, the upper bounds grows super-exponentially depending on the dimension n. Note that the right-hand side is approximately a tower of c's of height n.

Another Ramsey style proposition is introduced by Kanamori and McAloon [4]. A function $C : [X]^n \to \mathbb{N}$ is called *regressive* if $C(\bar{x}) < \min(\bar{x})$ for all $\bar{x} \in [X]^n$ such that $\min(\bar{x}) > 0$. A set $H \subseteq X$ is said to be *min-homogeneous* for C if, for all $s, t \in [H]^n$, $C(s) = C(t)$ holds whenever $\min(s) = \min(t)$. The fact that, given a regressive function C, such a min-homogeneous set H with $card(H) \geq k$ always exists is denoted by

$$X \to (k)^n_{reg}. \tag{2}$$

Put $KM := (\forall n, k)(\exists \ell)[\ell \to (k)^n_{reg}]$. KM is proved to be equivalent to PH, hence independent from Peano Arithmetic as well. Moreover, if we let PH^n and KM^n be the two propositions with fixed dimension n, then they are also equivalent. The following theorem summarises the main results in Paris and Harrington [2] and Kanamori and McAloon [4].

Theorem 2 ([2,4]). *Let $n > 0$.*

1. *(In $I\Sigma_1$) PH, KM, and the 1-consistency of PA are equivalent.*
2. *(In $I\Sigma_1$) PH^{n+1}, KM^{n+1}, and the 1-consistency of $I\Sigma_n$ are equivalent.*
3. *PH^n and KM^n are provable in $I\Sigma_n$.*
4. *PH^{n+1} and KM^{n+1} are $I\Sigma_n$-independent.*

The 1-consistency of a theory T is the assertion saying every T-provable Σ^0_1 sentence is true. We remark that the proofs in [2,4] are done model-theoretically.

A parametrised version of PH based on the growth rate of a number-theoretic function $f : \mathbb{N} \to \mathbb{N}$ is discussed in McAloon [5]. It results from replacing relative largeness with f-largeness, i.e. $card(H) \geq f(\min(H))$. We write $X \to^*_f (k)^n_c$ instead of (1) when the f-largeness holds. PH_f is then defined by

$$PH_f := (\forall n, c, k)(\exists \ell)[\ell \to^*_f (k)^n_c].$$

PH^n_f is defined with fixed dimension n. Let R^*_f denote the Skolem function associated with PH_f: $R^*_f(n, c, k)$ is the least ℓ such that $\ell \to^*_f (k)^n_c$.

A parametrised version of KM based on the growth rate of a function is given in Kanamori and McAloon [4]. A function $C : [X]^n \to \mathbb{N}$ is called f-*regressive* if for all $\bar{x} \in [X]^n$ such that $f(\min(\bar{x})) > 0$ we have $C(\bar{x}) < f(\min(\bar{x}))$. We write $X \to (k)^n_{f\text{-}reg}$ instead of (2) to denote that, given a f-regressive function

C, there exists always such a min-homogeneous set H for C with $card(H) \geq k$. KM_f is then denoted by

$$\mathrm{KM}_f :\equiv (\forall n, k)(\exists \ell)[\ell \to (k)^n_{f\text{-}reg}].$$

KM_f^n is defined with fixed dimension n. Let $R(\mu)_f$ denote the Skolem-function associated with KM_f: $R(\mu)_f(n, k)$ is the least ℓ such that $\ell \to (k)^n_{f\text{-}reg}$.

PH_f and KM_f are true for any function $f : \mathbb{N} \to \mathbb{N}$. They easily follow from König's lemma and the infinite version of Ramsey's theorem, cf. [4]. Note that both turn into the finite Ramsey theorem when f is a constant function. This and the PA-independence of PH and KM indicate that their PA-provability might depend on the growth rate of f. Indeed, Weiermann [6] showed that the PA-provability of PH_f depends on the growth rate of f by giving an interesting categorisation using Schwichtenberg-Buchholz-Wainer's fast-growing hierarchy [7,8,9].

In this paper, we introduce a way to establish the same result with respect to KM_f by revealing some relationship between KM_f and PH_f. Our results are based on some known results from [3,4] and an unpublished one by J. Paris. Let us first finish with preliminaries before abusing more undefined notions.

Peano Arithmetic (PA), is the first-order theory in the language with constants $0, 1, +, \cdot, <$, axioms defining the properties of these primitive notions, and the induction scheme for all formulae. A function $f : \mathbb{N}^d \to \mathbb{N}$ is provably recursive in PA if $f(\bar{m}) = n$ just holds when $\mathrm{PA} \vdash F(\bar{m}, n)$ for some formula $F(\bar{x}, y)$ which is Δ_1 in PA satisfying $\mathrm{PA} \vdash \forall \bar{x} \exists y\, F(\bar{x}, y)$. Here $\bar{x}$ stands for $x_1, ..., x_d$. $I\Sigma_n$ (resp. $I\Delta_0$) stands for Peano Arithmetic with the induction scheme for Σ_n (resp. Δ_0) formulae. (exp) denotes the totality of the exponential function: $\forall x \exists y\, (2^x = y)$.

2   The Fast-Growing Hierarchy and the Main Theorems

We start with recalling Schwichtenberg-Buchholz-Wainer's fast-growing Hierarchy $(F_\alpha)_{\alpha \leq \varepsilon_0}$. Given a function $f : \mathbb{N} \to \mathbb{N}$ and natural number d, f^d denotes the d-th iteration of f, i.e. $f^0(x) := x$ and $f^{d+1}(x) := f(f^d(x))$. Then the fast-growing hierarchy is defined as follows:

$$F_0(x) := x + 1, \quad F_{\alpha+1}(x) := F_\alpha^{x+1}(x), \quad \text{and} \quad F_\lambda(x) := F_{\lambda[x]}(x)$$

Here $\cdot [\cdot] : \varepsilon_0 \times \mathbb{N} \to \varepsilon_0$ is a fixed assignment of fundamental sequences to ordinals below ε_0 based on the Cantor Normal Form. $(\gamma + \omega^\lambda)[x] := \gamma + \omega^{\lambda[x]}$, $(\gamma + \omega^{\beta+1})[x] := \gamma + \omega^\beta \cdot x$, where $\alpha_0(x) := x$, $\alpha_{d+1}(x) := \alpha^{\alpha_d(x)}$ and $\alpha_d := \alpha_d(1)$. For technical reasons, we put $\varepsilon_0[x] := \omega_x$, $(\beta + 1)[x] := \beta$ and $0[x] := 0$.

Note that $F_1(x) = 2x$, $F_2(x) = 2^x \cdot x$, and $F_3(x) \geq 2_x(x)$. The function hierarchy $(F_\alpha)_{\alpha < \varepsilon_0}$ plays an important role in proof theory as the following theorem shows the close relationship to PA, which is a folklore in proof theory. We say a function f *captures* or *eventually dominates* another function g if there is an m such that $g(i) \leq f(i)$ for any $i \geq m$.

Theorem 3 ([7,8,9]). *Let $d > 0$. Then*

1. $\mathrm{PA} \vdash \forall x \exists y [F_\alpha(x) = y]$ *iff* $\alpha < \varepsilon_0$.
2. *Let f be a Σ_1-definable function. Then* PA *proves the totality of f if and only if f is primitive recursive in F_α for some $\alpha < \varepsilon_0$.*
3. F_{ε_0} *eventually dominates all* PA*-provably recursive functions.*

Based on Theorem 3, Weiermann [6] gave a classification of the provability threshold of PH using the following parameter functions:

$$|x| = \lfloor \log_2(x+1) \rfloor, \quad |x|_{d+1} = ||x|_d|, \quad \text{and} \quad \log^* x := \min\{d \,|\, |x|_d \leq 2\}.$$

For convenience, we put $|x|_0 = x$ and $\log_2 0 := 0$. For an unbounded function $f : \mathbb{N} \longrightarrow \mathbb{N}$ we denote by f^{-1} the inverse of f, i.e. $f^{-1}(x) := \min\{y : f(y) > x\}$. Note that $f^{-1}(x) \leq y$ iff $x < f(y)$. Given $\alpha \leq \varepsilon_0$, define

$$f_\alpha(i) := |i|_{F_\alpha^{-1}(i)}.$$

Then the PA-provability of PH_{f_α} corresponds exactly to the relationship between PA and the ordinal ε_0.

Theorem 4 (Weiermann [6])

1. $R^*_{\log^*}$ *is provably total in* $I\Sigma_1$, *hence* $I\Delta_0 + (\exp) \vdash \mathrm{PH}_{\log^*}$.
2. *For all $d \in \mathbb{N}$, the totality of $R^*_{|\cdot|_d}$ is not provable in* PA, *hence* $\mathrm{PH}_{|\cdot|_d}$ *is* PA*-independent.*
3. *The totality of $R^*_{f_\alpha}$ is provable in* PA *iff $\alpha < \varepsilon_0$, hence* PH_{f_α} *is* PA*-provable iff $\alpha < \varepsilon_0$.*

Proof. The provability part follows easily from Erdős and Rado's upper bound for the standard Ramsey theorem in Theorem 1. For the unprovability part, Weiermann constructed some interesting colouring functions which are refinements of the ordinal-based colourings by Loebl and Nešetřil [10]. In fact, he showed

- $R^*_{|\cdot|_{n-2}}(n+1, 3_n(n+k+3), k^{(n)}) \geq F_{\omega_{n-1}(k)}(k-1)$ for any $n \geq 2$, $k \geq 4$,
- $R^*_{f_{\varepsilon_0}}(n+1, 3_n(2n+3), n^{(n)}) \geq F_{\varepsilon_0}(n-2)$ for any $n \geq 4$,

where $k^{(i)} := k + 3 + 3^2 + \cdots + 3^i$.   □

Remark 5. Characterisation of the PA-provability of some propositions using parameter functions in the form of f_α is first introduced in Arai [11].

Corollary 6 (Weiermann [6]). *There exist two primitive recursive functions $p, q : \mathbb{N} \to \mathbb{N}$ such that*

$$R^*_{id}(n, p(n), q(n)) \geq F_{\varepsilon_0}(n-3)$$

for any $n \geq 5$.

Our main theorems say that the same classification of parameter functions for KM_f holds:

1. $R(\mu)_{\log^*}$ is provably total in $I\Sigma_1$, hence $I\Delta_0 + (\exp) \vdash KM_{\log^*}$.
2. For all $d \in \mathbb{N}$, the totality of $R(\mu)_{|\cdot|_d}$ is not provable in PA, hence $KM_{|\cdot|_d}$ is PA-independent.
3. The totality of $R(\mu)_{f_\alpha}$ is provable in PA iff $\alpha < \varepsilon_0$, hence KM_{f_α} is PA-provable iff $\alpha < \varepsilon_0$.

The proofs will be divided into Theorem 8, Theorem 12, and Theorem 13.

3   Provability

We start with a simple proof of the provability of $KM_{\log^*}$ and KM_{f_α} with $\alpha < \varepsilon_0$.

Lemma 7. *Given an unbounded function* $f : \mathbb{N} \to \mathbb{N}$, *set* $f' := |\cdot|_{f^{-1}(\cdot)}$. *Then there is a primitive recursive function* $p : \mathbb{N}^2 \to \mathbb{N}$ *satisfying*

$$R_{f'}(n,k) \leq 2_n(f(p(n,k))) + f(p(n,k))$$

for all $n, k > 0$.

Proof. We may assume w.l.o.g. that $f(x) \geq x$ for any $x \in \mathbb{N}$. Given $n, k > 0$, assume $n \leq k$ and let $p(n,k)$ be the least $x > n$ such that

$$L(n,k) := (c) * (c^{n-1}) * \cdots * (c^2) * (c(k-n)+1) < 2_n(c),$$

where $c := f(x)$. The existence of such an x can be seen by noting that the right-hand side is approximately a tower of c of height $n+1$ while the left-hand side is approximately of height n. Indeed we can show for $c \geq 2^{2(n+k)}$ that $L(n,k) \leq c_n(n) \leq 2_n(c)$. That is, $p(n,k) := 4^{n+k}$ is a good candidate.
   Put $m := 2_n(f(p(n,k))) + f(p(n,k))$. Then $m \leq 2_{n+1}(f(p(n,k)))$. Indeed, $2_n(i) + i \leq 2_{n+1}(i)$ for all n, i. Let $C : [m]^n \to \mathbb{N}$ be any f'-regressive function. Define a new f'-regressive function

$$D : [f(p(n,k)), m]^n \to \mathbb{N}$$

by restricting C. Note that, for any $y \in [f(p(n,k)), m]$, it holds

$$f'(y) \leq |2_{n+1}(f(p(n,k)))|_{f^{-1}(f(p(n,k)))} = |2_{n+1}(f(p(n,k)))|_{p(n,k)+1} < f(p(n,k)).$$

This implies that $\mathrm{Im}(D) \subseteq f(p(n,k))$. By Theorem 1, there is a D-homogeneous set Y, hence min-homogeneous for C such that $card(Y) \geq k$.   □

It is now obvious to see the provability part. Note that $\log^*(n) \leq Exp_s^{-1}(n)$, where Exp_s denotes the super-exponential function defined by $Exp_s(n) := 2_n(1)$.

Theorem 8. *Let* $\alpha < \varepsilon_0$.

1. $KM_{\log^*}$ *is provable in* $I\Delta_0 + (\exp)$.
2. KM_{f_α} *is provable in* PA.

Proof. Both claims follow from Lemma 7 together with Theorem 3.   □

4   Independence

We now turn our attention to the independent proof. We start with recalling a lemma.

Lemma 9 (Kanamori and McAloon [4]). *Let $I \subseteq \mathbb{N}$. If $C : [I]^n \to \mathbb{N}$ is regressive, then $H \subseteq I$ is min-homogeneous for C iff every $Y \subseteq H$ of cardinality $n+1$ is min-homogeneous for C.*

The following lemma is slightly modified from Lemma 3.3 in [4]. Given a function $f : \mathbb{N} \to \mathbb{N}$, the function $|f|$ is defined by $|f|(i) = |f(i)|$. Given $x \in \mathbb{N}$, note that any $y < x$ can be represented as $(y_0, \dots, y_{d-1})_2$ in binary notation, where $d = |x|$. Put $C(\bar{x}) = (C_0(\bar{x}), \dots, C_{d-1}(\bar{x}))_2$, where $\bar{x} \in [x, y]^n$ and $d = |f(\min(\bar{x}))|$.

Lemma 10. *If $C : [\ell]^n \to y$ is f-regressive and $f(i) \leq i$, then there is a regressive function C' such that*

- $C' : [\ell]^{n+1} \to y$ is $(2|f| + 1)$-regressive, and
- if H' is min-homogeneous for C', then $H = H' - (7 \cup \{\max(H')\})$ is min-homogeneous for C.

Proof. Define C' on $[x, y]^{n+1}$ as follows:

- $C'(x_0, \dots, x_n) = 0$ if either $x_0 < 7$, or $\{x_0, \dots, x_n\}$ is min-homogeneous for C;
- $C'(x_0, \dots, x_n) = 2i + C_i(x_0, \dots, x_{n-1}) + 1$, otherwise, where $i < |f(x_0)|$ is the least such that $\{x_0, \dots, x_n\}$ is not min-homogeneous for C_i.

Then C' is $2|f| + 1$-regressive and even regressive because $2|f(x_0)| + 1 \leq 2|x_0| + 1 \leq x_0$ for any $x_0 \geq 7$. Suppose H' is min-homogeneous for C' and H is as described. If $C' \restriction [H]^{n+1} = \{0\}$, then we are done by the previous lemma. Suppose on the contrary that there were $x_0 < \cdots < x_n$ all in H such that $C'(x_0, \dots, x_n) = 2i + C_i(x_0, \dots, x_{n-1}) + 1$. Given any $\bar{s}, \bar{t} \in [\{x_0, \dots, x_n\}]^n$ with $\min(\bar{s}) = \min(\bar{t}) = x_0$, note that

$$C'(\bar{s} \cup \{\max(H')\}) = C'(x_0, \dots, x_n) = C'(\bar{t} \cup \{\max(H')\})$$

by the min-homogeneity of H'. But then, $C_i(\bar{s}) = C_i(\bar{t})$, so that $\{x_0, \dots, x_n\}$ were min-homogeneous for C_i after all, which contradicts the assumption.   □

Now we are going to apply Lemma 10 iteratively. Define a sequence of functions as follows:

$$g_0(i) := i \quad \text{and} \quad g_{m+1}(i) := 2 \cdot |g_m(i)| + 1$$

Lemma 11. *Let $n, m \geq 1$. Then, in $I\Sigma_1$, $\mathrm{KM}_{g_m}^{n+m}$ implies KM^n.*

Proof. Assume $\mathrm{KM}_{g_m}^{n+m}$. Let k be given, then by assumption we can find ℓ satisfying

$$\ell \to (k + m + 7)_{g_m\text{-}reg}^{n+m}.$$

We claim $\ell \to (k)^n_{reg}$ holds. Suppose $C : [\ell]^n \to \ell$ is regressive. By applying the previous lemma m times, we get a g_m-regressive function $C' : [\ell]^{n+m} \to \ell$. Then there is a set $H' \subseteq \ell$, min-homogeneous for C', such that $card(H') \geq k + m + 7$, so

$$H = H' - 7 \cup \{\text{the last } m \text{ elements of } H'\}$$

is min-homogeneous for C and $card(H) \geq k$.   □

Theorem 12. *Let d be a natural number.*

1. In $I\Sigma_1$, $KM_{|\cdot|_d}$ implies KM.
2. $KM_{|\cdot|_d}$ is PA-independent, i.e. $R(\mu)_{|\cdot|_d}$ is not provably total in PA.

Proof. Note just that $KM_{g_{d+1}}$ follows from $KM_{|\cdot|_d}$. Therefore, $R(\mu)_{|\cdot|_d}$ cannot be captured by F_{ε_0}.   □

It remains to prove the PA-independence of $KM_{f_{\varepsilon_0}}$. Here we introduce two ways to prove it. The first one is based on an unpublished, combinatorial proof[1] by J. Paris of $(KM^n \to PH^n)$ by showing the following:

$$I\Delta_0 + (\exp) \vdash (KM^n \to KM^n_{Exp_2}) \wedge (KM^n_{Exp_2} \to PH^n)$$

for any $n \geq 2$, where $Exp_2(x) := 2^{2^x}$.

The second one is based on a refinement of Paris' proof. Indeed, we will give a direct, combinatorial proof of $(KM^n \to PH^n)$. The proof is not just a composition of two proofs given by Paris, but a real refinement of it.

The point here is that both proofs given by Paris and us are purely combinatorial and hence provide us with a primitive recursive function $p_1 : \mathbb{N}^3 \to \mathbb{N}$ satisfying

$$R(\mu)(n, p_1(n, c, k)) \geq R^*(n, c, k) \tag{3}$$

for any $n \geq 2$ and for any $c, k \geq 0$. And we can use this fact to prove the PA-independence of KM_{f_α}. We first show how to make use of (3) and then present a sketch of our refined proof in Appendix.

Theorem 13. $KM_{f_{\varepsilon_0}}$ *is PA-independent.*

Proof. From the proof of Lemma 11, we know

$$R(\mu)_{|\cdot|_d}(n + d + 1, k + d + 8) \geq R(\mu)_{g_{d+1}}(n + d + 1, k + d + 8) \geq R(\mu)(n, k)$$

for any positive n, d, k. Using this and (3), it follows that

$$R(\mu)_{|\cdot|_d}(n + d + 1, p_1(n, c, k) + d + 8) \geq R^*(n, c, k),$$

and hence

$$R(\mu)_{|\cdot|_d}(n + d + 1, p_1(n, p(n), q(n)) + d + 8) \geq F_{\varepsilon_0}(n - 3)$$

---

[1] This is mentioned in Kanamori-McAloon [4], page 39, and we got a copy of the proof.

for any $n \geq 5$, where p, q are some primitive recursive functions satisfying Corollary 6. This indicates the existence of a primitive recursive function $p' : \mathbb{N} \to \mathbb{N}$ such that, putting $d := n$,

$$R(\mu)_{|\cdot|_n}(2n + 1, p'(n)) \geq F_{\varepsilon_0}(n - 3) \tag{4}$$

for any $n \geq 5$. We use this fact to show that PA cannot prove the totality of $R(\mu)_{f_{\varepsilon_0}}$. Let $n \geq 5$ and p' be a primitive recursive function satisfying (4). We claim

$$R(\mu)_{f_{\varepsilon_0}}(2n + 1, p'(n)) > F_{\varepsilon_0}(n - 3),$$

which implies that $R_{f_{\varepsilon_0}}$ cannot be provably recursive in PA.

Assume otherwise. Note first that by (4), there is some $|\cdot|_n$-regressive function $G : [F_{\varepsilon_0}(n - 3) - 1]^{2n+1} \to \mathbb{N}$ which has no min-homogeneous set of cardinality $p'(n)$. On the other hand, G is f_{ε_0}-regressive. In fact, for all $i < F_{\varepsilon_0}(n - 3)$ it holds that $F_{\varepsilon_0}^{-1}(i) \leq n - 3$, hence $|i|_n \leq |i|_{n-3} \leq |i|_{F_{\varepsilon_0}^{-1}(i)}$. This means G should have a min-homogeneous set of cardinality $p'(n)$, which is not possible.   □

Finally, Theorem 8, Theorem 12, and Theorem 13 establish that PH_f and KM_f behave in the same manner with respect to the provability in PA.

Theorem 14. *Let* $f_\alpha(i) := |i|_{F_\alpha^{-1}(i)}$, *where* $\alpha \leq \varepsilon_0$. *Then*

1. *Both* $KM_{\log^*}$ *and* $PH_{\log^*}$ *are PA-provable.*
2. *For all* $d \in \mathbb{N}$, *both* $KM_{|\cdot|_d}$ *and* $PH_{|\cdot|_d}$ *are PA-independent.*
3. KM_{f_α} *(resp.* PH_{f_α}) *is PA-provable if and only if* $\alpha < \varepsilon_0$.

5 Conclusion

What we have shown in this paper is about the *global* relationship between two propositions KM_f and PH_f with respect to the provability in PA. Contrary to the same behaviour in the global level, they behave differently in the local level as shown by Carlucci, Lee, and Weiermann [12] where it is shown that

$$I\Sigma_n \vdash KM_{|\cdot|_d}^{n+1} \text{ if and only if } d \geq n,$$

while Weiermann [13] showed that

$$I\Sigma_n \vdash PH_{|\cdot|_d}^{n+1} \text{ if and only if } d \geq n + 1.$$

In Kojman, Lee, Omri, and Weiermann [14] one can see some refinements of the results about Kanamori-McAloon principle in case $n = 1$.

The methodology used in current paper is different from that of [12] although the results in the latter are refinements of those in this paper. Here we investigated the global relationship between PH and KM in a direct way, while the results in [12] and [13] are compared indirectly. We also refer to Bovykin [15] where related results on the Kanamori-McAloon principle are introduced using a model-theoretic approach à la Paris and Harrington [2].

References

1. Ramsey, F.P.: On a problem of formal logic. Proc. London Math. Soc. 30(2), 264–286 (1929)
2. Paris, J., Harrington, L.: A mathematical incompleteness in peano arithmetic. In: Barwise, J. (ed.) Handbook of Mathematical Logic. Studies in Logic and the Foundations of Mathematics, vol. 90, pp. 1133–1142. North-Holland, Amsterdam (1977)
3. Erdös, P., Rado, R.: Combinatorial theorems on classifications of subsets of a given set. Proc. London Math. Soc. 2(3), 417–439 (1952)
4. Kanamori, A., McAloon, K.: On Gödel incompleteness and finite combinatorics. Ann. Pure Appl. Logic 33(1), 23–41 (1987)
5. McAloon, K.: Progressions transfinies de théories axiomatiques, formes combinatoires du théorème d'incomplétude et fonctions récursives a croissance rapide. In: McAloon, K. (ed.) Modèles de l'Arithmetique, France, Paris. Asterique, vol. 73, pp. 41–58. Société Mathématique de France, Paris (1980)
6. Weiermann, A.: A classification of rapidly growing Ramsey functions. Proc. Am. Math. Soc. 132(2), 553–561 (2004)
7. Schwichtenberg, H.: Eine Klassifikation der ε_0-rekursiven Funktionen. Z. Math. Logik Grundlagen Math. 17, 61–74 (1971)
8. Buchholz, W., Wainer, S.: Provably computable functions and the fast growing hierarchy. In: Simpson, S. (ed.) Logic and Combinatorics. The AMS-IMS-SIAM joint summer conference in 1985. Contemporary Mathematics Series, pp. 179–198. Amer. Math. Soc., Providence (1987)
9. Fairtlough, M., Wainer, S.S.: Hierarchies of provably recursive functions. In: Handbook of proof theory. Stud. Logic Found. Math., vol. 137, pp. 149–207. North-Holland, Amsterdam (1998)
10. Loebl, M., Nešetřil, J.: An unprovable Ramsey-type theorem. Proc. Amer. Math. Soc. 116(3), 819–824 (1992)
11. Arai, T.: On the slowly well orderedness of ε_0. Math. Log. Q. 48(1), 125–130 (2002)
12. Carlucci, L., Lee, G., Weiermann, A.: Classifying the phase transition threshold for regressive ramsey functions (preprint)
13. Weiermann, A.: Analytic combinatorics, proof-theoretic ordinals, and phase transitions for independence results. Ann. Pure Appl. Logic 136(1-2), 189–218 (2005)
14. Kojman, M., Lee, G., Omri, E., Weiermann, A.: Sharp thresholds for the phase transition between primitive recursive and ackermannian ramsey numbers. J. Comb. Theory, Ser. A 115(6), 1036–1055 (2008)
15. Bovykin, A.: Several proofs of PA-unprovability. In: Logic and its applications. Contemp. Math., vol. 380, pp. 29–43. Amer. Math. Soc., Providence (2005)
16. Lee, G.: Phase Transitions in Axiomatic Thought. PhD thesis, University of Münster, Germany (2005)

Appendix

Here we give a combinatorial proof of KM^n implies PH^n. The basic idea is getting min-homogeneous sets of very large cardinality such that some fine thinning can be chosen whose every two elements lie sufficiently far away from each other. This is also one of the basic ideas of Paris' original proof. We shall demand somewhat more, and this will be achieved by using the following lemma for the construction of such sets. The following lemma from [4] is crucial. We present the original proof again because the proof itself plays an important role later.

Lemma 15 (Kanamori and McAloon [4]). *There are three regressive functions $\eta_1, \eta_2, \eta_3 : [\mathbb{N}]^2 \to \mathbb{N}$ such that whenever H' is min-homogeneous for all of them, then*
$$H = H' \setminus \{\text{the last three elements of } H'\}$$
has the property that $x < y$ both in H implies $x^x \leq y$.

Proof. Define $\eta_1, \eta_2, \eta_3 : [\mathbb{N}]^2 \to \mathbb{N}$ by:

$$\eta_1(x,y) = \begin{cases} 0 & \text{if } 2x \leq y, \\ y \dot- x & \text{otherwise,} \end{cases}$$

$$\eta_2(x,y) = \begin{cases} 0 & \text{if } x^2 \leq y, \\ u & \text{otherwise, where } u \cdot x \leq y < (u+1) \cdot x, \end{cases}$$

$$\eta_3(x,y) = \begin{cases} 0 & \text{if } x^x \leq y, \\ v & \text{otherwise, where } x^v \leq y < x^{v+1}. \end{cases}$$

Suppose that H' is as hypothesised, and let $z_1 < z_2 < z_3$ be the last three elements of H'. If $x < y$ are both in $H' \setminus \{z_3\}$, then since $\eta_1(x,y) = \eta_1(x,z_3)$, clearly we must have $\eta_1(x,y) = 0$. Hence, η_1 on $[H' \setminus \{z_3\}]^2$ is constantly 0.

Next, assume that $x < y$ are both in $H' \setminus \{z_2, z_3\}$ and $\eta_2(x,y) = u > 0$. then $u \cdot x \leq y < z_2 < u \cdot x + x$ by min-homogeneity, and so we have $u \cdot x + x \leq y + x \leq y + y \leq z_2$ by the previous paragraph. But this leads to the contradiction $z_2 < z_2$. Hence, η_2 on $[H' \setminus \{z_2, z_3\}]^2$ is constantly 0.

Finally, we can iterate the argument to show that η_3 on $[H' \setminus \{z_1, z_2, z_3\}]^2$ is constantly 0, and so the proof is complete. $\qquad\square$

Theorem 16 (In $I\Sigma_1$). KM^n *implies* PH^n *for all* $n \geq 2$.

Proof. For a better readability we prove the claim for $n = 3$. This case is general enough, and a proof for arbitrary n needs so many symbols that it would be difficult to see the ground idea of the proof. A general proof is presented in the author's PhD thesis [16].

Assume KM^n. Given z, k, put $\ell := 21z + 22$. Then we can find m and y satisfying $m \to (\ell + 7)_4^3$ and $[x,y] \to (m)_{reg}^3$, where x is chosen such that $x \geq \max\{7, z, k, M\}$ and M is so large that any $i \geq M$ satisfies

$$2|i| + 1 < i, \quad 2^{2^{|i|_3}} < i, \quad \text{and} \quad |i|_3 > 2. \tag{5}$$

How large x should be will become obvious during the proof below. Notice that the existence of m is guaranteed by the finite Ramsey theorem, and y can be chosen such that $y \to (m + x)^3_{reg}$. We claim $y \to^* (k)^3_z$. Indeed, we are going to show $[x, y] \to^* (k)^3_z$.

Let a function $f : [x, y]^3 \to z$ be given. We will define a regressive function $g : [x, y]^3 \to y$ such that a fine thinning of a min-homogeneous set for g leads to a f-homogeneous set of cardinality k. We need some preparation. Let α, β, γ be numbers from $[x, y]$ satisfying $\alpha < \beta < \gamma$. Below we will consider finite sequences as functions with finite domains.

– First construct a finite sequence Q_γ of length at most $(\gamma - x)$ by letting $Q_\gamma(0) := x$, $Q_\gamma(1) := x + 1$, and, assuming $Q_\gamma(i - 1)$ is defined, $Q_\gamma(i)$ be the least t such that $Q_\gamma(i - 1) < t < \gamma$ and

$$\forall j, p < i\, [\, j < p \to f(Q_\gamma(j), Q_\gamma(p), t) = f(Q_\gamma(j), Q_\gamma(p), \gamma)\,]\,.$$

If there is no such t, then the construction stops. Now put

$$R_{\gamma\alpha}(j, p) := f(Q_\gamma(j), Q_\gamma(p), \gamma)$$

for $j, p \in \alpha \cap \mathrm{dom}(Q_\gamma)$, with $j < p$. Notice that $Q_\gamma \restriction \alpha$ can be regained from f and $R_{\gamma\alpha}$, i.e., γ is not necessary.

– If $\beta \in \mathrm{dom}(Q_\gamma)$, construct another sequence $P_{\gamma\beta}$ by letting $P_{\gamma\beta}(0) := x$, and, assuming $P_{\gamma\beta}(i - 1)$ is defined, $P_{\gamma\beta}(i) :=$ be the least t such that $P_{\gamma\beta}(i - 1) < t$, $t \in \mathrm{Im}(Q_\gamma \restriction \beta)$, and

$$\forall j < i(f(P_{\gamma\beta}(j), t, \gamma) = f(P_{\gamma\beta}(j), Q_\gamma(\beta), \gamma))\,.$$

If there is no such t, then the construction stops. Now put

$$S_{\gamma\beta\alpha}(j) := f(P_{\gamma\beta}(j), Q_\gamma(\beta), \gamma)$$

for $j \in \alpha \cap \mathrm{dom}(P_{\gamma\beta})$. Notice that $P_{\gamma\beta} \restriction \alpha$ can be regained from f, $S_{\gamma\beta\alpha}$ and γ.

– Let η_j, $j = 1, 2, 3$, be the three regressive functions from Lemma 15. Applying Lemma 10, define $\bar{\eta}_j : [x, y]^3 \to 2|y| + 1$, $j = 1, 2, 3$, such that, if $\bar{H}$ min-homogeneous for all $\bar{\eta}_j$, then $H := \bar{H} - \{\text{the last element of } \bar{H}\}$ is min-homogeneous for all η_j. (Note that $x \geq 7$.) Define a function h as follows:[2]

$$h(\alpha, \beta, \gamma) := \begin{cases} 0 & \text{if } \bar{\eta}_j(\alpha, \beta, \gamma) = 0 \text{ for each } j \in \{1, 2, 3\}, \\ j & \text{if } j \text{ is the least one s.t. } \bar{\eta}_j(\alpha, \beta, \gamma) \neq 0. \end{cases}$$

---

[2] Here is the place where a refinement of Paris' original proof takes place.

We can now define g on $[x, y]^3$:

$$g(\alpha, \beta, \gamma) := \begin{cases} \bar{\eta}_j(\alpha, \beta, \gamma) \text{ if } h(\alpha, \beta, \gamma) = j > 0, \\ 0 \qquad \text{otherwise and } \neg(x \le |\alpha|_3 < |\beta|_3 < |\gamma|_3), \\ \langle R_{|\gamma|_3|\alpha|_3}, |\gamma|_3 \pmod{2z|\alpha|_3}\rangle_2 \\ \qquad \text{otherwise and } |\beta|_3 \notin \mathrm{dom}(Q_{|\gamma|_3}), \\ \langle R_{|\gamma|_3|\alpha|_3}, S_{|\gamma|_3|\beta|_3|\alpha|_3}, |\gamma|_3 \pmod{2z|\alpha|_3}\rangle_3 \\ \qquad \text{otherwise.} \end{cases}$$

Here we assume that R and S are coded as natural numbers by suitable encoding functions $\langle -, -\rangle_2$ and $\langle -, -, -\rangle_3$ satisfying the following:

$$\langle R_{\gamma\alpha}, \gamma \pmod{2z\alpha}\rangle_2, \langle R_{\gamma\alpha}, S_{\gamma\beta\alpha}, \gamma \pmod{2z\alpha}\rangle_3 \le 2^{2^{\alpha}} \tag{6}$$

for all $\alpha \ge x$. Notice that $\mathrm{dom}(R_{\gamma\alpha}) \subseteq \alpha \times \alpha$, $\mathrm{dom}(S_{\gamma\beta\alpha}) \subseteq \alpha$ and $\mathrm{Im}(R_{\gamma\alpha}) \cup \mathrm{Im}(S_{\gamma\beta\alpha}) \subseteq z \le x$.

This is true if x is large enough. From now on we assume that (6) is always satisfied for any $\alpha \ge x$. This implies that g is regressive:

$$g(\alpha, \beta, \gamma) \le \max\{\bar{\eta}_j(\alpha, \beta, \gamma), 2^{2^{|\alpha|_3}}\} < \alpha.$$

Notice that $\alpha \ge x \ge M$.

Let X_0 be min-homogeneous for g and homogeneous for h such that $\mathrm{card}(X_0) \ge \ell + 7$. Define X_1 and X by

$$X_1 := X_0 - \{\text{the last four elements of } X_0\},$$
$$X := X_1 - \{\text{the first three elements of } X_1\},$$

then $\mathrm{card}(X) \ge \ell$. Let also Y' be the set of every third element of X and Y the set of every second element of Y', i.e. Y is the set of every 6th element of X, so $\mathrm{card}(Y) \ge \ell/7 > 3z + 3$. Now we show a series of claims.

Claim 1: $h \upharpoonright [X_1]^3$ is the constant function with value 0.

Proof of Claim 1: Let $a < b < c < d$ be the last four elements of X_0 and assume $h \upharpoonright [X_1]^3 = 1$. Then $h \upharpoonright [X_0]^3 = 1$ and $g \upharpoonright [X_0]^3 = \bar{\eta}_1 \upharpoonright [X_0]^3$. It follows that X_0 is min-homogeneous for $\bar{\eta}_1$, so $X_0 \setminus \{d\}$ is min-homogeneous for η_1. By the proof of Lemma 15 $\eta_1 \upharpoonright [X_0 \setminus \{c, d\}]^2 = 0$. Hence $\bar{\eta}_1 \upharpoonright [X_0 \setminus \{c, d\}]^3 = 0$ contradicting $h \upharpoonright [X_0]^3 = 1$. Therefore, $h \upharpoonright [X_0]^3 \ne 1$ and $\bar{\eta}_1 \upharpoonright [X_0]^3 = 0$ because it is a constant function. In particular, X_0 is min-homogeneous for $\bar{\eta}_1$, and so $\eta_1 \upharpoonright [X_0 \setminus \{c, d\}]^2 = 0$.

Finally, we can iterate the same argument to show that $\eta_2 \upharpoonright [X_0 \setminus \{b, c, d\}]^2 = 0$ and $\eta_3 \upharpoonright [X_0 \setminus \{a, b, c, d\}]^2 = 0$. It follows that $h \upharpoonright [X_1]^3 \notin \{1, 2, 3\}$, and so we should have $h \upharpoonright [X_1]^3 = 0$.   q.e.d.

Claim 2: $g \upharpoonright [X]^3 > 0$.

Proof of Claim 2: By Lemma 10 and Lemma 15, for all $\alpha < \beta \in X_1$, $2^{\alpha} < \beta$, and hence $|\alpha|_3 < |\beta|_3$ if $|\alpha|_3 > 2$. Because there are three elements from X_1 which are smaller than $\min(X)$, we also have $x \le |\alpha|_3$ for all $\alpha \in X$.   q.e.d.

Claim 3: Let $\alpha < \beta < \delta < \gamma$ be from Y'. Then $z|\alpha|_3 < |\delta|_3 - |\beta|_3$.

Proof of Claim 3: Let $\tau < \rho \in X$ such that $\alpha < \tau < \rho < \beta < \delta$. Because $g(\alpha, \tau, \rho) = g(\alpha, \tau, \beta) = g(\alpha, \tau, \delta)$ by min-homogeneity of $X \subseteq X_0$ we have

$$|\rho|_3 \ (\text{mod } 2z|\alpha|_3) = |\beta|_3 \ (\text{mod } 2z|\alpha|_3) = |\delta|_3 \ (\text{mod } 2z|\alpha|_3).$$

Then for some $n_1 < n_2 \in \mathbb{N}$, $|\beta|_3 = |\rho|_3 + 2z|\alpha|_3 \cdot n_1$ and $|\delta|_3 = |\rho|_3 + 2z|\alpha|_3 \cdot n_2$. Therefore, $|\delta|_3 - |\beta|_3 > z|\alpha|_3$.   q.e.d.

Claim 4: Let $\alpha < \beta < \delta < \gamma$ be from Y'. Then $|\beta|_3 \in \text{dom}(Q_{|\gamma|_3})$, $Q_{|\gamma|_3}(|\beta|_3) < |\delta|_3$, $Q_{|\gamma|_3} \upharpoonright |\beta|_3 = Q_{|\delta|_3} \upharpoonright |\beta|_3$, and $\text{dom}(R_{|\gamma|_3|\alpha|_3}) = |\alpha|_3$.

Proof of Claim 4: Let $\tau, \rho \in X$ such that $\beta < \tau < \rho < \delta$. Because $g(\beta, \tau, \rho) = g(\beta, \tau, \gamma)$ by min-homogeneity of $X \subseteq X_0$ we have $R_{|\rho|_3|\beta|_3} = R_{|\gamma|_3|\beta|_3}$ and $Q_{|\rho|_3} \upharpoonright |\beta|_3 = Q_{|\gamma|_3} \upharpoonright |\beta|_3$. Therefore, for each $j < p \in |\beta|_3 \cap \text{dom}(Q_{|\rho|_3}) = |\beta|_3 \cap \text{dom}(Q_{|\gamma|_3})$ it holds that

$$f(Q_{|\rho|_3}(j), Q_{|\rho|_3}(p), |\rho|_3) = f(Q_{|\gamma|_3}(j), Q_{|\gamma|_3}(p), |\gamma|_3).$$

Let $\mu := |\beta|_3 \cap \text{dom}(Q_{|\rho|_3})$. Then $\mu = |\beta|_3$, since otherwise it would follow that $Q_{|\gamma|_3}(\mu)$ is defined. Note that $|\rho|_3 < |\gamma|_3$. This contradicts the definition of μ. Here we used the fact that $Q_{|\rho|_3} \upharpoonright \mu = Q_{|\gamma|_3} \upharpoonright \mu$. In the same way, we can show $|\beta|_3 \in \text{dom}(Q_{|\gamma|_3})$ and $Q_{|\gamma|_3}(|\beta|_3) \leq |\rho|_3 < |\delta|_3$. Replacing $|\gamma|_3$ with $|\delta|_3$, we get $Q_{|\delta|_3} \upharpoonright |\beta|_3 = Q_{|\gamma|_3} \upharpoonright |\beta|_3$. This implies that $\text{dom}(R_{|\gamma|_3|\alpha|_3}) = |\alpha|_3$.   q.e.d.

Claim 5: Let $\alpha < \beta < \delta < \gamma < \eta$ be from Y. Then $\text{dom}(S_{|\gamma|_3|\beta|_3|\alpha|_3}) = |\alpha|_3$ and $P_{|\gamma|_3|\beta|_3}(|\alpha|_3) < |\beta|_3$.

Proof of Claim 5: Let $\tau \in Y'$ be such that $\alpha < \tau < \beta$. Then $g(\alpha, \tau, \gamma) = g(\alpha, \beta, \gamma)$ by min-homogeneity of $Y \subseteq X_0$. By the same arguments as above, we can show $S_{|\gamma|_3|\tau|_3|\alpha|_3} = S_{|\gamma|_3|\beta|_3|\alpha|_3}$, i.e., $\mu := |\alpha|_3 \cap \text{dom}(P_{|\gamma|_3|\tau|_3}) = |\alpha|_3 \cap \text{dom}(P_{|\gamma|_3|\beta|_3})$ and for each j, u, with $j < \mu$,

$$f(P_{|\gamma|_3|\tau|_3}(j), Q_{|\gamma|_3}(|\tau|_3), |\gamma|_3) = f(P_{|\gamma|_3|\beta|_3}(j), Q_{|\gamma|_3}(|\beta|_3), |\gamma|_3).$$

As in the proof of Claim 4, we can show that $\mu = |\alpha|_3$ and $P_{|\gamma|_3|\beta|_3}(|\alpha|_3) \leq Q_{|\gamma|_3}(|\tau|_3) < |\beta|_3$.   q.e.d.

Claim 6: Let $\alpha < \beta < \delta < \gamma < \eta$ be from Y. Then $P_{|\gamma|_3|\delta|_3}(|\alpha|_3) < |\beta|_3$ and $P_{|\delta|_3|\beta|_3} \upharpoonright |\alpha|_3 = P_{|\eta|_3|\gamma|_3} \upharpoonright |\alpha|_3$.

Proof of Claim 6: Let τ be as above. Replacing β with δ above, it holds that $P_{|\gamma|_3|\delta|_3}(|\alpha|_3) \leq Q_{|\gamma|_3}(|\tau|_3) < |\beta|_3$. Then $P_{|\delta|_3|\beta|_3} \upharpoonright |\alpha|_3 = P_{|\eta|_3|\gamma|_3} \upharpoonright |\alpha|_3$ follows directly from $g(\alpha, \beta, \delta) = g(\alpha, \gamma, \eta)$.   q.e.d.

Let $\alpha_0, \alpha_1, \ldots, \alpha_p$ enumerate all the elements of Y in the ascending order. Then choose a set, for each $i < [p/3] - 1$,

$$Z_i \subseteq \text{Im}(P_{|\alpha_p|_3|\alpha_{p-1}|_3} \upharpoonright [\,|\alpha_{3i}|_3, |\alpha_{3i+1}|_3))$$

such that $|Z_i| \geq \frac{|\alpha_{3i+1}|_3 - |\alpha_{3i}|_3}{z}$ and the function $t \mapsto f(t, Q_{|\alpha_p|_3}(|\alpha_{p-1}|_3), |\alpha_p|_3)$ is constant on Z_i, with a constant value, say $c_i < z$. This is possible because $P_{|\alpha_p|_3|\alpha_{p-1}|_3}$ is strictly increasing. On the other hand, $[p/3] - 1 > z$ implies $c_{i_0} = c_{i_1}$ for some i_0, i_1, with $i_0 < i_1 < [p/3] - 1$. We claim $Z_{i_0} \cup Z_{i_1}$ is a large homogeneous set for f. Note first that

$$|Z_{i_0} \cup Z_{i_1}| > |Z_{i_1}| \geq \frac{|\alpha_{3i_1+1}|_3 - |\alpha_{3i_1}|_3}{z} > |\alpha_{3i_1-1}|_3 > x > k \, .$$

Given $u, v, w \in Z_{i_0} \cup Z_{i_1}$ such that $u < v < w$, we have

$$f(u, v, w) = f(u, v, |\alpha_p|_3) = f(u, Q_{|\alpha_p|_3}(|\alpha_{p-1}|_3), |\alpha_p|_3)$$

because $\{u, v, w\} \subseteq \text{Im}(Q_{|\alpha_p|_3})$. On the other hand,

$$f(u, Q_{|\alpha_p|_3}(|\alpha_{p-1}|_3), |\alpha_p|_3) = f(u', Q_{|\alpha_p|_3}(|\alpha_{p-1}|_3), |\alpha_p|_3)$$

for all $u' \in Z_{i_0} \cup Z_{i_1}$. Therefore, $Z_{i_0} \cup Z_{i_1}$ is f-homogeneous. The relative largeness follows because

$$\min(Z_{i_0} \cup Z_{i_1}) \leq \min(Z_{i_0}) < |\alpha_{3i_0+2}|_3 < \frac{|\alpha_{3i_1+1}|_3 - |\alpha_{3i_1}|_3}{z} \leq |Z_{i_1}| < |Z_{i_0} \cup Z_{i_1}| \, .$$

This completes the proof.   □

The First Order Theories of the Medvedev and Muchnik Lattices

Andrew Lewis[1],[*], André Nies[2],[**], and Andrea Sorbi[3],[* * *]

[1] University of Siena, 53100 Siena, Italy
andy@aemlewis.co.uk
http://aemlewis.co.uk/
[2] Department of Computer Science, University of Auckland, New Zealand
andre@cs.auckland.ac.nz
http://www.cs.auckland.ac.nz/~nies/
[3] University of Siena, 53100 Siena, Italy
sorbi@unisi.it
http://www.dsmi.unisi.it/~sorbi/

Abstract. We show that the first order theories of the Medevdev lattice and the Muchnik lattice are both computably isomorphic to the third order theory of the natural numbers.

1 Introduction

A major theme in the study of computability theoretic reducibilities has been the question of how complicated the first order theories of the corresponding degree structures are. A computability theoretic reducibility is usually a preordering relation on sets of numbers, or on number-theoretic functions. If $\leq_r$ is a reducibility (i.e. a preordering relation) on, say, functions, then $f \leq_r g$, with f, g functions, usually has an arithmetical definition. Therefore, if $(\mathcal{P}, \leq_r)$ is the degree structure corresponding to $\leq_r$, then first order statements about the poset $(\mathcal{P}, \leq_r)$ can be translated into second order arithmetical statements, allowing for quantification over functions. This usually establishes that $\mathrm{Th}(\mathcal{P}, \leq_r) \leq_1 \mathrm{Th}_2(\mathbb{N})$. (Here $\leq_1$ denotes 1–1 reducibility, $\mathrm{Th}(\mathcal{P}, \leq_r)$ denotes the set of first order sentences in the language with equality of partial orders that are true in the poset $\langle \mathcal{P}, \leq_r \rangle$, and by $\mathrm{Th}_n(\mathbb{N})$ we denote the set of n-th order arithmetical sentences that are true of the natural numbers $\mathbb{N}$: precise definitions for $n = 2, 3$ will be given later. $\mathrm{Th}_n(\mathbb{N})$ is usually called the *n-th order theory* of $\mathbb{N}$.)

For instance, if one considers the Turing degrees $\mathfrak{D}_T = (\mathcal{D}_T, \leq_T)$, it immediately follows from the above that $\mathrm{Th}(\mathfrak{D}_T) \leq_1 \mathrm{Th}_2(\mathbb{N})$. On the other hand a

---

[*] This research was supported by a Marie Curie Intra-European Fellowship, Contract MEIF-CT-2005-023657, within the 6th European Community Framework Programme. Current address of the first author: Department of Pure Mathematics, School of Mathematics, University of Leeds, Leeds, LS2 9JT, U. K.

[**] Partially supported by the Marsden Fund of New Zealand, grant no. 03-UOA-130.

[* * *] Partially supported by the NSFC Grand International Joint Project *New Directions in Theory and Applications of Models of Computation*, No. 60310213.

K. Ambos-Spies, B. Löwe, and W. Merkle (Eds.): CiE 2009, LNCS 5635, pp. 324–331, 2009.
© Springer-Verlag Berlin Heidelberg 2009

classical result due to Simpson, [8] (see also [9]), shows that $\mathrm{Th}_2(\mathbb{N}) \leq_m \mathrm{Th}(\mathfrak{D}_T)$. Thus the first order theory of the Turing degrees $\mathfrak{D}_T = (\mathcal{D}_T, \leq_T)$ is as complicated as it can be, i.e. computably isomorphic to second order arithmetic $\mathrm{Th}_2(\mathbb{N})$. For an updated survey on this subject and related topics, we refer the reader to the recent survey by R. Shore, [7].

An interesting, although not much studied, computability theoretic reducibility, is Medvedev reducibility. Here the story is completely different, since Medvedev reducibility is a preordering relation defined on sets of functions. Therefore we need quantification over sets of functions to express first order statements about the corresponding degree structure, which is a bounded distributive lattice called the Medvedev lattice. This suggests that in order to find an upper bound for the complexity of the first order theory of the Medvedev lattice, one has to turn to third order arithmetic. The purpose of this note is to show that third order arithmetic is indeed the exact level: we will show that the first order theories of the Medvedev lattice, and of its nonuniform version called the Muchnik lattice, are in fact computably isomorphic to third order arithmetic.

2   Basics

We briefly review basic definitions concerning the Medvedev lattice and the Muchnik lattice. For more detail the reader is referred to [6], and [10].

A *mass problem* is a subset of $\mathbb{N}^{\mathbb{N}}$. On mass problems one can define the following preordering relation: $\mathcal{A} \leq_M \mathcal{B}$ if there is a Turing functional Ψ such that for all $f \in \mathcal{B}$, $\Psi(f)$ is total, and $\Psi(f) \in \mathcal{A}$. The relation $\leq_M$ induces an equivalence relation on mass problems: $\mathcal{A} \equiv_M \mathcal{B}$ if $\mathcal{A} \leq_M \mathcal{B}$ and $\mathcal{B} \leq_M \mathcal{A}$. The equivalence class of $\mathcal{A}$ is denoted by $\deg_M(\mathcal{A})$ and is called the *Medvedev degree* of $\mathcal{A}$ (or, following Medvedev [3], the *degree of difficulty* of $\mathcal{A}$). The collection of all Medvedev degrees is denoted by $\mathcal{M}$, partially ordered by $\deg_M(\mathcal{A}) \leq_M \deg_M(\mathcal{B})$ if $\mathcal{A} \leq_M \mathcal{B}$. Note that there is a smallest Medvedev degree $\mathbf{0}$, namely the degree of any mass problem containing a computable function. There is also a largest degree $\mathbf{1}$, the degree of the empty mass problem. For functions f and g, as usual we define the function $f \oplus g$ by $f \oplus g(2x) = f(x)$ and $f \oplus g(2x+1) = g(x)$. Let $n \hat{\ } \mathcal{A} = \{n \hat{\ } f : f \in \mathcal{A}\}$, where $n \hat{\ } f$ is the function such that $n \hat{\ } f(0) = n$, and for $x > 0$ $n \hat{\ } f(x) = f(x-1)$. The *join* operation

$$\mathcal{A} \vee \mathcal{B} = \{f \oplus g : f \in \mathcal{A} \wedge g \in \mathcal{B}\},$$

and the *meet* operation

$$\mathcal{A} \wedge \mathcal{B} = 0 \hat{\ } \mathcal{A} \cup 1 \hat{\ } \mathcal{B}$$

on mass problems originate well defined operations on Medvedev degrees that make $\mathcal{M}$ a bounded distributive lattice $\mathfrak{M} = (\mathcal{M}, \vee, \wedge, \mathbf{0}, \mathbf{1})$, called the *Medvedev lattice*. Henceforth, when talking about the first order theory of the Medvedev lattice, denoted $\mathrm{Th}(\mathfrak{M})$, we will refer to $\mathrm{Th}(\mathcal{M}, \leq_M)$. Clearly $\mathrm{Th}(\mathcal{M}, \leq_M) \equiv \mathrm{Th}(\mathcal{M}, \vee, \wedge, \mathbf{0}, \mathbf{1})$, where the symbol $\equiv$ denotes computable isomorphism.

One can consider a nonuniform variant of the Medvedev lattice, the *Muchnik lattice* $\mathfrak{M}_w = (\mathcal{M}_w, \leq_w)$, introduced and studied in [4]. This is the structure resulting from the reduction relation on mass problems defined by

$$\mathcal{A} \leq_w \mathcal{B} \Leftrightarrow (\forall g \in \mathcal{B})(\exists f \in \mathcal{A})[f \leq_T g],$$

where $\leq_T$ denotes Turing reducibility. Again, $\leq_w$ generates an equivalence relation $\equiv_w$ on mass problems. The equivalence class of $\mathcal{A}$ is called the *Muchnik degree* of $\mathcal{A}$, denoted by $\deg_w(\mathcal{A})$. The above displayed operations on mass problems turn $\mathcal{M}_w$ into a lattice too, denoted by $\mathfrak{M}_w = (\mathcal{M}_w, \vee, \wedge, \mathbf{0}_w, \mathbf{1}_w)$, where $\mathbf{0}_w$ is the Muchnik degree of any mass problem containing a computable function, and $\mathbf{1}_w = \deg_w(\emptyset)$. The first order theory of the Muchnik lattice, in the language of partial orders, will be denoted by $\text{Th}(\mathfrak{M}_w)$.

It is well known that the Turing degrees can be embedded into both $\mathcal{M}$ and $\mathcal{M}_w$. Indeed, the mappings

$$i(\deg_T(A)) = \deg_M(\{c_A\}),$$
$$i_w(\deg_T(A)) = \deg_w(\{c_A\})$$

(where, given a set A, we denote by c_A its characteristic function), are well defined embeddings of $(\mathcal{D}_T, \leq_T)$ into $(\mathcal{M}, \leq_M)$ and $(\mathcal{M}_w, \leq_w)$, respectively. Moreover, i and i_w preserve least element, and the join operation. Henceforth, we will often identify the Turing degrees with the range of i, or i_w, according to the case. Thus, we say that a Medvedev degree (respectively, a Muchnik degree) $\mathbf{X}$ is a Turing degree if it is in the range of i (respectively, i_w). It is easy to see that $\mathbf{X} \in \mathfrak{M}$ (respectively, $\mathbf{X} \in \mathfrak{M}_w$) is a Turing degree if and only if $\mathbf{X} = \deg(\{f\})$ (respectively, $\mathbf{X} = \deg_w(\{f\})$) for some function f. When thinking of a Turing degree $\mathbf{X}$ within $\mathfrak{M}$, or $\mathfrak{M}_w$, we will always choose a mass problem that is a singleton as a representative of $\mathbf{X}$.

Lemma 1. *The Turing degrees are first order definable in both $(\mathcal{M}, \leq_M)$ and $(\mathcal{M}_w, \leq_w)$ via the formula*

$$\varphi(u) =^{def} \exists v \, [u < v \wedge \forall w \, [u < w \rightarrow v \leq w]] \, .$$

Proof. See [1]. □

It is perhaps worth observing that the Medvedev lattice and the Muchnik lattice are not elementarily equivalent:

Theorem 1. $\text{Th}(\mathfrak{M}) \neq \text{Th}(\mathfrak{M}_w)$.

Proof. We exhibit an explicit first order difference. Let

$$\mathbf{0}' = \deg_M(\{f : f \text{ non computable}\}),$$
$$\mathbf{0}'_w = \deg_w(\{f : f \text{ non computable}\}).$$

Notice that $\mathbf{0}'$ and $\mathbf{0}'_w$ are definable in the respective structures by the same first order formula, expressing that $\mathbf{0}'$ is the least element amongst the nonzero

Medvedev degrees, and $\mathbf{0}'_w$ is the least element amongst the nonzero Muchnik degrees. (Notice that $\mathbf{0}'$ is the element v witnessing the existential quantifier in the above formula $\varphi(u)$ when u is interpreted as the least Turing degree in the Medvedev lattice; similarly $\mathbf{0}'_w$ is the element v witnessing the existential quantifier in the above formula $\varphi(u)$ when u is interpreted as the least Turing degree in the Muchnik lattice.) It is now easy to notice an elementary difference between the Medvedev and the Muchnik lattice, as it can be shown that $\mathbf{0}'$ is meet-irreducible in $\mathfrak{M}$: this follows from the characterization of meet-irreducible elements of $\mathfrak{M}$ given in [1], see also [10, Theorem 5.1]. On the other hand, let f be a function of minimal Turing degree, and let $\mathbf{A} = \deg_w(\{f\})$, $\mathbf{B} = \deg_w(\{g : f \not\leq_T g\})$. Then, in $\mathfrak{M}_w$, $\mathbf{0}'_w = \mathbf{A} \wedge \mathbf{B}$, i.e. $\mathbf{0}'_w$ is meet-reducible.   □

3   The Complexity of the First Order Theory

We will show that the first order theories of the Medevdev lattice and the Muchnik lattice are both computably isomorphic to third order arithmetic.

3.1   Some Logical Systems

We now introduce second and third order arithmetic and some useful related logical systems.

Third order arithmetic. Third order arithmetic is the logical system defined as follows. The language, with equality, consists of: The basic symbols $+, \times, 0, 1, <$ of elementary arithmetic; first order variables $x_0, x_1, \ldots$ (for numbers); second order variables $p_0, p_1, \ldots$ (for unary functions on numbers); third order variables $X_0, X_1, \ldots$ (for sets of functions, i.e. mass problems). Terms and formulas are built up as usual, but similarly to function symbols, second order variables are allowed to form terms: thus if t is a term and p is a second order variable, then $p(t)$ is a term; if p is a second order variable and X is a third order variable then $p \in X$ is allowed as an atomic formula. Finally, we are allowed quantification also on second order variables, and on third order variables. Sentences are formulas in which all variables are quantified. A sentence is *true* if its standard interpretation in the natural numbers is true (with first order variables being interpreted by numbers; second order variables being interpreted by unary functions from $\mathbb{N}$ to $\mathbb{N}$; third order variables being interpreted by mass problems; the symbol $\in$, here and in the following systems, is interpreted as membership). The collection of all true sentences, under this interpretation, is called *third order arithmetic*, denoted by $\mathrm{Th}_3(\mathbb{N})$. Notice that by limiting ourselves to adding to elementary arithmetic only variables for functions, we get what is known as *second order arithmetic*, denoted by $\mathrm{Th}_2(\mathbb{N})$.

Second order theory of the real numbers. The second order theory of the field $\mathbb{R}$ of the real numbers is the logical system (with equality) defined as follows. The language, with equality, consists of the basic symbols $+, \times, 0, 1, <$. We have first order variables $r_0, r_1, \ldots$ (for real numbers); second order variables $X_0, X_1, \ldots$

(for sets of real numbers). Terms, atomic formulas and formulas are built in the usual way, where we regard $r \in Y$ as an atomic formula if r is a first order variable, and X is a second order variable. Quantification on both first and second order variables is allowed. Sentences are formulas in which all variables are quantified. A sentence is *true* if the standard interpretation of the sentence in the field of real numbers is true (where first order variables are interpreted by real numbers; second order variables are interpreted by sets of real numbers). By the *second order theory of the field* $\mathbb{R}$, denoted by $\mathrm{Th}_2(\mathbb{R})$, we mean the collection of all such true sentences.

We are now ready to give a useful, although simple, characterization of third order arithmetic $\mathrm{Th}_3(\mathbb{N})$.

Lemma 2. $\mathrm{Th}_3(\mathbb{N})$ *is computably isomorphic to* $\mathrm{Th}_2(\mathbb{R})$.

Proof. Let EA_1 be the logical system obtained from elementary arithmetic as follows. The language, with equality, consists of the basic elementary symbols of arithmetic $+, \times, 0, 1, <$; we have first order *numerical variables* $x_0, x_1, \ldots$, and in addition we have first order variables of a different sort, $r_0, r_1, \ldots$, called *real variables*. Then the system EA_1 is obtained by taking all sentences which are true under interpreting numerical variables with numbers, real variables with real numbers, and interpreting $+, \times, 0, 1, <$ accordingly. This system is sometimes known as *elementary analysis*. It is known (see for instance [6, Theorem 16.XIII], for a proof) that $\mathrm{Th}_2(\mathbb{N}) \equiv EA_1$. Let now EA_2 be the logical system obtained by adding to the language of EA_1 second order variables $R_0, R_1, \ldots$ (for sets of reals); and by adding atomic formulas of the form $r \in R$, where r is a real variable and R is a second order variable. Then EA_2 is the collection of all sentences in this language that are true under the additional interpretation of second order variables as sets of real numbers. Following up the argument in Rogers, [6], it is now easy to show that $\mathrm{Th}_3(\mathbb{N}) \equiv EA_2$. It is then sufficient to show that $EA_2 \equiv \mathrm{Th}_2(\mathbb{R})$. Indeed, $EA_2 \leq_1 \mathrm{Th}_2(\mathbb{R})$ follows from the fact that $\mathbb{N} \subseteq \mathbb{R}$ is second-order definable in the field of real numbers, being the smallest inductive set. On the other hand, it is clear that $\mathrm{Th}_2(\mathbb{R}) \leq_1 EA_2$. $\square$

Lemma 3. *Let* $\mathfrak{A} \subseteq \mathfrak{D}_T$ *be an antichain, let* $\mathfrak{B} \subseteq \mathfrak{A}$, *and via the embedding of the Turing degrees into* $\mathfrak{M}$ *directly regard* $\mathfrak{A}$ *as a subset of* $\mathfrak{M}$. *For every* $\mathbf{X} \in \mathfrak{A}$ *let* $f_\mathbf{X}$ *be a function such that* $\mathbf{X} = \deg_M(\{f_\mathbf{X}\})$. *Let* $\mathbf{C}$ *be the Medvedev degree of the mass problem* $\mathcal{C} = \{f_\mathbf{Y} : \mathbf{Y} \in \mathfrak{B}\}$. *Then*

$$(\forall \mathbf{X} \in \mathfrak{A})[\mathbf{C} \leq_M \mathbf{X} \Leftrightarrow \mathbf{X} \in \mathfrak{B}].$$

A similar result applies to the Muchnik lattice. In this latter case, we of course regard each $\mathbf{X} \in \mathfrak{A}$ *as the Muchnik degree* $\deg_w(\{f_\mathbf{X}\})$ *of some function* $f_\mathbf{X}$, *and we work with* $\leq_w$ *instead of* $\leq_M$.

Proof. Suppose we work with the Medvedev lattice, and let $\mathfrak{A} \subseteq \mathfrak{D}_T$ be an antichain, viewed as an antichain in $\mathfrak{M}$ via the embedding of the Turing degrees. Let $\mathfrak{B} \subseteq \mathfrak{A}$, $f_\mathbf{X}$, $\mathcal{C}$ and $\mathbf{C}$ be defined as in the statement of the lemma.

If $\mathbf{C} \leq_M \mathbf{X}$ then $\mathcal{C} \leq_M \{f_{\mathbf{X}}\}$, which implies that $f_{\mathbf{X}} \in \mathcal{C}$, hence $\mathbf{X} \in \mathfrak{B}$. For the other direction, if $\mathbf{X} \in \mathfrak{B}$ then $f_{\mathbf{X}} \in \mathcal{C}$, implying $\mathcal{C} \leq_M \{f_{\mathbf{X}}\}$, i.e. $\mathbf{C} \leq_M \mathbf{X}$. The case of the Muchnik lattice is similar. $\qquad\square$

3.2 The Complexity of the Theory

We now show that the first order theories of the Medvedev lattice and the Muchnik lattice have the same m-degree as $\mathrm{Th}_3(\mathbb{N})$.

One direction is trivial:

Lemma 4. $\mathrm{Th}(\mathfrak{M}), \mathrm{Th}(\mathfrak{M}_w) \leq_m \mathrm{Th}_3(\mathbb{N})$.

Proof. This follows from the fact that Turing reducibility is arithmetically definable, and thus the first order theories of $\mathfrak{M}$ and $\mathfrak{M}_w$ can be interpreted in third order arithmetic. $\qquad\square$

For the converse, we first need a computability theoretic result. All unexplained computability notions which are used in this section can be found in [2], see in particular Chapter V. Following [2], we say that a *tree* T is a function from binary strings to binary strings such that for every binary string σ, $T(\sigma\hat{\ }0)$ and $T(\sigma\hat{\ }1)$ are incomparable extensions of $T(\sigma)$. Here the symbol $\hat{\ }$ denotes concatenation of strings, and if $d \in \{0,1\}$ we often identify d with the string $\langle d \rangle$. If σ is a string and n is a number then we let $\sigma\hat{\ }n = \sigma\hat{\ }\langle n \rangle$: a similar convention holds of $n\hat{\ }\sigma$. The length of a string σ is denoted by $|\sigma|$. A tree T is *computable* if T is computable as a function. We also say that T' is a *subtree of* T if $\mathrm{range}(T') \subseteq \mathrm{range}(T)$. Given a tree T, the collection of all infinite paths in T will be denoted by $[T]$.

Lemma 5. *There is a tree T such that, for any Turing degrees $\mathbf{x}, \mathbf{y}, \mathbf{z}$ of distinct paths of T, the following hold:*

(i) $\mathbf{x}$ is minimal;
(ii) $\mathbf{x} \not\leq \mathbf{y} \vee \mathbf{z}$.

Proof. Given any tree T and any σ, σ' let $\mathrm{Ext}(T, \sigma)(\sigma') = T(\sigma\hat{\ }\sigma')$: if $d \in \{0,1\}$ then $\mathrm{Ext}(T,d) = \mathrm{Ext}(T, \langle d \rangle)$. For every computable tree T and every n let $\mathrm{Min}(T, n)$ be a computable subtree of T such that if $A \in [\mathrm{Min}(T, n)]$ then

$$\varphi_n^A \text{ total} \Rightarrow [\varphi_n^A \text{ computable} \vee A \leq_T \varphi_n^A] :$$

see for instance [2] for the details of the construction of $\mathrm{Min}(T, n)$ starting from T. For any computable trees T_0, T_1, T_2 and any n, let for each $i \leq 2$,

$$\mathrm{Diag}_n^i(T_0, T_1, T_2) = \hat{T}_i$$

for some computable $\hat{T}_i \subseteq T_i$ such that if $A_0 \in \hat{T}_0$, $A_1 \in \hat{T}_1$ and $A_2 \in \hat{T}_2$ then $A_0 \neq \varphi_n^{A_1 \oplus A_2}$. That $\mathrm{Diag}_n^i(T_0, T_1, T_2)$ exists can be seen as follows: Let T_0, T_1, T_2 and n be given. We distinguish the following cases:

Case 1. $(\exists x)(\exists\rho_1)(\exists\rho_2)(\forall\tau_1 \supseteq \rho_1)(\forall\tau_2 \supseteq \rho_2)[\varphi_n^{T_1(\tau_1)\oplus T_2(\tau_2)}(x)\uparrow]$: in this case choose ρ_1 and ρ_2 and define

$$\hat{T}_0 = T_0 \qquad \hat{T}_1 = \text{Ext}(T_1,\rho_1) \qquad \hat{T}_2 = \text{Ext}(T_2,\rho_2).$$

Case 2. Otherwise, we can find strings τ_1 and τ_2 such that

$$(\forall x < |T_0(0)|, |T_0(1)|)[\varphi_n^{T_1(\tau_1)\oplus T_2(\tau_2)}(x)\downarrow].$$

Since $T_0(0) \neq T_0(1)$ we can choose $j \in \{0,1\}$ such that

$$T_0(j)(x) \neq \varphi_n^{T_1(\tau_1)\oplus T_2(\tau_2)}(x),$$

for some $x < |T_0(0)|, |T_0(1)|$, and define

$$\hat{T}_0 = \text{Ext}(T_0,j) \qquad \hat{T}_1 = \text{Ext}(T_1,\tau_1) \qquad \hat{T}_2 = \text{Ext}(T_2,\tau_2).$$

We now define T which satisfies the hypothesis of the lemma in stages, defining $T(\sigma)$ for all σ of length n at stage n. For each σ we also define an auxiliary value T_σ.

Stage 0. Let $T(\lambda) = \lambda$ and define $T_\lambda = \text{Id}$.

Stage $n + 1$. Let $\{T_i^0 : i < 2^{n+1}\}$ be the set of all values $\text{Ext}(T_\sigma, d)$ such that σ is of length n and $d \in \{0,1\}$ and for each $i < 2^{n+1}$ let $\sigma_i = \sigma\hat{\ }d$ for σ and d such that $T_i^0 = \text{Ext}(T_\sigma, d)$. Let r be the number of triples (k,l,m) with $k,l,m < 2^{n+1}$ and k,l,m pairwise distinct.

Step (1). Fixing any order on the set of all such triples, proceed as follows for each such triple in turn. For the jth triple (k,l,m), given T_k^{j-1}, T_l^{j-1} and T_m^{j-1}, let $T_k^j = \text{Diag}_n^0(T_k^{j-1}, T_l^{j-1}, T_m^{j-1})$, $T_l^j = \text{Diag}_n^1(T_k^{j-1}, T_l^{j-1}, T_m^{j-1})$ and $T_m^j = \text{Diag}_n^2(T_k^{j-1}, T_l^{j-1}, T_m^{j-1})$. For each $i < 2^{n+1}$ such that $i \notin \{k,l,m\}$ define $T_i^j = T_i^{j-1}$.

Step (2). For each $i < 2^{n+1}$, define $T_{\sigma_i} = \text{Min}(T_i^r, n)$ and $T(\sigma_i) = T_{\sigma_i}(\lambda)$.   □

Lemma 6. $\text{Th}_2(\mathbb{R})$ *can be interpreted in both* $\text{Th}(\mathfrak{M})$ *and* $\text{Th}(\mathfrak{M}_w)$.

Proof. Again the proof is given for $\mathfrak{M}$, but *mutatis mutandis* it works for $\mathfrak{M}_w$ too. By the usual coding methods, see e.g. [5], the ordered field $\mathbb{R}$ can be first-order defined in a symmetric graph $\langle V, E\rangle$, where we may assume $V = 2^{\mathbb{N}}$, the Cantor space. Since T as in Lemma 5 is homeomorphic to $2^{\mathbb{N}}$, we may assume that in fact V is the set of paths of T. We can now obtain a coding scheme $\mathbf{R}_{A,B}$ to code with two appropriate parameters A, B a copy of the ordered field $\mathbb{R}$ into $\mathfrak{M}$. Let $\mathfrak{B}$ be the collection of Turing degrees of the paths of T (viewed inside $\mathfrak{M}$). The parameter A picks up $\mathfrak{B}$ among the minimal Turing degrees, that are first order definable in $\mathfrak{M}$, via Lemma 3. The parameter B picks the edge relation

$$\{\mathbf{x} \vee \mathbf{y} : Exy\},$$

for $x, y \in V$, obtained by applying Lemma 3 to the antichain

$$\{\mathbf{x} \vee \mathbf{y} : \mathbf{x} \neq \mathbf{y} \text{ and } \mathbf{x}, \mathbf{y} \in \mathfrak{B}\}.$$

Applying Lemma 3, we may now quantify over subsets of the coded copy of $\mathbb{R}$. It is clear how to translate each second order sentence Φ in the language of $\mathrm{Th}_2(\mathbb{R})$ into a formula $\widehat{\Phi}_{A,B}$ with parameters A, B, according to this coding scheme of $\mathbb{R}$ into $\mathfrak{M}$.

We obtain a correctness condition on parameters, $\alpha(A, B)$, saying that the coded model $\mathbf{R}_{A,B}$ is isomorphic to $\mathbb{R}$, by requiring the second order axioms of a complete ordered field (i.e. each bounded nonempty subset has a supremum). So

$$\Phi \in \mathrm{Th}_2(\mathbb{R}) \Leftrightarrow \mathfrak{M} \models (\exists A, B)[\alpha(A, B) \wedge \widehat{\Phi}_{A,B}].$$

This shows the claim.   □

Theorem 2. $\mathrm{Th}(\mathfrak{M}), \mathrm{Th}(\mathfrak{M}_w) \equiv \mathrm{Th}_3(\mathbb{N})$.

Proof. By Lemma 4 and by the fact that theories are cylinders (see [6]), we get $\mathrm{Th}(\mathfrak{M}), \mathrm{Th}(\mathfrak{M}_w) \leq_1 \mathrm{Th}_3(\mathbb{N})$. On the other hand, by Lemma 6, we get $\mathrm{Th}_2(\mathbb{R}) \leq_1 \mathrm{Th}(\mathfrak{M}), \mathrm{Th}(\mathfrak{M}_w)$, and thus by Lemma 2, $\mathrm{Th}_3(\mathbb{N}) \leq_1 \mathrm{Th}(\mathfrak{M}), \mathrm{Th}(\mathfrak{M}_w)$.   □

Note added in proof. The result reported in Theorem 2 has been independently obtained also by Paul Shafer, at Cornell University.

References

1. Dyment, E.: Certain properties of the Medvedev lattice. Mathematics of the USSR Sbornik 30, 321–340 (1976) (English Translation)
2. Lerman, M.: Degrees of Unsolvability. Perspectives in Mathematical Logic. Springer, Heidelberg (1983)
3. Medevdev, Y.T.: Degrees of difficulty of the mass problems. Dokl. Nauk. SSSR 104, 501–504 (1955)
4. Muchnik, A.: On strong and weak reducibility of algorithmic problems. Sibirskii Matematicheskii Zhurnal 4, 1328–1341 (1963) (Russian)
5. Nies, A.: Undecidable fragments of elementary theories. Algebra Universalis 35, 8–33 (1996)
6. Rogers Jr., H.: Theory of Recursive Functions and Effective Computability. McGraw-Hill, New York (1967)
7. Shore, R.A.: Degree structures: local and global investigations. Bull. Symbolic Logic 12(3), 369–389 (2006)
8. Simpson, S.G.: First order theory of the degrees of recursive unsolvability. Ann. of Math. 105, 121–139 (1977)
9. Slaman, T.A., Woodin, W.H.: Definability in the Turing degrees. Illinois J. Math. 30, 320–334 (1986)
10. Sorbi, A.: The Medvedev lattice of degrees of difficulty. In: Cooper, S.B., Slaman, T.A., Wainer, S.S. (eds.) Computability, Enumerability, Unsolvability - Directions in Recursion theory. London Mathematical Society Lecture Notes Series, pp. 289–312. Cambridge University Press, New York (1996)

Infima of $d.r.e.$ Degrees

Jiang Liu, Shengling Wang, and Guohua Wu

Division of Mathematical Sciences
School of Physical and Mathematical Sciences
Nanyang Technological University
Singapore 637371, Singapore

Abstract. Lachlan observed that the infimum of two r.e. degrees considered in the r.e. degres coincides with the one considered in the Δ_2^0 degrees. It is not true anymore for the d.r.e. degrees. Kaddah proved in [7] that there are $d.r.e.$ degrees $\mathbf{a}, \mathbf{b}, \mathbf{c}$ and a 3-$r.e.$ degree $\mathbf{x}$ such that $\mathbf{a}$ is the infimum of $\mathbf{b}, \mathbf{c}$ in the d.r.e. degrees, but not in the 3-r.e. degrees, as $\mathbf{a} < \mathbf{x} < \mathbf{b}, \mathbf{c}$. In this paper, we extend Kaddah's result by showing that such a infima difference occurs densely in the r.e. degrees.

1 Introduction

In [3], Cooper, Harrington, Lachlan, Lempp and Soare constructed an incomplete maximal d.r.e. degrees, and hence proved that the d.r.e. degrees are not dense. After this, several weak density of various unusual properties of the d.c.e. degrees have been established by Cooper, Lempp, and Watson in [4], Arslanov, Lempp and Shore in [1]; LaForte in [10], Ding and Qian in [5]; and by Wu more recently in [12]. Our result in this paper follows this line and we will prove that Kaddah's construction of two d.r.e. degrees with an infimum in the d.r.e. degrees but fails to be an infimum in the 3-r.e. degrees can be implemented in any proper interval of the r.e. degrees. We will explain in detail how delayed permissions are used in the construction, which is a core component in the density argument.

For any sets $A, B \subseteq \omega$, the supremum of the Turing degrees of A and B always exists and equals to the Turing degree of $A \oplus B$. However, the infimum of the Turing degrees of A and B may not exist as proved by Kleene-Post in [8]. Later, Lachlan proved in [9] that such examples of A and B can be r.e. Lachlan also pointed out in this paper that for any r.e. sets A and B, the infimum of the Turing degrees of these two sets, considered in the r.e. degrees, if exists, is the same as the one when considered in the Δ_2^0 degrees. One natural question is whether such a coincidence is true when A and B are not r.e.. Kaddah proved in [7] that the "r.e." condition on A and B cannot be omitted. She actually proved that there are $d.r.e.$ degrees $\mathbf{b}, \mathbf{c}$ having the infimum in the d.r.e. degrees different from the one in the 3-r.e. degrees, an extremely strong result. In particular, she proved that there are $d.r.e.$ degrees $\mathbf{a}, \mathbf{b}, \mathbf{c}$ and a 3-$r.e.$ degree $\mathbf{x}$ such that $\mathbf{a}$ is the infimum of $\mathbf{b}, \mathbf{c}$ in the d.r.e. degrees, but $\mathbf{a} < \mathbf{x} < \mathbf{b}, \mathbf{c}$. In this paper, we extend Kaddah's result by showing that such a difference can happen everywhere in the r.e. degrees.

K. Ambos-Spies, B. Löwe, and W. Merkle (Eds.): CiE 2009, LNCS 5635, pp. 332–341, 2009.
© Springer-Verlag Berlin Heidelberg 2009

Theorem 1. *Given r.e. degrees* $\mathbf{u} < \mathbf{v}$, *there are d.r.e. degrees* $\mathbf{a}, \mathbf{b_1}, \mathbf{b_2}$ *and a 3-r.e. degree* $\mathbf{x}$ *between* $\mathbf{u}$ *and* $\mathbf{v}$ *such that* $\mathbf{a} < \mathbf{x} < \mathbf{b_1}, \mathbf{b_2}$ *and* $\mathbf{b_1}$ *and* $\mathbf{b_2}$ *have infimum* $\mathbf{a}$ *in the d.r.e. degrees.*

Our notation and terminology are standard and generally follow Soare [11].

2   Requirements

In this section, we list the requirements needed to prove Theorem 1, and describe the strategy for satisfying each requirement. Given U and V with $U <_T V$, we will construct three *d.r.e.* sets A, B_1, B_2, a 3-*r.e.* set X and auxiliary functionals Γ_1, Γ_2 and Δ_j satisfying the following requirements:

$\mathcal{G}$: $A, B_1, B_2, X \leq_T V$;
$\mathcal{R}$: $X = \Gamma_1^{B_1, A, U} = \Gamma_2^{B_2, A, U}$;
$\mathcal{P}_n$: $X \neq \Phi_n^{A, U}$;
$\mathcal{N}_n$: $\Phi_e^{B_1, A, U} = \Phi_e^{B_2, A, U} = D_j \Rightarrow D_j = \Delta_j^{A, U}$, where $n = \langle e, j \rangle$ and $\{D_j : j \in \omega\}$ is some standard listing of all the differences of *r.e.* sets.

2.1   The $\mathcal{G}$ and $\mathcal{R}$-Strategies

The $\mathcal{G}$ and $\mathcal{R}$ requirements are both global. To satisfy $\mathcal{G}$, we apply the delayed permitting argument, which was first introduced by Sacks in his famous density theorem.

For the $\mathcal{R}$-requirement, we apply the usual coding strategy. That is, we enumerate two functional axioms Γ_i with $i = 1, 2$ following the rules below:

(1) When $\Gamma_i^{B_i, A, U}(x)$ is first defined at a stage s, its use $\gamma_i(x)[s]$ is selected as a fresh number.
(2) If x enters or exits X at stage t after $\Gamma_i^{B_i, A, U}(x)$ is defined, then $\Gamma_i^{B_i, A, U}(x)[t]$ must be undefined or recovered to ensure that it is finally defined agreeing with $X(x)$. In our construction, we put or, extract a number less than or equal to $\gamma_i(x)[t]$ from $B_i \cup A$.
(3) For every x, there is a stage s_x such that after s_x, $\Gamma_i^{B_i, A, U}(x)$ can never be undefined.
(4) $\gamma_i(x)$ is increasing with respect to x.

2.2   A $\mathcal{P}$-Strategy

The basic idea of satisfying a $\mathcal{P}$-requirement is the standard Friedberg-Muchnik argument. That is, we first pick a witness x for $\mathcal{P}$, and wait for $\Phi^{A, U}(x) \downarrow = 0$. After we see that $\Phi^{A, U}(x) \downarrow = 0$, we put x into X and impose a restraint on $A \lceil \varphi(x)$. The problem is that after $\Phi^{A, U}(x) \downarrow = 0$, the computation $\Phi^{A, U}(x)$ may be destroyed by $U \lceil \varphi(x)$ even though we put restraint $A \lceil \varphi(x)$. To get out of this dilemma, we will impose "indirect" restraint on U by threatening $V \leq_T U$ via Ψ.

Let α be a $\mathcal{P}$-strategy. α will run cycles j for $j \in \omega$. α starts cycle 0 first. Each cycle j can start only cycle $j + 1$, however stop any cycle j' with $j' > j$. All the cycles of α define a functional Ψ_α jointly. The cycle j is responsible for the definition of $\Psi_\alpha^U(j)$.

Given $j \in \omega$, α runs cycle j as follows:

(1) Pick x_j as a fresh number.
(2) Wait for a stage s_0 such that $\Phi^{A,U}(x_j) \downarrow [s_0] = 0$.
(3) Preserve $A\lceil \varphi(x_j)[s_0]$ from other strategies.
(4) Set $\Psi_\alpha^U(j)[s_0] = V(j)[s_0]$ with use $\psi_\alpha(j) = \varphi(x_j)[s_0]$ and start cycle $j + 1$ simultaneously.
(5) Wait for $U\lceil \varphi(x_j)[s_0]$ or $V(j)$ to change.

   (a) If $U\lceil \varphi(x_j)[s_0]$ changes first, then cancel all cycles $j' > j$ and drop the A-restraint of cycle j to 0. Go back to step 2.
   (b) If $V(j)$ changes first, then stop cycles $j' > j$, and go to step 6.

(6) Put x_j into X and wait for $U\lceil \varphi(x_j)[s_0]$ to change.
(7) Define $\Psi_\alpha^U(j) = V(j) = 1$ with use 0, and start cycle $j + 1$.

α has two sorts of outcomes:

(j, f) : There is a stage s after which no new cycle runs.
   (*So some cycle j_0 waits forever at step 2 or 6. If α waits forever at step 2 then $\Phi^{A,U}(x_{j_0}) \downarrow= 0$ is not true. If α waits for ever at step 6 then $\Phi^{A,U}(x_{j_0}) \downarrow= 0$ and $x_{j_0} \in X$. In any case, x_{j_0} is a witness for the requirement $\mathcal{P}$.*)
(j, ∞) : Some cycle j_0 runs infinitely often, but no cycle $j' < j$ does so.
   (*It must be true that cycle j_0 goes from step 5 to 2 infinitely often. Thus $\Phi^{A,U}(x_{j_0})$ diverges. So x_{j_0} is clearly a witness for the requirement $\mathcal{P}$.*)

Note that it is impossible that, there are stages s_j for all $j \in \omega$ such that no cycle j runs after stage s_j but there are infinitely many stages at which some cycle runs, as otherwise, we would be able to prove that $V \leq_T U$ via Ψ.

Now, we consider the interaction between strategies $\mathcal{P}$ and $\mathcal{G}(\mathcal{R})$. When α puts a number x_j into X at step 6, if $\Gamma_i^{B_i,A,U}(x_j)$ is defined then we must make it undefined. To do this, we will put its use $\gamma_i(x_j)$ into B_i. But for $B_i \leq_T V$, we have to get a V-permitting. In this construction the V-permitting is realized via a $V(j)$-change. So it suffices to modify step 6 as follows:

(6') Put x_j into X, $\gamma_i(x_j)$ into B_i and wait for $U\lceil \varphi(x_j)[s_0]$ to change.

For the convenience of description, we introduce some notions here. Say that a j-cycle acts if it chooses a fresh number x_j as its attacker at step 1 or, it changes the value of $X(x_j)$ by enumerating x_j into X at step 6'. We also say that j-cycle is active at stage s if at this stage, when α is visited, α is running j-cycle, except the situation that j-cycle is just started at stage s.

2.3   An $\mathcal{N}$-Strategy

An $\mathcal{N}_e$-strategy, β say, is devoted to the construction of a partial functional Δ_β such that if $\Phi_e^{B_1,A,U} = \Phi_e^{B_2,A,U} = D_j$ then $\Delta_\beta^{A,U}$ is well-defined and computes D_j correctly. For simplicity, we will omit the subscript in the basic strategy, and it will be indicated if needed. First define the agreement function as follows:

$$l(\beta, s) = \max\{y < s : \Phi^{B_1,A,U}\lceil y \downarrow [s] = \Phi^{B_2,A,U}\lceil y \downarrow [s] = D_s\lceil y)\},$$
$$m(\beta, s) = \max\{0, l(\beta, t) : t < s \ \& \ t \text{ is a } \beta\text{-stage}\}.$$

Say a stage s is β-expansionary if $s = 0$ or s is an β-stage such that $l(\beta, s) > m(\beta, s)$.

If neither the U-changes nor the V-permission is involved, then the basic strategy for this requirement is exactly the same as the one introduced by Kaddah [7]: After $\Delta^{A,U}(z)$ was defined, we allowed the computations $\Phi^{B_i,A,U}(z)$'s to be destroyed simultaneously (by $\mathcal{P}$-strategies) instead of preserving one of them as in Lachlan's minimal pair construction. As a consequence, as we are constructing B_i d.r.e., we are allowed to remove numbers from one of them to recover a computation to a previous one, and force a disagreement between D and $\Phi^{B_i,A,U}$ for some $i \in \{0, 1\}$ at an argument z.

(1) Define $\Delta^{A,U}(z)[s_0] = D_{s_0}(z)$ with $\delta(z) = s_0$ at some β-expansionary stage s_0. (*We assume that $D_{s_0}(z) = 0$ here, mainly because it is the most complicated situation.*)

(2) Some $\mathcal{P}_e$-strategy with lower priority enumerates a small number x into X_{s_1} at stage $s_1 > s_0$, and for the sake of the global requirement $\mathcal{R}$, we put $\gamma_1(x)[s_1]$ into $B_1[s_1]$ and $\gamma_2(x)[s_1]$ into B_{2,s_1}. The computations $\Phi^{B_i,A,U}(z)[s_0]$ could be *injured* by these enumerations.

(3) At stage $s_2 > s_1$, z enters D_{s_2}.

(4) Then at some β-expansionary stage $s_3 > s_2$, we force that $D(z) \neq \Phi^{B_1,A,U}(z)$ by extracting $\gamma_1(x)[s_1]$ from B_{1,s_3} to recover the computation $\Phi^{B_1,A,U}(z)$ to $\Phi^{B_1,A,U}(z)[s_0]$. To make sure that Γ_1 and Γ_2 well-defined and compute X correctly, we need to extract x from X, and put a number, the current $\gamma_2(x)[s_3]$ into B_2 to undefine $\Gamma_2(x)$. For the sake of consistency between the $\mathcal{N}$-strategies, we also enumerate s_1 into A.

(5) At stage $s_4 > s_3$, z goes out of D_{s_4}. *Since D is d.r.e., z cannot enter D later.*

(6) At a β-expansionary stage $s_5 > s_4$, $\Phi^{B_1,A,U}(z)[s_5] = \Phi^{B_2,A,U}(z)[s_5] = D_{s_5}(z) = 0$ again. Now, we are ready to force that $\Phi^{B_2,A,U}(z) \neq D(z)$ by extracting $\gamma_2(x)[s_3]$ from B_2, and s_1 from A (to recover the computation $\Phi^{B_2,A,U}(z)$ to $\Phi^{B_2,A,U}(z)[s_3]$) and enumerating x into X. To correct $\Gamma_1^{B_1,A,U}(x)$, we put a number less than $\gamma_1(x)[s_1]$, $\gamma_1(x)[s_1] - 1$ say, into B_{1,s_5}. Again, for the sake of consistency between the $\mathcal{N}$-strategies, we also enumerate s_3 into A.

Now we consider the U and V-changes involved in such a process. Note that step 4 and 6 assume that $U\lceil\varphi(B_1, A, U; z)[s_0]$ and $U\lceil\varphi(B_1, A, U; z)[s_3]$ don't change after stage s_0 and s_3 respectively. If $U\lceil\varphi(B_1, A, U; z)[s_0]$ changes then it can correct $\Delta^{A,U}(z)[s_0]$. Consequently, this U-change allow us to correct $\Delta^{A,U}(z)[s_0]$ (this U-change may damage our plan of making a disagreement at z). If the change happens after step 4, we will deal with this by threatening $V \leq_L U$ via a functional Θ. That is, we will make infinitely many attempts to satisfy $\mathcal{N}$ as above by an infinite sequence of cycles.

Every cycle k will run as above, in addition, insert a step before step 4:

(4°) Set $\Theta^U(l) = V(l)[s_3]$ with use $\theta(l) = \delta(z)[s_3]$. Start $(l+1)$-cycle simultaneously, wait for $V(l)$-change (i.e. V-permitting), then stop the cycles $l' > l$ and go to step 5.

Once some cycle finds a $U\lceil\theta(l)$-change after stage s_3, we will stop all cycles $l' > l$ and go back to step 1 as the $U\lceil\theta(l)$-change can correct $\Delta^{A,U}(z)[s_0]$.

A similar modification applies to step 6. We attach an additional step 6° before step 6 so as to run the disagreement strategy and pass through another V-permitting argument.

To combine two disagreement strategies, we need two V-permissions, and as in [4], we arrange the basic $\mathcal{N}$-strategy in $\omega \times \omega$ many cycles (k, l) where $k, l \in \omega$. Their priority is arranged by the lexicographical ordering. $(0, 0)$-cycle starts first, and each (k, l)-cycle can start cycles $(k, l+1)$ or $(k+1, 0)$ and stop, or cancel cycles (k', l') for $(k, l) < (k', l')$. Each (k, l)-cycle can define $\Theta_k^U(l)$ and $\Xi^U(k)$. At each stage, only one cycle can do so.

(k, l)-cycle runs as follows:

(1) Wait for a β-expansionary stage, s say.

(2) If $\Delta^{A,U}(z)[s] = D_s(z)$ for all z such that $\Delta^{A,U}(z) \downarrow [s]$, then go to (3). Otherwise, let z be the least one such that $\Delta^{A,U}(z) \downarrow [s] \neq D_s(z)$, and then go to step 4.

(3) Define $\Delta^{A,U}(z)[s] = D_s(z)$ for all $z < l(\beta, s)$ that $\Delta^{A,U}(z) \uparrow [s]$ with $\delta(z) = s$, then go back to step 1.

(4) Restrain $A\lceil\delta(z)[s]$, $B_1\lceil\delta(z)[s]$ and $B_2\lceil\delta(z)[s]$ from other strategies. Go to step 5.

(5) Set $\Theta_k^U(l) = V_s(l)$ with $\theta_k(l)[s] = \delta(z)[s]$ $(\delta(z)[s] < s)$. Start $(k, l+1)$-cycle, and simultaneously, wait for $U\lceil\theta_k(l)[s]$ or $V(l)$ to change.

(6.1) If $U\lceil\theta_k(l)[s]$ changes first then cancel cycles $(k', l') > (k, l)$, drop the A, B_1, B_2-restraints of (k, l)-cycle to 0, undefine $\Delta^{A,U}(z)[s]$ and go back to step 1.

(6.2) If $V(l)$ changes first at stage $t > s$ then we just remove $\gamma_1(x)[s_0]$ from B_1 and x from X_t, put s_0 into A and $\gamma_2(x)[t]$, where $s_0 < s$ is the stage at which x is enumerated into X. Now stop cycles $(k', l') > (k, l)$ and wait for a bigger β-expansionary stage s'.

(7) Set $\Xi^U(k) = V_{s'}(k)$ with $\xi(k)[s] = s'$. Start $(k+1, 0)$-cycle, and simultaneously, wait for $U\lceil\xi(k)[s]$ or $V(k)$ to change.

(7.1) If $U\lceil\xi(k)[s']$ changes first then cancel cycles $(k', l') > (k, l)$, and redefine $\Xi^U(k) = V(k)$ with $\xi(k)[s] = s'$.

(7.2) If $V(k)$ changes first at stage $t' > s'$, then we just remove s_0 from A, remove $\gamma_2(x)[t]$ from B_2 and, put x into X, $\gamma_1(x)[s_0] - 1$ into B_1, and also t into A.

(8) Wait for $U[\xi(k)[s']$ to change. If this change happens then $\Xi^U(k)$ is undefined, and we can define it again, which will compute $V(k)$ correctly. Start cycle $(k + 1, 0)$.

Similar to the $\mathcal{P}$-strategies, β has two possible outcomes:

(k, l, ∞): Some cycle (k, l) (the leftmost one) runs infinitely often, then $\Delta^{A,U}$ is totally defined and computes D correctly.

(k, l, f): Some cycle (k, l) waits at step 1 eventually, then a disagreement appears between $\Phi^{B_1,A,U}$ and D, or between $\Phi^{B_2,A,U}$ and D.

3 Construction

First of all, the priority of the requirements are arranged as follows:

$$\mathcal{G} < \mathcal{R} < \mathcal{N}_0 < \mathcal{P}_0 < \mathcal{N}_1 < \mathcal{P}_1 < \cdots < \mathcal{N}_n < \mathcal{P}_n < \cdots,$$

where each $\mathcal{X} < \mathcal{Y}$ mean that, as usual, $\mathcal{X}$ has higher priority than $\mathcal{Y}$.

Each $\mathcal{N}$-strategy has outcomes $(k, l, \infty), (k, l, f), k, l \in \omega$, in order type ω^2. Each $\mathcal{P}$-strategy has outcomes $(j, f), (j, \infty)$ for $j \in \omega$. These outcomes are listed in lexicographic order. The priority tree, T, is constructed recursively by the outcomes of the strategies corresponding to the requirements and grows downwards. The requirements $\mathcal{G}$ and $\mathcal{R}$ are both global, and hence their outcomes are not on T. The full construction is given as follows. The construction will proceed by stages. Especially, at odd stages, we define Γ_i's; at even stages, approximate the true path. For the sake of clarifying what a number enters B_i, we require that the $\gamma_i(-)$-uses are defined by odd numbers.

Construction

Stage 0: Initialize all nodes on T, and let $A = B_1 = B_2 = X = \Gamma_1 = \Gamma_2 = \emptyset$. Let $\sigma_0 = \lambda$.

Stage $s = 2n + 1 > 0$: Define $\Gamma_1^{B_1,A,U}(x) = \Gamma_2^{B_2,A,U}(x) = X_s(x)$ for least $x < s$ for which Γ's are not defined, with fresh uses $\gamma_1(x)[s]$ and $\gamma_2(x)[s]$.

Stage $s = 2n > 0$:

Substage 0: Let $\sigma_s(0) = \lambda$ the root node.

Substage t: Given $\zeta = \sigma_s[t]$. First initialize all the nodes $>_L \zeta$. If $t = s$ then define $\sigma_s = \zeta$ and initialize all the nodes with lower priority than σ_s. If $t < s$, take action for ζ and define $\sigma_s(t+1)$ depending on which requirement ζ works for. In any case, for $\mathcal{N}$-strategies on σ_s, extend the definition of the associated Δs in a standard way, to compute the corresponding D.

Case 1. If $\zeta = \alpha$ is a $\mathcal{P}$-strategy, there are three subcases.

  $\alpha 1$. If α has no cycle started, then start cycle 0 and choose a fresh number $x_{\alpha,0}$ as its attacker. Define $\sigma_s = \alpha^\frown(0, f)$, initialize all the nodes with lower priority.

  $\alpha 2$. If $\alpha 1$ fails, let j be the largest active cycle at the last ζ-stage if any. Now, implement the delayed permitting (if any) as follows:

  $\alpha 2.1$ If U has a change below the restraint of a (least) cycle $j' \leq j$, then define $\sigma_s(t + 1)$ be $\alpha^\frown(j', \infty)$ and then go to the next substage, if cycle j' has not received the V-permission on j' so far; and define $\sigma_s(t + 1)$ as $\alpha^\frown(j' + 1, f)$, if cycle j' received a V-permission on j' before. In the latter case, redefine $\psi_\alpha(j')$ as $\psi_\alpha(j' - 1)[s]$ (j' is now in V) and define $x_{\alpha,j'+1}$ as a new number. Let $\sigma_s = \alpha^\frown(j' + 1, f)$, initialize all the nodes with priority lower than σ_s and go to the next stage.

  $\alpha 2.2$ $\alpha 2.1$ fails and V has a change on a (least) number $j' \leq j$ between the last α-stage and stage $s + 1$. Then let cycle j' act at this stage as follows: enumerate $x_{\alpha,j'}$ into X and $\gamma_i(x_{\alpha,j'})$ into B_i. Initialize all the nodes with priority lower than or equal to $\alpha^\frown(j', f)$ and go to the next stage. Let $\sigma_s = \alpha^\frown(j', f)$.

  $\alpha 3$. Neither $\alpha 1$ nor $\alpha 2$ is true. Let α be in j-cycle. Check whether $\Phi_{e(\alpha)}^{A,U}(x_{\alpha,j})[s + 1] \downarrow = 0$. If yes, then set $\Psi_\alpha^U(j) = V(j)$ with use $\psi_\alpha(x_{\alpha,j})$ equal to $\varphi_{e(\alpha)}(x_{\alpha,j})[s]$. Start $(j + 1)$-cycle by choosing a fresh number $x_{\alpha,j+1}$ as the attacker and define $\sigma_s = \alpha^\frown(j + 1, f)$. Otherwise, define $\sigma_s(t + 1)$ as $\alpha^\frown(j, f)$ and go to the next substage.

Case 2. $\zeta = \beta$ is an $\mathcal{N}$-strategy. Let (k, l) be the active cycle with the lowest priority which is not initialized since the last β-stage. There are three subcases.

  $\beta 1$. U changes below some restraint of a cycle $(k', l') \leq_L (k, l)$, or V changes and gives a V-permission for a cycle $(k', l') \leq_L (k, l)$. Without loss of generality, let (k', l') be such a cycle with the highest priority.

  $\beta 1.1$ If U has a change below the restraint of a cycle $(k', l') <_L (k, l)$. Without loss of generality, we assume (k', l') is the one with the highest priority, then see whether (k', l') has received some V-permissions. If no V-permission on k' and l' occurs, then we define $\sigma_s(t+1)$ as $\beta^\frown(k', l', \infty)$ or $\beta^\frown(k', l', f)$ according to whether $s + 1$ is β-expansionary (with respect to cycle (k', l')), and go to the next substage. If V-permission on l' occurs before, but not on k', then define σ_s as $\beta^\frown(k', l' + 1, \infty)$ (we start cycle $(k', l'+1)$, and stop stage $s+1$). If V-permission on k', occurs before, then we define σ_s as $\beta^\frown(k' + 1, 0, \infty)$ (we start cycle $(k' + 1, 0)$, and stop stage $s + 1$).

  $\beta 1.2$ $\beta 1.1$ fails and cycle (k', l') receives a V-permission. There are two cases:

  Case 1. $V(l')$ changes, then remove the corresponding $\gamma_1(x)$ (old) from B_1, x from X, enumerate the current $\gamma_2(x)$ into B_2, and enumerate $s_{x,1}$ into A, where $s_{x,1}$ is the stage when x is enumerated into X.

Case 2. $V(k')$ changes, then remove the corresponding $\gamma_2(x)$ from B_2 (the one being enumerated into B_2 when the cycle received the $V(l')$-permission), $s_{x,1}$ out A, and enumerate x into X, and $s_{x,2}$ into A, where $s_{x,2}$ is the stage when the $V(l')$-permission was performed.

$\beta 2.$ $\beta 1$ fails. Check whether $s+1$ is a β-expansionary stage (with respect to cycle (k,l)).

If no, then let $\sigma_s(t+1)$ be $\beta^\frown(k,l,f)$ and go to the next substage.

If yes, then check whether cycle (k,l) has received and performed the V-permission on l. If yes, then we set $\Xi^U(k) = V(k)[s]$ with $\xi(k)[s] = s$. Start cycle $(k+1,0)$, and let σ_s be $\beta^\frown(k+1,0,\infty)$. Initialize all the nodes with priority lower than $\beta^\frown(k+1,0,\infty)$ and go to the next stage. Otherwise, check whether $\Delta_\beta^{A,U}$ computes $D_{j(\beta)}$ correctly. If yes again, then extend the definition of $\Delta_\beta^{A,U}$ correspondingly, and let $\sigma_s(t+1)$ be $\beta^\frown(k,l,\infty)$ and go to the next substage. If no, there is some (least) z with $\Delta_\beta^{A,U}(z) \neq D_{j(\beta)}(z)$, cycle (k,l) acts by setting $\Theta_k^U(l) = V(l)[s]$ with $\theta_k(l)[s] = \delta_\beta(z)[s]$ (note that $\delta_\beta 9z)[s] < s$). Start cycle $(k,l+1)$, and let σ_s be $\beta^\frown(k,l+1,\infty)$. Initialize all the nodes with priority lower than $\beta^\frown(k,l+1,\infty)$ and go to the next stage.

This completes stage s and hence the whole construction.

4 Verification

We now verify that the constructed sets A, B, C and X satisfy all of the requirements, and completes the proof of Theorem 1.

Let $TP = \liminf_s \sigma_{2s}$ be the true path of the construction. We first prove that TP is infinite and then verify that the construction given above satisfies all the requirements.

The following lemma is standard.

Lemma 1. *Let σ be any node on TP. Then*

(1) σ *can only be initialized finitely often.*

(2) σ *has an outcome $\mathcal{O}$ such that $\sigma^\frown\mathcal{O}$ is on TP.*

(3) σ *can initialize strategies below $\sigma^\frown\mathcal{O}$ at most finitely often.*

Furthermore, if σ is a $\mathcal{P}$ or $\mathcal{N}$-strategy, then the corresponding $\mathcal{P}$ or $\mathcal{N}$-requirement is satisfied.

We now prove that all the constructed sets are reducible to V.

Lemma 2. $A, B_1, B_2 \leq_T V \oplus U$.

Proof. By our construction, for a given strategy σ, we implement the V-permitting only when σ is visited again even though such a V-change happens before σ is visited. We demonstrate the way of reducing A to $U \oplus V$, and the other reductions share the same idea, even though, sometimes, a little bit more complicated.

To show that $A \leq_T V \oplus U$, fix a number $n \in \omega$. We will show that we can $V \oplus U$-recursively determine whether $n \in A$ or not. In the course of the construction, only $\mathcal{N}$-strategies can put or extract numbers from A. Furthermore, if n can be enumerated into A, then it must be that some $\mathcal{P}$-strategy, α say, enumerates a number x into X, and $\gamma_i(x)[n-1]$ into $B_{i,n}$ at stage n, and later this x is extracted from X by some $\mathcal{N}$-strategy $\beta < \alpha$ when n is enumerated into A. So if at stage n, no number is enumerated into X, then $n \notin A$. Otherwise, assume x is enumerated into X at stage n by some α, and $\gamma_i(x)[n-1]$ into $B_{i,n}$ simultaneously. Find out the highest priority $\mathcal{N}$-strategy $\beta \subset \alpha$ (if any) with $\beta^\frown \mathcal{O} \subseteq \alpha$ such that for some z with $\Delta^{A,U}_{\beta^\frown \mathcal{O}}(z)$ defined, the computations $\Phi^{B_1,A,U}_{e(\beta)}(z)$ and $\Phi^{B_2,A,U}_{e(\beta)}(z)$ are injured by the enumeration of $\gamma_i(x)$-uses. If this β does not exist, then n will not be in A.

Now we assume the existence of such $\mathcal{N}$-strategy $\beta \subset \alpha$. For the convenience in the following proof, we introduce the following notions. Given $\sigma \in T$, the *permission-bound at stage* s, $b_s(\sigma)$, is defined as

$$\max(\{k+1, l+1, j+1 : \exists \tau \leq_L \sigma \ (\text{ active } \tau^\frown(k,l) \leq_L \sigma \text{ or active} \atop \tau^\frown(j,-) \leq_L \sigma, - \in \{\infty, f\})\}),$$

where "active" means that it is not initialized, cancelled or stopped at stage s. Let $S(\sigma_s)$ be the least stage such that $V_{S(\sigma_s)}\lceil b_s(\sigma) = V \lceil b_s(\sigma)$. Obviously, $S(\sigma_s)$ is V-computable.

To $U \oplus V$-recursively find the stage s such that $n \in A \iff n \in A_s$, we need the following crucial claim (core of the density argument), which can be proved by induction:

Claim: Assume that a $\mathcal{P}$-strategy σ runs cycle j at stage s_0 and

$$s_1 \geq \max\{s_0, S(\sigma_s), j+1\}$$

is a U-true stage. If σ does not run cycle j at stage s_1, then after stage s_1, if σ runs cycle j again, at stage s_2 say, cycle j must have been cancelled or initialized between stage s_0 and s_2. A similar result is true for an $\mathcal{N}$-strategy σ with (k,l) cycles instead.

This claim says that the permission can be delayed, but cannot be delayed unlimited.

Now we go back to the proof of $A \leq_T V \oplus U$. Suppose that $\sigma = \beta^\frown(k,l,-) \subseteq \alpha$ for some $k, l \in \omega$. Now if $l \in V_n$ or $l \notin V$, then $n \notin A$. Assume l enters V at stage $t > n$, let $t' > t$ be the least U-true stage greater than both t and $S(\sigma_t)$. t' is $U \oplus V$-computable, and so it is V-computable. At stage t', if σ is not visited, or if σ is visited but does not run cycle (k,l), then by the claim, $n \in A$ iff $n \in A_{t'}$. If σ runs cycle (k,l) at stage t', it is easy to see that $n \in A$ is possible only when $n \in A_{t'}$, after stage t', n cannot get V-permission to enter A. Now by a similar discussion on k, we can have a corresponding stage t'', $U \oplus V$-computable, after which n cannot be removed from A if $n \in A_{t'}$. In this case, $n \in A$ iff $n \in A_{t''}$. Therefore $A \leq_T U \oplus V$.

By a similar argument, we can show that both B_1 and B_2, are reducible to $V \oplus U$.

We complete the proof of Theorem 1 by showing the following lemma.

Lemma 3. $\Gamma_1^{B_1,A,U}$ *and* $\Gamma_2^{B_2,A,U}$ *are well-defined. Furthermore,* $X = \Gamma_1^{B_1,A,U} = \Gamma_2^{B_2,A,U}$. *As a consequence,* $X \leq_T V$.

Acknowledgement. Wu is partially supported by a research grant RG58/06 from Nanyang Technological University.

References

1. Arslanov, M.M., Lempp, S., Shore, R.A.: On isolating r.e. and isolated d.r.e. degrees. In: Cooper, S.B., Slaman, T.A., Wainer, S.S. (eds.) Computability, Enumerability, Unsolvability, pp. 61–80 (1996)
2. Cenzer, D., LaForte, F., Wu, G.: The nonisolating degrees are nowhere dense in the computably enumeralbe degrees (to appear)
3. Cooper, S.B., Harrington, L., Lachlan, A.H., Lempp, S., Soare, R.I.: The d.r.e. degrees are not dense. Ann. Pure Appl. Logic 55, 125–151 (1991)
4. Cooper, S.B., Lempp, S., Watson, P.: Weak density and cupping in the d-r.e. degrees. Israel J. Math. 67, 137–152 (1989)
5. Ding, D., Qian, L.: Isolated d.r.e. degrees are dense in r.e. degree structure. Arch. Math. Logic 36, 1–10 (1996)
6. Fejer, P.A.: The density of the nonbranching degrees. Ann. Pure Appl. Logic 24, 113–130 (1983)
7. Kaddah, D.: Infima in the d.r.e. degrees. Ann. Pure Appl. Logic 62, 207–263 (1993)
8. Kleene, S.C., Post, E.L.: The upper semi-lattice of degrees of recursive unsolvability. Ann. of Math. 59, 379–407 (1954)
9. Lachlan, A.H.: Lower bounds for pairs of recursively enumerable degrees. Proc. London Math. Soc. 16, 537–569 (1966)
10. LaForte, G.L.: The isolated d.r.e. degrees are dense in the r.e. degrees. Math. Logic Quart. 42, 83–103 (1996)
11. Soare, R.I.: Recursively Enumerable Sets and Degrees. Perspectives in Mathematical Logic. Springer, Heidelberg (1987)
12. Wu, G.: On the density of the pseudo-isolated degrees. In: Proceedings of London Mathematical Society, vol. 88, pp. 273–288 (2004)

A Divergence Formula for Randomness and Dimension

Jack H. Lutz*

Department of Computer Science, Iowa State University, Ames, IA 50011 USA
lutz@cs.iastate.edu

Abstract. If S is an infinite sequence over a finite alphabet Σ and β is a probability measure on Σ, then the *dimension* of S with respect to β, written $\dim^\beta(S)$, is a constructive version of Billingsley dimension that coincides with the (constructive Hausdorff) dimension $\dim(S)$ when β is the uniform probability measure. This paper shows that $\dim^\beta(S)$ and its dual $\text{Dim}^\beta(S)$, the *strong dimension* of S with respect to β, can be used in conjunction with randomness to measure the similarity of two probability measures α and β on Σ. Specifically, we prove that the *divergence formula*

$$\dim^\beta(R) = \text{Dim}^\beta(R) = \frac{\mathcal{H}(\alpha)}{\mathcal{H}(\alpha) + \mathcal{D}(\alpha||\beta)}$$

holds whenever α and β are computable, positive probability measures on Σ and $R \in \Sigma^\infty$ is random with respect to α. In this formula, $\mathcal{H}(\alpha)$ is the Shannon entropy of α, and $\mathcal{D}(\alpha||\beta)$ is the Kullback-Leibler divergence between α and β.

1 Introduction

The constructive dimension $\dim(S)$ and the constructive strong dimension $\text{Dim}(S)$ of an infinite sequence S over a finite alphabet Σ are constructive versions of the two most important classical fractal dimensions, namely, Hausdorff dimension [7] and packing dimension [19,18], respectively. These two constructive dimensions, which were introduced in [11,1], have been shown to have the useful characterizations

$$\dim(S) = \liminf_{w \to S} \frac{\text{K}(w)}{|w| \log |\Sigma|} \tag{1.1}$$

and

$$\text{Dim}(S) = \limsup_{w \to S} \frac{\text{K}(w)}{|w| \log |\Sigma|}, \tag{1.2}$$

* This research was supported in part by National Science Foundation Grants 9988483, 0344187, 0652569, and 0728806 and by the Spanish Ministry of Education and Science (MEC) and the European Regional Development Fund (ERDF) under project TIN2005-08832-C03-02.

K. Ambos-Spies, B. Löwe, and W. Merkle (Eds.): CiE 2009, LNCS 5635, pp. 342–351, 2009.
© Springer-Verlag Berlin Heidelberg 2009

where the logarithm is base-2 [15,1]. In these equations, $K(w)$ is the Kolmogorov complexity of the prefix w of S, i.e., the *length in bits of the shortest program* that prints the string w. (See section 2.4 or [9] for details.) The numerators in these equations are thus the *algorithmic information content* of w, while the denominators are the "naive" information content of w, also in bits. We thus understand (1.1) and (1.2) to say that $\dim(S)$ and $\mathrm{Dim}(S)$ are the lower and upper *information densities* of the sequence S. These constructive dimensions and their analogs at other levels of effectivity have been investigated extensively in recent years [8].

The constructive dimensions $\dim(S)$ and $\mathrm{Dim}(S)$ have recently been generalized to incorporate a probability measure ν on the sequence space Σ^∞ as a parameter [13]. Specifically, for each such ν and each sequence $S \in \Sigma^\infty$, we now have the constructive dimension $\dim^\nu(S)$ and the constructive strong dimension $\mathrm{Dim}^\nu(S)$ of S with respect to ν. (The first of these is a constructive version of Billingsley dimension [2].) When ν is the uniform probability measure on Σ^∞, we have $\dim^\nu(S) = \dim(S)$ and $\mathrm{Dim}^\nu(S) = \mathrm{Dim}(S)$. A more interesting example occurs when ν is the product measure generated by a nonuniform probability measure β on the alphabet Σ. In this case, $\dim^\nu(S)$ and $\mathrm{Dim}^\nu(S)$, which we write as $\dim^\beta(S)$ and $\mathrm{Dim}^\beta(S)$, are again the lower and upper information densities of S, but these densities are now measured with respect to unequal letter costs. Specifically, it was shown in [13] that

$$\dim^\beta(S) = \liminf_{w \to S} \frac{K(w)}{\mathcal{I}_\beta(w)} \qquad (1.3)$$

and

$$\mathrm{Dim}^\beta(S) = \limsup_{w \to S} \frac{K(w)}{\mathcal{I}_\beta(w)}, \qquad (1.4)$$

where

$$\mathcal{I}_\beta(w) = \sum_{i=0}^{|w|-1} \log \frac{1}{\beta(w[i])}$$

is the Shannon self-information of w with respect to β. These unequal letter costs $\log(1/\beta(a))$ for $a \in \Sigma$ can in fact be useful. For example, the complete analysis of the dimensions of individual points in self-similar fractals given by [13] requires these constructive dimensions with a particular choice of the probability measure β on Σ.

In this paper we show how to use the constructive dimensions $\dim^\beta(S)$ and $\mathrm{Dim}^\beta(S)$ in conjunction with randomness to measure the degree to which two probability measures on Σ are similar. To see why this might be possible, we note that the inequalities

$$0 \le \dim^\beta(S) \le \mathrm{Dim}^\beta(S) \le 1$$

hold for all β and S and that the maximum values

$$\dim^\beta(R) = \mathrm{Dim}^\beta(R) = 1 \qquad (1.5)$$

are achieved if (but not only if) the sequence R is random with respect to β. It is thus reasonable to hope that, if R is random with respect to some other probability measure α on Σ, then $\dim^\beta(R)$ and $\mathrm{Dim}^\beta(R)$ will take on values whose closeness to 1 reflects the degree to which α is similar to β. This is indeed the case. Our main theorem says that the *divergence formula*

$$\dim^\beta(R) = \mathrm{Dim}^\beta(R) = \frac{\mathcal{H}(\alpha)}{\mathcal{H}(\alpha) + \mathcal{D}(\alpha||\beta)} \tag{1.6}$$

holds whenever α and β are computable, positive probability measures on Σ and $R \in \Sigma^\infty$ is random with respect to α. In this formula, $\mathcal{H}(\alpha)$ is the Shannon entropy of α, and $\mathcal{D}(\alpha||\beta)$ is the Kullback-Leibler divergence between α and β. When $\alpha = \beta$, the Kullback-Leibler divergence $\mathcal{D}(\alpha||\beta)$ is 0, so (1.6) coincides with (1.5). When α and β are dissimilar, the Kullback-Leibler divergence $\mathcal{D}(\alpha||\beta)$ is large, so the right-hand side of (1.6) is small. Hence the divergence formula tells us that, when R is α-random, $\dim^\beta(R) = \mathrm{Dim}^\beta(R)$ is a quantity in $[0,1]$ whose closeness to 1 is an indicator of the similarity between α and β.

In the full version of this paper [12] we prove that the divergence formula also holds for the finite-state β-dimensions [5] of sequences that are α-normal in the sense of Borel [3]. This proof is more challenging, but it uses ideas from the present proof.

2 Preliminaries

2.1 Notation and Setting

Throughout this paper we work in a finite alphabet $\Sigma = \{0, 1, \ldots, k-1\}$, where $k \geq 2$. We write Σ^* for the set of (finite) *strings* over Σ and Σ^∞ for the set of (infinite) *sequences* over Σ. We write $|w|$ for the length of a string w and λ for the empty string. For $w \in \Sigma^*$ and $0 \leq i < |w|$, $w[i]$ is the ith symbol in w. Similarly, for $S \in \Sigma^\infty$ and $i \in \mathbb{N}$ ($= \{0, 1, 2, \ldots\}$), $S[i]$ is the ith symbol in S. Note that the leftmost symbol in a string or sequence is the 0th symbol. For $a \in \Sigma$ and $w \in \Sigma^*$, we write $\#(a, w)$ for the number of a's in w.

A *prefix* of a string or sequence $x \in \Sigma^* \cup \Sigma^\infty$ is a string $w \in \Sigma^*$ for which there exists a string or sequence $y \in \Sigma^* \cup \Sigma^\infty$ such that $x = wy$. In this case we write $w \sqsubseteq x$. The equation $\lim_{w \to S} f(w) = L$ means that, for all $\epsilon > 0$, for all sufficiently long prefixes $w \sqsubseteq S$, $|f(w) - L| < \epsilon$. We also use the limit inferior,

$$\liminf_{w \to S} f(w) = \lim_{w \to S} \inf \{f(x) \mid w \sqsubseteq x \sqsubseteq S\},$$

and the limit superior

$$\limsup_{w \to S} f(w) = \lim_{w \to S} \sup \{f(x) \mid w \sqsubseteq x \sqsubseteq S\}.$$

2.2   Probability Measures, Gales, and Shannon Information

A *probability measure* on Σ is a function $\alpha : \Sigma \to [0,1]$ such that $\sum_{a \in \Sigma} \alpha(a) = 1$. A probability measure α on Σ is *positive* if $\alpha(a) > 0$ for every $\alpha \in \Sigma$. A probability measure α on Σ is *rational* if $\alpha(a) \in \mathbb{Q}$ (i.e., $\alpha(a)$ is a rational number) for every $a \in \Sigma$.

A *probability measure* on Σ^∞ is a function $\nu : \Sigma^* \to [0,1]$ such that $\nu(\lambda) = 1$ and, for all $w \in \Sigma^*$, $\nu(w) = \sum_{a \in \Sigma} \nu(wa)$. (Intuitively, $\nu(w)$ is the probability that $w \sqsubseteq S$ when the sequence $S \in \Sigma^\infty$ is "chosen according to ν.") Each probability measure α on Σ naturally induces the probability measure α on Σ^∞ defined by

$$\alpha(w) = \prod_{i=0}^{|w|-1} \alpha(w[i]) \tag{2.1}$$

for all $w \in \Sigma^*$.

We reserve the symbol μ for the *uniform probability measure* on Σ, i.e.,

$$\mu(a) = \frac{1}{k} \text{ for all } a \in \Sigma,$$

and also for the *uniform probability measure* on Σ^∞, i.e.,

$$\mu(w) = k^{-|w|} \text{ for all } w \in \Sigma^*.$$

If α is a probability measure on Σ, then a sequence $S \in \Sigma^\infty$ has *asymptotic frequency* α, and we write $S \in \text{FREQ}^\alpha$, if $\lim_{w \to S} \frac{\#(a,w)}{|w|} = \alpha(a)$ holds for all $a \in \Sigma$. Such sequences S are also said to be α-1-*normal* in the sense of Borel [3].

If α is a probability measure on Σ and $s \in [0, \infty)$, then an s-α-*gale* is a function $d : \Sigma^* \to [0, \infty)$ satisfying

$$d(w) = \sum_{a \in \Sigma} d(wa)\alpha(a)^s \tag{2.2}$$

for all $w \in \Sigma^*$. A 1-α-gale is also called an α-*martingale*. When $\alpha = \mu$, we omit it from this terminology, so an s-μ-gale is called an s-*gale*, and a μ-martingale is called a *martingale*.

We frequently use the following simple fact without explicit citation.

Observation 2.1 *Let α and β be positive probability measures on Σ, and let $s, t \in [0, \infty)$. If $d : \Sigma^* \to [0, \infty)$ is an s-α-gale, then the function $\tilde{d} : \Sigma^* \to [0, \infty)$ defined by*

$$\tilde{d}(w) = \frac{\alpha(w)^s}{\beta(w)^t} d(w)$$

is a t-β-gale.

Intuitively, an s-α-gale is a strategy for betting on the successive symbols in a sequence $S \in \Sigma^\infty$. For each prefix $w \sqsubseteq S$, $d(w)$ denotes the amount of capital (money) that the gale d has after betting on the symbols in w. If $s = 1$, then the right-hand side of (2.2) is the conditional expectation of $d(wa)$, given that w has occurred, so (2.2) says that the payoffs are fair. If $s < 1$, then (2.2) says that the payoffs are unfair.

Let d be a gale, and let $S \in \Sigma^\infty$. Then d *succeeds* on S if $\limsup_{w \to S} d(w) = \infty$, and d *succeeds strongly* on S if $\liminf_{w \to S} d(w) = \infty$. The *success set* of d is the set $S^\infty[d]$ of all sequences on which d succeeds, and the *strong success set* of d is the set $S^\infty_{\mathrm{str}}[d]$ of all sequences on which d succeeds strongly.

The k-ary *Shannon entropy* of a probability measure α on Σ is

$$\mathcal{H}_k(\alpha) = \sum_{a \in \Sigma} \alpha(a) \log_k \frac{1}{\alpha(a)},$$

where $0 \log_k \frac{1}{0} = 0$. The k-ary *Kullback-Leibler divergence* between two probability measures α and β on Σ is

$$\mathcal{D}_k(\alpha \| \beta) = \sum_{a \in \Sigma} \alpha(a) \log_k \frac{\alpha(a)}{\beta(a)}.$$

When $k = 2$ in $\mathcal{H}_k$, $\mathcal{D}_k$, or $\log_k$, we omit it from the notation. The Kullback-Leibler divergence is used to quantify how "far apart" the two probability measures α and β are. The *Shannon self-information* of a string $w \in \Sigma^*$ with respect to a probability measure β on Σ is

$$\mathcal{I}_\beta(w) = \log \frac{1}{\beta(w)} = \sum_{i=0}^{|w|-1} \log \frac{1}{\beta(w[i])}.$$

Discussions of $\mathcal{H}(\alpha)$, $\mathcal{D}(\alpha \| \beta)$, $\mathcal{I}_\beta(w)$ and their properties may be found in any good text on information theory, e.g., [4].

2.3   Hausdorff, Packing, and Billingsley Dimensions

Given a probability measure β on Σ, each set $X \subseteq \Sigma^\infty$ has a *Hausdorff dimension* $\dim(X)$, a *packing dimension* $\mathrm{Dim}(X)$, a *Billingsley dimension* $\dim^\beta(X)$, and a *strong Billingsley dimension* $\mathrm{Dim}^\beta(X)$, all of which are real numbers in the interval $[0, 1]$. In this paper we are not concerned with the original definitions of these classical dimensions, but rather in their recent characterizations (which may be taken as definitions) in terms of gales.

Notation. For each probability measure β on Σ and each set $X \subseteq \Sigma^\infty$, let $\mathcal{G}^\beta(X)$ (respectively, $\mathcal{G}^{\beta,\mathrm{str}}(X)$) be the set of all $s \in [0, \infty)$ such that there is a β-s-gale d satisfying $X \subseteq S^\infty[d]$ (respectively, $X \subseteq S^\infty_{\mathrm{str}}[d]$).

Theorem 2.2 (gale characterizations of classical fractal dimensions)
Let β be a probability measure on Σ, and let $X \subseteq \Sigma^\infty$.

1. *[10]* $\dim(X) = \inf \mathcal{G}^\mu(X)$.   3. *[13]* $\dim^\beta(X) = \inf \mathcal{G}^\beta(X)$.
2. *[1]* $\mathrm{Dim}(X) = \inf \mathcal{G}^{\mu,\mathrm{str}}(X)$.   4. *[13]* $\mathrm{Dim}^\beta(X) = \inf \mathcal{G}^{\beta,\mathrm{str}}(X)$.

2.4   Randomness and Constructive Dimensions

Randomness and constructive dimensions are defined by imposing computability constraints on gales.

A real-valued function $f : \Sigma^* \to \mathbb{R}$ is *computable* if there is a computable, rational-valued function $\hat{f} : \Sigma^* \times \mathbb{N} \to \mathbb{Q}$ such that, for all $w \in \Sigma^*$ and $r \in \mathbb{N}$,

$$|\hat{f}(w,r) - f(w)| \leq 2^{-r}.$$

A real-valued function $f : \Sigma^* \to \mathbb{R}$ is *constructive*, or *lower semicomputable*, if there is a computable, rational-valued function $\hat{f} : \Sigma^* \times \mathbb{N} \to \mathbb{Q}$ such that

(i) for all $w \in \Sigma^*$ and $t \in \mathbb{N}$, $\hat{f}(w,t) \leq \hat{f}(w,t+1) < f(w)$, and
(ii) for all $w \in \Sigma^*$, $f(w) = \lim_{t \to \infty} \hat{f}(w,t)$.

The first successful definition of the randomness of individual sequences $S \in \Sigma^\infty$ was formulated by Martin-Löf [14]. Many characterizations (equivalent definitions) of randomness are now known, of which the following is the most pertinent.

Theorem 2.3 (Schnorr [16,17]). *Let α be a probability measure on Σ. A sequence $S \in \Sigma^\infty$ is* random *with respect to α (or, briefly, α-random) if there is no constructive α-martingale that succeeds on S.*

Motivated by Theorem 2.2, we now define the constructive dimensions.

Notation. We define the sets $\mathcal{G}^\beta_{\mathrm{constr}}(X)$ and $\mathcal{G}^{\beta,\mathrm{str}}_{\mathrm{constr}}(X)$ to be like the sets $\mathcal{G}^\beta(X)$ and $\mathcal{G}^{\beta,\mathrm{constr}}(X)$ of section 2.3, except that the β-s-gales are now required to be constructive.

Definition. Let β be a probability measure on Σ, let $X \subseteq \Sigma^\infty$, and let $S \in \Sigma^\infty$.

1. [11] The *constructive dimension* of X is $\mathrm{cdim}(X) = \inf \mathcal{G}^\mu_{\mathrm{constr}}(X)$.
2. [1] The *constructive strong dimension* of X is $\mathrm{cDim}(X) = \inf \mathcal{G}^{\mu,\mathrm{str}}_{\mathrm{constr}}(X)$.
3. [13] The *constructive β-dimension* of X is $\mathrm{cdim}^\beta(X) = \inf \mathcal{G}^\beta_{\mathrm{constr}}(X)$.
4. [13] The *constructive strong β-dimension* of X is $\mathrm{cDim}^\beta(X) = \inf \mathcal{G}^{\beta,\mathrm{str}}_{\mathrm{constr}}(X)$.
5. [11] The *dimension* of S is $\dim(S) = \mathrm{cdim}(\{S\})$.
6. [1] The *strong dimension* of S is $\mathrm{Dim}(S) = \mathrm{cDim}(\{S\})$.
7. [13] The *β-dimension* of S is $\dim^\beta(S) = \mathrm{cdim}^\beta(\{S\})$.
8. [13] The *strong β-dimension* of S is $\mathrm{Dim}^\beta(S) = \mathrm{cDim}^\beta(\{S\})$.

It is clear that definitions 1, 2, 5, and 6 above are the special case $\beta = \mu$ of definitions 3, 4, 7, and 8, respectively. It is known that $\mathrm{cdim}^\beta(X) = \sup_{S \in X} \dim^\beta(S)$ and that $\mathrm{cDim}^\beta(X) = \sup_{S \in X} \mathrm{Dim}^\beta(S)$ [13]. Constructive dimensions are thus investigated in terms of the dimensions of individual sequences. Since one does not discuss the classical dimension of an individual sequence (because the dimensions of section 2.3 are all zero for singleton, or even countable, sets), no confusion results from the notation $\dim(S)$, $\mathrm{Dim}(S)$, $\dim^\beta(S)$, and $\mathrm{Dim}^\beta(S)$.

The *Kolmogorov complexity* $\mathrm{K}(w)$ of a string $w \in \Sigma^*$ is the minimum length of a program $\pi \in \{0,1\}^*$ for which $U(\pi) = w$, where U is a fixed universal self-delimiting Turing machine [9].

Theorem 2.4. *Let β be a probability measure on Σ, and let $S \in \Sigma^\infty$.*

1. [15] $\dim(S) = \liminf_{w \to S} \frac{\mathrm{K}(w)}{|w| \log k}$.   *3. [13]* $\dim^\beta(S) = \liminf_{w \to S} \frac{\mathrm{K}(w)}{\mathcal{I}_\beta(w)}$.

2. [1] $\mathrm{Dim}(S) = \limsup_{w \to S} \frac{\mathrm{K}(w)}{|w| \log k}$.   *4. [13]* $\mathrm{Dim}^\beta(S) = \limsup_{w \to S} \frac{\mathrm{K}(w)}{\mathcal{I}_\beta(w)}$.

3   Divergence Formula for Randomness and Constructive Dimensions

This section proves the divergence formula for α-randomness, constructive β-dimension, and constructive strong β-dimension. The key point here is that the Kolmogorov complexity characterizations of these β-dimensions reviewed in section 2.4 immediately imply the following fact.

Lemma 3.1. *If α and β are computable, positive probability measures on Σ, then, for all $S \in \Sigma^\infty$,*

$$\liminf_{w \to S} \frac{\mathcal{I}_\alpha(w)}{\mathcal{I}_\beta(w)} \leq \frac{\dim^\beta(S)}{\dim^\alpha(S)} \leq \limsup_{w \to S} \frac{\mathcal{I}_\alpha(w)}{\mathcal{I}_\beta(w)},$$

and

$$\liminf_{w \to S} \frac{\mathcal{I}_\alpha(w)}{\mathcal{I}_\beta(w)} \leq \frac{\mathrm{Dim}^\beta(S)}{\mathrm{Dim}^\alpha(S)} \leq \limsup_{w \to S} \frac{\mathcal{I}_\alpha(w)}{\mathcal{I}_\beta(w)}.$$

It is interesting to note that Kolmogorov complexity's only role in the proof of our main theorem is to be cancelled from the numerators and denominators of fractions when deriving Lemma 3.1 from Theorem 2.4. This cameo appearance makes our proof significantly easier than our first proof, which did not use Kolmogorov complexity.

The following lemma is crucial to our argument.

Lemma 3.2 (frequency divergence lemma). *If α and β are positive probability measures on Σ, then, for all $S \in \mathrm{FREQ}^\alpha$,*

$$\mathcal{I}_\beta(w) = (\mathcal{H}(\alpha) + \mathcal{D}(\alpha||\beta))|w| + o(|w|)$$

as $w \to S$.

Proof. Assume the hypothesis, and let $S \in \mathrm{FREQ}^\alpha$. Then, as $w \to S$, we have

$$
\begin{aligned}
\mathcal{I}_\beta(w) &= \sum_{i=0}^{|w|-1} \log \frac{1}{\beta(w[i])} \\
&= \sum_{a \in \Sigma} \#(a, w) \log \frac{1}{\beta(a)} \\
&= |w| \sum_{a \in \Sigma} \mathrm{freq}_a(w) \log \frac{1}{\beta(a)} \\
&= |w| \sum_{a \in \Sigma} (\alpha(a) + o(1)) \log \frac{1}{\beta(a)} \\
&= |w| \sum_{a \in \Sigma} \alpha(a) \log \frac{1}{\beta(a)} + o(|w|) \\
&= |w| \sum_{a \in \Sigma} \left(\alpha(a) \log \frac{1}{\alpha(a)} + \alpha(a) \log \frac{\alpha(a)}{\beta(a)} \right) + o(|w|) \\
&= (\mathcal{H}(\alpha) + \mathcal{D}(\alpha||\beta))|w| + o(|w|).
\end{aligned}
$$

The next lemma gives a simple relationship between the constructive β-dimension and the constructive dimension of any sequence that is α-1-normal.

Lemma 3.3. *If α and β are computable, positive probability measures on Σ, then, for all $S \in \mathrm{FREQ}^\alpha$,*

$$
\dim^\beta(S) = \frac{\dim(S)}{\mathcal{H}_k(\alpha) + \mathcal{D}_k(\alpha||\beta)},
$$

and

$$
\mathrm{Dim}^\beta(S) = \frac{\mathrm{Dim}(S)}{\mathcal{H}_k(\alpha) + \mathcal{D}_k(\alpha||\beta)}.
$$

Proof. Let α, β, and S be as given. By the frequency divergence lemma, we have

$$
\begin{aligned}
\frac{\mathcal{I}_\mu(w)}{\mathcal{I}_\beta(w)} &= \frac{|w| \log k}{(\mathcal{H}(\alpha) + \mathcal{D}(\alpha||\beta))|w| + o(|w|)} \\
&= \frac{\log k}{\mathcal{H}(\alpha) + \mathcal{D}(\alpha||\beta) + o(1)} \\
&= \frac{\log k}{\mathcal{H}(\alpha) + \mathcal{D}(\alpha||\beta)} + o(1) \\
&= \frac{1}{\mathcal{H}_k(\alpha) + \mathcal{D}_k(\alpha||\beta)} + o(1)
\end{aligned}
$$

as $w \to S$. The present lemma follows from this and Lemma 3.1.

We now recall the following constructive strengthening of a 1949 theorem of Eggleston [6].

Theorem 3.4 ([11,1]). *If α is a computable probability measure on Σ, then, for every α-random sequence $R \in \Sigma^\infty$,*

$$\dim(R) = \text{Dim}(R) = \mathcal{H}_k(\alpha).$$

The main result of this section is now clear.

Theorem 3.5 (divergence formula for randomness and constructive dimensions). *If α and β are computable, positive probability measures on Σ, then, for every α-random sequence $R \in \Sigma^\infty$,*

$$\dim^\beta(R) = \text{Dim}^\beta(R) = \frac{\mathcal{H}(\alpha)}{\mathcal{H}(\alpha) + \mathcal{D}(\alpha||\beta)}.$$

Proof. This follows immediately from Lemma 3.3 and Theorem 3.4.

We note that $\mathcal{D}(\alpha||\mu) = \log k - \mathcal{H}(\alpha)$, so Theorem 3.4 is the case $\beta = \mu$ of Theorem 3.5.

Acknowledgments. I thank Xiaoyang Gu and Elvira Mayordomo for useful discussions.

References

1. Athreya, K.B., Hitchcock, J.M., Lutz, J.H., Mayordomo, E.: Effective strong dimension, algorithmic information, and computational complexity. SIAM Journal on Computing 37, 671–705 (2007)
2. Billingsley, P.: Hausdorff dimension in probability theory. Illinois Journal of Mathematics 4, 187–209 (1960)
3. Borel, E.: Sur les probabilités dénombrables et leurs applications arithmétiques. Rend. Circ. Mat. Palermo 27, 247–271 (1909)
4. Cover, T.M., Thomas, J.A.: Elements of Information Theory, 2nd edn. John Wiley & Sons, Inc., Chichester (2006)
5. Dai, J.J., Lathrop, J.I., Lutz, J.H., Mayordomo, E.: Finite-state dimension. Theoretical Computer Science 310, 1–33 (2004)
6. Eggleston, H.: The fractional dimension of a set defined by decimal properties. Quarterly Journal of Mathematics 20, 31–36 (1949)
7. Hausdorff, F.: Dimension und äusseres Mass. Mathematische Annalen 79, 157–179 (1919) (English translation)
8. Hitchcock, J.M.: Effective Fractal Dimension Bibliography (October 2008), http://www.cs.uwyo.edu/~jhitchco/bib/dim.shtml
9. Li, M., Vitányi, P.M.B.: An Introduction to Kolmogorov Complexity and its Applications, 2nd edn. Springer, Berlin (1997)
10. Lutz, J.H.: Dimension in complexity classes. SIAM Journal on Computing 32, 1236–1259 (2003)
11. Lutz, J.H.: The dimensions of individual strings and sequences. Information and Computation 187, 49–79 (2003)
12. Lutz, J.H.: A divergence formula for randomness and dimension. Technical Report cs.CC/0811.1825, Computing Research Repository (2008)

13. Lutz, J.H., Mayordomo, E.: Dimensions of points in self-similar fractals. SIAM Journal on Computing 38, 1080–1112 (2008)
14. Martin-Löf, P.: The definition of random sequences. Information and Control 9, 602–619 (1966)
15. Mayordomo, E.: A Kolmogorov complexity characterization of constructive Hausdorff dimension. Information Processing Letters 84(1), 1–3 (2002)
16. Schnorr, C.P.: A unified approach to the definition of random sequences. Mathematical Systems Theory 5, 246–258 (1971)
17. Schnorr, C.P.: A survey of the theory of random sequences. In: Butts, R.E., Hintikka, J. (eds.) Basic Problems in Methodology and Linguistics, pp. 193–210. D. Reidel, Dordrecht (1977)
18. Sullivan, D.: Entropy, Hausdorff measures old and new, and limit sets of geometrically finite Kleinian groups. Acta Mathematica 153, 259–277 (1984)
19. Tricot, C.: Two definitions of fractional dimension. In: Mathematical Proceedings of the Cambridge Philosophical Society, vol. 91, pp. 57–74 (1982)

On Ladner's Result for a Class of Real Machines with Restricted Use of Constants

Klaus Meer

BTU Cottbus
Konrad-Wachsmann-Allee 1
D-03046 Cottbus, Germany
meer@informatik.tu-cottbus.de

Abstract. We study the question whether there are analogues of Ladner's result in the computational model of Blum, Shub and Smale. It is known that in the complex and the additive BSS model a pure analogue holds, i.e. there are non-complete problems in NP \ P assuming NP ≠ P. In the (full) real number model only a non-uniform version is known. We define a new variant which seems relatively close to the full real number model. In this variant inputs can be treated as in the full model whereas real machine constants can be used in a restricted way only. Our main result shows that in this restricted model Ladner's result holds. Our techniques analyze a class P/const that has been known previously to be crucial for this kind of results. By topological arguments relying on the polyhedral structure of certain sets of machine constants we show that this class coincides with the new restricted version of $P_\mathbb{R}$, thus implying Ladner's result.

Keywords: Complexity, real number model, diagonal problems.

1 Introduction

Ladner's result [7] states that assuming P ≠ NP there exist problems in NP \ P which are not NP-complete. The respective question has been studied in other models as well. For Valiant's complexity classes VP and VNP a Ladner like result was shown in [3]. In the computational model introduced by Blum, Shub, and Smale several similar results have been proved. For the BSS model over the complex numbers Ladner's result holds analogously [8]. In [1] the authors analyzed in how far this result can be extended to further structures. They showed that a class P/const is of major importance for such questions. P/const was introduced by Michaux [9]. It denotes all decision problems that can be solved non-uniformly by a machine $M(x, c)$ that has a constant number k of non-rational machine constants $c \in \mathbb{R}$ and runs in polynomial time. M solves the problem up to input dimension n by taking for each n non-uniformly a potentially new set of constants $c^{(n)} \in \mathbb{R}^k$. For a more precise definition see below. It turns out that if P = P/const in a structure in which quantifier elimination is possible, then Ladner's result holds in this structure. The latter is the case for $\mathbb{C}$ as well as for $\{0, 1\}^*$. Over the reals it is unknown (and in fact unlikely) whether $P_\mathbb{R} = P_\mathbb{R}/\text{const}$. The currently

K. Ambos-Spies, B. Löwe, and W. Merkle (Eds.): CiE 2009, LNCS 5635, pp. 352–361, 2009.
© Springer-Verlag Berlin Heidelberg 2009

strongest known version of a Ladner like result in the full BSS model over $\mathbb{R}$ is the following non-uniform statement: If $\mathrm{NP}_{\mathbb{R}} \not\subseteq \mathrm{P}_{\mathbb{R}}/\mathrm{const}$ there are problems in $\mathrm{NP}_{\mathbb{R}} \setminus \mathrm{P}_{\mathbb{R}}/\mathrm{const}$ which are not $\mathrm{NP}_{\mathbb{R}}$-complete [1]. Chapuis and Koiran [4] give a deep model-theoretic analysis of real complexity classes of the form $\mathcal{C}/\mathrm{const}$ relating them to the notion of saturation. As by-product of their investigations they obtain Ladner's result for the reals with order and addition.

The present paper continues this line of research. In order to come closer to a complete analogue of Ladner's result in the full real number model we introduce a new variant of it that has to the best of our knowledge not been studied so far. This variant has full access to the input data but limited access to real machine constants. More precisely, an algorithm that uses machine constants $c_1, \ldots, c_k \in \mathbb{R}$ and gets an input $x \in \mathbb{R}^n$ is allowed to perform any operation among $\{+, -, *\}$ if an x_i is involved as operand. Two operands involving some of the c_i's can only be combined by $+$ or $-$. We call this model real number model with restricted use of constants. Its deterministic and non-deterministic polynomial time complexity classes are denoted by $\mathrm{P}_{\mathbb{R}}^{\mathrm{rc}}$ and $\mathrm{NP}_{\mathbb{R}}^{\mathrm{rc}}$. More details are given in Section 2. In a certain sense this variant swaps the role inputs and machine constants are playing in the linear BSS model [2]. Whereas in the latter all computed results depend linearly on the inputs but arbitrarily on machine constants, in the new variant it is just the opposite. However, this analogy is not complete since in our variant the degrees of intermediate results as polynomials in the input components can still grow exponentially in the number of computation steps, something that is not true in the linear BSS model with respect to the constants.

It turns out that the new variant is somehow closer to the full BSS model than other variants studied so far. For example, we show that some $\mathrm{NP}_{\mathbb{R}}$-complete problems are as well complete in $\mathrm{NP}_{\mathbb{R}}^{\mathrm{rc}}$.[1] The main result of the paper will be an analogue of Ladner's theorem in the restricted model. Towards this aim we start from the results in [1]. The main task is to analyze the corresponding class $\mathrm{P}_{\mathbb{R}}^{\mathrm{rc}}/\mathrm{const}$ and show on the one side that the diagonalization technique developped in [1] can be applied to $\mathrm{P}_{\mathbb{R}}^{\mathrm{rc}}/\mathrm{const}$ as well. On the other side the more difficult part will be to show that $\mathrm{P}_{\mathbb{R}}^{\mathrm{rc}}/\mathrm{const}$ actually is contained - and thus equal to - the class $\mathrm{P}_{\mathbb{R}}^{\mathrm{rc}}$. This implies Ladner's result.

The paper is organized as follows. Section 2 recalls previous results including necessary definitions of the models and complexity classes we are interested in. We introduce the restricted variant of the real number model. It is then shown that the $\mathrm{NP}_{\mathbb{R}}$-complete problem of deciding solvability of a system of real polynomial equations as well belongs to $\mathrm{NP}_{\mathbb{R}}^{\mathrm{rc}}$ and is complete in this class under $\mathrm{FP}_{\mathbb{R}}^{\mathrm{rc}}$-reductions. The diagonalization technique from [1] is then argued to work for $\mathrm{P}_{\mathbb{R}}^{\mathrm{rc}}/\mathrm{const}$ as well. In Section 3 the main result, namely the relation $\mathrm{P}_{\mathbb{R}}^{\mathrm{rc}}/\mathrm{const} \subseteq \mathrm{P}_{\mathbb{R}}^{\mathrm{rc}}$ is established. It implies the analogue of Ladner's theorem for $\mathrm{NP}_{\mathbb{R}}^{\mathrm{rc}}$. The paper finishes with some discussions.

---

[1] Though this is true as well for Koiran's weak model its different cost measure implies dramatic differences to the full model because $P \neq NP$ has been proved in the former [5].

2   Basic Notions and First Results

We suppose the reader to be familiar with the BSS model over the reals [2]. In this paper we deal with a variant of it which results from restricting the way in which algorithms are allowed to use real machine constants.

Definition 1. *a) Let A be a BSS algorithm using $c_1, \ldots c_k$ as its non-rational machine constants. An intermediate result computed by A on inputs x from some $\mathbb{R}^n$ is called* marked *if it as function of (x, c) depends on at least one of the c_i's, otherwise it is* unmarked.

b) The real number model with restricted use of constants is the variant of the full BSS model in which the following condition is imposed on usual BSS algorithms. A restricted machine *can perform the operations $\{+, -, *\}$ if at least one operand is unmarked. If both operands are marked, it can perform $+$ and $-$ only. Tests of the form 'is $x \geq 0$?' can be performed on both kind of operands.*

c) The cost of a restricted algorithm is defined as in the full model, i.e. each operation counts one. We denote the resulting analogues of $P_\mathbb{R}$ and $NP_\mathbb{R}$ by $P_\mathbb{R}^{rc}$ and $NP_\mathbb{R}^{rc}$, respectively.

A few remarks should clarify the power and the limits of the restricted model. We do not include divisions by unmarked operands for sake of simplicity. This would not change anything significantly. The condition on how to differentiate between marked and unmarked intermediate results guarantees a certain control on how real machine constants influence the results. More precisely, such an algorithm can compute arbitrarily with input components, whereas machine constants only occur linearly. Thus, each value computed by a restricted algorithm on an input $x \in \mathbb{R}^n$ is of form $\sum_{i=1}^{k} p_i(x) \cdot c_i + q(x)$. Here, the p_i's and q are polynomials in the input with rational coefficients. Contrary to for example the weak model introduced by Koiran the degrees of these polynomials still can grow exponentially in polynomial time. For rational machine constants no condition applies, i.e. they can be used arbitrarily. As a potential limit of the restricted model note that when composing machines constants of the first might become inputs of the second algorithm. Thus it is unclear whether reductions can be composed in general. Nevertheless, as we shall see below this problem has no significant impact for our pourposes since there exist $NP_\mathbb{R}^{rc}$-complete problems.

A typical example of a problem solvable efficiently by an algorithm in $P_\mathbb{R}^{rc}$ is the following: Given a polynomial $f \in \mathbb{R}[x_1, \ldots, x_n]$ and a $y \in \mathbb{R}^n$, is $f(y) = 0$? Here, f is given by the coefficients of its monomials. Both these coefficients and y are inputs to the problem, so the normal evaluation algorithm does not need additional constants. All of its operations can be performed by a restricted algorithm. Arguments of this kind shall be used again below to prove existence of complete problems. Another example takes as input a polynomial f as above and asks whether f has a real zero all of whose components can be found among the components of a real zero of a fixed real polynomial g. This problem is in

$NP_\mathbb{R}^{rc}$. Guess a zero y of g and a zero z of f such that all of z's components are among those of y. Evaluation of f in z is done as above, evaluation of g in y needs g's coefficients as machine constants. However, they only occur restrictedly in the evaluation procedure.

The study of transfer results of the P versus NP question between different structures led Michaux to the definition of the complexity class P/const [9]. It is a non-uniformly defined class which allows a uniform polynomial time machine to use non-uniformly different choices of machine constants for different input dimensions. Below we change a bit Michaux' original definition with respect to two technical details. We split rational uniform constants and potentially real constants that might be changed for each dimension. And we require the real machine constants to be bounded in absolute value by 1. Both changes do not affect the results from [1] as we are going to show later.

Definition 2. *A basic machine over $\mathbb{R}$ in the BSS-setting is a BSS-machine M with rational constants and with two blocks of parameters. One block x stands for a concrete input instance and takes values in $\mathbb{R}^\infty$, the other one c represents real constants used by the machine and has values in some $\mathbb{R}^k \cap [-1,1]^k$ ($k \in \mathbb{N}$ fixed for M). Here $\mathbb{R}^\infty := \bigcup_{n \geq 1} \mathbb{R}^n$ denotes the set of all finite sequences of real numbers.*

Basic machines for the restricted model are defined similarly.

The above definition of a basic machine intends to split the discrete skeleton of an original BSS machine from its real machine constants. That is done by regarding those constants as a second block of parameters. Fixing c we get back a usual BSS machine $M(\bullet, c)$ that uses c as its constants for all input instances x. If below we speak about the machine's constants we refer to the potentially real ones only.

Definition 3. *(cf. [9]) A problem L is in class $P_\mathbb{R}^{rc}/const$ if and only if there exists a restricted polynomial time basic machine M and for every $n \in \mathbb{N}$ a tuple $c^{(n)} \in [-1,1]^k$ of real constants for M such that $M(\bullet, c^{(n)})$ decides L upto size n.*

Lemma 1. *It is $P_\mathbb{R}^{rc} \subseteq P_\mathbb{R}^{rc}/const$.*

In a number of computational models the class P/const has strong relations to Ladner's result. In [1] the authors show for a number of structures including $\mathbb{R}, \mathbb{C}$ and finite structures that if NP $\not\subseteq$ P/const there are problems in NP\ P/const that are not NP-complete. Michaux proved in [9] that for recursively saturated structures P = P/const. As consequence the result mentioned above reproves Ladner's result both for the Turing model [7] and the BSS model over $\mathbb{C}$ [8]. Over the reals the currently strongest Ladner like theorem is the following.

Theorem 1. *[1] Suppose $NP_\mathbb{R} \not\subseteq P_\mathbb{R}/const$. Then there exist problems in $NP_\mathbb{R} \setminus P_\mathbb{R}/const$ not being $NP_\mathbb{R}$-complete.*

Concerning a further analysis of $P_\mathbb{R}/const$ and its relation to uniform classes, in particular to $P_\mathbb{R}$, not much is known. Chapuis and Koiran [4] argue that already

$P_{\mathbb{R}}/1 = P_{\mathbb{R}}$ is unlikely; here, $P_{\mathbb{R}}/1$ is defined by means of basic machines which use a finite number of uniform and a single non-uniform machine constant only. The main result below shows that the same line of arguments used in [1] extends to the restricted model as well. The way it is obtained might also shed new light on what can be expected for the full model. Some ideas are discussed in the conclusion section.

2.1   Complete Problems; Diagonalization between $P_{\mathbb{R}}^{rc}$ and $NP_{\mathbb{R}}^{rc}$

An indication for that the restricted model is not too exotic is to show that it shares complete problems with the original BSS model. Beside showing this the result given below is also needed with respect to the quantifier elimination procedure in the next subsection. The main point here is that such an elimination can be performed as well by a restricted machine. Towards this goal it is necessary to establish completeness of problems that do not involve non-rational constants.

Theorem 2. *The feasibility problem FEAS which asks for solvability of a system of real polynomial equations of degree at most two is $NP_{\mathbb{R}}^{rc}$-complete in the BSS model with restricted use of machine constants.*

Proof. The problem $FEAS$ belongs to $NP_{\mathbb{R}}^{rc}$. Given such a system by coding the coefficients of each monomial of the single polynomial equations a verification algorithm guesses as usual a potential real solution and evaluates all polynomials in the guessed point. Since both the coefficients and the potential solution are inputs for this algorithm it does not need any real constants and thus can be performed by making restricted use of constants.

For $NP_{\mathbb{R}}^{rc}$-completeness we only have to check the usual $NP_{\mathbb{R}}$-completeness proof of $FEAS$ from [2]. Here, the reduction of a computation of an $NP_{\mathbb{R}}$-machine M on an input x to the feasibility question of a polynomial system takes x as input. It then introduces for each step new variables which in a suitable way reflect the computation done in this step. To make the argument as easy as possible we can require that a (restricted) reduction algorithm at the very beginning introduces for each real constant c_i which M is using a new variable z_i together with the equation $c_i = z_i$ and then only uses z_i. The resulting polynomial system is constructed without additional real constants beside the c_i's. Moreover this construction algorithm uses the c_i's only in the demanded restricted way.   □

The next result says that an analogue variant of Theorem 1 is as well true in the restricted model.

Theorem 3. *Suppose $FEAS \not\subseteq P_{\mathbb{R}}^{rc}/const$ (and thus $NP_{\mathbb{R}}^{rc} \not\subseteq P_{\mathbb{R}}^{rc}/const$). Then there exist problems in $NP_{\mathbb{R}}^{rc} \setminus P_{\mathbb{R}}^{rc}/const$ not being $NP_{\mathbb{R}}^{rc}$-complete under restricted polynomial time reductions.*

We postpone the proof to the full paper and just indicate the main arguments why it is true. We adapt the proof of a corresponding result in [1]. Two properties have to be checked: first, one can describe the existence of error dimensions for

$P_{\mathbb{R}}^{rc}$/const- and reduction-machines by a (quantified) first-order formula in the theory of the reals. Secondly, the special form of this formula allows quantifier elimination by means of a restricted machine.

3 Proof of the Main Results

In this section we shall show $P_{\mathbb{R}}^{rc}$/const $= P_{\mathbb{R}}^{rc}$ thus implying Ladner's result in the BSS model with restricted use of constants. Since the proof is a bit involved let us first outline the main ideas. Let $L \in P_{\mathbb{R}}^{rc}$/const and M be a corresponding linear basic machine establishing this membership. Suppose M uses k machine constants. The overall goal is to find by means of a uniform algorithm for each dimension n a suitable choice of constants that could be used by M to decide $L \cap \mathbb{R}^{\leq n}$. We suppose that there is no uniform choice of one set of constants working for M; otherwise $L \in P_{\mathbb{R}}^{rc}$ is a trivial consequence.

Let $E_n \subseteq [-1,1]^k$ denote the suitable constants working for $L \cap \mathbb{R}^{\leq n}$ when used by M. Thus, for $x \in \mathbb{R}^{\leq n}$ we ideally would like to find efficiently and uniformly a $c \in E_n$ and let then run M on (x,c).[2] However, we shall not compute such a c. Instead we shall show that there are three vectors $c^*, d^*, e^* \in \mathbb{R}^k$ such that for all dimensions $n \in \mathbb{N}$ if we move a short step (depending on n) from c^* into the direction of d^* and afterwards a short step into the direction e^* we end up in E_n. This argument is relying on the topological structure of the E_n's. We are then left with deciding an assertion of the following structure:

$\forall n \in \mathbb{N} \; \exists \epsilon_1 > 0 \; \forall \mu_1 \in (0, \epsilon_1) \exists \epsilon_2 > 0 \; \forall \mu_2 \in (0, \epsilon_2)$ M works correctly on $\mathbb{R}^{\leq n}$ using as constants the vector $c^* + \mu_1 \cdot d^* + \mu_2 \cdot e^*$.

In a second (easy) step we show that in the model with restricted use of constants statements of the above form can be decided efficiently.

3.1 Finding the Correct Directions

For $L \in P_{\mathbb{R}}^{rc}$/const and M a suitable basic machine using k constants define

$$E_n := \{c \in [-1,1]^k | M(\bullet, c) \text{ correctly decides } L \cap \mathbb{R}^{\leq n}\} \subseteq [-1,1]^k .$$

M witnesses membership of L in $P_{\mathbb{R}}^{rc}$/const, so $E_n \neq \emptyset \; \forall n \in \mathbb{N}$. Without loss of generality suppose $\bigcap_{n=1}^{\infty} E_n = \emptyset$, otherwise $L \in P_{\mathbb{R}}^{rc}$ follows. The following facts about the sets E_n hold:

- For all $n \in \mathbb{N}$ it is $E_{n+1} \subseteq E_n$;
- none of the E_n is finite since otherwise there exists an index m with $E_m = \emptyset$;

---

[2] This is actually done in [4] to show that in the additive BSS model $P_{\mathbb{R}}^{add}$/const $= P_{\mathbb{R}}^{add}$.

- each E_n is a finite union of convex sets. Each of these convex sets is a (potentially infinite) intersection of open or closed halfspaces. This is true since for every n membership in E_n can be expressed by a formula resulting from writing down the behaviour of M on inputs from $\mathbb{R}^{\leq n}$ when using a correct vector of constants from E_n. All computed intermediate results are of the form $\sum_{i=1}^{k} p_i(x) \cdot c_i + q(x)$, where p_i, q are polynomials. Thus, the tests produce for each x an open or closed halfspace with respect to suitable choices of c. If finitely many x are branched along a path only this results in a polyedron, otherwise we obtain an infinite intersection of such halfspaces;
- nestedness and being a finite union of convex sets hold as well for the closures $\overline{E_n}$;
- each $\overline{E_n}$ has a finite number of connected components. This follows from the fact that the E_n are finite unions of convex and thus connected sets. Note that $\overline{E_n}$ is obtained from E_n by relaxing strict inequalities to non-strict ones.

Now apply as in [4] König's infinity lemma (cf. [6]) to the infinite tree which has as its nodes on level n the connected components of E_n. The nestedness of the E_n (and thus of its connected components) implies that there is an $s \geq 1, s \leq k$ and a nested sequence of convex sets in $[-1, 1]^k \cap E_n$ each of dimension s such that they have empty intersection. The existence of $s \geq 1$ follows because the E_n are non-finite. We denote this nested sequence of sets again by $\{E_n\}_n$. As first main result we show

Theorem 4. *For the family $\{E_n\}_n$ of convex connected sets as defined above there exist vectors $c^*, d^*, e^* \in \mathbb{R}^k$ such that*

$$\forall n \in \mathbb{N} \; \exists \epsilon_1 > 0 \; \forall \mu_1 \in (0, \epsilon_1) \; \exists \epsilon_2 > 0 \; \forall \mu_2 \in (0, \epsilon_2) : \quad c^* + \mu_1 \cdot d^* + \mu_2 \cdot e^* \in E_n \;.$$

Proof. Let $\{E_n\}_n$ be as above and $\overline{E_n}$ the closure of E_n. Both are of dimension $s \geq 1$. Since each $\overline{E_n}$ is an intersection of halfspaces it is contained in an affine subspace of dimension $s \geq 1$. Now E_n is convex and therefore contains a polyhedron of dimension s. Let S be the affine subspace of dimension s generated by $\overline{E_1}$. Nestedness of the E_n implies $\bigcup_{n=1}^{\infty} \overline{E_n} \subseteq S$. The geometric idea underlying the construction of c^*, d^*, e^* is as follows. We look for a point $c^* \in \bigcap_{n=1}^{\infty} \overline{E_n}$ such that there is a direction d^* leading from c^* into $\overline{E_n}$ for each $n \in \mathbb{N}$. This d^* is found by considering iteratively facets of decreasing dimension of certain polyhedra contained in suitable subsets of $E_n \cap S$. The vector e^* then points towards the interior E_n^0 of E_n in S. Thus moving a short enough step from c^* into the direction d^* and subsequently towards e^* a suitable choice for the machine constants for inputs of dimension at most n is found.

Let $P_n^{(s)}$ be a closed s-dimensional polyhedron in E_n and let $F_n^{(s-1)}$ denote an $s - 1$-dimensional facet of $P_n^{(s)}$. Let $e_n^{(s-1)}$ be a normal vector of this facet pointing to $P_n^{(s)}$. Thus $F_n^{(s-1)}$ together with $e_n^{(s-1)}$ generate S. The sequence $\{e_n^{(s-1)}\}_n$ of vectors of length 1 in $\mathbb{R}^k$ is bounded and thus has a condensation point $e^{(s-1)} \in \mathbb{R}^k$. Let $S^{(s-1)}$ denote the $s-1$-dimensional affine subspace defined as orthogonal complement of $e^{(s-1)}$ in S. Without loss of generality suppose that $e^{(s-1)}$ points for each n from $S^{(s-1)}$ towards a side where parts of E_n are located. Due to nestedness of the E_n and the fact that $S^{(s-1)} \cap \overline{E_n}$ is $s-1$-dimensional for all n at least one among the vectors $e^{(s-1)}$ and $-e^{(s-1)}$ must satisfy this condition. Then for all n and for all points x in the interior of $S^{(s-1)} \cap \overline{E_n}$ (the interior with respect to $S^{(s-1)}$) the point $x + \mu \cdot e^{(s-1)}$ for sufficiently small $\mu > 0$ lies in E_n. The vector e^* in the statement is chosen to be $e^{(s-1)}$.

We repeat the above argument iteratively, now starting with $S^{(s-1)}$ instead of S and considering a family of the $s-1$-dimensional polyhedra $P_n^{(s-1)} \subset S^{(s-1)} \cap \overline{E_n}$ as well as their $s-2$-dimensional facets $F_n^{(s-2)}$. This way affine subspaces $S^{(s-2)}, \ldots, S^{(1)}$ and corresponding vectors $e^{(s-2)}, \ldots, e^{(1)} \in \mathbb{R}^k$ are obtained such that $e^{(i)}$ is orthogonal to $S^{(i)}$, each $S^{(i)} \cap \overline{E_n}$ is a polyhedron of dimension i and for sufficiently small $\mu > 0$ and a point x in the interior (with respect to $S^{(i)}$) of $S^{(i)} \cap \overline{E_n}$ the point $x + \mu \cdot e^*$ belongs to E_n. The final step in the construction defines c^* and d^*. Since $S^{(1)} \cap \overline{E_n}$ is a closed 1-dimensional polyhedron for all $n \in \mathbb{N}$ and since $S^{(1)} \cap \overline{E_{n+1}} \subseteq S^{(1)} \cap \overline{E_n}$ there is a point $c^* \in \bigcap_{n \geq 1} S^{(1)} \cap \overline{E_n}$. Define d^* such that it points from c^* into $S^{(1)} \cap \overline{E_n}$, i.e. for all $n \in \mathbb{N}$ and sufficiently small $\mu > 0$ it is $c^* + \mu \cdot d^* \in \overline{E_n}$.

Altogether we have shown the existence of points $c^*, d^*, e^* \in \mathbb{R}^k$ satisfying

$$\forall n \in \mathbb{N} \; \exists \epsilon_1 > 0 \; \forall \mu_1 \in (0, \epsilon_1) \; \exists \epsilon_2 > 0 \; \forall \mu_2 \in (0, \epsilon_2) \; c^* + \mu_1 \cdot d^* + \mu_2 \cdot e^* \in E_n \ .$$

$\square$

Note that in the above construction one cannot stop after having found $S^{(s-1)}$ and $e^{(s-1)}$. It is in general not the case that choosing a point c in $S^{(s-1)} \cap \bigcap_{n \geq 1} \overline{E_n}$ and an arbitrary direction e from c into $S^{(s-1)}$ will result in the required property. We do not know how the sequence of the $\overline{E_n}$'s might contract.

Note as well that since we cannot for given n determine how small ϵ_1 and ϵ_2 have to be chosen the above construction does not give an obvious efficient way to compute points in E_n.

3.2 Efficient Elimination of $\exists \forall \exists \forall$ Quantifiers

In order to show $P_\mathbb{R}^{rc}/\mathrm{const} = P_\mathbb{R}^{rc}$ we are left with deciding the following question: Let $M(\bullet, c)$ be a basic machine in the restricted model using k machine constants and witnessing $L \in P_\mathbb{R}^{rc}/\mathrm{const}$. Given $x \in \mathbb{R}^n$ what is M's result on $M(x, c^* + \mu_1 \cdot d^* + \mu_2 \cdot e^*)$ for sufficiently small μ_1, μ_2? Here, we have to obey the order in which μ_1 and μ_2 tend to 0. That is for all sufficiently small $\mu_1 > 0$ M has to work the same way when μ_2 tends to 0.

Answering this question efficiently is easy due to the linear structure in which the coefficients of a restricted machine occur.

Theorem 5. *Let M be a restricted basic machine running within a polynomial time bound $t(n)$ and using k machine constants. Suppose there are points*

$c^*, d^*, e^* \in \mathbb{R}^k$ such that for all $n \in \mathbb{N}$ the following condition is satisfied: $\exists \epsilon_1 > 0 \; \forall \mu_1 \in (0, \epsilon_1) \exists \epsilon_2 > 0 \; \forall \mu_2 \in (0, \epsilon_2) \; M(x, c^* + \mu_1 \cdot d^* + \mu_2 \cdot e^*)$ decides correctly the n-dimensional part $L \cap \mathbb{R}^{\leq n}$ of a language L. Then $L \in P_{\mathbb{R}}^{\mathrm{rc}}$.

Proof. Let M be as in the statement, $x \in \mathbb{R}^n$ an input. A polynomial time restricted machine M' for L works as follows. The constants which M' uses are c^*, d^*, and e^*. It tries to simulate M on x by computing independently of the constants which M uses the form of each test M performs. Then M' evaluates the result of such a test assuming that M uses as its constants $c^* + \mu_1 \cdot d^* + \mu_2 \cdot e^*$ for small enough μ_1, μ_2. More precisely, any test which M performs when using as constants a vector $c \in \mathbb{R}^k$ has the form

$$(*) \qquad \sum_{i=1}^{k} p_i(x) \cdot c_i + q(x) \geq 0 \; ?$$

Here, the p_i and q are polynomials in the input x. They can be computed by M' without use of constants by simulating M symbolically, i.e. without actually performing those operations involving components of c. Machine M' only keeps track of which coefficient a c_i finally has in the representation $(*)$. Using constants c^*, d^*, e^* machine M' can as well compute the terms $T_1 := \sum_{i=1}^{k} p_i(x) \cdot c_i^* + q(x)$, $T_2 := \sum_{i=1}^{k} p_i(x) \cdot d_i^*$, $T_3 := \sum_{i=1}^{k} p_i(x) \cdot e_i^*$. For deciding the result of the above test for M and suitable μ_i the new machine tests one after another whether $T_1 = 0, T_2 = 0$ and $T_3 = 0$. The sign of the first T_i for which the test gives a result $\neq 0$ gives as well the answer for M's behaviour. If all $T_i = 0$, then M follows the 'yes' branch. In the same way M' simulates the entire computation of M. Clearly, the running time of M' is linear in the running time of M.   □

Given the results of this and the previous section it follows:

Theorem 6. *Assuming* $FEAS \notin P_{\mathbb{R}}^{\mathrm{rc}}$ *for the* $NP_{\mathbb{R}}^{\mathrm{rc}}$-*complete problem FEAS in the BSS model with restricted use of constants there are non-complete problems in* $NP_{\mathbb{R}}^{\mathrm{rc}} \setminus P_{\mathbb{R}}^{\mathrm{rc}}$.

Proof. Follows from Theorems 3 , 4 and 5.   □

4   Conclusions

We have extended the list of computational models for which Ladner's theorem holds by a new variant of the BSS model over the reals. The latter allows to compute as in the full model with input values but has a limited access to the machine constants only. Real constants can be used freely in additions and subtractions, but must not be multiplied with each other. This model seems closer to the full BSS model than the linear variants. It has as well solvability of systems of polynomial equations as $NP_{\mathbb{R}}^{\mathrm{rc}}$-complete problem and its P versus NP question seems not easy to be solved either.

The proof of Ladner's result in this model once more relies on showing that the class $P_{\mathbb{R}}^{rc}/const$ is equal to $P_{\mathbb{R}}^{rc}$. Establishing this equality however is much more involved than for the additive BSS model over $\mathbb{R}$ or the BSS model over $\mathbb{C}$. It relies on the topological structure of the set of suitable constants. We believe the proof technique to be more interesting than the result itself. In particular, techniques which allow to replace in certain situations machine constants by others are important in many arguments concerning real number complexity theory. It seems interesting to study whether similar ideas might help for dealing with the question in the full real number BSS model as well. A promising idea could be to consider a new variant of Michaux' class $P_{\mathbb{R}}/const$ which for example requires by definition convexity of the sets E_n of suitable constants. However, then it is unclear whether $P_{\mathbb{R}}$ is captured by this new non-uniform class. More generally, the diagonalization technique used for $P_{\mathbb{R}}/const$ in [1] and for $P_{\mathbb{R}}^{rc}/const$ above allows some degree of freedom as to how to define $P_{\mathbb{R}}/const$. This means that we can put some additional conditions onto the set of constants that we allow for a fixed dimension to work. To make the diagonalization work there are basically two aspects to take into account. First, the resulting class has to contain $P_{\mathbb{R}}$. Secondly, the conditions we put on the constants have to be definable semi-algebraically without additional real constants. An example of the latter occurs in our definition of $P_{\mathbb{R}}^{rc}/const$ in that we require the constants to be bounded by 1 in absolute value. Other properties are conceivable and might lead to new results.

References

1. Ben-David, S., Meer, K., Michaux, C.: A note on non-complete problems in $NP_{\mathbf{R}}$. J. Complexity 16(1), 324–332 (2000)
2. Blum, L., Cucker, F., Shub, M., Smale, S.: Complexity and real computation. Springer, New York (1998)
3. Bürgisser, P.: On the structure of Valiant's complexity classes. In: Meinel, C., Morvan, M. (eds.) STACS 1998. LNCS, vol. 1373, pp. 194–204. Springer, Heidelberg (1998)
4. Chapuis, O., Koiran, P.: Saturation and stability in the theory of computation over the reals. Ann. Pure Appl. Logic 99(1-3), 1–49 (1999)
5. Cucker, F., Shub, M., Smale, S.: Separation of complexity classes in Koiran's weak model. Theoret. Comput. Sci. 133(1), 3–14 (1994); Selected papers of the Workshop on Continuous Algorithms and Complexity (Barcelona, 1993)
6. Diestel, R.: Graph theory, 3rd edn. Graduate Texts in Mathematics, vol. 173. Springer, Berlin (2005)
7. Ladner, R.E.: On the structure of polynomial time reducibility. J. Assoc. Comput. Mach. 22, 155–171 (1975)
8. Malajovich, G., Meer, K.: On the structure of $NP_{\mathbb{C}}$. SIAM J. Comput. 28(1), 27–35 (1999)
9. Michaux, C.: $P \neq NP$ over the nonstandard reals implies $P \neq NP$ over $\mathbf{R}$. Theoret. Comput. Sci. 133(1), 95–104 (1994)

0″-Categorical Completely Decomposable Torsion-Free Abelian Groups

Alexander G. Melnikov

The University of Auckland, New Zealand
vokinlem@bk.ru

Abstract. We show that every homogeneous completely decomposable torsion-free abelian group is 0″-categorical. We give a description of effective categoricity for some natural class of torsion-free abelian groups. In particular, we give examples of 0″-categorical but not 0′-categorical torsion-free abelian groups.

Keywords: Computable model theory, torsion-free abelian groups, computable categoricity.

One of the main topics in computable model theory is the study of isomorphisms of computable structures. In particular, one is interested in finding isomorphisms that are least or most complex in terms of the Turing degree. This paper is devoted to the investigation of complexity of isomorphisms in the class of computable infinite abelian groups.

An infinite structure $\mathcal{A}$ is *computable* if its domain is ω and its atomic diagram is a computable set. Computability of an abelian group is thus equivalent to its domain being ω and its operation being a computable function. If $\mathcal{A}$ is a computable structure and $\mathcal{B}$ is isomorphic to $\mathcal{A}$ then $\mathcal{A}$ is a *computable presentation* of $\mathcal{B}$. We say that a structure $\mathcal{A}$ is $0^{(n)}$-categorical if any two computable presentations of $\mathcal{A}$ are isomorphic via an isomorphism computable in $0^{(n)}$, where $0^{(n)}$ is the nth-jump of the computable degree. Thus, a structure is *computably categorical* if there is a computable isomorphism between any two computable presentations of the structure.

Goncharov, Nurtasin, Remmel and others have characterized computably categorical models in the classes of Boolean algebras, linear orders and abelian groups [4], [9], [10], [3]. For instance, a Boolean algebra is computably categorical iff it has finitely many atoms [5]. A linear order is computably categorical iff it has finitely many adjacencies [10]. A torsion-free abelian group is computably categorical iff its rank is finite [9]. In [7] computably categorical trees of finite height are characterized. There are also general sufficient conditions for structures to be computably categorical [6], [2].

When one studies $0^{(n)}$-categorical structures, where $n > 0$, the situation becomes more complex. In [8] McCoy studies $0'$-categorical linear orders and Boolean algebras. In [7], for any given $n > 0$, a tree is constructed that is $0^{(n+1)}$-categorical but not $0^{(n)}$-categorical. It is also not hard to see that there are

K. Ambos-Spies, B. Löwe, and W. Merkle (Eds.): CiE 2009, LNCS 5635, pp. 362–371, 2009.
© Springer-Verlag Berlin Heidelberg 2009

examples of linear orders that are $0^{(n+1)}$-categorical but not $0^{(n)}$-categorical. Similar examples can be constructed in the classes of Boolean algebras. However, we do not know examples of abelian groups that are $0^{(n+1)}$-categorical but not $0^{(n)}$-categorical for any given $n > 0$. In this paper we address this problem and give partial answers.

We restrict our attention to torsion-free abelian groups that are *completely decomposable*, i.e. can be presented as a direct sum of subgroups of the additive group of the rationals. We consider two subclasses of this class. There are groups that are in both classes, but in general these classes are not comparable under inclusion.

Examples of completely decomposable groups are the additive group of a vector space over Q and free abelian groups. These examples have all direct summands isomorphic. Such groups are called *homogeneous completely decomposable groups*. We prove the following result:

Theorem 1. *Every homogeneous completely decomposable torsion-free abelian group is $0″$-categorical.*

We do not know if the $0″$-bound is sharp.

Our second class of completely decomposable groups defined as follows. Let P be a set of prime numbers, and let

$$Q_P = \{\tfrac{m}{n} \in Q : p|n \text{ implies } p \in P, \text{ for all primes } p\}.$$

Clearly, Q_P is an additive subgroup of the group Q of the rationals. We say that a group $\mathcal{A}$ is *P-complete* if it is a direct sum of copies of Q_P. We say that $\mathcal{G}$ is a *divisibly extended P-complete* group if it can be written as $\mathcal{G} = \mathcal{A} \oplus \mathcal{C}$, where $\mathcal{A}$ is P-complete and $\mathcal{C}$ is divisible (i.e. $\mathcal{C}$ is a direct sum of copies of the rationals Q with addition). We prove the following theorem:

Theorem 2. *Let $\mathcal{CD}_P$ be the class of divisibly extended P-complete torsion-free abelian groups. Then we have the following properties:*

1. *For each $\mathcal{G} \in \mathcal{CD}_P$, $\mathcal{G}$ is $0″$-categorical.*
2. *If P is not the set of all primes, and the P-complete and divisible summands of $\mathcal{G} \in \mathcal{CD}_P$ are both of infinite rank, then $\mathcal{G}$ is not $0'$-categorical.*
3. *If $\mathcal{G}$ is a divisible group of infinite dimension, then $\mathcal{G}$ is $0'$-categorical, but not computably categorical.*
4. *Any group of a finite rank (not necessarily in $\mathcal{CD}_P$) is computably categorical.*

Case *3* of the theorem is a corollary of the main result of [9]. Case *4* follows immediately from the following observation. Consider isomorphic images of a finite basis in given computable copies of $\mathcal{G}$. We define a computable isomorphism as an extension of a map between bases to linear combinations.

This theorem gives us a description of effective categoricity in $\mathcal{CD}_P$. As a corollary we get the first examples of a torsion-free abelian groups that are $0″$-categorical but not $0'$-categorical. For instance, the group $F \oplus C \in \mathcal{CD}_P$, where F is a free group of infinite rank and C is a divisible group of infinite rank, has this property.

1 Basic Concepts

We use known definitions and facts from computability theory and the theory of abelian groups. We mention some of them in this sections. Standard references are [11] for computability and [12] for the theory of torsion-free abelian groups.

Let $\{\Phi_i^A\}_{i \in \omega}$ be a standard enumeration of all partial computable functions with an oracle for a set $A \subseteq \omega$. The set $\{x : \Phi_x^A(x) \downarrow\} = \{x : x \in W_x^A\}$ is called the *jump* of A and is denoted by A'. The jump operator can be iterated as follows. For a given $n \in \omega$ set $A^{(n)} = (A^{(n-1)})'$, where $A^{(0)} = A$. The jump operation is well defined on Turing degrees. The iteration of jumps induces a hierarchy, known as the arithmetical hierarchy:

$$X = \{x : \varphi(x, T)\} \leftrightarrow X \in \Sigma_n^0 \leftrightarrow X \text{ is c.e. in } 0^{(n-1)},$$

where φ is a Σ_n-formula over a computable relation T, and 0 is the Turing degree of $\emptyset$. We define Π_n^0-sets to be the complements of Σ_n^0 sets. The Arithmetical hierarchy is proper because $0^{(n)} \in \Sigma_n^0 \setminus \Sigma_{n-1}^0$ (see [11] for more details).

Recall that a group $\langle G, \cdot \rangle$ is *computable* if $G = \omega$ and the operation $\cdot$ is a computable function. If A is a computable structure and B is a c.e. substructure of A (i.e. the domain of B is c.e.), then B has a computable presentation (e.g. see [2]). We will use this fact without an explicit reference to it.

For abelian groups, that is the groups whose group operations are commutative, we write "$+$" instead of "$\cdot$", "0" instead of "1", and "$-$" for the inverse operation. Therefore na, $n \in \omega$, denotes the element $\underbrace{a + a + \ldots + a}_{n \text{ times}}$. A group is *torsion-free* if $na = 0$ implies $n = 0$ or $a = 0$, for all $a \in G$ and $n \in \omega$.

Recall that elements $g_0, \ldots, g_n$ of a torsion-free abelian group G are *linearly independent* if, for all $c_0, \ldots, c_n \in Z$, the equality $c_0 g_0 + c_1 g_1 + \ldots + c_n g_n = 0$ implies that $c_0 = c_1 = \ldots = c_n = 0$. An infinite set is *linearly independent* if every finite subset of this set is linearly independent. A maximal linearly independent set is a *basis*. All bases of G have the same cardinality. This cardinality is called the *rank* of G. It is not hard to see that torsion-free abelian groups of the rank 1 are exactly additive subgroups of Q.

Definition 1. An abelian group G is the *direct sum* of groups A_i, $i \in I$, written $G = \bigoplus_{i \in I} A_i$, if G can be presented as follows:

1. The domain consists of infinite sequences $(a_0, a_1, a_2, \ldots, a_i, \ldots)$, each $a_i \in A_i$, such that the set $\{i : a_i \neq 0\}$ is finite.

2. The operation $+$ is defined component-wise.

The groups A_i are called *direct summands* of G. It is easy to prove that $G \cong \bigoplus_{i \in I} A_i$, where $A_i \leqq G$, if and only if (1) $G = \sum_{i \in I} A_i$, i.e. $\{A_i : i \in I\}$ generates G, and (2) for all j we have $A_j \cap \sum_{i \in I, i \neq j} A_i = \{0\}$.

Definition 2. A torsion-free abelian group is called *completely decomposable* if G is a direct sum of additive subgroups of the rationals Q.

We write $n|g$ and say that g is divisible by n in G if there exists an $h \in G$ for which $nh = g$.

Definition 3. A subgroup A of G is called *pure* if for all $a \in A$, $G \models n|a$ implies $A \models n|a$. For any subset B of a *torsion-free* abelian group G we denote by $[B]$ the least pure subgroup of G that contains B.

Definition 4. For $g \in G$, $g \neq 0$, and prime number p, set

$$h_p(g) = \begin{cases} \max\{k : G \models p^k|g\}, \text{if this maximum exists,} \\ \infty, \text{otherwise.} \end{cases}$$

Let us fix the canonical listing of the prime numbers: $p_1 < p_2 < \ldots$. Then the *characteristic* of g is $\chi(g) = (h_{p_1}(g), h_{p_2}(g), \ldots)$.

We say that $\chi(g)$ is c.e. if the set $\{\langle i, j \rangle : j \leq h_{p_i}, h_{p_i} > 0\}$ is c.e. It is easy to see that if G is computable then $\chi(g)$ is c.e. for every $g \in G$.

Definition 5. Two characteristics, $(k_1, k_2, \ldots)$ and $(l_1, l_2, \ldots)$, are *equivalent*, written $(k_1, k_2, \ldots) \simeq (l_1, l_2, \ldots)$ if $k_n \neq l_n$ only for finitely many n, and k_n and l_n are finite for these n. The equivalence classes of this relation are called *types*.

For $g \in G$, $g \neq 0$, write $\mathbf{t}(g)$ to denote the type of g. If G has rank 1 then all non-zero elements of G have equivalent types. Hence we can define the type of G to be $\mathbf{t}(g)$, for a non-zero $g \in G$, and denote by $\mathbf{t}(G)$. The following theorem classifies groups of rank 1:

Theorem 3 (Baer [1]). *Let G and H be torsion-free abelian groups of rank 1. Then G and H are isomorphic if and only if $\mathbf{t}(G) = \mathbf{t}(H)$.*

Definition 6. A torsion-free abelian group G is called *homogeneous* if all non-zero elements of G have the same type.

For a completely decomposable homogeneous group $G = \bigoplus_{i \in I} G_i$, the pair $(card(I), \mathbf{t}(G))$ describes its isomorphism type.

2 Homogeneous Groups and P-Bases

In this section we always assume that P is a set of primes. Let $\hat{P} = \{p \notin P : p \text{ is prime}\}$. There is the well-known notion of a p-basis (see [12] for more details). We generalize this notion as follows:

Definition 7. A subset B of a group G is P-*independent* if for any $b_1, \ldots, b_k \in B$, any integers $m_1, \ldots, m_k$ and $p \in P$, $G \models p| \sum_i m_i b_i$ implies $p|m_i$ for all i. If $P = \emptyset$ then by P-*independence* we mean a linear independence. The maximal P-independent subset of G is called a P-*basis*. We say that a P-basis is *excellent* if it is a basis of G.

Every P-basis is a linearly independent set, but not necessary a maximal one. It follows from *Lemma 35.1* in [12] that a free abelian group of infinite rank contains a P-basis that is not excellent, even for a singleton $P = \{p\}$.

Recall that Q_P is an additive subgroup of the rationals consisting of fractions which denominators are not divisible by primes from $\hat{P}$. A group G is P-complete if $G = \bigoplus_{i \in I} G_i$ where $G_i \cong Q_P$, for all $i \in I$. Note that for any $r \in Q_P$ and any element g of a P-complete G, the product rg is well-defined.

We will write $\sum_{a \in A} r_a a$ for an infinite set of elements A of a group and rationals r_a. This will always mean that $r_a = 0$ for all but finitely many $a \in A$. We denote by $(A)_R$ the subgroup of G (if this subgroup exists) generated by $A \subset G$ over $R \leq Q$, i.e. $(A)_R = \{\sum_{a \in A} r_a a : r_a \in R\}$. If a set A is linearly independent then every element of $(A)_R$ has the unique presentation $\sum_{a \in A} r_a a$. Therefore $(A)_R = \bigoplus_{a \in A} Ra$, where $Ra = (\{a\})_R$.

Lemma 1. *Let G be a P-complete group and let B be an excellent $\hat{P}$-basis of G. Then B generates G over Q_P, i.e. $G = \bigoplus_{b \in B} Q_P b$.*

Proof. Suppose $g \in G$. By our assumption, B is a basis of G, therefore there exist integers m and $\{m_b : b \in B\}$ s.t. $mg = \sum_b m_b b$ with only finitely many $m_b \neq 0$. Let $m = pm'$ for $p \in \hat{P}$. By the definition of a $\hat{P}$-basis, $p|m_b$ for all $b \in B$ and the equality can be reduced until $m = \prod_{p \in P} p^{\lambda_p}$ for some $\lambda_p \geq 1$. Therefore, $g = \sum_b \frac{m_b}{m} b \in (B)_{Q_P}$ and $G = (B)_{Q_P}$.

Suppose $a, b \in G$. Let $\chi(a) \leq \chi(b)$ iff $h_p(a) \leq h_p(b)$ for all p (for any integer k, $k < \infty$). Let $G_\chi = \{g \in G : \chi \leq \chi(g)\}$ for a characteristic χ. We have $h_p(a) = h_p(-a)$ and $\inf(h_p(a), h_p(b)) \leq h_p(a + b)$, for all p. Therefore G_χ is a subgroup of G, for any χ.

Lemma 2. *Let G be a completely decomposable homogeneous group of type t and $\chi \in t$. A subgroup G_χ defined above is P-complete for $P = \{p : h_p = \infty \text{ in } \chi\}$.*

Proof. Let $G = \bigoplus_{i \in I} G_i$ be a decomposition of G where for each i, $G_i \leq Q$. Suppose $g_i \in G_i$ s.t. $\chi(g_i) = \chi$. Then $\{g_i : i \in I\} \subseteq G_\chi$ is a basis of G and, therefore, G_χ. If $g \in G_\chi$, then there is the unique irreducible equality $ng = \sum_{i \in I} m_i g_i$, with integer coefficients. Every direct summand G_i is a pure subgroup of G, thus $n| \sum_{i \in I} m_i g_i$ implies $n|m_i g_i$ for all $i \in I$ and $g = \sum_{i \in I} \frac{m_i}{n} g_i$ or $\sum_{i \in I} \frac{m_i'}{n_i} g_i$ after reduction. If there is some $p \in \hat{P}$ and $i \in I$ s.t. $m_i' \neq 0$ and $n_i = pn_i'$ then $h_p(\frac{m_i'}{n_i} g_i) = h_p(\frac{m_i'}{n_i} \frac{g_i}{p}) \leq h_p(\frac{g_i}{p}) < h_p(g_i)$, because $(m_i', p) = 1$ and $h_p(g_i)$ is finite. As a consequence of the definitions of direct sum and p-height we have $h_p(g) = \min\{h_p(\frac{m_i'}{n_i} g_i) : i \in I, m_i \neq 0\}$. Therefore $h_p(g) \leq h_p(\frac{m_i'}{n_i} g_i) < h_p(g_i)$. But $\chi(g_i) = \chi$. Thus $\chi(g) \not\geq \chi$ and $g \notin G_\chi$. This proves that any $g \in G_\chi$ can be expressed as $g = \sum_{i \in I} r_i g_i$, where for each i, $r_i \in Q_P$.

On the other hand, if $p \in P$ then for any integer k, $\frac{g_i}{p^k} \in G_i$ and $h_p(\frac{g_i}{p^k}) = h_p(g_i) = \infty$. Therefore $\chi(g) = \inf\{\chi(r_i g_i) : i \in I\} \geq \chi$ and $g \in G_\chi$, for any $g \in (\{g_i : i \in I\})_{Q_P}$.

We see that $G_\chi = \sum_{i \in I} Q_P g_i$ and, therefore, $G_\chi = \bigoplus_{i \in I} Q_P g_i$. This group is P-complete, and $\{g_i : i \in I\}$ is an excellent $\hat{P}$-basis of G_χ.

Let $Q(\chi)$ be a subgroup of Q s.t. $1 \in Q(\chi)$ and $\chi(1) = \chi$.

Proposition 1. *Let G be a completely decomposable homogeneous group of type t. Suppose B is an excellent $\hat{P}$-basis of G_χ, where $\chi \in t$ and $P = \{p : h_p = \infty$ in $\chi\}$. Then B generates G over $Q(\chi)$.*

Proof. For every $b \in B$ consider $[b] = [\{b\}]$. Let $\langle B \rangle = \sum_{b \in B}[b] \leqq G$.

1. $\chi(b) = \chi$ in G, for all $b \in B$, therefore $[b] = Q(\chi)b$. If not, then $\chi(b) > \chi$ and $b = pa$ for some $a \in G_\chi$ and $p \in \hat{P}$, this is a contradiction (B is a $\hat{P}$-basis of G_χ, and $p|1b$).

2. $\langle B \rangle = \bigoplus_{b \in B}[b]$. Indeed, B is a linear independent system in G_χ and, therefore, in G.

3. $G = \langle B \rangle$ (and thus $\langle B \rangle = [B]$). Let $g \in G$, $g \neq 0$. There are integers m, n $((m,n) = 1)$ s.t. $\chi(\frac{m}{n}g) = \chi$, because $\chi(g) \in t$. Then $\frac{m}{n}g \in G_\chi$ and $\frac{m}{n}g = \sum_{b \in B, r_b \in Q_P} r_b b$, by *Lemmas 1* and *2*. For all $b \in B$, $\chi(b) = \chi(\frac{m}{n}g) = \chi$. But $G \models m|\frac{m}{n}g$. Therefore there is $x_b \in [b]$ s.t. $mx_b = b$, for every $b \in B$. Then $g = \sum_{b \in B} nr_b x_b$, where $nr_b x_b \in [b]$.

Remark 1. Let G be as in the previous proposition, and let $G' \cong G$. Suppose B and B' are X-c.e. excellent $\hat{P}$-bases of G_χ and G'_χ respectively. If G and G' are computable, then there is an isomorphism $\varphi : G \to G'$ computable in X. If $\varphi : B \to B'$ is a bijection, set $\varphi(\sum_{b \in B} q_b b) = \sum_{b \in B} q_b \varphi(b)$, for all $g \in G$.

Rado proved the basis theorem for free abelian groups (see e.g. [12], *Lemma 15.3*). This theorem says that every "good" element of a free abelian group can be extended to a basis of this group that generates it over Z. We will need the following generalisation.

Proposition 2. *Let $k < n$ and suppose $a_1, \ldots, a_k \in Z^n = \bigoplus_{1 \leq i \leq n} Ze_i$ is a $\hat{P}$-independent set of elements. Then there exist $a_{k+1}, \ldots, a_n \in Z^n$ s.t. an extension $\{a_1, \ldots, a_n\}$ is an excellent $\hat{P}$-basis of Z^n.*

Proof. We can assume $\hat{P} \neq \emptyset$.

Let $k=1$. Set $a_1 = (\alpha_1, \ldots, \alpha_n) \in Z^n$ and suppose m is the greatest common divisor of α_i, $1 \leq i \leq n$, and $(m, \hat{P}) = 1$ (i.e. for all primes $p \notin P$, m is not divisible by p). Then $m = \sum_i m_i \alpha_i$ for integers m_i, $1 \leq i \leq n$. Look at the transformation $a_1 = (\alpha_1, \ldots, \alpha_n) \to (\alpha_1, \ldots, \alpha_n + m_t \alpha_t)$, $1 \leq t \leq n$. It can be done by multiplication of a string a_1 by a matrix $A_{tn} = A_{tn}(m_t)$ with integer coefficients which looks like the identity matrix everywhere but the unique position where $a_{t,n} = m_t$. Note that A_{tn}^{-1} is just A_{tn} with m_t replaced by $-m_t$.

Therefore there is a matrix $A = A_{1n} \ldots A_{n-1n}A_{n1}(\frac{\alpha_1}{m}) \ldots A_{nn-1}(\frac{\alpha_{n-1}}{m})$ with integer coefficients s.t. A^{-1} is a matrix of integers and $a_1 A = (0, 0, \ldots, m) = me_n$, where $\{e_1, \ldots, e_n\}$ is the standard basis of Z^n (we could use Euclid's algorithm instead of Bezout's identity to obtain this matrix A). Set $a_i = e_i A^{-1} \in Z^n$, $1 \leq i < n$. If there is $p \in \hat{P}$ and $b \in Z^n$ s.t. $pb = \sum_i m_i a_i$ then $pb' = pbA = \sum_{i < n} m_i e_i + m_n m e_n$ and $p|m_i$ for all i $((m, p) = 1)$.

Now let $1 < k < n$ and $a_1 = (\alpha_{11}, \ldots, \alpha_{1n}), \ldots, a_k = (\alpha_{k1}, \ldots, \alpha_{kn})$. Assume that the G.C.D. of the integers $\{\alpha_{ji} : 1 \leq i \leq n\}$ is d_j, $j \leq k$. There is a matrix

A s.t. $a_1A = (0, \ldots, d_1)$. Let $a'_j = a_jA$, $1 < j \leq k$. There are integer coefficients $t_2, \ldots, t_k$ s.t. for every j, $1 < j \leq k$, $a''_j = d_1 a'_j - t_j d_1 e_n$ has 0 at the n'th position in its vector representation.

Suppose there is $h \in Z^{n-1}$ and $p \in \hat{P}$ s.t. $ph = \sum_{1<i\leq k} m_i a''_i$. Then $ph' = phA^{-1} = \sum_i m_i (d_1 a'_i - t_i d_1 e_n) A^{-1} = d_1 \sum_i m_i a_i - (\sum_i m_i t_i) a_1$ and $p|m_i$, $1 < i \leq k$, by our assumption $(p, d_1) = 1$.

Then there are elements $a''_{k+1}, \ldots, a''_n$ s.t. $\{a''_j : 1 < j \leq n\}$ is an excellent $\hat{P}$-basis of Z^{n-1}. Let $a_j = a''_j A^{-1}$, $k < j \leq n$.

If there are $h \in Z^n$ and $p \in \hat{P}$ s.t. $ph = \sum_i m_i a_i$ then $phA = \sum_i m_i a_i A = m_1 d_1 e_n + \sum_{1<i\leq k} m_i a'_i + \sum_{k<i\leq n} m_i a''_i = m_1 d_1 e_n + \sum_{1<i\leq k} m_i(\frac{a''_i}{d_1} + t_i e_n) + \sum_{k<i\leq n} m_i a''_i = (\sum_{i\leq k} m_i t_i)e_n + \sum_{1<i\leq k} m_i \frac{a''_i}{d_1} + \sum_{k<i\leq n} m_i a''_i$, where $t_1 = d_1$. Evidently $Z^n = Z^{n-1} \oplus Z = (e_1, \ldots, e_{n-1})_Z \oplus (e_n)_Z$, therefore $p| \sum_{1<i\leq k} m_i \frac{a''_i}{d_1} + \sum_{k<i\leq n} m_i a''_i$, $p| \sum_{1<i\leq k} m_i a''_i + d_1 \sum_{k<i\leq n} m_i a''_i$. By our assumption $\{a''_j : 1 < j \leq n\}$ is an excellent $\hat{P}$-basis of Z^{n-1} and $(p, d_1) = 1$. Therefore $p|m_i$, for $1 < i \leq n$.

Finally, $p|ph - \sum_{1<i\leq n} m_i a_i = m_1 a_1$ and $p|m_1$.

Remark 2. If $\{a_1, \ldots, a_k\}$ is an excellent $\hat{P}$-basis of Z^n, then for each $1 \leq i \leq n$, $n_i e_i \in (a_1, \ldots, a_n)_Z$ for an integer n_i with a property $(n_i, \hat{P}) = 1$.

Now we are ready to prove the main result of this section:

Theorem 4. *Let* $G = \bigoplus_{i\in\omega} G_i = \bigoplus_{i\in\omega} Q_P g_i$ *be a P-complete group of the rank* ω. *If G is computable then it has a $0''$-c.e. excellent $\hat{P}$-basis.*

Proof. Suppose that G is computable. Consider the following procedure:

Step 0. Set $k(1) = 0$ and $B_1 = \emptyset$.

Step n. Suppose that $k(n)$ and $B_n = \{b_1, \ldots, b_{k(n)}\}$ have been defined. For n (the nth element of G) find a family $A_n = \{a_1, \ldots, a_{r(n)}\}$ such that $B_n \cup A_n \cup \{n\}$ is linearly dependent, and $B_n \cup A_n$ is $\hat{P}$-independent. Set $k(n+1) = k(n) + r(n)$, $B_{n+1} = B_n \cup A_n$ and $b_{k(n)+i} = a_i$ for $1 \leq i \leq r(n)$.

Lemma 3. *For all n, step n halts.*

Proof. We know that G is generated by $\{g_i : i \in \omega\}$ over Q_P, therefore there are generators $g_0, \ldots, g_l$ s.t. $(g, b_1, \ldots, b_{k(n)})_{Q_P} \subseteq (g_0, \ldots, g_l)_{Q_P}$. There is an integer M such that $\{Mb_1, \ldots, Mb_{k(n)}\} \subset (g_0, \ldots, g_l)_Z$, and $\frac{1}{M} \in Q_P$. Then there is a natural correspondence between b_i and an l-dimensional vector B_i of integers. Moreover, for each $p \notin P$, $G \models p| \sum_i m_i b_i$ iff $Z^{l+1} \models p| \sum_i m_i B_i$ ($h_p(g_j) = 0$ for all j, and $(M, p) = 1$). Therefore $B_1, \ldots, B_{k(n)}$ can be extended to an excellent $\hat{P}$-basis of Z^{l+1}, and it means that $b_1, \ldots, b_{k(n)}$ can be extended to a $\hat{P}$-independent set that generates $(g_0, \ldots, g_l)_{Q_P}$ over Q_P (by *Remark 2*). So there is an extension that generates g over Q_P and is $\hat{P}$-independent. Any such extension will satisfy conditions needed for the next step.

Lemma 4. *Let B be the set of elements enumerated by the procedure. Then it is an excellent $\hat{P}$-basis of G.*

Proof. Evidently B is a $\hat{P}$-independent set, because each finite subset is. It is a basis, because each element of g was forced to be in $(B)_{Q_P}$.

On the **step n** we search for elements that satisfy the following formulas:
$(\forall \overline{m})(\forall \overline{n})(\forall p)(p \in P \vee \neg(p| \sum_i m_i b_i + \sum_j n_j a_j) \vee \bigwedge_i p|m_i \bigwedge_j p|n_j)$,
$(\exists \overline{m})(\exists \overline{n})(\exists t)(t \neq 0 \wedge tg = \sum_i m_i b_i + \sum_j n_j a_j)$,
which are Π_2^0 and Σ_1^0 respectively (P is c.e.; to justify this choose $g \in G$ satisfying $h_p(g) = 0$ for all $p \notin P$). We see that $0''$ is enough.

Remark 3. If P is computable then a P-complete group G has $0'$-c.e. excellent $\hat{P}$-basis. Indeed, the property "to be $\hat{P}$-independent in G" can be expressed by a Π_1^0-formula in this case.

3   Categoricity

Proposition 3. *Every P-complete torsion-free abelian group is $0''$-categorical. If P is computable then the group is $0'$-categorical.*

Proof. By *Theorem 4*, *Remark 1* and *Remark 3*.

For instance, the group $Z^{<\omega}$ (the free abelian group of the rank ω) is $0'$-categorical.

We are ready to prove that every homogeneous completely decomposable torsion-free abelian group is $0''$-categorical.

Proof (Theorem 1). Let G be a computable homogeneous group of the type χ and let $G_\chi = \{g \in G : \chi(g) \geq \chi\}$ for $\chi \in \mathbf{t}$. Let $h \in G$ s.t. $\chi(h) = \chi$. By the definition, $\chi(g) \geq \chi$ iff $(\forall p\text{-prime})(\forall k)(p^k|h \to p^k|g)$, therefore G_χ is a Π_2^0-subgroup of G. We know that G_χ is P-complete for a Π_2^0 set of primes $P = \{p : p^\infty|h\}$. To build an excellent $\hat{P}$-basis of G_χ we relativize the proof of *Theorem 4*.

On the **step n** we do the following. First, we find the nth element $g \in G_\chi$ computably in $0''$. Then we find an extension of B_n in G_χ (this subgroup is Π_2^0) which is linear dependent together with g (this is Σ_1^0) and is a $\hat{P}$-independent system. "To be a $\hat{P}$-independent system in G_χ for $P \in \Pi_2^0$" can be expressed by a Π_3^0 - formula:

$$(\forall \overline{m})(\forall \overline{n})(\forall p)([p \notin P \wedge (\exists x)(x \in G_\chi \wedge px = \sum_i m_i b_i + \sum_j n_j a_j)] \to$$
$$\to \bigwedge_i p|m_i \bigwedge_j p|n_j).$$

But for all $p \notin P$ and $b_i, a_j \in G_\chi$, $(\exists x)(x \in G_\chi \wedge px = \sum_i m_i b_i + \sum_j n_j a_j)$ iff $(\exists k)(\exists y \in G)(h_p < k \wedge p^k y = \sum_i m_i b_i + \sum_j n_j a_j)$, where h_p is the p-th component of χ. Indeed, if there is $x \in G_\chi$ s.t. $px = \sum_i m_i b_i + \sum_j n_j a_j$ then there exists y with $p^{h_p+1}y = px$ $(h_p(x) \geq h_p)$ and we can set $k = h_p + 1$. If the right holds, then $px = p^k y$ for $x = p^{k-1}y$. Evidently $h_p(x) \geq (k-1) \geq h_p$ and $h_q(x) = h_q(p^k y) \geq h_q$, therefore $\chi(x) \geq \chi$ and $x \in G_\chi$.

Therefore we can write an equivalent Π_2^0-formula:

$$(\forall \overline{m})(\forall \overline{n})(\forall p)([p \notin P \wedge (\exists k)(\exists y \in G)(h_p < k \wedge p^k y = \sum_i m_i b_i + \sum_j n_j a_j)] \rightarrow$$
$$\bigwedge_i p|m_i \bigwedge_j p|n_j),$$

where $h_p < k$ iff $\neg(\exists h_1)(p^k h_1 = h)$ $(\chi(h) = \chi)$.

We see that $0''$ is enough to enumerate an excellent $\hat{P}$-basis of G_χ if G is computable. By *Remark 1*, there is a $0''$-computable isomorphism between any two computable presentations of G.

We now focus our attention on divisibly extended P-complete groups.

Proposition 4. *Let $G = R \bigoplus C$, where R is a P-complete group and C is a divisible group. Then G is $0''$-categorical.*

Proof. We can assume $\hat{P} \neq \emptyset$. We modify the proof of *Theorem 4* for $G/C = R$, where G is computable. We build an excellent $\hat{P}$-basis $B = \bigcup_n B_n$ of $G/C = R$ (more precisely, a preimage of it in G). We do it as in *Theorem 4*, but we consider all the elements of G modulo a subgroup C.

Note that for all $h \in G$ and any natural k, $G \models k|h$ iff $G/C \models k|\tilde{h}$, where $\tilde{h} = h + C$ is a coset of h in G/C. Indeed, $(\exists \tilde{a} \in G/C)(k\tilde{a} = \tilde{h}) \leftrightarrow (\exists a \in G)(\exists c \in C)(ka + c = k(a + \frac{c}{k}) = h) \leftrightarrow (\exists d \in G)(kd = h)$. Therefore on each step we *check in* G the following formulas:

$$(\forall \overline{m})(\forall \overline{n})(\forall p)(p \in P \vee \neg(p| \sum_i m_i b_i + \sum_j n_j a_j) \vee \bigwedge_i p|m_i \bigwedge_j p|n_j),$$

$$(\exists \overline{m})(\exists \overline{n})(\exists k)(\exists c)(k \neq 0 \wedge c \in C \wedge kg = \sum_i m_i a_i + \sum_j n_j a_j + c),$$

which are Π_2^0 (P is c.e.) and Σ_3^0 (C is Π_2^0) respectively. Therefore the procedure that enumerates B is $0''$-computable. We can build a basis D of C with an oracle $0''$ because C is Π_2^0-subgroup of G. It follows that $G/C \cong \bigoplus_{b \in B} Q_P \tilde{b}$. Therefore $G = \sum_{b \in B} Q_P b + \sum_{d \in D} Q d$. But $B \cup D$ is linearly independent, thus $G = \bigoplus_{b \in B} Q_P b \oplus \bigoplus_{d \in D} Q d$. We can extend a bijection between bases of two computable presentations of G to an isomorphism computably, as in *Remark 1*. Therefore G is $0''$-categorical.

Proposition 5. *Suppose $G = R \bigoplus C$, where R is a P-complete ($\hat{P} \neq \emptyset$) group of an infinite rank and C is a divisible group of an infinite rank. Then it is not $0'$-categorical.*

Proof. Let S be a Σ_2^0-complete set. Then there is a computable relation T s.t. $S = \{x : \exists^{<\infty} y T(x, y)\}$ ([11]). We build a presentation $A = \bigoplus_{x \in \omega} A_x$ of G. We start with a c.e. P-complete subgroup $\bigoplus_{x \in \omega} Q_P a_x$ of a computable group $\bigoplus_{x \in \omega} Q a_x$ with a computable basis $\{a_x : x \in \omega\}$ (if G is computable then P is c.e., see the proof of *Theorem 4*). Then we enumerate S and add p_j^k-roots, $j \leq k$, to a_x if $\exists^k y T(x, y)$. By Baer theorem (*Theorem 3*), we will have

$$A_x \cong \begin{cases} Q_P, & \text{if } x \in S, \\ Q, & \text{if } x \notin S. \end{cases}$$ Then A has a computable copy, moreover the image of the basis is computable. We identify A with this computable copy.

Now let $H = \bigoplus_{x \in \omega} H_x$ where $H_x \cong \begin{cases} Q_P, & \text{if } x \text{ is even,} \\ Q, & \text{else.} \end{cases}$

Suppose there is a $0'$-computable isomorphism $\varphi : H \to A$. Using $0'$ we can enumerate images of divisible elements of H which are divisible in A. Every divisible element $c \in A$ has the unique decomposition $c = \sum_{x \notin S} r_x a_x$. We can find this decomposition effectively. Every divisible $c \in A$ has a divisible preimage. Hence there is a $0'$-computable procedure that enumerates $\{a_x : x \notin S\}$ and $\overline{S}$. This is a contradiction.

Proof (Theorem 2). By *Propositions 4* and *5*.

References

1. Baer, R.: Abelian Groups Without Elements of Finite Order. Duke Math. J. 3, 68–122 (1937)
2. Ershov, Y.L., Goncharov, S.S.: Constructive Models, Novosibirsk, Nauchnaya Kniga (1999)
3. Goncharov, S.S.: Countable Boolean Algebras and Decidability. Siberian School of Algebra and Logic, Novosibirsk, Nauchnaya Kniga (1996)
4. Goncharov, S.S.: Autostability of Models and Abelian Groups. Algebra and Logic 19, 13–27 (1980) (English translation)
5. Goncharov, S.S.: The Problem of the Number of Non-Autoequivalent Constructivisations. Algebra and Logic 19, 401–414 (1980) (English translation)
6. Goncharov, S.S., Dzgoev, V.D.: Autostability of Models. Algebra and Logic 19(1), 45–58 (1980)
7. Lempp, S., McCoy, C., Miller, R., Solomon, R.: Computable Categoricity of Trees of Finite Height. Journal of Symbolic Logic (2005)
8. McCoy, C.: Categoricity in Boolean Algebras and Linear Orderings. Annals of Pure and Applied Logic 119, 85–120 (2003)
9. Nurtazin, A.T.: Computable Classes and Algebraic Criteria of Autostability. Summary of Scientific Schools, Math. Inst. SB USSRAS, Novosibirsk (1974)
10. Remmel, J.B.: Recursively Categorical Linear Orderings. Proc. Amer. Math. Soc. 83, 387–391 (1981)
11. Soare, R.I.: Recursively Enumerable Sets and Degrees. Springer, Heidelberg (1987)
12. Fuchs, L.: Infinite Abelian Groups, vol. I, II. Academic Press, London (1973)

Notes on the Jump of a Structure

Antonio Montalbán*

University of Chicago,Chicago, IL 60637, USA
antonio@math.uchicago.edu
http://www.math.uchicago.edu/~antonio

Abstract. We introduce the notions of a *complete set of computably infinitary* Π_n^0 *relations* on a structure, of the *jump of a structure*, and of *admitting nth jump inversion*.

Introduction

This paper is part of the study of the interactions between structural properties of a model and computational properties of its presentations. We concentrate on the Turing jumps of the presentations of a structure, and the main notion defined in this paper is the one of the jump of a structure. Even if the definitions and theorems of this paper are all new, many of the ideas were already in the air, but they were not concretely formulated.

We start by defining what it means for a set of relations in a structure to be a complete set of computably infinitary Π_n^0 relations. The idea is that a set of relations is computably-infinitary-Π_n^0 complete if it captures the whole Π_n^0 structural information of the structure. In the second section we look at classes of structures which have a finite complete set of computably infinitary Π_n^0 relations. In the first section, we also define the nth jump of a structure to be the structure together with a complete set of computably infinitary Π_n^0 relations. In the last section we look at the degree spectrum of the jump of a structure. We also define the notion of a structure $\mathcal{A}$ admitting nth jump inversion and prove it is equivalent to saying whenever $\mathcal{A}$ has an X-low$_n$ copy for some X, it also has an X-computable copy.

Throughout this paper we use $\mathcal{L}$ to denote a computable first order language. We use Π_n^c to denote the set of computably infinitary Π_n^0 $\mathcal{L}$-formulas (see [AK00, Ch 7] for background on this language), and we use $\Pi_n^{c,Z}$ to denote the class of Z-computably infinitary Π_n^0 $\mathcal{L}$-formulas.

1 Main Definitions

Definition 1. *Let $\mathcal{A}$ be an $\mathcal{L}$-structure. Let $\{P_0, P_1, ...\}$ be a finite or infinite set of uniformly Π_n^c relations on $\mathcal{A}$. That is, there is a c.e. list of Π_n^c-formulas defining the relations P_i on $\mathcal{A}$. We say that $\{P_0, P_1, ...\}$ is a complete set of*

---

* This research was partially supported by NSF grant DMS-0600824.

Π_n^c relations on $\mathcal{A}$ if every Π_n^c $\mathcal{L}$-formula $\psi(\bar{x})$ is equivalent to a $\Sigma_1^{c,0^{(n)}}$ ($\mathcal{L} \cup \{P_0, ...\}$)-formula, and there is a computable procedure to find this equivalent formula. In other words, $\{P_0, P_1, ...\}$ is a complete set of Π_n^c relations on $\mathcal{A}$ if for every Π_n^c $\mathcal{L}$-formula $\psi(\bar{x})$ we can uniformly produce a $0^{(n)}$-computable list $\varphi_0(\bar{x}), \varphi_1(\bar{x}), \varphi_2(\bar{x}), ...$ of finitary existential $\mathcal{L}$-formulas that may mention the relations $P_0, P_1, ...$ such that

$$\mathcal{A} \models \psi(\bar{x}) \iff \bigvee_i \varphi_i(\bar{x}).$$

Observe that if $\{P_0, P_1, ...\}$ is a complete set of Π_n^c relations on $\mathcal{A}$ then every Σ_{n+1}^c $\mathcal{L}$-formula is also equivalent to a $\Sigma_1^{c,0^{(n)}}$ ($\mathcal{L} \cup \{P_0, ...\}$)-formula.

Definition 2. If $\{P_0, P_1, ...\}$ is a complete set of Π_n^c relations on $\mathcal{A}$, we say that $(\mathcal{A}, P_0, P_1, ...)$ is an nth jump of $\mathcal{A}$ and write

$$\mathcal{A}^{(n)} = (\mathcal{A}, P_0, P_1, ...).$$

Note that being the nth jump of a structure is a property of isomorphism types of structures and not of presentations of structures.

The second thing to notice is that every structure always has an nth jump: Just let $\{P_0, P_1, ...\}$ be a computable list of all the Π_n^c relations on $\mathcal{A}$.

Observation 1. If $\mathcal{A}$ has a copy $\leq_T X$, then $\mathcal{A}^{(n)}$ has a copy $\leq_T X^{(n)}$.

Even though a structure might have many nth jumps, the following lemma show that, from a complexity viewpoint, all these jumps are the same.

Lemma 1. Let $P_0, P_1, ...$ and $R_0, R_1, ...$ be complete sets of Π_n^c relations on $\mathcal{A}$. For every $Y \geq_T 0^{(n)}$ we have that

$$(\mathcal{A}, R_0, R_1, ...) \text{ has copy } \leq_T Y \iff (\mathcal{A}, P_0, P_1, ...) \text{ has copy } \leq_T Y.$$

Proof. Fix a presentation of $\mathcal{A}$ and assume that Y can compute $\mathcal{A}$ and all the relations R_i uniformly. We will show that Y uniformly computes all the relations P_i. Since each P_i is defined by a Π_n^c formula, and this formula is equivalent to a $\Sigma_1^{c,0^{(n)}}$ ($\mathcal{L} \cup \{R_0, ...\}$)-formula, we have that P_i is c.e. in $Y \oplus 0^{(n)} \equiv_T Y$. The complement of P_i is Σ_n^c, and in particular Σ_{n+1}^c. Hence it is also equivalent to a $\Sigma_1^{c,0^{(n)}}$ ($\mathcal{L} \cup \{R_0, ...\}$)-formula and is also c.e. in Y. So P_i is computable in Y, uniformly in i. □

Recall that for non-trivial structures (using Knight's theorem)

$$degSp(\mathcal{A}) = \{X \in \mathcal{D} : \mathcal{A} \text{ has a copy } \leq_T X\},$$

where $\mathcal{D}$ is the set of Turing degrees. So, the lemma above can be restated as

$$degSp(\mathcal{A}, R_0, R_1, ...) \cap \mathcal{D}_{(\geq_T 0^{(n)})} = degSp(\mathcal{A}, P_0, P_1, ...) \cap \mathcal{D}_{(\geq_T 0^{(n)})}$$

where $\mathcal{D}_{(\geq_T 0^{(n)})}$ is the set of Turing degrees above $0^{(n)}$). Therefore, we have that the degree spectrum of $\mathcal{A}^{(n)}$ on the degrees $\geq_T 0^{(n)}$ is independent of the possible choices of $\mathcal{A}^{(n)}$.

2   Examples

We now turn into looking at examples of jumps of structures. We start with linear orderings.

Lemma 2. *Let $\mathcal{A} = (A, <)$ be a linear ordering and let*

$$Succ^{\mathcal{A}} = \{(a, b) \in A^2 : \not\exists c \ (a < c < b)\}.$$

Then

$$\mathcal{A}' = (A, <, Succ^{\mathcal{A}}).$$

Proof (Sketch of the proof). We need to show that $Succ^{\mathcal{A}}$ is Π_1^c-complete. We will show that every Σ_1^c formula is equivalent to a finitary universal formula that uses the predicate $Succ$, and that $0'$ can uniformly find this formula. By taking complements, we will then get that every Π_1^c formula is equivalent to a $\Sigma_1^{c,0'}$ $(\mathcal{L} \cup \{Succ\})$-formula.

Suppose that $\mathcal{A}$ has a first and a last element called $-\infty$ and ∞; the general case is very similar. First, we note that every finitary existential sentence ψ is equivalent to a sentence that says that there are least n many different elements in $\mathcal{A}$. Let $\psi_n(x, y)$ be the formula that says that there are at least n many different elements in between x and y. Observe that $\psi_n(x, y)$ is equivalent to a finitary universal formula over the Successor predicate:

$$\psi_n(x, y) \iff \bigwedge_{j<n} \not\exists z_1, ..., z_j \left(x = z_1 < ... < z_j = y \wedge (\bigwedge_{i=1}^{j-1}(Succ(z_i, z_{i+1}))) \right).$$

Second, if we have a formula with free variables $\psi(x_1, ..., x_k)$, we can write ψ as a disjunction, over all the permutations $(\tau_1, ..., \tau_k)$ of $(1, ..., k)$, of the formulas $\left(\bigwedge_{j<k} x_{\tau_j} < x_{\tau_{j+1}} \right) \wedge \psi(x_1, ..., x_k)$. Third, for every finitary existential formula $\psi(x_1, ..., x_k)$, we have that $x_1 < x_2 < ... < x_k \wedge \psi(x_1, ..., x_k)$ is equivalent to a finite disjunction of formulas of the form

$$x_1 < x_2 < ... < x_k \wedge \psi_{n_0}(-\infty, x_{\tau_1}) \wedge \psi_{n_1}(x_{\tau_1}, x_{\tau_2}) \wedge ... \wedge \psi_{n_k}(x_{\tau_k}, \infty),$$

for some $n_0, ..., n_k \in \omega$. Therefore, we have that any Σ_1^c formula $\psi(x_1, ..., x_k)$ is equivalent to the disjunction over all the permutations $(\tau_1, ..., \tau_k)$ of $(1, ..., k)$ of formulas of the form

$$x_{\tau_1} < ... < x_{\tau_k} \wedge \left(\bigvee_j \left(\psi_{n_0^j}(-\infty, x_{\tau_1}) \wedge \psi_{n_1^j}(x_{\tau_1}, x_{\tau_2}) \wedge ... \wedge \psi_{n_k^j}(x_{\tau_k}, \infty) \right) \right).$$

Fourth, it can be shown that in the infinite disjunction in the formula above all but finitely many of the disjuncts are redundant. Furthermore, $0'$ can find these finitely many disjuncts. $\square$

Notice that in all the linear orderings the same relation $Succ^{\mathcal{A}}$ that is Π_1^c complete. This motivates the following definition.

Definition 3. *Let $\mathcal{K}$ be a class of $\mathcal{L}$-structures. A c.e. set $\varphi_0, \varphi_1, \ldots$ of Π_n^c formulas is a complete set of Π_n^c formulas for $\mathcal{K}$, if for each structure $\mathcal{A} \in \mathcal{K}$ we have that $\{\varphi_0^{\mathcal{A}}, \varphi_1^{\mathcal{A}}, \ldots\}$ is a complete set of Π_n^c formulas on $\mathcal{A}$.*

The Boolean algebra predicates considered by Downey, Jockusch [DJ94], Thurber [Thu95], Knight and Stob [KS00] are exactly the ones needed to define the first four jumps of a Boolean algebra.

Lemma 3 (Harris, Montalbán [HM]). *Let $\mathcal{B}$ be a Boolean algebra.*

- $\mathcal{B}' = (\mathcal{B}, atom^{\mathcal{B}})$,
- $\mathcal{B}'' = (\mathcal{B}, atom^{\mathcal{B}}, inf^{\mathcal{B}}, atomless^{\mathcal{B}})$.
- $\mathcal{B}''' = (\mathcal{B}, atom^{\mathcal{B}}, inf^{\mathcal{B}}, atomless^{\mathcal{B}}, atomic^{\mathcal{B}}, 1\text{-}atom^{\mathcal{B}}, atominf^{\mathcal{B}})$.
- $\mathcal{B}^{(4)} = (\mathcal{B}, atom^{\mathcal{B}}, inf^{\mathcal{B}}, atomless^{\mathcal{B}}, atomic^{\mathcal{B}}, 1\text{-}atom^{\mathcal{B}}, atominf^{\mathcal{B}}, \sim\text{-}inf^{\mathcal{B}},$
  $\quad Int(\omega + \eta)^{\mathcal{B}}, infatomicless^{\mathcal{B}}, 1\text{-}atomless^{\mathcal{B}}, nomaxatomless^{\mathcal{B}})$.

Furthermore, for every n there is a finite set of Π_n^c formulas which are Π_n^c complete for the class of Boolean algebras. (Definitions of the relations above can be found in [KS00] and [HM].)

We note that not all the predicates in the four items above are Π_n^c for the corresponding n, but they are Boolean combinations of Π_n^c predicates. There is no problem relaxing our definition of nth jump to allow the predicates to be Boolean combinations of Π_n^c predicates so long they still generate all other Π_n^c predicates.

Proof. Harris and the author [HM] proved that the unary predicates R_σ for $\sigma \in \mathbf{BF}_n$ are Π_n^c complete for the class of Boolean algebras. They showed that, for $n \leq 4$, the relations in the four items mentioned above are Boolean combinations of R_σ for $\sigma \in \mathbf{BF}_n$ and vice versa.   □

The main lemmas in Downey, Jockusch [DJ94], Thurber [Thu95], and Knight, Stob [KS00] can now be stated as follows

Lemma 4. *Let X be any set, and $\mathcal{B}$ a Boolean algebra.*

1. *[DJ94] $\mathcal{B}$ has a copy $\leq_T X$ if and only if $\mathcal{B}'$ has a copy $\leq_T X'$.*
2. *[Thu95] $\mathcal{B}'$ has a copy $\leq_T X$ if and only if $\mathcal{B}''$ has a copy $\leq_T X'$.*
3. *[KS00] $\mathcal{B}''$ has a copy $\leq_T X$ if and only if $\mathcal{B}'''$ has a copy $\leq_T X'$.*
4. *[KS00] $\mathcal{B}'''$ has a copy $\leq_T X$ if and only if $\mathcal{B}^{(4)}$ has a copy $\leq_T X'$.*

Corollary 1 (Knight and Stob [KS00]). *Suppose $Y^{(4)} \leq_T X^{(4)}$. Every Y-computable Boolean algebra has a X-computable copy.*

Proof. Suppose $\mathcal{B}$ has a copy computable in Y. Then, by Observation 1, $\mathcal{B}^{(4)}$ has a copy computable in $Y^{(4)} \leq_T X^{(4)}$. Applying the four items of the previous lemma one at the time starting from the last one, we get that $\mathcal{B}$ has a X-computable copy.   □

3   Jump Inversions

The following theorem is a sort of jump inversion theorem for structures. The proof is just an applications of the ideas about generic copies of structures developed by Ash, Knight, Mennasse and Slaman [AKMS89] and Chisholm [Chi90]. Essentially we prove that the jump $\mathcal{A}'$ of a structure $\mathcal{A}$ can compute a 1-generic copy of $\mathcal{A}$.

Theorem 1. *If* $Y \geq_T 0'$ *and* $\mathcal{A}'$ *has a copy* $\leq_T Y$, *then for some* X *with* $X' \leq_T Y$, $\mathcal{A}$ *has a copy computable in* X.

Proof. We will build a copy $\mathcal{B}$ of $\mathcal{A}$ with domain $B = \{b_0, b_1, ...\}$. Let $D(\mathcal{B}) \in 2^\omega$ be the diagram of $\mathcal{B}$. So, for some list of atomic formulas ψ_i with variables among $x_0, x_1, ...$ we have that $D(\mathcal{B}) = 1$ if and only if $\mathcal{B} \models \psi_i$ where x_j is interpreted as b_j. Assume that ψ_i only uses variables among $x_0, ..., x_i$. Therefore, to know the first n bits of $D(\mathcal{B})$ we only need to use the atomic relations among $b_0, ..., b_n$. For each $\sigma \in 2^n$, let $\psi_\sigma(x_0, ..., x_n)$ be the formula $\bigwedge_{i:\sigma(i)=1} \psi_i \wedge \bigwedge_{i:\sigma(i)=0} \neg\psi_i$. So, we have that $\mathcal{B} \models \psi_\sigma(b_0, ..., b_n) \iff \sigma \subseteq D(\mathcal{B})$.

We will build a bijection $F \colon B \to A$ and then define $\mathcal{B}$ by pulling back the structure of $\mathcal{A}$, and we will let $X = D(\mathcal{B})$. We need to define F computably in Y and we will also make sure that Y computes the Turing jump of $D(\mathcal{B})$. At each stage s we define a finite one-to-one partial map $p_s \colon B \to A$ with domain $\{b_0, ..., b_{n_s}\}$, and then we will let $F = \bigcup_s p_s$. Given a finite one-to-one partial map p that maps $b_0, ..., b_n$ to $a_0, ..., a_n$, let $D(p)$ be the $\sigma \in 2^n$ such that $\mathcal{A} \models \psi_\sigma(a_0, ..., a_n)$. Note that $D(\mathcal{B}) = \bigcup_n D(p_n)$.

Construction:

- Let p_0 map b_0 to a_0.
- At stage $s+1 = 2e$ extend p_s to p_{s+1} in any way so that b_e is in the domain of p_{s+1} and a_e is in the image.
- At stage $s+1 = 2e+1$ we want to decide the jump of $D(\mathcal{B})$. Suppose p_s maps $b_0, ..., b_{n_s}$ to $a_0, ..., a_{n_s}$. Using Y, decide whether there exists $q \supseteq p_s$ such that $\{e\}^{D(q)}(e) \downarrow$. Note that Y can decide this because, since it computes $\mathcal{A}'$, it knows whether

$$\mathcal{A} \models \bigvee_{\sigma \supseteq p_s, \{e\}^\sigma(e)\downarrow} \exists\bar{y}\ \psi_\sigma(a_0, ..., a_{n_s}, \bar{y}).$$

If the answer is positive, Y can search for witnesses σ and $\bar{y}$ and use them to define p_{s+1} adding $\bar{y}$ to the range of p_s. In this case Y knows that $e \in D(\mathcal{B})'$. Otherwise, we let $p_{s+1} = p_s$ and Y knows that $e \notin D(\mathcal{B})'$.

We have build $\mathcal{B} \cong \mathcal{A}$ so that $D(\mathcal{B})' \leq_T Y$ as wanted.   □

Corollary 2. *For every structure* $\mathcal{A}$,

$$degSp(\mathcal{A}') \cap \mathcal{D}_{(\geq_T 0')} = \{X' : X \in degSp(\mathcal{A})\}.$$

Proof. That $\{X' : X \in degSp(\mathcal{A})\} \subseteq degSp(\mathcal{A}') \cap \mathcal{D}_{(\geq_T 0')}$ follows from Observation 1. That $degSp(\mathcal{A}') \cap \mathcal{D}_{(\geq_T 0')} \subseteq \{X' : X \in degSp(\mathcal{A})\}$ follows from the previous theorem. □

Now we consider a stronger version of jump inversion.

Definition 4. *We say that $\mathcal{A}$ admits nth jump inversion if for every set X, we have that*

$$\mathcal{A}^{(n)} \text{ has copy } \leq_T X^{(n)} \iff \mathcal{A} \text{ has copy } \leq_T X.$$

Observation 2. *If $\mathcal{A}$ admits jump inversion, then*

$$degSp(\mathcal{A}) = \{X : X' \in degSp(\mathcal{A}')\}.$$

Structures which admit nth jump inversion have, in some sense, already been considered in the literature before. Lemma 3 above shows that Boolean algebras admit 4th jump inversion. In [HM], Harris and the author asked the following question (stated in a different way): does every Boolean algebra admit nth jump inversion? It follows from Theorem 2 below that this question is equivalent to the well-known question of whether for every X, every X-low$_n$ Boolean algebra has an X-computable copy. For linear orderings, the following results are known. Downey [DK92] proved that every linear ordering of the form $(\mathbb{Q}+2+\mathbb{Q})\cdot\mathcal{A}$ admits jump inversion. Ash [Ash91] showed that linear orderings of the form $\omega^n \cdot \mathcal{A}$ admit $2n$th-jump inversion. Kach and the author [KM] then used these results to prove that all the linear orderings with finitely many descending cuts admit nth jump inversion for every n. Graphs which admit αth jump inversion have been used in [GHK+05] and [CFG+] to show that there exists Δ^0_α-categorical structures which are not relatively Δ^0_α-categorical, lifting earlier results of Goncharov and others on computably categorical structures.

Theorem 2. *Let $\mathcal{A}$ be an $\mathcal{L}$-structure. The following are equivalent.*

1. *$\mathcal{A}$ admits nth jump inversion;*
2. *For every sets X, Y with $X^{(n)} \equiv_T Y^{(n)}$, we have that*

$$\mathcal{A} \text{ has copy } \leq_T X \iff \mathcal{A} \text{ has copy } \leq_T Y.$$

Proof. To prove that (1) implies (2), we have that if $\mathcal{A}$ has a copy computable in X, then $\mathcal{A}^{(n)}$ has a copy computable in $X^{(n)} \equiv_T Y^{(n)}$, and hence by (1) $\mathcal{A}$ has a copy computable in Y.

For the other direction, suppose that $\mathcal{A}^{(n)}$ has a $Y^{(n)}$-computable copy. Then, using n iterations of Theorem 1, for some X with $X^{(n)} \equiv_T Y^{(n)}$, $\mathcal{A}$ has a copy computable in X. But, then, by (2), $\mathcal{A}$ has a Y-computable copy. □

4   Questions

1. What are other examples of structures with finite sets of complete Π^c_n relations?
2. What are other examples of structures that admit jump inversion?
3. Is there a structural characterization of the structures that admit jump inversion?

References

AK00.   Ash, C.J., Knight, J.: Computable Structures and the Hyperarithmetical Hierarchy. Elsevier Science, Amsterdam (2000)

AKMS89.   Ash, C., Knight, J., Manasse, M., Slaman, T.: Generic copies of countable structures. Ann. Pure Appl. Logic 42(3), 195–205 (1989)

Ash91.   Ash, C.J.: A construction for recursive linear orderings. J. Symbolic Logic 56(2), 673–683 (1991)

CFG+.   Chisholm, J., Fokina, E., Goncharov, S., Knight, J., Miller, S.: Intrinsic bounds on complexity at limit levels. Journal of Symbolic Logic (to appear)

Chi90.   Chisholm, J.: Effective model theory vs. recursive model theory. J. Symbolic Logic 55(3), 1168–1191 (1990)

DJ94.   Downey, R., Jockusch, C.G.: Every low Boolean algebra is isomorphic to a recursive one. Proc. Amer. Math. Soc. 122(3), 871–880 (1994)

DK92.   Downey, R., Knight, J.F.: Orderings with αth jump degree $\mathbf{0}^{(\alpha)}$. Proc. Amer. Math. Soc. 114(2), 545–552 (1992)

GHK+05.   Goncharov, S., Harizanov, V., Knight, J., McCoy, C., Miller, R., Solomon, R.: Enumerations in computable structure theory. Ann. Pure Appl. Logic 136(3), 219–246 (2005)

HM.   Harris, K., Montalbán, A.: On the n-back-and-forth types of Boolean algebras (submitted for publication)

KM.   Kach, A., Montalbán, A.: Linear orders with finitely many descending cuts (in preparation)

KS00.   Knight, J.F., Stob, M.: Computable Boolean algebras. J. Symbolic Logic 65(4), 1605–1623 (2000)

Thu95.   Thurber, J.J.: Every low$_2$ Boolean algebra has a recursive copy. Proc. Amer. Math. Soc. 123(12), 3859–3866 (1995)

A General Representation Theorem for Probability Functions Satisfying Spectrum Exchangeability

J.B. Paris and A. Vencovská*

School of Mathematics
University of Manchester
Manchester M13 9PL, UK
jeff.paris@manchester.ac.uk, alena.vencovska@manchester.ac.uk

Abstract. In the context of polyadic inductive logic we give a general representation theorem applicable to all probability functions satisfying Spectrum Exchangeability as the weighted difference of two probability functions satisfying Spectrum Exchangeability and Language Invariance.

Keywords: Polyadic Inductive Logic, Spectrum Exchangeability, Language Invariance.

Introduction

The aim of this short note is to give a general representation theorem for those probability functions satisfying the Principle of Spectrum Exchangeability (within the context of polyadic inductive logic) as the weighted difference of two probability functions satisfying Spectrum Exchangeability *and* Language Invariance. The value of this result is that probability functions satisfying these latter two properties have themselves a rather straightforward de Finetti style representation theorem, see [3], [4], [5], from which various other properties of this class of probability functions can be derived, for example a version of 'Instantial Relevance', see [6].

Our framework and context, polyadic inductive logic, have previously been explained in some detail in a number of earlier papers, in particular [8], [2], and [5], though for the sake of completeness we shall briefly outline them again here.

We work with a first order language L containing finitely many relation symbols $P_1, P_2, \ldots, P_q$ with arities $r_1, r_2, \ldots, r_q$ respectively, countably many constants $a_1, a_2, a_3, \ldots$ and no function symbols nor equality. The intention here (as reflected in (P3) below) is that these constants exhaust the universe. Let SL, $QFSL$, respectively, denote the sentences and quantifier free sentences of L. Since our results are only of interest when L is not entirely unary we assume that some $r_i \geq 2$. Throughout we shall use $b_1, b_2, \ldots$ to denote distinct constants a_i from L.

A function $w : SL \to [0, 1]$ is a *probability function* on L if it satisfies that for all $\theta, \phi, \exists x \, \psi(x) \in SL$:

---

* Supported by a UK Engineering and Physical Sciences Research Council (EPSRC) Research Assistantship.

K. Ambos-Spies, B. Löwe, and W. Merkle (Eds.): CiE 2009, LNCS 5635, pp. 379–388, 2009.

(P1) If $\vDash \theta$ then $w(\theta) = 1$,

(P2) If $\vDash \neg(\theta \wedge \phi)$ then $w(\theta \vee \phi) = w(\theta) + w(\phi)$,

(P3) $w(\exists x\,\psi(x)) = \lim_{m\to\infty} w(\bigvee_{i=1}^{m} \psi(a_i))$.

By a theorem of Gaifman any probability function defined on $QFSL$ (i.e. satisfying (P1) and (P2) for $\theta, \phi \in QFSL$) extends uniquely to a probability function on L. Hence we can limit our considerations to probability functions defined just on $QFSL$. By the Disjunctive Normal Form Theorem it then follows that w is determined simply by its values on the *state descriptions*, that is sentences of the form $\Theta(b_1, b_2, \ldots, b_m)$ where

$$\Theta(b_1, b_2, \ldots, b_m) = \bigwedge_{s=1}^{q} \bigwedge_{i_1, i_2, \ldots, i_{r_s} \in \{1, \ldots, m\}} \pm P_s(b_{i_1}, b_{i_2}, \ldots, b_{i_{r_s}}) \quad .(1)$$

where $\pm P$ stands for P or $\neg P$ respectively. We use $SD(m)$ to denote the set of such state descriptions ($b_1, \ldots, b_m$ being implicit in the context).

We are interested in w that satisfy various rationality principles. In this paper, we shall be considering just two of them, Spectrum Exchangeability (Sx) and Language Invariance (Li).

In order to explain Sx we first need to introduce some notation. Given a state description $\Theta(b_1, b_2, \ldots, b_m)$ we say that b_i, b_j are *indistinguishable* mod Θ, written $b_i \sim_\Theta b_j$, if for any relation $P(x_1, x_2, \ldots, x_r)$ of L and any $t_1, \ldots, t_r \in \{1, \ldots, m\}$, the sentence $P(b_{t_1}, b_{t_2}, \ldots, b_{t_r})$ appears positively as a conjunct in $\Theta(b_1, b_2, \ldots, b_m)$ if and only if $P(b_{s_1}, b_{s_2}, \ldots, b_{s_r})$ also appears positively as a conjunct in $\Theta(b_1, b_2, \ldots, b_m)$ where $\langle b_{s_1}, b_{s_2}, \ldots, b_{s_r} \rangle$ is the result of replacing any number of occurrences of b_i in $\langle b_{t_1}, b_{t_2}, \ldots, b_{t_r} \rangle$ by b_j or vice-versa. Clearly $\sim_\Theta$ is an equivalence relation.

Define the *spectrum* of Θ, denoted $\mathcal{S}(\Theta)$, to be the multiset of sizes of the (non-empty) equivalence classes with respect to $\sim_\Theta$. The number of these equivalence classes is referred to as the *length* of the spectrum.

The Spectrum Exchangeability Principle, Sx

If $\Theta(b_1, b_2, \ldots, b_m), \Phi(c_1, c_2, \ldots, c_m)$ are state descriptions and $\mathcal{S}(\Theta) = \mathcal{S}(\Phi)$ then $w(\Theta) = w(\Phi)$.

The other principle which we shall be concerned with here is that of Language Invariance which asserts that w should be consistently extendable to any larger language (of the type which we are considering).

Language Invariance, Li[1]

Let w be a probability function on L satisfying Sx. We say that w satisfies *Language Invariance* if there exist a class of probability functions $w_\mathcal{L}$ satisfying Sx,

---

[1] Here we include the requirement of Sx in our definition of Li. The terminology is not entirely uniform and on some earlier occasions (for example [7]) 'Language Invariance' has simply asserted the existence of such a class of probability functions $w_\mathcal{L}$ without any requirement that they satisfy Sx. In the context of this paper however the current formulation appears most appropriate since it seems entirely reasonable to demand that these $w_\mathcal{L}$ also satisfy the same basic properties as w.

ine for each finite predicate language $\mathcal{L}$ such that whenever $\mathcal{L}'$ is a sublanguage of $\mathcal{L}$ then w restricted to $S\mathcal{L}'$ equals $w_{\mathcal{L}'}$ $(w_{\mathcal{L}} \restriction S\mathcal{L} = w_{\mathcal{L}'})$ and $w_L = w$.

Homogenous and heterogenous probability functions. Let w be a probability function satisfying Sx. We say that w is *homogenous* if for all k

$$\lim_{m \to \infty} \sum_{\substack{\Theta \in SD(m) \\ |\mathcal{S}(\Theta)| \leq k}} w(\Theta(b_1, \cdots, b_m)) = 0.$$

For a natural number $t \geq 1$, w is $\leq t$-*heterogenous* if $w(\Theta) = 0$ whenever $\mathcal{S}(\Theta)| > t$, and w is t-*heterogenous* if moreover

$$\lim_{m \to \infty} \sum_{\substack{\Theta \in SD(m) \\ |\mathcal{S}(\Theta)| < t}} w(\Theta(b_1, \cdots, b_m)) = 0.$$

Essentially then if w is t-heterogeneous it asserts that there are always exactly t distinguishable individuals in the population whilst if w is homogeneous it asserts that there are infinitely many distinguishable individuals.

In [8] it is shown that any probability function w satisfying Sx can be expressed uniquely[2]) as a weighted sum of a homogenous probability function and t-heterogenous probability functions:

$$w = \sum_{t=0}^{\infty} \eta_t w^{[t]} \tag{2}$$

for some non-negative $\eta_0, \eta_1, \cdots$ with $\sum_{t=0}^{\infty} \eta_t = 1$, $w^{[0]}$ homogenous and $w^{[t]}$ t-heterogenous for $t > 0$.

Note that for a $\leq t$-*heterogenous* function w, all η_k with $k > t$ are equal to 0.

The probability functions u^{p}. Let

$$\mathbb{B} = \{ \langle x_0, x_1, x_2, \ldots \rangle \mid x_1 \geq x_2 \geq \ldots \geq 0,\ x_0 \geq 0,\ \sum_{i=0}^{\infty} x_i = 1 \}$$

with the standard weak product topology inherited from $[0,1]^{\infty}$. Let

$$\boldsymbol{p} = \langle p_0, p_1, p_2, \ldots \rangle \in \mathbb{B}.$$

p_i should be thought of as the probability of picking 'colour' i (from some urn, with replacement).

For a state description $\Theta(b_1, b_2, \ldots, b_m)$ and colours

$$\boldsymbol{c} = \langle c_1, c_2, \ldots, c_m \rangle \in \{0, 1, 2, \ldots\}^m$$

we define probabilities $j^{\boldsymbol{p}}(\Theta(b_1, b_2, \ldots, b_m), \boldsymbol{c})$ inductively as follows:

---

[2] The η_t are unique and provided $\eta_t \neq 0$, so is $w^{[t]}$ for each t.

Set $j^{\boldsymbol{p}}(\top, \emptyset) = 1$. Suppose that at stage m we have defined the probability $j^{\boldsymbol{p}}(\Theta(b_1, b_2, \ldots, b_m), \boldsymbol{c})$. Pick colour c_{m+1} from $0, 1, 2, \ldots$ according to the probabilities $p_0, p_1, p_2, \ldots$ and let

$$\boldsymbol{c}^+ = \langle c_1, \ldots, , c_m, c_{m+1} \rangle.$$

If c_{m+1} is the same as an earlier colour, c_j say, with $c_j \neq 0$ extend $\Theta(b_1, b_2, \ldots, b_m)$ to the unique state description $\Theta^+(b_1, b_2, \ldots, b_m, b_{m+1})$ for which $b_j \sim_{\Theta^+} b_{m+1}$. (Notice this means that the equivalence classes mod Θ^+ are the same as those mod Θ except that $m + 1$ is added to the class containing j.) On the other hand if c_{m+1} is 0 or a previously unchosen colour then randomly (uniformly) choose $\Theta^+(b_1, b_2, \ldots, b_m, b_{m+1})$ extending $\Theta(b_1, b_2, \ldots, b_m)$ so that when $i, j \leq m$ are such that $c_i = c_j \neq 0$ then $b_i \sim_{\Theta^+} b_j$. Finally let $j^{\boldsymbol{p}}(\Theta^+, \boldsymbol{c}^+)$ be $j^{\boldsymbol{p}}(\Theta, \boldsymbol{c})$ times the probability as described of going from $\Theta, \boldsymbol{c}$ to $\Theta^+, \boldsymbol{c}^+$.

Having defined these $j^{\boldsymbol{p}}(\Theta, \boldsymbol{c})$ now set

$$u^{\boldsymbol{p}}(\Theta) = \sum_{\boldsymbol{c}} j^{\boldsymbol{p}}(\Theta, \boldsymbol{c}).$$

The functions $u^{\boldsymbol{p}}$ satisfy Sx and Li.

It is shown in [3] and [4] that probability functions w satisfying Sx and Li are precisely those of the form

$$\int_{\mathbb{B}} u^{\boldsymbol{p}} d\mu(\boldsymbol{p})$$

for some measure μ on $\mathbb{B}$. Theorem 6 in [5] shows that any homogenous function w has such a representation.

In this paper we show that any probability function satisfying merely Sx can be expressed as

$$G_+ w_+ - G_- w_-$$

for some $0 \leq G_+, G_- \leq 2$ and w_+, w_- satisfying Sx and Li.

The probability functions $v^{\boldsymbol{p}}$. Define $\mathbb{B}_t$ to be the space

$$\mathbb{B}_t = \{ \boldsymbol{y} = \langle y_1, \ldots, y_t \rangle \mid y_1 \geq \ldots \geq y_t > 0, \ \sum_{i=1}^{t} y_i = 1 \}$$

with the usual topology inherited from $\mathbb{R}^t$.

Let $\boldsymbol{p} = \langle p_1, p_2, \ldots, p_t \rangle \in \mathbb{B}_t$. For a state description $\Theta(b_1, b_2, \ldots, b_m)$ and

$$\boldsymbol{c} = \langle c_1, c_2, \ldots, c_m \rangle \in \{1, 2, \ldots, t\}^m$$

we define $l^{\boldsymbol{p}}(\Theta(b_1, b_2, \ldots, b_m), \boldsymbol{c})$ inductively as follows:

Set $l^{\boldsymbol{p}}(\top, \emptyset) = 1$. Suppose that at stage m we have defined the probability $l^{\boldsymbol{p}}(\Theta(b_1, b_2, \ldots, b_m), \boldsymbol{c})$. Pick colour c_{m+1} from $1, 2, \ldots, t$ according to the probabilities $p_1, p_2, \ldots, p_t$ and let

$$\boldsymbol{c}^+ = \langle c_1, \ldots, , c_m, c_{m+1} \rangle.$$

If c_{m+1} is the same as an earlier colour, c_j say, extend $\Theta(b_1, b_2, \ldots, b_m)$ to the unique state description $\Theta^+(b_1, b_2, \ldots, b_m, b_{m+1})$ for which $b_j \sim_{\Theta^+} b_{m+1}$. Otherwise extend randomly observing $b_i \sim_{\Theta_{k+1}} b_j$ when $i, j \leq m$ and $c_i = c_j$ so that the probability of Θ^+ being chosen is the ratio

the number of extensions of Θ_{c}^+ to an element of $SD(t)$ with spectrum of length t
the number of extensions of Θ_c to an element of $SD(t)$ with spectrum of length t

where Θ_c and Θ_c^+ are the state description resulting from Θ and Θ^+ respectively by restricting them to

$$\{b_i \mid \text{for all } j < i, \ c_i \neq c_j\}.$$

Finally let $l^{\boldsymbol{p}}(\Theta^+, \boldsymbol{c}^+)$ be $l^{\boldsymbol{p}}(\Theta, \boldsymbol{c})$ times the probability as described of going from $\Theta, \boldsymbol{c}$ to $\Theta^+, \boldsymbol{c}^+$.

As before, we now set

$$v^{\boldsymbol{p}}(\Theta) = \sum_c l^{\boldsymbol{p}}(\Theta, \boldsymbol{c}).$$

The functions $v^{\boldsymbol{p}}$ satisfy Sx and they are t-heterogeneous. Moreover, as shown in [5], t-heteogeneous probability functions w satisfying Sx are precisely those of the form

$$\int_{\mathbb{B}_t} v^{\boldsymbol{p}} d\mu(\boldsymbol{p})$$

for some measure μ on $\mathbb{B}_t$.

The Representation Theorem

Theorem 1. *Every probability function w satisfying Sx is of the form*

$$G_+ w_+ - G_- w_-$$

for some $0 \leq G_+, G_- \leq 2$ and probability functions w_+, w_- satisfying Sx and Language Invariance.

Let $\boldsymbol{p} = \langle p_1, p_2, \ldots, p_t \rangle \in \mathbb{B}_t$. Setting all other $p_i = 0$ gives an element $\boldsymbol{p}$ of $\mathbb{B}$, and via this association we may treat $\mathbb{B}_t$ as a subset of $\mathbb{B}$.

Let E_t be the set of all partitions of the set of colours $\{1, 2, \ldots, t\}$ and let $f = \{f_1, f_2, \ldots, f_m\} \in E_t$. We denote $m = |f|$, and for $f, g \in E_t$ we write $f \trianglelefteq g$ if g is a refinement of f. We take $\iota(t)$ (referred to as ι as long as t is fixed) to be the maximal partition in E_t consisting of all singletons.

Let $f(\boldsymbol{p})$ be the element of $\mathbb{B}_m$ formed by ordering the numbers

$$\sum_{j \in f_s} p_j, \ s = 1, 2, \ldots, m.$$

Note that $\iota(\boldsymbol{p}) = \boldsymbol{p}$. We will write s_f for $|SD(|f|)|$ and s_f^f for the number of those state descriptions from $SD(|f|)$ which have spectrum of length $|f|$, that is, those state descriptions which make all $|f|$ individuals distinguishable (this should really be $s^{|f|}_{|f|}$ in keeping with the earlier mentioned s^t_t, but we drop the $\|$ signs. Note that s^{ι}_{ι} is s^t_t). The following identity was originally proved by Jürgen Landes and appears as Theorem 12 of [1].

Lemma 2

$$u^{f(\boldsymbol{p})} = \sum_{g \trianglelefteq f} \frac{s^g_g}{s_f} v^{g(\boldsymbol{p})}.$$

Let A be the $|E_t \times E_t|$ matrix with entry on the fth row, gth column s^g_g/s_f for $g \trianglelefteq f$ and 0 otherwise. So $\boldsymbol{u}^T = A\boldsymbol{v}^T$. For $f \in E_t$ let the entry in the ιth row fth column of A^{-1} be a_f. Then for $g \triangleleft \iota$,

$$a_\iota = \frac{s_\iota}{s^{\iota}_{\iota}}, \tag{3}$$

$$\frac{a_g}{s_g} = -\sum_{g \triangleleft f} \frac{a_f}{s_f}, \tag{4}$$

$$v^{\iota(\boldsymbol{p})} = \sum_{f \in E_t} a_f u^{f(\boldsymbol{p})}. \tag{5}$$

From (3), (4) we see that

$$\frac{a_g}{s_g} = \frac{N_g a_\iota}{s_\iota} = \frac{N_g}{s^{\iota}_{\iota}}$$

where

$$N_g = \sum_{z:g \rightsquigarrow \iota} (-1)^{|z|}$$

and $z : g \rightsquigarrow \iota$ means that z is a path

$$g = f_0 \triangleleft f_1 \triangleleft f_2 \triangleleft \ldots \triangleleft f_k = \iota$$

(with $|z| = k$ in this case).
From this it follows from (5) that

$$v^{\boldsymbol{p}} = v^{\iota(\boldsymbol{p})} = \sum_{N_f > 0} \frac{N_f s_f u^{f(\boldsymbol{p})}}{s^{\iota}_{\iota}} - \sum_{N_f < 0} \frac{(-N_f) s_f u^{f(\boldsymbol{p})}}{s^{\iota}_{\iota}}. \tag{6}$$

Each of the summands in (6) is a linear combination of language invariant probability functions satisfying Sx, hence a constant multiple of a language invariant probability function satisfying Sx. Considering their values for a tautology shows the multiplicative factors to be

$$G^t_+ = \sum_{N_f > 0} \frac{N_f s_f}{s^{\iota}_{\iota}}, \qquad G^t_- = \sum_{N_f < 0} \frac{N_f s_f}{s^{\iota}_{\iota}}$$

Hence (6) shows that for any $p \in \mathbb{B}_t$ there are two language invariant probability functions w_+, w_- such that

$$v^p = G_+^t w_+ - G_-^t w_-$$

and it also specifies these functions.

Using the representation (2) and taking integrals and finite sums proves the theorem for any $\leq t$-heterogeneous probability function (satisfying Sx of course).

Now consider an arbitrary probability function w satisfying Sx. By the representation (2) we have

$$w = \sum_{t=0}^{\infty} \eta_t w^{[t]}$$

with $w^{[0]}$ homogeneous, the $w^{[t]}$ t-heterogeneous for $t > 0$ and $\sum_{t=0}^{\infty} \eta_t = 1$. For each $t > 0$ we have

$$w^{[t]} = G_+^t w_+^{[t]} - G_-^t w_-^{[t]}$$

so

$$w = \eta_0 w^{[0]} + \sum_{t=1}^{\infty} \eta_t (G_+^t w_+^{[t]} - G_-^t w_-^{[t]}).$$

To separate these positive and negative factors each into a constant times a language invariant probability function we require that

$$\sum_{t=1}^{\infty} \eta_t G_-^t < \infty. \tag{7}$$

To show this we need to prove that

$$\sum_{t>0} \sum_{\substack{f \in E_t \\ N_f < 0}} \eta_t \frac{(-N_f)s_f}{s_{\iota(t)}^{\iota(t)}} < \infty.$$

Clearly for this it would be enough if the

$$\sum_{\substack{f \in E_t \\ N_f < 0}} \frac{(-N_f)s_f}{s_{\iota(t)}^{\iota(t)}} \tag{8}$$

were bounded. Note that if this is the case then the corresponding sum with $N_f > 0$ can be at most 1 more (because $w_+^{[t]}(\top) = w_-^{[t]}(\top) = 1$).

Lemma 3. *For $f \in E_t$, $|f| = k$ we have $|N_f| \leq 2^{(t-k)^2}$.*

Proof. For $0 < r \leq t - k$ let

$$M_f^r = |\{ z : f \rightsquigarrow \iota \,|\, |z| = r \}|.$$

We illustrate a method of coding the elements of this set, in the first case with a simple example. Suppose $r = 3$,

$$f = \{\{1, 2, 3, 4\}, \{5, 6, 7, 8\}.$$

Let h be the map with domain $\{1, 2, 3, 5, 6, 7\}$ which sends 1 and 3 to 1, 6 to 2 and 2,5,7 to 3. Then h codes the $z_h : f \rightsquigarrow \iota$ given by successively dividing the sets in f at the numbers mapped to 1 by h, then those mapped to 2, then those mapped to 3, i.e. z_h is the path

$$f = \{\, \{1, 2, 3, 4\}, \{5, 6, 7, 8\}\, \},$$

$$\{\, \{1\}, \{2, 3\}, \{4\}, \{5, 6, 7, 8\}\}\, \},$$

$$\{\, \{1\}, \{2, 3\}, \{4\}, \{5, 6\}, \{7, 8\}\}\, \},$$

$$\{\, \{1\}, \{2\}, \{3\}, \{4\}, \{5\}, \{6\}, \{7\}, \{8\}\}\, \} = \iota.$$

Of course we assumed here that the original partition of f and all subsequent partitionings kept the elements in the order 1,2,3,4,5,6,7,8. In general this would not be the case. However any other such path can be obtained from this one by taking a permutation of the elements in the sets in f.

Some paths are thus obtained through more than one combination of a mapping and a permutation but we do have that

$$M_f^r \leq J(t - k, r) \prod_{i=1}^{k} (|f_i|!)$$

where $J(t - k, r)$ is the number of maps from a set with $t - k$ elements onto one with r elements and $f = \{f_1, f_2, \ldots, f_k\}$.

Consequently,

$$|N_f| \leq \sum_{r=1}^{t-k} |M_f^r| \leq (t - k)(t - k)^{t-k}(t - k)! \leq 2^{(t-k)^2},$$

as required.

Let $\left\{ {t \atop k} \right\}$ denote a Stirling Number of the second kind, that is the number of partitions of $\{1, 2, \ldots, t\}$ into k non-empty sets.

Lemma 4

$$\sum_{k=1}^{t-1} \left\{ {t \atop k} \right\} 4^{-k(t-k)} \leq 1/2.$$

Proof. The proof is by induction of t. If $t = 1$ then the result holds, so suppose it holds below $t \geq 2$. Then using the identity

$$\left\{ {t \atop k} \right\} = \left\{ {t-1 \atop k-1} \right\} + k \left\{ {t-1 \atop k} \right\}$$

and the inductive hypothesis, and dropping zero terms, we obtain that

$$\sum_{k=1}^{t-1} \left\{ {t \atop k} \right\} 4^{-k(t-k)} = \sum_{k=2}^{t-1} \left\{ {t-1 \atop k-1} \right\} 4^{-k(t-k)} + \sum_{k=1}^{t-1} k \left\{ {t-1 \atop k} \right\} 4^{-k(t-k)}$$

$$= \sum_{k=1}^{t-2} \left\{ {t-1 \atop k} \right\} 4^{-(k+1)(t-1-k)} + \sum_{k=1}^{t-1} k \left\{ {t-1 \atop k} \right\} 4^{-k(t-1-k)-k}$$

$$= \sum_{k=1}^{t-2} \left\{ {t-1 \atop k} \right\} 4^{-k(t-1-k)} \cdot 4^{-(t-1-k)} + \sum_{k=1}^{t-2} \left\{ {t-1 \atop k} \right\} 4^{-k(t-1-k)} \cdot \frac{k}{4^k}$$

$$+ (t-1) \left\{ {t-1 \atop t-1} \right\} 4^{-(t-1)}$$

$$\leq (1/2)4^{-1} + (1/2)4^{-1} + (t-1)4^{-(t-1)}$$

$$\leq 1/8 + 1/8 + 1/4 = 1/2$$

We can now conclude the proof of the main theorem. By considering the 'worst case' where L has a single binary relation (recall that L is not entirely unary) we have that

$$\frac{s_{L(t)}}{s^{L(t)}_{L(t)}} \leq 8/7 \quad \text{for all } t$$

and if $f \in E_t$ has k classes then

$$\frac{s_f}{s_{L(t)}} \leq \frac{2^{k^2}}{2^{t^2}}.$$

Consequently

$$\sum_{\substack{f \in E_t \\ N_f < 0}} \frac{(-N_f)s_f}{s^{L(t)}_{L(t)}} = \sum_{\substack{f \in E_t \\ N_f < 0}} \frac{(-N_f)s_f}{s_{L(t)}} \frac{s_{L(t)}}{s^{L(t)}_{L(t)}} \leq \sum_{k=1}^{t-1} \left\{ {t \atop k} \right\} 2^{(t-k)^2} \frac{2^{k^2}}{2^{t^2}} (8/7)$$

$$= (8/7) \sum_{k=1}^{t-1} \left\{ {t \atop k} \right\} 4^{-k(t-k)} \leq 1,$$

which concludes the proof of the Representation Theorem.

The following corollary is an immediate consequence of the main representation theorem from [5].

Corollary 5. *For any probability function on L satisfying Spectrum Exchangeability there are measures μ_+ and μ_- on $\mathbb{B}$ and $0 \leq G_+, G_- \leq 2$ such that*

$$w = \int_{\mathbb{B}} G_+ u^p d\mu_+ - \int_{\mathbb{B}} G_- u^p d\mu_-. \tag{9}$$

Note that a converse to Corollary 5 also holds: For any choice of G_+, G_-, μ_+, μ_-, if the right hand side of (9) defines a probability function *then* it will satisfy

Spectrum Exchangeability. This then gives an interesting insight into why some probability functions w satisfying Spectrum Exchangeability fail to satisfy Language Invariance. If we had $G_- = 0$ (so $G_+ = 1$) in (9) we could, as explained in [4], extend w to larger languages simply by taking the natural extensions of the u^p in (9). However in the case $G_- > 0$ this device can fail because the resulting function on the right hand side of (9) gives negative values when these natural extensions of the u^p are substituted.

References

1. Landes, J.: The Principle of Spectrum Exchangeability within Inductive Logic, Ph.D. Thesis, University of Manchester (2009)
2. Landes, J., Paris, J.B., Vencovská, A.: Some aspects of Polyadic Inductive Logic. Studia Logica 90, 3–16 (2008)
3. Landes, J., Paris, J.B., Vencovská, A.: A characterization of the Language Invariant Families satisfying Spectrum Exchangeability in Polyadic Inductive Logic. In: Proceeding of the IPM Logic Conference, Tehran (2007); to appear in a special edition of the Annals of Pure and Applied Logic
4. Landes, J., Paris, J.B., Vencovská, A.: Language Invariance and Spectrum Exchangeability in Inductive Logic. In: Mellouli, K. (ed.) ECSQARU 2007. LNCS (LNAI), vol. 4724, pp. 151–160. Springer, Heidelberg (2007)
5. Landes, J., Paris, J.B., Vencovská, A.: Representation Theorems for probability functions satisfying Spectrum Exchangeability in Inductive Logic. The International Journal of Approximate Reasoning (submitted)
6. Landes, J., Paris, J.B., Vencovská, A.: Instantial Relevance in Polyadic Inductive Logic. In: Ramanujam, R., Sarukkai, S. (eds.) Logic and Its Applications. LNCS, vol. 5378, pp. 162–169. Springer, Heidelberg (2009)
7. Nix, C.J., Paris, J.B.: A Continuum of inductive methods arising from a generalized principle of instantial relevance. Journal of Philosophical Logic 35(1), 83–115 (2006)
8. Nix, C.J., Paris, J.B.: A note on Binary Inductive Logic. Journal of Philosophical Logic 36(6), 735–771 (2007)

Stability under Strategy Switching

Soumya Paul, R. Ramanujam, and Sunil Simon

The Institute of Mathematical Sciences
C.I.T. Campus, Chennai 600 113, India
{soumya,jam,sunils}@imsc.res.in

Abstract. We suggest that a *process-like* notion of strategy is relevant
in the context of interactions in systems of self-interested agents. In this
view, strategies are not plans formulated by rational agents considering all possible futures and (mutually recursively) taking into account
strategies employed by other players. Instead, they are partial; players
start with a set of potential strategies and dynamically switch between
them. This necessitates some means in the model for players to access
each others' strategies, and we suggest a syntax by which players' rationale for such switching may be specified and structurally composed.
In such a model one can ask a stability question: given a game arena
and a strategy specification, whether players eventually settle down to
strategies without further switching. We show that this problem can be
algorithmically solved using automata theoretic methods.

Keywords: Graphical games, strategy specifications, strategy switching.

1 Overview

Consider the game of cricket[1]. A bowler, starting on his run-up, considers: Should
I bowl on his off-side or leg-side? Should I bowl a short-pitch ball? Should I bowl
a slower one? Since he mis-hit the last bouncer I bowled to him, should I bowl
one again? The batsman, on his part, considers as he takes his stance: If he
bowls on my legs, should I pelt him for a boundary and reveal my strength off
that flank? Or should I play it safe and settle for a single? I have already hit two
boundaries in this over; if I hit him for too many runs, will he be taken off the
attack?

In an ideal world, both bowler and batsman would have perfect information
not only about each other's prowess but also about the nature of the pitch,
and would play optimal mixed strategies, since they could go through all the
reasoning above before a single ball is ever bowled. We could compute equilibria
and predict rational cricket play.

Not only is the actual game far from ideal, it is also more interesting. If we
are interested in predicting, in addition to outcomes, also how the play is likely

---

[1] Wikipedia-level understanding of cricket http://en.wikipedia.org/wiki/Cricket is
enough to understand the points being made here, though some knowledge of cricket
would surely help.

K. Ambos-Spies, B. Löwe, and W. Merkle (Eds.): CiE 2009, LNCS 5635, pp. 389–398, 2009.
© Springer-Verlag Berlin Heidelberg 2009

to progress (at some partial play), we need to correspondingly look not just at *which* strategies are available to players, but also how they *select* a strategy from among many. Such considerations naturally lead to partial strategies, and the notion of *switching* between (partial) strategies.

In such a view, a player enters the game arena with information on the game structure and on other players' skills, as well as an initial set of possible strategies to employ. As the play progresses, she makes observations and accordingly revises strategies, switches from one to another, perhaps even devises new strategies that she hadn't considered before. The dynamics of such interaction eventually leads to some strategies being eliminated, and some becoming stable.

Such considerations can be entirely eliminated by taking into account all possible futures while strategizing. However, such omniscient strategizing may be impossible, even in principle, for finitary agents (who have access only to finite resources). Dynamical system models of social interaction and negotiations have for long considered such switching behaviour ([7], [5]), and we suggest that such a consideration is relevant for computational models as well.

What questions can one study in such a model? Since the model describes dynamics, it is best suited to address questions that relate to eventual patterns in game evolution dictated by the dynamics in the model. For instance, since some strategies may simply get eliminated in the course of play, eventual game evolution may get restricted, and one can ask: "Does the play finally settle down to some subset of the entire arena?", "Can a player ensure certain objectives using a strategy which does not involve switching between a set of strategies?"

Another interesting question is, "Given a sub-arena of the game, is the strategy of a player *live* in that subset?" A strategy is live if for every history in the subset, the action it specifies is present in the set. Such questions are especially relevant in the context of bargaining and negotiations, as evidenced in many political contexts.

In this work, we look at algorithmic issues concerning the above questions. We give a simple but expressive syntax for specifying and composing strategies. We then show that in the case of bounded memory strategies, these questions can be algorithmically solved. At the heart of these questions lie the issue of *liveness* of a strategy which we formalise in the next section. However what is emphasized is the need and possibility of computational models that include process aspects of strategizing and application of algorithmic tools on them, rather than a study of the complexity of determining stability under switching.

Related Work

Dynamic learning has been extensively studied in game theory: for instance, Young ([15],[16]) considers a model in which each player chooses an optimal strategy based on a sample of information about what other players have done in the past. Similar analyses have been carried out in the context of cooperative game theory as well: here players decide dynamically which coalition to join. One asks how coalition structures change over time, and which coalition players will eventually arrive at ([2]). Evolutionary game theory ([14]) studies how players observe payoffs of other players in their neighbourhood and accordingly change strategies to maximise fitness.

Our work is located in the logical foundations of game theory, and hence employs logical descriptions of strategies and algorithms to answer questions. Modal logics have been used in various ways to reason about games. Notable among these is the work on alternating temporal logic (ATL) [1], a logic where assertions are made on outcomes a coalition of players can ensure. Various extensions of ATL ([10],[11]) has been proposed to incorporate knowledge of players and strategies explicitly into the logic. In [8,9] van Benthem uses dynamic logic to describe games as well as strategies. [4] presents a complete axiomatisation of a logic describing both games and strategies in a dynamic logic framework where assertions are made about atomic strategies. [6] studies a logic in which not only are games structured, but so also are strategies.

Somewhat different in approach, and yet closely related is the work of De Vos and Vermeir ([12],[13]) in which the authors present a framework for decision making with circumstance dependent preferences and decisions (OCLP). It allows decisions that comprise of multiple alternatives which become available only when a choice between them is forced.

Due to space restrictions, detailed proofs have been omitted[2].

2   Preliminaries

We are interested in looking at infinite duration games. We first introduce extensive form games which constitutes our game models.

2.1   Extensive Form Games

Let $N = \{1, \ldots, n\}$ be the set of players. For each $i \in N$, let A_i be a finite set of actions, which represent the moves of the players. We assume that the action sets of the players are mutually disjoint, i.e., $A_i \cap A_j = \emptyset$ for $i \neq j$. Let $A = A_1 \times \cdots \times A_n$ denote the set of action tuples and $\widetilde{A} = A_1 \cup \cdots \cup A_n$ denote the set of actions of all the players. For any action tuple $\bar{a} = (a_1, \ldots, a_n) \in A$, we write $a \in \bar{a}$ if $a = a_i$ for some $1 \leq i \leq n$.

An extensive form game is a tree $\mathcal{T} = (T, \Rightarrow, t_0)$ where $T \subseteq A^*$ is a prefix-closed set called the set of nodes or game positions. The initial game position or the root of $\mathcal{T}$ is $t_0 = \epsilon$ (the empty word) and the edge relation is $\Rightarrow \subseteq T \times T$. A play in the game is just a path in $\mathcal{T}$ starting at t_0. For technical convenience we assume that all plays are infinite, i.e. for all $t \in T$, $\exists t'$ such that $t \Rightarrow t'$.

Strictly speaking, a game consists of a game tree along with *winning conditions* for the players. As we shall see later the winning conditions in our case will be some properties of the game model which are fairly general. Assuming that the outcomes and payoffs of the game arise from a fixed finite set, they can be coded up using propositions in our logical framework on the lines of [3]. Our main focus in this exposition is the strategies of players rather than the winning conditions themselves.

[2] For a full version see http://www.imsc.res.in/~soumya/Files/Stability.pdf

2.2   Strategies

A strategy for player i tells her at each game position, which action to choose. Given the game tree $\mathcal{T} = (T, \Rightarrow, t_0)$, a strategy μ for a player i is a function $\mu : T \rightarrow A_i$.

For a history $\bar{a}_1 \ldots \bar{a}_k$ of the game, a strategy for player i after the history $\bar{a}_1 \ldots \bar{a}_k$ is a function $\mu[\bar{a}_1 \ldots \bar{a}_k] : \{\bar{a}_1 \ldots \bar{a}_k u \in \mathcal{T}\} \rightarrow A_i$ where $u \in A^*$. Thus $\mu[\epsilon]$ is a strategy for the entire game and we denote it by μ itself. The function $\mu[\bar{a}_1 \ldots \bar{a}_k]$ may be viewed as a subtree $\mathcal{T}^{\mu[\bar{a}_1 \ldots \bar{a}_k]} = (T', \Rightarrow', t'_0)$ of $\mathcal{T}$ with root t'_0 such that $t'_0 = \bar{a}_1 \ldots \bar{a}_k \in T'$ and

- For any node $t = \bar{a}_1 \ldots \bar{a}_l \in T'$, $(l \geq k)$ if $\mu[\bar{a}_1 \ldots \bar{a}_k](t) = a$ then the children of t in $\mathcal{T}^{\mu[\bar{a}_1 \ldots \bar{a}_k]}$ are exactly those nodes $t\bar{a} \in \mathcal{T}$ such that the ith component of $\bar{a}$, $\bar{a}(i)$ is equal to a.

We shall call such a subtree $\mathcal{T}^{\mu[\bar{a}_1 \ldots \bar{a}_k]}$, a strategy tree for the strategy $\mu[\bar{a}_1 \ldots \bar{a}_k]$. Note that the values of $\mu[\bar{a}_1 \ldots \bar{a}_k]$ at positions, $t \notin T'$ does not affect the outcome of a play conforming to $\mu[\bar{a}_1 \ldots \bar{a}_k]$. Hence, we can interpret the semantics of a strategy in terms of its strategy tree without any loss of generality. We shall also use the terms 'strategy' and 'strategy tree' interchangeably. Let $\Omega_i(t)$ denote the set of all strategies of player i after history t in $\mathcal{T}$ and let $\Omega_i = \cup_{t \in \mathcal{T}} \Omega_i(t)$. Note that the set of strategies is infinite for any game $\mathcal{T}$.

Composition of Strategies: Let $\mu_1, \mu_2 \in \Omega_i$. Suppose player i starts playing the game $\mathcal{T}$ with strategy μ_1 and after k rounds ($k \geq 0$), she decides to use the strategy μ_2 for the rest of the game. The resulting prescription is also a strategy μ (say) in the set of strategies of player i, that is, $\mu \in \Omega_i$. In a sense μ may be viewed as a composition of the strategies μ_1 and μ_2. We denote the strategy μ by $\mu_1^k \mu_2$.

The strategy tree $\mathcal{T}^{\mu_1^k \mu_2}$ for the strategy is obtained by taking $\mathcal{T}^{\mu_1}$ and removing all the nodes with height greater than or equal to $k+1$, resulting in a tree of height k, and pasting $\mathcal{T}^{\mu_2[\bar{a}_1 \ldots \bar{a}_k]}$ at each leaf node $\bar{a}_1 \ldots \bar{a}_k$ of this resulting tree.

2.3   Partial Strategies

Given $\mathcal{T} = (T, \Rightarrow, t_0)$, a history $\bar{a}_1 \ldots \bar{a}_k \in \mathcal{T}$, a partial strategy $\sigma[\bar{a}_1 \ldots \bar{a}_k]$ for player i after this history is a partial function

$$\sigma[\bar{a}_1 \ldots \bar{a}_k] : \{\bar{a}_1 \ldots \bar{a}_k u \in \mathcal{T}\} \rightharpoonup A_i$$

where $u \in A^*$, with the interpretation that if σ is not defined for some history $\bar{a}_1 \ldots \bar{a}_k u \in \mathcal{T}$, the player may play any available action there. The strategy $\sigma[\epsilon]$ is identified with the strategy σ for the entire game.

The strategy tree $\mathcal{T}^{\sigma[\bar{a}_1 \ldots \bar{a}_k]} = (T', \Rightarrow', t'_0)$ is again a subtree of $\mathcal{T}$ with root $t'_0 = \bar{a}_1 \ldots \bar{a}_k \in T'$ and for any node $t = \bar{a}_1 \ldots \bar{a}_l \in T'$ $(l \geq k)$, if $\sigma[\bar{a}_1 \ldots \bar{a}_k](t) = a$, then the children of t are exactly those nodes $t\bar{a} \in \mathcal{T}$ such that the ith component of $\bar{a}$, $\bar{a}(i)$ is equal to a. On the other hand if $\sigma[\bar{a}_1 \ldots \bar{a}_k]$ is undefined on t, then the children of t are $\{t\bar{a} \mid t\bar{a} \in \mathcal{T}\}$, i.e., all the nodes that are the children of the node t in the game tree $\mathcal{T}$ itself.

We let $\Sigma_i(t)$ denote the set of all partial strategies of player i after history t in $\mathcal{T}$ and let $\Sigma_i = \cup_{t \in \mathcal{T}} \Sigma_i(t)$ denote the set of all partial strategies of player i.

A partial strategy may be viewed as a set of total strategies. Given the strategy tree $\mathcal{T}_G^\sigma$ for a partial strategy σ for player i we obtain a set of trees $\widetilde{\mathcal{T}}_G^\sigma$ of total strategies as follows. $\mathcal{T} = (T, \Rightarrow, t_0) \in \widetilde{\mathcal{T}}_G^\sigma$ if and only if $t_0 = \epsilon$ and

- If $\bar{a}_1 \ldots \bar{a}_k \in T$ then $\bar{a}_1 \ldots \bar{a}_{k+1} \in T$ if and only if $\bar{a}_1 \ldots \bar{a}_{k+1} \in \mathcal{T}_G^\sigma$ and for all $\bar{a}_1 \ldots \bar{a}_{k+1}$, $\bar{a}_1 \ldots \bar{a}'_{k+1} \in T$, $\bar{a}_{k+1}(i) = \bar{a}'_{k+1}(i)$.

For any history $\bar{a}_1 \ldots \bar{a}_k$, the set $\widetilde{\mathcal{T}}_G^{\sigma[\bar{a}_1 \ldots \bar{a}_k]}$ of total strategy trees for the partial strategy $\sigma[\bar{a}_1 \ldots \bar{a}_k]$ of player i may be defined similarly.

It is convenient to define the maps $\mathcal{PT}_i$ and $\mathcal{TP}_i$ for all $i \in N$. $\mathcal{PT}_i : \Sigma_i \rightarrow 2^{\Omega_i}$, such that $\mathcal{PT}_i(\mathcal{T}_G^{\sigma[\bar{a}_1 \ldots \bar{a}_k]}) = \widetilde{\mathcal{T}}_G^{\sigma[\bar{a}_1 \ldots \bar{a}_k]}$. And $\mathcal{TP}_i : 2^{\Omega_i} \rightarrow \Sigma_i$, such that given a set $\widetilde{\mathcal{T}}_G^{\mu[\bar{a}_1 \ldots \bar{a}_k]}$ of total strategy trees of player i, $\mathcal{TP}_i(\widetilde{\mathcal{T}}_G^{\mu[\bar{a}_1 \ldots \bar{a}_k]})$ is the partial strategy tree $(T, \Rightarrow, t_0)$ such that $t_0 = \bar{a}_1 \ldots \bar{a}_k$ and

- $t \in T$ if and only if $t \in T$ for some $\mathcal{T} \in \widetilde{\mathcal{T}}_G^{\mu[\bar{a}_1 \ldots \bar{a}_k]}$
- $\Rightarrow = \bigcup_{\mathcal{T} \in \widetilde{\mathcal{T}}_G^{\mu[\bar{a}_1 \ldots \bar{a}_k]}} \{\Rightarrow \in \mathcal{T}\}$.

2.4 Relevant Questions

Given the above notion of partial strategies, it makes sense to talk about what it means to compose several (usually simple) strategies to obtain another (more complex) strategy. A player will start out with a set (possibly finite) of *elementary* or *atomic* strategies, and as the game progresses, combine them to obtain new strategies. Switching from one strategy to another is based on certain observable properties of the game. Strategies thus generated may not be present in her initial set of strategies.

Given a region of the game arena, to check whether a player's strategy eventually becomes stable with respect to switching, we need to be able to first check whether a strategy is *live* in the region. For a subtree $\mathcal{T}'$ of $\mathcal{T}$ and a partial strategy σ_i of player i, we say σ_i is *live* in $\mathcal{T}'$ if $\forall t \in \mathcal{T}'$ the following condition holds:

- if $\sigma_i(t)$ is defined and $\sigma_i(t) = a$ then $\exists t' = t\bar{a} \in T$ such that $\bar{a}(i) = a$.

Given a game $\mathcal{T}$, natural questions of interest include:

- Given a subtree $\mathcal{T}'$ of $\mathcal{T}$ and a partial strategy σ_i, is σ_i live in $\mathcal{T}'$?
- Is it the case that a given strategy σ_i eventually becomes not live ?
- Find the set of all partial strategies which are live in a substructure.

Note that here we assume that every strategy is equally viable and switching between strategies does not involve any overhead. A model where different strategies have different *costs* would be interesting to study in its own right.

To solve these questions algorithmically and to subsequently address the stability issue, we need to present partial strategies and game trees in a finite manner. Below we show how this can be achieved.

3   Strategy Specifications

We present a syntax to specify partial strategies and their composition in a structural manner. We crucially use a construct which allows players to play the game with a strategy σ_1 up to some point and then switch to a strategy σ_2.

Syntax: The strategy set Π_i of player i is obtained by combining her atomic strategies as follows:

$$\Pi_i ::= \sigma \in \Sigma_i \mid \pi_1 \cup \pi_2 \mid \pi_1 \cap \pi_2 \mid \pi_1{}^\frown\pi_2 \mid (\pi_1 + \pi_2) \mid \psi?\pi$$

Using the *test operator* $\psi?\pi$, a player checks whether an observable condition ψ holds and then decides on a strategy. We think of these conditions as past time formulas of a simple tense logic over an atomic set of observables.

In the atomic case, σ simply denotes a partial strategy. The intuitive meaning of the operators are given as:

- $\pi_1 \cup \pi_2$ means that the player plays according to the strategy π_1 or the strategy π_2.
- $\pi_1 \cap \pi_2$ means that if at a history $t \in \mathcal{T}$, π_1 is defined then the player plays according to π_1; else if π_2 is defined at t then the player plays according to π_2. If both π_1 and π_2 are defined at t then the moves that π_1 and π_2 specify at t must be the same (we call such a pair π_1 and π_2, compatible). Henceforth, we shall use the $\cap$ operator only for compatible pairs of strategies.
- $\pi_1{}^\frown\pi_2$ means that the player plays according to the strategy π_1 and then after some history, switches to playing according to π_2. The position at which she makes the switch is not fixed in advance.
- $(\pi_1 + \pi_2)$ says that at every point, the player can choose to follow either π_1 or π_2.
- $\psi?\pi$ says at every history, the player tests if the property ψ holds of that history. If it does then she plays according to π.

Example: In the cricket example, let the bowler's set of atomic strategies be given as $\Sigma_{bowler} = \{\sigma_{short}, \sigma_{good}, \sigma_{outside-off}, \sigma_{legs}\}$ which corresponds to bowling a short-pitch, good length, off-side and leg-side ball respectively.

Let $p_{(short,sixer)}$ be the observable which says that the outcome of a short ball is a sixer. Then the following specification says that the bowler keeps bowling short balls till he is hit for a sixer after which he changes to good-length deliveries.

- $\neg \diamondsuit(p_{(short,sixer)}?(\sigma_{short})) \cup \diamondsuit(p_{(short,sixer)}?(\sigma_{good}))$

The specification $\sigma_{short}{}^\frown\sigma_{good}{}^\frown\sigma_{legs}$ for the bowler says that he starts by bowling short-pitch balls and after some point he switches to bowling at the batsman's legs and again switches to bowling good-length balls.

Semantics: Formally, given the game tree $\mathcal{T} = (T, \Rightarrow, t_0)$, the semantics of a strategy specification $\pi \in \Pi_i$ is a function $[\![\cdot]\!]_\mathcal{T} : \Pi_i \times T \to 2^{\Omega_i}$. That is, each specification at a node t of the game tree is associated with a set of total strategy trees after history t.

For any $t = \bar{a}_1 \ldots \bar{a}_k \in T$, $[\![\cdot]\!]_\mathcal{T}$ is defined inductively as follows:

- $[\![\sigma, (\bar{a}_1 \ldots \bar{a}_k)]\!]_\mathcal{T} = \mathcal{PT}_i(\mathcal{T}^{\sigma[\bar{a}_1 \ldots \bar{a}_k]})$.
- $[\![\pi_1 \cup \pi_2, (\bar{a}_1 \ldots \bar{a}_k)]\!]_\mathcal{T} = [\![\pi_1, (\bar{a}_1 \ldots \bar{a}_k)]\!]_\mathcal{T} \cup [\![\pi_2, (\bar{a}_1 \ldots \bar{a}_k)]\!]_\mathcal{T}$.
- $[\![\pi_1 \cap \pi_2, (\bar{a}_1 \ldots \bar{a}_k)]\!]_\mathcal{T} = [\![\pi_1, (\bar{a}_1 \ldots \bar{a}_k)]\!]_\mathcal{T} \cap [\![\pi_2, (\bar{a}_1 \ldots \bar{a}_k)]\!]_\mathcal{T}$.
- $[\![\pi_1 {}^\frown \pi_2, (\bar{a}_1 \ldots \bar{a}_k)]\!]_\mathcal{T} = \bigcup_{l \geq k} [\![[\![\pi_1, (\bar{a}_1 \ldots \bar{a}_k)]\!]_\mathcal{T} {}^\frown (\pi_2, l)]\!]_\mathcal{T}$

  where $[\![[\![\pi_1, (\bar{a}_1 \ldots \bar{a}_k)]\!]_\mathcal{T} {}^\frown (\pi_2, l)]\!]_\mathcal{T}$ is defined as follows: For every tree $\mathcal{T} \in [\![\pi_1, (\bar{a}_1 \ldots \bar{a}_k)]\!]_\mathcal{T}$, prune the tree $\mathcal{T}$ at depth l and call it $\mathcal{T}_l$. Then $[\![[\![\pi_1, (\bar{a}_1 \ldots \bar{a}_k)]\!]_\mathcal{T} {}^\frown (\pi_2, l)]\!]_\mathcal{T}$ is the set of trees got by appending to every leaf node $\bar{a}_1 \ldots \bar{a}_l$ of such trees $\mathcal{T}_l$, the trees in $[\![\pi_2, (\bar{a}_1 \ldots \bar{a}_l)]\!]_\mathcal{T}$.

- $[\![(\pi_1 + \pi_2), (\bar{a}_1 \ldots \bar{a}_k)]\!]_\mathcal{T} = \bigcup_{k_1, k_2, \ldots} [\![[\![\pi_1, (\bar{a}_1 \ldots \bar{a}_k)]\!]_\mathcal{T} {}^\frown (\pi_2, k_1)]\!]_\mathcal{T} {}^\frown (\pi_1, k_2)]\!]_\mathcal{T} \cdots$

  where $k \leq k_1 \leq k_2 \ldots$.

- $[\![\psi?\pi, (\bar{a}_1 \ldots \bar{a}_k)]\!]_\mathcal{T}$: $[\![\psi?\pi, (\bar{a}_1 \ldots \bar{a}_k)]\!]_\mathcal{T}$ is obtained from $[\![\pi, (\bar{a}_1 \ldots \bar{a}_k)]\!]_\mathcal{T}$ and $\mathcal{T}$ as follows. Let $\mathcal{TP}_i([\![\pi, (\bar{a}_1 \ldots \bar{a}_k)]\!]_\mathcal{T}) = \mathcal{T}^{\pi[\bar{a}_1 \ldots \bar{a}_k]}$ be the partial strategy tree of $\pi[\bar{a}_1 \ldots \bar{a}_k]$. Then $[\![\psi?\pi, (\bar{a}_1 \ldots \bar{a}_k)]\!]_\mathcal{T}$ is a set of trees such that the following holds. $\mathcal{T} \in [\![\psi?\pi, (\bar{a}_1 \ldots \bar{a}_k)]\!]_\mathcal{T}$ if and only if:
  - $\bar{a}_1 \ldots \bar{a}_k \in \mathcal{T}$.
  - If $\bar{a}_1 \ldots \bar{a}_l \in \mathcal{T}$ and ψ holds at $\bar{a}_1 \ldots \bar{a}_l$ then $\bar{a}_1 \ldots \bar{a}_{l+1} \in \mathcal{T}$ if and only if $\bar{a}_1 \ldots \bar{a}_{l+1} \in \mathcal{T}^{\pi[\bar{a}_1 \ldots \bar{a}_k]}$ and for all $\bar{a}_1 \ldots \bar{a}_{l+1}, \bar{a}_1 \ldots \bar{a}'_{l+1} \in \mathcal{T}, \bar{a}_{l+1}(i) = \bar{a}'_{l+1}(i)$. If $\bar{a}_1 \ldots \bar{a}_l \in \mathcal{T}$ and ψ does not hold at $\bar{a}_1 \ldots \bar{a}_l$ then $\bar{a}_1 \ldots \bar{a}_{l+1} \in \mathcal{T}$ if and only if $\bar{a}_1 \ldots \bar{a}_{l+1} \in \mathcal{T}$ and for all $\bar{a}_1 \ldots \bar{a}_{l+1}, \bar{a}_1 \ldots \bar{a}'_{l+1} \in \mathcal{T}, \bar{a}_{l+1}(i) = \bar{a}'_{l+1}(i)$.

4 Finite Presentation of Games and Strategies

For algorithmic analysis, we need to present the infinite game in a finite fashion. In this paper, we assume that the game is presented as a finite graph. The extensive form game is just the unfolding of this graph.

Game Arena: The game arena is a finite graph $G = (W, \to, w_0)$ where W is a finite set of game positions, $w_0 \in W$ is the initial position and $\to : W \times A \to W$, is the set of edges. For $w \in W$, let $w_\to = \{\bar{a} \mid w \xrightarrow{\bar{a}} w' \text{ for some } w' \in W\}$. For technical convenience, we assume that for all $w \in W$, $w_\to \neq \emptyset$. The infinite extensive form game tree corresponding to G is obtained by the *tree-unfolding* of G.

For a word on notation, given an arena G and a strategy specification π, we denote the function $[\![\cdot]\!]_{\mathcal{T}_G}$ by just $[\![\cdot]\!]_G$.

Finite State Transducers and Bounded Memory Strategies: A finite state transducer (FST) over the input alphabet A and output alphabet A_i is a tuple $\mathcal{A} = (Q, \to, I, f)$ where Q is a finite set of states, $I \subseteq Q$ is the set of initial states, $\rightarrow: Q \times A \rightarrow 2^Q$ is the transition function and $f : Q \rightarrow A_i$ is the output function.

The semantics of strategy specifications is presented with respect to the set of all strategies. For algorithmic concerns we restrict our attention to bounded memory strategies. As we will see later, strategy specifications can only enforce bounded memory strategies.

A strategy σ of player i is said to be **bounded memory** if there exists an FST $\mathcal{A} = (Q, \rightarrow, I, f)$ where the set of states Q is the *memory* of σ, I is the initial memory, $\rightarrow$ is the *memory update function* and f is the *action output function* such that the following is true. When $\bar{a}_1 \ldots \bar{a}_{k-1}$ is a play and the sequence $q_0, q_1, \ldots, q_k$ is determined by $q_0 \in I$ and $q_i \xrightarrow{\bar{a}_i} q_{i+1}$ then $\sigma(\bar{a}_1 \ldots \bar{a}_{k-1}) = f(q_k)$. The intuition is that the FST faithfully reflects the outputs of the strategy σ.

Given a strategy μ of player i, a **run** of an FST $\mathcal{A}$ on $\mathcal{T}_G^\mu$ is a Q labelled tree $(T, \Rightarrow, t_0, \chi)$. The labelling function $\chi : T \rightarrow Q$ is defined as: $\chi(t_0) = q_0 \in I$ and if $\bar{a}_1 \ldots \bar{a}_k \Rightarrow \bar{a}_1 \ldots \bar{a}_{k+1}$ then $\chi(\bar{a}_1 \ldots \bar{a}_k) \in \rightarrow (\chi(\bar{a}_1 \ldots \bar{a}_k), \bar{a}_{k+1})$.

We say that μ is accepted by $\mathcal{A}$ if there is a run χ of $\mathcal{A}$ on $\mathcal{T}_G^\mu$ satisfied the condition: $\forall t = \bar{a}_1 \ldots \bar{a}_k \in \mathcal{T}_G^\mu, \bar{a}(i) = f(\chi(t))$. The language of $\mathcal{A}$, $\mathcal{L}(\mathcal{A}) = \{\mu \mid \mu$ is accepted by $\mathcal{A}\}$.

The following lemma relates strategy specifications to finite state transducers.

Lemma 4.1. *Given game arena G, a player $i \in N$ and a strategy specification $\pi \in \Pi_i$, where all the atomic strategies mentioned in π are bounded memory, we can construct an FST $\mathcal{A}_\pi$ such that for all $\mu \in \Omega_i$ we have $\mu \in \llbracket \pi \rrbracket_G$ iff $\mu \in \mathcal{L}(\mathcal{A}_\pi)$.*

5   Stability

Call a strategy π **switch-free** if it does not have any of the $\frown$ or the $+$ construct. Given a strategy $\pi \in \Pi_i$ of player i, the set of substrategies of π, S_π are just the subformulae of π. Let $SF(S_\pi)$ be the set of switch-free strategies of S_π. Note that $SF(S_\pi)$ is a finite set for a given π.

Given a game arena G and strategy specifications of the players, we may ask whether there exists some subarena of G that the game settles down to if the players play according to their strategy specifications. This subarena is in some sense the equilibrium states of the game. It is also meaningful to ask if the game settles down to such an equilibrium subarena, then whether the strategy of a particular player attains stability with respect to switching.

Let $G = (W, \rightarrow, w_0)$ be the game arena, $\pi \in \Pi_i$ and $\mathcal{A}_\pi = (Q, \rightarrow, I, f, \lambda)$ be the FST for π. We define the restriction of G with respect to $\mathcal{A}_\pi$ as $G \restriction \mathcal{A}_\pi = (W', \rightarrow', w_0')$ where $W' = W \times Q$, $w_0' = \{w_0\} \times I$ and $(w_1, q_1) \xrightarrow{\bar{a}}' (w_2, q_2)$ iff $w_1 \xrightarrow{\bar{a}} w_2$, $q \xrightarrow{\bar{a}} q_2$ and $f(q_1) = \bar{a}(i)$.

Theorem 5.1. *Given a game arena $G = (W, \rightarrow, w_0)$ with a valuation of the observables on W, a subarena R of G and strategy specifications $\pi_1, \ldots, \pi_n$ for players 1 to n, the question, "Do all plays conforming to these specifications eventually settle down to R?" is decidable.*

Proof. Construct the graph $G_\pi = (\cdots((G{\upharpoonright}\mathcal{A}_{\pi_1}){\upharpoonright}\mathcal{A}_{\pi_2}\cdots){\upharpoonright}\mathcal{A}_{\pi_n}) = (W_\pi, \rightarrow_\pi, w_\pi)$. Let $F \subseteq G_\pi = (W', \rightarrow')$ such that $W' = \{(w, q_1, \ldots, q_n) \mid w \in R, q_1 \in Q_{\pi_1}, \ldots, q_n \in Q_{\pi_n}\}$ where $Q_{\pi_1}, \ldots, Q_{\pi_n}$ are the state sets of the FST's $\mathcal{A}_{\pi_1}, \ldots, \mathcal{A}_{\pi_n}$ respectively. Let $\rightarrow' = \rightarrow_\pi \cap(W' \times W')$.

1. Check if F is a maximal connected component in G_π. If so proceed to step 2, else output a 'NO'.
2. Check if all paths strating at all initial nodes $w' \in w_\pi$ reach F and output a 'YES'. Otherwise, output a 'NO'.

Theorem 5.2. *Given a game arena $G = (W, \rightarrow, w_0)$ with a valuation of the observables on W, a subarena R of G and strategy specifications $\pi_1, \ldots, \pi_n$ for players 1 to n, the question, "If all plays conforming to these specifications converge to R, does the strategy of player i become eventually stable with respect to switching?" is decidable in time $\mathcal{O}(m^m \cdot p \cdot 2^{np})$ where m is the size of the arena G and p is the maximum length of a specification formula π_1 or $\ldots$ or π_n.*

Proof. We first check if all the plays settle down to R. But in doing so we also have to keep track of all the strategy-switches ($\frown$'s) of player i along these plays. Because given a subformula of the form $\pi \frown \pi'$, once the player has switched to strategy π' she cannot play π later. We do this by an inductive procedure by first indexing all the subformulae of the specification π_i of player i and then augmenting the FST's with an output so that at each point they output the indices of only those subformulae that are still relevant at that point. We also keep track of the states each of the FST's are in when the plays reach R. Having done so, we check, by constructing FST's for each relevant switch free substrategy of π_i, whether the play stays inside R if player i plays according to that substrategy, given that all the FST's start at the states in which they were on reaching R.

6 Discussion

The framework presented here is intended only as an initial step of a research programme that studies computational models of social interaction. It is to be noted that the presentation of game arenas as graphs may be inappropriate for many contexts, and it may be more natural to define games by rules.

Moreover, the assumption of a fixed finite set of players may also be unrealistic for models of social dynamics. The notion of strategy switching is demonstrated quite naturally in a framework which consists of a population of players and a neighbourhood model. In such a set up, the players are parts of different neighbourhoods and can observe the outcomes within their neighbourhoods. Such a structure besides giving a rationale to the players for playing certain strategies and switching between them, would also model various game-theoretic and social scenarios more concretely. We hope that the study of formal models of dynamics in interaction will lead to new questions for games and computations.

References

1. Alur, R., Henzinger, T.A., Kupferman, O.: Alternating-time temporal logic. Journal of ACM 49(5), 672–713 (2002)
2. Arnold, T., Schwalbe, U.: Dynamic coalition formation and the core. Journal of Economic Behavior and Organization 49, 363–380 (2002)
3. Bonanno, G.: Modal logic and game theory: Two alternative approaches. Risk Decision and Policy 7, 309–324 (2002)
4. Ghosh, S.: Strategies made explicit in dynamic game logic. Logic and the Foundations of Game and Decision Theory (2008)
5. Horst, U.: Dynamic systems of social interactions. In: NSF/CEME Mathematical Economics Conference at Berkeley (2005)
6. Ramanujam, R., Simon, S.: Dynamic logic on games with structured strategies. In: Proceedings of the Conference on Principles of Knowledge Representation and Reasoning, pp. 49–58 (2008)
7. Skyrms, B., Pemantle, R.: A dynamic model of social network formation. Proceedings of the National Academy of Sciences 97(16), 9340–9346 (2000)
8. van Benthem, J.: Games in dynamic epistemic logic. Bulletin of Economic Research 53(4), 219–248 (2001)
9. van Benthem, J.: Extensive games as process models. Journal of Logic Language and Information 11, 289–313 (2002)
10. van der Hoek, W., Jamroga, W., Wooldridge, M.: A logic for strategic reasoning. In: Proceedings of AAMAS, pp. 157–164 (2005)
11. van der Hoek, W., Wooldridge, M.: Cooperation, knowledge, and time: Alternating-time temporal epistemic logic and its applications. Studia Logica 75(1), 125–157 (2003)
12. Vos, M.D., Vermeir, D.: A logic for modeling decision making with dynamic preferences. In: Brewka, G., Moniz Pereira, L., Ojeda-Aciego, M., de Guzmán, I.P. (eds.) JELIA 2000. LNCS (LNAI), vol. 1919, pp. 391–406. Springer, Heidelberg (2000)
13. Vos, M.D., Vermeir, D.: Dynamic decision-making in logic programming and game theory. In: McKay, B., Slaney, J.K. (eds.) Canadian AI 2002. LNCS, vol. 2557, pp. 36–47. Springer, Heidelberg (2002)
14. Weibull, J.W.: Evolutionary Game Theory. MIT Press, Cambridge (1997)
15. Young, H.P.: The evolution of conventions. In: Econometrica, vol. 61, pp. 57–84. Blackwell Publishing, Malden (1993)
16. Young, H.P.: The diffusion of innovations in social networks. Economics Working Paper Archive 437, The Johns Hopkins University, Department of Economics (May 2000)

Computational Heuristics for Simplifying a Biological Model

Ion Petre, Andrzej Mizera, and Ralph-Johan Back

Department of Information Technologies, Åbo Akademi University
Computational Biomodeling Laboratory, Turku Centre for Computer Science
FIN-20520 Turku, Finland
{ipetre,amizera,backrj}@abo.fi

Abstract. Computational biomodelers adopt either of the following approaches: build rich, as complete as possible models in an effort to obtain very realistic models, or on the contrary, build as simple as possible models focusing only on the core aspects of the process, in an effort to obtain a model that is easier to analyze, fit, and validate. When the latter strategy is adopted, the aspects that are left outside the models are very often up to the subjective options of the modeler. We discuss in this paper a heuristic method to simplify an already fit model in such a way that the numerical fit to the experimental data is not lost. We focus in particular on eliminating some of the variables of the model and the reactions they take part in, while also modifying some of the remaining reactions. We illustrate the method on a computational model for the eukaryotic heat shock response. We also discuss the limitations of this method.

Keywords: Model reduction, heat shock response, mathematical model.

1 Introduction

When designing a new molecular model for some biological process or network, the choice one has to make early on in the modeling process is whether to strive for a rich model, capturing many details, or on the contrary, to focus on a more abstract model, capturing only a few, main actors of interest. The choice is not obvious and depends heavily on the goals of the modeling project. On one hand, a rich model has the potential of being more realistic but it leads to a more complex mathematical model that may be difficult to fit to experimental data, to analyze, and ultimately may be less apt to answer to biological queries. On the other hand, a less finely grained molecular model leads to a smaller mathematical model (in terms of the number of variables and equations) that may be easier to work with, but it pays a price in ignoring a number of details. A main difficulty in choosing between a rich and a simplified molecular model is that the potential cost of starting off with a rich model only becomes transparent at a latter stage, in the process of analyzing the corresponding mathematical model. Moreover, in the case of choosing a simplified model, the selection of the aspects to be ignored in the model is left up to the subjective choice of the modeler. We

K. Ambos-Spies, B. Löwe, and W. Merkle (Eds.): CiE 2009, LNCS 5635, pp. 399–408, 2009.
© Springer-Verlag Berlin Heidelberg 2009

discuss in this paper an intermediate approach where we start with a (potentially large, rich) model that has already been fit and validated against experimental data and we aim to simplify it in such a way that its numerical behavior remains largely unchanged. In this way, the simplified model is the result of a systematic, numerical analysis of the larger model that preserves its validation. We illustrate the approach on a computational model for the eukaryotic heat shock response and discuss the biological relevance of the simplifications we operate on the model. We also discuss the strong dependency of this approach on the numerical setup of the model; we show that our approach in the case of the heat shock response model is robust to some changes in the numerical values of the parameters, but it is sensitive to others.

2   The Heat Shock Response Model

The heat shock response is a well-conserved defence mechanism across all eukaryotic cells that enables them to survive under conditions of elevated temperatures. When exposed to heat shock, proteins inside cells tend to misfold. In turn, as an effect of their hydrophobic core being exposed, misfolded proteins form bigger and bigger aggregates with disastrous consequences for the cell, see [1]. In order to survive, the cell has to immediately react by increasing the level of chaperons (proteins that assist other proteins in the process of folding or refolding). Once the heat shock is removed, the defence mechanism is turned off and the cell eventually re-establishes the original level of chaperons, see [7,11,17].

The heat shock response has been intensively investigated in recent years for at least three main reasons. First, as a well-conserved mechanism in all eukaryotes, it is considered a promising candidate for investigating the engineering principles of gene regulatory networks, see [3,4,8,18]. Second, heat shock proteins (hsp) act as main components in a large number of cellular processes such as signaling, regulation and inflammation, see [6,16]. Moreover, their contribution to the resilience of cancer cells makes them an attractive target for cancer treatment, see [2,9,10,19].

We consider in this paper the molecular model proposed in [14] for the eukaryotic heat shock response. This model consists of only the minimum number of components that any regulatory network must contain: an activation mechanism and a feedback mechanism. Moreover, the model consists of only well-documented reactions, without using any hypothetical, unknown cellular mechanism. The control over the cellular defence mechanism against protein misfolding is implemented through the regulation of the transactivation of the hsp-encoding gene. The transcription of the gene is activated by heat shock factors (hsf) which trimerize (the trimerization includes a transient dimerization phase) and in this form bind to the heat shock element (hse), which is the promoter of the hsp-encoding gene. Once the hsf trimer is bound to the specific DNA sequence, the gene is transactivated and the transcription and translation take place. As a result, new hsp molecules are eventually synthesized. When the level of hsp is high enough, the synthesis is switched off by the following

Table 1. The reactions of the heat shock response model of [14]

(i) $2\,\mathsf{hsf} \leftrightarrows \mathsf{hsf}_2$	(x) $\mathsf{prot} \to \mathsf{mfp}$
(ii) $\mathsf{hsf} + \mathsf{hsf}_2 \leftrightarrows \mathsf{hsf}_3$	(xi) $\mathsf{hsp} + \mathsf{mfp} \leftrightarrows \mathsf{hsp}\!:\!\mathsf{mfp}$
(iii) $\mathsf{hsf}_3 + \mathsf{hse} \leftrightarrows \mathsf{hsf}_3\!:\!\mathsf{hse}$	(xii) $\mathsf{hsp}\!:\!\mathsf{mfp} \to \mathsf{hsp} + \mathsf{prot}$
(iv) $\mathsf{hsf}_3\!:\!\mathsf{hse} \to \mathsf{hsf}_3\!:\!\mathsf{hse} + \mathsf{mhsp}$	(xiii) $\mathsf{hsf} \to \mathsf{mhsf}$
(v) $\mathsf{hsp} + \mathsf{hsf} \leftrightarrows \mathsf{hsp}\!:\!\mathsf{hsf}$	(xiv) $\mathsf{hsp} \to \mathsf{mhsp}$
(vi) $\mathsf{hsp} + \mathsf{hsf}_2 \to \mathsf{hsp}\!:\!\mathsf{hsf} + \mathsf{hsf}$	(xv) $\mathsf{hsp} + \mathsf{mhsf} \leftrightarrows \mathsf{hsp}\!:\!\mathsf{mhsf}$
(vii) $\mathsf{hsp} + \mathsf{hsf}_3 \to \mathsf{hsp}\!:\!\mathsf{hsf} + 2\,\mathsf{hsf}$	(xvi) $\mathsf{hsp}\!:\!\mathsf{mhsf} \to \mathsf{hsp} + \mathsf{hsf}$
(viii) $\mathsf{hsp} + \mathsf{hsf}_3\!:\!\mathsf{hse} \to \mathsf{hsp}\!:\!\mathsf{hsf} + 2\,\mathsf{hsf} + \mathsf{hse}$	(xvii) $\mathsf{hsp} + \mathsf{mhsp} \leftrightarrows \mathsf{hsp}\!:\!\mathsf{mhsp}$
(ix) $\mathsf{hsp} \to \emptyset$	(xviii) $\mathsf{hsp}\!:\!\mathsf{mhsp} \to 2\,\mathsf{hsp}$

mechanism: hsp bind to free hsf as well as break the hsf trimers (both free and those bound to DNA). This turns off DNA transcription and blocks the forming of new hsf trimers. The whole defense mechanism is turned on again when, as a result of raised temperature, the proteins (prot) in the cell begin misfolding again. To counteract, the heat shock proteins become involved in refolding them and they free the hsf, which in turn trimerize and activate the synthesis of hsp, etc. What drives the heat shock response is the race to keep under control the level of misfolded proteins, in such a way that they are not able to accumulate, form aggregates, and eventually lead to cell death. The model consists of the molecular reactions in Table 1.

When designing this molecular model, several criteria were followed, see [14], including that only well-documented reactions should be included and that the model should explicitly consider the temperature-induced protein misfolding as the trigger of the response. The model was also designed in such a way that is consistent with itself and with the kinetic principles of biochemistry. E.g., although hsf dimers are not experimentally detectable, they should be included in the model to account as a transient step in the formation of hsf trimers. Also, since hsp and hsf are themselves proteins, they should be subject to temperature-induced misfolding just like the regular proteins prot. Moreover, the refolding of mhsf and mhsp is controlled by the same kinetic constants as the refolding of mfp. The proper folding of newly synthesized hsp is assisted by chaperons as in the case of most proteins, see [1]. The degradation of hsf, prot, and mfp was on the other hand not included in the model so that intricate compensating mechanisms of protein synthesis could be ignored, see [14].

The mathematical model associated with the molecular model in Table 1 is in terms of ordinary differential equations and it is obtained by assuming for all reactions the law of mass-action. The reasons for this choice is so that the explicit contribution of each reaction to the overall behavior could be followed. Let us denote the reactants occurring in the model according to the convention in Table 2(a). We use $\kappa \in \mathbb{R}_+^{25}$ to denote the vector with all reaction rate constants as its components, see Table 2(b): $\kappa = (k_1^+, k_1^-, k_2^+, k_2^-, k_3^+, k_3^-, k_4, k_5^+, k_5^-, k_6, k_7, k_8, k_9, \phi(T), k_{11}^+, k_{11}^-, k_{12}, \phi(T), \phi(T), k_{11}^+, k_{11}^-, k_{12}, k_{11}^+, k_{11}^-, k_{12})$.

The corresponding mathematical model consists of the following differential equations:

$$dX_1/dt = -k_2^+\, X_1\, X_2 + k_2^-\, X_3 - k_5^+\, X_1\, X_7 + k_5^-\, X_9 + 2\, k_8\, X_4\, X_7 + k_6\, X_2\, X_7$$
$$\qquad -\varphi(T)\, X_1 + k_{14}\, X_{10} + 2\, k_7\, X_3\, X_7 - 2\, k_1^+\, X_1^2 + 2\, k_1^-\, X_2 \tag{1}$$
$$dX_2/dt = -k_2^+\, X_1\, X_2 + k_2^+\, X_3 - k_6\, X_2\, X_7 + k_1^+\, X_1^2 - k_1^-\, X_2 \tag{2}$$
$$dX_3/dt = -k_3^+\, X_3\, X_6 + k_2^+\, X_1\, X_2 - k_2^-\, X_3 + k_3^-\, X_4 - k_7\, X_3\, X_7 \tag{3}$$
$$dX_4/dt = k_3^+\, X_3\, X_6 - k_3^-\, X_4 - k_8\, X_4\, X_7 \tag{4}$$
$$dX_5/dt = \varphi(T)\, X_1 - k_{13}^+\, X_5\, X_7 + k_{13}^-\, X_{10} \tag{5}$$
$$dX_6/dt = -k_3^+\, X_3\, X_6 + k_3^-\, X_4 + k_8\, X_4\, X_7 \tag{6}$$
$$dX_7/dt = -k_5^+\, X_1\, X_7 + k_5^-\, X_9 - k_{11}^+\, X_7\, X_{14} + k_{11}^-\, X_{12} - k_8\, X_4\, X_7 - k_6\, X_2\, X_7$$
$$\qquad -k_{13}^+\, X_5\, X_7 + (k_{13}^- + k_{14})\, X_{10} - (\varphi(T) + k_9)\, X_7 - k_{15}^+\, X_7\, X_8$$
$$\qquad -k_7\, X_3\, X_7 + (k_{15}^- + 2\, k_{16})\, X_{11} + k_{12}\, X_{12} \tag{7}$$
$$dX_8/dt = k_4\, X_4 + \varphi(T)\, X_7 - k_{15}^+\, X_7\, X_8 + k_{15}^-\, X_{11} \tag{8}$$
$$dX_9/dt = k_5^+\, X_1\, X_7 - k_5^-\, X_9 + k_8\, X_4\, X_7 + k_6\, X_2\, X_7 + k_7\, X_3\, X_7 \tag{9}$$
$$dX_{10}/dt = k_{13}^+\, X_5\, X_7 - (k_{13}^- + k_{14})\, X_{10} \tag{10}$$
$$dX_{11}/dt = k_{15}^+\, X_7\, X_8 - (k_{15}^- + k_{16})\, X_{11} \tag{11}$$
$$dX_{12}/dt = k_{11}^+\, X_7\, X_{14} - (k_{11}^- + k_{12})\, X_{12} \tag{12}$$
$$dX_{13}/dt = k_{12}\, X_{12} - \varphi(T)\, X_{13} \tag{13}$$
$$dX_{14}/dt = -k_{11}^+\, X_7\, X_{14} + k_{11}^-\, X_{12} + \varphi(T)\, X_{13} \tag{14}$$

The rate coefficient of protein misfolding $\varphi(T)$ with respect to temperature T has been investigated experimentally in [12,13], and a mathematical expression describing the relation has been proposed in [11]. After adapting this formula in [11] to the time unit of our mathematical model (second), we obtain the following misfolding rate coefficient:

$$\varphi(T) = \left(1 - \frac{0.4}{e^{T-37}}\right) \cdot 1.4^{T-37} \cdot 1.45 \cdot 10^{-5}\ s^{-1}, \tag{15}$$

where T is the numerical value of the temperature of the environment in Celsius degrees. The formula is valid for $37 \leq T \leq 45$.

For the numerical fit of the model, data of [7] on DNA binding at $42°C$ was used to relate it to hsf$_3$: hse. Moreover, the initial values of the model were sought so that they give a steady state of the model at $37°C$. This latter restriction was imposed since the heat shock response is absent at $37°C$. Once suitable numerical values for the parameters were found, the model was subjected to a number of other validation tests. For a detailed discussion on the fit and the validation of the model we refer to [14] and [15]. The final numerical setup of the model is shown in Tables 2(a) and 2(b).

3   Simplifying the Model

We discuss in this section a series of numerical observations leading to several simplifications we can operate on our model, without changing its numerical

Table 2. (a) The list of variables in the mathematical model, their initial concentration values and their concentration values in one of the steady states of the system, for $T = 42$. Note that the initial state of the model is a steady state for $T = 37$. All concentrations are in $\frac{\#}{\text{cell}}$, where $\#$ denotes the number of molecules. The values should be interpreted as an average of a population of cells. [14,15]; (b) The numerical values of parameters for the fitted model [14,15].

Metabolite	Variable	Initial conc.
hsf	X_1	0.67
hsf$_2$	X_2	$8.73 \cdot 10^{-4}$
hsf$_3$	X_3	$1.22 \cdot 10^{-4}$
hsf$_3$: hse	X_4	3
hse	X_5	30
hsp	X_6	766.92
hsp : hsf	X_7	1403.26
hsp : mfp	X_8	71.65
prot	X_9	$1.14915 \cdot 10^8$
mfp	X_{10}	517.32
mhsf	X_{11}	$3.01 \cdot 10^{-6}$
mhsp	X_{12}	0.02
hsp : mhsf	X_{13}	$4.17 \cdot 10^{-7}$
hsp : mhsp	X_{14}	$2.24 \cdot 10^{-3}$

Constant	Reaction	Nr. value	Unit
k_1^+	(i), forward	3.49	$\frac{\text{cell}}{\# \cdot s}$
k_1^-	(i), backward	0.19	s^{-1}
k_2^+	(ii), forward	1.07	$\frac{\text{cell}}{\# \cdot s}$
k_2^-	(ii), backward	10^{-9}	s^{-1}
k_3^+	(iii), forward	0.17	$\frac{\text{cell}}{\# \cdot s}$
k_3^-	(iii), backward	$1.21 \cdot 10^{-6}$	s^{-1}
k_4	(iv)	$8.3 \cdot 10^{-3}$	s^{-1}
k_5^+	(v), forward	9.74	$\frac{\text{cell}}{\# \cdot s}$
k_5^-	(v), backward	3.56	s^{-1}
k_6	(vi)	2.33	$\frac{\text{cell}}{\# \cdot s}$
k_7	(vii)	$4.31 \cdot 10^{-5}$	$\frac{\text{cell}}{\# \cdot s}$
k_8	(viii)	$2.73 \cdot 10^{-7}$	$\frac{\text{cell}}{\# \cdot s}$
k_9	(ix)	$3.2 \cdot 10^{-5}$	s^{-1}
k_{11}^+	(xi), forward	$3.32 \cdot 10^{-3}$	$\frac{\text{cell}}{\# \cdot s}$
k_{11}^-	(xi), backward	4.44	s^{-1}
k_{12}	(xii)	13.94	s^{-1}

(a)   (b)

behavior, in particular without losing its experimental fit and validation. We then discuss the extent to which these simplifications are dependent on the numerical values of our parameters.

The first observation is that the variables mhsf and hsp: mhsf both assume negligible numerical values throughout numerical simulations for temperatures from $37°C$ to $45°C$. Even when their initial values are increased to higher values, e.g. to 100 each, their numerical convergence towards their steady state values is very fast. Moreover, if the increase in the initial values of mhsf and hsp: mhsf is so that the total amount of hsf and of hsp remain unchanged, then the experimental fit and validation of the model remain largely unchanged. The reason for this behavior is that the reactions having mhsf as a product, i.e. reactions (xiii) and the reverse reaction (xv) have a negligible flux rate, primarily due to the small kinetic rate constant of the protein misfolding law, see (15). Consequently, the reaction producing hsp: mhsf, i.e. reaction (xv), also has negligible flux rate. On the other hand, the reactions having mhsf and hsp: mhsf as reactants reach much higher flux rates because of larger kinetic constants and high levels of hsp, a co-reactant in reaction (xv). We decide then to eliminate both mhsf and hsp: mhsf from the model, along with the reactions where they take part in, i.e., reactions (xiii), (xv), and (xvi).

Note now that the situation is somewhat similar for hsf, hsf$_2$ and hsf$_3$: they all assume small (albeit not negligible) values throughout numerical simulations. There is however a crucial difference which points to their significance for the model: when increasing the initial level of hsf$_3$, even in such a way that the total level of hsf is unchanged, the fit to the DNA binding experimental data of [7] is drastically changed.

The observation that the flux of the hsf misfolding reaction is negligible was the main rationale behind eliminating mhsf and hsp: mhsf from the model. This leads to the observation that the flux of the hsp misfolding reaction, leading to the formation of mhsp is also negligible. The case of mhsp is however different because it is also the end product of reaction (iv). Moreover, mhsp plays a central role in our model, being the source of all induced hsp through reactions (iv), (xvii) and (xviii). The numerical values assumed by mhsp throughout simulations for temperatures between $37°C$ and $45°C$ are small, but not negligible. They are however negligible relative to the total level of hsp. Moreover, the numerical convergence of mhsp towards its steady state value is very fast, even in the case when the initial level of mhsp is increased several folds. This points to the observation that mhsp plays the role of a transient state towards hsp, having a very high turnover rate. As such, it could be eliminated from the model if only mhsp were replaced in reaction (iv) with hsp. Consequently, we eliminate mhsp from the model, along with reactions (xiv), (xvii) and (xviii). At the same time, we replace reaction (iv) with

(iv') hsf$_3$: hse $\rightarrow$ hsf$_3$: hse + hsp

The simplified molecular model has only 10 variables and 12 reactions, compared to 14 variables and 18 reactions in the initial model. The numerical simulations of the simplified model for temperatures between $37°C$ and $45°C$ are indistinguishable from those of the initial model.

Regarding the biological relevance, the simplified model differs from the initial model in ignoring the misfolded form of hsf and hsp, as well as ignoring that newly synthesized proteins often need chaperons to form their native fold. Excluding the misfolding of hsf and hsp is reasonable because the numerical levels of misfolded hsf and hsp are negligible with respect to the level of mfp and thus, their competition for the chaperon resources of the cell is insignificant. Excluding the role of chaperons in assisting the formation of the native fold of newly synthesized proteins is justified by the high speed of the reaction, relative to the speed of the other reactions in our model. As such, the complex chaperon - newly synthesized protein is a very fast transient stage in the model and can be ignored.

It should be noted that the simplifications we have made on the model are based on numerical arguments and so, in principle, they are dependant on the numerical values of the parameters of the model. To test the robustness of the model reductions against changes in the numerical setup of the model, we perform several tests. In each test, we either change the initial values of some variables, or we change the values of some kinetic rate constants. For each new numerical setup we set the initial values of all variables to their steady state values at $37°C$, similarly as done in [15] (to underline that the heat shock response is missing at $37°C$). Finally, we compare the numerical behavior of the model with that of its simplified version obtained as above, for temperatures between $37°C$ and $45°C$.

We first consider a numerical setup where the total level of hsf is increased by 1000 to a value of around 2400. In a second test, we increase both the total level of hsf by 1000 and the total level of hse by 100. In both tests, the numerical behaviors of the models and those of their simplified versions are undistinguishable. In a third test, we increase the total level of hsp by 1000. When estimating the steady state values of the model at $37°C$, we note that they are identical with those of the initial model, summarized in Table 2(a). This raises an intriguing problem of independent interest: is the steady state of the model independent of the initial total level of hsp?

A test where the complex chaperon–misfolded protein is made more unstable by increasing the kinetic rate constant k_{11}^- to 25 yields a numerically equivalent simplified model. In a final test, we decrease the value of the kinetic rate constant c_{12} of the refolding reaction (xii) from almost 14 to 1. In this way, we induce a great increase in the values of misfolded proteins of all types to test whether eliminating mhsf and mhsp is still possible in this context. It turns out that eliminating mhsf and hsp: mhsf is possible and yields a numerically equivalent simplified model. On the other hand, eliminating mhsp and hsp: mhsp changes the behavior of the model pronouncedly. E.g., mfp peaks at a lower value showing that the simplified model, where hsp is not subject to misfolding, is more efficient in fighting off the accumulation of mfp. A main reason why the elimination of misfolded hsp fails is because, unlike in the previous tests, the change in the refolding rate is not accounted for when setting the initial values of the variables to the steady state values at $37°C$, since the refolding reaction has a negligible flux at that temperature. At $42°C$ however, protein refolding, in particular that of mhsp, becomes very important and removing it from the model makes a big difference.

4   Discussion

Having simple biomodels is very important for being able to analyze their mathematical properties and for their integration into larger models. In the case of the heat shock response, adding the phosphorylation of hsf in all of its homo- and hetero-polymers, along with its influence on gene transcription leads to a combinatorial explosion in the number of variables of the model. As such, decreasing the number of variables, in particular the elimination of mhsf and hsp: mhsf reduces the difficulty of the problem.

Several aspects contribute to the model simplification succeeding in a given numerical setup. The most important is that we eliminate variables that have a fast numerical convergence to their steady state values. This procedure is often referred to as a time-separation principle. A factor here is the flux rate of the reactions producing certain variables of the model. If the total flux contributing to producing a given variable remains very small, then that variable will converge fast to its steady state value and it can be eliminated from the model. There are at least two reasons why a flux rate can be small: a small kinetic constant, or much higher kinetic constant in reactions using some of the same reactants. In the context of the heat shock response model, one more factor plays an important role: the condition that the initial values of all variables are a steady state of the model at $37°C$. It turns out that the model has an interesting property, formulated as a theorem in the appendix: the steady state values of most of its variables are independent of the temperature. In this way, even at higher temperature, several of the variables of the model start from their steady state values and witness only minor numerical disturbances before returning to the same values.

The model simplification discussed in this paper is dependant on the numerical setup of the model: on the initial values of the variables and on the numerical values of the kinetic constants. Even if the initial and the simplified models appear to be numerically equivalent in one particular setup, they may be very different in other setups. To evaluate the robustness of the model simplifications, one should compare the two models in several numerical setups, spanning the domain of expected values for the model parameters. Some of the simplifications may turn out to be robust against numerical variations, as it is the case with eliminating hsf and mhsf in the heat shock model, while others may be valid only in special numerical setups.

The main difficulty in designing a simple biomodel is that the decision to exclude variables and reactions from the model is most often done at the early stage of considering the molecular basis of the model. At that stage however it is crucial to ensure that all aspects of potential interest are included in the model. Appreciating the potentially insignificant contribution of some of the aspects is very difficult at that stage, without having first a well-validated numerical setup for the model. The approach we have discussed in this paper takes an intermediate view: one may start with a rich model that is first numerically fit and validated against experimental data and then it is subjected to a numerical analysis to identify the components that can be eliminated without changing the numerical behavior of the model. In this way, the result is a model that remains faithful to the biological data and soundly identifies those aspects of the biological reality that have insignificant contribution to the overall behavior.

Acknowledgments. This work has been partially supported by projects 108421 and 203667 (to I.P.) from Academy of Finland. The numerical simulations and estimations discussed in this paper were performed with Copasi, see [5].

References

1. Alberts, B., Johnson, A., Lewis, J., Raff, M., Roberts, K., Walter, P.: Essential Cell Biology, 2nd edn. Garland Science (2004)
2. Ciocca, D.R., Calderwood, S.K.: Heat shock proteins in cancer: diagnostic, prognostic, predictive, and treatment implications. Cell Stress and Chaperones 10(2), 86–103 (2005)

3. El-Samad, H., Kurata, H., Doyle, J., Gross, C.A., Khamash, M.: Surviving heat shock: control strategies for robustness and performance. PNAS 102(8), 2736–2741 (2005)

4. El-Samad, H., Prajna, S., Papachristodoulu, A., Khamash, M., Doyle, J.: Model validation and robust stability analysis of the bacterial heat shock response using sostools. In: Proceedings of the 42nd IEEE Conference on Decision and Control, pp. 3766–3741 (2003)

5. Hoops, S., Sahle, S., Gauges, R., Lee, C., Pahle, J., Simus, N., Singhal, M., Xu, L., Mendes, P., Kummer, U.: Copasi – a COmplex PAthway SImulator. Bioinformatics 22(24), 3067–3074 (2006)

6. Kampinga, H.K.: Thermotolerance in mammalian cells: protein denaturation and aggregation, and stress proteins. J. Cell Science 104, 11–17 (1993)

7. Kline, M.P., Morimoto, R.I.: Repression of the heat shock factor 1 transcriptional activation domain is modulated by constitutive phosphorylation. Molecular and Cellular Biology 17(4), 2107–2115 (1997)

8. Kurata, H., El-Samad, H., Yi, T.M., Khamash, M., Doyle, J.: Feedback regulation of the heat shock response in e.coli. In: Proceedings of the 40th IEEE Conference on Decision and Control, pp. 837–842 (2001)

9. Liu, B., DeFilippo, A.M., Li, Z.: Overcomming immune toerance to cancer by heat shock protein vaccines. Molecular cancer therapeutics 1, 1147–1151 (2002)

10. Lukacs, K.V., Pardo, O.E., Colston, M.J., Geddes, D.M., Eric WFW Alton: Heat shock proteins in cancer therapy. In: Habib (ed.) Cancer Gene Therapy: Past Achievements and Future Challenges, pp. 363–368 (2000)

11. Peper, A., Grimbergent, C.A., Spaan, J.A.E., Souren, J.E.M., van Wijk, R.: A mathematical model of the hsp70 regulation in the cell. Int. J. Hyperthermia 14, 97–124 (1997)

12. Lepock, J.R., Frey, H.E., Ritchie, K.P.: Protein denaturation in intact hepatocytes and isolated cellular organelles during heat shock. The Journal of Cell Biology 122(6), 1267–1276 (1993)

13. Lepock, J.R., Frey, H.E., Rodahl, A.M., Kruuv, J.: Thermal analysis of chl v79 cells using differential scanning calorimetry: Implications for hyperthermic cell killing and the heat shock response. Journal of Cellular Physiology 137(1), 14–24 (1988)

14. Petre, I., Mizera, A., Hyder, C.L., Mikhailov, A., Eriksson, J.E., Sistonen, L., Back, R.-J.: A new mathematical model for the heat shock response. In: Kok, J. (ed.) Algorithmic bioprocesses, Natural Computing. Springer, Heidelberg (2008)

15. Petre, I., Hyder, C.L., Mizera, A., Mikhailov, A., Eriksson, J.E., Sistonen, L., Back, R.-J.: A simple mathematical model for the eukaryotic heat shock response (manuscript, 2009)

16. Graham Pockley, A.: Heat shock proteins as regulators of the immune response. The Lancet 362(9382), 469–476 (2003)

17. Rieger, T.R., Morimoto, R.I., Hatzimanikatis, V.: Mathematical modeling of the eukaryotic heat shock response: Dynamics of the hsp70 promoter. Biophysical Journal 88(3), 1646–1658 (2005)

18. Tomlin, C.J., Axelrod, J.D.: Understanding biology by reverse engineering the control. PNAS 102(12), 4219–4220 (2005)

19. Workman, P., de Billy, E.: Putting the heat on cancer. Nature Medicine 13(12), 1415–1417 (2007)

Appendix

The next theorem formulates an interesting property of the heat shock response model. We formulate the property for the simplified model of the heat shock response.

Theorem 1. *Let* $c^1 = (c_1^1, c_2^1, c_3^1, \ldots, c_{10}^1)$ *be a steady state of the system at temperature* T_1 *and* $c^2 = (c_1^2, c_2^2, c_3^2, \ldots, c_{10}^2)$ *a steady state at temperature* T_2, *where* c_i^1 *and* c_i^2 *for* $i = 1, \ldots, 10$ *are steady state concentrations of metabolite* X_i *at temperatures* T_1 *and* T_2 *respectively. Then* $c = (c_1^1, \ldots, c_7^1, c_8^2, c_9^2, c_{10}^2)$ *is a steady state of the system at temperature* T_2.

Proof. Let c^1 and c^2 be steady states at temperatures T_1 and T_2, respectively. Further, let us split the system of differential equations (1)-(10) into two subsystems: one containing equations (1)-(7) and the other consisting of equations (8)-(10). Equation (6) is the only one in the first subsystem with right-hand side containing functions defined by the second subsystem, i.e. $X_8(t)$, $X_9(t)$ and $X_{10}(t)$, and can be by (9) rewritten in the following form:

$$dX_6/dt = k_4\,X_4 - k_5^+\,X_1\,X_6 + k_5^-\,X_7 - k_8\,X_4\,X_6 - k_6\,X_2\,X_6$$
$$-k_7\,X_3\,X_6 - k_9\,X_6 - dX_8/dt. \tag{16}$$

When considering the steady states, the left-hand sides of (1)-(10) are set to 0 and in consequence equation (16) can be written as

$$0 = k_4\,X_4 - k_5^+\,X_1\,X_6 + k_5^-\,X_7 - k_8\,X_4\,X_6 - k_6\,X_2\,X_6 - k_7\,X_3\,X_6 - k_9\,X_6.$$

This algebraic relation does not contain any of functions $X_8(t)$, $X_9(t)$ or $X_{10}(t)$ and hence the steady state algebraic relations of subsystem (1)-(7) become independent of them. As a consequence, the relations do not contain temperature as a parameter and are the same both for T_1 and T_2. Since the same equations have the same solutions, it follows that $c = (c_1^1, \ldots, c_7^1, c_8^2, c_9^2, c_{10}^2)$ is a steady state of the whole system at temperature T_2.

The biological significance of Theorem 1 deserves some comments. Even though the cell approaches similar steady state levels regardless of the temperature values, the *time* it takes to arrive in a certain neighborhood of the steady state is longer for higher temperature values. Even if one starts in the steady state, the *effort* required of the cell is higher for higher temperatures: the fluxes of all reactions are higher for higher temperatures. The intuitive reason for this is that the misfolding rate is vastly accelerated for higher temperatures, eventually accelerating all other reactions.

Functions Definable by Arithmetic Circuits

Ian Pratt-Hartmann[1] and Ivo Düntsch[2]

[1] School of Computer Science, University of Manchester,
Manchester M13 9PL, U.K.
ipratt@cs.man.ac.uk
[2] Department of Computer Science, Brock University,
St. Catharines, ON, L2S 3A1, Canada
duentsch@brocku.ca

Abstract. An *arithmetic circuit* is a labelled, directed, acyclic graph specifying a cascade of arithmetic and logical operations to be performed on sets of non-negative integers. In this paper, we consider the definability of functions from tuples of sets of non-negative integers to sets of non-negative integers by means of arithmetic circuits. We prove two negative results: the first shows, roughly, that a function is not circuit-definable if it has an infinite range and sub-linear growth; the second shows, roughly, that a function is not circuit-definable if it has a finite range and fails to converge on certain 'sparse' chains under inclusion. We observe that various functions of interest fall under these descriptions.

Keywords: Arithmetic circuit, integer expression, complex algebra, expressive power.

1   Introduction

An *arithmetic circuit* (McKenzie and Wagner [7,8]) is a labelled, directed, acyclic graph specifying a cascade of arithmetic and logical operations to be performed on sets of non-negative integers. (Henceforth, we refer to non-negative integers simply as *numbers*). Each node in this graph evaluates to a set of numbers, representing a stage of the computation performed by the circuit. Nodes without predecessors in the graph are called *input nodes*, and their labels indicate the sets of numbers to which they evaluate. Nodes with predecessors in the graph are called *arithmetic gates*, and their labels indicate operations to be performed on the values of their immediate predecessors; the results of these operations are then taken to be the values of the arithmetic gates in question. Multiple edges are allowed. One of the nodes in the graph (usually, a node with no successors) is designated as the *circuit output*; the set of numbers to which it evaluates is taken to be the value of the circuit as a whole.

We allow input nodes to be labelled with any of the constant symbols $\{1\}$, $\{0\}$, $\emptyset$, $\mathbb{N}$, or with one of an infinite stock of variables. The constants denote subsets of $\mathbb{N}$ in the conventional way, and the variables are taken to range over the power set of $\mathbb{N}$. Similarly, we allow arithmetic gates to be labelled with any of the symbols $+$, $\bullet$, $^{-}$, $\cap$ or $\cup$, denoting an operation on sets of numbers.

K. Ambos-Spies, B. Löwe, and W. Merkle (Eds.): CiE 2009, LNCS 5635, pp. 409–418, 2009.

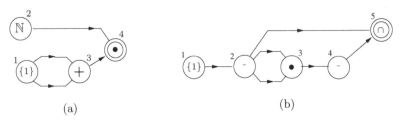

(a)   (b)

Fig. 1. Arithmetic circuits defining: (a) the set of even numbers; (b) the set of primes. The integers next to the nodes are for reference only.

The symbols $^{-}$, $\cap$, $\cup$ have the obvious Boolean interpretations (with $^{-}$ denoting complementation in $\mathbb{N}$), while $+$ and $\bullet$ denote the result of lifting addition and multiplication to the algebra of sets, thus:

$$s + t = \{i + j | i \in s \text{ and } j \in t\}; \qquad s \bullet t = \{i \cdot j | i \in s \text{ and } j \in t\}.$$

We assume that arithmetic gates have the appropriate number of immediate predecessors (1 or 2) for their labels.

Fig. 1 shows two examples of arithmetic circuits, where the output gate is indicated by the double circle. In Fig. 1a, Node 1 evaluates to $\{1\}$, and Node 2 to $\mathbb{N}$; hence, Node 3 evaluates to $\{1\} + \{1\} = \{2\}$, and Node 4, the output of the circuit, to $\{2\} \bullet \mathbb{N}$, i.e. the set of even numbers. The circuit of Fig. 1b functions similarly: Node 2 evaluates to $\{0\} \cup \{n \in \mathbb{N} \mid n \geq 2\}$, and Node 3 to $\{0\} \cup \{n \in \mathbb{N} \mid n \text{ is composite}\}$; hence, Node 4 evaluates to the set of numbers which are either prime or equal to 1, and Node 5, the output of the circuit, to the set of primes. We say that the circuits of Fig. 1a and Fig. 1b *define*, respectively, the set of even numbers and the set of primes. Any arithmetic circuit with no variable inputs defines a set of numbers in this way. Likewise, any arithmetic circuit with one or more variable inputs defines a function from (tuples of) sets of numbers to sets of numbers. The question naturally arises as to which sets and functions are definable by arithmetic circuits.

The study of arithmetic circuits originates in Stockmeyer and Meyer [10], who studied *integer expressions*—in effect, arithmetic circuits in the form of trees, with no variables and no multiplication gates. (Integer expressions are essentially the same as star-free regular expressions over a 1-element alphabet, where the integer n stands for the string of length n). The *membership* problem for integer expressions is as follows: given a number and an integer expression, determine whether that number is in the set defined by that integer expression. The *non-emptiness problem* is as follows: given an integer expression, determine whether the set of numbers it defines is non-empty. Stockmeyer and Meyer showed that both these problems are PSPACE-complete.

The corresponding membership and non-emptiness problems for variable-free arithmetic circuits—i.e., circuits featuring, additionally, the $\bullet$-operator, but still with no variable inputs—are discussed by McKenzie and Wagner [7]. It is easy to see that these problems are reducible to one another; however, the question of their decidability is currently still open. The complexity of the membership problem (and related problems) for variable-free arithmetic circuits with operators chosen from various proper subsets of $\{\cup, \cap, ^-, +, \bullet\}$ are studied in Glaßer et al. [1,2], building on the work of Meyer and Stockmeyer, op. cit., McKenzie and Wagner, op. cit. and Yang [12]. A somewhat different complexity-theoretic profile is obtained if arithmetic circuits are taken to compute over sets of integers (possibly negative), as shown in Travers [11].

Arithmetic circuits with variable inputs give rise to different complexity-theoretic problems. Jeż and Okhotin [6] consider systems of equations $\phi_i(\boldsymbol{x}) = \psi_i(\boldsymbol{x})$ ($1 \leq i \leq n$) where the ϕ_i and ψ_i are arithmetic circuits with a tuple of variable inputs $\boldsymbol{x}$, but without the multiplication or complementation operators. It is shown, inter alia, that all recursively enumerable sets are representable as the least solutions of such systems of equations. Further, it is shown in Jeż and Okhotin [5] that, when consideration is restricted to so-called resolved sets of equations (i.e. ϕ_i is the variable x_i, where $\boldsymbol{x} = x_1, \ldots, x_n$), the family of sets representable as least solutions of such equations is included in EXPTIME, and moreover contains some EXPTIME-hard sets.

In the present paper, we consider the expressive power of arithmetic circuits. In particular, we ask: which functions (from tuples of sets of numbers to sets of numbers) are definable by arithmetic circuits? We obtain two negative results: the first shows, roughly, that a function is not circuit-definable if it has an infinite range and sub-linear growth; the second shows, roughly, that a function is not circuit-definable if it has a finite range and fails to converge on certain 'sparse' chains under inclusion. We observe that various functions of interest fall under these descriptions.

2   Preliminaries

The full complex algebra of $\mathbb{N}$ is the structure $\mathfrak{Cm} = \{2^{\mathbb{N}}, \cap, \cup, ^-, \emptyset, \mathbb{N}, +, \{0\}, \bullet, \{1\}\}$, with $+$ and $\bullet$ as defined above. Let V be an infinite set of variables. A complex arithmetic term (or simply term) is a member of the smallest set C of expressions satisfying the conditions: (i) $V \subseteq C$; (ii) the constants $\{0\}$, $\{1\}$, $\emptyset$ and $\mathbb{N}$ are in C; and (iii) if σ, τ are in C, then so are $\sigma \cap \tau$, $\sigma \cup \tau$, $\overline{\sigma}$, $\sigma + \tau$ and $\sigma \bullet \tau$. If $\boldsymbol{x} = x_1, \ldots, x_n$ is a non-empty tuple of variables, and τ a term all of whose variables are among the $\boldsymbol{x}$, we optionally write τ as $\tau(\boldsymbol{x})$ to denote the order of variables. The algebra of terms $\tau(\boldsymbol{x})$ is denoted $C[\boldsymbol{x}]$. Likewise, the algebra of variable-free terms is denoted $C[\ ]$.

An interpretation is a function $\iota : V \to 2^{\mathbb{N}}$ mapping variables to sets of numbers. For any tuple of variables $\boldsymbol{x}$ (possibly empty), ι is extended homomorphically to a function $C[\boldsymbol{x}] \to \mathfrak{Cm}$. For the sake of readability, if τ is a term in $C[\boldsymbol{x}]$, and ι an interpretation mapping the tuple of variables $\boldsymbol{x}$ to the tuple of sets of numbers $\boldsymbol{s}$, then we denote $\iota(\tau)$ by $\tau(\boldsymbol{s})$. In particular, if τ is variable-free, the set $\iota(\tau)$ (which is independent of ι) is denoted $\tau(\ )$. If $\boldsymbol{x}$ is an n-tuple of variables ($n > 0$), the function defined by a term $\tau \in C[\boldsymbol{x}]$ is the function $\boldsymbol{s} \mapsto \tau(\boldsymbol{s})$. Any function $f : (2^{\mathbb{N}})^n \to 2^{\mathbb{N}}$ which can be written in this way is said to be *term-definable*. Likewise, if τ is a variable-free term, the set *defined by τ* is the set of numbers $\tau(\ )$. Any set $s \subseteq \mathbb{N}$ which can be written in this way is said to be *term-definable*.

For the purposes of investigating term-definability, there is no difference between complex arithmetic terms and arithmetic circuits, as described in Section 1. Thus, for example, the circuits of Fig. 1 correspond to the respective terms

$$\tau_e = (\{1\} + \{1\}) \bullet \mathbb{N} \qquad \tau_p = \overline{\{1\}} \cap (\overline{\{1\} \bullet \{1\}}). \qquad (1)$$

For the sake of familiarity, therefore, we speak of *circuits* in preference to *terms*, referring to term-definable functions as *circuit-definable* functions, and similarly for term-definable sets.

3   Circuit-Definable Sets

We observed in Section 1 that it is not known whether the membership problem for arithmetic circuits is decidable. However, for any *fixed* arithmetic circuit $\tau \in C[\ ]$, the problem of determining membership in the set $\tau(\ )$ is easily seen to be decidable, and in fact to have relatively low computational complexity.

Define the binary operation $\circ$ on sets of numbers by

$$s \circ t = \{i \cdot j \mid i \in s \setminus \{0\}, j \in t \setminus \{0\}\}.$$

Thus, $\circ$ is a modified form of set multiplication. Let us define the set of *modified circuits* C° in exactly the same way as C, except that the multiplication operator $\bullet$ is replaced by the modified multiplication operator $\circ$.

If $\boldsymbol{x}$ is a tuple of variables (possibly empty), the algebra of modified circuits $\tau(\boldsymbol{x})$ will be denoted $C^\circ[\boldsymbol{x}]$. The concept of a circuit's *defining* a function or set is carried over to modified circuits in the obvious way.

Theorem 1. *A set is definable by an arithmetic circuit if and only if it is definable by a modified circuit.*

Proof. Define $g : C[\ ] \rightarrow C^\circ[\ ]$ recursively: (*i*) if $\tau \in \{\{0\}, \{1\}, \emptyset, \mathbb{N}\}$, then $g(\tau) = \tau$; (*ii*) $g(\overline{\tau}) = \overline{g(\tau)}$; (*iii*) if $\star \in \{\cap, \cup, +\}$, then $g(\sigma \star \tau) = g(\sigma) \star g(\tau)$; (*iv*)

$$g(\sigma \bullet \tau) = \begin{cases} (g(\sigma) \circ g(\tau)) \cup \{0\} & \text{if } 0 \in \tau(\ ) \text{ and } \sigma(\ ) \neq \emptyset \\ (g(\sigma) \circ g(\tau)) \cup \{0\} & \text{if } 0 \in \sigma(\ ) \text{ and } \tau(\ ) \neq \emptyset \\ g(\sigma) \circ g(\tau) & \text{otherwise.} \end{cases}$$

Routine induction shows that, for all $\tau \in C[\ ]$, $\tau(\ ) = g(\tau)(\ )$. Hence, every set definable by an arithmetic circuit is definable by a modified circuit. The converse is proved similarly.

We remark that, since the conditions $0 \in \tau(\ )$ or $\tau(\ ) = \emptyset$ are not known to be decidable, g is not known to be computable. That is: we know that every circuit-definable set is definable by some modified circuit or other, but we do not know which one.

If m is a number, and s, t sets of numbers, to determine whether $m \in s \circ t$, it suffices to check whether m can be factored as $m = m_1 \cdot m_2$, *with m_1, m_2 bounded by m*, such that $m_1 \in s$ and $m_2 \in t$. (This is not true for $s \bullet t$.) We remind the reader that the language of *bounded arithmetic* is the first-order language over the signature $(+, \cdot, 1, 0)$, but with all quantification restricted to the forms $(\forall x \leq t)\varphi$ and $(\exists x \leq t)\varphi$, where t is a term (see, e.g. [3]).

Corollary 1. *Every circuit-definable set is definable by a formula of* bounded arithmetic.

It follows that every circuit-definable set is in the polynomial hierarchy, PH. It also follows that the characteristic function of every circuit-definable set is computable in deterministic linear time, or, equivalently, lies the Grzegorczyk class, $\mathcal{E}_*^2$ (Ritchie [9], Corollary, p. 153). Hence, circuit sets, regarded as languages over the alphabet $\{0, 1\}$, are certainly all context-sensitive. Note, however, that the set of primes, defined by the circuit in Fig. 1b, is known not to be context-free (Hartmanis and Shank [4]).

4   Circuit-Definable Functions

Many natural functions involving sets of numbers turn out to be circuit-definable. For example, the function

$$d(x) = \{n \in \mathbb{N} \mid \forall m \in x, n < m\} \tag{2}$$

is defined by the circuit $\tau_d(x) = \overline{x + \mathbb{N}}$ (but cf. Corollary 4 below). Likewise,

$$\mu(x) = \begin{cases} \{\min(x)\} & \text{if } x \neq \emptyset \\ \emptyset & \text{otherwise} \end{cases}$$

is defined by the circuit $\tau_\mu(x) = (\tau_d(x + \{1\})) \cap x$ (but cf. Corollary 3 below).

Boolean-valued functions may be defined by arithmetic circuits, employing some encoding of truth values as sets of numbers. The precise encoding chosen is not important: we write $\top$ for $\{0\}$ and $\bot$ for $\emptyset$. For example, the function

$$S(x) = \begin{cases} \top & \text{if } x \text{ is a singleton} \\ \bot & \text{otherwise} \end{cases} \tag{3}$$

is defined by the circuit $\tau_S(x) = (x \bullet \{0\}) \cap \overline{((x \cap \overline{\tau_\mu(x)}) \bullet \{0\})}$.

Definition by cases is also possible: if the functions $F, G, H : (2^{\mathbb{N}})^n \to 2^{\mathbb{N}}$ are defined by the circuits $\rho(\boldsymbol{x})$, $\sigma(\boldsymbol{x})$, $\tau(\boldsymbol{x})$, respectively, then the function

$$\boldsymbol{x} \mapsto \begin{cases} G(\boldsymbol{x}) & \text{if } F(\boldsymbol{x}) \neq \emptyset \\ H(\boldsymbol{x}) & \text{otherwise} \end{cases}$$

is defined by the circuit $(((\rho(\boldsymbol{x}) \bullet \{0\}) + \mathbb{N}) \cap \sigma(\boldsymbol{x})) \cup (\overline{((\rho(\boldsymbol{x}) \bullet \{0\}) + \mathbb{N})} \cap \tau(\boldsymbol{x}))$.

If $f : \mathbb{N}^n \to \mathbb{N}$ is a function from tuples of numbers to numbers, we say that f is *circuit-definable* if there exists a circuit-definable function $F : (2^{\mathbb{N}})^n \to 2^{\mathbb{N}}$ such that, for all $m_1, \ldots, m_n$, $F(\{m_1\}, \ldots, \{m_n\}) = \{f(m_1, \ldots, m_n)\}$. For example, given a fixed number $p > 1$, the function $n \mapsto (n \bmod p)$ is defined, in this sense, by the circuit

$$\tau_{\bmod p}(x) = \bigcup_{0 \le k < p} \left(\left(\left(x \cap \left(\left(\{p\} \bullet \mathbb{N} \right) + \{k\} \right) \right) \bullet \{0\} \right) + \{k\} \right),$$

where, for $m > 1$, $\{m\}$ abbreviates $\{1\} + \cdots + \{1\}$ (m times). Notice that, when discussing the circuit-definability of a function $f : \mathbb{N}^n \to \mathbb{N}$, we do not care what values the defining circuit takes on non-singleton inputs.

We now proceed to establish some simple results on functions which are not circuit-definable. We employ the following notation. If $s \subseteq \mathbb{N}$ and $m \in \mathbb{N}$, denote by $s_{|m}$ the set $s \cap [0, m]$. If $\boldsymbol{s} = s_1, \ldots, s_n$ is a tuple of elements of $2^{\mathbb{N}}$, write $\boldsymbol{s}_{|m}$ for $(s_1)_{|m}, \ldots, (s_n)_{|m}$.

Lemma 1. *Let $\tau \in C^\circ[\boldsymbol{x}]$ with $\boldsymbol{x}$ of arity n, let $m \in \mathbb{N}$, and let $\boldsymbol{s}, \boldsymbol{t} \in (2^{\mathbb{N}})^n$. If $\boldsymbol{s}_{|m} = \boldsymbol{t}_{|m}$, then, $\tau(\boldsymbol{s})_{|m} = \tau(\boldsymbol{t})_{|m}$.*

Proof. Straightforward induction on the structure of $\tau(\boldsymbol{x})$.

Note that Lemma 1 fails if the condition $\tau \in C^\circ[\boldsymbol{x}]$ is replaced by $\tau \in C[\boldsymbol{x}]$.

Let $\boldsymbol{x}$ be an n-tuple of variables ($n \ge 0$). A *circuit-condition (in $\boldsymbol{x}$)* is an expression α of the form

$$\bigwedge_{1 \le i \le a} \sigma_i(\boldsymbol{x}) = \emptyset \wedge \bigwedge_{1 \le i \le b} \tau_i(\boldsymbol{x}) \ne \emptyset \wedge \bigwedge_{1 \le i \le c} 0 \in \psi_i(\boldsymbol{x}) \wedge \bigwedge_{1 \le i \le d} 0 \notin \varphi_i(\boldsymbol{x}), \quad (4)$$

where $a, b, c, d \ge 0$, and the σ_i, τ_i, ψ_i and φ_i are circuits in $C[\boldsymbol{x}]$. (The empty conjunction with $a = b = c = d = 0$ is allowed.) When writing circuit-conditions, we silently re-order conjuncts and remove duplicates as necessary. If we wish to specify a particular order of the variables in α, we write $\alpha(\boldsymbol{x})$. If $\boldsymbol{s} \in (2^{\mathbb{N}})^n$, we take $\alpha(\boldsymbol{s})$ to be the Boolean value (either $\top$ or $\bot$) computed by substituting $\boldsymbol{s}$ for $\boldsymbol{x}$ in (4), in the obvious way. Two circuit-conditions $\alpha(\boldsymbol{x})$ and $\alpha'(\boldsymbol{x})$ are *disjoint* if, for all $\boldsymbol{s}$, one of $\alpha(\boldsymbol{s})$ or $\alpha'(\boldsymbol{s})$ is $\bot$, and a k-tuple of circuit-conditions $\alpha_1(\boldsymbol{x}), \ldots, \alpha_k(\boldsymbol{x})$ is *jointly exhaustive* if, for all $\boldsymbol{s}$, one of $\alpha_1(\boldsymbol{s})$, $\ldots$, $\alpha_k(\boldsymbol{s})$ is $\top$.

A *circuit-clause (in $\boldsymbol{x}$)* is an expression γ of the form $\alpha \to \tau$, where α is a circuit-condition in $\boldsymbol{x}$ and τ a circuit in $\boldsymbol{x}$; in this case, α is called the *condition of γ*. A *circuit-ensemble (in $\boldsymbol{x}$)* is a finite set $\varepsilon = \{\gamma_1, \ldots, \gamma_k\}$ of circuit-clauses in $\boldsymbol{x}$ such that the respective conditions $\alpha_1(\boldsymbol{x})$, $\ldots$, $\alpha_k(\boldsymbol{x})$ are pairwise disjoint and jointly exhaustive. Again, we write $\gamma(\boldsymbol{x})$, $\varepsilon(\boldsymbol{x})$ *etc.* to indicate the order of the variables; and if $\boldsymbol{s}$ is an n-tuple of elements of $2^{\mathbb{N}}$, we take $\varepsilon(\boldsymbol{s})$ to be the set $\tau(\boldsymbol{s})$, where $\alpha \to \tau$ is the unique clause of ε such that $\alpha(\boldsymbol{s}) = \top$. If all the circuits mentioned in ε (including those occurring in the conditions of the clauses) are modified circuits (i.e. are in $C^\circ[\boldsymbol{x}]$), then we say that ε is a *modified* circuit-ensemble. We denote the set of all modified circuit-ensembles in $\boldsymbol{x}$ by $E^\circ[\boldsymbol{x}]$. Let $\tau(\boldsymbol{x})$ be a circuit and $\varepsilon(\boldsymbol{x})$ a circuit-ensemble: we say $\tau(\boldsymbol{x})$ is *equivalent* to $\varepsilon(\boldsymbol{x})$ if, for all $\boldsymbol{s}$, $\tau(\boldsymbol{s}) = \varepsilon(\boldsymbol{s})$.

Lemma 2. *For every $\tau \in C[\boldsymbol{x}]$, there exists an $\varepsilon \in E^\circ[\boldsymbol{x}]$ such that $\tau(\boldsymbol{x})$ is equivalent to $\varepsilon(\boldsymbol{x})$.*

Proof. We construct an equivalent modified circuit-ensemble ε by induction on the structure of τ. We illustrate with the case $\sigma(\boldsymbol{x}) \bullet \tau(\boldsymbol{x})$. Suppose that $\sigma(\boldsymbol{x})$ is equivalent to a modified circuit-ensemble $\{\alpha_i \to \sigma_i \mid 1 \leq i \leq l\}$ and $\tau(\boldsymbol{x})$ is equivalent to a modified circuit-ensemble $\{\beta_j \to \tau_j \mid 1 \leq j \leq m\}$. Then it is easy to see that the set of clauses of the forms

$$\alpha_i \wedge \beta_j \wedge \sigma_i \neq \emptyset \wedge 0 \in \tau_j \to (\sigma_i \circ \tau_j) \cup \{0\}$$
$$\alpha_i \wedge \beta_j \wedge 0 \in \sigma_i \wedge \tau_j \neq \emptyset \wedge 0 \notin \tau_j \to (\sigma_i \circ \tau_j) \cup \{0\}$$
$$\alpha_i \wedge \beta_j \wedge \sigma_i \neq \emptyset \wedge 0 \notin \sigma_i \wedge \tau_j \neq \emptyset \wedge 0 \notin \tau_j \to \sigma_i \circ \tau_j$$
$$\alpha_i \wedge \beta_j \wedge \sigma_i = \emptyset \wedge \tau_j \neq \emptyset \to \emptyset$$
$$\alpha_i \wedge \beta_j \wedge \tau_j = \emptyset \to \emptyset,$$

where i and j take all values in the ranges $0 \leq i \leq l$ and $0 \leq j \leq m$, is a modified circuit-ensemble equivalent to $\sigma(\boldsymbol{x}) \bullet \tau(\boldsymbol{x})$. The other cases are similar.

In the sequel, we use the notation $\emptyset$ to denote the tuple $\emptyset, \ldots, \emptyset$ whose arity will be clear from context.

Lemma 3. *For any $\tau \in C[\boldsymbol{x}]$, with $\boldsymbol{x}$ of arity n, there exist $k > 0$ and $\tau_1, \ldots, \tau_k \in C[\ ]$ with the following property: for all $m \in \mathbb{N}$ and all $\boldsymbol{s} \in (2^{\mathbb{N}})^n$ with $\boldsymbol{s}_{|m} = \emptyset$, there exists i $(1 \leq i \leq k)$ such that $\tau(\boldsymbol{s})_{|m} = \tau_i(\ )_{|m}$.*

Proof. By Lemma 2, let $\tau(\boldsymbol{x})$ be equivalent to $\varepsilon(\boldsymbol{x}) \in E^\circ[\boldsymbol{x}]$, and let $\varepsilon(\boldsymbol{x}) = \{\alpha_i(\boldsymbol{x}) \to \tau_i'(\boldsymbol{x}) \mid 1 \leq i \leq k\}$. Let $\tau_1, \ldots, \tau_k$ be obtained by replacing all variables in $\tau_1', \ldots, \tau_k'$ with the constant $\emptyset$. Fix any $m \in \mathbb{N}$ and $\boldsymbol{s} \in (2^{\mathbb{N}})^n$ with $\boldsymbol{s}_{|m} = \emptyset$. Let i be the unique integer $(1 \leq i \leq k)$ such that $\alpha_i(\boldsymbol{s}) = \top$. Then, $\tau(\boldsymbol{s})_{|m} = \tau_i'(\boldsymbol{s})_{|m} = \tau_i'(\emptyset)_{|m}$, by Lemma 1 (since $\boldsymbol{s}_{|m} = \emptyset$). Thus, $\tau(\boldsymbol{s})_{|m} = \tau_i(\ )_{|m}$.

Lemma 4. *For any $\tau \in C[\boldsymbol{x}]$, with $\boldsymbol{x}$ of arity n, the family of sets $\{\tau(\boldsymbol{s})_{|m} \mid m \geq 0, \boldsymbol{s} \in (2^{\mathbb{N}})^n$ s.t. $\boldsymbol{s}_{|m} = \emptyset\}$ is the union of finitely many chains under inclusion.*

Proof. Let $\tau_1, \ldots, \tau_k$ be as in Lemma 3. Every $\{\tau_i(\ )_{|m}\}_{m \geq 0}$ is such a chain.

This shows that a function (from numbers to numbers) is not circuit-definable if it has an infinite range and sub-linear growth:

Theorem 2. *Let $f : \mathbb{N} \to \mathbb{N}$ be a function. If $\{f(n) \mid n \in \mathbb{N}, f(n) < n\}$ is infinite, then f is not circuit-definable.*

Proof. Suppose, for contradiction, that $\tau(x)$ defines f. Consider an infinite collection M of numbers m such that the values $f(m + 1)$ are pairwise distinct, with $f(m + 1) \leq m$. Then

$$\{\{f(m+1)\} \mid m \in M\} \subseteq \{\tau(\{n\})_{|m} \mid 0 \leq m < n\}$$
$$\subseteq \{\tau(\boldsymbol{s})_{|m} \mid m \geq 0, \boldsymbol{s} \in 2^{\mathbb{N}} \text{ s.t. } \boldsymbol{s}_{|m} = \emptyset\}.$$

But the infinite set of singletons $\{\{f(m+1)\} \mid m \in M\}$ is certainly not included in the union of any finite collection of chains, contrary to Lemma 4.

Corollary 2. *If $0 < \alpha < 1$, then the functions $f(n) = \lceil \alpha n \rceil$ and $f(n) = \lceil n^\alpha \rceil$ are not circuit-definable. If $1 < \beta$, then neither are:*

$$f(n) = \begin{cases} n-1 & \text{if } n > 0 \\ 0 & \text{otherwise} \end{cases} \qquad f(n) = \begin{cases} \lceil \log_\beta n \rceil & \text{if } n > 0 \\ 0 & \text{otherwise.} \end{cases}$$

Note that Theorem 2 fails if the condition that $\{f(n) \mid n \in \mathbb{N}, \ f(n) < n\}$ is infinite is replaced by the condition that $\{n \in \mathbb{N} \mid f(n) < n\}$ is infinite. For example, we have already seen that the function $n \mapsto (n \bmod p)$ is circuit-definable, for all $p > 1$. Thus, arithmetic circuits can compute remainders (for fixed divisors), but not quotients.

As a further corollary, we see that, while the circuit τ_μ given above computes minima, no arithmetic circuit computes maxima:

Corollary 3. *If $F : 2^\mathbb{N} \to 2^\mathbb{N}$ has the property that, for all finite, non-empty $s \subseteq \mathbb{N}$, $F(s) = \{\max(s)\}$, then F is not circuit-definable.*

Proof. Suppose that F is defined by a circuit $\tau(x)$. We observed above that the function d given in (2) is defined by the circuit τ_d. Then the circuit $\tau(\tau_d(x))$ maps any singleton $\{n\}$ $(n > 0)$ to $\{n-1\}$, contradicting Theorem 2.

We remark on another consequence of Lemma 2, contrasting with the circuit-definability of the function $d(x)$ given in (2).

Corollary 4. *Denote $\{n \in \mathbb{N} \mid \exists m \in x, n < m\}$ by $\downarrow (x)$. There is no arithmetic circuit $\tau(x)$ such that, for all finite s, $\tau(s) = \downarrow (s)$.*

Proof. Suppose otherwise. By Lemma 2, let $\varepsilon(x)$ be a modified circuit-ensemble such that $\varepsilon(s) = \downarrow (s)$ for all finite s. For all even numbers i, define $s_i = \{m \in \mathbb{N} \mid m \text{ odd and } m < i\}$. Note that, for i, j even with $i < j$, we have $(s_i)_{|i} = (s_j)_{|i}$, but $(\downarrow (s_i))_{|i} = [0, i-1] \neq [0, i] = (\downarrow (s_j))_{|i}$. Since the number of clauses of ε is finite, we can find even numbers i, j, with $i < j$, such that s_i and s_j satisfy (the condition of) the same clause of ε. But then, by Lemma 1, $\varepsilon(s_i)_{|i} = \varepsilon(s_j)_{|i}$, contradicting the fact that $(\downarrow (s_i))_{|i} \neq (\downarrow (s_j))_{|i}$.

Continuing with functions of sets of numbers, we now show that no such function is circuit-definable if it has a finite range and fails to converge on certain 'sparse' chains under inclusion. Let s be a finite, non-empty set of numbers, t a set of numbers, and m a number. We write $s \sqsubseteq_m t$ if $m \geq \max(s)$ and $s_{|m} = t_{|m}$. In that case, we may think of the (possibly empty) interval $[\max(s) + 1, m]$ as a 'buffer' following the largest element of s, in which no element of t occurs.

Theorem 3. *Let $F : 2^\mathbb{N} \to 2^\mathbb{N}$ be a function with finite range. And suppose that, for all finite, non-empty $s \subseteq \mathbb{N}$ and all $m \geq \max(s)$, there exists $t \subseteq \mathbb{N}$ for which $s \sqsubseteq_m t$ and $F(t) \neq F(s)$. Then F is not circuit-definable.*

Proof. First, since the range of F is finite, fix a number q such that, if $F(s) \neq F(s')$, then $F(s)$ and $F(s')$ differ on some number less than or equal to q.

Suppose, for contradiction, that F is defined by a circuit $\tau(x)$; and, by Lemma 2, let $\varepsilon(x) = \{\alpha_i(x) \to \tau_i'(x) \mid 1 \leq i \leq k\}$ be a modified circuit-ensemble equivalent to $\tau(x)$. Let $\sigma_1(x), \ldots, \sigma_p(x)$ be a list, in some (arbitrary) order, of all the (modified) circuits $\sigma(x)$ such that $\sigma(x)$ occurs in a conjunct $\sigma = \emptyset$, $\sigma \neq \emptyset$, $0 \in \sigma$ or $0 \notin \sigma$ of any of the α_i. If $1 \leq j \leq p$, we write $\beta_j(x)$ for the condition $\sigma_j(x) \neq \emptyset$. If $s \subseteq \mathbb{N}$, let us say that the *profile* of s is the tuple of Boolean values $\pi(s) = \langle \beta_1(s), \ldots, \beta_p(s) \rangle$. We shall write $s \preceq t$ if (i) $s \cap \{0\} = t \cap \{0\}$, and (ii) for all j $(1 \leq j \leq p)$, $\beta_j(s) = \top$ implies $\beta_j(t) = \top$.

Now select a $\preceq$-maximal finite, non-empty set s. That is: if u is finite with $s \preceq u$, then $\pi(s) = \pi(u)$. This must be possible, since the number of β_j is finite. For all j $(1 \leq j \leq p)$, let m_j be some element of $\sigma_j(s)$ if $\beta_j(s) = \top$, and 0 otherwise. Let $m = \max(s \cup \{m_j \mid 1 \leq j \leq p\} \cup \{q\})$. By the hypothesis of the lemma, let t be a subset of $\mathbb{N}$ such that $s \sqsubseteq_m t$ and $F(s) \neq F(t)$. Since $s \sqsubseteq_m t$, we have $s_{|m} = t_{|m}$. Then $\beta_j(s) = \top \Rightarrow m_j \in \sigma_j(s)_{|m} \Rightarrow m_j \in \sigma_j(t)_{|m}$ (by Lemma 1) $\Rightarrow \beta_j(t) = \top$. Moreover, we certainly have $s \cap \{0\} = t \cap \{0\}$. On the other hand, since $F(s) \neq F(t)$, it follows that s and t have different profiles. For otherwise, s and t would satisfy the same $\alpha_j(x)$ $(1 \leq j \leq k)$, whence $F(s) = \tau_j'(s)$ and $F(t) = \tau_j'(t)$. But $s_{|m} = t_{|m}$ would then imply $\tau_j'(s)_{|m} = \tau_j'(t)_{|m}$ (by Lemma 1), contradicting the fact that $F(s)$ and $F(t)$ disagree on some number less than or equal to $q \leq m$. Thus, there exists h $(1 \leq h \leq p)$ such that $\beta_h(t) = \top$ and $\beta_h(s) = \bot$. Let m_h be a number in $\beta_h(t)$, let $m^* = \max(m, m_h)$, and let $u = t_{|m^*}$. Then u is finite, $s \preceq u$ (by the same argument as for $s \preceq t$), and $\beta_h(u) = \top \neq \beta_h(s)$. This contradicts the $\preceq$-maximality of s among finite sets.

Corollary 5. *The function* $F_{\mathrm{fin}} : 2^{\mathbb{N}} \to 2^{\mathbb{N}}$ *such that* $F_{\mathrm{fin}}(s) = \top$ *if s is finite, and* $F_{\mathrm{fin}}(s) = \bot$ *otherwise, is not circuit-definable. Further, no circuit-definable function* $F : 2^{\mathbb{N}} \to 2^{\mathbb{N}}$ *satisfies any of the following conditions for all finite (non-empty)* $s \subseteq \mathbb{N}$:

$$F(s) = \begin{cases} \top & \text{if } |s| \text{ is even} \\ \bot & \text{otherwise;} \end{cases} \quad F(s) = \begin{cases} \top & \text{if } \max(s) \text{ even} \\ \bot & \text{otherwise;} \end{cases} \quad F(s) = \begin{cases} \top & \text{if } \sum s \text{ even} \\ \bot & \text{otherwise.} \end{cases}$$

Proof. By considering the function $s \mapsto F(s) \bullet \{0\}$ if necessary, we may assume without loss of generality that the range of F is finite. Now apply Theorem 3.

Observe that the circuit $(\tau_e \cap x) \bullet \{0\}$, where τ_e is given in (1), defines a function F such that, for all finite s, $F(s) = \top$ if $\prod s$ is even, and $F(s) = \bot$ otherwise!

Corollary 6. *No circuit-definable function* $F : 2^{\mathbb{N}} \to 2^{\mathbb{N}}$ *satisfies the condition* $F(s) = \{\sum s\}$ *for all finite* $s \subseteq \mathbb{N}$. *No circuit-definable function* $F : 2^{\mathbb{N}} \to 2^{\mathbb{N}}$ *satisfies the condition* $F(s) = \{|s|\}$ *for all finite* $s \subseteq \mathbb{N}$.

Proof. Suppose, for contradiction, that the circuit $\tau(x)$ computes the sum of any finite argument s. Then the circuit $(\tau_e \cap \tau(x)) \bullet \{0\}$, where τ_e is given in (1), violates Corollary 5. The case of $|s|$ is handled similarly.

Acknowledgement

The second author is supported by NSERC. Both authors gratefully acknowledge the support of the EPSRC (grant ref. EP/F069154), and would like to thank Alasdair Urquhart for acquainting them with arithmetic circuits.

References

1. Glaßer, C., Herr, K., Reitwießner, C., Travers, S., Waldherr, M.: Equivalence problems for circuits over sets of natural numbers. In: Diekert, V., Volkov, M.V., Voronkov, A. (eds.) CSR 2007. LNCS, vol. 4649, pp. 127–138. Springer, Heidelberg (2007)
2. Glaßer, C., Reitwießner, C., Travers, S., Waldherr, M.: Satisfiability of algebraic circuits over sets of natural numbers. In: Arvind, V., Prasad, S. (eds.) FSTTCS 2007. LNCS, vol. 4855, pp. 253–264. Springer, Heidelberg (2007)
3. Harrow, K.: The bounded arithmetic hierarchy. Information and Control 36(1), 102–117 (1978)
4. Hartmanis, J., Shank, H.: On the recognition of primes by automata. Journal of the Association for Computing Machinery 15(3), 382–389 (1968)
5. Jeż, A., Okhotin, A.: Complexity of solutions of equations over sets of natural numbers. In: Albers, S., Weil, P. (eds.) Proceedings of STACS 2008, Internationales Begegnungs- und Forschungszentrum für Informatik, Schloß Dagstuhl. Dagstuhl Seminar Proceedings, vol. 08001, pp. 373–384 (2008)
6. Jeż, A., Okhotin, A.: On the computational completeness of equations over sets of natural numbers. In: Aceto, L., Damgård, I., Goldberg, L.A., Halldórsson, M.M., Ingólfsdóttir, A., Walukiewicz, I. (eds.) ICALP 2008, Part II. LNCS, vol. 5126, pp. 63–74. Springer, Heidelberg (2008)
7. McKenzie, P., Wagner, K.: The complexity of membership problems for circuits over sets of natural numbers. In: Alt, H., Habib, M. (eds.) STACS 2003. LNCS, vol. 2607, pp. 571–582. Springer, Heidelberg (2003)
8. McKenzie, P., Wagner, K.: The complexity of membership problems for circuits over sets of natural numbers. Computational Complexity 16(3), 211–244 (2007)
9. Ritchie, R.: Classes of predictably computable functions. Transactions of the American Mathematical Society 106, 139–173 (1963)
10. Stockmeyer, L., Meyer, A.: Word problems requiring exponential time (preliminary report). In: Proceedings of the Fifth Annual ACM Symposium on Theory of Computing, ACM Digital Library, pp. 1–9 (1973)
11. Travers, S.: The complexity of membership problems for circuits over sets of integers. Theoretical Computer Science 369, 211–229 (2006)
12. Yang, K.: Integer circuit evaluation is PSPACE-complete. In: Proceedings, 15th IEEE Conference on Computational Complexity, pp. 204–211. IEEE, Los Alamitos (2000)

Survey on Oblivious Routing Strategies

Harald Räcke

Computer Science Department
University of Warwick
harry@dcs.warwick.ac.uk

Abstract. We give a survey about recent advances in the design of oblivious routing algorithms. These routing algorithms choose their routing paths independent of the traffic in the network and they are therefore very well suited for distributed environments in which no central entitiy exist that could make routing decisions based on the whole traffic pattern in the network.

1 Introduction

Routing protocols play a major role for the performance and usability of massively parallel systems or large-scale communication networks like the Internet. A fundamental component in a routing protocol is a path selection strategy. Given a routing request consisting of a source-target pair $(s, t) \in V \times V$ from a network $G = (V, E)$, this strategy determines a path from the source s to the target t along the edges of the network.

In general, the strategy could base its path selection on a variety of information, like e.g. the paths that have previously been chosen for other requests, or à priori knowledge of the request pattern, if available. A path selection that uses this information is called *adaptive* as it can dynamically adapt its routing decisions to the current traffic in the system. This approach promises the best possible quality of the computed routing paths but has the severe disadvantage that it may be difficult to implement in a distributed scenario. The reason for this is that all non-local information that is used for making a routing decision has also to be distributed to the node making this decision. This requires extra communication which also needs to be taken into account when judging the performance of a path selection strategy.

Oblivious routing is a totally different approach. For an oblivious algorithm the route chosen for a request may not depend on any other request, which means that routing decisions are completely independent from the current traffic in the network. In fact, for an oblivious algorithm a routing path may only depend on the source node, the target node, and on some random bits for the case of randomized routing algorithms. Therefore, such an algorithm can be implemented very easily via routing tables that store at each node $v \in V$, a probability distribution over paths for each possible destination node. Alternatively, we can view a randomized oblivious routing algorithm as specifying a unit flow between every source target node in the network, while a deterministic oblivious algorithm specifies a single routing path for every pair.

K. Ambos-Spies, B. Löwe, and W. Merkle (Eds.): CiE 2009, LNCS 5635, pp. 419–429, 2009.

Because of this simplicity *obliviousness* is a very desirable feature for a routing algorithm. However, it is important to quantify the possible loss in performance that results from restricting yourself to oblivious routing algorithms instead of considering their adaptive counterparts.

In the following we review different results from the theory of oblivious routing algorithms. These results show that for many combinations of cost-measures and for many networks oblivious routing proves a powerful concept that provides performance guarantees similar to that of adaptive routing.

The model. We model the network as a edge-weighted graph $G = (V, E)$ with node set V. We use n to denote the cardinality of V, i.e., $|V| = n$. A weight function $c : E \to \mathbb{R}_0^+$ describes for an edge $e \in E$ the link-capacity of the edge. We assume that the weight function c is normalized, i.e., the minimum nonzero capacity of a link is 1. We denote the maximum capacity of a network link with $c_{\max}$.

A randomized oblivious routing scheme consists of a probability distribution over s-t paths for each source-target pair s, t. Equivalently, such a probability distribution can be viewed as a unit flow between s and t.

For simplifying the following discussion we will often assume that the routing algorithm that uses the output of the oblivious path selection strategy may route fractionally, i.e., a routing request of demand d between s and t may be fulfilled via a flow of value d between s and t, and is not restricted to use only a single path.

When using fractional routing we can neglect individual routing requests and only need the total demand between every source-target pair in order to evaluate the different performance metrics of the path selection strategy. This is done via a *demand matrix* D which is an $n \times n$ nonnegative matrix where the diagonal entries are 0.

For a given routing algorithm and a demand matrix we define the (absolute) load of a link as the total amount of data transmitted by the link. The relative load of a link is defined to be its load divided by its capacity. Finally, we define the *congestion* to be the maximum over the relative loads of all links in the network. We define the *dilation* as the length of the longest path that experiences non-zero traffic. Finally, we define the *total load* as the sum of the absolute loads over all network links.

Cost-measures. There exist several different (sometimes conflicting) objectives that characterize a good path-system – each of which may be more or less important, depending on the specific application scenario.

The first requirement is that one usually aims at having short routing paths. This leads either to the minimization of the *dilation* or the *total load* (where the total load can be viewed as minimizing the average path length of messages). Both these cost-measures correspond to minimizing the latency in the network.

However, only focusing on these distance-based cost-measures may create serious bottlenecks in the network and, hence, may adversely affect the throughput of the system. In order to enable a high throughput the communication load has to be distributed evenly among all network resources. This corresponds to minimizing the *congestion*. This is usually the main objective considered when analyzing virtual circuit routing algorithms.

In packet routing if one is given a set of packets and for each packet a path from its source to its destination, the routing time that is required to *schedule* the packets to their destinations is close to $C + D$ for many protocols [19,20], where C denotes the congestion and D denotes the dilation of the path system. Hence, for packet routing applications *congestion+dilation* is the cost-measure that should be minimized.

The following results are stated either in absolute terms or are formulated in a competitive/approximate framework. A result of the first type would, e.g. be that *"any permutation in a hypercube can be routed by an oblivious algorithm with congestion at most $O(\log n)$"* while the approximate framework gives results that compare the performance of the oblivious algorithm to the performance of an optimal algorithm, i.e., it gives result of the form *"any routing instance in a mesh can be routed with congestion at most $O(\log n) \cdot C_{\mathrm{opt}}$, where C_{opt} denotes the optimal congestion that can be obtained for the instance"*. Usually results of the second type are more desirable. For these, the factor between the performance of the oblivious scheme and an optimal path selection is called the *competitive ratio*.

2 Survey of Some Results

2.1 Deterministic Oblivious Routing

One of the first results about oblivious routing is a result by Borodin and Hopcroft [7] that essentially states that randomization is required in order to obtain a good performance. They show that given an (unweighted, i.e., $\forall e \in E : c(e) = 1$) network with n nodes and maximum degree Δ, together with a deterministic oblivious routing scheme, there exists a permutation routing instance such that the congestion induced by the oblivious routing scheme is at least $\Omega(\sqrt{n}/\Delta^{3/2})$. This means, that when using deterministic oblivious path selection in a network with small degree, then a packet routing algorithm that uses this path selection will require at least $\Omega(\sqrt{n}/\Delta^{3/2})$ for routing a permutation. However, there are small-degree networks like the hypercube in which any permutation can be routed in only $O(\log n)$ steps. Hence, when evaluating oblivious routing strategies in a competitive framework, this leads to a large competitive ratio.

This result was improved by Kaklamanis et al [15] to a lower bound of $\Omega(\sqrt{n}/\Delta)$, which is in particular important for networks with non-constant degree like the hypercube.

2.2 Cost Measure Congestion+Dilation

Because of their simple implementations, much effort has been made to design oblivious routing algorithms for specific network topologies. Valiant and Brebner [26] were the first to perform a worst case theoretical analysis for oblivious routing on the hypercube. They design a randomized packet routing algorithm that routes any permutation in $O(\log n)$ steps. As a path-selection strategy they use a randomized oblivious strategy that is known as "Valiant's trick". Instead of routing directly from a source node s to the target t, route first to a random intermediate destination x chosen uniformly among the nodes of the hypercube. The sub-paths between s-x and x-t use greedy dimension-order routing. This trick distributes the load evenly among the network resources, and the expected load received by any edge is only $O(1)$.

To see this consider only the load that is generated by routing from sources to intermediate destinations. Every source distributes its load in a fractional way on 2^d different routing paths (possible choices of intermediate nodes), where d denotes the dimension of the hypercube. In total there are 2^{2d} paths each of length at most d. Because of the symmetry in the hypercube each of the $d \cdot 2^d/2$ edges has the same number of paths passing through it. Therefore, the number of paths passing through an edge is at most $\frac{d \cdot 2^{2d}}{d \cdot 2^d/2} = 2^{d+1}$. Each path carries an (expected) load of at most $1/2^d$ and hence the expected load of an edge is constant.

A disadvantage of this technique is that the path length increase from d to up to $2d$. Möcking [28] has introduced a different way of choosing the intermediate nodes that guarantees that path length are bounded by d. Note that in these works finding a good path system is only one part of the routing scheme, as their is also a packet scheduling mechanism required that decides in which order to forward packets in the network. Further, note that in the above analysis we only considered the expected load on any edge. When the routing is not fractional one can use Chernoff bounds and obtains that with high probability the maximum load of an edge is logarithmic.

Following the results by Valiant and Brebner, their technique has been applied to a variety of other networks like fat trees, expanders, Caley graphs, shuffle exchange network, De Bruijn network etc. [21,25]. A more general version of Valiants trick is given by Kolman and Scheideler [16]. They introduce the flow number F of a network which is defined as follows. Define a multicommodity flow problem in which the demand between a node u and node v is defined as $\Delta(u)\Delta(v)/\Delta(V)$, where for a node $x \in V$, $\Delta(x)$ denotes the (weighted) degree of x and $\Delta(V)$ is defined as $\Delta(V) := \sum_{x \in V} \Delta(x)$. The flow number $F(G)$ is the minimum of $\max(\text{Congestion}(\mathcal{S}), \text{Dilation}(\mathcal{S}))$ taken over all feasible solutions $\mathcal{S}$. They show that by using Valiant's trick any *balanced multicommodity flow problem* can be routed obliviously with congestion and dilation at most $O(F)$ (A balanced multicommodity flow problem is an instance in which the total traffic sent and received at any node is less than the total capacity of edges adjacent to the node).

When applying Valiant's trick to the problem of say permutation routing on a 2-dimensional grid of n nodes one obtains a path system with congestion at most $O(\sqrt{n})$ and dilation $O(\sqrt{n})$. Then one can use packet scheduling algorithms to route packets in nearly $O(\sqrt{n})$ steps. This is optimal in the sense that their exist permutations that require this many steps. However, for certain inputs like permutations that send packets to direct neighbors this bound may be very weak

Scheideler gives an oblivious path-selection scheme for 2-dimensional meshes that obtains congestion $O(\log n \cdot C_{opt})$ and dilation $O(D_{opt})$, where C_{opt} and D_{opt} denote the optimum congestion and optimum dilation, respectively, that can be obtained for the given instance.

A different algorithm with similar guarantees is given by Busch et al. [9]. This latter algorithm is also generalized to higher dimensional meshes. The same set of authors also provide oblivious routing algorithms that approximate the sum of congestion and dilation for so-called *geometric networks* [8] that are used to model wireless networks.

2.3   Cost-Measure Congestion

When considering virtual circuit routing the most important cost-measure is the congestion of the network links because this has a high impact on the throughput. Of course, all bounds from the previous section also hold for this cost-measure. There are very simple (undirected) networks in which no oblivious algorithm can obtain a good competitive ratio with respect to the sum of congestion and dilation (see [23]).

For a long time it was unclear whether it is possible to obtain a small competitive ratio for any network with an oblivious routing scheme when only considering the cost-measure congestion. In [17] Maggs et al. introduced a new approach for data management on grids that is based on a hierarchical decomposition of the grid into smaller and smaller sub-grids. This hierarchical decomposition corresponds to a decomposition tree. Maggs et al. show that a routing instance on the grid G can be efficiently simulated on the tree T so that the congestion on T is at most equal to the best possible congestion that could be obtained on G. In a second step they show that the grid can simulate a routing on T while only experiencing a logarithmic increase in congestion. This bi-simulation approach also gives an oblivious routing algorithm for the grid with a logarithmic competitive ratio with respect to congestion.

This result in itself is not an improvement over the algorithm by Scheideler mentioned in the previous section as it obtains the same competitive ratio with respect to congestion and does not give a bound on the path length. However, the strong relationship between the grid and its decomposition tree allows it to solve arbitrary problems (that aim at minimizing the congestion) first on the decomposition tree, and then to transfer this solution to the grid with only a modest loss in performance. They use this to solve a complex data management problem, which in particular generalizes the virtual circuit routing problem. They apply their decomposition technique to many other networks like e.g. expanders, hypercubic networks (i.e. shuffle exchange network, De Bruijn network, Benes network etc.), and Caley graphs [17,27,29].

Vöcking [27] conjectured that their framework can be extended to arbitrary undirected graphs by finding a suitable hierarchical decomposition. In 2002 Räcke [22] proved that this conjecture is true. He showed that for any graph G there exists a decomposition tree T that allows a bi-simulation between G and T according to the framework of Maggs et al. such that the increase in congestion when transferring a solution from T to G is only $O(\log^3 n)$. This gives for any graph an oblivious routing algorithm with competitive ratio $O(\log^3 n)$.

One disadvantage of his approach was that his result was non-constructive in the sense that only an exponential time algorithm was given for constructing the routing scheme. This issue was subsequently addressed by Azar et al. [2] who show that the optimum oblivious routing scheme, i.e., the scheme that guarantees the best possible competitive ratio, can be constructed in polynomial time by using a linear program. The method by Azar et al. does not give the possibility to derive general bounds on the competitive ratio for certain types of graphs. It can only give the best oblivious algorithm and the corresponding competitive ratio for the graph in question.

Another disadvantage of [2] was that it did not give a polynomial time construction of the hierarchy used in [22], which has proven to be useful in many other applications (see e.g. [1,11,18]). A polynomial time algorithm for this problem was independently given by Bienkowski et al. [6] and Harrelson et al. [13]. Whereas the first result shows a slightly weaker competitive ratio for the constructed hierarchy than the non-constructive result in the original paper [22], the latter paper by Harrelson, Hildrum and Rao has improved the competitive ratio to $O(\log^2 n \log \log n)$ using a technique originally introduced by Seymour. A slight caveat of both these polynomial time constructions is that they are only pseudo-polynomial because their running time is polynomial in the maximum edge-capacity $c_{\max}$ which may be super-polynomial in n.

In 2008 Räcke [24] introduced a routing scheme that is not based on a single decomposition tree that then is simulated in a complicated way on the graph G but instead is simply based on a convex combination of different trees in which the leaf nodes correspond to nodes in G. For each tree $T = (V_t, E_t)$ in the convex combination there exist embedding functions $m_V : V_t \rightarrow V$ and $m_E : E_t \rightarrow E$. The function m_V maps nodes of T to nodes of G (and in particular it induces a bijection between leafs of T and nodes of G). The function m_E maps an edge $e_t = (u_t, v_t)$ of T to a path $P_{u,v}$ in G between the corresponding end-points $u = m_V(u_t)$ and $v = m_V(v_t)$.

The oblivious routing algorithm then is very simple. For routing traffic from a source s to a target node t the traffic is first split among the different trees (according to the weight of the convex multiplier). Then, in each tree the unique path between the leaf nodes corresponding to s and t is chosen. In order to obtain paths in G the tree-paths are then mapped to G via the edge-mapping functions m_E for each tree.

It is shown that there exists a convex combination of trees such that the corresponding oblivious routing scheme has a competitive ratio of $O(\log n)$ with respect to congestion. This bound is optimal as it is not possible to obtain a sub-logarithmic competitive ratio even on the 2-dimensional grid (see [17,5]).

2.4   Cost-Measure Total Communication Load

Designing an oblivious routing algorithm that minimizes the total communication load is straightforward as one only needs to apply shortest path routing which is oblivious to the traffic in the network. However, in this section we want to give a very brief overview over algorithms that are not only oblivious but they are *tree based*, i.e., there structure is very similar to the oblivious strategies that obtain the optimal competitive ratio of $O(\log n)$ for the cost-measure congestion.

These algorithm are well known under the concept of probabilistically approximating metrics by tree metrics which was introduced by Bartal in 1996 [3]. Call a tree T a *decomposition tree* for a graph G if there is a bijection between the leaf nodes of T and the nodes of G. Bartal shows how to compute for a given graph G a convex combination of trees such that a) for each tree T the distance between two leafs is larger than the distance of the corresponding nodes in the original graph, and b) the *expected tree distance* between two nodes when choosing a random tree is not much larger (only a factor $O(\log^2 n)$) than their distance in the graph.

One can easily define node-mapping function m_V and edge-mapping functions n_E for the trees such that the oblivious routing scheme based on these functions (see the description of tree-based routing at the end of Section) obtains a competitive ratio of $O(\log^2 n)$ with respect to the total communication load. Of course, the main application for these trees is not routing but they allow to solve any problem for which the objective function can be expressed as a sum of distances on trees. The tree solution can then be transferred to the real graph with only a small increase in approximation guarantee.

This approach of reducing a distance-related problem on a general undirected graph to a tree network has been very valuable for many optimization and online problems on graphs. The original competitive ratio of $O(\log^2 n)$ has been first improved by Bartal [4] to $O(\log n \log\log n)$ and later by Fakcharoenphol, Rao and Talwar [12] to $O(\log n)$ which is optimal.

3   Connections between Congestion Based and Distance Based Strategies

Interestingly, the optimum strategies for the cost measure congestion and the tree based strategies for the cost measure total communication load are closely related. In [10] Charikar et al. showed how to obtain a probabilistic approximation of metrics by tree metrics via the Minimum Communication Cost Tree Problem which is defined as follows.

We are given an undirected graph in which every edge $e \in E$ has an associated *length* $\ell(e)$, and we use $\mathrm{dist}(u,v)$ to denote the resulting shortest path distance between two nodes $u, v \in V$. Furthermore, we are given a demand function $d : V \times V \to \mathbb{R}_0^+$ that specifies an amount of traffic that has to be sent between u and v.

The goal is to route the demands in a *tree-like fashion* while minimizing the total traffic. Formally, the task is to construct a *decomposition tree* $T = (V_t, E_t)$, that minimizes

$$\mathrm{cost}(T) = \sum_{(u,v)} \mathrm{dist}_T(u,v) \cdot d(u,v),$$

where $\text{dist}_T(u, v)$ denotes the distance when connecting u and v via the tree. This is defined as follows. We define the length of a tree edge $e_t = (u_t, v_t)$ as the length of the corresponding path $m_E(e_t)$ in G (Here m_E denotes the edge-mapping function that maps tree edges to paths in G. See the description at the end of Section). Then, the tree-distance $\text{dist}_T(u, v)$ between graph nodes u and v is given by the shortest path distance between the corresponding leaf-nodes in T.

We can also write the above cost in a different way. A tree edge $e_t = (u_t, v_t)$ partitions the leaf nodes of the tree and, hence, the nodes of the graph, into two disjoint sets V_{u_t} and V_{v_t}. Let $d(e_t) := \sum_{u \in V_{u_t}, v \in V_{v_t}} d(u, v)$ denote the total demand that has to cross the corresponding cut. All this traffic has to be forwarded via the path $m_E(e_t)$.

We define the *load* $\text{load}_T(e)$ that is induced on an edge $e \in E$ by tree T (and its embedding) as

$$\text{load}_T(e) := \sum_{e_t \in E_t : e \in m_E(e_t)} d(e_t),$$

which is the total traffic that goes over e if the demand is routed via T along the chosen path system. With these definitions we can write the cost of a decomposition tree for a Minimum Communication Cost Tree instance as

$$\text{cost}(T) = \sum_{e \in E} \text{load}_T(e) \cdot \ell(e).$$

We will use the following result about the Minimum Communication Cost Tree Problem which is due to Fakcharoenphol, Rao and Talwar [12], and states that routing along a tree only costs a logarithmic factor more than shortest path routing.

Theorem 1 (Fakcharoenphol et al. [12]). *Given an instance for the Minimum Communication Cost Tree Problem, a solution with cost $O(\log n) \cdot \sum_{u,v} d(u, v) \cdot \text{dist}(u, v)$ can be computed in polynomial time.*

3.1   Finding Good Trees for Distance Based Routing

Charikar et al. formulated the problem of finding the best convex combination of decomposition tree for distance based routing as the following linear program:

$$
\begin{aligned}
\text{minimize} \quad & \delta \\
\text{subject to:} \quad \forall u, v \quad & \sum_i \lambda_i \cdot \text{dist}_{T_i}(u, v) \leq \delta \cdot \text{dist}(u, v) \\
& \sum_i \lambda_i \geq 1 \\
\forall i \quad & \lambda_i \geq 0
\end{aligned}
\qquad (\text{Primal 1})
$$

This gives a convex combination of decomposition trees such that the expected stretch of any pair (u, v) is at most δ. The dual of this linear program is the following.

$$\text{maximize} \quad \beta$$

$$\text{subject to:} \quad \forall i \quad \sum_{u,v} d(u,v) \cdot \text{dist}_{T_i}(e) \;\geq\; \beta$$
$$\sum_e d(u,v) \cdot \text{dist}(u,v) \;\leq\; 1 \qquad \text{(DUAL 1)}$$
$$\forall u,v \qquad\qquad\qquad d(u,v) \;\geq\; 0$$

This asks for a demand between the vertices in the graph such that the total traffic $\sum_{u,v} d(u,v)\, \text{dist}(u,v)$ when routing these demands along shortest path in the graph is at most 1, while the cost when routing the demands along any decomposition tree is large (at least β). However, Theorem says that this is not possible if $\beta \geq \omega(\log n)$ as the theorem will always find a violated constraint.

3.2   Finding Good Trees for Congestion Minimization

For finding good tree decompositions for routing with respect to the cost-measure congestion a very similar approach can be used [24]. The problem can be formulated as the following linear program. Here we define the demand $d(u,v)$ between two nodes as the capacity of the edge between them ($d(u,v) = 0$ if there is no edge). Then $d(e_t)$ is equal to the capacity of the cut that is induced by e_t. This means it is an upper bound on the demand that has to cross the cut (or the tree edge e_t) in any feasible routing instance (i.e., an instance that can be solved with congestion at most 1).

$$\text{minimize} \quad \delta$$

$$\text{subject to:} \quad \forall e \in E \quad \sum_i \lambda_i \cdot \text{load}_{T_i}(e) \;\leq\; \delta \cdot c(e)$$
$$\sum_i \lambda_i \;\geq\; 1 \qquad \text{(PRIMAL 2)}$$
$$\forall i \qquad\qquad\qquad \lambda_i \;\geq\; 0$$

Observe that $\sum_i \lambda_i \cdot \text{load}_{T_i}(e)$ is an upper bound on the traffic that has to be forwarded by graph edge e (in a feasible instance). Therefore, the above LP minimizes the factor by which the edge is *overloaded*. The dual of this LP is as follows.

$$\text{maximize} \quad \beta$$

$$\text{subject to:} \quad \forall i \quad \sum_e \ell(e) \cdot \text{load}_{T_i}(e) \;\geq\; \beta$$
$$\sum_e \ell(e) \cdot c(e) \;\leq\; 1 \qquad \text{(DUAL 2)}$$
$$\forall e \in E \qquad\qquad \ell(e) \;\geq\; 0$$

This LP asks for a length assignment to the edges of the graph that fulfills certain properties. However, observe that the term $\sum_e \ell(e) \cdot \text{load}_{T_i}(e)$ denotes simply the cost for routing along the tree T_i in a Minimum Communication Cost Tree problem as shown in the previous section. We have set the demand to $d(u,v) = \ell((u,v))$ for every edge (u,v). This means the cost for shortest path routing is given by $\sum_{u,v} d(u,v)\,\text{dist}(u,v) = \sum_{e=(u,v)} c(e)\,\text{dist}(u,v) \le \sum_{e=(u,v)} c(e)\ell(e) \le 1$ as the last equation is a constraint in the linear program. Therefore, Theorem gives us a tree for which the routing cost is at most $O(\log n)$. Hence, the solution to the LP will have $\beta = O(\log n)$ as otherwise we would be able to find a violated constraint.

4   Conclusions

We have given a survey about oblivious routing algorithms. There is a remarkable similarity between strategies for congestion minimization in oblivious routing and the technique of approximating arbitrary metrics by tree metrics which can be viewed as giving a tree based oblivious algorithm for minimizing the total communication load. It seems to be an important open question whether this relationship can be extended to find tree based oblivious routing algorithm for other cost measures apart from congestion or total load.

One interesting cost-measure would be sum-of-squares where we take the sum of squares of all edge-loads in the network. This is interesting as it corresponds to minimizing average latency for networks with linear latency functions. So far only very limited results exist for this cost-measure (see [14]).

References

1. Alon, N., Awerbuch, B., Azar, Y., Buchbinder, N., Naor, J.S.: A general approach to online network optimization problems. In: Proc. of the 15th SODA, pp. 577–586 (2004)
2. Azar, Y., Cohen, E., Fiat, A., Kaplan, H., Räcke, H.: Optimal oblivious routing in polynomial time. In: Proc. of the 35th STOC, pp. 383–388 (2003)
3. Bartal, Y.: Probabilistic approximations of metric spaces and its algorithmic applications. In: Proc. of the 37th FOCS, pp. 184–193 (1996)
4. Bartal, Y.: On approximating arbitrary metrics by tree metrics. In: Proc. of the 30th STOC, pp. 161–168 (1998)
5. Bartal, Y., Leonardi, S.: On-line routing in all-optical networks. Theor. Comput. Sci. 221(1-2), 19–39 (1999); Also In: Degano, P., Gorrieri, R., Marchetti Spaccamela, A. (eds.) ICALP 1997. LNCS, vol. 1256, pp. 516–526. Springer, Heidelberg (1997)
6. Bienkowski, M., Korzeniowski, M., Räcke, H.: A practical algorithm for constructing oblivious routing schemes. In: Proc. of the 15th SPAA, pp. 24–33 (2003)
7. Borodin, A., Hopcroft, J.E.: Routing, merging and sorting on parallel models of computation. J. Comput. Syst. Sci. 30(1), 130–145 (1985)
8. Busch, C., Magdon-Ismail, M., Xi, J.: Oblivious routing on geometric networks. In: Proc. of the 17th SPAA, pp. 316–324 (2005)
9. Busch, C., Magdon-Ismail, M., Xi, J.: Optimal oblivious path selection on the mesh. In: Proc. of the 20th IPDPS, pp. 82–91 (2005)
10. Charikar, M., Chekuri, C., Goel, A., Guha, S., Plotkin, S.A.: Approximating a finite metric by a small number of tree metrics. In: Proc. of the 39th FOCS, pp. 379–388 (1998)

11. Chekuri, C., Khanna, S., Shepherd, B.: The all-or-nothing multicommodity flow problem. In: Proc. of the 36th STOC, pp. 156–165 (2004)

12. Fakcharoenphol, J., Rao, S.B., Talwar, K.: A tight bound on approximating arbitrary metrics by tree metrics. In: Proc. of the 35th STOC, pp. 448–455 (2003)

13. Harrelson, C., Hildrum, K., Rao, S.B.: A polynomial-time tree decomposition to minimize congestion. In: Proc. of the 15th SPAA, pp. 34–43 (2003)

14. Harsha, P., Hayes, T.P., Narayanan, H., Räcke, H., Radhakrishnan, J.: Minimizing average latency in oblivious routing. In: Proc. of the 19th SODA, pp. 200–207 (2008)

15. Kaklamanis, C., Krizanc, D., Tsantilas, T.: Tight bounds for oblivious routing in the hypercube. In: Proc. of the 2nd SPAA, pp. 31–36 (1990)

16. Kolman, P., Scheideler, C.: Improved bounds for the unsplittable flow problem. J. Algorithms 61(1), 20–44 (2006)

17. Maggs, B.M., Meyer auf der Heide, F., Vöcking, B., Westermann, M.: Exploiting locality for networks of limited bandwidth. In: Proc. of the 38th FOCS, pp. 284–293 (1997)

18. Maggs, B.M., Miller, G.L., Parekh, O., Ravi, R., Woo, S.L.M.: Finding effective support-tree preconditioners. In: Proc. of the 17th SPAA, pp. 176–185 (2005)

19. Meyer auf der Heide, F., Vöcking, B.: A packet routing protocol for arbitrary networks. In: Mayr, E.W., Puech, C. (eds.) STACS 1995. LNCS, vol. 900, pp. 291–302. Springer, Heidelberg (1995)

20. Ostrovsky, R., Rabani, Y.: Universal $O(\text{congestion} + \text{dilation} + \log^{1+\epsilon} N)$ local control packet switching algorithms. In: Proc. of the 29th STOC, pp. 644–653 (1997)

21. Rabin, M.O.: Efficient dispersal of information for security, load balancing and fault tolerance. J. ACM 36, 335–348 (1989)

22. Räcke, H.: Minimizing congestion in general networks. In: Proc. of the 43rd FOCS, pp. 43–52 (2002)

23. Räcke, H.: Data Management and Routing in General Networks. PhD thesis, Universität Paderborn (2003)

24. Räcke, H.: Optimal hierarchical decompositions for congestion minimization in networks. In: Proc. of the 40th STOC, pp. 255–264 (2008)

25. Upfal, E.: Efficient schemes for parallel communication. J. ACM 31(3), 507–517 (1984)

26. Valiant, L.G., Brebner, G.J.: Universal schemes for parallel communication. In: Proc. of the 13th STOC, pp. 263–277 (1981)

27. Vöcking, B.: Static and Dynamic Data Management in Networks. PhD thesis, Universität Paderborn (December 1998)

28. Vöcking, B.: Almost optimal permutation routing on hypercubes. In: Proc. of the 33rd STOC, pp. 530–539 (2001)

29. Westermann, M.: Caching in Networks: Non-Uniform Algorithms and Memory Capacity Constraints. PhD thesis, Universität Paderborn (November 2000)

An Approach to the Engineering of Cellular Models Based on P Systems

Francisco J. Romero-Campero and Natalio Krasnogor

Automated Scheduling, Optimisation and Planning Research Group
School of Computer Science, Jubilee Campus, University of Nottingham
Nottingham NG8 1BB, United Kingdom
{fxc,nxk}@cs.nott.ac.uk

Extended Abstract

Living cells assembled into colonies or tissues communicate using *complex systems*. These systems consist in the interaction between many molecular species distributed over many compartments. Among the different cellular processes used by cells to monitor their environment and respond accordingly, gene regulatory networks, rather than individual genes, are responsible for the information processing and orchestration of the appropriate response [16].

In this respect, *synthetic biology* has emerged recently as a novel discipline aiming at unravelling the design principles in gene regulatory systems by synthetically engineering transcriptional networks which perform a specific and prefixed task [2]. Formal modelling and analysis are key methodologies used in the field to engineer, assess and compare different genetic designs or devices.

In order to model cellular systems in colonies or tissues one requires a formalism able to represent the following relevant features:

- Single cells should be described as the *elementary units* in the system. Nevertheless, they cannot be represented as homogeneous points as they exhibit *complex structures* containing different compartments where specific molecular species interact according to particular reactions.
- The molecular interactions taking place in cellular systems are inherently *discrete and stochastic processes*. This is a key feature of cellular systems that needs to be taken into account when describing their dynamics [9].
- It has been postulated that gene regulatory networks are organised in a *modular* manner in such a way that cellular processes arise from the orchestrated interactions between different genetic transcriptional units that can be considered separable modules [1].
- Spatial and geometric information must be represented in the system in order to describe processes involving *pattern formation*.

In this work we review recent advances in the use of the computational paradigm *membrane computing* or *P systems* as a formal methodology in *synthetic biology* for the specification and analysis on cellular system models according to the previously presented points.

K. Ambos-Spies, B. Löwe, and W. Merkle (Eds.): CiE 2009, LNCS 5635, pp. 430–436, 2009.
© Springer-Verlag Berlin Heidelberg 2009

Stochastic P systems and modularity

P systems were introduced by G. Păun in 2000 as a computational abstraction of the structure and functioning of the *living cell* [11]. The main constituents of a generic P system are the following:

- A set of membranes representing compartments arranged in a hierarchical manner (membranes can contain other membranes) all of them embedded within a single membrane called *skin* that defines the system.
- Multisets of objects specifying molecular species distributed over the regions or compartments defined by membranes.
- Rewriting rules on multisets of objects associated specifically with each compartment and describing molecular interactions which determine the computation or evolution of the system.

For a detailed description we refer to [12]. Originally, the rewriting rules were applied according to a maximally parallel and nondeterministic strategy. More specifically, all the objects in every membrane that can evolve according to a rewriting rule must do so in a single step. If an object can evolve according to more than one rewriting rule the rule that is actually applied is chosen nondeterministically [11,12]. A considerable part of the research in this field has focused on the study of the computational universality and efficiency of the different proposed variants. Nevertheless, there is an emerging application of P systems as a framework for the specification and analysis of cellular systems [4,6,7,10,13,14,15,17,18,19,21,22].

In the following we will discuss *stochastic P systems*, a variant of the generic P system that satisfies the requirements presented previously to model cellular systems.

The original maximally parallel and nondeterministic semantics for the evolution of P systems was proved not suitable for reproducing the dynamics of cellular systems since it does not capture the different rates at which molecular interactions take place. In order to solve this problem stochastic semantics based on *Gillespie's theory of stochastic kinetics* [8] were introduced in the generic model to define *stochastic P system* [14]. This variant of P systems differs from the generic model in that a *stochastic kinetic constant* is associated specifically with each rewriting rule on multisets of objects. In particular, *boundary rules* [5] of the following form are used:

$$r : \; u\,[\,v\,]_i \xrightarrow{\;c\;} u'\,[\,v'\,]_i$$

where u, v, u', v' are multisets of objects that are replaced simultaneously outside and inside the corresponding compartment i represented with square brackets. The *kinetic stochastic constant* c associated with each rule is used to compute its propensity by multiplying it with the number of distinct possible combinations of objects in the left hand side of the rule. The sum of the propensities of all the rules is taken as the parameter of an exponential distribution which determines the time elapsed between rule applications. The rule that is applied is chosen according to a multinomial distribution with parameters the normalised values of the rule propensities [8,14].

Stochastic P systems has been compared with other computational paradigms, like Petri nets and process algebra, that have also been adapted for their application to the modelling of cellular systems. Their similarities/differences and advantages/disadvantages were discussed [4,20].

As mention previously modularity is an interesting feature to study in cellular systems. Modularity can be explicitly represented in stochastic P systems by allowing the specification of the sets of rewriting rules associated with compartments as a combination of *instantiated P system modules* [21,22]. A P system module is a set of rewriting rules over multisets of variables representing objects, Var_O, whose stochastic kinetic constants are also variables Var_C. A module is identified with a name, Mod and is specified as $Mod(Var_O, Var_C)$. Then the instantiation of such module Mod with specific objects O_0, constants C_0 is specified as $Mod(O_0, C_0)$ and its rules are obtained from the rules in $Mod(Var_O, Var_C)$ by replacing all the occurrences of the variables Var_O and Var_C with the specific objects O_0 and stochastic kinetic constants C_0.

Note that the set of rewriting rules associated with a module Mod can be obtained by applying set union to simpler modules $Mod_1, \ldots, Mod_n$. In this way a nested hierarchical modular structure can be obtained.

Lattice Population P systems

Finally, a spatially distributed colony or tissue of cells can be represented using *lattice population P systems* introduced in [22]. Spatial and geometric information is introduced in P systems by using a finite regular geometrical *lattice*. The different types of cells in a colony or tissue are represented by individual stochastic P systems $\Pi_1, \ldots, \Pi_n$. Copies of these P systems are then distributed over the points of the geometrical lattice to represent the spatial distribution of cells in the colony or tissue. In this way each position $\mathbf{p}$ in the lattice has a specific P system associated with it that will be denoted $\Pi(\mathbf{p})$. The P systems distributed over the lattice can exchange objects using *translocation rules* of the following form which describe molecular processes like passive diffusion or active transport of molecular species.

$$r : [\, u\,]_l \overset{\mathbf{v}}{\leftrightharpoons} [\ ] \overset{c}{\longrightarrow} [\ ]_l \overset{\mathbf{v}}{\leftrightharpoons} [\, u\,]$$

where u is a multiset of objects representing the molecular species that are to be diffused or transported, $\mathbf{v}$ is a vector and c is the corresponding kinetic stochastic constant. When rule r is applied in the P system located in position $\mathbf{p}$, $\Pi(\mathbf{p})$ the objects u are translocated from $\Pi(\mathbf{p})$ to the P system located in position $\mathbf{p} + \mathbf{v}$, $\Pi(\mathbf{p} + \mathbf{v})$.

In order to illustrate our approach we use the following example. Our genetic design consists of three colonies of bacteria arranged in a specific spatial distribution. These colonies of bacteria posses special genetic circuits that produce the temporal expression of a particular gene in a specific colony.

Table 1. The library of modules used in our example consisting in constitutive ($UnReg$), positive ($PosReg$) and negative ($NegReg$) gene regulation as well as protein degradation (Deg) and diffusion ($Diff$)

Module	Rules
$UnReg(\{G, P\}, \{c_1\})$	$r_1 : [\, G \,]_b \xrightarrow{c_1} [\, G + P \,]_b$
$PosReg(\{A, G, P\}, \{c_1, c_2, c_3\})$	$r_1 : [\, A + G \,]_b \xrightarrow{c_1} [\, A.G \,]_b$ $r_2 : [\, A.G \,]_b \xrightarrow{c_2} [\, A + G \,]_b$ $r_3 : [\, A.G \,]_b \xrightarrow{c_3} [\, A.G + P \,]_b$
$NegReg(\{R, G\}, \{c_1, c_2\})$	$r_1 : [\, R + G \,]_b \xrightarrow{c_1} [\, R.G \,]_b$ $r_2 : [\, R.G \,]_b \xrightarrow{c_2} [\, R + G \,]_b$
$Deg(\{P\}, \{c_1\})$	$r_1 : [\, P \,]_b \xrightarrow{c_1} [\;]_b$
$Diff(\{P\}, \{c_1\})$	$r_1 : [\, P \,]_b \overset{(1,0)}{\leftrightsquigarrow} [\;] \xrightarrow{c_1} [\;]_b \overset{(1,0)}{\leftrightsquigarrow} [\, P \,]$ $r_2 : [\, P \,]_b \overset{(-1,0)}{\leftrightsquigarrow} [\;] \xrightarrow{c_1} [\;]_b \overset{(-1,0)}{\leftrightsquigarrow} [\, P \,]$ $r_3 : [\, P \,]_b \overset{(0,1)}{\leftrightsquigarrow} [\;] \xrightarrow{c_1} [\;]_b \overset{(0,1)}{\leftrightsquigarrow} [\, P \,]$ $r_4 : [\, P \,]_b \overset{(0,-1)}{\leftrightsquigarrow} [\;] \xrightarrow{c_1} [\;]_b \overset{(0,-1)}{\leftrightsquigarrow} [\, P \,]$

The three different types of bacteria forming the three colonies are specified by the P systems Π_1, Π_2 and Π_3, see Table 2. These P systems consist of a single membrane or compartment representing a bacterium. The corresponding set of rules associated with each P system is obtained as a combination of the P systems modules in Table 1. These modules represent the constitutive ($UnReg$), positive ($PosReg$) and negative ($NegReg$) regulation of genes as well as protein degradation (Deg) and diffusion ($Diff$). An additional P system Π_0 is used to represent a subvolume of the media between the different colonies.

The P systems from Table 2 are distributed over a rectangular lattice as shown in Figure 1. The leftmost colony consists of bacteria of type 1 represented by Π_1. These bacteria express constitutively $gene1$ whose protein product diffuses freely across the whole system. In the rightmost colony, described by Π_3, $gene2$ is regulated positively by $prot1$ producing the freely diffusible $prot2$. Finally, the colony in the centre of the lattice, specified by Π_2, possesses a genetic circuit according to which $prot1$ regulates positively $gene3$ whereas $prot2$ represses it.

Figure 2 shows the evolution over time of the molecules of $prot1, prot2$ and $prot3$ in the bacterium at position $(4, 2)$ in the lattice presented in Figure 1. It can be observed that at around 50 minutes $prot1$ is present in the compartment producing the activation of $gene3$ and the accumulation of $prot3$. When $prot1$ reaches the rightmost colony, $prot2$ starts to be produced and it diffuses across the system. After a delay of around 300 minutes $prot2$ reaches our bacterium at position $(4, 2)$ and represses the expression of $prot3$. Therefore our engineered design consisting in specific gene regulatory circuits and in a particular spatial

Table 2. The four different P systems, Π_0, Π_1, Π_2 and Π_3 representing a subvolume from the media, a bacterium from colony one, two and three respectively

P System	Modules [a]
Π_0	$Diff(\{prot1\}, \{0.01\})$ $Diff(\{prot2\}, \{0.01\})$
Π_1	$UnReg(\{gene1, prot1\}, \{10\})$ $Deg(\{prot1\}, \{0.0015\})$ $Deg(\{prot2\}, \{0.0015\})$ $Diff(\{prot1\}, \{0.01\})$ $Diff(\{prot2\}, \{0.01\})$
Π_2	$PosReg(\{prot1, gene3\}, \{1, 5, 5\})$ $NegReg(\{prot2, gene3\}, \{1, 0.001\})$ $Deg(\{prot1\}, \{0.0015\})$ $Deg(\{prot2\}, \{0.0015\})$ $Deg(\{prot3\}, \{0.03\})$ $Diff(\{prot1\}, \{0.01\})$ $Diff(\{prot2\}, \{0.01\})$
Π_3	$PosReg(\{prot1, gene2\}, \{1, 0.01, 10\})$ $Deg(\{prot1\}, \{0.0015\})$ $Deg(\{prot2\}, \{0.0015\})$ $Diff(\{prot1\}, \{0.01\})$ $Diff(\{prot2\}, \{0.01\})$

[a] The units of the constants associated with the modules are min^{-1}

Π_1	Π_1	Π_0	Π_0	Π_2	Π_2	Π_0	Π_0	Π_3	Π_3
(0,4)	(1,4)	(2,4)	(3,4)	(4,4)	(5,4)	(6,4)	(7,4)	(8,4)	(9,4)
Π_1	Π_1	Π_0	Π_0	Π_2	Π_2	Π_0	Π_0	Π_3	Π_3
(0,3)	(1,3)	(2,3)	(3,3)	(4,3)	(5,3)	(6,3)	(7,3)	(8,3)	(9,3)
Π_1	Π_1	Π_0	Π_0	Π_2	Π_2	Π_0	Π_0	Π_3	Π_3
(0,2)	(1,2)	(2,2)	(3,2)	(4,2)	(5,2)	(6,2)	(7,2)	(8,2)	(9,2)
Π_1	Π_1	Π_0	Π_0	Π_2	Π_2	Π_0	Π_0	Π_3	Π_3
(0,1)	(1,1)	(2,1)	(3,1)	(4,1)	(5,1)	(6,1)	(7,1)	(8,1)	(9,1)
Π_1	Π_1	Π_0	Π_0	Π_2	Π_2	Π_0	Π_0	Π_3	Π_3
(0,0)	(1,0)	(2,0)	(3,0)	(4,0)	(5,0)	(6,0)	(7,0)	(8,0)	(9,0)

Fig. 1. Spatial distribution of the three different bacterial colonies over a rectangular lattice

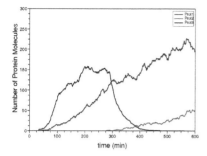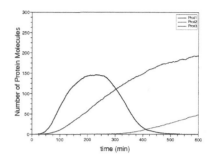

Fig. 2. Number of molecules of *prot1*, *prot2* and *prot3* in the bacterium at position (4,2) in single simulation (left) and the average over 100 simulations

distribution produces a transient expression of around 250 minutes of *prot3* in the colony at the center of our system.

Acknowledgements

We would like to acknowledge EPSRC grant EP/E017215/1 and BBSRC grant BB/F01855X/1.

References

1. Alon, U.: Network motifs: theory and experimental approaches. Nature Reviews Genetics 8, 450–461 (2007)
2. Andrianantoandro, E., Basu, S., Karig, D., Weiss, R.: Synthetic biology: new engineering rules for an emerging discipline. Molecular Systems biology 2, 2006.0028 (2006)
3. Bernardini, F., Gheorghe, M., Romero-Campero, F.J., Walkinshaw, N.: A Hybrid Approach to Modeling Biological Systems. In: Eleftherakis, G., Kefalas, P., Păun, G., Rozenberg, G., Salomaa, A. (eds.) WMC 2007. LNCS, vol. 4860, pp. 138–159. Springer, Heidelberg (2007)
4. Bernardini, F., Gheorghe, M., Krasnogor, N.: Quorum Sensing P Systems. Theoretical Computer Science 371(1-2), 20–33 (2007)
5. Bernardini, F., Manca, V.: Dynamical aspects of P systems. BioSystems 70(2), 85–93 (2003)
6. Besozzi, D., Cazzaniga, P., Pescini, D., Mauri, G., Colombo, S., Martegani, E.: Modeling and stochastic simulation of the Ras/cAMP/PKA pathway in the yeast Saccharomyces cerevisiae evidences a key regulatory function for intracellular guanine nucleotides pools. Journal of Biotechnology 133(3), 377–385 (2008)
7. Bianco, L., Fontana, F., Manca, V.: P systems with reaction maps. International Journal of Foundations of Computer Science 17(1), 27–48 (2006)
8. Gillespie, D.T.: Stochastic Simulation of Chemical Kinetics. Annu. Rev. Phys. Chem. 58, 35–55 (2007)

9. Kaern, M., Elston, T.C., Blake, W.J., Collins, J.J.: Stochasticity in gene expression: from theories to phenotypes. Nature Reviews Genetics 6, 451–464 (2005)

10. Krasnogor, N., Gheorghe, M., Terrazas, G., Diggle, S., Williams, P., Camara, M.: An appealing computational mechanism drawn from bacterial quorum sensing. Bulletin of the EATCS 85, 135–148 (2005)

11. Păun, G.: Computing with Membranes. Journal of Computer and System Sciences 61(1), 108–143 (2000)

12. Păun, G.: Membrane Computing: An Introduction. Springer, Heidelberg (2002)

13. Păun, A., Pérez-Jiménez, M.J., Romero-Campero, F.J.: Modeling signal transduction using P systems. In: Hoogeboom, H.J., Păun, G., Rozenberg, G., Salomaa, A. (eds.) WMC 2006. LNCS, vol. 4361, pp. 100–122. Springer, Heidelberg (2006)

14. Pérez-Jiménez, M.J., Romero-Campero, F.J.: P Systems, a new computational modelling tool for systems biology. In: Priami, C., Plotkin, G. (eds.) Transactions on Computational Systems Biology VI. LNCS (LNBI), vol. 4220, pp. 176–197. Springer, Heidelberg (2006)

15. Pescini, D., Besozzi, D., Mauri, G., Zandron, C.: Dynamical probabilistic P systems. International Journal of Foundations of Computer Science 17(1), 183–204 (2006)

16. Ptashne, M., Gann, A.: Genes and Signals. Cold Spring Harbor Laboratory Press (2002)

17. Romero-Campero, F.J., Pérez-Jiménez, M.J.: Modelling gene expression control using P systems: the Lac Operon, a case study. BioSystems 91(3), 438–457 (2008)

18. Romero-Campero, F.J., Pérez-Jiménez, M.J.: A model of the quorum sensing system in Vibrio fischeri using P systems. Artificial Life 14(1), 1–15 (2008)

19. Păun, G., Romero-Campero, F.J.: Membrane Computing as a Modeling Framework. Cellular Systems Case Studies. In: Bernardo, M., Degano, P., Zavattaro, G. (eds.) SFM 2008. LNCS, vol. 5016, pp. 168–214. Springer, Heidelberg (2008)

20. Romero-Campero, F.J., Gheorghe, M., Ciobanu, G., Auld, J.M., Pérez-Jiménez, M.J.: Cellular modelling using P systems and process algebra. Progress in Natural Science 17(4), 375–383 (2007)

21. Romero-Campero, F.J., Cao, H., Cámara, M., Krasnogor, N.: Structure and Parameter Estimation for Cell Systems Biology Models. In: Proc. of the Genetic and Evolutionary Computation Conference, Atlanta, USA, July 12 - 16, pp. 331–338 (2008)

22. Romero-Campero, F.J., Twycross, J., Cámara, M., Bennett, M., Gheorghe, M. Krasnogor, N.: Modular Assembly of Cell Systems Biology Models Using P Systems. International Journal of Foundations of Computer Science (2008) (in press)

Decidability of Sub-theories of Polynomials over a Finite Field[*]

Alla Sirokofskich

Hausdorff Research Institute for Mathematics
Poppelsdorfer Allee 45, D-53115, Bonn, Germany
Department of Mathematics
University of Crete, 714 09 Heraklion, Greece
`asirokof@math.uoc.gr`

Abstract. Let $\mathbb{F}_q$ be a finite field with q elements. We produce an (effective) elimination of quantifiers for the structure of the set of polynomials, $\mathbb{F}_q[t]$, of one variable, in the language which contains symbols for addition, multiplication by t, inequalities of degrees, divisibility of degrees by a positive integer and, for each $d \in \mathbb{F}_q[t]$, a symbol for divisibility by d. We discuss the possibility of extending our results to the structure which results if one includes a predicate for the relation "x is a power of t".

1 Introduction

In what follows $\mathbb{F}_q$ is a finite field with $q = p^n$, p a prime; $\mathbb{F}_q[t]$ is the ring of polynomials over $\mathbb{F}_q$ in the variable t. By $\mathbb{N}$ we denote the set of positive integers and by $\mathbb{N}_0$ the set of non-negative integers. In what follows $+$ denotes regular addition in $\mathbb{F}_q[t]$ and f_t is a one placed functional symbol interpreted by $f_t(x) = tx$ (in other words, we allow multiplication by t). The constant symbols 0 and 1 are interpreted in the usual way. We work in the language

Definition 1

$$L = \{+, 0, 1, f_t\} \cup \{|_\alpha : \ \alpha \in \mathbb{F}_q[t]\} \cup \ \{D_<\} \cup \{D_n : n \in \mathbb{N}\}$$

where

$$D_<(\omega_1, \omega_2) \ stands \ for \ "deg \ \omega_1 < deg \ \omega_2",$$

$$D_n(\omega) \ stands \ for \ "n|deg \ \omega",$$

$$|_\alpha(\omega) \ stands \ for \ " \ \exists x(x \cdot \alpha = \omega)".$$

We consider the structure $\mathcal{A}$ with universe $\mathbb{F}_q[t]$ in the language L, where the symbols are interpreted as above. We show that the first-order theory of $\mathcal{A}$ admits

---

[*] Supported by the Trimester Program on Diophantine Equations, January - April 2009.

K. Ambos-Spies, B. Löwe, and W. Merkle (Eds.): CiE 2009, LNCS 5635, pp. 437–446, 2009.
© Springer-Verlag Berlin Heidelberg 2009

elimination of quantifiers, i.e., each first-order formula of L is equivalent in $\mathcal{A}$ to a quantifier-free formula. The elimination is constructive. As a consequence we obtain that the first-order theory of $\mathcal{A}$ is decidable, that is, there is an algorithm which, given any formula of L, decides whether that is true or not in $\mathcal{A}$. Our main Theorem is

Theorem 1. *The theory of the structure $\mathcal{A}$ in the language L admits elimination of quantifiers and is decidable.*

Since Goedel's Incompleteness Theorem which asserts undecidability of the ring-theory of the rational integers, many researchers have investigated various rings of interest from the point of view of decidability of their theories. In [9] R. Robinson proved that the theory of a ring of polynomials $A[t]$ of the variable t in the language of rings, augmented by a symbol for t, is undecidable. Following the negative answer to 'Hilbert's Tenth Problem', Denef in [1] and [2] showed that the existential theory of $A[t]$ is undecidable, if A is a domain. In consequence decidability can be a property of theories weaker, only, than the ring theory of $A[t]$. The situation is analogous to the ring of integers: Since no general algorithms can exist for the ring theory of $\mathbb{Z}$, one can look into sub-theories that correspond to structures on $\mathbb{Z}$ weaker than the ring structure. Two examples are: (a) (L. Lipshitz in [3]) the existential theory of $\mathbb{Z}$ in the language of addition and divisibility is decidable (but the full first order theory is undecidable) and (b) (A. Semenov in [10] and [11]) the elementary theory of addition and the function $n \to 2^n$ over $\mathbb{Z}$ is decidable. Th. Pheidas proved a result analogous to those of Lipshitz in (a) for polynomials in one variable over a field with decidable existential theory (in his Ph. D. Thesis) - but the similar problem for polynomials in two variables has an undecidable existential theory. Th. Pheidas and K. Zahidi in [6] showed that the theory of the structure $(\mathbb{F}_q[t]; +; x \to x^p; f_t; 0, 1)$ is model complete and therefore decidable ($x \to x^p$ is the Frobenius function). For surveys on relevant decidability questions and results the reader may consult [4], [5], [6], [7] and [8].

Our results provide a mild strengthening of the analogue, for polynomials over finite fields, of the decidability of 'Presburger Arithmetic' (which is, essentially, the theory of addition and order) for $\mathbb{N}$.

1.1   A List of Open Problems

1. Presently we do not have any estimate for the complexity of the decision algorithm. The existential theory of the structure $\mathcal{A}$ is already exponential and NP-hard since it contains the problem of dynamic programming over polynomials. At the moment it is unclear what the complexity of the whole theory is.

2. Does the similar problem for polynomial rings $F[t]$ have a similar answer (decidability) for any field F with a decidable theory?

2   Analogue of Presburger Arithmetic in $\mathbb{F}_q[t]$

By $\wedge, \vee, \neg$ we mean the usual logical connectives and deg x stands for the degree of the polynomial x. In what follows, addition, multiplication and degree are meant in $\mathbb{F}_q[t]$.

Consider any quantifier free formula $\psi(\bar{x})$ in L, where $\bar{x} = (x_1, \ldots, x_n)$. Then $\psi(\bar{x})$ is equivalent to a quantifier-free formula in disjunctive-normal form with literals among the following relations:

$$D_<(\omega_1, \omega_2), \ |_c(\omega), \ D_n(\omega), \ \omega = 0$$

and their negations, where $\omega, \omega_1, \omega_2$ are terms of the language L with variables among $x_1, \ldots, x_n$. The following negations can be eliminated:

- $\neg D_<(\omega_1, \omega_2)$ is equivalent to $D_<(\omega_2, \omega_1) \vee [D_<(\omega_1, t \cdot \omega_2) \wedge D_<(\omega_2, t \cdot \omega_1)]$.
- $\neg D_n(\omega)$ is equivalent to a finite disjunction of $D_n(t^i \omega)$ for $1 \leq i < n$.
- $\not|_c(\omega)$ can be replaced by

$$\bigvee_{r \neq 0, deg(r) < deg(c)} |_c(\omega + r).$$

- $\omega \neq 0$ is equivalent to $D_<(0, \omega)$, (recall that $\deg(0) = -\infty$).

This can be summarized in the next Proposition.

Proposition 1. *Every existential formula of L is equivalent to a finite disjunction of formulas of the form*

$$\sigma(\bar{\omega}) : \ \sigma_0 \wedge \exists \bar{x} = (x_1, \ldots, x_n)\sigma_1 \wedge \sigma_2 \wedge \sigma_3 \wedge \sigma_4 \tag{1}$$

where σ_0 is an open formula with parameters $\bar{\omega} = (\omega_1, \ldots, \omega_k)$,

$$\sigma_1(\bar{x}, \bar{\omega}) : \ \bigwedge_i f_i(\bar{x}) = h_i(\bar{\omega}) \ , \tag{2}$$

$$\sigma_2(\bar{x}, \bar{\omega}) : \ \bigwedge_\rho D_<(\pi_{1,\rho}(\bar{x}, \bar{\omega}), \pi_{2,\rho}(\bar{x}, \bar{\omega})) \ , \tag{3}$$

$$\sigma_3(\bar{x}, \bar{\omega}) : \ \bigwedge_\lambda |_{c_\lambda}(\chi_\lambda(\bar{x}, \bar{\omega})) \ , \tag{4}$$

$$\sigma_4(\bar{x}, \bar{\omega}) : \ \bigwedge_\xi D_{n_\xi}(g_\xi(\bar{x}, \bar{\omega})) \ , \tag{5}$$

*where
each index among i, ρ, λ, ξ ranges over a finite set, $n_\xi \in \mathbb{N}$, each of f_i, h_i, $\pi_{1,\rho}$, $\pi_{2,\rho}$, χ_λ, g_ξ is a degree-one polynomial of the indicated variables over $\mathbb{F}_q[t]$, and each f_i is a homogeneous polynomial.*

$D_=(X, Y)$ is an abbreviation for the formula $D_<(X, tY) \land D_<(Y, tX)$. Also $D_\leq(X, Y)$ stands for the formula $D_<(X, Y) \lor D_=(X, Y)$.

Definition 2. *Let* $X, Y, Z \in \mathbb{F}[t]$, *with* $\deg(X) = \deg(Y) = \deg(Z)$. *We define the depth of the cancellation in the sum* $X + Y$ *to be*

$$dc(X + Y) = \deg(Y) - \deg(X + Y).$$

We say that X *fits better into* Y *than into* Z, *if* $dc(X + Y) > dc(X + Z)$.

We continue with several facts about the depth of the cancellation. Let

$$a_1 x = \sum_{i \leq k} u_i t^i, \qquad \omega_1 = \sum_{i \leq k} v_i t^i, \qquad \omega_2 = \sum_{i \leq k} w_i t^i,$$

with $u_i, v_i, w_i \in \mathbb{F}_q$. Assume that there is some $\lambda \leq k$ such that $u_i = -v_i$ for all $i \geq \lambda$. Let λ_1 be the least such λ. If $\lambda_1 \geq 1$, then the degree of $a_1 x + \omega_1$ is $\lambda_1 - 1$ and thus $dc(a_1 x + \omega_1) = k - \lambda_1 + 1$. Note that in case $\lambda_1 = 0$, then $a_1 x = -\omega_1$ and the degree of $a_1 x + \omega_1$ is $-\infty$.

Assume that $dc(a_1 x + \omega_1) > 0$. Consider any ω_2 with the properties $\deg(a_1 x) = \deg(\omega_2)$ and $dc(a_1 x + \omega_1) < dc(a_1 x + \omega_2)$. The crucial observation is that for any i such that $\forall j \geq i(u_j = -w_j)$, we have that i should be greater than λ_1. Therefore $dc(a_1 x + \omega_1) > dc(\omega_2 + (-\omega_1))$. Thus $\deg(a_1 x + \omega_1) < \deg(\omega_2 - \omega_1)$.

For the sake of completeness we list several facts for the relation of the form $D_<(a_1 x + \omega_1, a_2 x + \omega_2)$, where $a_i \in \mathbb{F}_q[t] \setminus \{0\}$, ω_i are parameters and x is a variable.

Lemma 1. *The relation* $D_=(a_1 x + \omega_1, a_1 x + \omega_2)$ *is equivalent to the disjunction of*

(1.1) $D_<(a_1 x, \omega_1) \land D_<(a_1 x, \omega_2) \land D_=(\omega_1, \omega_2)$,

(1.2) $D_<(\omega_1, a_1 x) \land D_<(\omega_2, a_1 x)$,

(1.3) $D_=(a_1 x + \omega_1, \omega_1) \land D_=(a_1 x, \omega_1) \land D_<(\omega_2, \omega_1)$,

(1.4) $D_=(a_1 x + \omega_1, \omega_1) \land D_=(a_1 x + \omega_2, \omega_2) \land D_=(a_1 x, \omega_1) \land D_=(\omega_1, \omega_2)$,

(1.5) $D_<(a_1 x + \omega_1, \omega_1) \land D_<(a_1 x + \omega_2, \omega_2) \land D_\leq(\omega_1 - \omega_2, a_1 x + \omega_1) \land D_=(a_1 x, \omega_1) \land$
$\quad\ D_=(\omega_1, \omega_2) \land D_\leq(\omega_1 - \omega_2, a_1 x + \omega_2)$,

(1.6) $D_=(a_1 x + \omega_2, \omega_2) \land D_=(a_1 x, \omega_2) \land D_<(\omega_1, \omega_2)$.

Proof. "$\Leftarrow$"

• Assume that (1.1) holds. Then $D_=(a_1 x + \omega_1, \omega_1)$ and $D_=(a_1 x + \omega_2, \omega_2)$ therefore $D_=(a_1 x + \omega_1, a_1 x + \omega_2)$ holds.

• Assume that (1.2) holds. Then $D_=(a_1 x + \omega_1, a_1 x)$ and $D_=(a_1 x + \omega_2, a_1 x)$ therefore $D_=(a_1 x + \omega_1, a_1 x + \omega_2)$ holds.

• Assume that (1.3) holds. Then $D_=(a_1 x + \omega_1, a_1 x)$ and $D_=(a_1 x + \omega_2, a_1 x)$ therefore $D_=(a_1 x + \omega_1, a_1 x + \omega_2)$ holds.

• Assume that (1.4) holds. Then it is obvious that $D_=(a_1 x + \omega_1, a_1 x + \omega_2)$ holds true.

• Assume that (1.5) holds. Following the notation given after Definition 2, let λ_1 be as defined and λ_2 be the least λ such that $u_i = -w_i$ for all $i \geq \lambda$. Note that if $\lambda_1 < \lambda_2$, then $\deg(a_1 x + w_1) < \deg(w_1 - w_2)$ and this contradicts the assumption. Similarly if $\lambda_2 < \lambda_1$, we have that $\deg(a_1 x + w_2) < \deg(w_1 - w_2)$ and this also contradicts the assumption. Thus $\lambda_1 = \lambda_2$, therefore we have that $D_=(a_1 x + w_1, a_1 x + w_2)$ holds.

• Assume that (1.6) holds. Then $D_=(a_1 x + w_1, a_1 x)$ and $D_=(a_1 x + w_2, a_1 x)$, therefore $D_=(a_1 x + w_1, a_1 x + w_2)$ holds.

"⇒" Assume that $D_=(a_1 x + w_1, a_1 x + w_2)$ holds. We examine all possible linear orderings of the set $\{a_1 x, w_1, w_2\}$.

• Let $D_<(a_1 x, w_2)$. The cases $D_=(a_1 x, w_1)$ and $D_<(w_1, a_1 x)$ are impossible. If $D_<(a_1 x, w_1)$, then (1.1) holds.

• Let $D_<(w_2, a_1 x)$. Then either $D_<(w_1, a_1 x)$, thus (1.2) holds, or $D_=(w_1, a_1 x)$ and $\deg(a_1 x + w_1) = \deg(w_1)$ i.e., (1.3) holds.

• Let $D_=(w_2, a_1 x)$. The case $D_<(a_1 x, w_1)$ is impossible. If $D_<(w_1, a_1 x)$, then (1.6) holds. If $D_=(w_1, a_1 x)$, then we $dc(a_1 x + w_1) = dc(a_1 x + w_2)$. If both depths are zero, then (1.4) holds. If the depths are non-zero, then we have that $v_i = w_i$, for all $i \geq \lambda_1 = \lambda_2$. Note that v_i, w_i might be equal and for some $i < \lambda_1$, i.e., $\deg(w_2 - w_1) \leq \lambda_1 - 1 = \lambda_2 - 1$. Therefore (1.5) holds.   □

Lemma 2. *For* $k \in \mathbb{N}$ *and* $X, Y \in \mathbb{F}_q[t]$, *we define* $D_{<_k}(X, Y)$ *to be* $D_<(t^{k-1} X, Y)$. *With this notation the formula* $D_{<_k}(a_1 x + w_1, a_1 x + w_2)$ *is equivalent to the disjunction of*

(2.1) $D_<(a_1 x, w_1) \wedge D_<(a_1 x, w_2) \wedge D_{<_k}(w_1, w_2)$,

(2.2) $D_<(w_1, a_1 x) \wedge D_{<_k}(a_1 x, w_2)$,

(2.3) $D_\leq(a_1 x + w_1, w_1) \wedge D_=(a_1 x, w_1) \wedge D_<(w_1, w_2) \wedge D_{<_k}(a_1 x + w_1, w_2)$,

(2.4) $D_\leq(a_1 x + w_1, w_1) \wedge D_=(a_1 x, w_1) \wedge D_<(w_2, w_1) \wedge D_{<_k}(a_1 x + w_1, w_1)$,

(2.5) $D_<(a_1 x + w_1, w_1) \wedge D_=(a_1 x, w_1) \wedge D_=(a_1 x, w_2) \wedge D_{<_k}(a_1 x + w_1, w_2 - w_1)$.

Proof. "⇐"

• Assume that (2.1) holds. Then $D_=(a_1 x + w_1, w_1)$ and $D_=(a_1 x + w_2, w_2)$, therefore $D_{<_k}(a_1 x + w_1, a_1 x + w_2)$ holds.

• Assume that (2.2) holds. Then $D_=(a_1 x + w_1, a_1 x)$, $D_<(a_1 x, w_2)$, $k \geq 1$ and $D_=(a_1 x + w_2, w_2)$, therefore $D_{<_k}(a_1 x + w_1, a_1 x + w_2)$ holds.

• Assume that (2.3) holds. Then for the reasons given above, we have that $D_{<_k}(a_1 x + w_1, a_1 x + w_2)$ holds.

• Assume that (2.4) holds. Then $D_=(a_1 x + w_2, w_1)$ and $D_=(a_1 x, w_1)$, therefore $D_{<_k}(a_1 x + w_1, a_1 x + w_2)$ holds.

• Assume that (2.5) holds. Then we have that there is a cancellation in the sum $a_1 x + w_1$. Also the cancellation, if there is any, in the sum $w_2 + (-w_1)$ is smaller from the former one. Thus the cancellation (if there is) in the sum $a_1 x + w_2$ is smaller than the cancellation in the sum $a_1 x + w_1$. Therefore $D_{<_k}(a_1 x + w_1, a_1 x + w_2)$ holds.

"⇒" Assume that $D_{<_k}(a_1 x + w_1, a_1 x + w_2)$ holds.

• Let $D_<(a_1x, \omega_2)$. If $D_=(a_1x, \omega_1)$, then (2.3) holds. If $D_<(a_1x, \omega_1)$, then (2.1) holds. If $D_<(\omega_1, a_1x)$, then (2.2) holds.

• Let $D_<(\omega_2, a_1x)$. Then we must have a cancellation at least of depth k in the sum $a_1x + \omega_1$, i.e., $\deg(a_1x + \omega_1) \leq \deg(\omega_1) + k$, i.e., (2.4) holds.

• Let $D_=(\omega_2, a_1x)$. Then we must have a cancellation in the sum $a_1x + \omega_1$ of at least depth k plus the depth of cancellation in the sum $a_1x + \omega_2$, i.e., (2.5) holds.

Lemma 3. For $k \in \mathbb{N}$ and $X, Y \in \mathbb{F}_q[t]$, we define $D_{<k}(X, Y)$ to be $D_<(X, Yt^k)$. With this notation the formula $D_{<k}(a_1x + \omega_1, a_1x + \omega_2)$ is equivalent to the disjunction of

(3.1) $D_<(a_1x, \omega_2) \wedge D_{<k}(a_1x + \omega_1, \omega_2)$,

(3.2) $D_\leq(\omega_1, a_1x) \wedge D_<(\omega_2, a_1x)$,

(3.3) $D_<(a_1x, \omega_1) \wedge D_<(\omega_2, a_1x) \wedge D_{<k}(\omega_1, a_1x)$,

(3.4) $D_=(a_1x, \omega_2) \wedge D_<(a_1x, \omega_1) \wedge D_{<k}(\omega_1, a_1x + \omega_2)$,

(3.5) $D_=(a_1x, \omega_2) \wedge D_<(\omega_1, a_1x) \wedge [D_{<k}((\omega_2, a_1x + \omega_2)]$,

(4.6) $D_=(a_1x, \omega_2) \wedge D_=(\omega_1, a_1x) \wedge D_\leq(a_1x + \omega_1, a_1x + \omega_2)$,

(4.7) $D_=(a_1x, \omega_2) \wedge D_=(\omega_1, \omega_2) \wedge D_=(a_1x + \omega_1, \omega_2 - \omega_1) \wedge \Big[\bigvee_{i=1}^{k-1} D_=(a_1x + \omega_2, t^i(\omega_2 - \omega_1)) \Big]$.

The purpose of the above Lemmas is to show that when the coefficients of x in the relation $D_<(a_1x + \omega_1, a_2x + \omega_2)$ are the same, then this relation is equivalent to a disjunction of relations of the form $D_<$, where we have at most one appearance of x in each relation $D_<$. Our next goal is to deal with the relation $D_<(a_1x + \omega_1, a_2x + \omega_2)$, where the coefficients of x are not the same.

Lemma 4. Consider the relation $D_<(a_1x + \omega_1, a_2x + \omega_2)$, with $a_1 \neq a_2$. Then it is equivalent to the disjunction of

(4.1) $D_<(a_1, a_2) \wedge D_{<k_1}(a_1a_2x + a_2\omega_1, a_1a_2x + a_1\omega_2)$,

(4.2) $D_<(a_2, a_1) \wedge D_{<k_2}(a_1a_2x + a_2\omega_1, a_1a_2x + a_1\omega_2)$,

(4.3) $D_=(a_1, a_2) \wedge D_<(a_1a_2x + a_2\omega_1, a_1a_2x + a_1\omega_2)$,

where $k_1 = \deg(a_2) - \deg(a_1)$, $k_2 = \deg(a_1) - \deg(a_2) + 1$,

In order to proceed with the elimination of quantifiers, we need to prove one fact.

Proposition 2. Consider σ as given in Proposition 1 for $n = 1$ (i.e. $\bar{x} = x_1 = x$). Then there are quantifier-free formulae $\tilde{\sigma}_0, \tilde{\sigma}_1, \tilde{\sigma}_2, \tilde{\sigma}_3$ and $\tilde{\sigma}_4$ such that

$$\sigma_0 \wedge \exists x \, (\sigma_1 \wedge \sigma_2 \wedge \sigma_3 \wedge \sigma_4) \iff \bigvee (\tilde{\sigma}_0 \wedge \exists z \, (\tilde{\sigma}_1 \wedge \tilde{\sigma}_2 \wedge \tilde{\sigma}_3 \wedge \tilde{\sigma}_4)),$$

where $\tilde{\sigma}_0$ is a quantifier-free formula with parameters $\bar{\omega}$,

$$\tilde{\sigma}_1(z, \bar{\omega}) : \bigwedge_i z = \tilde{h}_i(\bar{\omega}) , \tag{6}$$

$$\tilde{\sigma}_2(z, \bar{\omega}) : \bigwedge_\rho D_<(z, \tilde{\pi}_{2,\rho}(\bar{\omega})) \wedge D_<(\tilde{\pi}'_{1,\rho}(\bar{\omega}), z), \tag{7}$$

$$\tilde{\sigma}_3(z) : \bigwedge_\lambda |_{c_\lambda}(\tilde{\chi}_\lambda(z)), \tag{8}$$

$$\tilde{\sigma}_4(z) : \bigwedge_\xi D_{n_\xi}(z) \tag{9}$$

where
each index among i, ρ, λ, ξ ranges over a finite set, each of $\tilde{h}_i, \tilde{\pi}_{2,\rho}, \tilde{\pi}'_{1,\rho}$ is a degree-one polynomial in the parameters $\bar{\omega}$ over $\mathbb{F}_q[t]$, each of $\tilde{\chi}_\lambda$ is a degree-one polynomial in the variable z over $\mathbb{F}_q[t]$.

Proof. Let σ be as in the hypothesis. We follow the notation as given in Proposition 1. According to the above Lemmas, we can assume that for every ρ in the formula σ_2, the coefficient of x is non-zero in exactly one of the polynomials $\pi_{1,\rho}, \pi_{2,\rho}$.

Consider A to be the set of all coefficients of x in σ. Let a' be the least common multiple of all coefficients of x in σ. Let a be the least element in $\mathbb{F}_q[t]$ such that $a'|a$ and $n_\xi|deg(\frac{a}{b})$, for all n_ξ given in σ_4 and for all $b \in A$. Next we modify σ in the following way.

- By multiplying suitably, we arrange the coefficient of x in the terms $f_i(x)$ to be a. Thus we may assume that $f_i(x) = ax$, for all i.
- Consider any relation of the form $|_{c_\lambda}(\chi_\lambda(x, \bar{\omega}))$ and let a_1 be the coefficient of x. Then

$$|_{c_\lambda}(\chi_\lambda(x, \bar{\omega})) \text{ if and only if } |_{\frac{a \cdot c_\lambda}{a_1}}(\frac{a}{a_1}\chi_\lambda(x, \bar{\omega})).$$

Therefore we may assume that $\chi_\lambda(x, \bar{\omega}) = ax + \chi'_\lambda(\bar{\omega})$.
- Consider any relation of the form $D_{n_\xi}(g_\xi(x, \bar{\omega}))$ and let a_1 be the coefficient of x. Then

$$D_{n_\xi}(g_\xi(x, \bar{\omega})) \text{ if and only if } D_{n_\xi}(\frac{a}{a_1}g_\xi(x, \bar{\omega})),$$

because $deg(\frac{a}{a_1}g_\xi(x, \bar{\omega})) = deg(\frac{a}{a_1}) + deg(g_\xi(x, \bar{\omega}))$ and $n_\xi|deg(\frac{a}{a_1})$ Therefore we may assume that $g_\xi(x, \bar{\omega}) = ax + g'_\xi(\bar{\omega})$.
- Consider any relation of the form $D_<(\pi_{1,\rho}(x, \bar{\omega}), \pi_{2,\rho}(x, \bar{\omega}))$. As we mentioned before, due to Lemmas 1 -4 for every ρ exactly one of the polynomials $\pi_{1,\rho}, \pi_{2,\rho}$ has a non-trivial appearance of x. Let a_1 be the non-zero coefficient of x. Then

$$D_<(\pi_{1,\rho}(x, \bar{\omega}), \pi_{2,\rho}(x, \bar{\omega})) \text{ if and only if } D_<(\frac{a}{a_1}\pi_{1,\rho}(x, \bar{\omega}), \frac{a}{a_1}\pi_{2,\rho}(x, \bar{\omega})).$$

Therefore we may assume that either $\pi_{1,\rho}(x, \bar{\omega}) = ax + \pi'_{1,\rho}(\bar{\omega})$, $\pi_{2,\rho}(x, \bar{\omega}) = \pi'_{2,\rho}(\bar{\omega})$, or $\pi_{1,\rho}(x, \bar{\omega}) = \pi'_{1,\rho}(\bar{\omega})$, $\pi_{2,\rho}(x, \bar{\omega}) = ax + \pi'_{2,\rho}(\bar{\omega})$.

We take a disjunction over all possible total orderings of the degrees of the terms $ax, ax + \pi'_{1,\rho}(\bar{\omega}), ax + \pi'_{2,\rho}(\bar{\omega}), \pi'_{1,\rho}(\bar{\omega}), \pi'_{2,\rho}(\bar{\omega}), ax + \chi'_\lambda(\bar{\omega}), ax + g'_\xi(\bar{\omega})$ that occur in σ. Since the existential quantifier $\exists x$ distributes over $\vee$ we may assume, without loss of generality, that σ_2 implies such an ordering. Let T be a term of lowest degree (according to this ordering), in which x occurs non-trivially. Clearly, T must be of the form $ax + u(\bar{\omega})$ where u is a term of L in which x does not occur. We perform the change of variables $z = ax + u$ and we substitute each occurrence of ax in the above terms by the resulting value of ax, $z - u$. We adjoin in σ_3 the divisibility $|_a(z - u)$. In detail,

- each formula of the form $ax = h_i(\bar{\omega})$ is replaced by $z = \tilde{h}(\bar{\omega})$, where $\tilde{h}(\bar{\omega}) = h_i(\bar{\omega}) + u(\bar{\omega})$,
- each formula of the form $|_c(ax + \chi'_\lambda(\bar{\omega}))$ is replaced by $\bigvee_r |_c(z+r) \wedge |_c(\chi'(\bar{\omega}) - u(\bar{\omega}) - r))$, where r runs over all polynomials with degree less then $\deg(c)$,
- each formula of the form $D_<(ax + \pi'_{1,\rho}(\bar{\omega}), \pi'_{2,\rho}(\bar{\omega})) \wedge D_<(ax + u(\bar{\omega}), ax + \pi'_{1,\rho}(\bar{\omega}))$ is replaced by $D_\le(\pi'_{1,\rho}(\bar{\omega}) - u(\bar{\omega}), z) \wedge D_<(z, \pi'_{2,\rho}(\bar{\omega}))$,
- each formula of the form $D_<(ax + \pi'_{1,\rho}(\bar{\omega}), \pi'_{2,\rho}(\bar{\omega})) \wedge D_=(ax + u(\bar{\omega}), ax + \pi'_{1,\rho}(\bar{\omega}))$ is replaced by $D_<(z, \pi'_{2,\rho}(\bar{\omega})) \wedge D_\le(\pi'_{1,\rho}(\bar{\omega}) - u(\bar{\omega}), z)$,
- each formula of the form $D_<(\pi'_{1,\rho}(\bar{\omega}), ax + \pi'_{2,\rho}(\bar{\omega})) \wedge D_<(ax + u(\bar{\omega}), ax + \pi'_{2,\rho}(\bar{\omega}))$ is replaced by $D_<(z, \pi'_{2,\rho}(\bar{\omega}) - u(\bar{\omega})) \wedge D_<(\pi'_{1,\rho}(\bar{\omega}), \pi'_{2,\rho}(\bar{\omega})) - u(\bar{\omega})$
- each formula of the form $D_<(\pi'_{1,\rho}(\bar{\omega}), ax + \pi'_{2,\rho}(\bar{\omega})) \wedge D_=(ax + u(\bar{\omega}), ax + \pi'_{2,\rho}(\bar{\omega}))$ is replaced by $D_<(\pi'_{1,\rho}(\bar{\omega}), z) \wedge D_\le(\pi'_{2,\rho}(\bar{\omega}) - u(\bar{\omega}), z)$,
- each formula of the form $D_n(ax + g'_\xi(\bar{\omega})) \wedge D_<(ax + u(\bar{\omega}), ax + g'_\xi(\bar{\omega}))$ is replaced by $D_n(g'_\xi(\bar{\omega}) - u(\bar{\omega})) \wedge D_<(z, g'_\xi(\bar{\omega}) - u(\bar{\omega}))$,
- each formula of the form $D_n(ax + g'_\xi(\bar{\omega})) \wedge D_=(ax + u(\bar{\omega}), ax + g'_\xi(\bar{\omega}))$ is replaced by $D_n(z) \wedge D_\le(g'_\xi(\bar{\omega}) - u(\bar{\omega}), z)$.

This completes the proof of the separation of x from $\bar{\omega}$.   ⊏

We are ready to eliminate the existential quantifiers over the variables $\bar{x}$ in the existential formula σ of Proposition 1.

Theorem 2. *Every formula σ of L is equivalent over $\mathbb{F}_q[t]$ to an open formula of L.*

Proof. Let σ be as in Proposition 1. If σ_1 is not void (i.e. equivalent to $1 = 1$) then solve for one of the variables in terms of the remaining ones over $\mathbb{F}_q(t)$, substitute each occurrence of it by the value implied by the equations and adjoin the corresponding divisibility to σ_3 as indicated in the proof of Proposition 2. Iterate until there are no equations. Hence we assume that σ_1 is void.

According to Proposition 1 we assume that $\sigma_2 \wedge \sigma_3 \wedge \sigma_4$ has the form indicated in Proposition 2, with respect to the variable x_n.

In order to achieve the elimination of x_n, we separate the variable x_n from the rest of the variables $x_1, \ldots, x_{n-1}$, by considering $x_1, \ldots, x_{n-1}, \omega_1, \ldots, \omega_m$ as parameters. Thus after applying Proposition 2 to σ, we may assume from the beginning that each σ_i (as given in Proposition 1) is already in separated form with $\bar{x} = x_n$ and that the coefficient of every nontrivial appearance of x_n in σ is equal to 1.

Let $x_1, \ldots, x_{n-1}$ and $\bar{\omega}$ be given. First, we observe that we may substitute the relations of σ_4 by only one divisibility $D_{n_{\xi_0}}(x_n)$, where n_{ξ_0} is the least common multiple of all n_ξ appearing in σ_4.

Case 1: There is no upper bound for the degree of x_n. Then x_n can be eliminated if and only if the conditions for the Generalized Chinese Theorem hold for σ_3.

Case 2: There is an upper bound for the degree of x_n. Now note that

$$D_<(x_n, \theta_1(x_1, \ldots, x_{n-1}, \bar{\omega})) \wedge D_<(x_n, \theta_2(x_1, \ldots, x_{n-1}, \bar{\omega})) \iff$$

$$[D_<(\theta_1(x_1, \ldots, x_{n-1}, \bar{\omega}), \theta_2(x_1, \ldots, x_{n-1}, \bar{\omega})) \wedge D_<(x_n, \theta_1(x_1, \ldots, x_{n-1}, \bar{\omega}))] \vee$$

$$[D_<(\theta_2(x_1, \ldots, x_{n-1}, \bar{\omega}), t\theta_1(x_1, \ldots, x_{n-1}, \bar{\omega})) \wedge D_<(x_n, \theta_2(x_1, \ldots, x_{n-1}, \bar{\omega}))].$$

Let $\theta_{m_2}(x_1, \ldots, x_{n-1}, \bar{\omega})$ be such that its degree is the least upper bound for the degree of x_n. Using the Generalized Chinese Theorem, we check if the system of divisibilities of σ_3 has some solution $x_n \in \mathbb{F}_q[t]$. If it does, then there is a solution $x_n \in \mathbb{F}_q[t]$ such that $D_{n_{\xi_0}}(x_n)$.

Case 2(a): Assume that there is no $\theta(x_1, \ldots, x_{n-1}, \bar{\omega})$ such that $D_<(\theta(x_1, \ldots, x_{n-1}, \bar{\omega}), x_n)$. Then x_n should be a constant polynomial, i.e., there is an elimination of x_n.

Case 2(b): There are $\theta_1(x_1, \ldots, x_{n-1}, \bar{\omega})$ and $\theta_2(x_1, \ldots, x_{n-1}, \bar{\omega})$ such that

$$D_<(\theta_1(x_1, \ldots, x_{n-1}, \bar{\omega}), x_n) \wedge D_<(x_n, \theta_2(x_1, \ldots, x_{n-1}, \bar{\omega})).$$

Let $\theta_{m_2}(x_1, \ldots, x_{n-1}, \bar{\omega})$ be as defined above and $\theta_{m_1}(x_1, \ldots, x_{n-1}, \bar{\omega})$ such that its degree is the least lower bound for the degree of x_n. For simplicity we denote $\theta_{m_1}(x_1, \ldots, x_{n-1}, \bar{\omega}), \theta_{m_2}(x_1, \ldots, x_{n-1}, \bar{\omega})$ by $\theta_{m_1}, \theta_{m_2}$ respectively. We repeat the previous algorithm to decide if there exists a x_n that satisfies $\sigma_3 \wedge D_{n_{\xi_0}}(x_n)$. If there is such $x_n \in \mathbb{F}_q[t]$, then let d be the least positive integer with the property: if x_n is a solution of $\sigma_3 \wedge D_{n_{\xi_0}}(x_n)$, then the next solution of $\sigma_3 \wedge D_{n_{\xi_0}}(x_n)$ is of degree $\deg(x_n) + d$. Such d exists due to the Generalized Chinese Theorem. Thus

$$\exists x_n(\sigma_2 \wedge \sigma_3 \wedge \sigma_4) \iff \bigvee_{i=0}^{d-1} [D_d(t^{d-i}\theta_{m_1}) \wedge D_<(t^{d-i}\theta_{m_1}, \theta_{m_2})].$$

Thus by induction on n we obtain the required statement of elimination.   □

An Enrichment for $(\mathbb{F}_q[t]; +; |_a; P; f_t; 0, 1)$

We start by augmenting the language of the structure $\mathcal{A}$ to a language L_P.

Definition 3. *Let q and t be given. We define the language*

$$L_P = L \cup \{P\}$$

where the predicate $P(\omega)$ stands for "ω is a power of t".

This extension of a language L is an analogue to the extension of Presburger arithmetic by the relation "x is a power of 2", which, over $\mathbb{N}$, has a decidable theory, as mentioned in the Introduction.

Currently, we are investigating the theory of $\mathbb{F}_q[t]$ in L_P from the point of view of decidability. Our results so far indicate that this theory may be model-complete.

Acknowledgments

The author would like to thank Th. Pheidas and the referees for helpful suggestions.

References

1. Denef, J.: The diophantine problem for polynomial rings and fields of rational functions. Transactions of the American Mathematical Society 242, 391–399 (1978)
2. Denef, J.: The diophantine problem for polynomial rings of positive characteristic Logic Colloquium, vol. 78, pp. 131–145. North Holland, Amsterdam (1984)
3. Lipshitz, L.: The diophantine problem for addition and divisibility. Transactions of the American Mathematical Society 235, 271–283 (1978)
4. Pheidas, T.: Extensions of Hilbert's Tenth Problem. The Journal of Symbolic Logic 59(2), 372–397 (1994)
5. Pheidas, T., Zahidi, K.: Undecidability of existential theories of rings and fields: A survey. Contemporary Mathematics 270, 49–106 (2000)
6. Pheidas, T., Zahidi, K.: Elimination theory for addition and the Frobenius map in polynomial rings. The Journal of Symbolic Logic 69(4), 1006–1026 (2004)
7. Pheidas, T., Zahidi, K.: Analogues of Hilbert's tenth problem. In: Chatzidakis, Z., Macpherson, D., Pillay, A., Wilkie, A. (eds.) Model theory with Applications to Algebra and Analysis. London Math Soc. Lecture Note Series, Nr. 350, vol. 2, pp. 207–236 (2008)
8. Poonen, B.: Undecidability in Number Theory. Notices A.M.S. 55(3), 344–350 (2008)
9. Robinson, R.: Undecidable rings. Transactions of the American Mathematical Society 70, 137–159 (1951)
10. Semenov, A.: Logical theories of one-place functions on the set of natural numbers Math. USSR Izvestija 22, 587–618 (1984)
11. Semenov, A.: On the definability of arithmetic in its fragments. Soviet Math Dokl. 25, 300–303 (1982)

Chaitin Ω Numbers and Halting Problems

Kohtaro Tadaki

Research and Development Initiative, Chuo University
CREST, JST
1–13–27 Kasuga, Bunkyo-ku, Tokyo 112-8551, Japan
tadaki@kc.chuo-u.ac.jp

Abstract. Chaitin [G. J. Chaitin, *J. Assoc. Comput. Mach.*, vol. 22, pp. 329–340, 1975] introduced his Ω number as a concrete example of random real. The real Ω is defined as the probability that an optimal computer halts, where the optimal computer is a universal decoding algorithm used to define the notion of program-size complexity. Chaitin showed Ω to be random by discovering the property that the first n bits of the base-two expansion of Ω solve the halting problem of the optimal computer for all binary inputs of length at most n. In the present paper we investigate this property from various aspects. It is known that the base-two expansion of Ω and the halting problem are Turing equivalent. We consider elaborations of both the Turing reductions which constitute the Turing equivalence. These elaborations can be seen as a variant of the weak truth-table reduction, where a computable bound on the use function is explicitly specified. We thus consider the relative computational power between the base-two expansion of Ω and the halting problem by imposing the restriction to finite size on both the problems.

Keywords: algorithmic information theory, Chaitin Ω number, halting problem, Turing reduction, algorithmic randomness, program-size complexity.

1 Introduction

Algorithmic information theory (AIT, for short) is a framework for applying information-theoretic and probabilistic ideas to recursive function theory. One of the primary concepts of AIT is the *program-size complexity* (or *Kolmogorov complexity*) $H(s)$ of a finite binary string s, which is defined as the length of the shortest binary input for a universal decoding algorithm U, called an *optimal computer*, to output s. By the definition, $H(s)$ can be thought of as the information content of the individual finite binary string s. In fact, AIT has precisely the formal properties of classical information theory (see Chaitin [2]). In particular, the notion of program-size complexity plays a crucial role in characterizing the *randomness* of an infinite binary string, or equivalently, a real. In [2] Chaitin introduced the halting probability Ω_U as an example of random real. His Ω_U is defined as the probability that the optimal computer U halts, and plays a central role in the metamathematical development of AIT. The real Ω_U is shown to be random, based on the following fact:

. Ambos-Spies, B. Löwe, and W. Merkle (Eds.): CiE 2009, LNCS 5635, pp. 447–456, 2009.
) Springer-Verlag Berlin Heidelberg 2009

Fact 1 (Chaitin [2]). *The first n bits of the base-two expansion of Ω_U solve the halting problem of U for inputs of length at most n.* □

In this paper, we first consider the following converse problem:

Problem 1. *For every positive integer n, if n and the list of all halting inputs for U of length at most n are given, can the first n bits of the base-two expansion of Ω_U be calculated ?* □

As a result of this paper, we can answer this problem negatively. In this paper, however, we consider more general problems in the following forms. Let V and W be arbitrary optimal computers.

Problem 2. *Find a succinct equivalent characterization of a total recursive function $f\colon \mathbb{N}^+ \to \mathbb{N}$ which satisfies the condition: For all $n \in \mathbb{N}^+$, if n and the list of all halting inputs for V of length at most n are given, then the first $n - f(n) - O(1)$ bits of the base-two expansion of Ω_W can be calculated.* □

Problem 3. *Find a succinct equivalent characterization of a total recursive function $f\colon \mathbb{N}^+ \to \mathbb{N}$ which satisfies the condition: For infinitely many $n \in \mathbb{N}^+$, if n and the list of all halting inputs for V of length at most n are given, then the first $n - f(n) - O(1)$ bits of the base-two expansion of Ω_W can be calculated.* □

Here $\mathbb{N}^+$ denotes the set of positive integers and $\mathbb{N} = \{0\} \cup \mathbb{N}^+$. Theorem 4 and Theorem 10 below are two of the main results of this paper. On the one hand, Theorem 4 gives to Problem 2 a solution that the total recursive function f must satisfy $\sum_{n=1}^{\infty} 2^{-f(n)} < \infty$, which is the Kraft inequality in essence. Note that the condition $\sum_{n=1}^{\infty} 2^{-f(n)} < \infty$ holds for $f(n) = \lfloor (1 + \varepsilon) \log_2 n \rfloor$ with an arbitrary computable real $\varepsilon > 0$, while this condition does not hold for $f(n) = \lfloor \log_2 n \rfloor$. On the other hand, Theorem 10 gives to Problem 3 a solution that the total recursive function f must not be bounded to the above. Theorem 10 also results in Corollary 2 below, which refutes Problem 1 completely.

It is also important to consider whether the bound n on the length of halting inputs given in Fact 1 is tight or not. We consider this problem in the following form:

Problem 4. *Find a succinct equivalent characterization of a total recursive function $f\colon \mathbb{N}^+ \to \mathbb{N}$ which satisfies the condition: For all $n \in \mathbb{N}^+$, if n and the first n bits of the base-two expansion of Ω_V are given, then the list of all halting inputs for W of length at most $n + f(n) - O(1)$ can be calculated.* □

Theorem 11, which is one of the main results of this paper, gives to Problem 4 a solution that the total recursive function f must be bounded to the above. Thus, we see that the bound n on the length of halting inputs given in Fact 1 is tight up to an additive constant.

It is well known that the base-two expansion of Ω_U and the halting problem of U are Turing equivalent, i.e., $\Omega_U \equiv_T \operatorname{dom} U$ holds, where $\operatorname{dom} U$ denotes the domain of definition of U. This paper investigates an elaboration of the Turing equivalence. For example, consider the Turing reduction $\Omega_U \leq_T \operatorname{dom} U$, which partly constitutes the Turing equivalence $\Omega_U \equiv_T \operatorname{dom} U$. The Turing reduction can be equivalent to the condition that there exists an oracle deterministic Turing machine M such that, for all $n \in \mathbb{N}^+$,

$$M^{\operatorname{dom} U}(n) = \Omega_U \restriction_n, \tag{1}$$

where $\Omega_U \restriction_n$ denotes the first n bits of the base-two expansion of Ω_U. Let $g: \mathbb{N}^+ \to \mathbb{N}$ and $h: \mathbb{N}^+ \to \mathbb{N}$ be total recursive functions. Then the condition (1) can be elaborated to the condition that there exists an oracle deterministic Turing machine M such that, for all $n \in \mathbb{N}^+$,

$$M^{\operatorname{dom} U \restriction_{g(n)}}(n) = \Omega_U \restriction_{h(n)}, \tag{2}$$

where $\operatorname{dom} U \restriction_{g(n)}$ denotes the set of all strings in $\operatorname{dom} U$ of length at most $g(n)$. This elaboration allows us to consider the asymptotic behavior of h which satisfies the condition (2), for a given g. We might regard g as the degree of the relaxation of the restrictions on the computational resource (i.e., on the oracle $\operatorname{dom} U$) and h as the difficulty of the problem to solve. Thus, even in the context of computability theory, we can deal with the notion of asymptotic behavior in a manner like in computational complexity theory in some sense. Note also that the condition (2) can be seen as a variant of the weak truth-table reduction of the function $\Omega_U \restriction_{h(n)}$ of n to $\operatorname{dom} U$, where a computable bound on the use of $M^{\operatorname{dom} U}(n)$ is explicitly specified by the function g. Theorem 4, a solution to Problem 2, is obtained as a result of the investigation in this line, and gives the upper bound of the function h in the case of $g(n) = n$.

The other Turing reduction $\operatorname{dom} U \leq_T \Omega_U$, which constitutes $\Omega_U \equiv_T \operatorname{dom} U$, is also elaborated in the same manner as above to lead to Theorem 11, a solution to Problem 4.

Thus, in this paper, we study the relationship between the base-two expansion of Ω and the halting problem of an optimal computer using a more rigorous and insightful notion than the notion of Turing equivalence. The paper is organized as follows. We begin in Section 2 with some preliminaries to AIT. We then present Theorems 4, 10, and 11 in Sections 3, 4, and 5, respectively. Due to the 10-page limit, we omit some proofs, in particular, the proof of Theorem 10. A full paper which describes all the proofs and other related results is in preparation.

2   Preliminaries

We start with some notation about numbers and strings which will be used in this paper. $\#S$ is the cardinality of S for any set S. $\mathbb{N} = \{0, 1, 2, 3, \dots\}$ is the set of natural numbers, and $\mathbb{N}^+$ is the set of positive integers. $\mathbb{Z}$ is the set of integers, and $\mathbb{Q}$ is the set of rationals. $\mathbb{R}$ is the set of reals. Normally, $O(1)$ denotes any function $f: \mathbb{N}^+ \to \mathbb{R}$ such that there is $C \in \mathbb{R}$ with the property that $|f(n)| \leq C$ for all $n \in \mathbb{N}^+$.

$\{0,1\}^* = \{\lambda, 0, 1, 00, 01, 10, 11, 000, \ldots\}$ is the set of finite binary strings where λ denotes the *empty string*, and $\{0,1\}^*$ is ordered as indicated. We identify any string in $\{0,1\}^*$ with a natural number in this order, i.e., we consider $\varphi\colon \{0,1\}^* \to \mathbb{N}$ such that $\varphi(s) = 1s - 1$ where the concatenation $1s$ of strings 1 and s is regarded as a dyadic integer, and then we identify s with $\varphi(s)$. For any $s \in \{0,1\}^*$, $|s|$ is the *length* of s. A subset S of $\{0,1\}^*$ is called *prefix-free* if no string in S is a prefix of another string in S. For any subset S of $\{0,1\}^*$ and any $n \in \mathbb{Z}$, we denote by $S{\upharpoonright}_n$ the set $\{s \in S \mid |s| \le n\}$. Note that $S{\upharpoonright}_n = \emptyset$ for every subset S of $\{0,1\}^*$ and every negative integer $n \in \mathbb{Z}$. For any partial function f, the domain of definition of f is denoted by $\operatorname{dom} f$. We write "r.e." instead of "recursively enumerable."

Let α be an arbitrary real. For any $n \in \mathbb{N}^+$, we denote by $\alpha{\upharpoonright}_n \in \{0,1\}^*$ the first n bits of the base-two expansion of $\alpha - \lfloor \alpha \rfloor$ with infinitely many zeros, where $\lfloor \alpha \rfloor$ is the greatest integer less than or equal to α. For example, in the case of $\alpha = 5/8$, $\alpha{\upharpoonright}_6 = 101000$. On the other hand, for any non-positive integer $n \in \mathbb{Z}$, we set $\alpha{\upharpoonright}_n = \lambda$. A real α is called *r.e.* if there exists a total recursive function $f\colon \mathbb{N}^+ \to \mathbb{Q}$ such that $f(n) \le \alpha$ for all $n \in \mathbb{N}^+$ and $\lim_{n\to\infty} f(n) = \alpha$. An r.e. real is also called a *left-computable* real.

2.1 Algorithmic Information Theory

In the following we concisely review some definitions and results of algorithmic information theory [2,3]. A *computer* is a partial recursive function $C\colon \{0,1\}^* \to \{0,1\}^*$ such that $\operatorname{dom} C$ is a prefix-free set. For each computer C and each $s \in \{0,1\}^*$, $H_C(s)$ is defined by $H_C(s) = \min\{|p| \mid p \in \{0,1\}^* \,\&\, C(p) = s\}$ (may be ∞). A computer U is said to be *optimal* if for each computer C there exists $d \in \mathbb{N}$ with the following property; if $p \in \operatorname{dom} C$, then there is $q \in \operatorname{dom} U$ for which $U(q) = C(p)$ and $|q| \le |p| + d$. It is easy to see that there exists an optimal computer. We choose a particular optimal computer U as the standard one for use, and define $H(s)$ as $H_U(s)$, which is referred to as the *program-size complexity* of s or the *Kolmogorov complexity* of s. It follows that for every computer C there exists $d \in \mathbb{N}$ such that, for every $s \in \{0,1\}^*$,

$$H(s) \le H_C(s) + d. \tag{3}$$

Based on this we can show that, for every partial recursive function $\Psi\colon \{0,1\}^* \to \{0,1\}^*$, there exists $d \in \mathbb{N}$ such that, for every $s \in \operatorname{dom} \Psi$,

$$H(\Psi(s)) \le H(s) + d. \tag{4}$$

For any $s \in \{0,1\}^*$, we define s^* as $\min\{p \in \{0,1\}^* \mid U(p) = s\}$, i.e., the first element in the ordered set $\{0,1\}^*$ of all strings p such that $U(p) = s$. Then, $|s^*| = H(s)$ for every $s \in \{0,1\}^*$. For any $s, t \in \{0,1\}^*$, we define $H(s,t)$ as $H(b(s,t))$ where $b\colon \{0,1\}^* \times \{0,1\}^* \to \{0,1\}^*$ is a particular bijective total recursive function. Note also that, for every $n \in \mathbb{N}$, $H(n)$ is $H(\text{the } n\text{th element of } \{0,1\}^*)$.

Definition 1 (Chaitin Ω number, Chaitin [2]). *For any optimal computer V, the halting probability Ω_V of V is defined by*

$$\Omega_V = \sum_{p \in \text{dom}\, V} 2^{-|p|}. \qquad \square$$

For every optimal computer V, since $\text{dom}\, V$ is prefix-free, Ω_V converges and $0 < \Omega_V \le 1$. For any $\alpha \in \mathbb{R}$, we say that α is *weakly Chaitin random* if there exists $c \in \mathbb{N}$ such that $n - c \le H(\alpha\!\restriction_n)$ for all $n \in \mathbb{N}^+$ [2,3]. Based on Fact 1, Chaitin [2] showed that Ω_V is weakly Chaitin random for every optimal computer V. Therefore $0 < \Omega_V < 1$ for every optimal computer V. For any $\alpha \in \mathbb{R}$, we say that α is *Chaitin random* if $\lim_{n \to \infty} H(\alpha\!\restriction_n) - n = \infty$ [3]. We can then show the following theorem (see Chaitin [3] for the proof and historical detail).

Theorem 1. *For every $\alpha \in \mathbb{R}$, α is weakly Chaitin random if and only if α is Chaitin random.* $\qquad \square$

Miller and Yu [7] recently strengthened Theorem 1 to the following form.

Theorem 2 (Ample Excess Lemma, Miller and Yu [7]). *For every $\alpha \in \mathbb{R}$, α is weakly Chaitin random if and only if $\sum_{n=1}^{\infty} 2^{n-H(\alpha\!\restriction_n)} < \infty$.* $\qquad \square$

The following is an important result on random r.e. reals.

Theorem 3 (Calude, et al. [1], Kučera and Slaman [6]). *For every $\alpha \in (0,1)$, α is r.e. and weakly Chaitin random if and only if there exists an optimal computer V such that $\alpha = \Omega_V$.* $\qquad \square$

Elaboration I of the Turing Reduction $\Omega_U \le_T \text{dom}\, U$

Theorem 4 (main result I). *Let V and W be optimal computers, and let $f \colon \mathbb{N}^+ \to \mathbb{N}$ be a total recursive function. Then the following two conditions are equivalent:*

i) There exist an oracle deterministic Turing machine M and $c \in \mathbb{N}$ such that, for all $n \in \mathbb{N}^+$, $M^{\text{dom}\, V\restriction_n}(n) = \Omega_W\!\restriction_{n-f(n)-c}$.

ii) $\sum_{n=1}^{\infty} 2^{-f(n)} < \infty$. $\qquad \square$

Theorem 4 follows from Theorem 5 and Theorem 6 below, and Theorem 3.

Theorem 5. *Let α be an r.e. real, and let V be an optimal computer. For every total recursive function $f \colon \mathbb{N}^+ \to \mathbb{N}$, if $\sum_{n=1}^{\infty} 2^{-f(n)} < \infty$, then there exist an oracle deterministic Turing machine M and $c \in \mathbb{N}$ such that, for all $n \in \mathbb{N}^+$, $M^{\text{dom}\, V\restriction_n}(n) = \alpha\!\restriction_{n-f(n)-c}$.* $\qquad \square$

Theorem 6. *Let α be a real which is weakly Chaitin random, and let V be an optimal computer. For every total recursive function $f \colon \mathbb{N}^+ \to \mathbb{N}$, if there exists an oracle deterministic Turing machine M such that, for all $n \in \mathbb{N}^+$, $M^{\text{dom}\, V\restriction_n}(n) = \alpha\!\restriction_{n-f(n)}$, then $\sum_{n=1}^{\infty} 2^{-f(n)} < \infty$.* $\qquad \square$

The proofs of Theorem 5 and Theorem 6 are given in the next two subsections, respectively.

3.1   The Proof of Theorem 5

In order to prove Theorem 5, we need Theorem 7 and Corollary 1 below.

Theorem 7 (Kraft-Chaitin Theorem, Chaitin [2]). *Let $f \colon \mathbb{N}^+ \to \mathbb{N}$ be a total recursive function such that $\sum_{n=1}^{\infty} 2^{-f(n)} \leq 1$. Then there exists a total recursive function $g \colon \mathbb{N}^+ \to \{0,1\}^*$ such that (i) g is an injection, (ii) the set $\{\, g(n) \mid n \in \mathbb{N}^+\}$ is prefix-free, and (iii) $|g(n)| = f(n)$ for all $n \in \mathbb{N}^+$.* ⬜

Let M be a deterministic Turing machine with the input and output alphabet $\{0,1\}$, and let C be a computer. We say that M *computes* C if the following holds: for every $p \in \{0,1\}^*$, when M starts with the input p, (i) M halts and outputs $C(p)$ if $p \in \operatorname{dom} C$; (ii) M does not halt forever otherwise. We use this convention on the computation of a computer by a deterministic Turing machine throughout the rest of this paper. Thus, we exclude the possibility that there is $p \in \{0,1\}^*$ such that, when M starts with the input p, M halts but $p \notin \operatorname{dom} C$. For any $p \in \{0,1\}^*$, we denote the running time of M on the input p by $T_M(p)$ (may be ∞). Thus, $T_M(p) \in \mathbb{N}$ for every $p \in \operatorname{dom} C$ if M computes C.

Theorem 8. *Let V be an optimal computer. Then, for every computer C there exists $d \in \mathbb{N}$ such that, for every $p \in \{0,1\}^*$, if p and the list of all halting inputs for V of length at most $|p| + d$ are given, then the halting problem of the input p for C can be solved.*

Proof. Let M be a deterministic Turing machine which computes a computer C. We consider the computer D such that (i) $\operatorname{dom} D = \operatorname{dom} C$ and (ii) $D(p) = T_M(p)$ for every $p \in \operatorname{dom} C$. Recall here that we identify $\{0,1\}^*$ with $\mathbb{N}$. It is easy to see that such a computer D exists. Then, since V is an optimal computer from the definition of optimality there exists $d \in \mathbb{N}$ with the following property: if $p \in \operatorname{dom} D$, then there is $q \in \operatorname{dom} V$ for which $V(q) = D(p)$ and $|q| \leq |p| + d$.

Given $p \in \{0,1\}^*$ and the list $\{q_1, \ldots, q_L\}$ of all halting inputs for V of length at most $|p| + d$, one first calculates the finite set $S = \{\, V(q_i) \mid i = 1, \ldots, L \,\}$, and then calculates $T_{\max} = \max S$ where S is regarded as a subset of $\mathbb{N}$. One then simulates the computation of M with the input p until at most the time step $T_{\max}$. In the simulation, if M halts until at most the time step $T_{\max}$, one knows that $p \in \operatorname{dom} C$. On the other hand, note that if $p \in \operatorname{dom} C$ then there is $q \in \operatorname{dom} V$ such that $V(q) = T_M(p)$ and $|q| \leq |p| + d$, and therefore $q \in \{q_1, \ldots, q_L\}$ and $T_M(p) \leq T_{\max}$. Thus, in the simulation, if M does not yet halt at the time step $T_{\max}$, one knows that M does not halt forever and therefore $p \notin \operatorname{dom} C$. ⬜

As a corollary of Theorem 8 we obtain the following.

Corollary 1. *Let V be an optimal computer. Then, for every computer C there exist an oracle deterministic Turing machine M and $d \in \mathbb{N}$ such that, for all $n \in \mathbb{N}^+$, $M^{\operatorname{dom} V \upharpoonright_{n+d}}(n) = \operatorname{dom} C \upharpoonright_n$, where the finite subset $\operatorname{dom} C \upharpoonright_n$ of $\{0,1\}^*$ is represented as a finite binary string in a certain format.* ⬜

Based on Theorem 7 and Corollary 1, Theorem 5 is proved as follows.

Proof (of Theorem 5). Let α be an r.e. real, and let V be an optimal computer. For an arbitrary total recursive function $f\colon \mathbb{N}^+ \to \mathbb{N}$, assume that $\sum_{n=1}^{\infty} 2^{-f(n)} < \infty$. In the case of $\alpha \in \mathbb{Q}$, the result is obvious. Thus, in what follows, we assume that $\alpha \notin \mathbb{Q}$ and therefore the base-two expansion of $\alpha - \lfloor \alpha \rfloor$ is unique and contains infinitely many ones.

Since $\sum_{n=1}^{\infty} 2^{-f(n)} < \infty$, there exists $d_0 \in \mathbb{N}$ such that $\sum_{n=1}^{\infty} 2^{-f(n)-d_0} \le 1$. Hence, by the Kraft-Chaitin Theorem, i.e., Theorem 7, there exists a total recursive function $g\colon \mathbb{N}^+ \to \{0,1\}^*$ such that (i) the function g is an injection, (ii) the set $\{ g(n) \mid n \in \mathbb{N}^+ \}$ is prefix-free, and (iii) $|g(n)| = f(n) + d_0$ for all $n \in \mathbb{N}^+$. On the other hand, since α is r.e., there exists a total recursive function $h\colon \mathbb{N}^+ \to \mathbb{Q}$ such that $h(k) \le \alpha$ for all $k \in \mathbb{N}^+$ and $\lim_{k\to\infty} h(k) = \alpha$.

Now, let us consider the following computer C. For each $n \in \mathbb{N}^+$, $p, s \in \{0,1\}^*$ and $l \in \mathbb{N}$ such that $U(p) = l$, $g(n)ps \in \operatorname{dom} C$ if and only if (i) $|g(n)ps| = n - l$ and (ii) $0.s < h(k) - \lfloor \alpha \rfloor$ for some $k \in \mathbb{N}^+$. It is easy to see that such a computer C exists. Then, by Corollary 1, there exist an oracle deterministic Turing machine M and $d \in \mathbb{N}$ such that, for all $n \in \mathbb{N}^+$, $M^{\operatorname{dom} V \restriction_{n+d}}(n) = \operatorname{dom} C \restriction_n$, where the finite subset $\operatorname{dom} C \restriction_n$ of $\{0,1\}^*$ is represented as a finite binary string in a certain format. We then see that, for every $n \in \mathbb{N}^+$ and $s \in \{0,1\}^*$ such that $|s| = n - |g(n)| - d - |d^*|$,

$$g(n)d^*s \in \operatorname{dom} C \text{ if and only if } s \le \alpha \restriction_{n-|g(n)|-d-|d^*|}, \tag{5}$$

where s and $\alpha \restriction_{n-|g(n)|-d-|d^*|}$ are regarded as a dyadic integer. Then, by the following procedure, we see that there exist an oracle deterministic Turing machine M_1 and $c \in \mathbb{N}$ such that, for all $n \in \mathbb{N}^+$, $M_1^{\operatorname{dom} V \restriction_n}(n) = \alpha \restriction_{n-f(n)-c}$. Note here that $|g(n)| = f(n) + d_0$ for all $n \in \mathbb{N}^+$ and also $H(d) = |d^*|$.

Given n and $\operatorname{dom} V \restriction_n$ with $n > d$, one first checks whether $n - |g(n)| - H(d) \le 0$ holds. If this holds then one outputs λ. If this does not hold, one then calculates the finite set $\operatorname{dom} C \restriction_{n-d}$ by simulating the computation of M with the input $n - d$ and the oracle $\operatorname{dom} V \restriction_n$. Then, based on (5), one determines $\alpha \restriction_{n-|g(n)|-d-H(d)}$ by checking whether $g(n)d^*s \in \operatorname{dom} C$ holds or not for each $s \in \{0,1\}^*$ with $|s| = n - |g(n)| - d - H(d)$. This is possible since $|g(n)d^*s| = n - d$ for every $s \in \{0,1\}^*$ with $|s| = n - |g(n)| - d - H(d)$. Finally, one outputs $\alpha \restriction_{n-|g(n)|-d-H(d)}$. $\qquad\square$

8.2   The Proof of Theorem 6

In order to prove Theorem 6, we need Theorem 9 below and the Ample Excess Lemma (i.e., Theorem 2).

Let M be an arbitrary deterministic Turing machine with the input and output alphabet $\{0,1\}$. We define $L_M = \min\{ |p| \mid p \in \{0,1\}^* \ \& \ M \text{ halts on input } p\}$ (may be ∞). For any $n \ge L_M$, we define I_M^n as the set of all halting inputs p for M with $|p| \le n$ which take longest to halt in the computation of M, i.e., as the set $\{ p \in \{0,1\}^* \mid |p| \le n \ \& \ T_M(p) = T_M^n \}$ where T_M^n is the maximum running time of M on all halting inputs of length at most n. We can slightly strengthen the result presented in Chaitin [3] to obtain the following (see Note in Section 8.1 of Chaitin [3]).

Theorem 9. *Let V be an optimal computer, and let M be a deterministic Turing machine which computes V. Then $n = H(n,p) + O(1) = H(p) + O(1)$ for all (n,p) with $n \geq L_M$ and $p \in I_M^n$.*   □

Proof (of Theorem 6). Let α be a real which is weakly Chaitin random. Let V be an optimal computer, and let M be a deterministic Turing machine which computes V. For each $n \geq L_M$, we choose a particular p_n from I_M^n. For an arbitrary total recursive function $f: \mathbb{N}^+ \to \mathbb{N}$, assume that there exists an oracle deterministic Turing machine M_0 such that, for all $n \in \mathbb{N}^+$, $M_0^{\text{dom } V \upharpoonright_n}(n) = \alpha \upharpoonright_{n-f(n)}$. Note that, given (n, p_n) with $n \geq L_M$, one can calculate the finite set $\text{dom } V \upharpoonright_n$ by simulating the computation of M with the input q until at most the time step $T_M(p_n)$, for each $q \in \{0,1\}^*$ with $|q| \leq n$. This can be possible because $T_M(p_n) = T_M^n$ for every $n \geq L_M$. Thus, we see that there exists a partial recursive function $\Psi: \mathbb{N} \times \{0,1\}^* \to \{0,1\}^*$ such that, for all $n \geq L_M$, $\Psi(n, p_n) = \alpha \upharpoonright_{n-f(n)}$. It follows from (4) that $H(\alpha \upharpoonright_{n-f(n)}) \leq H(n, p_n) + O(1)$ for all $n \geq L_M$. Thus, by Theorem 9 we have

$$H(\alpha \upharpoonright_{n-f(n)}) \leq n + O(1) \tag{6}$$

for all $n \in \mathbb{N}^+$.

In the case where the function $n - f(n)$ of n is bounded to the above, there exists $c \in \mathbb{N}$ such that, for every $n \in \mathbb{N}^+$, $-f(n) \leq c - n$, and therefore $\sum_{n=1}^{\infty} 2^{-f(n)} \leq 2^c$. Thus, in what follows, we assume that the function $n - f(n)$ of n is not bounded to the above.

We define a function $g: \mathbb{N}^+ \to \mathbb{Z}$ by $g(n) = \max\{k - f(k) \mid 1 \leq k \leq n\}$. It follows that the function g is non-decreasing and $\lim_{n \to \infty} g(n) = \infty$. Thus we can choose an enumeration $n_1, n_2, n_3, \ldots$ of the countably infinite set $\{n \in \mathbb{N}^+ \mid n \geq 2 \ \& \ 0 \leq g(n-1) < g(n)\}$ with $n_j < n_{j+1}$. It is then easy to see that $g(n_j) = n_j - f(n_j)$ and $1 \leq n_j - f(n_j) < n_{j+1} - f(n_{j+1})$ hold for all j. On the other hand, since α is weakly Chaitin random, using the Ample Excess Lemma i.e., Theorem 2, we have $\sum_{n=1}^{\infty} 2^{n-H(\alpha \upharpoonright_n)} < \infty$. Thus, using (6) we see that

$$\sum_{j=1}^{\infty} 2^{-f(n_j)} \leq \sum_{j=1}^{\infty} 2^{n_j - f(n_j) - H(\alpha \upharpoonright_{n_j - f(n_j)}) + O(1)} \leq \sum_{n=1}^{\infty} 2^{n - H(\alpha \upharpoonright_n) + O(1)} < \infty. \tag{7}$$

On the other hand, it is easy to see that (i) $g(n) \geq n - f(n)$ for every $n \in \mathbb{N}^+$ and (ii) $g(n) = g(n_j)$ for every j and n with $n_j \leq n < n_{j+1}$. Thus, for each $k \geq 2$, it is shown that

$$\sum_{n=n_1}^{n_k - 1} 2^{-f(n)} \leq \sum_{n=n_1}^{n_k - 1} 2^{g(n)-n} = \sum_{j=1}^{k-1} \sum_{n=n_j}^{n_{j+1}-1} 2^{g(n)-n} = \sum_{j=1}^{k-1} 2^{g(n_j)} \sum_{n=n_j}^{n_{j+1}-1} 2^{-n}$$

$$= \sum_{j=1}^{k-1} 2^{n_j - f(n_j)} 2^{-n_j + 1} \left(1 - 2^{-n_{j+1} + n_j}\right) < 2 \sum_{j=1}^{k-1} 2^{-f(n_j)}.$$

Thus, using (7) we see that $\lim_{k \to \infty} \sum_{n=n_1}^{n_k - 1} 2^{-f(n)} < \infty$. Since $2^{-f(n)} > 0$ for all $n \in \mathbb{N}^+$ and $\lim_{j \to \infty} n_j = \infty$, we have $\sum_{n=1}^{\infty} 2^{-f(n)} < \infty$.   □

4 Elaboration II of the Turing Reduction $\Omega_U \leq_T \operatorname{dom} U$

Theorem 10 (main result II). *Let V and W be optimal computers, and let $f \colon \mathbb{N}^+ \to \mathbb{N}$ be a total recursive function. Then the following two conditions are equivalent:*

(i) There exist an oracle deterministic Turing machine M and $c \in \mathbb{N}$ such that, for infinitely many $n \in \mathbb{N}^+$, $M^{\operatorname{dom} V \restriction_n}(n) = \Omega_W \restriction_{n-f(n)-c}$.

(ii) The function f is not bounded to the above. $\square$

In a similar manner to the proof of Theorem 4 we can prove Theorem 10. The implication (ii) $\Rightarrow$ (i) of Theorem 10 follows from Lemma 1 below and Corollary 1. On the other hand, the implication (i) $\Rightarrow$ (ii) of Theorem 10 follows from Theorem 9 and Theorem 1. By setting $f(n) = 0$ and $W = V$ in Theorem 10, we obtain the following.

Corollary 2. *Let V be an optimal computer. Then, for every $c \in \mathbb{N}$, there does not exist an oracle deterministic Turing machine M such that, for infinitely many $n \in \mathbb{N}^+$, $M^{\operatorname{dom} V \restriction_{n+c}}(n) = \Omega_V \restriction_n$.* $\square$

Elaboration of the Turing Reduction $\operatorname{dom} U \leq_T \Omega_U$

Theorem 11 (main result III). *Let V and W be optimal computers, and let $f \colon \mathbb{N}^+ \to \mathbb{N}$ be a total recursive function. Then the following two conditions are equivalent:*

(i) There exist an oracle deterministic Turing machine M and $c \in \mathbb{N}$ such that, for all $n \in \mathbb{N}^+$, $M^{\{\Omega_V \restriction_n\}}(n) = \operatorname{dom} W \restriction_{n+f(n)-c}$, where the finite subset $\operatorname{dom} W \restriction_{n+f(n)-c}$ of $\{0,1\}^*$ is represented as a finite binary string in a certain format.

(ii) The function f is bounded to the above. $\square$

The implication (ii) $\Rightarrow$ (i) of Theorem 11 follows immediately from Fact 1 and Corollary 1. On the other hand, in order to prove the implication (i) $\Rightarrow$ (ii) of Theorem 11, we need the following lemma first.

Lemma 1. *Let $f \colon \mathbb{N}^+ \to \mathbb{N}$ be a total recursive function. If the function f is not bounded to the above, then the function $f(n) - H(n)$ of $n \in \mathbb{N}^+$ is not bounded to the above.*

Proof. Contrarily, assume that there exists $c \in \mathbb{N}$ such that, for every $n \in \mathbb{N}^+$, $f(n) \leq H(n) + c$. Then, since f is not bounded to the above, it is easy to see that there exists a total recursive function $\Psi \colon \mathbb{N}^+ \to \mathbb{N}^+$ such that, for every $k \in \mathbb{N}^+$, $k \leq H(\Psi(k))$. It follows from (4) that $k \leq H(k) + O(1)$ for all $k \in \mathbb{N}^+$. On the other hand, using (3) we can show that $H(k) \leq 2\log_2 k + O(1)$ for all $k \in \mathbb{N}^+$. Thus we have $k \leq 2\log_2 k + O(1)$ for all $k \in \mathbb{N}^+$. However, we have a contradiction on letting $k \to \infty$ in this inequality, and the result follows. $\square$

Proof (of (i) ⇒ (ii) of Theorem 11). Let V and W be optimal computers. For an arbitrary total recursive function $f\colon \mathbb{N}^+ \to \mathbb{N}$, assume that there exist an oracle deterministic Turing machine M and $c \in \mathbb{N}$ such that, for all $n \in \mathbb{N}^+$, $M^{\{\Omega_V \restriction n\}}(n) = \operatorname{dom} W \restriction_{n+f(n)-c}$. Then, by considering the following procedure, we see that $n + f(n) < H(\Omega_V \restriction n) + O(1)$ for all $n \in \mathbb{N}^+$.

Given $\Omega_V \restriction n$, one first calculates the finite set $\operatorname{dom} W \restriction_{n+f(n)-c}$ by simulating the computation of M with the input n and the oracle $\Omega_V \restriction n$. Then, by calculating the set $\{\, W(p) \mid p \in \operatorname{dom} W \restriction_{n+f(n)-c} \,\}$ and picking any one finite binary string s which is not in this set, one can obtain $s \in \{0,1\}^*$ such that $n+f(n)-c < H_W(s)$.

Thus, there exists a partial recursive function $\Psi\colon \{0,1\}^* \to \{0,1\}^*$ such that, for all $n \in \mathbb{N}^+$, $n + f(n) - c < H_W(\Psi(\Omega_V \restriction n))$. It follows from the optimality of W and (4) that $n + f(n) < H(\Omega_V \restriction n) + O(1)$ for all $n \in \mathbb{N}^+$, as desired. On the other hand, using (3) we can show that $H(s) \le |s| + H(|s|) + O(1)$ for all $s \in \{0,1\}^*$. Therefore we have $f(n) < H(n) + O(1)$ for all $n \in \mathbb{N}^+$. Thus, it follows from Lemma 1 that f is bounded to the above.   ◻

Acknowledgments. The proof of Theorem 8 was originally based on the computation history of M with input p in place of $T_M(p)$. Inspired by the suggestion of the anonymous referee to use $(p, T_M(p))$ instead, we can reach the current form of the proof using $T_M(p)$. In addition, the proof of the implication (i) ⇒ (ii) of Theorem 11 is significantly shortened by the suggestion of the same referee. Thus the author is grateful to the referee for the insightful suggestions. This work was supported by KAKENHI, Grant-in-Aid for Scientific Research (C) (20540134), by SCOPE from the Ministry of Internal Affairs and Communications of Japan, and by CREST from Japan Science and Technology Agency.

References

1. Calude, C.S., Hertling, P.H., Khoussainov, B., Wang, Y.: Recursively enumerable reals and Chaitin Ω numbers. Theoret. Comput. Sci. 255, 125–149 (2001)
2. Chaitin, G.J.: A theory of program size formally identical to information theory. J. Assoc. Comput. Mach. 22, 329–340 (1975)
3. Chaitin, G.J.: Algorithmic Information Theory. Cambridge University Press, Cambridge (1987)
4. Chaitin, G.J.: Program-size complexity computes the halting problem. Bulletin of the European Association for Theoretical Computer Science 57, 198 (1995)
5. Downey, R.G., Hirschfeldt, D.R.: Algorithmic Randomness and Complexity. Springer, Heidelberg (to appear)
6. Kučera, A., Slaman, T.A.: Randomness and recursive enumerability. SIAM J. Comput. 31(1), 199–211 (2001)
7. Miller, J., Yu, L.: On initial segment complexity and degrees of randomness. Trans. Amer. Math. Soc. 360, 3193–3210 (2008)
8. Nies, A.: Computability and Randomness. Oxford University Press, New York (2009)
9. Solovay, R.M.: Draft of a paper (or series of papers) on Chaitin's work.. done for the most part during the period of September–December 1974, IBM Thomas J. Watson Research Center, Yorktown Heights, New York, pp. 215 (May 1975) (unpublished manuscript)

Bayesian Data Integration and Enrichment Analysis for Predicting Gene Function in Malaria

Philip M.R. Tedder[1], James R. Bradford[2], Chris J. Needham[3],
Glenn A. McConkey[1], Andrew J. Bulpitt[3], and David R. Westhead[1]

[1] Faculty of Biological Sciences, University of Leeds LS2 9JT, UK
[2] Applied Computational Biology and Bioinformatics,
Paterson Institute for Cancer Research,
The University of Manchester, Manchester, M20 4BX, UK
[3] School of Computing, University of Leeds, Leeds, LS2 9JT, UK

Abstract. Malaria is one of the world's most deadly diseases and is caused by the parasite *Plasmodium falciparum*. Sixty percent of *P. falciparum* genes have no known function and therefore new methods of gene function prediction are needed. To address this problem, we train a naïve Bayes classifier on multiple sources of data and subsequently apply a modified version of the Gene Set Enrichment Analysis Algorithm to predict gene function in *P. falciparum*. To define gene function, we exploit the hierarchical structure of the Gene Ontology, specifically using the Biological Process category. We demonstrate the value of integrating multiple data sources by achieving accurate predictions on genes that cannot be annotated using simple sequence similarity based methods.

Keywords: *Plasmodium falciparum*, malaria, gene function prediction, Bayes classifier, heterogeneous data sources, machine learning.

Introduction

The speed and decreased cost of DNA sequencing has led, in the past decade, to the genome sequences of hundreds of organisms being revealed [1]. Annotating the function of each gene by conventional laboratory techniques for even just a single genome would take decades and would be exceedingly expensive. Therefore, there is an urgent need for tools that can either automate the prediction of gene function, or speed up the process by suggesting the function of particular genes, which can then be tested experimentally.

The most established and reliable methods for annotating genes of unknown function are based on sequence similarity based methods such as BLAST [2], but unfortunately in most genomes such methods still leave a large proportion of the genome with no known function. The process of annotating the remaining genes of unknown function has been aided in the past decade by using large, genome wide data sets. However, these datasets are often noisy and incomplete leading to problems of low reliability as well as limited coverage. To try and assuage these

Ambos-Spies, B. Löwe, and W. Merkle (Eds.): CiE 2009, LNCS 5635, pp. 457–466, 2009.

problems, multiple data sets have often been combined using machine learning techniques such as neural networks, support vector machines and Bayesian networks to create synergistic predictions that have produced encouraging results [4]. Owing to the paucity of large, genome wide data sets, most gene function prediction programs so far have been restricted to model organisms, particularly the yeast *Saccharomyces cerevisiae*.

The most common annotation database used by gene function prediction programs is the Gene Ontology (GO) database [3]. The GO database seperates a gene's annotation into three distinct classes: molecular function, biological process and cellular location. Each class contains thousands of annotations, so even just predicting the most specific annotations for a gene leads to over 10^{12} possible functional classes making computing all possible annotations intractable. Most gene function prediction programs attempt to solve this problem by inferring gene function from other genes using a "guilt by association" approach [4] which uses the fact that genes that share a similar function are likely to show similar patterns of results in genome wide data sets.

Plasmodium falciparum is the causative organism of the most deadly form of human malaria. Its genome was sequenced in 2002 [5] but 60% of the genes could not be assigned a function at that time. Despite the large number of unannotated genes, *in silico* investigation of *P. falciparum* genes has been limited. To date, the only program that uses different genome wide data sets and a machine learning method to produce a combined gene function prediction is that of Plasmodraft [6]. However, Plasmodraft only uses three types of data and a simple machine learning method (a "guilt by association" k-nearest neighbour approach). There still remains a vital need for a gene function prediction tool which brings together the disparate data sets available for *P. falciparum*, and uses them synergistically via sophisticated machine learning methods to produce putative annotations. Equally, the computational complexity of gene function description and prediction calls for the development of new approaches. Here we present a novel approach to these computational issues and an illustrate it by application to *P. falciparum*.

2   Methods

In this study we limited our focus to predicting biological process, as several previous studies have suggested that non-sequence similarity based methods are most suited to predicting this aspect of gene function. The functions of *P. falciparum* genes were defined using the annotations of the GO database. This is a controlled vocabulary where annotations are organised as directed acyclic graph (DAG) with more specific annotations inheriting annotations (sometimes from multiple parents) from more general terms higher up the graph.

We solve the computability problem of gene function prediction by using the DAG structure of GO to reduce the number of functional classes by gathering similar terms. It is important to balance the need for the most detailed prediction possible with the demand for sufficient training data, which leads to prediction f fewer and broader functional classes. Here we approach this issue by clustering enes into groups of similar function using the semantic similarity of GO terms nd gene annotations.

.1 Deriving Clusters of Genes of Similar Function

'he semantic similarity between two GO terms was measured using the method f Resnik [7], first applied to GO by Lord and co-workers [8]. This measure is ased on the idea of the probability of the minimum subsumer (p_{ms}) as defined a equation 1 for the GO terms (t_1, t_2):

$$p_{ms}(t_1, t_2) = \min_{t \in A(t_1, t_2)} \{p(t)\} \tag{1}$$

'ere p(t) is the proportion of all GO annotated *P. falciparum* genes that are nnotated with this particular GO term and A(t_1, t_2) is the set of parental O terms shared by t_1 and t_2. The probability of the minimum subsumer was lculated not only for the most specific terms annotated to each gene product ut their ancestors as well. From this, the semantic similarity score between the vo terms can then be calculated using equation 2:

$$S(t_1, t_2) = -\ln p_{ms}(t_1, t_2). \tag{2}$$

or our purposes, this similarity score was normalised between zero and one by iving by the highest similarity score possible in that category.

Gene function similarity was then measured using these semantic similarity ores by adapting the method of Wang and co-workers [9]. For two genes G1 ad G2 annotated with GO term sets T1 and T2 respectively, where T1= $\{t_i^{(1)},$ =1..n\} and T2= $\{t_j^{(2)}$, j=1..m\}, the gene function similarity of the two genes is efined in equation 3:

$$(G_1, G_2) = \frac{1}{(n+m)} \left(\sum_{i=1}^{n} \max_{t^{(2)} \in T2} \{S\left(t_i^{(1)}, t^{(2)}\right)\} + \sum_{j=1}^{m} \max_{t^{(1)} \in T1} \{S\left(t_j^{(2)}, t^{(1)}\right)\} \right) \tag{3}$$

he above gene function similarity measure was used to cluster genes, using .e clique finding algorithm DFMAX [10], into a set of clusters at each level of nctional specificity above a defined threshold H, where H = 0.10-0.80 at 0.10 crements.

A two stage process was then adopted to predict gene function. First, a naïve ayes method was used to train a binary classifier predicting whether pairs of nes belong to the same functional cluster. Then, individual gene functions were 'edicted by using enrichment analysis (see section 2.4) on a list of gene pairs nked by likelihood of sharing cluster membership.

2.2   Data

For each specificity level, the Bayes classifier was trained on data consisting of a set of gene pairs, positive if the two genes belong to the same cluster or negative otherwise. The conditional probability distributions for eight different data sources (all of which have been shown to be able to predict gene function) were then learnt separately. These eight features consist of five continuous variables: co-expression in Agilent and Affymetrix microarray data, co-expression in mass spectrometry data, percentage of different domains shared between the two genes and the distance between the two genes; and three binary variables: whether or not the two proteins interact, whether a gene fusion has occurred between any of the genes' orthologs and whether they share an identical pattern of gene conservation. The details of how the data for each method was collated and formatted for input into the Bayes classifier is outlined below.

Sequence domains. The domain annotations for each gene were extracted from the InterPro database [11] using InterPro scan [12]. Then, the proportion of different domains shared by each gene pair was calculated.

Gene neighbourhood. Gene chromosomal location was downloaded from PlasmoDB [13] and the distance between a pair of genes was calculated and then normalized by dividing by the total length of the chromosome on which the genes resided. Genes on different chromosome were deemed to have a distance of 1.

Gene conservation. Gene conservation data for all organisms was downloaded from OrthoMCL-DB [14]. The gene conservation profiles were represented as binary strings corresponding to each of the 55 organisms in OrthoMCL: one indicating presence of a gene in a given organism, zero otherwise. Genes with orthologs present in fewer than two organisms were ignored, as were genes sharing identical profiles with greater than twenty other genes.

Gene fusion. Gene fusion data was based on the gene fusion data of Date and Stoeckert [15] where 163 genomes were mined for incidences of gene fusion events.

Agilent microarray. Agilent microarray data consisted of expression profile across 48 individual one hour time-points from the intra-erythrocytic developmental cycle in the HB3 strain [16] of *P. falciparum*.

Affymetrix microarray. Affymetrix *P. falciparum* microarray data consisted of expression profiles across 38 experiments from Young and co-workers [17] and Le Roch and co-workers [18].

Co-expression between a pair of genes for both microarray methods was quantified by calculating the Pearson correlation coefficient.

Mass spectrometry. The peptide counts from 26 *Plasmodium* mass spectrometry experiments were used [19][20][21][22]. This data is unsuitable for correlation based measures of similarity so the Manhattan distance measure was used.

Protein-Protein interactions. Interaction data for *P. falciparum* was downloaded from the Intact database [23] which for *P. falciparum* mostly comprised the yeast two-hybrid data of LaCount and co-workers [24]. Two proteins were deemed not to interact if their interaction was not indicated in the Intact database.

.3   Bayes Classifier

A combined prediction using all 8 data types was then produced using a naïve Bayes classifier (which assumes the data sets to be independent). This was implemented using the Bayesian Network Toolbox for Matlab [25] with the Gaussian distribution used to model the distribution of the continuous variables and the maximum likelihood algorithm used to learn the conditional probability distributions. At each level, the classifier was trained on a set of all gene pairs with the class variable being whether or not the two genes belong to the same cluster and the 8 dependent variables being the methods similarity score for the pair. The number of functional classes at each level is shown in tables 1 and 2.

.4   Gene Set Enrichment

To predict individual gene function, the gene is first compared with all genes in clusters using the Bayes classifier to assign probabilities of shared cluster membership. This list is then ranked by probability, on the hypothesis that genes with the same function (cluster) as the unknown gene would be enriched towards the top of the list. To measure this enrichment score (ES) we adapt an algorithm originally designed to perform a similar such calculation for a group of genes in microarray data, the Gene Set Enrichment Analysis algorithm [26], employing the same procedure at each level of functional specificity, as described below.

1. For the set of N genes, rank the list of all genes (g_j) in descending order of probability (r_j) of shared cluster membership with the query gene.

2. For each cluster C (containing S genes) step down through the list and at every step evaluate the fraction of genes in cluster C and the fraction of genes not in cluster C up to a given position i in N using the equations below:

$$P_{hit}(S,i) = \sum_{j \leq i,\, g_j \in C} \frac{r_j^p}{N_R}, \ where\ N_R = \sum_{g_j \in C} r_j^p \qquad (4)$$

$$P_{miss}(S,i) = \sum_{j \leq i,\, g_i \notin C} \frac{1}{(N-S)} \qquad (5)$$

3. The ES is then:

$$ES = \max_i \{P_{hit}(S,i) - P_{miss}(S,i)\} \qquad (6)$$

This reduces to the standard Kolmogorov-Smirnov statistic when p=0 (here we used p=1). An ES close to zero corresponds to a random distribution of genes from the cluster in the ranking, and a high ES corresponds to possible cluster membership. The ES is then normalised (NES) to take into account the size of each cluster C by comparing it with the set of random ES scores: $NES=ES/ES_{Rmean}$, where ES_{Rmean} is the mean ES score of 1000 random assignments of genes to cluster C from N.

For the query gene, the NES score was calculated for all clusters at each specificity level, the cluster with the highest NES score was then taken as the prediction of the gene's function for that specificity level.

To evaluate the gene function prediction program, a leave one out cross validation procedure was performed on all *P. falciparum* genes with biological process GO annotations that could not be inferred via a simple BLAST search. GO annotations that had been inferred without any manual intervention were also excluded. In each round of cross validation, the training set comprised all gene pairs except those that included the left out gene. A succesful prediction of a gene's function was defined as the correct annotation cluster having one of the three highest NES scores for that specificity level. Coverage was defined as the number of genes for a which a prediction was produced by a method divided by the total number of genes used in the evaluation.

3   Results

Accuracy and coverage of all the methods are displayed in tables 1 and 2 and figure 1. A higher number of predictions (coverage) were produced at the more general specificity levels as there were more genes clustered at these levels and therefore more data available to produce a prediction. There are also fewer clusters at more general levels, making it easier for the program to classify the gene in the correct cluster. As genes whose function could be predicted by a simple BLAST search were removed some of the methods produced very few predictions. The gene conservation and protein-protein interaction methods produce the least predictions, often failing to produce any correct predictions for a particular specificity level. Gene fusion, whose coverage was still comparatively low produced very accurate predictions (accuracy > 0.95 for all specificity levels). As would be expected the sequenced based method of domains produces high accuracy predictions and at a much higher coverage than the previous three methods. Interestingly, the domain accuracy is lowest for the intermediate specificity level and highest at the most general and specific levels. The three methods that measure the co-expression of the query gene with other *P. falciparum* genes - mass spectrometry, Affymetrix and Agilent microarray all produce encouraging results especially at more general levels. The performance of gene neighbourhood in predicting gene function is also encouraging especially at the more general specificity levels. The combined classifier has the largest coverage but is not the most accurate.

4   Discussion

The gene conservation and protein-protein interaction methods produced by far the least predictions, with these predictions often being of low quality. The

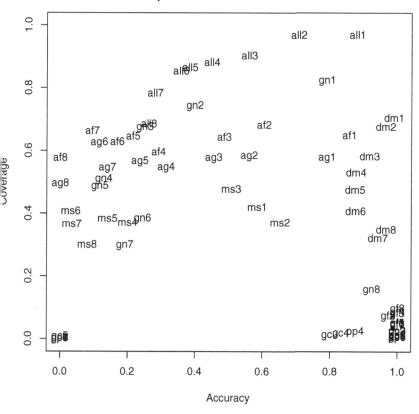

Fig. 1. Accuracy versus coverage of gene function predictions for different prediction methods at the different specificity levels, the digit at the end is the specificity level of the prediction (1 = most general, 8 = most specific), abbreviations of each method are defined below: gn = gene neighbourhood, gf = gene fusion, gc = gene conservation, pp = protein-protein interactions, af = Affymetrix, ag =Agilent, ms = mass spectrometry, all =combined classifier

method of gene conservation relies on two genes being co-conserved across several species. *P. falciparum* has a relatively unique life cycle and as such has many genes with specific functions not seen in many other organisms. This lack of orthologous processes and therefore genes may well explain the paucity of predictions produced by gene conservation. The protein-protein interaction data for *P. falciparum* is largely from the yeast two-hybrid experiment of LaCount [24] and the quality of this data has previously been questioned [27]. Only recently has mass spectrometry been used for gene function prediction, but these results suggest that as mass spectrometry methods improve this may become as useful and established as microarray data is for predicting gene function.

Table 1. Accuracy: Proportion of genes whose correct annotation was predicted in the top 3 clusters by each method for each specificity level (1 = most general, 8 = most specific)

	Specificity level							
	1	2	3	4	5	6	7	8
Number of classes	10	20	29	43	54	63	66	49
Agilent microarray	0.79	0.56	0.46	0.31	0.24	0.12	0.14	0.00
Affymetrix microarray	0.86	0.60	0.49	0.29	0.22	0.17	0.10	0.00
Mass spectrometry	0.58	0.65	0.51	0.20	0.14	0.03	0.03	0.08
Gene conservation	1.00	1.00	0.80	0.83	0.00	0.00	0.00	0.00
Gene neighbourhood	0.79	0.40	0.25	0.13	0.12	0.24	0.19	0.92
Gene fusion	1.00	0.97	1.00	1.00	1.00	1.00	1.00	1.00
Protein-Protein interaction	0.00	1.00	1.00	0.88	1.00	1.00	0.00	0.00
Domain	0.99	0.96	0.92	0.88	0.87	0.88	0.94	0.97
All	0.88	0.71	0.56	0.45	0.38	0.36	0.28	0.26

Table 2. Coverage: Proportion of total number of genes for which a prediction is produced by each method for each specificity level (1 = most general, 8 = most specific)

	Specificity level							
	1	2	3	4	5	6	7	8
Number of classes	10	20	29	43	54	63	66	49
Agilent microarray	0.58	0.58	0.58	0.55	0.57	0.63	0.54	0.49
Affymetrix microarray	0.65	0.68	0.64	0.60	0.65	0.63	0.66	0.58
Mass spectrometry	0.42	0.37	0.48	0.37	0.38	0.41	0.37	0.30
Gene conservation	0.01	0.02	0.01	0.02	0.01	0.00	0.00	0.00
Gene neighbourhood	0.82	0.74	0.67	0.51	0.49	0.38	0.30	0.16
Gene fusion	0.05	0.07	0.08	0.09	0.05	0.05	0.04	0.10
Protein-Protein interaction	0.00	0.01	0.02	0.02	0.01	0.00	0.00	0.00
Domain	0.71	0.68	0.58	0.53	0.48	0.41	0.32	0.35
All	0.97	0.97	0.90	0.88	0.87	0.86	0.78	0.69

Combining a Bayesian classifier and our enrichment analysis algorithm, we have produced predictions of gene function for a large proportion of *P. falciparum* genes that can not be predicted from the most widely used method of inferring gene function, that of a BLAST sequence alignment. Several methods produce highly encouraging results with the combined classifier producing predictions for a large proportion of the *P. falciparum* genome. An example of a prediction produced for a gene of unknown function is that of the gene with identifier PF10_0351. The gene was predicted to be involved in invasion of the host cell (GO:0030260) owing to the gene being highly correlated in the Agilent, Affymetrix and mass spectrometry methods with genes involved in invasion of red blood cells and because it also found next to genes involved in invasion in the *P. falciparum* genome. This prediction of function is corroborated by the gene containing a signal peptide normally associated with gene ivolved in invasion. This example demonstrates the ability of the gene function prediction program to produce putative predictions of a gene's function which hen can be experimentally verified.

References

1. Liolios, K., et al.: The Genomes On Line Database (GOLD) in 2007: status of genomic and metagenomic projects and their associated metadata. Nucleic Acids Res. 36, D475–D479 (2008)
2. Altschul, S.F., et al.: Gapped BLAST and PSI-BLAST: a new generation of protein database search programs. Nucleic Acids Res. 25, 3389–3402 (1997)
3. Ashburner, M., et al.: Gene ontology: tool for the unification of biology. The Gene Ontology Consortium. Nat. Genet. 25, 25–29 (2000)
4. Pena-Castillo, L., et al.: A critical assessment of Mus musculus gene function prediction using integrated genomic evidence. Genome Biol. 9(suppl. 1), S2 (2008)
5. Gardner, M.J., et al.: Genome sequence of the human malaria parasite Plasmodium falciparum. Nature 419, 498–511 (2002)
6. Brehelin, L., et al.: PlasmoDraft: a database of Plasmodium falciparum gene function predictions based on postgenomic data. BMC Bioinformatics 9, 440 (2008)
7. Resnik, P.: Semantic similarity in a taxonomy: An information-based measure and its application to problems of ambiguity in natural language. Journal of Artificial Intelligence Research 11, 95–130 (1999)
8. Lord, P.W., et al.: Investigating semantic similarity measures across the Gene Ontology: the relationship between sequence and annotation. Bioinformatics 19, 1275–1283 (2003)
9. Wang, J.Z., et al.: A new method to measure the semantic similarity of GO terms. Bioinformatics 23, 1274–1281 (2007)
10. dfmax.c, ftp://dimacs.rutgers.edu/pub/challenge/graph/solvers
11. Mulder, N.J., et al.: New developments in the InterPro database. Nucleic Acids Res. 35, D224–D228 (2007)
12. Quevillon, E., et al.: InterProScan: protein domains identifier. Nucleic Acids Res. 33, W116–W120 (2005)
13. Stoeckert Jr., C.J., et al.: PlasmoDB v5: new looks, new genomes. Trends Parasitol 22, 543–546 (2006)
14. Chen, F., et al.: OrthoMCL-DB: querying a comprehensive multi-species collection of ortholog groups. Nucleic Acids Res. 34, D363–D368 (2006)
15. Date, S.V., Stoeckert Jr., C.J.: Computational modeling of the Plasmodium falciparum interactome reveals protein function on a genome-wide scale. Genome Res. 16, 542–549 (2006)
16. Llinas, M., et al.: Comparative whole genome transcriptome analysis of three Plasmodium falciparum strains. Nucleic Acids Res. 34, 1166–1173 (2006)
17. Young, J.A., et al.: The Plasmodium falciparum sexual development transcriptome: a microarray analysis using ontology-based pattern identification. Mol. Biochem. Parasitol 143, 67–79 (2005)
18. Le Roch, K.G., et al.: Discovery of gene function by expression profiling of the malaria parasite life cycle. Science 301, 1503–1508 (2003)
19. Florens, L., et al.: A proteomic view of the Plasmodium falciparum life cycle. Nature 419, 520–526 (2002)

20. Lasonder, E., et al.: Analysis of the Plasmodium falciparum proteome by high-accuracy mass spectrometry. Nature 419, 537–542 (2002)
21. Khan, S.M., et al.: Proteome analysis of separated male and female gametocytes reveals novel sex-specific Plasmodium biology. Cell 121, 675–687 (2005)
22. Le Roch, K.G., et al.: Global analysis of transcript and protein levels across the Plasmodium falciparum life cycle. Genome Res. 14, 2308–2318 (2004)
23. Hermjakob, H., et al.: IntAct: an open source molecular interaction database. Nucleic Acids Res. 32, D452–D455 (2004)
24. LaCount, D.J., et al.: A protein interaction network of the malaria parasite Plasmodium falciparum. Nature 438, 103–107 (2005)
25. Murphy, K.P.: The Bayes Net Toolbox for Matlab. Computing Science and Statistics 33 (2001)
26. Subramanian, A., et al.: Gene set enrichment analysis: a knowledge-based approach for interpreting genome-wide expression profiles. Proc. Natl. Acad. Sci. U. S. A. 102, 15545–15550 (2005)
27. Wuchty, S., Ipsaro, J.J.: A draft of protein interactions in the malaria parasite P falciparum. J. Proteome Res. 6, 1461–1470 (2007)

Dialectica Interpretation with Fine Computational Control

Trifon Trifonov*

Mathematics Institute, University of Munich, Germany
trifonov@math.lmu.de

Abstract. Computational proof interpretations enrich the logical meaning of formula connectives and quantifiers with algorithmic relevance and
allow to extract the construction contained in a proof. Berger showed
that quantifiers can be selectively declared irrelevant for the modified realisability interpretation, thus removing unnecessary computation from
the extracted program [1]. Hernest adapted the uniform quantifiers to
Gödel's Dialectica interpretation [2] and later demonstrated together
with the author how they can be refined [5]. The present paper gives a
further extension, in which the computational meaning can be controlled
separately for every component of the Dialectica interpretation. Apart
from enriching the possibilities to remove redundancies in extracted programs, this finer approach also allows to independently switch off the
postive or negative algorithmic contribution of whole formulas.

Introduction

Constructive proofs contain two sorts of information: an algorithm computing a
witness for the conclusion and logical reasoning for its correctness. The modified
realisability interpretation [6] provides us with means to automatically separate
computation from verification. This distinction is entirely determined by the
proof and more precisely by the computational relevance of the formulas in the
nodes of the proof tree. Generally, to modify the computational content, we
might need to rephrase the entire proof. However, Berger [1] showed how we can
selectively remove content by marking quantifiers as computationally irrelevant
(or uniform). This effectively corresponds to converting all introductions and
eliminations of such quantifiers from possibly computational to purely logical.

A natural question to consider is whether such optimisations are possible
when extracting from proofs making use of classical principles. Berger already
showed in his paper [1] how uniform quantifiers can optimise programs, which
are extracted from a proof from contradiction. This was possible, because the
extraction method used (refined A-translation) applies modified realisability on
a translated proof.

* The author gratefully acknowledges financial support by MATHLOGAPS (MEST-
CT-2004-504029), a Marie Curie Early Stage Training Site.

Ambos-Spies, B. Löwe, and W. Merkle (Eds.): CiE 2009, LNCS 5635, pp. 467–477, 2009.
Springer-Verlag Berlin Heidelberg 2009

The context is different when we extract from non-constructive proofs using Gödel's functional (Dialectica) interpretation [3]. It can be viewed as an extension of realisability that allows tracking of terms, which are used to instantiate universal premises of implications. In a natural deduction setting this is achieved by keeping the negative computational content of all open assumptions. In addition, a case distinction over the quantifier-free Dialectica translation of an assumption formula is required whenever an assumption is used more than once. A careful analysis of the extracted programs shows that because of the extensive witness collection of Dialectica, undesired computations appear even more often than with modified realisability. Hernest [2] was the first to give a possible adaptation of uniform quantifiers to this setting. He showed that disabling tracking of certain instantiations of universal premises should be done carefully, as it could introduce quantifiers in the otherwise quantifier-free translation and make the proof uninterpretable.

An alternative approach for cleaning extracted terms is to remove the computational meaning of whole subformulas. Hernest and Oliva [4] followed this path for Gödel's original interpretation and showed that in a linear logic setting one can independently switch off positive and negative contributions of a whole formula. Oliva presented a further example, in which it is necessary to switch off *only* the negative content of a formula in order to still be able to interpret the proof. This example could not be directly simulated with Hernest's extension and naturally led to the idea of expressing positive and negative uniform quantifications independently of each other [5].

Even if the refined uniform quantifiers are more expressive, they seem insufficient to fully remove negative content of a subformula, as in Hernest and Oliva's hybrid interpretation. The reason is that negative content is generated in two different ways in Dialectica: from instantiations of universal quantifiers and from positive content of premises in implications. In the present paper we show a variant of Gödel's interpretation with uniform annotations of quantifiers and implications that allows to separately omit every computational component of the proof. We will show how the defined extension allows to completely discard positive or negative content of a formula.

2   Negative Arithmetic

We work in a restriction of Heyting Arithmetic with finite types (denoted HA^ω in [8]) to the language of $\to$ and $\forall$. We refer to the resulting system as Negative Arithmetic (NA).

Definition 1. *Types* (ρ, σ), *(object) terms* (s, t) *and formulas* (A, B) *are defined as follows:*

$$\rho, \sigma \quad ::= \quad \mathsf{B} \mid \mathsf{N} \mid \rho \Rightarrow \sigma \mid \rho \times \sigma$$

$$s, t \quad ::= \quad x^\rho \mid (\lambda_{x^\rho} t^\sigma)^{\rho \Rightarrow \sigma} \mid (s^{\rho \Rightarrow \sigma} t^\rho)^\sigma \mid \langle s^\rho, t^\sigma \rangle^{\rho \times \sigma} \mid (t^{\rho \times \sigma} {\llcorner})^\rho \mid (t^{\rho \times \sigma} {\lrcorner})^\sigma \mid$$
$$\mathsf{tt}^\mathsf{B} \mid \mathsf{ff}^\mathsf{B} \mid 0^\mathsf{N} \mid \mathsf{S}^{\mathsf{N} \Rightarrow \mathsf{N}} \mid \mathcal{C}^{\mathsf{B} \Rightarrow \sigma \Rightarrow \sigma \Rightarrow \sigma} \mid \mathcal{R}^{\mathsf{N} \Rightarrow \sigma \Rightarrow (\mathsf{N} \Rightarrow \sigma \Rightarrow \sigma) \Rightarrow \sigma}$$

$$A, B \quad ::= \quad \mathsf{at}(t^\mathsf{B}) \mid A \to B \mid \forall_{x^\rho} A$$

The base types of booleans B and natural numbers N are equipped with the usual constructors and structural recursor constants. Here x denotes a typed object variable. The sets of free variables $\mathsf{FV}(t)$, $\mathsf{FV}(A)$ are defined inductively as usual. Substitution of terms for object variables $s\,[x := t]$, $A\,[x := t]$ is always assumed to be capture-free with respect to abstraction and quantification.

The operational semantics of object terms is given by the usual β-reduction rules and computation rules for the recursor constants:

$$\langle s, t \rangle \llcorner \mapsto s \qquad\qquad \mathcal{C}\,\mathsf{tt}\,t_1\,t_2 \mapsto t_1 \qquad \mathcal{R}\,0\,s\,t \quad \mapsto s$$
$$\langle s, t \rangle \lrcorner \mapsto t \qquad\qquad \mathcal{C}\,\mathsf{ff}\,t_1\,t_2 \mapsto t_2 \qquad \mathcal{R}\,(\mathsf{S}n)\,s\,t \mapsto t\,n\,(\mathcal{R}\,n\,s\,t)$$
$$(\lambda_x s)t \mapsto s\,[x := t]$$

We express derivations in a natural deduction system with a similar syntax to that of object terms to stress the Curry-Howard correspondence. Proof terms are typed by their conclusion formulas and are built from *assumption variables*.

Definition 2. Proof terms (M, N) of NA *are defined as follows:*

$$
\begin{aligned}
M, N \quad ::= \quad & u^A \mid (\lambda_{u^A} M^B)^{A \to B} \mid (M^{A \to B} N^A)^B \mid \\
(*) \qquad & (\lambda_{x^\rho} M^{A(x)})^{\forall_{x^\rho} A(x)} \mid (M^{\forall_{x^\rho} A(x)} t)^{A(t)} \mid \\
& \mathsf{AxT} : \mathsf{at}(\mathsf{tt}) \mid \mathsf{Cases}^{A(b)} : \forall_{b^\mathsf{B}} \big(A(\mathsf{tt}) \to A(\mathsf{ff}) \to A(b) \big) \mid \\
& \mathsf{Ind}^{A(n)} : \forall_{n^\mathsf{N}} \big(A(0) \to \forall_{n^\mathsf{N}} (A(n) \to A(\mathsf{S}n)) \to A(n) \big)
\end{aligned}
$$

with the usual variable condition $(*)$ *that the object variable x does not occur free in any of the open assumptions of M. The sets of free variables $\mathsf{FV}(M)$ and free (open) assumption variables $\mathsf{FA}(M)$ as well as capture-free substitutions $M\,[x := t]$ and $M\,[u := N]$ are defined inductively as usual.*

The *truth axiom* AxT defines the logical meaning of $\mathsf{at}(\cdot)$ and allows us to consider any boolean valued function defined in our term system as a decidable predicate. When we write for example $n = m$, we actually mean $\mathsf{at}(\mathsf{Eq}\,n\,m)$, where $\mathsf{Eq}^{\mathsf{N} \Rightarrow \mathsf{N} \Rightarrow \mathsf{B}}$ is a term defining the decidable equality for natural numbers.

In our negative language, defining falsity as $\mathsf{F} := \mathsf{at}(\mathsf{ff})$ already gives us the full power of classical logic. In particular, for a formula A we can prove *Ex falso quodlibet* (efq): $\vdash \mathsf{F} \to A$ and *Stability*: $\vdash ((A \to \mathsf{F}) \to \mathsf{F}) \to A$ by meta induction on A, using AxT and Cases for the base case. We will thus use the abbreviations $\neg A := A \to \mathsf{F}$ and $\tilde{\exists}_{x^\rho} A := \neg \forall_{x^\rho} \neg A$.

The term system in consideration is essentially Gödel's T and is well known to be strongly normalising and confluent. Without loss of generality we assume that all considered object terms are in normal form. Note that we make no such assumption for proof terms in order to allow their modular structure to be transferred to the extracted programs.

Notation. For technical convenience we will use ε for denoting a special *nulltype*. By abuse of notation we also use ε to denote all terms of nulltype. We stipulate that the following simplifications are always carried out implicitly:

$$\rho \times \varepsilon \rightsquigarrow \rho, \quad t^{\rho \times \varepsilon}{}_{\llcorner} \rightsquigarrow t, \quad \langle t, \varepsilon \rangle \rightsquigarrow t, \quad \rho \Rightarrow \varepsilon \rightsquigarrow \varepsilon, \quad \lambda_x \varepsilon \rightsquigarrow \varepsilon, \quad \varepsilon t \rightsquigarrow \varepsilon$$

$$\varepsilon \times \rho \rightsquigarrow \rho, \quad t^{\varepsilon \times \rho}{}_{\lrcorner} \rightsquigarrow t, \quad \langle \varepsilon, t \rangle \rightsquigarrow t, \quad \varepsilon \Rightarrow \rho \rightsquigarrow \rho, \quad \lambda_{x^\varepsilon} t \rightsquigarrow t, \quad t \varepsilon \rightsquigarrow t \qquad (\varepsilon)$$

$$\forall_{x^\varepsilon} A \rightsquigarrow A, \quad M\varepsilon \rightsquigarrow M$$

3   The Dialectica Interpretation

In this section we present the Dialectica interpretation in our natural deduction setting, following the treatment in [7]. We translate[1] each formula A to a quantifier-free formula $|A|^x_y$, connecting a realising variable $x : \tau^+(A)$ and a challenging variable $y : \tau^-(A)$. We refer to the types $\tau^+(A)$ and $\tau^-(A)$ as positive and negative computational types of A. The Dialectica interpretation starts from a proof M and produces a witnessing term t, with y not among its free variables, together with a verifying proof of $\forall_y |A|^t_y$. Table 1 summarises the formal definition of the interpretation. If $\tau^+(A)$ or $\tau^-(A)$ are ε, then we omit them from the translation, i.e., we write $|A|_y$ or $|A|^x$.

Table 1. Dialectica interpretation

A	$\tau^+(A)$	$\tau^-(A)$	$|A|^r_s$
$at(b)$	ε	ε	$at(b)$
$B \to C$	$(\tau^+(B) \Rightarrow \tau^+(C)) \times$ $(\tau^+(B) \Rightarrow \tau^-(C) \Rightarrow \tau^-(B))$	$\tau^+(B) \times \tau^-(C)$	$\|B\|^{s_{\llcorner}}_{r_{\lrcorner}(s_{\llcorner})(s_{\lrcorner})} \to \|C\|^{r_{\llcorner}(s_{\llcorner})}_{s_{\lrcorner}}$
$\forall_{x^\rho} B(x)$	$\rho \Rightarrow \tau^+(B)$	$\rho \times \tau^-(B)$	$\|B(s_{\llcorner})\|^{r(s_{\llcorner})}_{s_{\lrcorner}}$

Theorem 1 (Soundness). *Let M be a proof of A from assumptions $u^{C_i}_i$. Let $x^{\tau^+(C_i)}_i$ and $y^{\tau^-(A)}_A$ be fresh variables, associated uniquely with the corresponding assumption variables u_i and the conclusion formula A. Then there are terms $[\![M]\!]^+ : \tau^+(A)$ and $[\![M]\!]^-_i : \tau^-(C_i)$ with $y_A \notin \mathsf{FV}([\![M]\!]^+)$ and a proof M^* of $|A|^{[\![M]\!]^+}_{y_A}$ from assumptions $u^*_i : |C_i|^{x_i}_{[\![M]\!]^-_i}$. Moreover, $\mathsf{FV}(M^*) \subseteq \mathsf{FV}(M) \cup \{x_i, y_A\}$.*

Proof. The proof goes by induction on M. We will present only some of the cases, for complete treatment the reader is referred to [7]. For technical convenience we assume that the axioms Cases and Ind are always applied to a sufficient number of arguments.

Case u^A. Set $[\![M]\!]^- := y$ and $[\![M]\!]^+ := x$.

Case $N^{A \to B}_1 N^A_2$. By IH we have proofs of

$$|A \to B|^{[\![N_1]\!]^+}_z \text{ from assumptions } |C_i|^{x_i}_{[\![N_1]\!]^-_i},$$

$$|A|^{[\![N_2]\!]^+}_w \text{ from assumptions } |C_i|^{x_i}_{[\![N_2]\!]^-_i}.$$

---

[1] Paulo Oliva introduced $|A|^x_y$ as a common notation for different functional interpretations. In our presentation $|A|^x_y$ denotes only the Dialectica translation.

Here we substitute $\xi_1 = \left[z := \left\langle [\![N_2]\!]^+, y \right\rangle\right]$ and $\xi_2 = \left[w := [\![N_1]\!]^+ \lrcorner [\![N_2]\!]^+ y\right]$. However, if the two subproofs share assumption variables, we need that $|C_i|_{[\![M]\!]_i^-}^{x_i}$ implies $|C_i|_{[\![N_j]\!]_i^- \xi_2}^{x_i}$ for $j = 1, 2$ in order to be able to use the IH. At this point we take advantage of the quantifier-free Dialectica translation and consider the following case distinction:

$$t_1 \overset{C_i}{\bowtie} t_2 := \begin{cases} t_{3-j}, & \text{if } u_i : C_i \notin \mathsf{FA}(N_j), \\ \mathbf{if}\ |C_i|_{t_1}^{x_i}\ \mathbf{then}\ t_2\ \mathbf{else}\ t_1, & \text{otherwise.} \end{cases}$$

We will show that $|C_i|_{t_1 \overset{C_i}{\bowtie} t_2}^{x_i}$ implies $|C_i|_{t_j}^{x}$ for $j = 1, 2$ and $u_i \in \mathsf{FA}(N_1) \cap \mathsf{FA}(N_2)$ by case distinction. If $|C_i|_{t_1}^{x}$ then $t_1 \overset{C_i}{\bowtie} t_2 = t_2$ and we are done. If $|C_i|_{t_1}^{x} \to \mathsf{F}$, then $t_1 \overset{C_i}{\bowtie} t_2 = t_1$. Hence, we can derive F and finish by using efq. Finally, we set $[\![M]\!]^+ := [\![N_1]\!]^+ \lrcorner [\![N_2]\!]^+$, $[\![M]\!]_i^- := \left([\![N_1]\!]_i^- \xi_1\right) \overset{C_i}{\bowtie} \left([\![N_2]\!]_i^- \xi_2\right)$.

Case Cases $b\ N_1\ N_2$. It is easy to check that we can set $[\![M]\!]_i^- := [\![N_1]\!]_i^- \overset{C_i}{\bowtie} [\![N_2]\!]_i^-$ and $[\![M]\!]^+ := \mathbf{if}\ b\ \mathbf{then}\ [\![N_1]\!]^+\ \mathbf{else}\ [\![N_2]\!]^+$.

Case $\mathsf{Ind}^{A(n)}\ n\ N_1\ N_2$. With $\xi = \left[z := \left\langle n, [\![M]\!]^+, y \right\rangle\right]$, we set:

$$[\![M]\!]^+ := \mathcal{R}\, n\, [\![N_1]\!]^+\, (\lambda_n [\![N_2]\!]^+ n_\llcorner),$$

$$[\![M]\!]_i^- := \mathcal{R}\, n\, (\lambda_y [\![N_1]\!]_i^-)\, \left(\lambda_{n,p,y}([\![N_2]\!]_i^- \xi) \overset{i}{\bowtie} \left(p([\![N_2]\!]^+ n \lrcorner [\![M]\!]^+ y)\right)\right)\, y.$$

Uniform Annotations

Since we would like to be able to switch on and off the computational contribution of different components of the formulas, let us examine positive and negative types in more detail. For each type component we define a *uniformity flag* of the connective as shown on Figure 1. The flag specifies that the corresponding part of the computational type is removed; multiple flags can appear on the same connective. We refer to the system with such enriched logical language as NA^U. We need to show how to extend the Dialectica translation $|A|_y^x$ to all flag combinations. For every type component there is a corresponding subterm, which needs to be omitted if the respective uniformity flag is present. Thus we will

$$\tau^+(\forall_x A) = \underbrace{\rho}_{\overset{+}{\forall}} \Rightarrow \tau^+(A) \qquad \tau^-(\forall_x A) = \underbrace{\rho}_{\overset{-}{\forall}} \times \tau^-(A)$$

$$\tau^+(A \to B) = \underbrace{(\tau^+(A) \Rightarrow \tau^+(B))}_{\overset{\#}{\to}} \times \underbrace{(\tau^+(A) \Rightarrow \tau^-(B) \Rightarrow \tau^-(A))}_{\overset{\pm}{\to} \quad \overset{=}{\to}}$$

$$\tau^-(A \to B) = \underbrace{\tau^+(A)}_{\overset{\to}{+}} \times \underbrace{\tau^-(B)}_{\overset{\to}{-}}$$

Fig. 1. Uniformity flags

Table 2. Interpretation of uniform annotations

Translation	Restriction	Redundant if
$\left\|\overset{+}{\forall}_x A\right\|^r_{x,u} := \|A\|^r_u$	$x \notin \mathsf{FV}(\llbracket M \rrbracket^+)$	$\tau^+(A) = \varepsilon$
$\left\|\bar{\forall}_x A\right\|^f_u := \forall_x \|A\|^{fx}_u$	$x \notin \mathsf{FV}(\llbracket M \rrbracket^-_i)$	$x \notin \mathsf{FV}(\llbracket M \rrbracket^-_i)$
$\left\|A \overset{\#}{\to} B\right\|^{r,g}_{x,u} := \|A\|^x_{gxu} \to \|B\|^r_u$	$x_u \notin \mathsf{FV}(\llbracket M \rrbracket^+)$	$\tau^+(A) = \varepsilon$ $\tau^+(B) = \varepsilon$
$\left\|A \overset{\pm}{\to} B\right\|^{f,g}_{x,u} := \|A\|^x_{gu} \to \|B\|^{fx}_u$	$x_u \notin \mathsf{FV}(\llbracket M \rrbracket^-_A)$	$\tau^+(A) = \varepsilon$ $\tau^-(A) = \varepsilon$
$\left\|A \overset{=}{\to} B\right\|^{f,g}_{x,u} := \|A\|^x_{gx} \to \|B\|^{fx}_u$	$y_A \notin \mathsf{FV}(\llbracket M \rrbracket^-_A)$	$\tau^-(B) = \varepsilon$ $\tau^-(A) = \varepsilon$
$\left\|A \underset{+}{\to} B\right\|^{f,g}_u := \forall_x \big(\|A\|^x_{gxu} \to \|B\|^{fx}_u\big)$	$x_u \notin \mathsf{FV}(\llbracket M \rrbracket^-_i)$	$\tau^+(A) = \varepsilon$
$\left\|A \underset{-}{\to} B\right\|^{f,g}_x := \forall_u \big(\|A\|^x_{gxu} \to \|B\|^{fx}_u\big)$	$y_A \notin \mathsf{FV}(\llbracket M \rrbracket^-_i)$	$\tau^-(B) = \varepsilon$

define the translation for every separate flag in a modular manner, since it can be extended for any combination of flags in a straightforward way.[2]

Naturally, in order to preserve interpretation soundness, uniform annotations come with restrictions on the corresponding introduction rules (cf. [1,2,5]). Here a condition is associated with every uniform flag and in order for a combination of flags to be applicable, the conjunction of all associated restrictions must hold. The interpretations of the connective flags together with the corresponding restrictions are presented in Table 2, where for clarity pairs of variables are used instead of projections of a variable of pair type. Table 3 denotes which parts of the extracted Dialectica terms are removed when a uniformity flag is used on the connective introduced or eliminated by the corresponding proof rule.

We will be interested in those proofs in $\mathrm{NA}^{\mathcal{U}}$ whose logical symbols are annotated in a way to preserve soundness.

Definition 3. *We call a proof* computationally correct *if:*

1. *every instance of a rule preserves annotations on non-principal connectives*
2. *every introduction conforms to the uniformity restrictions in Table 2*
3. *every assumption u^A with $\tau^-(A) \neq \varepsilon$ that is used more than once does not contain any of the flags $\bar{\forall}$, $\underset{+}{\to}$ and $\underset{-}{\to}$.*

---

[2] The four possible combinations of the universal quantifier flags $\forall, \overset{+}{\forall}, \bar{\forall}$ and $\overset{\pm}{\forall}$, were already presented in [5], where they were denoted $\forall_\pm, \forall_-, \forall_+$ and $\forall_\emptyset$ respectively.

Table 3. Extracted terms in NA^U

$\mathcal{P}$	$[\![\mathcal{P}]\!]^+$	$[\![\mathcal{P}]\!]_i^-$
$\lambda_u M$	$\langle \underbrace{\lambda_{x_u} [\![M]\!]^+}_{\overset{\#}{\rightarrow}}, \underbrace{\lambda_{x_u} \lambda_{y_B} [\![M]\!]_u^-}_{\overset{\pm}{\rightarrow}\ \overset{=}{\rightarrow}} \rangle$	$[\![M]\!]_i^- \underbrace{[x_u := y_{A \to B} {\llcorner}]}_{\overset{\rightarrow}{+}} \underbrace{[y_B := y_{A \to B} {\lrcorner}]}_{\overset{\rightarrow}{-}}$
MN	$\underbrace{([\![M]\!]^+ {\llcorner}) [\![N]\!]^+}_{\overset{\#}{\rightarrow}}$	$[\![M]\!]_i^- \left[y_{A \to B} := \langle \underbrace{[\![N]\!]^+}_{\overset{\rightarrow}{+}}, \underbrace{y_B}_{\overset{\rightarrow}{-}} \rangle \right]$ $\overset{i}{\bowtie}$ $[\![N]\!]_i^- \left[y_A := \underbrace{([\![M]\!]^+ {\lrcorner}) [\![N]\!]^+}_{\overset{\pm}{\rightarrow}\ \overset{=}{\rightarrow}} y_B \right]$
$\lambda_x M$	$\underbrace{\lambda_x [\![M]\!]^+}_{\overset{\forall}{+}}$	$[\![M]\!]_i^- \underbrace{[x := y_{\forall_x A} {\llcorner}]}_{\overset{\forall}{-}} [y_A := y_{\forall_x A} {\lrcorner}]$
Mt	$[\![M]\!]^+ \underbrace{t}_{\overset{\forall}{+}}$	$[\![M]\!]_i^- \left[y_{\forall_x A} := \langle \underbrace{t}_{\overset{\forall}{-}}, y_A \rangle \right]$

Condition *1* specifies that uniform annotations are treated as a part of the formula: once introduced, a connective keeps its flags until eliminated. Condition ? ensures that we do not discard computationally relevant components, while condition *3* ensures that whenever we need to make a case distinction on a translation $|A|_y^x$, it is quantifier-free and hence decidable.

Theorem 2 (Soundness for NA^U). *The Dialectica interpretation is sound for every computationally correct proof in NA^U.*

Proof. The proof is a modification of the original soundness argument (Theorem ?), where the non-trivial changes come in the treatment of introduction rules. We will present only the case of implication introduction; for the universal quantifier the reader is referred to [5].

Case $\lambda_{u^A} N^B$. By IH we have a proof of $|B|_z^{[\![N]\!]^+}$ from assumptions $|C_i|_{[\![N]\!]_i^-}^{x_i}$ and $|A|_{[\![N]\!]_0^-}^{x_0}$. The flags $\overset{\#,\pm,=}{\longrightarrow}$ correspond to removing the respective three abstracted variables from the usual extracted term $[\![M]\!]^+ := \langle \lambda_{x_0} [\![N]\!]^+, \lambda_{x_0,z} [\![N]\!]_0^- \rangle$. Depending on which of the flags $\overset{}{\underset{+,-}{\longrightarrow}}$ are used, the variables x_0, z will not appear in any of the terms $[\![N]\!]_i^-$ for $i \geq 1$. Then we would have

$$|A|_{[\![N]\!]_0^-}^{x_0} \to |B|_z^{[\![N]\!]^+} \quad \text{from} \quad |C_i|_{[\![N]\!]_i^-}^{x_i}$$

and we can introduce a universal quantifier on x_0 or z (or both), because the quantified variable will not occur in the open assumptions.   □

5   Properties of Uniform Annotations

The uniform flags turn out to be sufficient to discard completely only the positive or only the negative content of a formula.

Theorem 3. *For every formula A in* NA *there exist decorated variants $A^\oplus$ and $A^\ominus$ in* NA^{U} *which remove positive and negative content of A respectively, while fully preserving the opposite content. Formally, we have:*

1. $\tau^+(A^\oplus) = \tau^-(A^\ominus) = \varepsilon,$ $\tau^-(A^\oplus) = \tau^-(A),$ $\tau^+(A^\ominus) = \tau^+(A)$
2. $\tau^-(|A^\oplus|_y) = \tau^+(|A^\ominus|^x) = \varepsilon,$ $\tau^-(|A^\ominus|^x) = \tau^-(A)$
3. $|A^\ominus|^x \leftrightarrow \forall_y |A|_y^x$ *and* $||A^\ominus|^x|_y = |A|_y^x$
4. *If* $|A^\oplus|_y$ *is provable then* $|A|_y^t$ *is provable for some term t with $y \notin \mathsf{FV}(t)$*

Proof. Intuitively, in $A^\ominus$ we use uniform flags to inductively discard *only* negative content and then apply this annotation for implication premises in $A^\oplus$:

$$(\mathrm{at}(t))^\oplus := \mathrm{at}(t) \qquad (\forall_z B)^\oplus := \forall_z B^\oplus \qquad (B \to C)^\oplus := B^\ominus \to C^\oplus$$

$$(\mathrm{at}(t))^\ominus := \mathrm{at}(t) \qquad (\forall_z B)^\ominus := \bar{\forall}_z B^\ominus \qquad (B \to C)^\ominus := B \xrightarrow[+,-]{} C$$

All claims are proved by induction on the definition above. The proofs of *1-3* are straightforward and *4* follows from the stronger claim:

Lemma 1. *If M is a computationally correct proof of $|A^\oplus|_y$ from assumptions $u_i : |C_i^\ominus|^{x_i}$, then there are terms r_i and t with all their variables in $\mathsf{FV}(C_i) \cup \mathsf{FV}(A) \cup \{x_i, y\}$, such that $y \notin \mathsf{FV}(t)$, for which there is a proof M^* of $|A|_y^t$ from assumptions $u_i^* : |C_i|_{r_i}^{x_i}$.*

Proof. Induction on the definition of $A^\oplus$.

Case $\mathrm{at}(t)$. We apply the Dialectica interpretation to the proof M and by the Soundness theorem obtain the terms r_i and a proof M^* of $\mathrm{at}(t)$ from assumptions $u_i^* : ||C_i^\ominus|^{x_i}|_{r_i}$, which by *3* are in fact $u_i^* : |C_i|_{r_i}^{x_i}$.

Case $\forall_z B^\oplus$. We apply the IH to the proof $M [y := \langle z, v \rangle]$ of $|\forall_z B^\oplus|_{z,v} = |B^\oplus|_v$ to obtain terms r_i', t' and a proof M' of $|B|_v^{t'}$. We set $t := \lambda_z t', r_i := r_i' \xi$ with $\xi = [z, v := y_{\llcorner}, y_{\lrcorner}]$ and obtain the proof $M^* := M' \xi$ of $|\forall_z B|_y^t$.

Case $B^\ominus \to C^\oplus$. We note that $|B^\ominus \to C^\oplus|_y = |B^\ominus|^{y_{\llcorner}} \to |C^\oplus|_{y_{\lrcorner}}$ and apply the IH to the proof $(M [y := \langle x_0, v \rangle] u_0)$ with $u_0 : |B^\ominus|^{x_0}$ a fresh assumption variable. We obtain terms r_i', t' and a proof M' of $|C|_v^{t'}$. We set $t := \langle \lambda_{x_0} t', \lambda_{x_0, v} r_0' \rangle$ and

$_i := r'_i \xi$ for $i > 0$ with $\xi = [x_0, v := y_{\llcorner}, y_{\lrcorner}]$ and obtain the proof $M^* := u_0^*(M'\xi)$ of $|B \to C|_y^t$.   □

Points 3 and 4 show that the effect of removing positive or negative content is achieved by "pushing" the content from the formula to its translation. The Dialectica translation of a formula with no uniform annotations is always quantifier-free and hence void of any computational meaning. However, the translation of a NA^U formula might itself have nonempty positive or negative computational type. Thus by applying the interpretation a second time we are able to recover the originally discarded content.

The definitions $A^\oplus$ and $A^\ominus$ take advantage only of the flags $\bar{\forall}$, $\underset{+}{\to}$ and $\underset{-}{\to}$. In fact, these are precisely the flags which introduce computational content into the translations and we will refer to them as *strong*. They are somehow similar to Berger's uniform existential quantifier $\{\exists_x\}$, which introduces content in the modified realisability translation. The rest of the flags can be used only to discard redundant parameters of functions. In this sense they are *weak* and are similar to Berger's uniform universal quantifier. Still, there is one very notable difference between the realisability and Dialectica setting, which we will illustrate by an example scenario. Consider the following formula:

$$D := \forall_x \forall_y (P(x, y) \to \exists_z Q(x, y, z)). \tag{1}$$

Here $\exists$ denotes a strong (constructive) existential quantifier, while P and Q denote quantifier-free formulas. The program extracted from a proof L of D would be of the form $f := \lambda_{x,y} r$, where r is a term computing a witness for z. Let us assume that in order to show $Q(x, y, z)$ we used that $P(x, y)$ holds for some y, but its value was never used in the construction for z itself. This means that $y \notin \mathsf{FV}(r)$, so we are allowed to mark the quantifier $\forall_y$ as uniform, changing the extracted program to $f' := \lambda_x r$.

Now assume that L is a part of a larger proof M. Assume also that having x is hard to compute any y for which $P(x, y)$, but is relatively easy to calculate for which $Q(x, y, z)$. If we do not use a uniform quantification over y, the program extracted from M would need to pass a value for y to the function

On the other hand, if we express the fact that y is redundant by marking its quantifier as uniform, then we would use the function f' instead and save expensive computations of values of y.

Let us consider the weak formulation of D and its corresponding classical logic proof L', which is to be interpreted by Dialectica:

$$D' := \forall_x \forall_y (P(x, y) \to \tilde{\exists}_z Q(x, y, z)). \tag{2}$$

By the same reasoning as before we should be able to mark $\overset{+}{\forall}_y$. However, if L' uses open assumptions, for example as a part of a larger classical proof, we might need to make use of the negative content for y, even if we have discarded the positive one. Thus, if we have a term t computing y, we can avoid passing it as parameter to the positive extracted content, but we will still need to insert it in the negative extracted content of any relevant open assumptions (see last row of Table 3). Therefore, in this case we would not save the expensive computation of y by using the flag $\overset{+}{\forall}$; we would be able to do so if we were able to discard *all* computational uses of y by marking $\overset{\pm}{\forall}$.

Similar scenarios can be constructed for all the weak flags. They demonstrate that removing a single use of some computational component might lead to cleaning up the extracted programs, but removing all of its uses is necessary for an efficiency improvement. Rephrased syntactically, the usage of weak flags by themselves, that is, without being able to apply the corresponding strong flags, can have only sanitary applications. Moreover, this situation is specific to the Dialectica interpretation, where we have duality of computational relevance.

6   Conclusion and Future Work

The Dialectica interpretation allows for a much richer set of uniform annotations than realisability. In fact, the presented approach can be extended to Heyting Arithmetic with finite types, which allows us to completely simulate the effect of modified realisability inside a Dialectica context, including Berger's uniform quantifiers. A similar result was already reported by Hernest and Oliva in a linear logic setting [4]: a hint that the newly introduced uniform annotations will probably be able to simulate their hybrid functional interpretation.

We have shown that through the annotations $A^{\oplus}$ and $A^{\ominus}$ we can completely remove positive or negative content of a formula in a given proof, *if the annotated proof is computationally correct*. A subject of future research will be to investigate in which cases is this practically possible; or when it is not, how can additional annotations be *automatically* introduced in an optimal way to repair computational correctness. Such a result would allow us to understand better the relation between removing computational content by a "deep annotation" such as $A^{\oplus} \to B$, versus a "shallow annotation", such as $A \xrightarrow{\#,\pm}{+} B$ (or $?_k A \multimap B$ in the linear logic setting of [4]).

References

1. Berger, U.: Uniform Heyting Arithmetic. Ann. Pure Appl. Logic 133(1-3), 125–148 (2005)
2. Hernest, M.-D.: Optimized programs from (non-constructive) proofs by the light (monotone) Dialectica interpretation. PhD thesis (2007)
3. Gödel, K.: Über eine bisher noch nicht benützte Erweiterung des finiten Standpunktes. Dialectica 12, 280–287 (1958)
4. Hernest, M.-D., Oliva, P.: Hybrid functional interpretations. In: Beckmann, A., Dimitracopoulos, C., Löwe, B. (eds.) CiE 2008. LNCS, vol. 5028, pp. 251–260. Springer, Heidelberg (2008)

Hernest, M.-D., Trifonov, T.: Light dialectica revisited. Submitted to APAL (2008), http://www.math.lmu.de/~trifonov/papers/ldrev.pdf

Kreisel, G.: Interpretation of analysis by means of constructive functionals of finite types. In: Heyting, A. (ed.) Constructivity in Mathematics, pp. 101–128. North-Holland Publishing Company, Amsterdam (1959)

Schwichtenberg, H.: Dialectica interpretation of well-founded induction. Mathematical Logic Quarterly 54(3), 229–239 (2008)

Troelstra, A.S.: Metamathematical Investigation of Intuitionistic Arithmetic and Analysis. Lecture Notes in Mathematics, vol. 344. Springer, Heidelberg (1973)

Algorithmic Minimal Sufficient Statistic Revisited

Nikolay Vereshchagin

Moscow State University, Leninskie gory 1,
Moscow 119991, Russia
ver@mccme.ru
http://lpcs.math.msu.su/~ver

Abstract. We express some criticism about the definition of an algorithmic sufficient statistic and, in particular, of an algorithmic minimal
sufficient statistic. We propose another definition, which might have better properties.

1 Introduction

Let x be a binary string. A finite set A containing x is called an (algorithmic)
sufficient statistic of x if the sum of Kolmogorov complexity of A and the log-
cardinality of A is close to Kolmogorov complexity $C(x)$ of x:

$$C(A) + \log_2 |A| \approx C(x). \tag{1}$$

Let A^* denote a minimal length description of A and i the index of x in the list
of all elements of A arranged lexicographically. The equality (1) means that the
two part description (A^*, i) of x is as concise as the minimal length code of x.

It turns out that A is a sufficient statistic of x iff $C(A|x) \approx 0$ and $C(x|A) \approx
\log |A|$. The former equality means that the information in A^* is a part of in-
formation in x. The latter equality means that x is a typical member of A: x
has no regularities that allow to describe x given A in a shorter way than just
by specifying its $\log |A|$-bit index in A. Thus A^* contains all useful information
present in x and i contains only accidental information (noise).

Sufficient statistics may also contain noise. For example, this happens if x
is a random string and $A = \{x\}$. Is it true that for all x there is a sufficient
statistic that contains no noise? To answer this question we can try to use the
notion of a minimal sufficient statistics defined in [3]. In this paper we argue that
(1) this notion is not well-defined for some x (although for some x the notion
is well-defined) and (2) even for those x for which the notion of a minimal
sufficient statistic is well-defined not every minimal sufficient statistic qualifies
for a "denoised version of x". We propose another definition of a (minimal)
sufficient statistic that might have better properties.

K. Ambos-Spies, B. Löwe, and W. Merkle (Eds.): CiE 2009, LNCS 5635, pp. 478–487, 2009.

Sufficient Statistics

et x be a given string of length n. The goal of algorithmic statistics is to
explain" x. As possible explanations we consider finite sets containing x. We
all any finite $A \ni x$ a *model for* x. Every model A corresponds to the sta-
stical hypothesis "x was obtained by selecting a random element of A". In
hich case is such a hypothesis plausible? As argued in [3,4,5], it is plausible
$C(x|A) \approx \log |A|$ and $C(A|x) \approx 0$ (we prefer to avoid rigorous definitions
p to a certain point; approximate equalities should be thought as equalities
p to an additive $O(\log n)$ term). In the expressions $C(x|A), C(A|x)$ the set A
understood as a finite object. More precisely, we fix any computable bijec-
on $A \mapsto [A]$ between finite sets of binary strings and binary strings and let
$(x|A) = C(x|[A]), C(A|x) = C([A]|x), C(A) = C([A])$.
As shown in [3,5] this is equivalent to saying that $C(A) + \log |A| \approx C(x)$.
deed, assume that A contains x and $C(A) \le n$. Then, given A, the string x can
e specified by its $\log |A|$-bit index in A. Recalling the symmetry of information
d omitting additive terms of order $O(\log n)$, we obtain

$$C(x) \le C(x) + C(A|x) = C(A) + C(x|A) \le C(A) + \log |A|.$$

ssume now that $C(x|A) \approx \log |A|$ and $C(A|x) \approx 0$. Then all inequalities here
ecome equalities and hence A is a sufficient statistic. Conversely, if $C(x) \approx$
$(A) + \log |A|$ then the left hand side and the right hand side in these inequalities
incide. Thus $C(x|A) \approx \log |A|$ and $C(A|x) \approx 0$.
The inequality

$$C(x) \le C(A) + \log |A| \tag{2}$$

hich is true up to an additive $O(\log n)$ term) has the following meaning. Con-
der the two part code (A^*, i) of x, consisting of the minimal program A^* for x
d the $\log |A|$-bit index of x in the list of all elements of A arranged lexicographi-
lly. The equality means that its total length $C(A) + \log |A|$ cannot exceed $C(x)$.
$C(A) + \log |A|$ is close to $C(x)$, then we call A a *sufficient statistic* of x. To make
is notion rigorous we have to specify what we mean by "closeness". In [3] this is
ecified as follows: fix a constant c and call A a sufficient statistic if

$$|(C(A) + \log |A|) - C(x)| \le c. \tag{3}$$

ore precisely, [3] uses prefix complexity K in place of plain complexity C. For
efix complexity the inequality (2) holds up to a constant error term. If we
oose c large enough then sufficient statistics exists, witnessed by $A = \{x\}$.
he paper [1] suggests to set $c = 0$ and to use $C(x|n)$ and $C(A|n)$ in place
$C(x)$ and $C(A)$ in the definition of a sufficient statistic. For such definition
fficient statistics might not exist.)
To avoid the discussion on how small c should be let us call $A \ni x$ a *c-sufficient*
atistic if (3) holds. The smaller c is the more sufficient A is. This notion is non-
cuous only for $c = O(\log n)$ as the inequality (2) holds only with logarithmic
ecision.

3   Minimal Sufficient Statistics

Naturally, we are interested in squeezing as much noise from the given string x as possible. What does it mean? Every sufficient statistic A identifies $\log |A|$ bits of noise in x. Thus a sufficient statistic with maximal $\log |A|$ (and hence minimal $C(A)$) identifies the maximal possible amount of noise in x. So we arrive at the notion of a minimal sufficient statistic: a sufficient statistic with minimal $C(A)$ is called a minimal sufficient statistic (MSS).

Is this notion well-defined? Recall that actually we only have the notion of a c-sufficient statistic (where c is either a parameter or a constant). That is, we have actually defined the notion of a minimal c-sufficient statistic. Is this a good notion? We argue that for some strings x it is not whatever the value of c is. There are strings x for which it is impossible to identify MSS in an intuitively appealing way. For those x the complexity of the minimal c-sufficient statistic decreases substantially, as c increases a little.

To present such strings we need to recall a theorem from [7]. Let S_x stand for the *structure set* of x:

$$S_x = \{(i,j) \mid \exists A \ni x, \ C(A) \leq i, \ \log |A| \leq j\}.$$

This set can be identified by either of its two "border line" functions:

$$h_x(i) = \min\{\log |A| \mid A \ni x, \ C(A) \leq i\}, \quad g_x(j) = \min\{C(A) \mid A \ni x, \ \log |A| \leq j\}$$

The function h_x is called the *Kolmogorov structure function* of x; for small i it might take infinite values due to lack of models of small complexity. In contrast the function g_x is total for all x.

As pointed out by Kolmogorov [4], the structure set S_x of every string x of length n and Kolmogorov complexity k has the following three properties (we state the properties in terms of the function g_x): (1) $g_x(0) = k + O(1)$ (witnessed by $A = \{x\}$). (2) $g_x(n) = O(\log n)$ (witnessed by $A = \{0,1\}^n$). (3) g_x in non increasing and $g_x(j+l) \geq g_x(j) - l - O(\log l)$ for every $j, l \in \mathbb{N}$.

For the proof of the last property see [5,7]. Properties (1) and (3) imply that $i + j \geq k - O(\log n)$ for every $(i,j) \in S_x$. Sufficient statistics correspond to those $(i,j) \in S_x$ with $i + j \approx k$. The line $i + j = k$ is therefore called *the sufficiency line*.

A result of [7, Remark IV.4] states that for every g that satisfies (1)–(3) there is x of length n and complexity close to k such that g_x is close to g.[1] More specifically, the following holds:

Theorem 1 ([7]). *Let g be any non-increasing function $g : \{0,\ldots,n\} \to \mathbb{N}$ such that $g(0) = k$, $g(n) = 0$ and such that $g(j+l) \geq g_x(j) - l$ for every $j, l \in \mathbb{N}$ with $j + l \leq n$. Then there is a string x of length n and complexity $k \pm \varepsilon$ such that $|g_x(j) - g(j)| \leq \varepsilon$ for all $j \leq n$. Here $\varepsilon = O(\log n + C(g))$ and $C(g)$ stands for the Kolmogorov complexity of the graph of g: $C(g) = C(\{\langle j, g(j)\rangle \mid 0 \leq j \leq n\})$.*

---

[1] Actually, [7] provides the description of possible shapes of S_x in terms of the Kolmogorov structure function h_x. We use here g_x instead of h_x, as in terms of g_x the description is easier to understand.

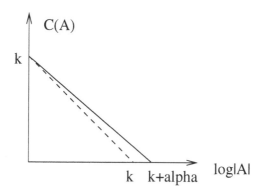

Fig. 1. The structure function of a string for which MSS is not well-defined

We are ready to present strings for which the notion of a MSS is not well-defined. Fix a large n and let $k = n/2$ and $g(j) = \max\{k - jk/(k+\alpha), 0\}$, where $= \alpha(k) \le k$ is a computable function of k with natural values. Then n, k, g satisfy all conditions of Theorem 1. Hence there is a string x of length n and complexity $k+O(\log n)$ with $g_x(j) = g(j)+O(\log n)$ (note that $C(g) = O(\log n)$). Its structure function is shown on Fig. 1. Choose α so that α/k is negligible compared to k) but α is not.

For very small j the graph of g_x is close to the sufficiency line and for $j = k+\alpha$ it is already at a large distance α from it. As j increases by one, the value $g(j) + j - C(x)$ increases by at most $\alpha/(k + \alpha) + O(\log n)$, which is negligible. Therefore, it is not clear where the graph of g_x leaves the sufficiency line. The complexity of the minimal c-sufficient statistic is $k - (c + O(\log n)) \cdot k/\alpha$ and decreases fast as a function of c.

Thus there are strings for which it is hard to identify the complexity of MSS. There is also another minor point regarding minimal sufficient statistics. Namely, there is a string x for which the complexity of minimal sufficient statistic is well-defined but not all MSS qualify as denoised versions of x. Namely, some of them have a weird structure function. What kind of structure set we expect of a denoised string? To answer this question consider the following example. Let y be a string, m a natural number and z a string of length $l(z) = m$ that is random relative to y. The latter means that $C(z|y) \ge m - \beta$ for a small β. Consider the string $x = \langle y, z \rangle$. Intuitively, z is a noise in x. In other words, we can say that x is obtained from x by removing m bits of noise. What is the relation between the structure set of x and that of y?

Theorem 2. *Assume that z is a string of length m with $C(z|y) \ge m-\beta$. Then for all $\ge m$ we have $g_x(j) = g_y(j-m)$ and for all $j \le m$ we have $g_x(j) = C(y)+m-j = (0) + m - j$. The equalities here hold up to $O(\log m + \log C(y) + \beta)$ term.*

Proof. In the proof we will ignore terms of order $O(\log m + \log C(y) + \beta)$.
The easy part is the equality $g_x(j) = C(y) + m - j$ for $j \le m$. Indeed, we have $g_x(m) \le C(y)$ witnessed by $A = \{\langle y, z' \rangle \mid l(z') = m\}$. On the other hand,

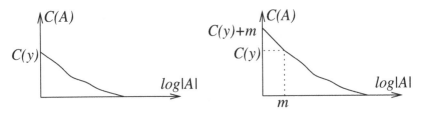

Fig. 2. Structure functions of y and x

$g_x(0) = C(x) = C(y) + C(z|y) = C(y) + m$. Thus $g_x(j)$ should have maximal possible rate of decrease on the segment $[0, m]$ to drop from $C(y) + m$ to $C(y)$.

Another easy part is the inequality $g_x(j) \leq g_y(j-m)$. Indeed, for every model A of y with $|A| \leq 2^{j-m}$ consider the model

$$A' = A \times \{0,1\}^m = \{\langle y', z' \rangle \mid y' \in A,\ l(z') = m\}$$

of cardinality at most 2^j. Its complexity is at most that of $|A|$, which proves $g_x(j) \leq g_y(j - m)$.

The tricky part is the inverse inequality $g_x(j) \geq g_y(j - m)$. Let A be a model for x with $|A| \leq 2^j$ and $C(A) = g_y(j)$. We need to show that there is a model of y of cardinality at most 2^{j-m} and of the same (or lower) complexity. We will prove it in a non-constructive way using a result from [7].

The first idea is to consider the projection of A: $\{y' \mid \langle y', z' \rangle \in A\}$. However this set may be as large as A itself. Reduce it as follows. Consider the yth section of A: $A_y = \{z' \mid \langle y, z' \rangle \in A\}$. Define i as the natural number such that $2^i \leq |A_y| < 2^{i+1}$. Let A' be the set of those y' whose y'th section has at least 2^i elements. Then by counting arguments we have $|A'| \leq 2^{j-i}$. If $i \geq m$, we are done. However, it might be not the case. To lower bound i, we will relate it to the conditional complexity of z given y and A. Indeed, we have $C(z|A, y) \leq i$ as z can be identified by its ordinal number in yth section of A. Hence we know that $\log |A'| \leq j - C(z|A, y)$. Now we will improve A' using a result of [7]:

Lemma 1 (Lemma A.4 in [7]). *For every $A' \ni y$ there is $A'' \ni y$ with* $C(A'') \leq C(A') - C(A'|y)$ *and* $\lfloor \log |A''| \rfloor = \lfloor \log |A'| \rfloor$.

By this lemma we get the inequality

$$g_y(j - C(z|A, y)) \leq C(A') - C(A'|y).$$

Note that

$$C(A') - C(A'|y) = I(y : A') \leq I(y : A) = C(A) - C(A|y),$$

as $C(A'|A)$ is negligible. Thus we have

$$g_y(j - C(z|A, y)) \leq C(A) - C(A|y).$$

We claim that by the property (3) of the structure set this inequality implies that $g_y(j - m) \leq C(A)$. Indeed, as $C(z|A, y) \leq m$ we have by property (3):

$$g_y(j - m) \leq m - C(z|A, y) + C(A) - C(A|y) \leq m + C(A) - C(z|y) = C(A).$$

In all the above inequalities, we need to be careful about the error term, as they
include sets, denoted by A or A', and thus the error term includes $O(\log C(A))$ or
$(\log C(A'))$. All the sets involved are either models of y or of x. W.l.o.g. we may
assume that their complexity is at most $C(x)+O(1)$. Indeed, there is no need to
consider models of y or x of larger complexity, as the models $\{y\}$ and $\{x\}$ have
the least possible cardinality and their complexity is at most $C(x)+O(1)$. Since
$(x) \leq C(y)+O(C(z|y)) \leq C(y)+O(m)$, the term $O(\log C(A))$ is absorbed by
the general error term.

This theorem answers our question: if y is obtained from x by removing m bits
of noise then we expect that g_y satisfy Theorem 2. Now we will show that there
are strings x as in Theorem 2 for which the notion of the MSS is well-defined
but the structure function of some minimal sufficient statistics does not satisfy
Theorem 2. The structure set of a finite set A of strings is defined as that of
[1]. It is not hard to see that if we switch to another computable bijection
$\mapsto [A]$ the value of $g_{[A]}(j)$ changes by an additive constant. Thus S_A and g_A
are well-defined for finite sets A.

Theorem 3. *For every k there is a string y of length $2k$ and Kolmogorov complexity $C(y) = k$ such that*

$$g_y(j) = \begin{cases} k, & \text{if } j \leq k, \\ 2k - j, & \text{if } k \leq j \leq 2k \end{cases}$$

*and hence for any z of length k and conditional complexity $C(z|y) = k$ the
structure function of the sting $x = \langle y, z \rangle$ is the following*

$$g_x(j) = \begin{cases} 2k - j, & \text{if } j \leq k, \\ k, & \text{if } k \leq j \leq 2k, \\ 3k - j, & \text{if } 2k \leq j \leq 3k. \end{cases}$$

*(See Fig. 3.) Moreover, for every such z the string $x = \langle y, z \rangle$ has a model B of
complexity $C(B) = k$ and log-cardinality $\log |B| = k$ such that $g_B(j) = k$ for all
$\leq 2k$. All equalities here hold up to $O(\log k)$ additive error term.*

The structure set of $x = \langle y, z \rangle$ clearly leaves the sufficiency line at the point
$= k$. Thus k is intuitively the complexity of minimal sufficient statistic and
both models $A = y \times \{0, 1\}^k$ and B are minimal sufficient statistics. The model A,
a finite object, is identical to y and hence the structure function of A coincides
with that of y. In contrast, the shape of the structure set of B is intuitively
incompatible with the hypothesis that B, as a finite object, is a denoised x.

Desired Properties of Sufficient Statistics and a New Definition

We have seen that there is a string x that has two very different minimal sufficient
statistics A and B. Recall the probabilistic notion of sufficient statistic [2]. In the

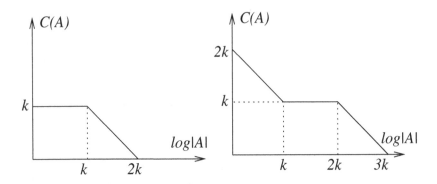

Fig. 3. Structure functions of y and x

probabilistic setting, we are given a parameter set Θ and for each $\theta \in \Theta$ we are given a probability distribution on a set X. For every probability distribution on Θ we thus obtain a probability distribution on $\Theta \times X$. A function $f : X - Y$ (where Y is any set) is called a sufficient statistic, if for every probability distribution on Θ, the random variables x and θ are independent relative to $f(x)$. That is, for all $a \in X$, $c \in \Theta$,

$$\mathrm{Prob}[\theta = c | x = a] = \mathrm{Prob}[\theta = c | f(x) = f(a)].$$

In other words, $x \to f(x) \to \theta$ is a Markov chain (for every probability distribution on Θ). We say that a sufficient statistic f is *less* than a sufficient statistic g if for some function h with probability 1 it holds $f(x) \equiv h(g(x))$. An easy observation is that there is always a sufficient statistic f that is less than any other sufficient statistic: $f(a)$ is equal to the function $c \mapsto \mathrm{Prob}[\theta = c | x = a]$. Such sufficient statistics are called minimal. Any two minimal sufficient statistic have the same distribution and by definition every minimal sufficient statistic is a function of every sufficient statistic. Is it possible to define a notion of an algorithmic sufficient statistic that has similar properties? More specifically, we wish it to have the following properties.

(1) If A is an (algorithmic) sufficient statistic of x and $\log|A| = m$ then the structure function of $y = A$ satisfies the equality of Theorem 2. In particular structure functions of every MSS A, B of x coincide.

(2) Assume that A is a MSS and B is a sufficient statistic of x. Then $C(A|B) \approx 0$

As the example of Theorem 3 demonstrates, the property (1) does not hold for the definitions of Sections 2 and 3, and we do not know whether (2) holds Now we propose an approach towards a definition that (hopefully) satisfies both (1) and (2). The main idea of the definition is as follows. As observed in [6] in order to have the same structure sets, the strings x, y should be equivalent in the following strong sense: there should exist short *total* programs p, q with $D(p, x) = y$ and $D(q, y) = x$ (where D is an optimal description mode in the definition of conditional Kolmogorov complexity). A program p is called *total* if $D(p, z)$ converges for *all* z.

Let $CT_D(x|y)$ stand for the minimal length of p such that p is total and $D(p, y) = x$. For the sequel we need that the conditional description mode D have the following property. For any other description mode D' there is a constant such that $CT_D(x|y) \leq CT_{D'}(x|y) + c$ for all x, y. (The existence of such a D straightforward.) Fixing such D we get the defintion of the *total* Kolmogorov complexity $CT(x|y)$. If both $CT(x|y), CT(y|x)$ are small then we will say that $, y$ are *strongly equivalent*. The following lemma is straightforward.

emma 2. *For all x, y we have $|g_x(j) - g_y(j)| \leq 2\max\{CT(x|y), CT(y|x)\} + O(1)$. (If x, y are strongly equivalent then their structure sets are close.)*

all A a *strongly sufficient statistic* of x if $CT(A|x) \approx 0$ and $C(x|A) \approx \log|A|$. lore specifically, call a model A of x an α, β-*strongly sufficient statistic* of x $CT(A|x) \leq \alpha$ and $C(x|A) \geq \log|A| - \beta$. The following theorem states that rongly sufficient statistics satisfy the property (1). It is a direct corollary of heorem 2 and Lemma 2.

heorem 4. *Assume that y is an α, β-strongly sufficient statistic of x and $g|y| = m$. Then for all $j \geq m$ we have $g_x(j) = g_y(j - m)$ and for all $j \leq m$ e have $g_x(j) = C(y) + m - j$. The equalities here hold up to a $O(\log C(y) + g\, m + \alpha + \beta)$ term.*

et us turn now to the second desired property of algorithmic sufficient statistics. e do not know whether (2) holds in the case when both A, B are strongly ifficient statistics. Actually, for strongly sufficient statistics it is more natural to quire that the property (2) hold in a stronger form: (2') Assume that A is a MSS id both A, B are strongly sufficient statistics of x. Then $CT(A|B) \approx 0$. Or, in an ven stronger form: (2") Assume that A is a minimal strongly sufficient statistic MSSS) of x and B is a strongly sufficient statistic of x. Then $CT(A|B) \approx 0$.

An interesting related question: (3) Is there always a strongly sufficient statis- c that is a MSS?

Of course, we should require that properties (2), (2') and (2") hold only for lose x for which the notion of MSS or MSSS is well-defined. Let us state the roperties in a formal way. To this end we introduce the notation $\Delta_x(A) = T(A|x) + \log|A| - C(x|A)$, which measures "the deficiency of strong sufficiency" a model A of x. In the case $x \notin A$ we let $\Delta_x(A) = \infty$. To avoid cumbersome itations we reduce generality and focus on strings x whose structure set is as Theorem 3. In this case the properties (2') and (3) read as follows: (2') For l models A, B of x,

$$CT(A|B) = O(|C(A) - k| + \Delta T_x(A) + \Delta T_x(B) + \log k).$$

) Is there always a model A of x such that $CT(A|x) = O(\log k)$, $\log|A| = + O(\log k)$ and $C(x|A) = k + O(\log k)$.

It is not clear how to formulate property (2") even in the case of strings x tisfying Theorem 3 (the knowledge of g_x does not help).

We are only able to prove (2') in the case when both A, B are MSS. By a sult of [7], in this case $C(A|B) \approx 0$ (see Theorem 5 below). Thus our result strengthens this result of [7] in the case when both A, B are strongly sufficient statistics (actually we need only that A is strong).

Let us present the mentioned result of [7]. Recalling that the notion of MSS is not well-defined, the reader should not expect a simple formulation. Let $d(u, v)$ stand for $\max\{C(u|v), C(v|u)\}$ (a sort of algorithmic distance between u and v).

Theorem 5 (Theorem V.4(iii) from [7]). *Let N^i stand for the number of strings of complexity at most i.* [2] *For all $A \ni x$ and i, either $d(N^i, A) \leq C(A) - i$ or there is $T \ni x$ such that $\log |T| + C(T) \leq \log |A| + C(A)$ and $C(T) \leq i - d(N^i, A)$, where all inequalities hold up to $O(\log(|A| + C(A)))$ additive term.*

Theorem 6. *There is a function $\gamma = O(\log n)$ of n such that the following holds. Assume that we are given a string x of length n and natural number $i \leq n$ and $\varepsilon < \delta \leq n$ such that the complexity of every $\varepsilon + \gamma$-sufficient statistic of x is greater than $i - \delta$. Then for every ε-sufficient statistics A, B of x of complexity at most $i + \varepsilon$, we have $CT(A|B) \leq 2 \cdot CT(A|x) + \varepsilon + 2\delta + \gamma$.*

Let us see what this statement yields for the string $x = \langle y, z \rangle$ from Theorem 3. Let $i = k$ and $\varepsilon = 100 \log k$, say. Then the assumption of Theorem 6 holds for $\delta = O(\log k)$ and thus $CT(A|B) \leq 2 \cdot CT(A|x) + O(\log k)$ for all $100 \log k$-sufficient B, A of complexity at most $k + 100 \log k$.

Proof. Fix models A, B as in Theorem 6. We claim that if $\gamma = c \log n$ and c is large enough constant, then the assumption of Theorem 6 implies $d(B, A) \leq 2\delta - O(\log n)$. Indeed, we have $K(A) + \log |A| = O(n)$. Therefore all the inequalities of Theorem 5 hold with $O(\log n)$ precision. Thus for some constant c, by Theorem 5 we have $d(N^i, A) \leq \varepsilon + c \log n$ (in the first case) or we have a T with $C(T) - \log |T| \leq i + \varepsilon + c \log n$ and $d(N^i, A) \leq i - C(T) + c \log n$ (in the second case). Let $\gamma = c \log n$. The assumption of Theorem 6 then implies that in the second case $C(T) > i - \delta$ and hence $d(N^i, A) < \delta + c \log n$. Thus anyway we have $d(N^i, A) \leq \delta + c \log n$. The same arguments apply to B and therefore $d(A, B) \leq 2\delta + O(\log n)$.

In the course of the proof, we will neglect terms of order $O(\log n)$. They will be absorbed by γ in the final upper bound of $CT(A|B)$ (we may increase γ).

Let p be a total program witnessing $CT(A|x)$. We will prove that there are many $x' \in B$ with $x' \in p(x') = A$ (otherwise $C(x|B)$ would be smaller than assumed). We will then consider all A' such that there are many $x' \in B$ with $x' \in p(x') = A'$. We will then identify A given B in few bits by its ordinal number among all such A's.

Let $D = \{x' \in B \mid x' \in p(x') = A\}$. Obviously, D is a model of x with

$$C(D|B) \leq C(A|B) + l(p) \leq 2\delta + l(p).$$

Therefore

$$C(x|B) \leq C(D|B) + \log |D| \leq \log |D| + 2\delta + l(p).$$

---

[2] Actually, the authors of [7] use prefix complexity in place of the plain complexity. It is easy to verify that Theorem V.4(iii) holds for plain complexity as well.

On the other hand, $C(x|B) \geq \log|B| - \varepsilon$, hence $\log|D| \geq \log|B| - \varepsilon - 2\delta - l(p)$. Consider now all A' such that

$$\log|\{x' \in B \mid x' \in p(x') = A'\}| \geq \log|B| - \varepsilon - 2\delta - l(p).$$

These A' are pairwise disjoint and each of them has at least $|B|/2^{\varepsilon+2\delta+l(p)}$ elements of B. Thus there are at most $2^{\varepsilon+2\delta+l(p)}$ different such A's. Given B and p, ε, δ we are able to find the list of all A's. The program that maps B to the list of A's is obviously total. Therefore there is a total program of $\varepsilon + 2\delta + 2l(p)$ bits that maps B to A and $CT(A|B) \leq \varepsilon + 2\delta + 2l(p)$.

Another interesting related question is whether the following holds: (4) *Merging strongly sufficient statistics:* If A, B are strongly sufficient statistics for x then x has a strongly sufficient statistic D with $\log|D| \approx \log|A| + \log|B| - \log|A \cap B|$.

It is not hard to see that (4) implies (2"). Indeed, as merging A and B cannot result in a strongly sufficient statistic larger than A we have $\log|B| \approx \log|A \cap B|$. Thus to prove that $CT(A|B)$ is negligible, we can argue as in the last part of the proof of Theorem 6.

References

Antunes, L., Fortnow, L.: Sophistication revisited. In: Baeten, J.C.M., Lenstra, J.K., Parrow, J., Woeginger, G.J. (eds.) ICALP 2003. LNCS, vol. 2719, pp. 267–277. Springer, Heidelberg (2003)

Cover, T.M., Thomas, J.A.: Elements of Information Theory. Wiley, New York (1991)

Gács, P., Tromp, J., Vitányi, P.M.B.: Algorithmic statistics. IEEE Trans. Inform. Th. 47(6), 2443–2463 (2001)

Kolmogorov, A.N.: Talk at the Information Theory Symposium in Tallinn, Estonia (1974)

Shen, A.K.: Discussion on Kolmogorov complexity and statistical analysis. The Computer Journal 42(4), 340–342 (1999)

Shen, A.K.: Personal communication (2002)

Vereshchagin, N.K., Vitányi, P.M.B.: Kolmogorov's structure functions and model selection. IEEE Trans. Information Theory 50(12), 3265–3290 (2004)

A Computation of the Maximal Order Type of the Term Ordering on Finite Multisets

Andreas Weiermann[*]

University of Ghent,
Vakgroep Zuivere Wiskunde en Computeralgebra,
Krijgslaan 281 Gebouw S22,
9000 Ghent, Belgium
Andreas.Weiermann@ugent.be

Abstract. We give a sharpening of a recent result of Aschenbrenner and Pong about the maximal order type of the term ordering on the finite multisets over a wpo. Moreover we discuss an approach to compute maximal order types of well-partial orders which are related to tree embeddings.

Keywords: well-quasi orderings, well-partial orderings, maximal order types, ordinals, term orderings, finite multisets.

1 Introduction

A *well-partial order* (wpo) is a partial order $\langle X, \leq_X \rangle$ such that for all infinite sequences $\{x_i\}_{i=0}^{\infty}$ of elements in X there exist natural numbers i, j such that $i < j$ and $x_i \leq_X x_j$. There are lots of examples for wpo's known, for example well-orders or wpo's resulting from Higman's Lemma and Kruskal's theorem. Well-partial orderings (and similarly well-quasi orderings) play an important role in logic, mathematics and computer science, since they form a convenient tool for proving termination of algorithms. For example the well-foundedness of syntactic termination orderings like the recursive path ordering follows easily by an appeal to Kruskal's theorem [6]. Famous applications of wpo's in logic are provided for example, by Ehrenfeucht's well-foundedness proof for Skolem's ordering on extended polynomials [7] or the termination proof for decision procedures related to relevance logic [18].

In mathematics wpo's are, for example, used to show termination of the algorithm which computes Gröbner bases in polynomial rings [2]. An obvious question related to such termination proofs for algorithms is the question about resulting running times of the algorithms.

From the folklore of subrecursive hierarchy theory it is known that as a rule of thumb such running times are bounded by α-descent recursive functions where

---

[*] The author acknowledges gratefully funding by the John Templeton Foundation, the Fonds voor Wetenschappelijk Onderzoek Vlaanderen (FWO) and (via a joint project with Michel Rathjen) funding by the Royal Society (UK).

K. Ambos-Spies, B. Löwe, and W. Merkle (Eds.): CiE 2009, LNCS 5635, pp. 488–498, 2009.

is the maximal order type of the underlying wpo [3]. This is particularly interesting if the maximal order type in question is small.

Therefore let us explain maximal order types and parts of their history in more detail. An early but very fundamental result concerning maximal order types is the following result of de Jongh and Parikh [4]:

For every wpo $\langle X, \leq_X \rangle$ there exists a linear extension $\leq^+$ of $\leq_X$ on the same set X such that the order type is maximal possible under all such well-ordered extensions.

In this situation we put $o(X, \leq_X) := otype(X, \leq^+)$ and the ordinal $o(X, \leq_X)$ is then (for obvious reasons) called the *maximal order type* of $\langle X, \leq \rangle$. (In the sequel we write $\leq$ for $\leq_X$ when there is no danger of confusion.) The result by de Jongh and Parikh can be used to actually compute maximal order types using the following formula:

$$o(X, \leq) = \sup\{o(L_X(x), \leq) + 1 : x \in X\}$$

here $L_X(x) := \{y \in X : \neg x \leq y\}$.

In logic, and proof theory in particular, maximal order types are typically the invariants which determine the proof theoretic strength of an assertion that a given poset is a wpo. As a rule of thumb (see, for example, [13] for a paradigm) one (quite often) observes the following principle: If it is possible to calculate the maximal order type of a wpo X in terms of an ordinal notation system then the assertion that the poset X is a wpo is over ACA_0 equivalent to the assertion that the linearly ordered ordinal terms below the term representing $o(X)$ form well-order.

There are several results known about maximal order types, and important sources for maximal order types are still [4] and the (unpublished) Habilitationsschrift of Diana Schmidt [15].

Some rudiments of wpo-theory are as follows. Suppose we have given two posets X_0 and X_1. Then we can define induced partial orders on the disjoint union $X_0 \oplus X_1$ and the cartesian product $X_0 \otimes X_1$ in the natural way. Moreover the set X^* of finite sequences of elements over X can be partially ordered using the natural pointwise ordering induced on subsequences (Higman ordering). With $\oplus$ and $\otimes$ we denote the (commutative) natural sum and the (commutative) natural product of ordinals.

Theorem 1 (De Jongh and Parikh [4], Schmidt [15])

. If X_0 and X_1 are wpo's then $X_0 \oplus X_1$ and $X_0 \otimes X_1$ are wpo's, $o(X_0 \oplus X_1) = o(X_0) \oplus o(X_1)$ and $o(X_0 \otimes X_1) = o(X_0) \otimes o(X_1)$.
. If X is a wpo then X^* is a wpo and the following cases occur:

$$o(X^*) = \begin{cases} \omega^{\omega^{o(X)-1}} & \text{if } X \text{ is finite,} \\ \omega^{\omega^{o(X)+1}} & \text{if } o(X) = \epsilon + n \text{ where } \epsilon \text{ is an epsilon number} \\ & \text{and } n \text{ is finite,} \\ \omega^{\omega^{o(X)}} & \text{otherwise.} \end{cases}$$

Quite recently wpo's related to monomial ideals have been investigated in detail by M. Aschenbrenner and W. Y. Pong [1]. Among other things they provided bounds on maximal order type of the term ordering on finite multisets. The aim of this paper is to provide a precise formula for such a maximal order type.

We intend to use this specific result in future investigations on maximal order types emerging from well-partial orderings related to tree-embeddability relations which are based on a Friedman style gap condition [17].

To this end we recently developed (in a joint research project with Michael Rathjen) a very satisfying and general formula which predicts in all natural cases (at which we had a look at) good upper bounds for the maximal order type of a tree-based wpo under consideration [21].

To explain this formula informally let us recall briefly the definition of the collapsing function needed for a proof-theoretic analysis of ID_1 (see, for example [14] for an exposition). Let Ω denote the first uncountable ordinal and $\varepsilon_{\Omega+1}$ the first epsilon number above Ω. Then recall that any ordinal $\alpha < \varepsilon_{\Omega+1}$ can be described uniquely in terms of its Cantor normal form

$$\alpha = \Omega^{\alpha_1}\beta + \cdots + \Omega^{\alpha_n} \cdot \beta_n$$

where $\alpha_1 > \ldots > \alpha_n$ and $0 < \beta_1, \ldots, \beta_n < \Omega$. In this situation we define the countable subterms $K\alpha$ of α recursively via

$$K\alpha := K\alpha_1 \cup \ldots \cup K\alpha_n \cup \{\beta_1, \ldots, \beta_n\}$$

where $K0 := 0$. Let $AP = \{\omega^\delta : \delta \in ON\}$. We can then put

$$\vartheta\alpha := \min\{\beta \in AP : \beta \geq \max K\alpha \wedge \forall \gamma < \alpha(K\gamma < \beta \rightarrow \vartheta\gamma < \beta\} . \quad (1$$

One easily verifies $\vartheta\alpha < \Omega$ by induction on α using a cardinality argument. It is moreover easy to verify that then $\varepsilon_0 = \vartheta\Omega$ and $\Gamma_0 = \vartheta\Omega^2$ (see, for example [13].)

To explain the expected formula concerning wpo's let us consider a given explicit operator W which maps a (countable) wpo X to a (countable) wpo $W(X$ so that the elements of $W(X)$ can be described as generalized terms in which the variables are replaced by constants for the elements of X. We assume that the ordering between elements of $W(X)$ is induced effectively by the ordering from X. (This resembles Feferman's notion of effective relative categoricity [8] Girard's notion of denotation system [10] or Joyal's notion of analytic functor. The latter notion seems to contain Feferman's and Girard's notions, as indicated e.g. in [11].) In concrete situations W may for example stand for an iterated application of basic constructions like disjoint union and cartesian product, the set of finite sequences construction, the multiset construction, or a tree constructor and the like. We assume that for W we have an explicit knowledge of $o(W(X)$ such that $o(W(X)) = o(W(o(X)))$ and such that this equality can be proved using an effective reification (An example for this technique is, for example, given in [13] or [16]).

Using W we then build the set of W-constructor trees $T(W(Rec))$ as follows:

1. $\cdot \in T(W(Rec))$.
2. If (s_i) is a sequence of elements in $T(W(Rec))$ and $w((x_i))$ is a term from $W(X)$ then $\cdot(w((s_i))) \in T(W(Rec))$.

he embeddability relation $\trianglelefteq$ on $T(W(Rec))$ is defined recursively as follows:

1. $\cdot \trianglelefteq t$.
2. If $s \trianglelefteq t_i$ then $s \trianglelefteq \cdot(w((t_i)))$
3. If $w((s_i)) \le w'((t_j))$ mod $W(T(W(Rec)))$ is induced recursively by $\trianglelefteq$ then

$$\cdot(w((s_i))) \trianglelefteq \cdot(w'((t_j))) \ .$$

The general principle is now that

$$T(W(Rec)) \text{ is a wpo}$$

nd

$$o(\langle T(W(Rec)), \trianglelefteq \rangle) \le \vartheta o(W(\Omega)) \tag{2}$$

r $o(W(\Omega)) \in dom(\vartheta)$ with $o(W(\Omega)) \ge \Omega^3$. [Moreover the reverse inequality llows in many cases by direct inspection.]

The formula (2) is true for several natural examples which appear as suborerings of Friedman's FKT^n [17] (provided that when necessary the domain of is suitably extended, but discussing such matters is beyond the scope of the esent article). We believe that (2) will be the key property in finally analyzing iedman's FKT^n and we have already obtained far reaching applications.

In general (2) can be proved along the following general outline. (This outline pplies to all cases which we considered so far.)

roof ("Proof outline" for (2)). The inequality is proved by induction on $W(\Omega)$). Let $t = w((t_j)) \in T(W(Rec))$. We claim $o(L_{T(W(Rec))}(t)) < \vartheta o(W(\Omega))$ d may assume by induction hypothesis that

$$o(L_{T(W(Rec))}(t_j)) < \vartheta o(W(\Omega)) \ .$$

now $s \in L_{T(W(Rec))}(t)$ then there will be natural quasi-embedding putting s to a well-partial order $W'(Rec, (t_i))$ such that

$$o(W'(\Omega, (t_i))) < o(W(\Omega))$$

d such that

$$K(o(W'(\Omega, (t_i)))) \le \max \left(K(o(W(\Omega)) \cup \{o(L_{T(W(Rec))}(t_j)) : j\} \right)$$

is step uses the assumption that the maximal order type resulting from W can computed by an effective reification a la [13] or [16]. Therefore the definition ϑ yields

$$\vartheta(o(W'(\Omega, (t_i)))) < \vartheta(o(W(\Omega))) \ .$$

induction hypothesis

$$o(L_{T(W(Rec))}(t)) \le o(T(W'(\Omega, (t_i)))) \le \vartheta(o(W'(\Omega, (t_i))))$$

d we are done.   □

This proof outline can be used to prove (rigorously) several of the main results of the Habilitationsschrift of Diana Schmidt [15] in a short and uniform way, but there already have been lots of more applications (which exceed the realm of the usual Kruskal theorem).

Examples 1 (Rathjen and Weiermann)

1. If $W(X) = X^*$ then
   $o(\langle T(W(Rec)), \trianglelefteq \rangle) = \vartheta \Omega^\omega$ (since $o(\Omega^*) = \omega^{\omega^{\Omega+1}} = \Omega^\omega$).
2. If $W(X) = \bigotimes_{i<n} X$ then $o(\langle T(W(Rec)), \trianglelefteq \rangle) = \vartheta \Omega^n$ (since $o(\bigotimes_{i<n} \Omega) = \Omega^n$).
3. If $W(X) = (X^*)^*$ then $o(\langle T(W(Rec)), \trianglelefteq \rangle) = \vartheta \Omega^{\Omega^{\Omega^\omega}}$ (since $o((\Omega^*)^*) = \omega^{\omega^{\omega^{\Omega+1}}} = \Omega^{\Omega^{\Omega^\omega}}$).

Further examples arise from the *multiset construction*. Let $M'(X)$ be the set of finite multisets over X ordered by (cf., e.g., [19])

$$m \ll m' \iff (\forall x \in m \setminus m \cap m')(\exists y \in m' \setminus m \cap m')[x < y] .$$

Further let $B(X)$ be the set of binary (planar) trees labeled with elements from X ordered under homeomorphic embeddability.

Examples 2 (Rathjen and Weiermann)

1. If $W(X) = M'(X)$ then
   $o(\langle T(W(Rec)), \trianglelefteq \rangle) \leq \vartheta \Omega$ (since $o(M'(\Omega)) = \omega^\Omega = \Omega$).
2. If $W(X) = M'(X \otimes X)$ then $o(\langle T(W(Rec)), \trianglelefteq \rangle) = \vartheta \Omega^\Omega$ (since $o(M'(\Omega \otimes \Omega)) = \omega^{\Omega \otimes \Omega} = \Omega^\Omega$).
3. If $W(X) = B(X)$ then $o(\langle T(W(Rec)), \trianglelefteq \rangle) = \vartheta \varepsilon_{\Omega+1}$ (since $o(B(\Omega)) = \varepsilon_{\Omega+1}$).

Finally let $M(X)$ be the set of finite multisets over X but now according to [1] ordered by

$$m \leq^\circ m' \iff (\exists f : m \hookrightarrow m')(\forall x \in m)[x \leq f(x) \bmod X] . \tag{3}$$

The main result proved in this paper is then the following. Given α let

$$\alpha' := \begin{cases} \alpha + 1 \text{ if } \alpha \text{ is an epsilon number,} \\ \alpha \text{ if } \alpha \text{ is not an epsilon number.} \end{cases} \tag{4}$$

Theorem 2. If $o(X) = \omega^{\alpha_1} + \cdots + \omega^{\alpha_n} \geq \alpha_1 \geq \ldots \geq \alpha_n$ then

$$o(M(X)) = \omega^{\omega^{\alpha_1'} + \cdots + \omega^{\alpha_n'}} .$$

This result and the formula (2) lead in many natural cases to the correct maximal order types for wpo's resulting from non planar trees since, e.g., $o(M(\Omega)) = \Omega$ and $o(T(M(Rec)) = \vartheta \Omega^\omega$ [13]. Moreover since $o(M(\varepsilon_0 \otimes \Omega)) = \omega^{\varepsilon_0 \otimes \Omega}$ we find the result $o(T\binom{1}{\varepsilon_0}) = \vartheta(\Omega^{\varepsilon_0})$ from [20]. Here the class $T\binom{1}{\varepsilon_0}$ from [20] corresponds to $T(M(\varepsilon_0 \otimes \Omega))$ in the current setting.

Proof of the Main Theorem

Before proving our main result let us first recall a basic fact from wpo-theory which is useful for proving results on maximal order types.

Definition 1. *Let X, Y be two posets. A map $e : X \to Y$ is called a quasi-embedding if for all $x, x' \in X$ with $e(x) \le e(x')$ mod Y we have $x \le x'$ mod X.*

Lemma 1. *If X, Y are posets and $e : X \to Y$ is a quasi-embedding and Y is a wpo, then X is a wpo and $o(X) \le o(Y)$.*

Let us now come to our Main Theorem. Assume that $\langle X, \le \rangle$ is a partial ordering and that $M(X)$ is the set of finite multisets over X. Recall the definition of the term ordering $\le^\circ$ [cf. (3)] and the definition of the operation $\alpha \mapsto \alpha'$ [cf. (4)] from Sec. 1.

Main Theorem 1. *Let $\langle X, \le \rangle$ be a well-partial ordering with*

$$o(X) = \omega^{\alpha_1} + \cdots + \omega^{\alpha_n} \ge \alpha_1 \ge \ldots \ge \alpha_n$$

and let $M(X)$ be the set of finite multisets over X. Then

$$o(M(X), \le^\circ) = \omega^{\omega^{\alpha_1'} + \cdots + \omega^{\alpha_n'}} \ .$$

Proof. For $\alpha = \omega^{\alpha_1} + \cdots + \omega^{\alpha_n} \ge \alpha_1 \ge \ldots \ge \alpha_n$ we write $\hat{\alpha} := \omega^{\alpha_1'} + \cdots + \omega^{\alpha_n'}$. Then $\widehat{\alpha \oplus \beta} = \hat{\alpha} \oplus \hat{\beta}$. The proof of the inequality

$$o(M(X), \le^\circ) \le \hat{\alpha}$$

is very similar to the proof provided by Aschenbrenner and Pong [1]. For convenience of the reader we recall the whole argument and we fill in the modifications when needed. The proof is by induction on $\alpha := o(X)$. The case $\alpha = 0$ is trivial and we may assume that $\alpha > 0$. Now assume that $\alpha = \alpha_1 \oplus \alpha_2$ where $\alpha_1, \alpha_2 < \alpha$. Then X is a disjoint union $X_1 \cup X_2$ with $o(X_1) \le \alpha_1$ and $o(X_2) \le \alpha_2$. Then

$$o(M(X)) \le o(M(X_1) \otimes M(X_2)) = o(M(X_1) \otimes o(M(X_2))$$
$$\le \omega^{\hat{\alpha_1}} \otimes \omega^{\hat{\alpha_2}} = \omega^{\widehat{\alpha_1 \oplus \alpha_2}} = \omega^{\hat{\alpha}}$$

using the induction hypothesis. Now suppose that $\alpha = \omega^{\alpha_1}$. We may assume that $\alpha_1 > 0$.

It suffices to show that $o(L(w)) < \omega^{\hat{\alpha}}$ for all $w \in M(X)$. We show this by induction on the length of w. If the length of w is zero then we are done. Assume now that $w = [\![x_0, \ldots, x_{m-1}]\!]$ with $x_0, \ldots, x_{m-1} \in X$. Then there is a quasi-embedding $e : L(w) \to L_X(x_0) \oplus (X \otimes L(w'))$ where $w' = [\![x_1, \ldots, x_{m-1}]\!]$. To see this let $v = [\![y_0, \ldots, y_{n-1}]\!] \in M(X)$ such that $\neg w \le^\circ v$. Then either $x_0 \le y_i$ mod X for all i; or $y_i \ge x_0$ mod X for some i, so after reordering the ys we may assume that $x_0 \le y_0$ mod X and $\neg w' \le^\circ v'$ for $v' := [\![y_1, \ldots, y_{n-1}]\!]$. In the first case we put $e(v) := v \in M(L(x_0))$ and in the second case we put

$e(v) = \langle y_0, v' \rangle \in X \otimes L(w')$. It is easy to check that this is a quasi-embedding Hence

$$o(L(w)) \leq o(M(L(x_0))) \oplus (\alpha \otimes o(L(w')))$$

by the previous Lemma. Put $\gamma = o(L(x_0))$, then $\gamma < \alpha$ hence $\hat{\gamma} < \hat{\alpha}$ and $o(M(L(x_0))) \leq \omega^{\hat{\gamma}} < \omega^{\hat{\alpha}}$ by induction hypothesis on α. By induction hypothesis on w we have $\delta := o(L(w')) < \omega^{\hat{\alpha}}$. Hence it suffices to show that $\alpha \cdot \delta < \omega^{\hat{\alpha}}$. If $\alpha_1' = \alpha_1$ then $\alpha_1 < \omega^{\omega^{\alpha_1}}$ and $\omega^{\omega^{\alpha_1}}$ is closed under natural multiplication. Hence $\delta < \omega^{\omega^{\alpha_1}}$ yields $\alpha \otimes \delta < \omega^{\hat{\alpha}}$. If $\alpha_1' > \alpha_1$ then $\alpha = \omega^{\omega^{\alpha_1}} < \omega^{\omega^{\alpha_1'}}$ and $\omega^{\omega^{\alpha_1'}}$ is again closed under natural multiplication. Hence $\delta < \omega^{\hat{\alpha}}$ yields $\alpha \otimes \delta < \omega^{\hat{\alpha}}$. We are done in this case.

We now prove the other direction. Assume that $\alpha := o(X) = \omega^{\alpha_1} + \cdots + \omega^{\alpha_r}$ where $\omega^{\alpha_1} + \cdots + \omega^{\alpha_n} \geq \alpha_1 \geq \ldots \geq \alpha_n$. We claim that $o(M(X), \leq^\circ) \geq \omega^{\hat{\alpha}}$. For this it suffices to show that $o(M(X), \preceq^\circ) \geq \omega^{\hat{\alpha}}$ where $\preceq$ is a linear extension of $\leq$ on X having (maximal possible) order type $o(X)$. It is easy to see that $o(M(X), \leq^\circ) \geq o(M(X), \preceq^\circ)$ since any quasi-embedding $e : o(X) \to X$ gives rise to a corresponding quasi-embedding $\hat{e} : M(o(X)) \to M(X)$.

While proving

$$o(M(X), \preceq^\circ) \geq \omega^{\hat{\alpha}}$$

we identify X with $o(X)$, hence with α.

Assume first that α_1 is not an epsilon number. (This case will be considerably easier than the other.) We define a quasi-embedding $e : \omega^{\widehat{o(X)}} \to M(X)$ by induction on $o(X)$. Assume that $\beta < \omega^{\hat{\alpha}} = \omega^{\omega^{\alpha_1} + \omega^{\alpha_2'} + \cdots + \omega^{\alpha_n'}}$. Assume that

$$\beta = \omega^{\omega^{\alpha_1} + \beta_1} + \cdots + \omega^{\omega^{\alpha_1} + \beta_r} + \omega^{\beta_{r+1}} + \cdots + \omega^{\beta_{r+s}}$$

where β is written in Cantor normal form and $\beta_1, \ldots, \beta_r < \omega^{\alpha_2'} + \cdots + \omega^{\alpha_n}$ and $\beta_{r+1}, \ldots, \beta_{r+s} < \omega^{\alpha_1}$. Then $\omega^{\beta_1} + \cdots + \omega^{\beta_r} < \omega^{\omega^{\alpha_2'} + \cdots + \omega^{\alpha_n}}$. By induction hypothesis we may assume that there is a quasi-embedding

$$f : \omega^{\omega^{\alpha_2'} + \cdots + \omega^{\alpha_n'}} \to M(\omega^{\alpha_2} + \cdots + \omega^{\alpha_n}).$$

Assume that $f(\omega^{\beta_1} + \cdots + \omega^{\beta_r}) = [\![\delta_1, \ldots, \delta_k]\!]$.

Now put

$$e(\beta) = [\![\omega^{\alpha_1} + \delta_1, \ldots, \omega^{\alpha_1} + \delta_k]\!] \cup [\![\beta_{r+1}, \ldots, \beta_{r+s}]\!].$$

Then $e(\beta) \in M(X)$. We claim that e is indeed a quasi-embedding.

For, assume that $e(\beta) \leq^\circ e(\gamma)$ with $\beta, \gamma < \omega^{o(\hat{X})}$. Write

$$e(\beta) = [\![\omega^{\alpha_1} + \delta_1, \ldots, \omega^{\alpha_1} + \delta_k]\!] \cup [\![\beta_{r+1}, \ldots, \beta_{r+s}]\!]$$

as before. Assume that

$$\gamma = \omega^{\omega^{\alpha_1} + \gamma_1} + \cdots + \omega^{\omega^{\alpha_1} + \gamma_t} + \omega^{\gamma_{t+1}} + \cdots + \omega^{\gamma_{t+u}}$$

here γ is written in Cantor normal form and $\gamma_1, \ldots, \gamma_t < \omega^{\alpha_2'} + \cdots + \omega^{\alpha_n'}$ and $\gamma_{t+1}, \ldots, \gamma_{t+u} < \omega^{\alpha_1}$. Assume that $f(\omega^{\gamma_1} + \cdots + \omega^{\gamma_t}) = [\![\xi_1, \ldots, \xi_l]\!]$ and write

$$e(\gamma) = [\![\omega^{\alpha_1} + \xi_1, \ldots, \omega^{\alpha_1} + \xi_l]\!] \cup [\![\gamma_{t+1}, \ldots, \gamma_{t+u}]\!].$$

From $e(\beta) \leq^\diamond e(\gamma)$ we conclude

$$[\![\omega^{\alpha_1} + \delta_1, \ldots, \omega^{\alpha_1} + \delta_k]\!] \leq [\![\omega^{\alpha_1} + \xi_1, \ldots, \omega^{\alpha_1} + \xi_l]\!] \bmod (M(\omega^{\alpha_2} + \cdots + \omega^{\alpha_n}))$$

hence $f(\omega^{\beta_1} + \cdots + \omega^{\beta_r}) \leq f(\omega^{\gamma_1} + \cdots + \omega^{\gamma_t})$ thus

$$\omega^{\beta_1} + \cdots + \omega^{\beta_r} \leq \omega^{\gamma_1} + \cdots + \omega^{\gamma_t}.$$

Multiplication by $\omega^{\omega^{\alpha_1}}$ on the left yields

$$\omega^{\omega^{\alpha_1} + \beta_1} + \cdots + \omega^{\omega^{\alpha_1} + \beta_r} \leq \omega^{\omega^{\alpha_1} + \gamma_1} + \cdots + \omega^{\omega^{\alpha_1} + \gamma_t}.$$

If the inequality would be strict, then $\beta < \gamma$ would follow immediately. So assume that

$$\omega^{\omega^{\alpha_1} + \beta_1} + \cdots + \omega^{\omega^{\alpha_1} + \beta_r} = \omega^{\omega^{\alpha_1} + \gamma_1} + \cdots + \omega^{\omega^{\alpha_1} + \gamma_t}.$$

Then

$$[\![\beta_{r+1}, \ldots, \beta_{r+s}]\!] \leq [\![\gamma_{t+1}, \ldots, \gamma_{t+u}]\!] \bmod M(\omega^{\alpha_1}).$$

This yields $\omega^{\beta_{r+1}} + \cdots + \omega^{\beta_{r+s}} \leq \omega^{\gamma_{t+1}} + \cdots + \omega^{\gamma_{t+u}}$ hence $\beta \leq \gamma$.

Now we turn to the critical case that α_1 is an epsilon number.

Let us assume that $o(X) = \epsilon + \tau$ where $\epsilon = \alpha_1$ and $\tau = \omega^{\alpha_2} + \cdots + \omega^{\alpha_n}$. Then $\epsilon = \epsilon \cdot \omega$ and $\hat{\tau} = \omega^{\alpha_2'} + \cdots + \omega^{\alpha_n'}$. We define $e : \omega^{\epsilon \cdot \omega + \hat{\tau}} \to M(X)$ as follows. Pick $\beta < \omega^{\epsilon \cdot \omega + \hat{\tau}}$ and assume that

$$\beta = \omega^{\epsilon \cdot \omega + \beta_1} + \cdots + \omega^{\epsilon \cdot \omega + \beta_r} + \epsilon^s \cdot \beta_{r+1} + \cdots + \epsilon^0 \cdot \beta_{r+1+s+1}$$

where $\hat{\tau} > \beta_1 \geq \ldots \geq \beta_r$ and $0 < \beta_{r+1}$ and $\beta_{r+1}, \ldots, \beta_{r+1+s+1} < \epsilon$.

Then $\omega^{\beta_1} + \cdots + \omega^{\beta_r} < \omega^{\tau'}$. By induction hypothesis there exists a quasi-embedding $f : \omega^{\hat{\tau}} \to M(\tau)$. Let $f(\omega^{\beta_1} + \cdots + \omega^{\beta_r}) = [\![\delta_1, \ldots, \delta_k]\!]$.

In this situation we define

$$e(\beta) := [\![\epsilon + \delta_1, \ldots, \epsilon + \delta_k, \beta_{r+1}, \beta_{r+1} + \beta_{r+2}, \ldots, \beta_{r+1} + \beta_{r+2} + \cdots + \beta_n]\!].$$

Then $e(\beta) \in M(\epsilon + \tau)$. Assume now that $e(\beta) \leq^\diamond e(\gamma)$ where

$$\beta = \omega^{\epsilon \cdot \omega + \beta_1} + \cdots + \omega^{\epsilon \cdot \omega + \beta_r} + \epsilon^s \cdot \beta_{r+1} + \cdots + \epsilon^0 \cdot \beta_{r+1+s+1}$$

and $\tau > \beta_1 \geq \ldots \geq \beta_r$ and $0 < \beta_{r+1}$ and $\beta_{r+1}, \ldots, \beta_{r+1+s+1} < \epsilon$ and

$$\gamma = \omega^{\epsilon \cdot \omega + \gamma_1} + \cdots + \omega^{\epsilon \cdot \omega + \gamma_t} + \epsilon^u \cdot \gamma_{t+1} + \cdots + \epsilon^0 \cdot \gamma_{t+1+u+1}$$

where $\tau > \gamma_1 \geq \ldots \geq \gamma_t$ and $0 < \gamma_{t+1}, \ldots, \gamma_{t+1+u+1} < \epsilon$.

From $e(\beta) \leq e(\gamma)$ we obtain

$$[\![\epsilon + \delta_1, \ldots, \epsilon + \delta_k]\!] \leq [\![\epsilon + \xi_1, \ldots, \epsilon + \xi_l]\!] \bmod M(\epsilon + \tau)$$

hence $[\![\delta_1, \ldots, \delta_k]\!] \leq [\![\xi_1, \ldots, \xi_l]\!]$ mod $M(\tau)$ thus

$$f(\omega^{\beta_1} + \cdots + \omega^{\beta_r}) \leq f(\omega^{\gamma_1} + \cdots + \omega^{\gamma_t}) \text{ mod } M(\tau)$$

hence

$$\omega^{\beta_1} + \cdots + \omega^{\beta_r} \leq \omega^{\gamma_1} + \cdots + \omega^{\gamma_t}$$

and

$$\omega^{\epsilon \cdot \omega + \beta_1} + \cdots + \omega^{\epsilon \cdot \omega + \beta_r} \leq \omega^{\epsilon \cdot \omega + \gamma_1} + \cdots + \omega^{\epsilon \cdot \omega + \gamma_t} .$$

If strict inequality would hold then $\beta < \gamma$. So assume that

$$\omega^{\epsilon \cdot \omega + \beta_1} + \cdots + \omega^{\epsilon \cdot \omega + \beta_r} = \omega^{\epsilon \cdot \omega + \gamma_1} + \cdots + \omega^{\epsilon \cdot \omega + \gamma_t}.$$

Then

$$[\![\beta_{r+1}, \beta_{r+1}+\beta_{r+2}, \ldots, \beta_{r+1}+\cdots+\beta_{r+1+s+1}]\!] \leq^\diamond [\![\gamma_{t+1}, \ldots, \gamma_{t+1}+\cdots+\gamma_{t+1+u+1}]\!]$$

Then necessarily $s \leq u$ since we need an injection for the embeddability. If $s < u$
then $\beta < \gamma$. So assume that $s = u$ and that g is the injection witnessing

$$[\![\beta_{r+1}, \beta_{r+1}+\beta_{r+2}, \ldots, \beta_{r+1}+\cdots+\beta_{r+1+s+1}]\!] \leq^\diamond [\![\gamma_{t+1}, \ldots, \gamma_{t+1}+\cdots+\gamma_{t+1+u+1}]\!]$$

We claim that $\beta_{r+1} \leq \gamma_{t+1}$. Otherwise there would be an m such that $\beta_{r+1} + \cdots + \beta_{r+m} \leq \gamma_{t+1}$ under g. But then again $\beta_{r+1} \leq \gamma_{t+1}$. If now $\beta_{r+1} < \gamma_{t+1}$ then $\beta < \gamma$ and so assume that $\beta_{r+1} = \gamma_{t+1}$. Then the same g also witnesses

$$[\![\beta_{r+1}+\beta_{r+2}, \ldots, \beta_{r+1}+\cdots+\beta_{r+1+s+1}]\!] \leq [\![\gamma_{t+1}+\gamma_{t+2}, \ldots, \gamma_{t+1}+\cdots+\gamma_{t+1+u+1}]\!]$$

Hence as before we may assume that

$$\beta_{r+1} + \beta_{r+2} = \gamma_{t+1} + \gamma_{t+2}$$

hence $\beta_{r+2} = \gamma_{t+2}$. This provides an easy inductive argument for proving $\beta \leq \gamma$.

As an application of our main theorem we obtain (following the proof of Corollary
4.2 in [1]) a refinement of a result of van den Dries and Ehrlich [5].

Corollary 1. *Let Γ be an ordered abelian group and $S \subset \Gamma^{\geq 0}$ well-ordered of
order type $\alpha = o(S)$. Then the monoid generated by S in Γ is well-ordered of
order type less than or equal to $\omega^{\hat{\alpha}}$.*

Remarks. The calculation of $o(M(X))$ is not as simple as a corresponding calculation in [19]. There, to obtain a maximal linear extension, it was sufficient to
extend X to a (maximal) well-order and the induced multiset order was already
maximal. In this paper the extension from X to $o(X)$ leads only from an extension of the partial order $\leq^\diamond$ to another partial order $\preceq^\diamond$ which still has to be
linearized to produce the maximal linear extension. This phenomenon is familiar
from Schmidt's calculation of maximal order types of several tree embeddability

lations [15]. In her context it is also possible to construct (in a *modular* way) maximal linear extensions by first linearizing the sets of labels and then linearizing embeddability relations on trees with well-ordered sets of labels.

Acknowledgements. The author is grateful to Michael Rathjen and Antonio Montalban for inspiring discussions on the subject. The authors is grateful to the referees who provided helpful suggestions which led to an improved exposition.

References

1. Aschenbrenner, M., Pong, W.Y.: Orderings of monomial ideals. Fund. Math. 181(1), 27–74 (2004)
2. Becker, T., Weispfenning, V.: Gröbner bases. A computational approach to commutative algebra. In: Cooperation with Heinz Kredel. Graduate Texts in Mathematics, vol. 141. Springer, New York (1993)
3. Buchholz, W., Cichon, E.A., Weiermann, A.: A uniform approach to fundamental sequences and hierarchies. Math. Logic Quart. 40(2), 273–286 (1994)
4. De Jongh, D.H.J., Parikh, R.R.: Well-partial orderings and hierarchies. Nederl. Akad. Wetensch. Proc. Ser. A 80=Indag. Math. 39(3), 195–207 (1977)
5. Van den Dries, L., Ehrlich, P.: Fields of surreal numbers an exponentiation. Fund. Math. 168(2), 173–188 (2001); Erratum to [5] in Fund. Math. 167(3), 295–297 (2001)
6. Dershowitz, N.N., Jouannaud, J.P.: Rewrite systems. In: Handbook of Theoretical Computer Science, Part B, pp. 243–320. Elsevier, Amsterdam (1990)
7. Ehrenfeucht, A.: Polynomial functions with exponentiation are well-ordered. Algebra Universalis 3, 261–262 (1973)
8. Feferman, S.: Systems of predicative analysis. II. Representations of ordinals. J. Symbolic Logic 33, 193–220 (1968)
9. Gallier, J.H.: What's so special about Kruskal's theorem and the ordinal Γ_0? A survey of some results in proof theory. Annals of Pure and Applied Logic 53, 199–260 (1991)
9. Girard, J.Y.: Π_2^1-logic. I. Dilators. Ann. Math. Logic 21(2-3), 75–219 (1981)
. Hasegawa, R.: Two applications of analytic functors. Theories of types and proofs (Tokyo, 1997). Theoret. Comput. Sci. 272(1-2), 113–175 (2002)
2. Montalbán, A.: Computable linearizations of well-partial-orderings. Order 24(1), 39–48 (2007)
3. Rathjen, M., Weiermann, A.: Proof-theoretic investigations on Kruskal's theorem. Annals of Pure and Applied Logic 60, 49–88 (1993)
. Pohlers, W.: Proof Theory. The first step into Impredicativity. Springer, Berlin (2009)
. Schmidt, D.: Well-Partial Orderings and Their Maximal Order Types. Habilitationsschrift, Heidelberg (1979)
. Schütte, K., Simpson, S.G.: Ein in der reinen Zahlentheorie unbeweisbarer Satz über endliche Folgen von natürlichen Zahlen. Archiv für mathematische Logik und Grundlagenforschung 25, 75–89 (1985)
. Simpson, S.G.: Nonprovability of certain combinatorial properties of finite trees. In: Harvey Friedman's research on the foundations of mathematics. Stud. Logic Found. Math., vol. 117, pp. 87–117. North-Holland, Amsterdam (1985)

18. Urquhart, A.: The complexity of decision procedures in relevance logic. II. J. Symbolic Logic 64(4), 1774–1802 (1999)
19. Weiermann, A.: Proving termination for term rewriting systems. In: Kleine Büning H., Jäger, G., Börger, E., Richter, M.M. (eds.) CSL 1991. LNCS, vol. 626, pp. 419–428. Springer, Heidelberg (1992)
20. Weiermann, A.: An order-theoretic characterization of the Schütte-Veblen hierarchy. Math. Logic Quart. 39(3), 367–383 (1993)
21. Weiermann, A.: Well partial orderings and their strengths measured in maximal order types. In: Summary of a talk given at the conference on computability, reverse mathematics and combinatorics in Banff (2008)

On Generating Independent Random Strings

Marius Zimand*

Department of Computer and Information Sciences,
Towson University, Baltimore, MD, USA
http://triton.towson.edu/~mzimand

Abstract. It is shown that from two strings that are partially random and independent (in the sense of Kolmogorov complexity) it is possible to effectively construct polynomially many strings that are random and pairwise independent. If the two initial strings are random, then the above task can be performed in polynomial time. It is also possible to construct in polynomial time a random string, from two strings that have constant randomness rate.

Keywords: Kolmogorov complexity, random strings, independent strings, randomness extraction.

Introduction

his paper belongs to a line of research that investigates whether certain at-ibutes of randomness can be improved effectively. We focus on finite binary rings and we regard randomness from the point of view of Kolmogorov com-exity. Thus, the amount of randomness in a binary string x is given by $K(x)$, e Kolmogorov complexity of x and the randomness rate of x is defined as $(x)/|x|$, where $|x|$ is the length of x. Roughly speaking, a string x is considered be random if its randomness rate is approximately equal to 1. It is obvious that ndomness cannot be created from nothing (e.g., from the empty string). On the her hand, it might be possible that if we already possess some randomness, we n produce "better" randomness or "new" randomness. For the case when we art with *one* string x, it is known that there exists no computable function that oduces another string y with higher randomness rate (i.e., "better" random-ss), and it is also clear that there is no computable function that produces ew" randomness, by which we mean a string y that has non-constant Kol-ogorov complexity conditioned by x. In fact, Vereshchagin and Vyugin [VV02, h. 4] construct a string x with high Kolmogorov complexity so that any shorter ring that has small Kolmogorov complexity conditioned by x (in particular any ring effectively constructed from x) has small Kolmogorov complexity uncon-tionally. Therefore, we need to analyze what is achievable if we start with two more strings that have a certain amount of randomness and a certain degree independence. In this case, in certain circumstances, positive solutions exist. r example, Fortnow, Hitchcock, Pavan, Vinodchandran and Wang [FHP$^+$06]

The author is supported in part by NSF grant CCF 0634830.

Ambos-Spies, B. Löwe, and W. Merkle (Eds.): CiE 2009, LNCS 5635, pp. 499–508, 2009.
Springer-Verlag Berlin Heidelberg 2009

show that, for any σ there exists a constant ℓ and a polynomial-time procedure that from an input consisting of ℓ n-bit strings $x_1, \ldots, x_\ell$, each with Kolmogorov complexity at least σn, constructs an n-bit string with Kolmogorov complexity $\succeq n - \mathrm{dep}(x_1, \ldots, x_\ell)$ ($\mathrm{dep}(x_1, \ldots, x_\ell)$ measures the dependency of the input strings and is defined as $\sum_{i=1}^{\ell} K(x_i) - K(x_1 \ldots x_\ell)$; $\succeq$ means that the inequality holds within an error of $O(\log n)$).

In this paper we focus on the case when the input consists of *two* strings x and y of length n. We say that x and y have dependency at most $\alpha(n)$ if the complexity of each string does not decrease by more than $\alpha(n)$ when it is conditioned by the other string, i.e., if $K(x) - K(x \mid y) \le \alpha(n)$ and $K(y) - K(y \mid x) \le \alpha(n)$. The reader should have in mind the situation $\alpha(n) = O(\log n)$, in which case we say that x and y are independent (see [CZ08] for a discussion of independence for finite binary strings and infinite binary sequences). We address the following two questions:

Question 1. Given x and y with a certain amount of randomness and a certain degree of independence, is it possible to effectively/efficiently construct a string z that is random?

Question 2. (a more ambitious version of Question 1) Given x and y with certain amount of randomness and a certain degree of independence, is it possible to effectively/efficiently construct strings that are random and have small dependency with x, with y, and pairwise among themselves? How many such strings exhibiting "new" randomness can be produced?

A construction is *effective* if it can be done by a computable function, and it is *efficient* if it can be done by a polynomial-time computable function.

We first recall the well-known (and easy-to-prove) fact that if x and y are random and independent, then the string z obtained by bit-wise XOR-ing the bits of x and y is random and independent with x and with y. Our first result is an extension of the above fact.

Theorem 1. (Informal statement.) If x and y are random and have dependence at most $\alpha(n)$, then by doing simple arithmetic operations in the field $\mathrm{GF}[2^n]$ (which take polynomial time), it is possible to produce polynomially many strings $z_1, \ldots, z_{\mathrm{poly}(n)}$ of length n such that $K(z_i) \succeq n - \alpha(n)$ and the strings x, y, $z_1, \ldots, z_{\mathrm{poly}(n)}$ are pairwise at most $\approx \alpha(n)$-dependent, where $\approx$ ($\succeq$) means that the equality (resp., the inequality) is within an error of $O(\log n)$. In particular, if x and y are independent, then the output strings are random and together with the input strings form a collection of pairwise independent strings.

The problem is more complicated when the two input strings x and y have randomness rate significantly smaller than 1. In this case, our questions are related to randomness extractors, which have been studied extensively in computational complexity. A randomness extractor is a polynomial-time computable procedure that improves the quality of a defective source of randomness. A source of randomness is modeled by a distribution X on $\{0, 1\}^n$, for some n, and its defectiveness is modeled by the min-entropy of X (X has min-entropy k if 2^{-k}

e largest probability that X assigns to any string in $\{0,1\}^n$). There are sev-
al type of extractors; for us, multi-source extractors are of particular interest.
n ℓ-multisource extractor takes as input ℓ defective independent distributions
1 the set of n-bit strings and outputs a string whose induced distribution is
atistically close to the uniform distribution. The analogy between random-
ess extractors and our questions is quite direct: The number of sources of the
ctractor corresponds to the number of input strings and the min-entropy of
e sources corresponds to the Kolmogorov complexity of the input strings. For
$= 2$, the best multisource extractors are (a) the extractor given by Raz [Raz05]
ith one source having min-entropy $((1/2) + \alpha)n$ (for some small α) and the
cond source having min-entropy $\text{polylog}(n)$, and (b) the extractor given by
ourgain [Bou05] with both sources having min-entropy $((1/2) - \alpha)n$ (for some
nall α). Both these extractors are based on recent results in arithmetic com-
natorics. It appears that finding polynomial-time constructions achieving the
oals in Question 2 is difficult. If we settle for effective constructions, then pos-
ive solutions exist. In [Zim09], we have shown that there exists a computable
nction f such that if x and y have Kolmogorov complexity $s(n)$ and depen-
ency at most $\alpha(n)$, then $f(x, y)$ outputs a string z of length $m \approx s(n)/2$ such
at $K(z \mid x) \succeq m - \alpha(n)$ and $K(z \mid y) \succeq m - \alpha(n)$. Our second result extends
e methods from [Zim09] and shows that it is possible to effectively construct
olynomially many strings exhibiting "new" randomness.

heorem 2. (Informal statement.) For every function $O(\log n) \leq s(n) \leq n$, there
cists a computable function f such that if x and y have Kolmogorov complex-
y $s(n)$ and dependency at most $\alpha(n)$, then $f(x, y)$ outputs polynomially many
rings $z_1, \ldots, z_{\text{poly}(n)}$ of length $m \approx s(n)/3$ such that $K(z_i) \succeq m - \alpha(n)$ and the
rings $(x, y, z_1, \ldots, z_{\text{poly}(n)})$ are pairwise at most $\approx \alpha(n)$-dependent. In particu-
r, if x and y are independent, then the output strings are random and together
ith the input strings form a collection of pairwise independent strings.

or Question 1, we give a polynomial-time construction in case x and y have
near Kolmogorov complexity, i.e., $K(x) \geq \delta n$ and $K(y) \geq \delta n$, for a positive
onstant $\delta > 0$. The proof relies heavily on a recent result of Rao [Rao08], which
ows the existence of 2-source condensers. (A 2-source condenser is similar but
eaker than a 2-source extractor in that the condenser's output is only required
be statistically close to a distribution that has larger min-entropy rate than
at of its inputs, while the extractor's output is required to be statistically close
the uniform distribution.)

heorem 3. (Informal statement.) For every constant $\delta > 0$, there exists a
olynomial-time computable function f such that if x and y have Kolmogorov
mplexity δn and dependency at most $\alpha(n)$, then $f(x, y)$ outputs a string z of
ngth $m = \Omega(\delta n)$ and $K(z) \geq m - (\alpha(n) + \text{poly}(\log n))$.

e main proof technique is an extension of the method used in [Zim08] and
[Zim09]. It uses ideas from Fortnow et al. [FHP$^+$06], who showed that a
ulti-source extractor can also be used to extract Kolmogorov complexity. A key

element is the use of *balanced tables*, which are combinatorial objects similar t
2-source extractors. A balanced table is an N-by-N table whose cells are colore
with M colors in such a way that each sufficiently large rectangle inside the tabl
is colored in a balanced way, in the sense that all colors appear approximatel
the same number of times. The exact requirements for the balancing propert
are tailored according to their application. The type of balanced table required i
Theorem 2 is shown to exist using the probabilistic method and then constructe
using exhaustive search. This is why the transformation in Theorem 2 is onl
effective, and not polynomial-time computable. The existence of the type c
balanced table used in Theorem 3 is a direct consequence of Rao's 2-sourc
condenser.

The paper is structured as follows. Sections 1.1 and 1.2 introduce the nota
tion and the main concepts of Kolmogorov complexity. Section 1.3 is dedicate
to balanced tables. Theorem 1 and Theorem 2 are proved in Section 2, an
Theorem 3 is proved in Section 3.

Due to the space constraints of the proceedings, some proofs are omitted. Th
full version of the paper is available on the author's web page.

1.1   Preliminaries

$\mathbb{N}$ denotes the set of natural numbers. For $n \in \mathbb{N}$, $[n]$ denotes the set $\{1, 2, \ldots, n$
We work over the binary alphabet $\{0, 1\}$. A string is an element of $\{0, 1\}^*$. If
is a string, $|x|$ denotes its length. The cardinality of a finite set A is denoted $|A$
Let M be a standard Turing machine. For any string x, define the *Kolmogoro*
complexity of x with respect to M, as $K_M(x) = \min\{|p| \mid M(p) = x\}$. There is
universal Turing machine U such that for every machine M there is a constar
c such that for all x, $K_U(x) \le K_M(x) + c$. We fix such a universal machin
U and dropping the subscript, we let $K(x)$ denote the Kolmogorov complexit
of x with respect to U. For the concept of conditional Komogorov complexit
the underlying machine is a Turing machine that in addition to the read/wor
tape which in the initial state contains the input p, has a second tape containir
initially a string y, which is called the conditioning information. Given such
machine M, we define the Kolmogorov complexity of x conditioned by y wit
respect to M as $K_M(x \mid y) = \min\{|p| \mid M(p, y) = x\}$. Similarly to the abov
there exist universal machines of this type and a constant c and they satisfy t
relation similar to the one above, but for conditional complexity. We fix such
universal machine U, and dropping the subscript U, we let $K(x \mid y)$ denote tl
Kolmogorov complexity of x conditioned by y with respect to U. In this pape
the constants implied in the $O(\cdot)$ notation depend only on the universal machin

The Symmetry of Information Theorem (see [ZL70]) states that for all strin
x and y:

$$|(K(x) - K(x \mid y)) - (K(y) - K(y \mid x))| \le O(\log K(x) + \log K(y)).$$

In case the strings x and y have length n, it can be shown that

$$|(K(x) - K(x \mid y)) - (K(y) - K(y \mid x))| \le 2 \log n + O(1).$$

Sometimes we need to concatenate two strings a and b in a self-delimiting atter, i.e., in a way that allows to retrieve each one of them. A simple way to do is is by taking $a_1 a_1 a_2 a_2 \ldots a_n a_n 01 b$, where $a = a_1 \ldots a_n$, with each $a_i \in \{0, 1\}$. more efficient encoding is as follows. Let $|a|$ in binary notation be $c_1 c_2 \ldots c_k$. ote that $k = \lfloor \log |a| \rfloor + 1$. Then we define $\mathrm{concat}(a, b) = c_1 c_1 c_2 c_2 \ldots c_k c_k 01 a b$. ote that $|\mathrm{concat}(a, b)| = |a| + |b| + 2\lfloor \log |a| \rfloor + 4$.

2   Independent Strings

efinition 1. *(a) Two strings x and y are at most $\alpha(n)$-dependent if $K(x) - (x|y) \le \alpha(|x|)$ and $K(y) - K(y|x) \le \alpha(|y|)$.*

(b) The strings $(x_1, x_2, \ldots)$ are pairwise at most $\alpha(n)$-dependent, if for every $\ne j$, x_i and x_j are at most $\alpha(n)$-dependent.

3   Balanced Tables

table is a function $T : [N] \times [N] \to [M]$. In our applications, N and M are)wers of 2, i.e., $N = 2^n$ and $M = 2^m$. We identify $[N]$ with $\{0, 1\}^n$ and $[M]$ th $\{0, 1\}^m$. Henceforth, we assume this setting.

It is convenient to view such a function as a two dimensional table with N ws and N columns where each entry has a color from the set $[M]$. If B_1, B_2 e subsets of $[N]$, the $B_1 \times B_2$ rectangle of table T is the part of T comprised the rows in B_1 and the columns in B_2. If $A \subseteq \{0, 1\}^m$ and $(x, y) \in [N] \times [N]$, : say that the cell (x, y) is A-colored if $T(x, y) \in A$.

In our proofs, we need the various tables to be *balanced*, which, roughly speak-g, requires that in each sufficiently large rectangle $B_1 \times B_2$, all colors appear)proximately the same number of times.

One variant of this concept is given in the following definition.

efinition 2. *Let $k \in \mathbb{N}$. The table T is (S, n^k)-strongly balanced if for every ir of sets B_1 and B_2, where $B_1 \subseteq [N]$, $|B_1| \ge S$, $B_2 \subseteq [N]$, $|B_2| \ge S$, the llowing two inequalities hold:*

) For every $a \in [M]$,

$$|\{(x, y) \in B_1 \times B_2 \mid T(x, y) = a\}| \le \frac{2}{M} |B_1 \times B_2|,$$

) for every $(a, b) \in [M]^2$ and for every $(i, j) \in [n^k]^2$,

$$|\{(x, y) \in B_1 \times B_2 \mid T(x + i, y) = a \text{ and } T(x + j, y) = b\}| \le \frac{2}{M^2} |B_1 \times B_2|,$$

where addition is done modulo N.

sing the probabilistic method, we show that, under some settings for the pa-meters, strongly-balanced tables exist.

Lemma 1. *If $S^2 > 3M^2 \ln M + 6M^2 \cdot k \cdot \ln n + 6SM^2 + 6SM^2 + 6SM^2 \ln(N/S) - 3M^2$, then there exists an (S, n^k) - strongly balanced table.*

NOTE: The condition is satisfied if $M = o((1/\sqrt{n})S^{1/2})$.

The proof (omitted in this extended abstract) uses the probabilistic metho which does not indicate an efficient way to construct such tables. In our appl cation, we will build such tables by exhaustive search, an operation that can b done in EXPSPACE.

A weaker type of a balanced table can be constructed in polynomial-time using a recent result of Rao [Rao08]. We first recall the following definitions. Le X and Y be two probability distributions on $\{0, 1\}^n$. The distributions X an Y are ϵ-close if for every $A \subseteq \{0, 1\}^n$, $|\mathrm{Prob}(X \in A) - \mathrm{Prob}(Y \in A)| < \epsilon$. Th min-entropy of distribution X is $\max_{a \in \{0,1\}^n}(\log(1/\mathrm{Prob}(X = a)))$.

Fact 1. *[Rao08] For every $\delta > 0$, $\epsilon > 0$, there exists a constant c an a polynomial-time computable function $Ext : \{0, 1\}^n \times \{0, 1\}^n \to \{0, 1\}^m$ where $m = \Omega(\delta n)$, such that if X and Y are two independent random var ables taking values in $\{0, 1\}^n$ and following distributions over $\{0, 1\}^n$ with mi entropy at least δn, then $Ext(X, Y)$ is ϵ-close to a distribution with min-entrop $m - (\delta \log 1/\epsilon)^c$.*

Rao's result easily implies the existence of a polynomial-time table with a usef balancing property.

Lemma 2. *Let $\delta > 0$, $\epsilon > 0$ and let c be the constant and $Ext : \{0, 1\}^n$ $\{0, 1\}^n \to \{0, 1\}^m$ be the function from Fact 1, corresponding to these param eters. We identify $\{0, 1\}^n$ with $[N]$ and $\{0, 1\}^m$ with $[M]$ and view Ext as a $[N] \times [N]$ table colored with M colors. Then for every rectangle $B_1 \times B_2$ $[N] \times [N]$, where $|B_1| \geq 2^{\delta n}$ and $|B_2| \geq 2^{\delta n}$ and for every $A \subseteq [M]$, the numb of cells in $B_1 \times B_2$ that are A-colored is at most*

$$\left(\frac{|A|}{M} 2^{(\delta \log(1/\epsilon))^c} + \epsilon\right) \cdot |B_1 \times B_2|.$$

Proof. Let B_1 and B_2 be two subsets of $\{0, 1\}^n$ of size $\geq 2^{\delta n}$. Let X and be two independent random variables that follow the uniform distributions B_1, respectively B_2 and assume the value 0 on $\{0, 1\}^n - B_1$, respectively 0 $\{0, 1\}^n \mp B_2$. Since X and Y have min-entropy $\geq 2^{\delta n}$, it follows that $Ext(X, Y)$ is ϵ-close to a distribution Z on $\{0, 1\}^m$ that has min-entropy $m - (\delta \log 1/\epsilon)$ If $A \subseteq \{0, 1\}^m$, then Z assigns to A probability mass at most $\frac{|A|}{M} 2^{(\delta \log 1/\epsilon)}$ because it assigns to each element in $\{0, 1\}^m$ at most $2^{-(m-(\delta \log 1/\epsilon)^c)}$. Thu $Ext(X, Y)$ assigns to A probability mass at most $\frac{|A|}{M} 2^{(\delta \log 1/\epsilon)^c} + \epsilon$. This mea that the number of occurrences of A-colored cells in the $B_1 \times B_2$ rectangle bounded by $\left(\frac{|A|}{M} 2^{(\delta \log 1/\epsilon)^c} + \epsilon\right) \cdot |B_1 \times B_2|$.

Generating Multiple Random Independent Strings

The formal statement of Theorem 1 follows.

Theorem 1. *For every $k \in \mathbb{N}$, there is a polynomial-time computable function f that on input x_1, x_2, two strings of length n, outputs n^k strings $x_3, x_4, \ldots, x_{n^k+2}$, strings of length n, with the following property. For every sufficiently large n and for every function $\alpha(n)$, if x_1 and x_2 satisfy*

(i) $K(x_1) \geq n - \log n$,
(ii) $K(x_2) \geq n - \log n$, and
(iii) x_1 and x_2 are at most $\alpha(n)$-dependent,
then
(a) $K(x_i) \geq n - (\alpha(n) + (k + O(1)) \log n$, for every $i \in \{3, \ldots, n^k + 2\}$, and
(b) the strings $x_1, x_2, \ldots, x_{n^k+2}$ are pairwise at most $\alpha(n) + (3k + O(1)) \log n$-dependent.

Due to space constraints, we limit to present the function f which on inputs x_1 and x_2, outputs the strings $x_3, x_4, \ldots, x_{n^k+2}$ defined by:

$$x_3 = x_1 + 1 \cdot x_2, \ \ x_4 = x_1 + 2 \cdot x_2, \ldots, \ \ x_{n^k+2} = x_1 + n^k \cdot x_2,$$

where the arithmetic is done in the finite field $GF[2^n]$ and $1, 2, \ldots, n^k$ denote the first (in some canonical ordering) n^k non-zero elements of $GF[2^n]$. We next prove Theorem 2. The formal statement is as follows.

Theorem 2. *For every $k \in \mathbb{N}$, for every computable function $s(n)$ verifying $(k + 15) \log n < s(n) \leq n$ for every n, there exists a computable function f that, for every n, on input two strings x_1 and x_2 of length n, outputs n^k strings $x_3, x_4, \ldots, x_{n^k+2}$ of length $m = s(n)/3 - (2k+5) \log n$ with the following property. For every sufficiently large n and for every function $\alpha(n)$, if*

(i) $K(x_1) \geq s(n)$,
(ii) $K(x_2) \geq s(n)$ and
(iii) x_1 and x_2 are at most $\alpha(n)$ - dependent,
then
(a) $K(x_i) \geq m - (\alpha(n) + O(\log n))$, for every $i \in \{3, \ldots, n^k + 2\}$ and
(b) the strings in the set $\{x_1, x_2, \ldots, x_{n^k+2}\}$ are pairwise at most $\alpha(n) + (2k + O(1)) \log n$-dependent.

Proof. We fix n and let $N = 2^n$, $m = s(n)/3 - (2k + 5) \log n$, $M = 2^m$, $S = 2^{s(n)/3}$. We also take $t = \alpha(n) + 7 \log n$. The requirements of Lemma 1 are satisfied and therefore there exists a table $T : [N] \times [N] \to [M]$ that is (S, n^k)-strongly balanced. By brute force, we find the smallest (in some canonical sense) such table T. Note that the table T can be described with $\log n + O(1)$ bits. The function f outputs

$$x_3 = T(x_1 + 1, x_2), \ x_4 = T(x_1 + 2, x_2), \ldots, \ x_{n^k+2} = T(x_1 + n^k, x_2).$$

Claim 1. *For every $j \in \{3, \ldots, n^k + 2\}$, $K(x_j \mid x_1) \geq K(x_j) - (\alpha(n) + O(\log n))$ and $K(x_j \mid x_2) \geq K(x_j) - (\alpha(n) + O(\log n))$.*

Claim 2. *For every $i, j \in \{3, \ldots, n^k + 2\}$, $K(x_j \mid x_i) \geq K(x_j) - (\alpha(n) + (2k - O(1))) \log n$.*

Claim 1 is using ideas from the paper [Zim09] and its proof is omitted. We next prove Claim 2.

We fix two elements $i \neq j$ in $\{3, \ldots, k + 2\}$ and analyze $K(x_i | x_j)$. Let $t_1 = K(x_1)$ and $t_2 = K(x_2)$. From hypothesis, $t_1 \geq s(n)$ and $t_2 \geq s(n)$. We define $B_1 = \{u \in \{0, 1\}^n \mid K(u) \leq t_1\}$ and $B_2 = \{u \in \{0, 1\}^n \mid K(u) \leq t_2\}$. We have $S \leq |B_1| < 2^{t_1 + 1}$ and $S \leq |B_2| < 2^{t_2 + 1}$. ($B_1$ and B_2 have size larger than $S = 2^{2s(n)/3}$, because they contain the set $0^{s(n)/3}\{0, 1\}^{2s(n)/3}$.) Let $T_{i,j}^{-1}(x_i, x_j)$ denote the set of pairs $(u, v) \in [N] \times [N]$ such that $T(u + i, v) = x_i$ and $T(u + j, v) = x_j$. Note that $(x_1, x_2) \in T_{i,j}^{-1}(x_i, x_j) \cap (B_1 \times B_2)$. Since the table T is strongly balanced,

$$|T_{i,j}^{-1}(x_i, x_j) \cap (B_1 \times B_2)| \leq \frac{2}{2^{-2m}} 2^{t_1 + t_2 + 2} = 2^{t_1 + t_2 - 2m + 3}.$$

Note that $T_{i,j}^{-1}(x_i, x_j) \cap (B_1 \times B_2)$ can be effectively enumerated given x_i, x_j, j, and the table T. Thus $x_1 x_2$ can be described from $x_i x_j$, the rank of (x_1, x_2) in the above enumeration, i, j, and the table T. This implies that

$$K(x_1 x_2) \leq t_1 + t_2 - 2m + 3 + K(x_i x_j) + $$
$$+ 2k \log n + 2(\log k + \log \log n) + O(\log n)$$
$$\leq t_1 + t_2 - 2m + K(x_i x_j) + (2k + O(1)) \log n.$$

On the other hand, $K(x_1 x_2) \geq K(x_1) + K(x_2 \mid x_1) - O(\log n)$ and $K(x_2 \mid x_1) \geq K(x_2) - \alpha(n)$. Therefore,

$$K(x_1 x_2) \geq K(x_1) + K(x_2) - (\alpha(n) + O(\log n))$$
$$= t_1 + t_2 - (\alpha(n) + O(\log n)).$$

Combining the last two inequalities, we get that

$$t_1 + t_2 - (\alpha(n) + O(\log n)) \leq t_1 + t_2 - 2m + K(x_i x_j) + (2k + O(1)) \log n$$

which implies that

$$K(x_i x_j) \geq 2m - \alpha(n) - (2k + O(1)) \log n.$$

Therefore

$$K(x_j | x_i) \geq K(x_i x_j) - K(x_i) - O(\log n)$$
$$\geq (2m - \alpha(n) - (2k + O(1)) \log n) - (m + O(1)) - O(\log n)$$
$$= m - \alpha(n) - (2k + O(1)) \log n.$$

It follows that

$$K(x_j) - K(x_j \mid x_i) \leq (m + O(1)) - (m - \alpha(n) - (2k + O(1)) \log n) - O(\log n)$$
$$\leq \alpha(n) + (2k + O(1)) \log n.$$

Thus, x_j and x_i are at most $\alpha(n) + (2k + O(1)) \log n$-dependent.

Polynomial-Time Generation of One Random String

In this section we prove Theorem 3. The formal statement is as follows.

Theorem 3. *For every $\delta > 0$ and for every function $\alpha(n)$, there exists a constant c and a polynomial-time computable function $f : \{0,1\}^n \times \{0,1\}^n \to \{0,1\}^m$, where $m = \Omega(\delta n)$, with the following property. If n is sufficiently large and x and y are two strings of length n satisfying*

(i) $K(x) \geq \delta n$
(ii) $K(y) \geq \delta n$
(iii) x *and y are at most $\alpha(n)$ - dependent,*
then

$$K(f(x,y)) \geq m - (\alpha(n) + O((\log n)^c)).$$

Proof. Let $\epsilon = 1/(8n^{10} \cdot \alpha(n))$ and let c be the constant and $Ext : \{0,1\}^n \times \{0,1\}^n \to \{0,1\}^m$ be the function given by Lemma 2 for parameters $(\delta/2)$ and Let $t = \alpha(n) + 10\log n + ((\delta/2)\log 1/\epsilon)^c + 3 = \alpha(n) + O((\log n)^c)$.

The function f on input x and y returns $z = Ext(x,y)$. We show that $K(z) \geq m - t$.

Suppose $K(z) < m - t$.

Let $t_1 = K(x)$, $t_2 = K(y)$, $B_1 = \{u \in \{0,1\}^n \mid K(u) \leq t_1\}$, $B_2 = \{u \in \{0,1\}^n \mid K(u) \leq t_2\}$. From hypothesis, $t_1 \geq \delta n$ and $t_2 \geq \delta n$.

Note also that $2^{\delta n/2} \leq |B_1| \leq 2^{t_1+1}$ and $2^{\delta n/2} \leq |B_2| \leq 2^{t_2+1}$. (The sets B_1 and B_2 have size $\geq 2^{\delta n/2}$ because they contain $0^{n-\delta n/2}\{0,1\}^{\delta n/2}$.)

Let

$$A = \{v \in \{0,1\}^m \mid K(v) < m - t\}$$

We focus on the table defined by the function $Ext : [N] \times [N] \to [M]$, where, as usual, we have identified $\{0,1\}^n$ with $[N]$ and $\{0,1\}^m$ with $[M]$.

Let G be the subset of $B_1 \times B_2$ of cells in the rectangle $B_1 \times B_2$ that are A-colored.

Since $Ext(x,y) = z \in A$, $x \in B_1$ and $y \in B_2$, the cell (x,y) belongs to the rectangle $B_1 \times B_2$ and is A-colored. In other words, $x \in G$.

Taking into account Lemma 2, we can bound the size of G by

$$\left(\frac{|A|}{M} \cdot 2^{((\delta/2)\log 1/\epsilon)^c} + \epsilon\right)|B_1 \times B_2|$$

$$\leq 2^{t_1+t_2+2}\left(\frac{2^{m-t}}{2^m} \cdot 2^{((\delta/2)\log 1/\epsilon)^c} + 2^{-\log 1/\epsilon}\right)$$

$$= 2^{t_1+t_2-t+((\delta/2)\log 1/\epsilon)^c+2} + 2^{t_1+t_2-\log 1/\epsilon+2}$$

$$\leq 2^{t_1+t_2-(\alpha(n)+10\log n)}.$$

The last inequality follows from the choice of ϵ and t.

The set G can be enumerated if we are given t_1, t_2, δ and n (from which we can derive t and the table Ext), and every element in G can be described by its rank in the enumeration and by the information neeeded to perform the enumeration.

Since $x \in G$, it follows that

$$K(xy) \leq t_1 + t_2 - \alpha(n) - 10 \log n + 2(\log t_1 + \log t_2 + \log n) + O(1)$$
$$< t_1 + t_2 - \alpha(n) - 4 \log n$$
$$= K(x) + K(y) - \alpha(n) - 4 \log n.$$

We have used the fact that $t_1 \leq n + O(1)$ and $t_2 \leq n + O(2)$. By the Symmetry of Information Theorem,

$$K(xy) \geq K(y) + K(x \mid y) - 2 \log n - O(1).$$

Combining the last two inequalities, we get

$$K(x) - K(x \mid y) > \alpha(n) + \log n,$$

which contradicts the fact that x and y are at most $\alpha(n)$-dependent.

References

[Bou05]   Bourgain, J.: More on the sum-product phenomenon in prime fields and its applications. International Journal of Number Theory 1, 1–32 (2005)

[CZ08]   Calude, C., Zimand, M.: Algorithmically independent sequences. In: Ito, M., Toyama, M. (eds.) DLT 2008. LNCS, vol. 5257, pp. 183–195. Springer, Heidelberg (2008)

[FHP+06]   Fortnow, L., Hitchcock, J., Pavan, A., Vinodchandran, N.V., Wang, F.: Extracting Kolmogorov complexity with applications to dimension zero-one laws. In: Bugliesi, M., Preneel, B., Sassone, V., Wegener, I. (eds.) ICALP 2006. LNCS, vol. 4051, pp. 335–345. Springer, Heidelberg (2006)

[Rao08]   Rao, A.: A 2-source almost-extractor for linear entropy. In: Goel, A., Jansen, K., Rolim, J.D.P., Rubinfeld, R. (eds.) APPROX and RANDOM 2008. LNCS, vol. 5171, pp. 549–556. Springer, Heidelberg (2008)

[Raz05]   Raz, R.: Extractors with weak random seeds. In: Gabow, H.N., Fagin, R. (eds.) STOC, pp. 11–20. ACM, New York (2005)

[VV02]   Vereshchagin, N.K., Vyugin, M.V.: Independent minimum length programs to translate between given strings. Theor. Comput. Sci. 271(1-2), 131–143 (2002)

[Zim08]   Zimand, M.: Two sources are better than one for increasing the Kolmogorov complexity of infinite sequences. In: Hirsch, E.A., Razborov, A.A., Semenov, A.L., Slissenko, A. (eds.) Computer Science – Theory and Applications. LNCS, vol. 5010, pp. 326–338. Springer, Heidelberg (2008)

[Zim09]   Zimand, M.: Extracting the Kolmogorov complexity of strings and sequences from sources with limited independence. In: Proceedings 26th STACS, Freiburg, Germany, February 26–29 (2009)

[ZL70]   Zvonkin, A., Levin, L.: The complexity of finite objects and the development of the concepts of information and randomness by means of the theory of algorithms. Russian Mathematical Surveys 25(6), 83–124 (1970)

Author Index